THE MACMILLAN Dictionary *of* Military Biography

THE MACMILLAN Dictionary *of* Military Biography

Alan Axelrod and Charles Phillips

MACMILLAN • USA

MACMILLAN
A Simon & Schuster Macmillan Company
1633 Broadway
New York, NY 10019

Macmillan Publishing books may be purchased for business or sales promotional use. For information please write:
Special Markets Department, Macmillan Publishing USA, 1633 Broadway, New York, NY 10019.

Library of Congress Cataloging-in-Publication Data

Axelrod, Alan, 1952–
The Macmillan dictionary of military biography / Alan Axelrod and Charles Phillips.
p. cm.
Includes bibliographical references.
ISBN: 0-02-861994-3 (alk. paper)
1. Military biography—Dictionaries. I. Phillips, Charles, 1948–. II. Title.
U51.A94 1998
355'.0092'2—dc21
[B]

97-31779
CIP

Book design by Nick Anderson

Manufactured in the United States of America

10 9 8 7 6 5 4 3 2 1

List of Subjects

Abd-el-Kader 1
Abrams, Creighton Williams, Jr. 2
Adan, Avraham "Bren" 3
Aëtius, Flavius 4
Agesilaus II 5
Agricola, Gnaeus Julius 5
Agrippa, Marcus Vipsanius 6
Alanbrooke, Sir Alan Francis Brooke 7
Alexander I 8
Alexander III the Great 9
Alexius I Comnenus 11
Alfonso VIII 12
Allen, Ethan 13
Allenby, Edmund Henry Hynman, First Viscount 13
Anderson, William C. ("Bloody Bill") 14
Anson, George, Lord 15
Anza, Juan Bautista de 15
Arnim, Count Hans Georg of 16
Arnold, Henry Harley ("Hap") 17
Atatürk (Mustafa Kemal, Kemal Atatürk) 18
Atkinson, Henry 19
Attila the Hun 20
Auchinleck, Sir Claude John Ayre 21

Bainbridge, William 23
Barry, John 23
Belisarius 24
Berthier, Louis Alexandre 25
Blake, Robert 26
Blücher, Gebhard Leberecht von 27
Bolívar, Simón 28
Botha, Louis 30
Braddock, Edward 30
Bradley, Omar Nelson 31
Bragg, Braxton 33
Brant, Joseph 34
Breckinridge, John Cabell 35
Brunswick, Frederick William, Duke of 36
Budënny, Semën Mikhailovich 36
Bugeaud de la Piconnerie, Thomas Robert, Duke of Isly 38
Bülow, Friedrich Wilhelm von 38
Burgoyne, John 39
Burnside, Ambrose Everett 40
Byng, Julian Hedworth George, First Viscount Byng of Vimy 42

Caesar, Gaius Julius 44
Campbell, Sir Colin, First Baron Clyde 45
Canby, Edward Richard Spring 46
Carleton, Sir Guy 47
Carlson, Evans Fordyce 48
Carnot, Lazare Nicolas Marguerite 49
Cassius Longinus, Gaius 50
Cato, Marcus Porcius Cato ("Cato the Elder") 50
Cetshwayo 51
Chaffee, Adna Romanza, Jr. 51
Champlain, Samuel de 53

Chandragupta Maurya 54
Charlemagne (Karl Der Grosse, Carolus Magnus) 54
Charles XII 55
Chiang Kai-shek 57
Chu Teh (Zhu De) 60
Churchill, Sir Winston Spencer 61
Cimon 63
Cincinnatus, Lucius Quinctius 64
Clark, George Rogers 65
Clark, Mark Wayne 66
Clausewitz, Karl Maria von 67
Cochise 68
Coenus 69
Coligny, Gaspard II de 70
Condé, Louis II de Bourbon, Prince of 70
Constantine I the Great 72
Cornwallis, Charles, First Marquess 73
Cortés, Hérnan 74
Croesus 76
Cromwell, Oliver 76
Crook, George 79
Cumberland, William Augustus, Duke of 80
Cunningham, Sir Andrew Browne, First Viscount Cunningham of Hyndhope 81
Custer, George Armstrong 82
Cyrus II the Great 83

David 85
Dayan, Moshe 86
Decatur, Stephen 87
De Gaulle, Charles (André-Marie-Joseph) 88
Dewey, George 89
Diocletian (Gaius Aurelius Vaerius Diocletianus) 90
Dionysius I (Dionysius the Elder) 91
Doniphan, Alexander William 92
Dönitz, Karl 92
Doolittle, James ("Jimmy") Harold 93
Doria, Andrea 94
Du Pont, Samuel Francis 95

Edward I Longshanks 97
Edward III of Windsor 98
Edward, The Black Prince 98
Eisenhower, Dwight David 99
Eugene, Prince of Savoy-Carignan 100
Ewell, Richard Stoddert 102

Fabius Maximus Verrucosus, Quintus 104
Fairfax of Cameron, Sir Thomas Fairfax, 3d Baron 104
Farragut, David Glasgow 105
Feng, Yü-hsiang 106
Foch, Ferdinand 107
Foote, Andrew Hull 109
Forrest, Nathan Bedford 109
Franco (Bahamonde), Francisco (Paulin Hermenegildo Teódulo) 110
Frederick I (Frederick Barbarossa, "Red Beard") 112
Frederick William I 114
Frederick II 114
Frederick II the Great 116
French, Sir John Denton Pinkstone, First Earl of Ypres 117
Frunze, Mikhail Vasilyevich 118

Gage, Thomas 120
Galtieri, Leopoldo Fortunato 121
Gates, Horatio 122
Geiger, Roy Stanley 123
Genghis Khan (Temujin) 124
Geronimo 125
Giap, Vo Nguyen 126
Gibbon, John 127
Gneisenau, August Wilhelm Anton, Count Neihardt von 128
Gonzalo De Cordoba, Hernandez 129
Gordon, Charles George ("Chinese Gordon") 130
Göring, Hermann William 131
Grant, Ulysses Simpson 133
Grasse, Count François Joseph Paul de 136
Greene, Nathanael 137
Grierson, Benjamin Henry 138
Guderian, Heinz 139
Gustav II Adolf (Gustavus Adolphus) 141

Hadrian (Publius Aelius Hadrianus) 143
Haig, Alexander Meigs, Jr. 144
Haig, Douglas, First Earl 144
Halleck, Henry Wager 145

Halsey, William Frederick, Jr. 146
Hamilcar Barca 148
Hamilton, Henry 149
Hamilton, Sir Ian Standish Monteith 149
Hancock, Winfield Scott 150
Hannibal 151
Hannibal Barca 152
Hannibal Gisco 153
Hardee, William Joseph 153
Harrison, William Henry 154
Hasegawa, Yoshimichi 156
Hastings, Francis Rawdon-Hastings, Earl of Moira and First Marquess of (Lord Rawdon) 156
Henry II 157
Henry V 159
Heraclius 160
Hill, Ambrose Powell 162
Hill, Daniel Harvey 162
Hill, Sir Rowland, First Viscount 163
Hindenburg, Paul Ludwig Hans von 164
Hitler, Adolf 165
Ho Chi Minh 168
Hooker, Joseph 169
Houston, Sam(uel) 170
Howard, Oliver Otis 171
Howe, Richard, Earl 172
Howe, Sir William, Second Earl 173
Hull, Isaac 174
Hung Hsiu-ch'üan 175
Hunyadi, Jänos 176
Hussein, Saddam (Takriti) 177

Ivan III the Great (Ivan Vasilievich) 179

Jackson, Andrew 181
Jackson, Thomas Jonathan ("Stonewall") 183
James II 184
Jellicoe, John Rushworth, First Earl 185
Joan of Arc (Jeanne d'Arc) 186
Jodl, Alfred 187
Joffre, Joseph Jacques Césaire 188
John I Zimisces 189
John II Comnenus 189
John III (John Sobieski) 190
Johnston, Albert Sidney 191
Johnston, Joseph Eggleston 192
Jomini, Antoine Henri, Baron de 193
Jones, John Paul 194
Chief Joseph 195
Joubert, Petrus Jacobus 196
Juel, Niels 197
Junot, Jean Andoche, Duke of Abrantes 198
Justinian I the Great 199

Kalb, Johann (Baron de Kalb) 201
Karageorge (George Petrovich) 201
Kearny, Philip 202
Kearny, Stephen Watts 203
Keitel, Wilhelm 205
Kesselring, Albert von 206
Kimmel, Husband Edward 207
King, Ernest Joseph 208
Kinkaid, Thomas Cassin 209
Kirby Smith, Edmund 210
Kitchener, Horatio Herbert, First Earl Kitchener of Khartoum 211
Kléber, Jean-Baptiste 212
Kluck, Alexander von 213
Knox, Henry 213
Kolchak, Aleksandr Vasiliyevich 214
Kosciusko, Tadeusz Andrezj Bonawentura 215
Krueger, Walter 216
Kuropotkin, Aleksei Nikolaevich 218
Kutuzov, Mikhail Ilarionovich Golenischev, Prince of Smolensk 219

Lafayette, Marie Joseph Paul Yves Roch Gilbert du Motier, Marquis de 220
Lally, Thomas Arthur, Count de 221
Lannes, Jean, Duke of Montebello, Prince of Sievers 222
Lasalle, Count Antoine Charles Louis 223
La Trémoile, Louis II de, Viscount of Thouars 224
Lawrence, James 225
Lawrence, Stringer 225
Lawrence, T(homas) E(dward) 226
Lawton, Henry Ware 227
Lee, Charles 228
Lee, Fitzhugh 228
Lee, Henry ("Light Horse Harry") 230

Lee, Robert Edward 230
Lefebvre, Francis Joseph, Duke of Danzig 232
LeMay, Curtis Emerson 234
Lettow-Vorbeck, Paul Emil von 235
Liggett, Hunter 236
Lincoln, Abraham 237
Lin Piao 239
Li Shih-min (T'ai Tsung) 240
Li Tsung-jen 240
Logan, John Alexander 241
Longstreet, James 242
Lucas, John Porter 243
Ludendorff, Erich 244
Luxembourg, François Henri de Montmorency-Boutteville, Duke of 246
Lysander 247

MacArthur, Douglas 249
Macdonald, Jacques Etienne Joseph Alexandre, Duke of Taranto 251
MacDonough, Thomas 252
Machiavelli, Niccolò 253
Mackensen, August von 254
Mackenzie, Ranald Slidell 255
Magruder, John Bankhead ("Prince John") 256
Magsaysay, Ramón 257
Mahan, Dennis Hart 258
Mahmud of Ghazna 258
Majorianus, Julius Valerius (Majorian) 259
Mangin, Charles Marie Emmanuel 260
Mannerheim, Carl Gustav Emil von 260
Manstein, Erich von 261
Mao Tse-tung 262
Marcellus, Marcus Claudius 265
Marcus Aurelius (Antoninus) 265
Marion, Francis 266
Marius, Gaius 267
Marlborough, John Churchill, First Duke of 268
Marmont, Auguste Frederic Louis Viesse de, Duke of Ragusa 269
Marshall, George Catlett 271
Martinet, Jean 272
Masséna, André, Duke of Rivoli and Prince of Essling 273
Matsui, Iwane 274
Matthias I Corvinus 275
Maude, Sir Frederick Stanley 276
Maurice (Flavius Tiberius Mauricius) 276
Maximilian I, Duke and Elector of Bavaria 277
Mazepa (Mazeppa), Ivan Stepanovich 278
McClellan, George Brinton 278
McCulloch, Ben 280
McDowell, Irvin 280
McMorris, Charles Horatio 281
McNair, Lesley James 282
Meade, George Gordon 283
Medici, Giovanni de' ("The Invincible," "Giovanni of the Black Bands") 285
Merrill, Frank Dow 285
Middleton, Troy H. 286
Miles, Nelson Appleton 287
Minamoto, Yoshitsune (Ushiwaka) 289
Mitchell, William ("Billy") 290
Mitscher, Marc Andrew 291
Model, Walther 292
Mohammed II the Conqueror 293
Moltke, Count Helmuth Johannes Ludwig von 294
Moltke, Count Helmuth Karl Bernhard von 295
Moncey, Bon Adrien Jeannot de, Duke of Conegliano 297
Monck, George, First Duke of Albemarle 298
Monmouth, James Scott, Duke of 299
Montcalm, Louis-Joseph de Montcalm-Gozon, Marquis of 300
Montgomery, Sir Bernard Law, First Viscount Montgomery of Alamein 301
Montmorency, Anne, Duke of 303
Montrose, James Graham, Marquess of 304
Moreau, Jean Victor Marie 305
Morgan, John Hunt 306
Mortier, Edouard Adolphe Casimir Joseph, Duc de Treviso 306
Mountbatten, Louis Francis Albert Victor Nicholas, Earl Mountbatten of Burma 308
Murat, Joachim, King of Naples, Duke of Cleve and Berg 310
Murray, Lord George 311
Mussolini, Benito 312

Napier, Charles James 315
Napoleon I (Napoleon Bonaparte) 316
Narses 320
Navarro, Pedro, Count of Oliveto 320
Nelson, Horatio, Viscount 321
Nero, Gaius Claudius 323
Ney, Michel, Prince de la Moskova, Duc of Elchingen 323
Nicholas (Nikolai Nikolaevich Romanov), Grand Duke 325
Nimitz, Chester William 325
Nogi, Maresuke 327
Norstad, Lauris 327
Nurhachi 328

Oda, Nohunaga 329
Odaenathus, Septimius 330
Oku, Yasukata 330
Oldendorf, Jesse Bartlett 331
Ord, Edward Otho Cresap 332
Otto I the Great 332
Oudinot, Nicolas Charles, Duke of Reggio 334

Pappenheim, Count Gottfried Heinrich 336
Patch, Alexander McCarrell, Jr. 336
Pate, Randolph McCall 337
Patton, George Smith 338
Paulus, Friedrich von 340
Pericles 342
Perry, Matthew Calbraith 342
Perry, Oliver Hazard 343
Perseus 344
Pershing, John Joseph 345
Pétain, Henri Philippe 346
Peter I the Great 348
Philip II 349
Philip II Augustus 350
Philopoemen 350
Phraates IV 351
Piccolomini, Prince Ottavio 352
Pickett, George Edward 353
Pizarro, Francisco 354
Pompey the Great (Gnaeus Pompeius Magnus) 355
Popé 356
Porter, David 357
Porter, David Dixon 358
Potemkin, Prince Gregori Aleksandrovich 359
Powell, Colin Luther 360
Ptolemy I Soter 361
Pulaski, Count Kazimierz (Casimir) 362
Pyrrhus 363

Quantrill, William Clarke 365

Rabin, Yitzhak 366
Raeder, Erich 366
Richard I the Lion-Hearted 367
Richard III 368
Richelieu, Armand Jean du Plessis, Cardinal et Duc de 369
Ridgway, Matthew Bunker 370
Rochambeau, Jean Baptiste Donatien de Virneur, Count of 372
Rodney, George Brydges, Baron 373
Rommel, Erwin Johannes Eugen 374
Rundstedt (Karl Rudolf), Gerd von 375
Rupert, Prince, Count Palatine of the Rhine, Duke of Bavaria 376

Saint-Cyr, Laurent Gouvion, Count (Marquis of) 378
Saladin (Salah Ad-din Yusuf Ibn Ayyub, "Righteous of the Faith") 379
Samori Toure 380
Samsonov, Aleksandr Vasilievich 381
Samudragupta 381
Santa Anna, Antonio Lopez de 382
Saxe, Hermann Maurice, Count 383
Scharnhorst, Count Gerhard Johann David von 384
Schlieffen, Count Alfred von 385
Schwarzkopf, H(erbert) Norman 386
Scipio Africanus Major (Publius Cornelius Scipio) 387
Scott, Winfield 388
Severus (Lucius Septimius Severus Pertinax) 390
Shaka (Chaka, Tshaka) 391
Shapur I 392
Sharon, Ariel 392

Sheridan, Philip Henry 393
Sherman, William Tecumseh 394
Short, Walter Campbell 396
Sims, William Sowden 397
Sitting Bull (Tatanka Yotanka) 398
Slim, Sir William Joseph 399
Smuts, Jan Christiaan 400
Soult, Nicolas Jean de Dieu, Duke of Dalmatia 402
Stalin, Josef 403
Steuben, Baron Friedrich Wilhelm von 406
Stilwell, Joseph Warren 406
Stuart, James Ewell Brown (Jeb) 408
Stuyvesant, Peter 409
Suleiman I The Magnificent 410
Sulla, Lucius Cornelius 411

Takeda, Shingen (Harunobu) 413
Tamerlane (Timur, Timur the Lane) 413
Taylor, Zachary 414
Tecumseh 415
Tiglath-pileser III 416
Tigranes 417
Tilly, Count Johan Tserclaes 418
Tito, Josip Broz 418
Tojo, Hideki 420
Trajan (Marcus Ulpius Traianus) 421
Turenne, Henri de la Tour d'Auvergne, Viscount of 422

Vandegrift, Alexander Archer 425
Vandenburg, Hoyt Sanford 426
Van Fleet, James Alward 427
Vercingetorix 428
Vespasian (Titus Flavius Vespasianus) 428
Victor-Perrin, Claude, Duke of Belluno 429

Wainwright, Jonathan Mayhew IV 431
Walker, William 432
Wallenstein, Albert Eusebius von, Duke of Friedland and Mecklenburg, Prince of Sagan 433
Washington, George 434
Wayne, Anthony 437
Wellington, Arthur Wellesley, First Duke of 438
Westmoreland, William Childs 440
Weyler y Nicolau, Valeriano 441
Wilhelm II 442
William I the Conqueror 443
Wolfe, James 444
Wood, Leonard 445

Xerxes I 447

Yamamoto, Isoroku 448
Yamashita, Tomoyuki 450
Yang Chien 451
Yen Hsi-shan 451

Zhukov, Georgi Konstantinovich 453

Introduction

General George S. Patton once observed that, next to war, all other human endeavors pale into puny insignificance. Leaving aside issues of the morality and desirability of warfare, that statement rings with simple, if chilling, truth. What is more costly, in money, materials, effort, pain, and life, than war? What requires a greater degree of highly coordinated energy and intensely focused thought? What other human endeavor encompasses such a range of enterprise—from acts of darkest secrecy and utter squalor to the grandest mobilization of entire peoples, vast armies shaping and reshaping the world?

Nothing has more deeply stained the fabric of history than war. The Swiss historian Jean Jacques Babel has estimated that 5,500 years of recorded history present a meager total of 292 years of peace. This book is about some of the men (and one woman, **Joan of Arc**) who shaped at least 5,208 years of history. It is a narrative biographical encyclopedia of the lives of the most significant military leaders, generals, popular militant and guerrilla leaders, tribal chiefs, and emperors from 3500 B.C. (roughly the beginning of recorded history) to the present.

"Most significant" is, admittedly, a difficult phrase. Our chief criterion for including a figure in this book—for judging a commander "most significant"—is the consensus of recent military historians. Beyond this, while *The Macmillan Encyclopedia of Military Biography* is international in scope and covers a vast range of history, it is frankly and deliberately written from an American perspective, even at the risk of giving the military figures of American history disproportionate weight. In this, we have attempted to anticipate the likely interest of our readers.

Finally, the figures included here are, first and foremost, military commanders. They are not heroes of lower rank—the Paul Reveres and the Sergeant Yorks—but, for the most part, commanders of general officer or equivalent rank. Many of the military commanders treated here are also political figures—kings, dictators, even American presidents. After all, as **Karl von Clausewitz** wrote, war is politics by other means. Yet the overwhelming emphasis is on professional soldiers, not politicians.

Each entry begins with the figure's dates of birth and death, followed by a concise statement of the figure's significance to military history and a capsule chronology of the subject's military service. Following this information is an essay that aims to present the relevant chronology, but also a miniature portrait of character that usually ends with an evaluation of the figure's significance. While both authors are strong believers in the importance of social history—the actions of populations and "ordinary people"—we also cautiously subscribe to Thomas Carlyle's assertion that history is the biography of great men. Certainly, this is nowhere more true than in the case of warfare, an overwhelmingly male activity and an enterprise in which masses of men purposefully align themselves behind a handful of leaders, whom they suppose are "great men." We hope that what we have created is a series of concise biographies that are also illuminating as short histories of the human conflicts that have occupied so much of our collective time on earth.

Alan Axelrod
Charles Phillips

THE MACMILLAN Dictionary of Military Biography

ABD-EL-KADER

BORN: September 6, 1808
DIED: May 25/26, 1883
SERVICE: Resistance to the French conquest of Algeria (1830–47)

Unifier of Algerian tribes against French conquest

Abd-el-Kader (also spelled Abd-al-Kadir) was the third son of Mahi-el-Din, who led the Hashim tribe and the fierce Sufi fundamentalist Qadariyya sect. Abd-el-Kader became emir of Mascara following his father's death in 1832 and renewed the jihad (holy war) his father had waged against the French colonizers of Algeria.

Intelligent, charismatic, and adroit, Abd-el-Kader amassed his power with great skill, even going so far as to strike an alliance with the hated French in order to drive out Turkish forces during July 1834. As a result of this maneuver, he emerged as the principal native leader in the region by the end of the year. With the Turks ousted, he quickly turned against the French once again, achieving a brilliant victory at the Battle of La Macta on June 28, 1835. Frustratingly, however, this did not stop the sack of el-Kader's native Mascara by forces under French general Bertrand Clausel in December.

The next blow came at the Battle of Sikkah (July 6, 1836), when el-Kader was defeated by **General Thomas R. Bugeaud,** with whom he subsequently negotiated the terms of the Peace of the Tama. This treaty gave him French sanction of his authority over the disparate tribes of the region. It is an example of el-Kader's pragmatic brilliance, for even as he suffered defeat, and when the tribes were on the verge of deserting the cause, he managed to negotiate a surrender that actually strengthened his position among the native population.

As a government leader, el-Kader organized an efficient theocracy with authority over the native-held two thirds of Algeria. This allowed him not only to maintain, but to strengthen the allegiance of the tribes, which he rallied into new resistance. When the French stormed Constantine on October 13, 1837, capping a program of renewed expansion, el-Kader confidently declared a new jihad in November 1839. He organized a well-disciplined army of 40,000 men, but met with disaster in 1842 when he lost Tlemcen. In the long aftermath of this defeat, his great force was scattered by a mere two thousand troops under the Duke of Aumale at the Battle of Smala (May 10, 1843).

Abd-el-Kader's resourcefulness and reputation were such that, after retreating into Morocco, he was quickly able to reorganize a new army of 45,000 men. However, this new force suffered a decisive defeat at the hands of Bugeaud, who commanded a column of 7,500 at the Battle of the Isly River (August 14, 1844). Following this, el-Kader was formally declared an outlaw by the Moroccan government. This act did not stop el-Kader from exploiting a rebellion in the Dahra area, north of the Chelif River, as cover for his return to Algeria. He won a series of victories, culminating in Battle of Sidi Brahim (September 1845), after which, however, massed French forces drove him back into Morocco.

In Morocco, Abd-el-Kader became active in the always rebellious Rif region and was able to unite the tribes and stir them to renewed resistance. Learning of his activities, Moroccan government forces declared war on him. Initially, el-Kader enjoyed success in this latest effort at resistance, but a series of defeats sent him fleeing back to Algeria, where he subsequently surrendered (December 23, 1847) to the Duke of Aumale in exchange for safe conduct for himself and his family to the Levant. Aumale agreed, but French officials yielded to a public clamor for el-Kader's punishment. He was taken captive and sent to prison in France.

Abd-el-Kader languished in prison from 1848 to October 16, 1852, when he was released by order of Napoleon III, who allowed him to leave France. He lived for a time in Turkey and then moved with his family to Damascus in 1855. Five years later, he successfully intervened to save 1,800 Christians from a mob of Muslim fanatics, an action Napoleon III recognized by awarding to his nation's former enemy the Legion of Honor. Fêted by the Parisians, he remained in France from 1863 to 1865, then returned quietly to Damascus, where he lived in retirement until his death.

Although Abd-el-Kader ultimately failed to drive the French out of Algeria, his skill, determination, and charismatic ability to unite notoriously independent tribespeople did stave off conquest for some twenty years.

Intelligent, devout, physically hardy, and well-educated, Abd-el-Kader was a remarkable leader; he unified the Algerian tribes through a mixture of guile, diplomacy, religious devotion, and force. He also created a standing army of over 10,000 men, usually supplemented by local levies, and used his army and state to frustrate French designs in Algeria for almost two decades.

ABRAMS, Creighton Williams, Jr.

BORN: September 15, 1914

DIED: September 4, 1974

SERVICE: World War II (1941–45); Korean War (1950–53); Vietnam War (1964–73)

U.S. general best known for his service in the late phase of American operations during the Vietnam War

Abrams was a native of Springfield, Massachusetts, and received an appointment to West Point, where he was a superb student, graduating in 1936, eighteenth in a class of 276. Commissioned a second lieutenant of cavalry, he was promoted to captain in 1940 and transferred to armor. In Europe, during World War II, he brilliantly commanded the Thirty-Seventh Tank Battalion and, at the critical Battle of the Bulge, he led the column of Fourth Armored Division into Bastogne on December 6, 1944, relieving the 101st Airborne Division, which had been pinned down by a massive German offensive. **General George S. Patton,** in command of the U.S. Third Army and the master planner of the bold relief operation at Bastogne, was generous in his praise: "I'm supposed to be the best tank commander in the Army, but I have one peer—Abe Abrams." At the end of World War II, in 1945, Abrams was promoted to temporary colonel and subsequently received an appointment to the Armor School at Fort Knox, Kentucky, as its director. After serving in this post from 1946 to 1948, he was graduated from the Command and General Staff School in 1949. His next combat service came in the Korean War, when he served successively as chief of staff for the I, X, and IX Corps from June 1950 through July 1953. Following the cease-fire in 1953, Abrams was singled out for a staff post and attended the Army War College.

After receiving a promotion to brigadier general in February 1956, Abrams was named deputy assistant chief of staff for reserve components on the Army General Staff. He was awarded a second star in May 1960 and continued to serve in various command and staff posts, including commander of the Third Armored Division from 1960 to 1962. During September 1962, through May of the following year, he was given the sensitive and difficult task of leading federal troops dispatched to Mississippi to quell riots associated with the enforcement of civil rights legislation. In the late summer of 1963, Abrams was promoted to lieutenant general and given command of V Corps in Germany. A year later, in September 1964, he received his fourth star and was named army vice chief of staff.

Creighton Abrams came to widespread public attention as the Vietnam War escalated and he was named deputy commander of the U.S. Military Assistance Command Vietnam (MACV) in May 1967. His greatest challenge came as director of operations in northern South Vietnam during the so-called Tet

Offensive of January 30 through February 29, 1968. A massive Viet Cong attack on all fronts, the offensive was an attempt to steamroller South Vietnamese resistance in a single stroke. The Tet operation was perceived by many Americans on the homefront as a defeat for South Vietnamese and U.S. forces. Indeed, the Tet Offensive added great impetus to the growing American antiwar movement. Nevertheless, in military terms, the Tet Offensive was a costly failure for the North Vietnamese, and Abrams was named to succeed **General William Westmoreland** as commander of U.S. MACV. Abrams departed sharply from Westmoreland's strategy of search and destroy, adopting instead a plan of patrol and ambush that was better suited to the guerrilla nature of the conflict and was effective in disrupting the support and supply lines of the North Vietnamese forces. Abrams served as MACV commander during the long phase of "Vietnamization" (the transfer of the burden of the war from U.S. to South Vietnamese forces), and, during the presidential administration of Richard M. Nixon, he directed the reduction of American forces in Vietnam through July 1972, when he was appointed chief of staff of the army. He held this post until his death. In recognition of his service as a commander of armored forces, the U.S. Army named its principal tank, the M-1, in his honor.

ADAN, Avraham "Bren"

BORN: 1926

SERVICE: Israeli War of Independence (1948–49); Suez-Sinai War (1956); Six-Day War (1967); War of Attrition (1968–70); October War (1973)

Controversial Israeli military leader who advocated and helped create the "Bar Lev Line" of fortified observation posts along the troubled Suez in 1969

A native-born Israeli, Avraham Adan enlisted in 1943, at age seventeen, in the Palmach, the most elite of the forces that comprised the underground army of the Zionist settlers in Palestine. When the War of Independence began in earnest in 1948, the youthful Adan was given command of a company in the Negev Brigade, which operated in southern Palestine against the Egyptians. With the conclusion of the war, during March 5–10, 1949, Adan led a mechanized company to the Red Sea at Eliat, an operation that secured vital Israeli access to the Red Sea.

Adan left the army shortly after this operation, but was recalled to active duty during the Suez Crisis of 1956. Precipitated by Egypt's nationalization of the Suez Canal, the crisis touched off a war between Egypt and Israel, in which Adan commanded a tank battalion in the Battle of Abu Agheila (October 30th). While the outcome of that engagement was not favorable to the Israelis, Adan scored a signal success and was quickly named commander of an armored brigade. Eleven years later, during the Six-Day War of June 5–10, 1967, Adan served as deputy commander of Major General Avraham Yoffe's armored division task force in the Sinai. The task force played a key role in Israel's decisive victory.

After the Six-Day War, Adan presided over a board of army officers that recommended building a line of fortified observation posts, even under the Egyptian artillery barrages during the War of Attrition that followed the Six-Day War. The posts were designed to maintain Israeli control of the east bank of the Suez Canal. The recommendation was a controversial one and was bitterly opposed by many other Israeli officers, who were reluctant to expose forces so directly to enemy attack. However, the plan was finally approved by Israeli Chief of Staff Lieutenant General Chaim Bar-Lev, who assigned Adan to supervise construction of the so-called Bar Lev Line in 1969.

When the October War commenced with a surprise Egyptian assault across the Suez Canal on October 6–7, 1973, Adan was in command of a reserve division, which he mobilized within twenty-four hours, sending some of its elements into action in the Sinai less than forty-eight hours after the attack. On October 8, following poorly conceived orders issued by General Shmuel Gonen, commander of the Southern Command, based on inadequate intelligence, Adan launched a disastrous counterattack in the Kantara–Firdan sector. It became the worst defeat ever suffered by the Israeli Army. However, Adan was able to recover from the catastrophe and achieved victory at the Battle of the Chinese Farm on October 16–17. This success reestablished communication with the division commanded by **Ariel Sharon,** and, together, Adan's and Sharon's units penetrated deeply into Egyptian-held territory, all the way to the Gulf of Suez. This movement cut off two divisions of the Egyptian Third Army, enabling Adan to move against the canal itself. However, his initial efforts to seize that strategic waterway were repulsed, and the war ended in a cease-fire on October 24 with Egypt still in possession of the Suez Canal.

In the wake of the sobering October War, Adan bitterly and loudly disputed with fellow generals, especially Ariel Sharon and Shmuel Gonen, over blame for the relatively poor conduct of operations, especially during the first days of the war. Despite this rancorous situation, Adan was named to the prestigious post of military attaché in Washington, D.C. He served in this capacity until his retirement in 1977.

AËTIUS, Flavius

BORN: ca. 395

DIED: September 21, 454

SERVICE: Roman Civil Wars and Roman wars against the barbarians, 423–454

The last great general of imperial Rome

Flavius Aëtius was born at Durostorum (Silistra) in the Roman province of Moesia (now northern Bulgaria). His father, Caudentius, was a Scythian who served as a Roman *magister equitum* (cavalry commander). Caudentius was killed in a mutiny about 401, and Aëtius was delivered as a hostage to the court of Alaric the Goth. This would prove valuable training for his later campaigns against the barbarians; for Aëtius learned the language of this Germanic people and also absorbed a great deal of Alaric's military philosophy, including practical tactics.

Indeed, young Aëtius proved popular among the Goths, whose army he entered around 412. Subsequently, he recruited a force of Huns to march into Italy in aid of John (Iohannes), who, following the death of Emperor Honorius in 423, sought the Roman imperial throne. Although Aëtius arrived too late to help John, he reached an understanding with the sister of Honorius, Galla Placidia, in 425 and was appointed *magister equitum per Gallias*—cavalry commander for Gaul. In this capacity, he defeated Theodoric, the Visigoth king of Toulouse, at the Battle of Arles in 425. He next launched a series of vigorous campaigns throughout Gaul to neutralize the Franks and other Germanic tribes. Through his efforts, Roman authority was reestablished throughout the turbulent region by 430, except in Aquitaine, which remained in the hands of the Visigoths.

In 427, Aëtius returned briefly to Italy, where he was named *magister militum II*—vice-commander of the West. However, he fell under suspicion of having assassinated his superior and of having designs on the throne. He met demands for resignation in disgrace by leading forces loyal to him in a 432 invasion of Italy, only to be repulsed near Ravenna by Boniface, who perished in the battle and was therefore unable to press the pursuit. Aëtius fled to the Huns, among whom he recruited a large force in Pannonia (now western Hungary). With these troops at his back, he persuaded Galla Placidia to restore him to her good graces in 433.

It was a good thing for imperial Rome that she did so. For Aëtius led his forces against a Burgundian uprising around 434 and then forced a favorable peace with the Vandal king Gaiseric in 435. That same year, he successfully defended against a renewed invasion by Theodoric at Arles, and he crushed a Visigoth army at Narbonne. The years 437–439 saw additional fighting, culminating in an enduring treaty with Theodoric concluded in 442. The next year, Aëtius directed military operations that pushed the Burgundians from Worms to Savoy, and by about 445, he had subdued the Salian Franks led by Chlodian, whom he allowed to settle along the Somme.

The next major threat came from no less than **Attila,** whose forces were massed for an invasion of Gaul. Aëtius enlisted the aid of the Franks and the Visigoths (under Theodoric) and met Attila in battle at Châlons during June of 451. Victorious, Aëtius forced Attila to withdraw into eastern Europe.

The fate of many a Roman hero is not everlasting glory. During an audience with Emperor Valentian III, Aëtius was murdered on September 21, 454—most likely as a preemptive strike against any further imperial ambitions. It is believed that the assassination was instigated by Petronius Maximus, who later became emperor.

AGESILAUS II

BORN: 444 B.C.

DIED: 360 B.C.

SERVICE: Spartan–Persian War (400–387 B.C.); Spartan–Corinthian War (395–387 B.C.); Theban War of Independence (379–371 B.C.); Egyptian War (361–360 B.C.)

An expansionist Spartan ruler who commanded his armies with courage and great tactical skill

Born in Sparta, Agesilaus was the son of Archidamus II, who ruled Sparta as head of the Eurypontid house. A crafty politician, Agesilaus was able to succeed his brother Agis II about 398 B.C. after he enlisted the aid of **Lysander,** admiral of the Peloponnesian fleet, to persuade the citizens of Sparta to set aside Agis's son, Leotychidas.

In Asia Minor during 396 B.C., Agesilaus made peace with the satrap Tissaphernes, and, that accomplished, raided Phrygia (central Turkey). Next, he raised a force of cavalry and led it on raids throughout Phrygia and Lydia (east central Turkey) during 395 B.C. When he was recalled to Sparta in 394 B.C., Agesilaus traveled overland so that he could avenge the defeat of Lysander (who had been defeated and slain at Boeotia) by winning victory against Corinthian forces at Coronea. He laid siege to Corinth itself at the end of 394 B.C. and spent the next four years—through 390 B.C.—in a stalemated war with that city-state. The conflict was resolved by the arrival of an Athenian army under Iphicrates, which forced Agesilaus to break off his siege. He then turned to fighting the forces of the Boeotian League in Acarnania (southwestern Greece) in 389 B.C., which resulted in the favorable Peace of Antalcidas in 386 B.C. This brought about the independence of the city-states and gave Agesilaus the legal basis he needed to break up the Boeotian League as well as the union of Argos and Corinth during 385–382 B.C.

With rival states divided, Agesilaus successfully supported the Spartan commander Phoebidas's effort to take the citadel at Thebes in 382, and he personally led a lengthy campaign against Thebes during that city-state's war for independence (378–371 B.C.). He then launched an attack on the Mantinean region during 370 B.C., which stretched into a hard campaign of several years and included defensive operations against Epaminondas during 362 B.C.

Late in his career, in 361 B.C., Agesilaus led a mercenary Spartan force in the service of the Egyptian King Zedhor in an invasion of Palestine and Syria, only to shift his support to Zedhor's rival Nekhtharheb in a successful revolt. Agesilaus died at Cyrene while he was on his way back to Sparta.

The historian Xenophon celebrated Agesilaus II as a politician and, even more, as a great military commander. Although his guile is distasteful to many modern students of classical and ancient history, his remarkable record of military victory lends credence to Xenophon's high appraisal.

AGRICOLA, Gnaeus Julius

BORN: June 13, 37

DIED: August 23, 93

SERVICE: Roman conquest of Britain

The Roman general who conquered and then governed Britain

Agricola was born at Forum Julii (Fréjus) in Gallia Narbonensisi (the area of southern France and the

Rhone Valley). His father was a senator executed by Caligula. Agricola's first appointment came as tribune on the staff of Suetonius Paulinus in Britain during 59 to 61. He became quaestor in Asia during 64, and the people's tribune two years later. In 68, he was made praetor and joined the forces of **Vespasian** during the civil war of 68–69.

Agricola was given command of the XX ("Valeria Victrix") Legion serving in Britain during 70, and, when he returned to Rome in 73, he was elevated to patrician rank. The following year, Agricola was made governor of Aquitania (southern France), serving in that capacity until 77, when he was named consul and governor of Britain. He immediately set about the task of subduing northern Wales and the Island of Mona (Anglesey). He also conquered the Ordovices. Agricola acted next to enforce his conquests by constructing a fortress for the II Legion at Deva (modern Chester) during 78–79, then pressed his conquests northward, reaching the Tay River by 80. Along the northern reaches of Roman Britain, Agricola created a series of fortified outposts during 81–83. With these well established, he moved aggressively against the Caledonians beyond the River Forth (in modern Scotland) and destroyed their army in 84 in a battle at Mons Graupius.

Following his conquest of the Caledonians, Agricola built a large legionary fortress at Inchtuthil (near modern Dunkeld, Scotland), then returned to Rome, where he retired from public life. Never covetous of glory or power for its own sake, Agricola declined an appointment as proconsul of Asia.

Agricola was a brilliant and methodical military tactician, who knew how to retain the territory he conquered. He Romanized Britain deliberately and with much tolerance, permitting a high degree of self-rule among those conquered. He also fostered urbanization in the south of Britain. The great Roman historian Tacitus, Agricola's son-in-law, left a detailed account of his enlightened rule in Britain.

AGRIPPA, Marcus Vipsanius

BORN: ca. 63 B.C.

DIED: March 12, 12 B.C.

SERVICE: War of the Second Triumvirate (43–34 B.C.); War of Antony and Octavius (33–30 B.C.)

Roman general and ally of Octavius (later Emperor Augustus), who was instrumental in defeating Mark Antony at Actium in 31 B.C.

Agrippa was born in (or near) Rome and became, in childhood, the friend of Octavius. Agrippa was with Octavius in the spring of 44 B.C., when he returned to Rome following the assassination of **Julius Caesar** and became his right-hand man, raising an army in his behalf. It is likely that he fought at the battles of Philippi from October–November 42 B.C. and he was instrumental in Octavius's campaign against Lucius Antonius and Fulvia (Antony's sister) during 41 B.C. His victory at the Battle of Perusia (41 B.C.) was even more spectacular.

Following Perusia, Agrippa was named *praetor urbanis*—in effect, magistrate—of Gaul in 40 B.C. That same year, he was elevated to governor, serving until 37 B.C., when he was recalled following the defeat of Octavius by Sextus Pompeius. Undaunted, Agrippa supervised the creation of a new fleet and harbor at Naples and personally trained the crews to man his vessels. In 36 B.C., he led his fleet against the Pompeian forces at Mylae and Naulochus, winning splendid victories. He then took command of Octavius's campaign against the Illyrians during 35–34 B.C. and resumed command of the fleet during operations against Mark Antony. Agrippa was in personal command at the decisive victory of Actium (September 2, 31 B.C.), which resulted in the final defeat of Mark Antony.

With the defeat of Antony, Octavius became the Emperor Augustus, and Agrippa was elevated to consul during 28 and 27 B.C., functioning as the Emperor's chief lieutenant and troubleshooter, especially in provincial matters. It was Agrippa to whom Augustus entrusted his signet ring when he fell gravely ill in 23

B.C., and Agrippa took as his third wife, Julia, the daughter of Augustus.

Agrippa served in the East during 23–21 B.C. and in Gaul and Spain during 21–19 B.C.. Augustus bestowed upon his friend special powers as *tribunicia potestas* in 18 B.C. and sent him to the East as chief governor of the region. From 17 B.C. until his death in 12 B.C., Agrippa founded Roman cities throughout the East and brought peace between the Roman Empire and the Bosporian kingdom, as well as with Herod the Great. Agrippa was on a journey to Pannonia (western Hungary) when he succumbed to an illness and died.

Agrippa was invaluable in consolidating, under Augustus, Rome's power in the East, in Gaul, and in Spain. He also lavished his personal fortune on the construction of great public works for Rome itself, including the Pantheon and the first public baths.

ALANBROOKE, Sir Alan Francis Brooke

BORN: July 23, 1883
DIED: June 17, 1963
SERVICE: World War I (1914–18); World War II (1939–45)

British field marshal; one of World War II's great strategists and field commanders

Alanbrooke was born in Ireland, at Bagnéres de Bigorre, and was educated on the continent. After returning to England, he joined the army as a junior artillery officer (1902) and saw service on the Western Front during World War I, where he was given command of Canadian and Indian troops. In recognition of his superior aptitude as a leader and his brilliant grasp of tactics and strategy, Allanbrooke was assigned as an instructor at the Staff College, Camberley, serving in 1919 and from 1923 to 1927, when he was reassigned to the Imperial Defense College. He served at the college during 1927 and from 1932 to 1934. Two years later, he was appointed director of military training at the War Office, where he served during 1936–37. Concurrently with this assignment, he commanded the Antiaircraft Command and the Southern Command before and through the early months of World War II (1937–40).

Alanbrooke was given command of II Corps of the British Expeditionary Forces (BEF) and displayed great skill and presence of mind during the disastrous assault on the continent, the harrowing retreat to Dunkirk, and the evacuation back to England during May–June, 1940. It was in no small part because of the skill of Alanbrooke that the British forces were saved from annihilation in this battle.

Following Dunkirk, Alanbrooke was named commander in chief of home forces in July 1940, then succeeded Sir John Dill as chief of the Imperial General Staff in December 1941. In this position, Alanbrooke participated in the major strategic decisions of the war, including plans for Operation Overlord, which became the D-Day invasion. Although Alanbrooke lobbied vigorously to obtain command of the operation, that responsibility ultimately was given to **Dwight David Eisenhower.** Alanbrooke, the consummate professional, swallowed any disappointment he may have felt and gave his wholehearted cooperation to the American supreme allied commander and to other American officers. Indeed, among Alanbrooke's greatest gifts was a facility for working well with American as well as Soviet military and civilian leaders. At the Teheran Conference, from November 28 to December 1, 1943, Alanbrooke was instrumental in determining and coordinating the Allies' European strategy. He earned the respect of President Franklin D. Roosevelt, British prime minister **Winston Churchill,** and—perhaps most importantly—the friendship of Soviet premier **Josef Stalin.**

Alanbrooke was promoted to field marshal in 1944, and, in recognition of his distinguished war service, he was created a baron in 1945 and a viscount the following year. A year after the conclusion of World War II, in 1946, Alanbrooke resigned as chief of staff and retired from the military soon after. He died, in retirement, at his Hampshire estate in 1963.

One of the best British commanders of the war, his long tenure in staff posts has obscured his brilliance as a field commander. He was a gifted strategist, and he got on well with the British, Americans, and Russians, winning Stalin's friendship at the Teheran conference.

ALEXANDER I

BORN: December 23, 1777

DIED: December 1, 1825

SERVICE: Napoleonic Wars (1800–15); Persian War (1804–13); Turkish War (1806–12)

Czar of all the Russias from 1801 to 1825; credited with defeating Napoleon during his invasion of Russia in 1812

Son of future czar Paul I, Alexander was raised by his grandmother, the Empress Catherine the Great, who undertook the education of the young prince. Catherine, whose ambition was to "civilize" Russia by bringing to it the best of European culture—particularly the ideas of the French Enlightenment—hired the famed Swiss educator and republican zealot Frédéric-César de La Harpe as his principal tutor, instilling in the youth a passion for reform.

In 1793, when he was sixteen, Alexander was married to the fourteen-year-old Elisabeth of Baden-Durlach. Three years later, Alexander's father ascended the throne as Czar Paul I and began a disastrous reign marked by combined brutality and stupidity. A cabal of nobles assassinated him on March 24, 1801, whereupon Alexander I, now twenty-four, became czar.

The new czar began by acting on his zeal for reform, radically modernizing the archaic and bewildering Russian bureaucracy, putting in place the political machinery to bring an end to serfdom (which was not abolished, however, until 1861 under Czar Alexander II), outlawing torture as a method of interrogation and punishment, and extending land ownership rights beyond the gentry class. Alexander I formulated a bold plan to institute universal public education throughout Russia, and he founded three new universities. Like Catherine, he was a Francophile, and he greatly admired the reforms of Napoleon, especially the Code Napoléon, the all-encompassing legal reforms introduced into post-revolutionary France. Alexander emulated these in formulating a new system of law for his country, calling on the advice of the great Russian legal genius Mikhail Mikhailevich Speransky.

Despite his ambition for change, Alexander I was a mediocre administrator, and his domestic reforms, bold in conception, were half-hearted in execution. This was partly due to the restless monarch's apparent inability to focus on any single set of programs, but, increasingly, the domestic reforms fell victim to the czar's preoccupation with events in a Europe that was being engulfed by Napoleon. Like his grandmother, Alexander I saw Russia as part of the European community of nations. Accordingly, he sought and maintained alliances with Britain, Prussia, and Austria—alliances that drew his nation into the Napoleonic Wars, beginning 1804.

With high resolve, the czar sent his troops into battle in concert with the Austrians against Napoleon at Austerlitz and Moravia in 1805. The defeat of the czar's forces was decisive in both engagements, yet Napoleon gave no sign of an inclination to capitalize on his victories by invading Russia. Indeed, he made overtures of friendship to Alexander, who resisted out of mistrust of a man he had formerly greatly admired. But after another defeat, at Freidland, Alexander bowed to what he deemed the inevitable. On June 25, 1807, he met with Napoleon on a raft at Tilsit to discuss the terms of the surrender of Russia's ally, Prussia. The resulting treaty yielded an uneasy peace that was finally torn asunder by the French invasion of Russia on June 24, 1812.

Alexander, who had alienated certain of his subjects by having treated with Napoleon, now showed great skill in rallying his nation to what was called the Patriotic War. He put it to his people with stark simplicity: "Napoleon or I." In the end, it was a combination of the steadfastness of the czar's army, the vastness of the Russian land, and the unremitting rigors of the Russian winter that defeated the invasion. Napoleon managed to capture Moscow, but his supply lines were stretched beyond their limits, so, after burning the mostly evacuated city, he was forced to retreat through a nightmare of snow, ice, and starvation. The Russian campaign having broken his army, Napoleon was defeated at the Battle of Leipzig in October 1813. Alexander I marched into Paris the following year and was instrumental in restoring the Bourbon dynasty to France as Napoleon was sent into his first exile.

Hailed as the savior not only of Russia but of Europe itself, Alexander I organized the largest European congress up to that time, in Vienna, in 1814. After Napoleon returned from exile during the so-called Hundred Days, met defeat at Waterloo, and was exiled permanently to St. Helena in 1815, Alexander led the resulting peace conference in Paris. This same year, on September 26, Alexander I formulated and signed with Francis I of Austria (formerly Holy Roman Emperor Francis II), and Frederick William III, king of Prussia, the Holy Alliance. Framed on non-dogmatic religious principles, the alliance was intended to bring lasting peace to Europe, and it was subsequently signed by all but three of the European rulers (the British Prince Regent, the pope, and the Ottoman sultan). Yet, if anything, the Holy Alliance tended to provoke bitterness and rivalry. Worse, in the increasingly conservative political climate that set in after the Napoleonic Wars, the very name "Holy Alliance" came to be associated with the repressive policies of the original signatories.

Alexander I rapidly drifted away from reform. He had become the most powerful figure in continental Europe, yet he failed to capitalize on his new prestige and prominence. Disappointed and disillusioned by the failure of the Holy Alliance, he drifted into mysticism, becoming aloof and spending most of his time in prayer or in conversation with the visiting French visionary Barbara Krudener.

Alexander removed himself not only from military and international affairs, but from the domestic arena as well. He remained aloof from the growing revolutionary discontent among his subjects. History records that he died on December 1, 1825 of pneumonia, but a combination of Russian legend and tradition holds that he actually faked his death, escaped to Siberia, and lived anonymously as a hermit until he actually died 1864. Whether the czar died or exiled himself in 1825, revolutionary sentiment coalesced the following month into the Decembrist Revolt, Russia's first revolutionary movement, which attempted but failed to overthrow the government of Alexander's successor, Nicholas I.

ALEXANDER III THE GREAT

BORN: 356 B.C.

DIED: June 13, 323 B.C.

SERVICE: Fourth Sacred War (339–338 B.C.); Consolidation War (336 B.C.); Conquest of Persia (334–330 B.C.); Invasion of Central Asia (330–327 B.C.); Invasion of India (327–326 B.C.)

King of Macedonia; one of the greatest military conquerors of all time

The campaigns of Alexander the Great engulfed the Persian Empire, struck deep into the Indian subcontinent, and laid the foundation for the Greek empire of territorial kingdoms.

The son of **Philip II,** Alexander was born in Pella, Macedonia, and was tutored by no less a teacher than Aristotle. Alexander succeeded to the Macedonian throne in 336 B.C. and immediately turned his attention to matters of security at home, liquidating many of his rivals and consolidating his political power. Following this, in the spring of 334 B.C., he began the military expeditions that would occupy the rest of his life. Determined to liberate the Greek cities in Asia, he crossed the Hellespont with an army of about forty thousand men, conquered the Persian army, subdued his enemies in western Asia Minor, and in July 332 B.C. stormed the city of Tyre. Occupying Palestine and Phoenicia, he turned toward Egypt and subdued it between 332 and 331 B.C.

In 331 B.C., Alexander returned to Tyre, from which he marched his army across Mesopotamia to occupy Babylon. The following year, he entered Media to the north and captured its capital. Facing Persian Emperor Darius III and his Grand Army, Alexander was victorious at the epic battle of Gaugamela in 331 B.C. On the death of the Persian emperor in 330, Alexander assumed the title Basileus (great king).

In midsummer 330, Alexander set off for central Asia and through the next three years subjugated that vast region. In this case, he augmented military might with a diplomatic measure in order to secure what he had gained against continual rebellion and guerrilla actions. He married Roxane, a member of the Iranian nobility, and compelled ninety-one of his top lieutenants to marry Iranian women as well. For good measure, he conscripted 30,000 Iranian boys into his army and encompassed the Iranian cavalry into his own.

During the years of conquest in northeastern Iran and central Asia, Alexander's increasing despotism led to friction with Macedonian nobles and some Greeks. Philotas, the son of Parmenion, who had been Philip II's senior general, was the chief opponent of Alexander's new absolutist policies. As commander of the powerful cavalry, he was a dangerous rival, but, late in 330 B.C., Alexander engineered his elimination through a carefully staged coup d'etat and assassination. Yet another political foe, Cleitus, was killed by Alexander himself in a drunken brawl. Following this, Alexander decreed that his European subjects adopt the Asian custom of prostrating themselves before him. His Macedonian and Greek officers resisted, led by Callisthenes, a nephew of Aristotle who had been appointed official historian of the central Asian campaign. Callisthenes was executed on a charge of conspiracy, and despotic discipline was generally restored.

It is little wonder that Alexander had come to demand the tribute due a despot. He had conquered much of the ancient world and founded eight important cities named for himself—Alexandria—including the most important of all, in Egypt. Once he had restored internal discipline, Alexander set off on his last great campaign, into India. He invaded the Punjab in 327 B.C., then crossed the Indus River in 326, incorporating into his empire territories extending as far as the Hyphasis and lower Indus. The only significant challenge came from King Porus of Paurava, who offered battle at the river Hydaspes in 326 B.C. Alexander's army scored a stunning victory, boldly crossing the river in face of Porus's advancing army, which included large numbers of elephants.

Alexander ended his Indian campaign only after his long-suffering army, victorious in battle, but exhausted by the unremitting advance, refused to follow him farther. A crisis came in 324 B.C. when his officers, angered over the "mixed" army he had created of Macedonians and Iranians and resentful of the marriages they had been forced to make, nearly mutinied.

During the last year-and-a-half of his life, Alexander augmented his army with even more Iranian troops as he prepared to invade Arabia. While making preparations, he fell ill, probably of pernicious malaria, and died in Babylon on June 10, 323. He was thirty-three years old.

Without doubt, Alexander was a superb military leader with no peer among his many opponents. In addition to sheer audacity, his greatest talent as a commander was his ability to conduct lengthy campaigns over great distances, while preserving the ability to attack quickly and with overwhelming force. Alexander was a supremely versatile commander who freely combined a variety of different arms and successfully changed and adapted his tactics to meet the challenge of diverse enemies. Indeed, flexibility and responsiveness were the hallmarks of Alexander's generalship. He consistently understood how to exploit opportunity, foreseen or not. Beyond this, Alexander adhered to the principle of relentless pursuit. Tactical victory was never sufficient for him. His objective was to win the battle, then pursue and crush the opposing army. In this, as in almost all phases of combat, Alexander favored his cavalry, which he used so effectively that infantry was characteristically relegated to a secondary role. Yet his ability as a political administrator fell far short of his military prowess. Clearly, for him, the task of overseeing and managing his vast conquests was secondary to the process of conquest itself. He seems also to have lacked the character of a great civil leader. For example, he made official request to the Greek city-states for deification. To be sure, the elevation of great benefactors was not unheard of in Greek religion, but it was hardly something one asked for oneself. Furthermore, his Greek and Macedonian constituency forever resented Alexander's imposition of Iranian troops, nobles, and culture upon them. Generations of historians have speculated on the fate of Alexander had he lived long enough to rule as well as conquer.

ALEXIUS I COMNENUS

BORN: 1048

DIED: 1118

SERVICE: Byzantine Civil War (1071–81); Norman War (1081–85); Bogomil Revolt (1086–91); Revolt of Constantine Diogenes (1094); First Crusade (1097–1102); Norman War (1104–1108); Seljuk Turk War (1110–17)

Byzantine emperor who was also a fine military commander, skilled at the use of force combined with diplomacy

Alexius was born in Constantinople, the third son of John Comnenus and the nephew of Emperor Isaac I Comnenus. Alexius first saw combat against the Seljuk Turks during 1068–69 and 1070–71 during the reign of Romanus IV Diogenes, then fought against rebels during the insurrections that racked the reigns of Michael VII Ducas (1071–78) and Nicephorus III Botaniates (1078–81). In all of these campaigns, Alexius distinguished himself, and, on behalf of Nicephorus III Botaniates, he won a decisive victory over the rebel Nicephorus Briennus at the Battle of Calavryta in Thrace (1079). Despite this, Nicephorus III proved an inept ruler—a circumstance Alexius used to his advantage. In the free-for-all struggle for the throne that followed the triumph at Calavryta, Alexius defeated all comers, and, on April 4, 1081, was crowned the new emperor of Byzantium.

Almost immediately after ascending the throne, Alexius was faced with threats from the Normans, led by Robert Guiscard, and the Seljuks. Realizing that he could not successfully fight a two-front war, Alexius concluded a peace with the Seljuks, conceding them generous terms. This freed him up to concentrate on Guiscard, who controlled southern Italy. Alexius concluded an alliance with Venice, and sent a combined Byzantine and Venetian fleet to engage the forces of Robert Guiscard off the coast of Durazzo in 1081. While the Normans were defeated at sea, Alexius had to grant heavy concessions to Venice, and, on land, the Byzantine emperor did not fare well. Guiscard's cavalry dealt the Varangian Guard a severe blow in October 1081, and the Normans advanced across Greece. In 1082, an army under Bohemund laid siege to the Byzantine town of Larissa on the Vardar River. Here at last Alexius was able to halt the Norman advance by defeating Bohemund's forces. The war dragged on, however, until 1085, when Robert Guiscard died.

Within a year after the conclusion of the war with the Normans, Alexius was faced with a rebellion by the Bogomils. They found allies among the Petcheneg nomads, and Alexius turned to the Cumans for aid. At first the emperor's forces fared poorly against the combined armies of the Bogomils and Petchenegs. The most stunning defeat came in 1086 at the Battle of Durostorum, at which Alexius personally led the army. Yet it took four more years before the Bogomils and Petchenegs seriously threatened Constantinople (in 1090–91). Alexius again assumed personal command of his forces and this time succeeded in turning back the rebel assault on his capital at the momentous Battle of Mount Levunion (April 29, 1091). The Petchenegs were annihilated—a fact Alexius wielded as a club against the Serbs when they harassed the empire from the north in 1091–94.

In the midst of triumphs against the Bogomil-Petcheneg rebels, Alexius's Cuman allies rose against him in 1094, led by Constantine Diogenes, who claimed to be the son of Romanus IV. Alexius, however, was able to disperse Constantine's forces, and his army forced them across the Danube. On the eastern front, in the meantime, Alexius depended on diplomatic intrigue rather than military means to keep the Seljuks at bay. Through adroit political maneuvering, he kept the Seljuk tribes divided and quarreling among themselves, so that they posed no threat to the empire. Moreover, he concluded a treaty with the major Seljuk leader, Kilij Arslan, in 1091.

Ironically, during 1096–97, the arrival of the forces of the First Crusade—ostensibly Christian allies of the Byzantines—upset the balance Alexius had established in Asia Minor. Nevertheless, Alexius turned the Crusaders' presence to his advantage, judiciously supplying them with aid and supplies, and using them, from May 14–June 19, 1097, to capture Nicaea. This triumph, however, resulted in a jealous dispute between Alexius and the Crusaders, who left the Byzantines in order to continue their advance into Seljuk territory. For his part, Alexius used the momentum of the Crusade to capture Smyrna (Izmir), Ephesus, and Sardos (east of Smyrna) in addition to lesser cities throughout southwestern Asia Minor.

When the Crusaders took Antioch in 1098, which had formerly belonged to the Byzantine Empire, ties between Alexius and the Crusaders were completely severed, save for an alliance with Raymond of Toulouse. Among the Crusaders was Bohemund—whom Alexius had fought in 1082—who demanded that Alexius now assist him in conquering Turkish towns. Alexius, thoroughly alienated from the Crusader cause, declined and ejected Crusaders from various Byzantine territories. Using these rebuffs as pretext, Bohemund returned to Europe in 1105, raised a new army, and attacked at Durazzo in October 1107. Soundly defeated, he surrendered and yielded himself as vassal to Alexius, who granted him Antioch as a fief the following year.

Peace in the Byzantine world was never enduring. In 1110, the Seljuks invaded Anatolia in an effort to retake what Alexius had seized in 1097. At first, the Byzantine forces were repeatedly defeated; however, after some years of fighting, Alexius once again put himself at the head of his forces and led them to a monumental victory—the Battle of Philomelion in 1116. Beaten badly, the Seljuks sought peace. Two years later, Alexius died, leaving the empire far stronger than it was when he had assumed the throne. Still, the decay of the Byzantine Empire was inevitable and inexorable. The diplomatic and military triumphs of Alexius brought no more than fleeting peace.

ALFONSO VIII

BORN: 1155

DIED: 1214

SERVICE: Wars with Muslim Spain (1172–1212)

Monarch of Castile and León, Alfonso was one of the great warrior-kings of medieval Spain

Alfonso was the son of Sancho III, whom he succeeded to the Castilian throne when he was three years old. During his minority, Navarre repeatedly attempted to dominate Castile, but Alfonso's 1170 marriage to Eleanor the Younger of Aquitaine, daughter of England's King Henry II of England, brought Castile an ally that put an end to interference from Navarre. In 1179, Alfonso concluded the Pact of Cazorla, which divided Spain between him and Henry.

In 1188, Alfonso VIII secured the homage of his cousin Alfonso IX of León, with whom he put together a vast army and waged war against the Muslim Almohades, only to suffer a major defeat at the Battle of Alarcos in 1195. When Alfonso VIII and his forces limped back to Castile, he found his realm invaded by Navarre and León as well as by the Almohades, who had taken the offensive. Remarkably, Alfonso managed to rally his forces, who pushed all the invaders out of Castile. Nor did he remain in a defensive posture. Instead, Alfonso regrouped and prepared for an offensive against the Almohades. He managed to secure support throughout the Angevin Empire, from as far away as France and Italy. By 1212, he was once again in command of a huge army, which he marched into Andalusia. At the momentous Battle of Las Navas de Tolosa on July 16, 1212, Alfonso scored a victory that forever ended Almohad power in Spain. Over the course of the next four decades, the Muslims would be pushed back into the tiny kingdom of Granada and the Iberian peninsula emerged as a Christian-dominated region.

Although Alfonso VIII had triumphed at Las Navas de Tolosa, he never personally recovered from the stress and strain of his titanic struggle against the Muslims. Broken in health, he returned to Castile, where he died on October 6, 1214.

ALLEN, Ethan

BORN: 1738
DIED: February 11, 1789
SERVICE: French and Indian War (1754–63), American Revolution (1775–83)

Leader of the "Green Mountain Boys" in the American Revolution

Sources vary as to the exact date and location of Allen's birth. It was either in Roxbury, Connecticut on January 10, 1738 or Litchfield, Connecticut on the 21st. The early death of his father in 1755 cut short Allen's formal education, since he was required to support his family, but he read widely on his own. During the French and Indian War he served at Fort William Henry in 1757, then, in 1769, he acquired land in partnership with his four brothers in an area known as the New Hampshire Grants. Now the state of Vermont, the Grants were then an area disputed between New Hampshire and New York. After he settled in the Grants, Ethan Allen became leader of the Green Mountain Boys, a guerrilla force that, during 1774, waged war against New York settlers in the Grants to protect the holdings of settlers like himself.

The Green Mountain Boys were active when the American Revolution broke out, and Allen put them into service with Benedict Arnold in a successful campaign to capture British-held Fort Ticonderoga on May 10, 1775. Although he was beyond doubt a brilliant guerrilla leader, Ethan Allen was a cantankerous and disagreeable man who was never popular with his troops. They voted him out of command when the Green Mountain Boys joined the Continental Army, but Allen went on to participate in the ill-fated American invasion of Canada. During an assault on Montreal (September 25, 1775), Allen was taken prisoner and sent in chains to England. He was paroled in October 1776, then formally exchanged for a captured British officer on May 6, 1778, at which time he reported to **General George Washington** at Valley Forge, Pennsylvania. Although the commander-in-chief awarded him the brevet rank of colonel in the Continental Army, Allen saw no action in the regular army. Instead, he returned to Vermont, where he was made major general of the militia.

At this point during the Revolution, the status of Vermont was ambiguous. In 1777, it had declared itself a separate republic. The Continental Congress, however, did not recognize its independence, but neither did it act on Vermont's bid for statehood—despite Allen's efforts to stir Congressional interest. In 1779, along with his brother Ira (1751–1814), Ethan Allen approached the British commander at Quebec to obtain a guarantee of Vermont's independence in exchange for nominal status as a British colony. These negotiations came to nothing, and, after the Peace of Paris concluded the Revolution in 1783, he continued to lead the fight against New York's claims to Vermont. His death on February 11, 1789, came before Vermont was granted statehood on March 4, 1791.

ALLENBY, Edmund Henry Hynman, First Viscount

BORN: April 23, 1861
DIED: May 14, 1936
SERVICE: Second Anglo–Boer War (1899–1902); World War I (1914–18)

One of Britain's best field commanders in World War I, Allenby was in charge of operations in the Middle East

Allenby was born in Nottinghamshire and joined the Inniskilling Dragoons in 1882. With this unit, he served in the Bechuanaland (Botswana) expedition of 1884–85. At the turn of the century, during 1900–1902, he saw service in the Second Anglo–Boer War on the staff of Lord **Kitchener** and in the field. In 1902, he returned to England to command of the Fifth Lancers (1902–1905).

At the outbreak of World War I, Allenby was put in command of the cavalry division (later the cavalry corps) of the British Expeditionary Forces (BEF) during August–November 1914. He then commanded V Corps before receiving a promotion to command of the

Third Army in October 1915. He performed brilliantly in the Arras offensive during April 9–15, 1917, but his individualistic style and boldly aggressive methods were not popular with the British high command. He was removed from the Western Front, the war's principal theater, and sent to the Middle East in June 1917, as a replacement for Sir Archibald Murray as commander of British forces in Egypt. Allenby's arrival immediately boosted the morale of troops in this backwater of the war, and he set about with great energy to take the offensive by concentrating his forces against the Turkish inland flank. On October 31, he scored a signal victory in the Third Battle of Gaza, executing a surprise attack at Beersheba, then brushing aside Turkish resistance in a headlong advance on Jerusalem. His attack on the Turkish position at Junction Station during November 13–15 was a masterpiece that resulted in a Turkish rout and the capture of a crucial water supply. At last, on December 8, Allenby made an all-out assault on Turkish defenses in front of Jerusalem. After two days of combat, he entered the Holy City.

Unfortunately for Allenby, the demands of the European war drained off many of his troops at this point, and, during 1918, he was unable to continue his advance. His attempt to take Amman failed twice, during March 20–25 and April 30–May 3. However, after he belatedly received reinforcements, he resumed the offensive during July and August, using anti-Turkish Arab guerrilla troops as a diversion while he launched his main offensive near the coast. This achieved total surprise, and Allenby was able to smash the Turks' front line at Megiddo during September 19–21. Maintaining his momentum, he pursued the fleeing Turks from September 22 to October 30. During this pursuit, he took Damascus (October 1), Homs (Hims; October 16), and Aleppo (October 25).

Fighting in an often-neglected theater of the Great War, Allenby scored such sweeping victories in Palestine and Syria that Turkey withdrew from the war on October 30, 1918. That very month, Allenby was made a viscount and the following year was named high commissioner for Egypt. He served in this capacity until 1925. Even though he had fought for much of the war on a secondary front, Allenby was recognized as one of the great generals of World War I and one of England's most accomplished commanders.

ANDERSON, William C. ("Bloody Bill")

BORN: 1840

DIED: October 26, 1864

SERVICE: Civil War (1861–65)

Confederate guerrilla in U.S. Civil War

Missouri-born William C. Anderson was sprung from obscurity at the outbreak of the U.S. Civil War. After the death of one of his sisters in the collapse of a Kansas City prison for Southern sympathizers, he joined a partisan band from Clay County, Missouri, which later melded with **William Clarke Quantrill's** "Raiders." During raids against Union forces occupying Missouri, Anderson—along with Quantrill and other guerillas—quickly gained a reputation as a cold-blooded killer. After the Raiders' infamous August 21, 1863, attack on Lawrence, Kansas, Quantrill commissioned Anderson first lieutenant. The two soon fell to quarreling, however, and following an argument about the summary execution of one of Anderson's men, Anderson formed a renegade band of his own. "Bloody Bill," as he had become known, raided Union strongholds along the Missouri River and by late summer of 1864 had led his followers to join Confederate General Sterling "Old Pap" Price's cavalry in an incursion into southern Missouri. Union troops, belatedly reinforced, ultimately halted Price's march across the state outside St. Louis, but his progress sent shock waves through the Union command, which had earlier declared the area pacified.

Anderson's most infamous exploit was his raid on Centralia, Missouri, on September 27, 1864. After pillaging and burning the town, his 225 partisans stopped a noon train coming down the Wabash, St. Louis, and Pacific Railroad line. They murdered the engineer and fireman and then looted an estimated $3,000 from the express car. In the coaches, Anderson found twenty-five Union troops on furlough, disarmed them, lined them up, and killed all but one of them. He and his men then murdered several civilians who had tried to hide their valuables. Pursued later that day by 200 Union troops, Anderson's men mounted a devastating counterattack, killing some 116 of their pursuers. Among the Centralia raiders were two guerrillas later to become notorious postwar outlaws, Frank and Jesse James.

"Bloody Bill" Anderson died in an ambush by a Union militia company in Orrick, Missouri, on October 26, 1864. Union soldiers, outraged by the Centralia massacre and Anderson's reign of terror, decapitated him and stuck his severed head on a spiked telegraph pole in Richmond, Missouri.

ANSON, George, Lord

BORN: April 23, 1697
DIED: June 6, 1762
SERVICE: War of the Austrian Succession (1740–48)

The "Father of the British Navy"

Anson was born in Shugborough, Staffordshire, England, and joined the navy in 1712. His rise through the ranks was meteoric, and he became a captain in 1723. In 1740, he led a flotilla of six ships in a round-the-world voyage spanning September 1740 to June 1744. During this expedition, he captured the Manila galleon *Nuestra Señora de Cobadonga* in 1743 and sold it in Canton for £400,000. However, of the six vessels that undertook the circumnavigation, only the flagship *Centurion* completed the voyage and, at that, half of its crew perished from scurvy.

When Anson returned to England on June 15, 1744, he was hailed as a hero and given a post on the admiralty board, serving from 1745 until his death in 1762. In 1751, he was appointed First Lord of the Admiralty, from which post he completely reorganized the Royal Navy, fashioning it into the ever more formidable instrument it would become throughout the eighteenth century. Anson also created the British system of classifying ("rating") war ships, he was instrumental in a fundamental revision of the Articles of War in 1749, and he created the Royal Marines in 1755. Anson was at sea in 1747 and from 1756–57. His greatest feat was a battle with a French squadron off Cape Finisterre on May 14, 1747, where he captured six capital vessels and eight frigate escort ships.

Anson lacked the warmth, passion, and sheer dash of the Royal Navy's later—and far more famous—hero, Lord **Horatio Nelson.** However, it was his work that made possible the British navy that played such a key role in the creation of the Empire.

ANZA, Juan Bautista de

BORN: 1736
DIED: 1788
SERVICE: Campaigns against the Apaches (1759–88)

Spanish soldier, explorer, and governor of Spanish New Mexico

A highly competent, well-respected military officer, Juan Bautista de Anza is perhaps best known for opening an overland trail from the presidio of Tubac (Arizona) to California in 1774, after which he led a major colonizing expedition to California and founded the presidio and mission of San Francisco. He was born at the presidio of Fronteras, a hundred miles southeast of Tubac, the son of the presidio captain, a Basque descended from elite Spanish ranchers in Sonora. Anza entered military service at Fronteras in 1753, was promoted to lieutenant two years later, and distinguished himself in a campaign against the Apaches in 1759, which led the following year to his being named captain of the presidio of Tubac. He settled into his military duties—protecting the Jesuit missions and keeping the native populations of Seris, Apaches, and Papagos under control—and quickly distinguished himself as an able leader, skilled in battle, steady in command, and respected by his men.

In 1768, Franciscans took over Jesuit properties in Spanish America, and Anza soon met Father Francisco

Garces, who wanted to open a mission at the junction of the Gila and Colorado rivers. Anza accommodated him with a military expedition that opened the route from Tubac to Monterey in 1774 and established friendly relations with the Yuma Indians. In October of the following year, Anza trekked north to Monterey with 244 men, women, and children, returning to Sonora in 1776. Already promoted to colonel and made commandant of arms of the Province of Sonora back in 1771, he became governor of New Mexico by 1778. As head of the colony, one of his major jobs was to protect his subject from Indian attacks. The Apaches had always been a problem, and remained one. The Sonora-California route he had opened was itself closed after the Yuma Massacre of 1781. Nevertheless, in 1786 Anza managed to arrange a lasting peace with the Commanches and, at the same time, enter into an alliance with the Utes and Navajos in an attempt to control the Apaches.

Anza never gave up on the Apaches, continuously alternating between pursuing peace initiatives with them and launching military campaigns aimed at bringing them to terms. He died in Arizpe shortly after his Spanish soldiers had completed one such campaign.

ARNIM, Count Hans Georg of

BORN: 1582?

DIED: April 28, 1641

SERVICE: Russo–Swedish War (1613–17); Turco–Polish War (1617–21); Swedish–Polish War (1617–29); Thirty-Years' War (1618–48)

Skilled general in the service of the Saxons and the Austrians (Holy Roman Empire)

Arnim was born of noble lineage at Boitzenburg, Mecklenburg, into the fragmented world of German states and principalities. He trained as a professional soldier and entered the service of Sweden in 1613, fighting in Russia and Livonia. A dispute with the Swedish monarch Gustavus Adolphus in 1621 prompted Arnim to leave his service and to fight Turks in Poland, but he returned to the Swedes the following year and fought against the Poles. In 1626, he joined the army raised by **Albrecht Eusebius Wenzel von Wallenstein** in the service of the Holy Roman Empire and fought brilliantly in Silesia (1626–27) and Mecklenburg–Holstein (Sept-ember–October 1627). Arnim participated in Wallen-stein's long, failed siege of Stralsund during February 24–August 2, 1628, in the course of which he received a promotion to field marshal on April 21, 1628.

Arnim went to Poland in 1629 with imperial troops to fight Gustavus, achieving victory against him at the Battle of Sztum on June 17, 1629. However, religious differences with the Holy Roman Empire prompted the Lutheran Arnim to resign his commission in protest against the Edict of Restitution of March 29, 1629. He offered his services next to John George, Elector of Saxony, and was active in the creation of a "third party" to balance imperial power and the influence of France and Sweden. Arnim became field marshal and commander in chief of the Saxon army on June 21, 1631, and he led the Saxon forces at the First Battle of Breitenfeld on September 17, 1631, going on from this triumph to capture Prague in November 1631. During the spring of the following year, however, his former employer, Count Wallenstein, pushed him out of Bohemia. By this point, however, Arnim had become discontent with the policies of his Saxon masters, and he negotiated a new alliance with Wallenstein. This proved abortive when Wallenstein was assassinated on February 24, 1634, and the following year, after Saxony concluded the Peace of Prague with the Holy Roman Empire on May 30, Arnim resigned his Saxon commission.

Arnim rejected a commission in the French army and was arrested in 1637 by the Swedes, who took him to Stockholm. He managed to escape, however, and fled to Hamburg, where he sought and received reinstatement as an Imperial and Saxon lieutenant general in 1638. As he was preparing for operations against Sweden and France, he suddenly died on April 28, 1641.

ARNOLD, Henry Harley ("Hap")

BORN: June 25, 1886
DIED: January 15, 1950
SERVICE: World War I (1917–18); World War II (1941–45)

Instrumental in the creation of the U.S. Air Force

A native of Gladwyne, Pennsylvania, Arnold attended West Point, from which he graduated in 1907, and was commissioned an officer in the infantry. His first service was in the Philippines during 1907–1909, but he became passionately interested in flying and was a pioneer in military aviation. He transferred to the aeronautical section of the Signal Corps in April 1911 and received flight instruction from the Wright brothers themselves that June. A natural flier, Arnold won the Mackay Trophy for successfully completing the first reconnaissance flight in a heavier-than-air craft in October 1912.

Despite the efforts of Arnold and other military aviation enthusiasts, the army had little faith in the airplane, and Arnold, disappointed and disgusted, returned to the infantry in April 1913. Three years later, however, he was returned to the air service, where, promoted to captain in May 1916, he supervised the army's aviation training schools when the United States entered World War I. He functioned in this capacity from May 1917 through 1919. Despite sharp reductions in military funding during the years 1919–28, Arnold continued to work toward developing the Army Air Corps. He was sent to the army's Command and General Staff school, from which he graduated in 1929 with the rank of lieutenant colonel. In 1931, he was given command of the First Bomb Wing and the First Pursuit Wing at March Field, California, a post he held through February 1935. During July and August 1934, he led a flight of ten B-10 bombers on a round trip from Washington, D.C., to Fairbanks, Alaska, winning a second Mackay Trophy for his demonstration of the endurance of the modern bomber.

Promoted to brigadier general, Arnold took command of First Wing, GHQ Air Force, in February 1935, and became assistant chief of staff of the Air Corps in December of that year. With the death of General Oscar Westover in September 1938, Arnold was given a temporary promotion to major general and named chief of staff. He acted vigorously from this position to improve the combat readiness of the Air Corps, but was severely hampered by a shortage of funds and a lingering reluctance on the part of military planners to develop a fully effective air force. Nevertheless, in recognition of his effective advocacy of air power, Arnold became acting deputy chief of staff of the army for air matters in October 1940, and chief of the Air Corps after it was renamed the U.S. Army Air Forces in June 1941. This position carried with it a temporary promotion to lieutenant general, conferred shortly after the bombing of Pearl Harbor and America's entry into World War II.

In March 1942, Arnold was named commanding general of Army Air Forces, and the following year was given the temporary rank of general. Arnold served on the U.S. Joint Chiefs of Staff, playing a key role in shaping Allied strategy in the European and Pacific theaters. He organized the 20th Air Force in April 1944, to bomb Japan. The 20th answered directly to his command as a representative of the Joint Chiefs. This was a particularly significant step, since it took the Army Air Forces closer to becoming an independent service.

In December 1944, with generals **Dwight D. Eisenhower, Douglas MacArthur,** and **George C. Marshall,** Arnold was elevated to the rank of general of the army. He continued to command the Army Air Forces through the end of the war, retiring in March 1946. On September 18, 1947, thanks in large part to the foundation Arnold had laid, the Army Air Forces became an independent service: the U.S. Air Force. In recognition of the role he played, Arnold, in retirement, was named first general of the Air Force in May 1949. He died the following year on his ranch in Sonoma, California.

ATATÜRK
(Mustafa Kemal, Kemal Atatürk)

BORN: March 12, 1881

DIED: November 10, 1938

SERVICE: Turco–Italian War (1911–12); Balkan Wars (1912–13); World War I (1914–18); Greco–Turkish War (1920–22)

Turkish general and statesman; the father of modern Turkey

Atatürk almost singlehandedly transformed his nation from the backward-looking, Islamic state of its Ottoman past to a reform-minded, modern European country. Born in Salonika, Greece (at the time part of Ottoman-controlled Macedonia), Atatürk was the son of a lower-middle-class customs clerk married to a peasant wife. The early death of his father left the family in straitened circumstances, and Kemal endured a difficult, impoverished childhood. He was accepted at the state military school, moved on to the Senior Military School, and, finally, in 1899, enrolled in the Ottoman Military Academy in Istanbul. There, in addition to advanced military training, he read the likes of Rousseau, Voltaire, Hobbes, and other seminal social thinkers of the West. The result was a growing dissatisfaction with the moral and political bankruptcy of the ailing Ottoman empire.

At the age of twenty, Kemal was promoted to the General Staff College, where he and like-minded young officers founded the Vatan, a secret society with revolutionary leanings. When the Vatan failed to develop beyond a discussion group, Kemal joined the Committee on Union and Progress, which was aligned with the Young Turk movement. While he was not directly associated with the Young Turk Revolution of 1908, which forced Sultan Abd Al-Hamid II to restore the constitution and parliament of 1876 and rule more democratically, Kemal worked closely with many of the revolution's key leaders.

With the outbreak of World War I, Kemal was appalled by the sultan's alignment of the Ottoman Empire on the side of the Germans, whom he detested. Nevertheless, as an army officer, he set aside his personal views and led his troops with consummate skill on every Ottoman front during the war. At Gallipoli, beginning in April 1915, he held off the British and ANZAC forces for more than a month, earning him the name of "Savior of Istanbul." From there, he took control of the Second and Third Armies, stopping the advance of Russian forces south in the Caucasus in 1916. He was in command of the Seventh Army against the British at Aleppo when the war ended in 1918.

World War I was ruinous for the old Ottoman Empire. Long called by other nations the "Sick Man of Europe" because years of outmoded autocracy had led to hopeless internal decay, the Ottoman Empire loomed as a prize to be carved up by the victorious Allies. The terms of the armistice ending the war were harsh, and they included a secret agreement concerning the partition of the Ottoman Empire. For the time being, the Allies wished to maintain the semblance of Ottoman sovereignty, the better to exploit what Turkey had to offer, and many within the empire also hoped to maintain the sultanate under foreign regency. Kemal and others like him, however, wanted to create an independent state, sloughing off the rotten trappings of empire. Dispatched to Anatolia in 1919 to put down insurrection there, he instead organized the dissent into a coordinated movement against the numerous "foreign interests" that had invaded the country following World War I. He set up a provisional government in Anatolia, of which he was elected president, and fostered unified resistance to foreigners. In response, the sultan ordered a holy war against the Nationalists and in particular called for Kemal's execution.

In August 1920, the sultan signed the Treaty of Sèvres, effectively giving the Ottoman Empire to the Allies in return for his continued power over its remnants. It was a fatal blunder; for, with a stroke of the pen, the sultan consolidated vast popular support for Kemal's nationalism. Seizing the moment, Kemal marched his Nationalist Army against Istanbul. The Allies, seeking to hold their prize, looked to Turkey's perennial foe, Greece, for help. After eighteen months of costly combat, the Greek invaders were finally defeated in August 1922.

On November 1, 1922, the Grand National Assembly dissolved the sultanate of Mehmed VI.

Kemal's principal lieutenant, Ismet Inonu, negotiated the Treaty of Lausanne, concluded on July 24, 1923, whereby international recognition of the new republic of Turkey was achieved and Turkey regained not only Smyrna but also eastern Thrace and some of the Aegean islands. The new republic also resumed control of the Dardanelles, which had been internationalized under the Sèvres treaty. On October 29, 1923, Mustafa Kemal was elected president of the republic and a grateful nation bestowed upon him the title of Atatürk—literally, Father of the Turks.

Called president, Atatürk was a strongman and dictator, who outlawed all opposition, abolished rival political parties, and engineered repeated reelection until his death. All of this he did, however, for what he clearly considered was the good of his country. His so-called program of "Kemalisms" consisted of republicanism; secularism, whereby Islam was removed from official life and Turkish law; populism, which favored the common man; nationalism; statism, which revamped the ailing economy; and reformism, a more broadly cultural movement. Atatürk opened up the Turkish economy to the industrialized West, secularized the state, ended the Islamic suppression of women, and generally forced the Western way of life on his people, much as **Peter the Great** had done in Russia two centuries earlier.

Despite his stern and absolutist approach to reform, most of his countrymen welcomed the measures he introduced and were willing to overlook their president's personal failings, which included sexual indiscretion and alcoholism. Mustafa Kemal Atatürk died of cirrhosis of the liver on November 10, 1938.

ATKINSON, Henry

BORN: 1782

DIED: June 14, 1842

SERVICE: Black Hawk War (1832)

American general known for the military expeditions he led in the Yellowstone region

Henry Atkinson was born in North Carolina, the son of a tobacco grower, and served as an army officer from 1808 until his death.

From 1808 to 1813, he was a captain in the 3rd Infantry, stationed along the coast of the Gulf of Mexico. In April of 1813, he was transferred to the New York-Canada border region, and two years later was promoted from staff officer to the rank of colonel in the 6th Infantry. There he stayed the rest of his career.

Atkinson first came to the American West in 1819, when he was assigned to an expedition under orders to move troops northwest from St. Louis to the mouth of the Yellowstone River in eastern Montana. Steamboats carrying his men and supplies broke down, and he was forced to halt his troops just north of present-day Omaha, Nebraska. He threw up the rude Fort Atkinson, where, briefly, in 1820 he would become a brigadier general before army staff reductions cost him the rank. Five years later, he and an Indian agent named Benjamin O'Fallon led another expedition, 475 men strong, up the Missouri to the Yellowstone. This time they succeeded, reaching the stream's headwaters and negotiating peace treaties with more than a dozen Indian tribes.

Back in St. Louis, Atkinson spent his time supervising the construction and operation of the infantry school at Jefferson Barracks, and in 1827 he was called upon to prevent a war between miners and angry Winnebagoes in Wisconsin. He was in charge of the rag-tag troops that ran the famous Sauk war leader to ground in the Black Hawk War of 1832, and over the next decade he directed the removal of the Indian tribes of the Old Northwest beyond the Mississippi River. He died suddenly at age sixty at Jefferson Barracks, having secured "peace" and American military power in the Upper Mississippi and Missouri River valleys while all but destroying the societies and cultures of the area's Native Americans.

ATTILA THE HUN

DIED: 453
SERVICE: Wars with Rome (440s–452)

Celebrated conqueror whose name became a byword for rapacious and ruthless military action.

Called in his own time the "Scourge of God," Attila the Hun became, with his brother Bleda, joint chieftain of the Huns following the death of their uncle Ruas in 433. Ruas had led the Huns, a warlike tribe of Germanic nomads, to a position of such great power in central Europe that the Roman emperor Theodosius II was compelled to conclude a treaty with him whereby Rome paid an annual tribute and Ruas was made a general in the Roman army. The latter provision effectively ceded to the Huns sovereignty over the province of Panonia (modern Hungary). Following the death of Ruas, Attila and Bleda renewed the treaty with Theodosius, increasing the amount of the tribute to 700 pounds of gold annually. In return, the brothers gave the Roman Empire a half-dozen years of peace, as they prosecuted wars of conquest against Scythia, Media, and Persia.

During 441–43, the Vandal king Gaiseric offered Attila a large bounty to invade Rome's Eastern Empire. Attila readily violated the treaty with Theodosius and marched into Illyricum, leading his Huns into Moesia and Thrace and advancing to the walls of Constantinople itself. From those mighty walls, the city's defenders succeeded in repelling the invaders, and Attila turned away in order to destroy the imperial army of Aspar and ravage the Balkans. At last, a desperate Theodosius concluded another treaty in August 443, this one conceding an even greater tribute payment.

Two years after his triumph in the Balkans, Attila moved against his brother, assassinating him in order to become sole ruler of the Hun empire, which now stretched from southern Germany in the west to the Volga or Ural Rivers in the east and from the Baltic in the north to the Danube, Black Sea, and Caucasus in the south. Yet even that did not satisfy Attila's rapacity, and in 447, he led his armies against the beleaguered Eastern Empire. The capital, Constantinople, was thrown into panic because its walls, on which the defenders successfully relied during the first invasion, had been badly damaged by a recent earthquake. East Roman forces for once managed to divert Attila's advance, however, at the Battle of Utus. Ultimately, the imperial army was again defeated, but the action succeeded in sending Attila's Huns toward Greece and away from Constantinople.

Highly idealized depiction of Attila the Hun. From Charlotte M. Yonge, Pictorial History of the World's Great Nations, *1882.*

In Greece, Attila surrounded the fortress city of Thermopylae, but, as had happened previously outside of Constantinople, the Hun was unable to breach its walls. Yet, once more, Theodosius negotiated terms of peace, by which Attila extorted three times the original tribute amount as well as the cession of a fifty-mile-wide strip along the right bank of the Danube, from Singidunum (modern Belgrade) to Novae (Svistov, in modern Bulgaria).

Feeling that he had wrung from the Eastern Empire about all he could, Attila turned in 450 to the West. He was encouraged to invade the Western Empire by the Vandal leader Gaiseric, who wanted an ally against Rome. Moreover, one of two contending heirs to the Frankish throne also asked Attila for an alliance. Finally, the Hun sought vengeance for an insult he had suffered from Valentinian III, ruler of the Western Empire. Attila had sought the hand in marriage of Valentinian's sister Honoria, only to be rebuffed by the emperor. With his western alliances in place, Attila crossed the Rhine in 451 and attacked Gaul with a force reported to have numbered half a million—but which modern scholars believe to have been closer to 100,000.

Attila attempted to persuade the Visigoth chieftain Theodoric to join the battle, but the Roman general **Aetius** convinced Theodoric to ally the Visigoths with Rome. Thus Aetius was able to assemble against the Huns a coalition of imperial Roman forces, Visigoths, and others, principally the Alans, fickle kinsmen of the Huns. During May and June of 451, Attila besieged Orléans, which tottered on the verge of surrender when Aetius arrived to relieve it. Not only did he chase off Attila's bands, but he pursued the Hun's forces as far as the Catalaunian Plains, near Châlons-sur-Marne. There Aetius won a decisive victory at the Battle of Châlons during mid June 451. It was, in fact, a contest of momentous consequences, for a Hun victory would have spelled the end of Roman and Christian civilization and might well have signaled the commencement of the Asian domination of Europe. But Attila was, for the first time in his military career, badly defeated. Aetius permitted the remnant of his army to withdraw from the field and to retreat.

Undaunted, Attila once again demanded the hand of Honoria the following year, and, once again, he was rebuffed. In response, the Hun invaded Italy, destroying Aquileia and forcing the withdrawal of the people of Venetia, who, fleeing to islands off the Italian coast, contributed to the founding of Venice. Attila's hordes annihilated Padua and advanced on Minicio. General Aetius rushed back to Italy from Gaul, but he could do relatively little with the small force that marched with him.

Fortunately for the Romans, however, Attila received word that forces under command of one of his lieutenants in Illyricum had suffered a serious defeat. His own army in Italy was languishing in famine and pestilence. At this juncture, Pope Leo I visited Attila in camp. What happened inside Attila's tent is not known. Historians have speculated that Leo made a new offer of tribute. Others have suggested that Attila feared—and quite rightly—that the disease and starvation around him would destroy his army. Perhaps, as Catholic tradition has it, Attila was overawed by the holy majesty of the pontiff. For whatever reason or reasons, the Hun quickly withdrew from Italy.

The following year, Attila, wearied and discouraged, died. His vast empire almost immediately shattered into fragments as his sons squabbled over the throne, and as tributary tribes, including the Ostrogoths and the Gepidae, rebelled.

AUCHINLECK, Sir Claude John Ayre

BORN: 1884

DIED: March 23, 1981

SERVICE: World War I (1914–18); World War II (1939–45)

British field marshal most active in India

Auchinleck was the son of an army officer and, destined for a military career, was educated at Wellington and Sandhurst. He became an officer in the Indian army and saw service during World War I against Turkish forces in the Middle East. During the Great War, he rose rapidly through the ranks, becoming a lieutenant colonel by 1917. At the conclusion of the war, he was appointed to a teaching position at the Staff College, then returned to India. After attending the Imperial Defence College in 1927, he was assigned to command the First Battalion of the First Punjab Regiment during 1929–30, then taught at the Quetta Staff College during 1930–33. As commander of the Peshawar Brigade, he returned to the Northwest Frontier during 1933–36 to fight rebellious frontier tribesmen.

In 1936, Auchinleck became deputy chief of the General Staff at Indian Army headquarters in Simla, taking command of the Meerut District two years later. Promoted to major general in January 1940, he returned to England as commander of the ill-fated Anglo-French expeditionary force at Narvik, Norway. Auchinleck supervised the evacuation of the force in June and was returned to India to command all British forces there.

Auchinleck was named commander in chief of British forces in the Middle East in June 1941, but he soon lost the confidence of **Winston Churchill** because of his repeated failure to take the offensive. Auchinleck, who, after the Norway expedition, knew the folly of operating with unprepared forces, protested that he needed more time to forge his resources into an effective army. The fall of Tobruk in January 1942, was a severe blow that fatally undercut his position, even though he did stop **Rommel's** advance toward the Nile at El Alamein in June. However, in July, he was replaced by General Harold Alexander and returned to India, where he became commander in chief of operations there.

Despite his arguably indifferent performance in the Middle East, Auchinleck's achievements were recognized after the war by a promotion to field marshal in 1946.

B

BAINBRIDGE, William

BORN: May 7, 1774
DIED: July 27, 1833
SERVICE: Franco–American "Quasi-War" (1798–1800); Tripolitan War (1801–1805); War of 1812 (1812–15)

American naval officer who commanded the legendary frigate,* Constitution *("Old Ironsides")

William Bainbridge was born in Princeton, New Jersey, went to sea at fifteen, and became captain of a merchantman by 1790. He received a commission as an officer in the newly created U.S. Navy in 1798 during the so-called "Quasi-War" with France that followed the breakdown of Franco–American relations during the XYZ Affair. As a naval lieutenant, he commanded the schooner U.S.S. *Retaliation*, which was captured by French privateers in September. Briefly held prisoner in Guadeloupe, he was released and put in command of the brig *Norfolk*, from which he brilliantly led convoys of American merchantmen through blockades in the French West Indies. Promoted from master commandant to captain in 1800, he was given command of the frigate *George Washington*, assigned, during October 1800, to carry tribute money to the Barbary pirates in Algiers to ensure safe passage of American merchant vessels. When the pasha of Tripoli—one of the Barbary states—declared war on the United States, Bainbridge was given command of the U.S.S. *Philadelphia* and assigned to Commodore Edward Preble's squadron during the Tripolitan War (1801–05) against the so-called Barbary pirates. Unfortunately, in October 1803, the *Philadelphia* ran aground off Tripoli while pursuing an enemy vessel, and Bainbridge was compelled to surrender the vessel. He was imprisoned in Tripoli, gaining release with other American prisoners at the conclusion of the war in June 1805.

After his return to the United States, Bainbridge entered merchant service again, but was recalled to naval service in 1808 as captain of the frigate *President*, which he commanded through 1810. At the outbreak of the War of 1812, Bainbridge was given command of the Charleston Naval yard, then succeeded **Isaac Hull** as captain of the forty-four-gun frigate *Constitution*. He captured the British frigate *Java* off the coast of Brazil on December 29, 1812, and returned with it as a prize. Back in Charleston, Bainbridge supervised construction of a new frigate, the seventy-four-gun *Independence* and commanded her after the conclusion of the war. During the War of 1812, Bainbridge was an advocate of two important strategies that greatly increased the effectiveness of the grossly outnumbered and outgunned U.S. Navy: fleet dispersion and commerce raiding.

In addition to commanding the *Independence* after the War of 1812, Bainbridge was assigned important shore posts, including command of the Philadelphia Navy yard. In 1817, he created, at Charleston, the first school for U.S. Navy officers, then commanded the ship-of-the-line *Columbus* in the Mediterranean during 1820–21. During 1832–33, he was president of the Board of Naval Commissioners.

BARRY, John

BORN: 1745?
DIED: September 13, 1803
SERVICE: American Revolutionary War (1775–83); Franco–American "Quasi War" (1798–1800)

One of the key naval officers of the American Revolutionary War

A native of Ireland (County Wexford), Barry went to sea as a youth around 1756 and immigrated to Philadelphia by 1760. There he set up as a prosperous merchant captain and was active in the cause of American independence. After war broke out, he was commissioned a captain in the Continental Navy in March 1776, and assigned command of the thirty-two-gun U.S.S. *Lexington*. He captured H.M.S. *Edward* off the Virginia Capes on April 7, but was subsequently blockaded by the British in Delaware Bay. Undaunted, he took up arms and fought ashore at the battles of Trenton (December 26, 1776) and Princeton (January 3, 1777). Back at sea, Barry was forced to scuttle his ship, U.S.S. *Effingham*, to prevent her capture in September 1777. He made up for this loss by leading a boat party down the Delaware River, where he captured five British vessels, including the ten-gun H.M.S. *Alert* near Fort Penn on February 26, 1778.

Assigned command of the thirty-two-gun U.S.S. *Raleigh* in September 1778, he was outgunned and defeated by H.M.S. *Experiment* (fifty guns) and H.M.S. *Unicorn* (twenty guns) in Penobscot Bay during a series of engagements from September 24–27. As master of the *Alliance* (thirty-two guns), however, he captured many prizes in the Atlantic and Caribbean during a cruise that spanned 1781–83. Barry also fought the final naval action of the Revolution, engaging the H.M.S. *Sybill* on March 30, 1783. He broke off when other British ships approached.

Barry retired to merchant service, but was appointed senior captain in the newly created U.S. Navy in 1794 and given command of forty-four-gun frigate U.S.S. *United States*, then assumed command of the West Indies squadron during 1797–99. While the United States and France fought the undeclared "Quasi-War," Barry took the U. S. peace commissioners to Europe during fall 1799. After this, Barry's health declined, forcing his retirement from active command. He returned to Philadelphia in 1801 and died there two years later. In addition to his great skill and daring as a naval commander, Barry was instrumental in helping to shape the U.S. Navy and in training some of its future officers, the most prominent of whom was **Stephen Decatur**.

BELISARIUS

BORN: ca. 505

DIED: 565

SERVICE: Persian War (524–32); Vandal War (533–34); Italian (Gothic) War (534–54); Persian War (539–62); Bulgar invasion (559)

Able and loyal general to the Byzantine emperor Justinian I

Belisarius was a native Thracian who joined the army at a very young age and was entrusted with his first command in 530, when he led forces that would defeat the Persians at Daras. In April 531, he was not so fortunate, suffering defeat at the Battle of Callinicum (near Urfa), but, during the late summer and fall of that year, he successfully resisted additional Persian attacks at Sura. His bold defense brought the Persians to the peace table.

In 532, Belisarius was instrumental in the suppression of an uprising at Nika and the next year was given command of an expedition against the Vandals in North Africa. He established his staging area on the often-disputed island of Sicily, then launched an African invasion with 15,000 men early in September. At the Battle of Ad Decimum on September 13, he defeated the army of Gelimer and marched on Carthage, taking that city on the 15th. Late the next year, in December 533, he once again faced Gelimer (reinforced by his brother Tzazon) in battle at Tricamerum. Triumphant there, he ultimately captured the troublesome Vandal leader during March 534 and returned with him to Constantinople.

Hailed for triumphs, Belisarius was next given command of a modest expedition to Italy. In 535, with a force of 8,000, he laid successful siege to Palermo and went on once again to Africa in order to suppress a mutiny there during the spring of 536. That accomplished, he resumed his work in Italy, taking Naples during the summer of 536 and Rome on December 10.

Belisarius was called upon to defend Rome against the Goths, led by Vitiges, during 537–538. He managed to break the long siege and then took the offensive, in turn laying siege to Vitiges in Ravenna during 538–539, ultimately forcing his surrender. Yet, at the height of his effectiveness, Belisarius was suddenly recalled to Constantinople by Emperor **Justinian I,** who had grown jealous of his popularity and influence. He was redispatched to remote Mesopotamia to fight against the Persians in 542, and he succeeded in driving Persian forces out of Lazica (southeastern Turkey) by 544. As usual, he was quick to seize the initiative, and began a series of raids into Persia.

After a truce was concluded with the Persian leader Persiaos in 545, Belisarius was hurriedly recalled to Italy, where he failed to halt the recapture of Rome by the Goth Totila. However, during the next year, he recaptured the city and held it against Totila's repeated assaults. Yet Belisarius was unable to capitalize on his gains, since Justinian, again fearing too great a success for Belisarius, declined to send him badly needed reinforcements. Finally, he was recalled to Constantinople in 549, only to be called out of retirement in 554 to consummate the subjugation of southwest Spain. This accomplished, he again retired from active service—and was yet again recalled in 559 when a combined army of Bulgars and Slavs (Zabergan at Melanthius) invaded Moesia (northern Bulgaria) and Thrace.

For his pains, Justinian hastily imprisoned Belisarius on unfounded charges of treachery in 562; he was, however, quickly restored to favor in 563 and lived out the remainder of his life in comfort and with the gratitude of the emperor.

BERTHIER, Louis Alexandre

BORN: November 20, 1753

DIED: June 1, 1815

SERVICE: American Revolutionary War (1775–83); Wars of the French Revolution (1792–99); Napoleonic Wars (1800–15)

One of Napoleon's great marshals

Berthier was born in Versailles, the son of a geographical engineer. He followed his father's profession, receiving his military commission as an engineer on January 1, 1766. Four years later, he joined the army and during 1780–82, served with **General Jean Baptiste Rochambeau,** who led French land forces to aid France's new ally, the revolutionary republic of the United States, in its war for independence from Great Britain. Berthier rose quickly through the ranks, becoming a staff officer for Rochambeau and for the **Marquis de Lafayette** during 1791–92. However, his affiliation with Lafayette put him in disfavor with the new republican government of France, and he was dismissed from the army in August 1792.

The following year, Berthier nearly redeemed himself through his service against the Vendée rebels, but, still distrusted as a noble, he was removed after four months' service. By March 1795, Berthier had been reinstated yet again, with the rank of brigadier general, and was appointed chief of staff to General François E. C. Kellerman, commander of the Army of the Alps and Italy during the Wars of the French Revolution. When Bonaparte stepped into the picture, Berthier was promoted to major general and chief of staff to **Napoleon**'s Army of Italy on March 2, 1796, and he was at his commander's side throughout the triumphant Italian campaign. When Bonaparte left for Paris in December 1797, Berthier became commander of the Army of Italy. He occupied Rome and set up the Roman republic during the following February.

From July 1798–August 1799, Napoleon took up his Egyptian campaign, with Berthier again serving him as chief of staff. When his commander returned to France, Berthier accompanied him and was named minister of war after Eighteenth Brumaire (November 9, 1799)—Napoleon's successful overthrow of the French Directory and his rise to the Consulate that replaced it. Berthier was also accorded the title of commander of the Army of Reserve, but he actually served Napoleon once again as chief of staff. Berthier was slightly wounded at the Battle of Marengo on June 14, 1800.

On May 14, 1804, the faithful Berthier was appointed one of the original eighteen marshals of France and the following year became chief of staff for the Grande Armée. Emperor Napoleon further rewarded Berthier by creating him Prince de Neuchâtel in 1806. In some degree, this proved the high-water mark of his career. For, in August 1807, he resigned as minister of war in order to take direct command of the

Grande Armée during the commencement of the 1809 campaign. He was both indecisive and insufficiently aggressive—demonstrating that he was far better as a staff officer, a top-level subordinate, than as a commander in his own right. Fortunately for Berthier, his brilliant planning for the battle of Wagram (July 5–6, 1809) redeemed him in the emperor's eyes.

Berthier was instrumental in the disastrous Russian campaign of 1812; he played a critical role in holding the remnant of the army together during the long, tortured winter retreat from Moscow, and he went on to brilliant performance at the key battles of Lutzen (May 2, 1813), Bautzen (May 20–21), and Leipzig (October 16–18). Yet the years of continual service were wearing on the marshal. The turning point came, perhaps, when he was seriously and painfully wounded at Brienne on January 29, 1814. Morose and unsettled after this event, he joined the military faction calling for Napoleon's abdication on April 11 and sided with the Bourbons. When Napoleon returned from his first exile, during the "Hundred Days," it was Berthier who escorted Louis XVIII and his court out of Paris to Ghent. Berthier refused to join Napoleon. On June 1, 1815, he died in a mysterious fall from a window in Bamberg. His death was ruled accidental, the result of illness. However, many believe that he suffered assassination at the hands of die-hard supporters of Napoleon.

BLAKE, Robert

BORN: 1599

DIED: August 7, 1657

SERVICE: English Civil Wars (1642–51); First Anglo-Dutch War (1652–54); Anglo-Spanish War (1656–59)

As an audacious and innovative admiral, Blake was one of the fathers of the British navy

Blake was the eldest in a large Bridgewater, Somerset, family of twelve children. He was not initially destined for a career at sea, but took a bachelor of arts degree from Wadham College in 1618 and probably attended the Inns of Court, which suggests that he intended to become a lawyer. From 1625 to 1640, he was involved in his father's highly prosperous trading firm and was elected to Parliament for Bridgewater, but did not take his seat. A partisan of **Oliver Cromwell,** he joined the Parliamentary army, in which he distinguished himself at the failed defense of Bristol during July 15–26, 1643. Promoted to lieutenant colonel, he held Lyme Regis against Royalist forces in 1644, a feat for which he received a full colonelcy. Late in 1644, he captured Taunton and held it against repeated Royalist assaults through July 1645. Appointed governor of Taunton, he was again elected to Parliament from Bridgewater in 1645, but did not take his seat until the following year.

When the Second Civil War broke out in 1648, Blake organized forces in Somerset, but was then named one of three "generals at sea," along with Richard Deane and Edward Popham, in February 1649. His first command was of a squadron dispatched to blockade Prince **Rupert**'s ships at Kinsale, Ireland in April 1649. When Rupert broke through the blockade, Blake pursued him to the River Tagus in Portugal during September and October. Denied permission by Portuguese authorities to press the attack, Blake retaliated against the Brazil fleet by capturing six ships and burning another three. He then returned to England and once again set out after Rupert, managing to sink most of the Royalist flotilla off Cartagena, Spain, in November 1650. Blake went on to capture the Scilly Isles, an important base out of which Royalist privateers operated. He also participated in the siege and capture of Jersey. After this, he was made a member of the Council of State for the Commonwealth.

The next conflict the Commonwealth faced was war with the Dutch (1652–54), and Blake assumed command of the fleet in the English Channel. He boldly attacked a superior squadron at Goodwin Sands, taking two ships as prizes on May 29, 1652. When the formal declaration of war came on July 17, Blake set off to capture the Dutch North Sea herring fleet. A storm on August 3 prevented battle with the Dutch admiral Tromp, but, with William Penn (father of founder of Pennsylvania, also named William), he defeated Michael de Ruyter and Cornelius de Witt off the Kentish Knock on October 8. Tromp regrouped and defeated Blake on December 10 at Dungeness, driving him into the Thames. However, Blake then engaged Tromp in a three-day battle off Portland during February 28–March 2, 1653, in an attempt to fire on the fleet of Dutch merchantmen Tromp was escorting. The fierce battle ended inconclusively, and Blake was severely wounded. On June 12, 1653, Blake joined **George Monck** on the Gabbard Bank (off North Foreland) in the defeat of Tromp there.

In addition to battle command, Blake was instrumental in the reform and reorganization of the Parliamentary navy during 1652–53. He was among the authors of the *Fighting Instructions* issued in March 1653, and, ever the Cromwell adherent, he served briefly in the Barebones Parliament during 1653–54 before he assumed command of twenty-four ships in the Mediterranean for operations against pirates and corsairs. He succeeded in extracting a heavy indemnity from the Duke of Tuscany, but failed in his attempt to negotiate the release of English prisoners held in Tunis. Blake then attacked the Tunisian fleet at Porto Farina and destroyed the fortresses there early in 1655. The bey relented and released the prisoners.

When war broke out with Spain in 1656, Blake was dispatched to conduct operations off the coast of Cadiz and, if possible, to take the offensive and seize Gibraltar. Although he lacked sufficient forces to accomplish this, one of his captains, Richard Stayner, did capture the Spanish treasure fleet in September 1656, and, the next year, hearing of another treasure fleet lying at anchor in the fortified harbor at Santa Cruz de Tenerife in the Canaries, Blake organized another attack. On April 20, 1657, he hurled his vessels against the heavily fortified position, not only destroying the fleet, but demolishing the harbor forts as well. At this, the hour of his greatest naval triumph, Blake fell gravely ill. He returned with his fleet to England, but died a scant hour before his vessel entered Portsmouth harbor on August 7.

Parliament buried Blake with full military honors, interring him at Westminster Abbey. Following the Restoration, however, King Charles II ordered Blake's exhumation, and his remains were tossed into a common grave with those of other leading Cromwellians.

BLÜCHER, Gebhard Leberecht von

BORN: December 16, 1742

DIED: September 12, 1819

SERVICE: Seven Years' War (1756–63); Wars of the French Revolution (1792–99), Napoleonic Wars (1800–15)

"Alte Vorwärts"—Old Forward—Prussian field marshal who was one of Napoleon's ablest adversaries

Blücher was born at Rostock, Mecklenburg, and, despite his family's protests, entered the military early in 1757 as an enlisted man in the Swedish Morner cavalry regiment. His first service were three campaigns during the Seven Years' War *against* the Prussians. He was made a prisoner of war in the last of these engagements near Friedland in 1760, at which point he changed his allegiance and received a commission as cornet in the Eighth (Belling) Hussars, a Prussian regiment, and served with this unit throughout the rest of the Seven Years' War.

By 1773, posted to garrison duty, Blücher had become bored with army life and was also discouraged by his failure to achieve rapid promotion. He resigned his commission, married, and settled on a farm. Thirteen years later, in 1786, he asked to be recalled to the army and was promoted to the rank of major. With the outbreak of the Wars of the French Revolution, he came into his own as a bold and brilliant cavalry commander, serving with great distinction in the Rhineland. He was given command of the Eighth Hussars in March 1794, and, on May 28, trounced a French force at Landau. Discontent when Prussia made a separate peace with France in 1807, he loudly criticized Prussian policy and offered his services to **Frederick William,** Duke of Brunswick, who was still engaged in the conflict, which had by this time dissolved into the Napoleonic Wars. For Brunswick, Blücher led several cavalry charges at the battle of Auerstädt on October 14, 1806.

Bested by **Napoleon**'s forces, Blücher retreated and was captured near Lubeck late in 1806. He was soon released in exchange for the French commander Marshal Claude Victor and, now in his sixties, was appointed military governor of Pomerania in May 1807. After Napoleon engineered his removal from office late in 1811, the septuagenarian Blücher returned to active service (February 28, 1813), fighting Napoleon's forces at the Battle of Lutzen (May 1–2) and Bautzen (May 20). He triumphed over Marshal **Jacques Etienne Macdonald** at Katzbach on August 26, and made a vigorous advance toward Leipzig, which blocked Napoleon from using the city as a staging area and base. The Battle of Leipzig resulted in a major victory for the Allies during October 16–18.

Promoted to field marshal, Blücher led his forces across the Rhine to invade France on January 1, 1814, but was halted by Napoleon at Brienne on January 24. However, he badly defeated French forces at La Rothiere on February 1, only to be repulsed at Champaubert (near Reims) on February 10, Montmirail on February 11, Chateau-Thierry on February 12, and Vauchamps (near Montmirail) on February 14. Following this string of defeats, Blücher regrouped as Napoleon broke off to the south in order to fight the Austrian forces. Blücher then engaged Napoleon at Craonne on March 7 and was defeated, but he repulsed in turn Napoleon's attack at Laon two days later and, his way cleared, forged ahead toward Paris. His combined Prussian and Russian forces joined with other Allied troops against Marshals Marmont and Mortier at Montmartre on March 30. This brought an armistice the next day and, ultimately, Napoleon's first abdication and exile on April 6.

Tired and ailing, Blücher returned to his Silesian farmland. But, as was true for all of Europe, his rest would not be long. When Napoleon returned from exile during the Hundred Days, Blücher was recalled on March 8, 1815, to command the army in Belgium and was promptly defeated by Napoleon at Ligny on June 16, 1815. The superannuated commander was wounded and unhorsed, but nevertheless made good an escape from the battlefield and, acting on the recommendation of his chief of staff, Count Neithardt Gneisenau, decided to lead three-fourths of his army to Waterloo to reinforce **Wellington** there. A single corps would be left behind to engage the forces of Marshal Emmanuel de Grouchy at Wavre on June 18.

With the main portion of his army, Blücher attacked the left flank of the French near Placentoit, which relieved the pressure on Wellington and was instrumental in the final Allied victory at Waterloo that effectively ended the career of Napoleon Bonaparte.

In recognition of his services, Blücher was created Prince of Wahlstadt and at last retired to his farm at Kribolwitz, Silesia, where he died four years later. Blücher was not a brilliant strategist or a well-educated man. However, he was possessed of ample sound judgment, and he knew how to rely on his able staff officers. He was endowed with great courage and personal strength and had a rough charisma that consistently inspired and emboldened the troops under his command.

BOLÍVAR, Simón

BORN: July 24, 1783

DIED: December 17, 1830

SERVICE: Wars of Latin American Independence (1807–25)

A Latin American revolutionary leader instrumental in securing independence for Venezuela, Colombia, Ecuador, Peru, and Bolivia

Bolívar was born to a privileged Creole family in Caracas, where he was educated by private tutors, who, after the death of his father in 1789, largely raised him. Among these men was Simón Rodriguez, who instilled in Bolívar a passionate interest in Enlightenment thought, particularly the writings of Jean Jacques Rousseau. He traveled to Europe during 1799–1802 and met and married a Spanish noblewoman, who succumbed to yellow fever shortly after the couple returned to Venezuela. Heartbroken, Bolívar made a second trip to Europe in 1804 and drifted ever closer to revolutionary activism. He witnessed the self-coronation of **Napoleon** and briefly toured the eastern United States in 1807.

Emperor Napoleon deposed the Spanish Bourbons, thereby creating a political turmoil that made Spain's colonies ripe for rebellion. In July 1810, the ruling junta in Venezuela sent Bolívar to England to secure aid to further the cause of independence. Although this effort failed, Bolívar met the exiled Francisco Miranda, who had attempted a revolution in Venezuela in 1806. Inspired by Miranda, Bolívar returned to Venezuela in time to help compose the nation's July 5, 1811, declaration of independence.

Bolívar became a leader of the new republic's army and was put in command of the strategically critical fortress of Puerto Cabello. However, one of his trusted subordinates betrayed him to the royalists, and Miranda, overall commander of the Republican army, surrendered to Spanish General Juan Domingo Monteverde in July 1812. Bolívar fled to New Granada (present-day Colombia), where he personally raised a force of volunteers to liberate Venezuela. The new army invaded Venezuela during May 1813 and defeated the Spanish forces in a series of six bloody battles. On August 6, he took Caracas, where he was hailed as *el Libertador*.

Achieving independence was one thing; maintaining it, quite another. Bolívar's new government faced opposition from royalist factions among the Venezuelans, and a bitter civil war ensued. Once again, Bolívar assumed military command and defeated the counter-revolutionaries at Araure on December 5, 1813, at La Victoria in February 1814, at San Mateo in March and at Carabobo in May. However, he was outmatched and badly defeated at La Puerta in July. He was once again forced to find refuge in New Granada, where, yet again, he raised an army and immediately set about the liberation of Bogota. However, at the Battle of Santa Maria, his forces were crushed by Spanish troops under General Pablo Morillo. Following this defeat, Bolívar fled to Jamaica in 1815. Holing up there in exile, he wrote a fiery and eloquent manifesto, *La carta de Jamaica (Letter from Jamaica)*, then returned to Venezuela in December 1816. Meeting Morillo's forces once again, near Barcelona, on February 16, 1817, he defeated them, only to be defeated in turn at the Second Battle of La Puerta one year later, on March 15, 1818.

Following this setback, Bolívar regrouped in the remote Orinoco region, where, during 1818–19, he built a new army augmented by several thousand British and Irish veterans of the Napoleonic Wars. He united this force with other revolutionaries, including the troops of José Antonio Paez and Francisco de Paula Santander. They did not turn their attention immediately upon Venezuela, but decided instead to liberate New Granada. Bolívar left his headquarters at Ciudad Bolívar on June 11, 1819, leading 2,500 men in an epic and speedy march across the wilderness of Venezuela, fording rivers at flood stage and traversing the Andes via the Pisba Pass, which was considered impassable by anything so large as an army. Descending into the valley of the Sagamosa River (near Bucaramanga, in present-day Colombia) on July 6, they achieved total surprise and thereby defeated Spanish forces at the Battle of Boyaca on August 7, 1819. Bogata fell three days later, and Bolívar established a republic.

From his new position, President Bolívar convened the Angostura Congress, which resulted in the creation of Gran Colombia, a federation encompassing present-day Venezuela, Colombia, Panama, and Ecuador. The union was formalized at Cucuta in July 1821, and Bolívar became president of Gran Colombia. The following year, 1822, Bolívar met with the Argentine patriot José de San Martin at Guayaquil, Ecuador. No union with Argentina came, but, at this time, the revolutionary leader Antonio José de Sucre liberated Ecuador after victory at the Battle of Pichincha in May 1822. Subsequent victories in Peru—at Junin (August 1824) and Ayacucho (December 1824)—sent the remaining Spanish forces out of the so-called Bolívarian states: Venezuela, Colombia, Ecuador, Peru, and Bolivia.

Triumph over the Spanish did not bring peace to the Bolívarian states. Rivals for power stirred sporadic rebellion. Although Bolívar personally wrote a republican constitution for the state of Bolivia (as the liberated population of northern Peru decided to call their new nation), defiance of his authority was such that, in 1828, Bolívar assumed dictatorial powers over Gran Colombia. Instability persisted, and in the spring of 1830, discouraged, weary, and ill with tuberculosis, Bolívar stepped down as president of the nation. He died that winter.

Simón Bolívar died in the belief that he had failed. However, he had not only proven himself to be a remarkable military leader, he also became an enduringly influential political figure and thinker as well. His 1815 *Jamaica Letter* is still studied as an acute analysis of the Spanish–American character, and the *Angostura Discourse* of 1819 (the basis for the creation of Gran Colombia), as well as the Bolivian constitution of 1826, remain models of sound political thought and political ideals.

BOTHA, Louis

BORN: September 27, 1862

DIED: August 27, 1919

SERVICE: Second Great Boer War (1899–1902)

South African general who led Boer forces against the British

Botha was born near Greytown in Natal and, the son of a farming family, was given very little formal education. He was, however, intensely interested in politics and participated in the creation of the New Republic around Vryheid in 1884. He settled in that region, becoming field cornet (military leader) of Vryheid in 1894 and gaining election to the Volksraad (parliament) of Transvaal four years later. By this time, Botha had become a political moderate and pointedly abstained from the vote for war with Britain in October 1899. However, once war was declared, he quickly proved himself a highly able commander.

Botha rose swiftly to the position of second-in-command under General **Petrus J. "Piet" Joubert.** He performed brilliantly at the battles of Talana (October 20, 1899) and Ladysmith (October 30), both Boer victories. He was in charge of the siege of Ladysmith from November 2, 1899, through February 28, 1900, and assumed full command of Boer forces after Joubert was unhorsed and gravely injured on November 23. He checked Sir Redvers Buller's first three efforts to relieve Ladysmith at the battles of Colenso (December 15, 1899), Spion Kop (January 24, 1900), and Vaal Krantz (February 5–7), but was finally outflanked at Tugela River on February 17–18, and withdrew beyond Ladysmith. He took a stand in the Biggarsberg Mountains along the Natal–Orange Free state border, but was pushed out by Buller on May 14, 1900 and defeated by him in open battle at Bergendal on August 27—the final major engagement of the war.

Following the Boer defeat at Bergendal, the war took on the character of a guerilla conflict, centered primarily in the northeast Transvaal. Botha invaded Natal during September 1900 in cooperation with General **Jan Smuts**'s raid into the Cape Colony. He defeated forces under Lieutenant Colonel Hubert Gough at Blood River Poort on September 17. Botha was in turn badly defeated at Itala and Prospect, forts on the Transvaal–Zululand border, on September 26, 1901, but he won a narrow victory over G. E. Benson's rearguard at Bakenlaagte in the eastern Transvaal on October 30, 1901. Despite this, Botha's army entered the new year exhausted and ragged. Botha was among those who attended the peace conference at Vereeniging in May 1902 and signed the Treaty of Vereeniging on May 31, which ended the Boer War.

With the return of peace, Botha returned to politics and resumed his moderate stance. He labored with Jan Smuts for the unification of British and Boer factions that resulted in the creation of the Dominion of South Africa. Botha was elected the first prime minister of the Union of South Africa on May 31, 1901 and served in this office until his death.

BRADDOCK, Edward

BORN: 1695

DIED: 1755

SERVICE: War of the Austrian Succession (1740–48); French and Indian War (1754–63)

British general best known for having led the disastrous assault on Fort Duquesne during the French and Indian War

Born in London, Braddock was the son of a general and was destined by his parents for a military career. He was commissioned an ensign in the elite Coldstream Guards in 1710, but did not see battle service until the 1740s, when he fought in the War of the Austrian Succession (1740–48) at the Battle of Bergen op Zoom on September 18, 1747. In 1753, he was given command of the Fourteenth Regiment of Foot at Gibraltar, and the next year was promoted from colonel to major general. He was given command of British forces in America and landed at Hampton Roads, Virginia, in February 1755.

The force of regulars he led included the worst soldiers the British army fielded. It was not thought that fighting Indians and provincial French colonists in North America would be a terribly demanding task. After laying out a grand strategy for victory in the French and Indian War, Braddock united his regulars with Virginia militiamen in an expedition against the French stronghold of Fort Duquesne. This outpost, which had been built by the British and captured by the French, was located at the site of present-day Pittsburgh and controlled the strategically vital "forks of the Ohio"—the confluence of the Ohio, Allegheny, and Mononghela rivers. The nation that dominated here had control over the gateway to the vast western reaches of North America.

Braddock was a brave and determined officer, but he was also a hidebound military traditionalist, who refused to recognize the special exigencies of war in the American wilderness. He alienated would-be Indian allies as well as the "provincials," the colonial militia. Indeed, Braddock proved so offensive that colonial governors resisted collecting the war levies imposed on them and generally declined to cooperate with Braddock.

In March 1755, after many delays, Braddock marched two regiments of British regulars and a provincial detachment (under **George Washington**) out of Fort Cumberland, Maryland, bound for Fort Duquesne. This force of 2,500 men, loaded down with heavy equipment better suited to the open terrain in which European battles were fought, moved so slowly that French spies and snipers took a heavy toll on the advancing column. Finally, Washington advised Braddock to detach a "flying column" of 1,500 men to move more swiftly to an initial attack on Fort Duquesne. By July 7, the flying column set up a camp ten miles from their objective.

The size of Braddock's forces was enough to prompt thoughts of surrender in the mind of Fort Duquesne's commandant, Claude-Pierre Pécaudy de Contrecoeur. However, his subordinate, Captain Liénard de Beaujeu, persuaded him to take the offensive, and on July 9, 1775, 72 regulars of the French Marine, 146 Canadian militiamen, and 637 assorted Indian allies attacked Braddock's encampment. Even though the British troops outnumbered the attackers, panic seized Braddock's forces. The battle quickly degenerated into a rout. It was reported that many of the British regulars, in pathetic confusion, huddled in the road like flocks of sheep. General Braddock did much to distinguish himself in the attempt to rally his troops. In the course of the melee he had five horses shot from under him before he was mortally wounded himself. As he lay dying, he continued to observe the carnage around him. "Who would have thought it?" he said. They were his final words.

The Battle of the Wilderness was a terrible blow to the British. Of Braddock's 1,459 troops engaged, only 462 returned. The loss, attributable in large part to Braddock's lack of skill in an unfamiliar combat environment, nearly spelled early defeat for the British in the French and Indian War.

BRADLEY, Omar Nelson

BORN: February 12, 1893
DIED: April 8, 1981
SERVICE: World War I (1917–18); World War II (1941–45)

Called "The G.I. general" because of his rapport with the common soldier, Bradley was one of the most effective American generals of World War II

Bradley was Born in Clark, Missouri, and received an appointment to West Point, from which he graduated in 1915 with a commission as second lieutenant in the infantry. During 1915–18, he served in the West in a series of posts in Washington and Arizona and rose rapidly through the ranks, gaining promotion to major in June 1918. He was not shipped overseas during World War I.

At the end of the war, Bradley served as a military instructor at South Dakota State College (1919) and as an instructor at West Point from 1920 to 1924. He graduated from Infantry School at Fort Benning in 1925 and was posted to Hawaii from 1925 to 1928. In 1929, he graduated from the Command and General Staff School at Fort Leavenworth and was assigned as an instructor at the Infantry School during 1929–33. Bradley graduated from the Army War College in 1934 and became a tactical officer at West Point from 1934 to 1938, having been promoted to lieutenant colonel in June 1936.

In 1938, Bradley was appointed to the General Staff and promoted to brigadier general in February 1941. The following month, he was appointed commandant of the Infantry School, serving until February 1942, when he was put in command of the Eighty-Second Division (during March–June), and then the Twenty-Eighth Division (from June 1942 to January 1943). After serving briefly as deputy to General **Dwight D. Eisenhower** from January to March 1943, he was named to replace General **George S. Patton** as commander of II Corps. Bradley led the corps during the final stages of the Tunisian campaign, capturing Bizerto on May 8, and leading II Corps during the Sicilian campaign from July 10 to August 17, when he was transferred to England to participate in planning what would become the D-Day invasion of France.

As lieutenant general, Bradley was named commander of the First Army in January 1944, which was assigned the right wing position in the Allied D-Day landing at Normandy. Bradley planned and directed the key breakout at Normandy at Saint-Lô in July. The following month, he became commander of the 12th Army Group, consisting of the First Army (under General Courtney L. Hodges) and the Third Army (under General Patton). In this capacity, he directed the southern wing of the massive Allied advance across northern France, which spanned August to December 1944. At 1.3 million men, the 12th Army Group was the greatest force ever commanded by an American general. Promoted to four-star general in March 1945, Bradley continued in command of Twelfth Army Group during final operations in Germany from January to May, 1945.

Omar Bradley. Courtesy National Archives.

Following the war, Bradley was appointed head of the Veterans Administration, a post in which he served through December 1947. He then succeeded Eisenhower as Army Chief of Staff in February 1948 and became first chairman of the Joint Chiefs of Staff under the newly organized Defense Department during the Korean War. Bradley was promoted to general of the army in September 1950, and published his memoirs, *A Soldier's Story*, in 1951. He retired from the army in August 1953.

Bradley was not a brilliant tactician nor a great strategist. He was, however, a strong leader and, perhaps, the single most universally admired commander of World War II.

BRAGG, Braxton

BORN: March 22, 1817

DIED: September 27, 1876

SERVICE: Seminole War (1835–43); U.S.-Mexican War (1846–48); American Civil War (1861–65)

Passionate and flawed Confederate general during the American Civil War.

Braxton Bragg was born in Warrenton, North Carolina, and attended the U.S. Army Military Academy at West Point, from which he graduated in 1837 with a second lieutenant's commission in the Third Artillery. It was as an artilleryman that Bragg would prove most skilled throughout his military career.

Bragg's first service was in the First Seminole War, (1837–1841); then, like so many other commanders destined to make their mark during the Civil War, Bragg received his most extensive early battle experience in the U.S.–Mexican War, serving under **Zachary Taylor.** He was stationed first at Fort Brown (near present-day Brownsville, Texas), May 3–9, 1846, and was breveted to the rank of captain. At Monterrey, during September 20–24, he performed with distinction and was breveted major. Likewise at the battle of Buena Vista (February 23, 1847), Bragg performed with courage and skill, brilliantly deploying his artillery battery (using the "double shotting" technique, by which the effectiveness of his guns was virtually doubled) and playing a key role in Taylor's victory.

With the brevet rank of lieutenant colonel, Bragg resigned his commision in January 1856 and moved to Louisiana, where he established a plantation. Here he displayed great ability as an engineer, designing not only a drainage and levee system for his plantation, but for the entire state.

In February 1861, Bragg answered the call to service in the Louisiana militia, and entered the Confederate States Army as a brigadier general on March 7, 1861. In September, he was promoted to major general and took command of II Corps in the newly formed Army of Mississippi under General **Albert S. Johnston.** After leading the Confederate right flank at Shiloh, he was promoted to lieutenant general on April 12, 1862, a few days after the battle. Given command of the Army of Mississippi (subsequently renamed the Army of Tennessee) in June, Bragg invaded Kentucky from August–October, 1862. His intention was to bring this border state into the Confederate fold, but he narrowly failed to take Louisville when he moved his army to Lexington, thereby allowing Union general Don Carlos Buell to occupy Louisville. There the Union commander reinforced his army and was able to check Buell's invasion at the Battle of Perryville (Kentucky) on October 8, 1862.

Following his defeat in Kentucky, Buell withdrew from the state through the Cumberland Gap and fought four days against General William Rosecrans inconlusively at Stones River (also called Murfreesboro), Tennessee, starting December 31, 1862. Once again, Bragg was compelled to withdraw, but managed to maneuver out of Chattanooga early in September 1863, and defeat Rosecrans at the bloody Battle of Chickamauga on September 20. Union forces withdrew to Chattanooga, where Bragg held them under siege until General **Ulysses S. Grant** came to Rosecrans's relief and defeated Bragg at the battle of Chattanooga, November 23–25.

In December, after his loss to Grant, Bragg was replaced as commander of the Army of Tennessee by General **Joseph E. Johnston** and was appointed one of Jefferson Davis's military advisers. Early in 1865, Bragg personally raised a small force in North Carolina to defend against **Sherman**'s thrust to the north against Johnson. With General Johnston, Bragg surrendered to Sherman on April 26, 1865, and, with the formal conclusion of the war, returned to civil engineering, practicing in Texas.

No one ever questioned Bragg's patriotism and courage. However, along with the qualities of an able organizer, a skillful artillerist, and a fair degree of tactical acumen, he posessed a fatal lack of decisiveness and an irascible, irritable temperament that made him unpopular with his troops, peers, and superiors. Like Union general **George McClellan,** Bragg's tendency to overdeliberation prevented his successfully exploiting his military successes.

BRANT, Joseph

BORN: 1742

DIED: November 24, 1807

SERVICE: French and Indian War (1754–63); Pontiac's Rebellion (1763); American Revolutionary War (1775–83)

Mohawk war chief, skilled strategist, and valuable ally of the British during the French and Indian War and the American Revolution

Brant was born Thayendanegea in the Ohio Valley, the son of Tehowaghwengaraghkwin, a Mohawk chief. His father died soon after the boy's birth, and his mother, Margaret, remarried. Her new husband was soon killed in a raid, and Margaret married the Mohawk chief Nikus Brant, after which Thayendanegea called himself Joseph Brant.

Joseph Brant grew up in the Canajoharie region of upstate New York's Mohawk Valley. There he was befriended by Sir **William Johnson,** a British trader and land speculator who later maried Brant's sister, Molly. During this period, Brant was educated at the Anglican Mohawk mission school.

On September 8, 1755, when he was only thirteen years old, Brant served with distinction under Johnson at the Battle of Lake George during the French and Indian War. After this, Johnson sponsored Brant's further education at Moor's Charity School in Lebanon, Connecticut, where Brant became an interpreter for a missionary in 1763. That same year, Brant rejoined the French and Indian War during its "coda" phase, known as Pontiac's Rebellion. With other Iroquois (the Mohawks were members of the so-called Iroquois League) he fought on the side of the British against Pontiac.

Brant married the daughter of an Oneida chief in 1765, then turned to scholarly and religious pursuits, aiding the Anglican minister John Stewart in translating various religious works into Mohawk. He also served, in 1774, as secretary to William Johnson's nephew Guy Johnson, who was superintendent of Indian affairs for the British crown.

Intensely loyal to Britain, Brant was instrumental in persuading four of the six Iroquois League nations (all but the Oneidas and Tuscaroras) to join with British forces against the Americans when the Revolutionary War broke out. Serving the Mohawks as a war chief, Brant was made tribal spokesman and representative to the British. He received a captain's commission in the British provincial forces and sailed to England, where he was presented at court in 1775. A handsome man of supremely dignified appearance, Brant became the subject of fine portraits by the British painter George Romney and the London-based American expatriate Benjamin West.

Brant did not remain in England, however. He returned to America, where he fought at the Battle of the Cedars (on the St. Lawrence, west of Montreal), May 15–19, 1776, and, serving under the British general Barry St. Leger, fought at Oriskany (upstate New York) on August 6, 1777. Feared as a raider who terrorized the Mohawk Valley as well as portions of southern New York and northern Pennsylvania, he was present at the so-called Cherry Valley Massacre, November 11, 1778, but was noted as much for his efforts to restrain the brutal excesses of the warriors under his command as he was for his skill in deploying his forces during this action.

The Mohawk Valley raids Brant led were highly effective, and he was rewarded with a royal commission as colonel of Indians in 1779. During this year, Brant also managed to checkmate the efforts of Red Jacket, an influential Seneca chief, to make peace between the Iroquois and the Americans. During the spring and summer, Brant led additional New York raids, culminating in a triumph at Minisink on July 29, 1779.

Following the Minisink victory, Brant's objectives increasingly turned away from aiding the British and focused more narrowly on securing the lands of his people. He led the Mohawks west of the Niagara River and attempted to reach an agreement with the Americans. Failing this, he appealed to British Canada and secured a grant of land six miles wide on either side of the Grand River. With the conclusion of the war, Brant returned to England, where he solicited and obtained funds to indemnify the Iroquois for their extensive war losses. Using these moneys, he purchased additional lands in 1785 and dedicated the balance of his life to improving the lot of the Mohawks of the Grand River settlement. Brant founded the Old Mohawk Church, translated a variety of religious works, and struggled continually and successfully against a host of speculators who attempted to seize Mohawk lands. In 1792, Brant half-heartedly (and unsuccessfully) played the role of peacemaker during Little Turtle's War, fought in the Ohio Valley. He traveled to Philadelphia—then the capital of the United States—and was received by President **George Washington** and Secretary of War **Henry Knox.**

Brant was a formidable warrior and a brilliant tactician. A charismatic presence, he commanded the loyalty and respect of Indians and his British allies.

BRECKINRIDGE, John Cabell

BORN: January 21, 1821

DIED: May 17, 1875

SERVICE: U.S.–Mexican War (1846–48); Civil War (1861–65). Principal battles: Shiloh (1862); Stones River (1862–63); Chickamauga, Chattanooga (1863); New Market, Cold Harbor (both in Virginia), Nashville (1864)

Confederate general and the last secretary of war in the cabinet of Confederate president Jefferson Davis

Breckinridge (whose name is sometimes spelled Breckenridge) was born near Lexington, Kentucky. He graduated from Centre College in Danville, Kentucky, in 1839, and went on to study at Transylvania University in Lexington. He was admitted to the bar in 1841. After brief service as major of the Third Kentucky Volunteers in the U.S.–Mexican War during 1847, Breckinrdige entered politics, gaining election to the Kentucky legislature in 1849 and to the U.S. Congress in 1851. He served until March 1855, when he took office as vice president under President James Buchanan.

In 1860, Breckinridge unsuccessfully ran for president under the banner of a breakaway faction of the Democratic party. In the months immediately prior to the outbreak of the Civil War, he served as a U.S. senator (March–December 1861), even after he was appointed a brigadier general in the provisional army of the Confederate states in November 1861.

Breckinridge was given command of the reserve at Shiloh (April 6–7, 1862) and was promoted to major general in August 1862. Assigned to direct the defenses of the key Mississippi River Confederate stronghold of Vicksburg, he also effectively fortified Port Hudson, Mississippi. After this, he joined General **Braxton Bragg**'s Army of Tennessee and was appointed to divisional command under General **William Joseph Hardee** at the Battle of Stones River (December 31, 1862–January, 3, 1863).

Like many Civil War–era politicians, Breckinridge became a military commander. Unlike most of those political generals, however, he had a natural aptitude for command. On May 14, 1863, fighting under General **Joseph E. Johnston,** he distinguished himself at Jackson, Mississippi, and went on to divisional command under **D. H. Hill** in the Army of Tennessee. He performed well at bloody Chickamauga (September 19–20, 1863) and at Chattanooga (November 23–25), then, in 1864, was transferred to the Shenandoah Valley, where he scored a signal success at New Market on May 15.

During June 1–3, 1864, he fought at Cold Harbor under **Robert E. Lee,** serving as second in command to General Jubal Early. He was instrumental in the lightning raid on Washington, D.C., during July 11–12, then was assigned command of the Confederate Department of Southwest Virginia in September. He fought valiantly at Nashville (December 15–16, 1864) before Confederate president Jefferson Davis tapped him as the Confederacy's fifth—and last—secretary of war in February 1865.

Having occupied high elected office in the United States, Breckinridge feared that he would be prosecuted for treason following the Union victory. He fled to Cuba after the war and then to England, returning only after President Andrew Johnson declared amnesty in 1868. Breckinridge returned to the private practice of law in Kentucky and enjoyed financial success as a railroad investor. He died in Lexington, Kentucky.

Excelling in politics as well as military administration, Breckinridge was also a courageous and highly capable field commander—a natural soldier.

BRUNSWICK, Frederick William, Duke of

BORN: October 9, 1771
DIED: June 16, 1815
SERVICE: Napoleonic Wars (1800–15)

Known as "The Black Duke," Brunswick was a fierce warrior whose hatred of Napoleon drove his unremitting aggressiveness

Frederick William was the fourth son of Charles William and Augusta of Brunswick and inherited the title of duke upon his father's death in 1806, but could not claim his estates because the Treaty of Tilsit (July 7–9, 1807) dissolved his family's duchy. A dispossessed victim of Napoleon's aggression, Brunswick betook himself to Austria, where he agreed, on February 25, 1809, to raise a Brunswick corps of infantry and cavalry to fight Napoleon. His corps formed up at Náchod, Bohemia, on April 1. Clad in somber black uniforms, the unit was informally dubbed the Black Horde, and Brunswick was nicknamed the Black Duke.

Brunswick led his corps as part of the Austrian army in the invasion of Saxony during 1809 and fought at Zittau (May 30–31) and Borbitz (June 12). When Austria surrendered to Napoleon on July 12, Brunswick refused to relinquish his corps, but instead fought through to the north German coast, taking the town of Halberstadt on July 29 and reaching Brunswick on July 31, where he brilliantly outmaneuvered, at Oelper on August 1, the Westphalian forces that dogged him. Reaching Elsfleth on August 6, Brunswick and the Black Horde sailed for England, where he placed himself at the disposal of the British crown. A portion of his troops fought in British service in Portugal during 1810–1814, where they performed with great distinction. In the meantime, Brunswick lived in England until he was restored to his duchy in 1813. He then set about reorganizing the Black Horde, which he led in service of **Wellington**'s army at the Waterloo campaign of 1815. On June 16, at the Battle of Quatre Bras, Brunswick, always the daring commander at the head of his troops, received a mortal wound. Despite the loss of their commander, the Black Horde continued to fight and distinguished itself at the Battle of Waterloo proper two days later.

Although Brunswick was not a particularly distinguished tactician, he was an intensely zealous commander who was able to turn his intense personal hatred of Napoleon to effective military use. His troops adored him; and that he had trained them well is evidenced by their performance even in his absence, both in Portugal and, following his death, at Waterloo.

BUDËNNY, Semën Mikhailovich

BORN: April 25, 1883
DIED: October 27, 1973
SERVICE: Russo–Japanese War (1904–05); World War I (1914–18); Russian Civil War (1917–22); Russo–Polish War (1914–17); World War II (1939–45)

A Russian military commander of the old school, passionately devoted to the cavalry, Budënny performed with brilliance and dash during the Russian Civil War of 1917–22 and the Russo–Polish War of 1919–21, but wrought disaster during World War II

Budënny was born near Rostov-on-the-Don, the heart of the Don Cossack region, where he was steeped in the fighting traditions of the Cossacks. Although he received no formal education, Budënny was literate. He entered military service as an enlisted cavalryman in 1903 and participated in the Russo–Japanese War, although he saw no significant action. Despite this, Budënny's superiors appreciated the young man's potential, and he was enrolled in the St. Petersburg Riding

School of Imperial Cavalry in 1909. On graduation, he was promoted to sergeant-major, the highest noncommissioned rank in the czar's army.

Budënny served valiantly and extensively in World War I during 1914–17, receiving four St. George's Crosses for gallantry. With the outbreak of the Russian Revolution, Budënny, who was not a political revolutionary, nevertheless sided with the Bolsheviks. He was appointed to his division's soldier's soviet and formed a cavalry unit that successfully cleared Platovskaia of White (czarist) forces during February 1918. Budënny grew his unit to brigade proportions and, under heavy White onslaught, retreated to the north, where he joined forces with troops under **Josef Stalin** and Kliment Yefremovich Voroshilov. Budënny was assigned command of the cavalry division in Voroshilov's Tenth Army and was instrumental in retaking Tsaritsyn during January 1919.

As commander of the Cavalry Corps in June, Budënny deliberately disobeyed orders to avoid engaging the superior White cavalry forces around Orenburg and Voronezh. His bold gamble paid off; Budënny emerged victorious in October. Following this triumph, the Cavalry Corps was upgraded to the First Cavalry Army (known as the Konarmiya), and, leading it, Budënny defeated White forces in a critical battle at Bataysk, near Rostov in January 1920.

At this point, the Russian Civil War merged with the Russo–Polish War, and Budënny's army was dispatched to repulse Poles advancing into the Ukraine. On May 30, 1920, Budënny crossed the Dnieper River with 16,500 cavalry troopers and pushed through Polish lines at the Uman River by early June. In a remarkable 200-kilometer advance, Budënny engaged and smashed the Sixth Polish Army at Berdichev. Unfortunately, his further advance, on the city of L'vov, was poorly coordinated with the action of other units, and he soon found himself encircled. Retaining his sense of command, he managed to fight free by September and was dispatched to the south, where he and the First Cavalry Army played a key role in the conquest of the Crimea (October–January 1921).

Budënny's distinguished record prompted his elevation to assistant for cavalry to the commander in chief and gained him membership in the Revolutionary Military Council by 1923. Budënny's close association with Stalin (already political commissar by the time of the Polish War), helped secure great power and influence for Budënny, who set about reshaping the Red Army during the 1920s and 1930s. He was chiefly responsible for the army's anachronistic reliance on cavalry; there were no fewer than thirty cavalry divisions by 1938. The emphasis on cavalry retarded the mechnization of the Red Army, which put it at a severe disadvantage compared to the highly mechanized German army.

Budënny was appointed one of the first five Marshals of the Soviet Union in 1935 and, weathering Stalin's brutal purges of the military, rose to the position of deputy commissar of defense on the eve of World War II (1939). When **Adolf Hitler** violated his pact with Stalin and invaded the Soviet Union in June 1941, Budënny was named commander of the Soviet Reserve Army and stationed on the Southwest front with orders to defend Kiev. In July, facing overwhelming Nazi numbers, he urged retreat. Stalin would not hear of it, and, soon, Budënny's 600,000 men were surrounded by the invaders. Budënny escaped capture only because he was evacuated by air in September.

Following the Kiev debacle, Budënny was named to command along the North Caucasus Front from May–September, 1942, then was given full command of the Red Army cavalry beginning in May 1943. By this time, the grandiose post was largely political. Budënny was given no additional field or operational commands, but served as a member of the Party Central Committee from 1939 to 1952 and as a "candidate member" of the committee from 1952 until his death in 1973. This final honorary post reflected his enduring popularity as a hero of the Soviet Union despite his signal lack of success during World War II, which not only included failure in battle, but the even graver failure to prepare the Red Army adequately for twentieth-century warfare.

BUGEAUD DE LA PICONNERIE, Thomas Robert, Duke of Isly

BORN: October 15, 1784
DIED: June 10, 1849
SERVICE: Napoleonic Wars (1800–15); Conquest of Algeria (1830–49)

One of the most significant early commanders in the service of imperial France

Bugeaud was born at Limoges and, as a young man, enlisted in the *velites*, the light infantry of **Napoleon**'s Imperial Guard. His ability was quickly recognized, earning him a sublieutenant's commission in 1806. Within two years, he had distinguished himself at the sieges of Saragossa (December 20, 1808–February 20, 1809) and Pamplona. He proved particularly effective in operations against the guerrillas that harassed Napoleon's columns during 1808–1813.

With Napoleon's first abdication and exile to Elba in April 1814, Bugeaud sided with the restored Bourbons, only to rejoin the Napoleonic cause during the Hundred Days (April–June, 1815). After Napoleon's final defeat at Waterloo, Bugeaud was relieved of command and retired to his estate in Perigord, where he assumed the life of a gentleman farmer.

In 1830, with the outbreak of revolution, Bugeaud returned to the military and served Louis Philippe. He avidly accepted such unpopular assignments as jailer of the Duchess Du Berry in 1833 and, the following year, crushing the Paris riots. This last action was so brutally vigorous and effective that the government of Louis Philippe, fearing Bugeaud's continued presence in Paris would spark further riots, dispatched him to service in Algeria. What other military commanders would have regarded as bitter exile, Bugeaud turned to advantage. He launched an aggresive campaign against the formidable **Abd-el-Kader**, defeating him at the Battle of Sikkah on July 6, 1836, then confounded his Parisian critics by concluding an enlightened and generous armistice with the Algerian leader at the Tafna on May 30, 1837.

Bugeaud's handling of the Algerian military and civil situation earned him the governor generalship of that colony (December 29, 1840), and from this position he established an enduring military presence in Algeria. In contrast to the conventional reliance on massed shows of force, Bugeaud created light, highly mobile columns and sent them to raid enemy farmlands. In this way, he inexorably defeated Abd-el-Kader's Moroccan allies at the Isly by August 14, 1844.

Bugeaud ended his tenure as governor when he angrily resigned the post late in 1847 after the government repeatedly ignored his proposals for extended and decisive military colonization of the region. However, when the Revolution of 1848 threatened Louis Philippe, he rushed to suppress the violence. Although unsuccessful in saving his monarch's reign, Bugeaud continued to serve France. In 1849, he was assigned to command the Army of the Alps. However, Bugeaud succumbed to cholera in Paris during the epidemic of 1849 and never led his new troops.

While military historians agree that Bugeaud was a skilled and innovative commander, especially in his use of light, mobile forces to prosecute the ends of colonial government, many criticize his often brutal and unyielding conservatism. Such critics fail to recognize that his dealings with the Algerians were characteristically more enlightened—and far more generous—than what the colonial standards of the day dictated. A fine tactician, Bugeaud was a highly capable military administrator of colonial government.

BÜLOW, Friedrich Wilhelm von

BORN: 1755
DIED: 1816
SERVICE: French Revolutionary Wars (1792–99); Napoleonic Wars (1800–15)

An important Prussian general—Gebhard von Blücher's right-hand man—in service against Napoleon

Bülow was the older brother of another Prussian officer (better known as a military theorist than as a field commander), Heinrich Adam Dietrich von Bülow. Friedrich Wilhelm entered the Prussian army in the 1790s and participated in operations in the Rhineland during 1793 and 1794. He rose steadily through the ranks and, well into the Napoleonic Wars, was given command of Second East Prussian Fusilier Brigade during October 1806–July 1807.

Bülow became closely associated with Prince Louis Ferdinand, an ardent advocate of Prussian governmental and military reform. With the prince's help, Bülow gained promotion to major general and was given a key command in Pomerania during 1808–1811. In 1812, he was named provisional governor of East Prussia, and the following year was promoted to lieutenant general and commander of a division that fought at Mökern (April 5) and assaulted Halle (May 2). At the Battle of Luckau, on June 4, Bülow's forces defeated Marshal Nicholas Oudinot, and, on August 23, again engaged and defeated Oudinot's troops at Grossbeeren. The latter victory was achieved in defiance of the orders of his commander, Prince Johann Karl Bernadotte of Sweden, who did not want Bülow to attack.

Bülow succeeded in repulsing **Marshal Michel Ney**'s offensive at Dennewitz (September 6) and played a principal role in the critical Battle of Leipzig in October 1819. He then liberated Holland and Belgium from the French through the fall and winter of 1813–14. When Napoleon attacked Blücher at Laon on March 9, 1814, it was Bülow who checked the Frenchman's advance. After this, Blücher promoted his subordinate to *General der Infanterie* and made him commander of IV Corps.

Although Bülow did not participate in the pivotal Battle of Ligny on June 16, his corps was in the Prussian vanguard at Waterloo, where he successfully penetrated Napoleon's right wing on June 18, 1815. Bülow retired after the war, honored for his gallantry and enterprise, and widely regarded as Blücher's best subordinate.

BURGOYNE, John

BORN: 1722

DIED: June 4, 1792

SERVICE: Seven Years' War (1756–63); Spanish–Portuguese War (1761–63); American Revolution (1775–83)

Though infamous for the defeat he suffered at Saratoga, New York, during the American Revolution, Burgoyne was as capable a British commander as any of his age

Born in Lancashire, Burgoyne received a first-class education at Westminster School and retained a lifelong love of learning and letters. Indeed, his refinement and cultural tastes were such that, from early in his career, he earned the somewhat disparaging sobriquet "Gentleman Johnny." He was commissioned a cornet in Thirteenth Light Dragoons in 1740 and the following year was promoted to lieutenant. His impulsive elopement with the daughter of the Earl of Derby (who was the sister of Burgoyne's Westminster school chum) interrupted his military career. Financial hardship compelled him to sell his commission and prompted a move to France in 1746. In the fullness of time, however, Derby became reconciled with his daughter and son-in-law and exerted his considerable influence to gain for Burgoyne, in 1757, a captaincy in the Eleventh Dragoons during the Seven Years' War. The following year, Burgoyne saw action at Cherbourg and, in 1759, at Saint-Malo. An able administrator, he was largely responsible for raising two light-horse regiments, one of which he ably commanded beginning in August 1759.

Burgoyne was elected to Parliament in 1762, but that year was dispatched with his regiment to Portugal to counter the Spanish invasion there. It was during this action that he gained his first solid military reputation for dash and enterprise. On his return to London, he found himself welcomed as a hero and was quickly promoted colonel of Sixteenth Light Dragoons (1763).

In September 1774, with the rank of major general, he was assigned as subordinate to join **General Thomas Gage** to fight the American rebels in North America. His military acumen seemed to desert him, and he was dispatched back to England in November 1775.

However, as the situation grew increasingly critical in the colonies, Burgoyne was again sent to North America to reinforce Britain's Canadian forces. He scored a decisive victory against the Americans at the Battle of Trois-Rivieres on June 8, then returned to England during the winter.

The following year, Burgoyne returned to Canada, having formulated a plan to march down Lake Champlain and the Hudson River, where he would mass his forces with those of Colonel Barry St. Leger and General **Sir William Howe** in a grand strategy designed to sever New England from the rest of the colonies. Burgoyne successfully argued that New England was the source of the American Revolution; isolate it from the other colonies, and the rebellion would collapse.

Burgoyne successfully invaded New York, capturing Fort Ticonderoga during July 2–5. He then advanced southward, toward Albany. Now, however, his well-drilled and orderly columns were harassed by American irregulars, who fought using the unconventional tactics of the wilderness. Further hampering Burgoyne was a critical shortage of supply. Burgoyne dispatched one of his Hessian subordinates, Colonel Friedrich Baum, to forage. Baum's column was set upon at Bennington, Vermont, by American militamen under John Stark, who routed Baum on August 16.

Hungry and exhausted, Burgoyne's troops attacked Saratoga, only to be repulsed there at Freeman's Farm on September 19. A second Saratoga battle, at Bemis Heights on October 7, was even more decisive. Burgoyne's demoralized and under-supplied army was crushed. Burgoyne surrendered on October 17 and was made a prisoner of war.

After winning parole on condition that he return to England, Burgoyne was greeted with severe criticism, which effectively suspended his career until June 1782, when the liberal Whig party displaced the conservative Tories, and Burgoyne was named commander in chief in Ireland. He did not long occupy this post, however. In 1783, he retired from the military and turned instead to his first love, literature. He wrote with moderate success for the London stage until his death.

Burgoyne was as capable a commander as most, and, more than most, he combined enterprise with great regard for the welfare and safety of his troops. At his best during the latter phases of the Seven Years' War, he was ill-suited to combat in the American wilderness. His performance in North America suffered from his own inability to cope with unconventional combat conditions as well as from the failure of other British commanders, William Howe and Sir Henry Clinton, to coordinate with him and support his invasion of New York.

BURNSIDE, Ambrose Everett

BORN: May 23, 1824
DIED: September 13, 1881
SERVICE: Civil War (1861–65)

Popular but mediocre U.S. general in the Civil War

Burnside was born in Liberty, Indiana, and gained admission to the U.S. Army Military Academy at West Point, from which he graduated in 1847. Commissioned a second lieutenant in the artillery, Burnside, like many promising young officers of the period, soon found the army an unrewarding career and resigned his commission in 1853 to take up management of a gun factory in Bristol, Rhode Island. Burnside became an able gunsmith and designed an innovative breech-loading carbine in 1856, which nevertheless failed to save him from bankruptcy the following year. Following this failure, Burnside became an executive with the Illinois Central Railroad until the Civil War commenced in 1861. He returned to Rhode Island to command the First Rhode Island Volunteer Regiment in April, and by the First Battle of Bull Run (on July 21, 1861), he assumed brigade command. On August 6 he was promoted to brigadier general of volunteers and successfully mounted an amphibious expeditionary force in December, which he led against Confederate positions along the North Carolina coast during January 1862. Burnside took Roanoke Island on February 7 and New Bern on March 14. His unit sunk

Ambrose Burnside (seated at center) with Rhode Island regimental officers and staff. Courtesy Library of Congress.

Confederate ships riding at anchor in the Albemarle and Pamlico sounds. In recognition of his accomplishments, Burnside was promoted to major general of volunteers in March and, after his unit was redesignated IX Corps, he and his troops were transferred to regular federal service in the Army of the Potomac.

On September 14, 1862, Burnside commanded the right wing (I and IX Corps) at the Battle of South Mountain, and three days later was in command of the left wing at Antietam, a battle that should have gone to the Union but ended as a bloody draw. It was Burnside's deliberation, coupled with the over-caution of his commander, General **George B. McClellan,** that cost the victory. Nevertheless, President **Abraham Lincoln,** despairing of McClellan's lack of aggressiveness and enterprise, replaced him with Burnside as commander of the Army of the Potomac in November.

Burnside set about unwisely reorganizing the structure of the army into grandiose divisions of two corps each. These gave the Army of the Potomac an impressively streamlined appearance on paper, but soon proved unwieldy. Facing the great Confederate general **Robert E. Lee** at Fredericksburg on December 13, Burnside was badly defeated in a battle that cost 12,700 dead or wounded Union troops—even though Burnside's forces outnumbered Lee's by a considerable margin.

Like McClellan before him, Burnside was now replaced as commander of the Army of the Potomac on January 26, 1863. Burnside had accepted command of the Army of Potomac with great reluctance, for he was keenly aware of his own shortcomings. A man of lesser character might have resigned his commission after the Fredericksburg debacle, but Burnside accepted a more subordinate command. In charge of the Department of the Ohio in March, he successfully repulsed a raid into Ohio led by **John Hunt Morgan,** then took the offensive and swept through Tennessee, clearing the

Cumberland Gap of Confederate forces. He took Knoxville, Tennessee, and held it against the onslaught of General **James Longstreet** during September–December.

A poor strategic commander, Burnside was often an excellent subordinate and headed up IX Corps in General **George Meade**'s Army of the Potomac with great skill during the Wilderness campaign (April–May, 1864); however, at the siege of Petersburg during June 19–December 31, he suffered a bad defeat at the Battle of the Crater (July 20). The matter was serious enough to merit a court of inquiry, which found him culpable and prompted his resignation on the verge of the war's end, in April 1865. Burnside resumed railroad work, then gained election as governor of Rhode Island in 1866 (serving until 1869) and entered the U.S. Senate in 1874. He died in office in 1881.

Burnside was a handsome military figure whose characteristic mutton-chop whiskers became a popular fashion and have been known by a version of his name ever since: sideburns. He was extremely popular with his troops, for whom he had high regard. Often an able subordinate, he was not capable of independent command, and in this, unfortunately, was typical of many Union army general officers. **Ulysses S. Grant** evaluated Burnside frankly in his memoirs, calling him "an officer who was generally liked and respected. He was not, however, fitted to command an army. No one knew this better than himself."

BYNG, Julian Hedworth George, First Viscount Byng of Vimy

BORN: September 11, 1862

DIED: June 6, 1935

SERVICE: Mahdist War (1883–98); Second (Great) Anglo-Boer War (1899–1902); World War I (1914–18)

One of the best British generals of World War I

Byng was born into a privileged life at Wrotham Park, Barnet, the son of George Stevens Byng, Second Earl of Strafford, and was educated at Eton. He entered the militia, where he distinguished himself sufficiently to be gazetted into the elite Tenth Hussars in 1883. The following year, he was dispatched with his unit to the Sudan to fight the forces of the Mahdi there. By 1889, he had attained the rank of captain, and in 1894 graduated from the Staff College. Promoted to major in 1898, he was sent to South Africa in the fall of 1899 to fight in the Second Boer War. His first task was to form a unit of local horse soldiers under Sir Redvers Henry Buller in Natal. Leading these, Byng fought successfully at the Battle of the Tugela on February 17–18, 1900, and rode to the relief of Ladysmith on February 27. Gaining promotion to lieutenant colonel, Byng was given command of the South African Light Horse and was subsequently appointed to a position on **Lord Kitchener**'s staff.

By the end of the war in 1902, Byng was colonel of Tenth Hussars, and, two years later, was named to head the cavalry school at Netheravon. Elevated to brigade command in 1907, he was promoted to major general in 1909 and the following year assigned to command the East Anglia Territorial Division. Two years later, he was given overall command of Britain's forces in Egypt.

With the outbreak of World War I, Byng was recalled to Europe, where he led the Third Cavalry Division at the First Battle of Ypres during October 30–November 24, 1914. He commanded the Cavalry Corps on the western front until August 1915, when he headed IX Corps at Suvla Bay during the ill-fated Gallipoli campaign (August 1915–January 1916). As commander of the Canadian Corps in Flanders, he was instrumental in the capture of Vimy Ridge on April 9, 1917, the opening day of the Arras Offensive. Added to this triumph was his excellent performance as general and commander of Third Army—a post to which he was appointed in June and from which he planned and executed the Cambrai Offensive during November 20–December 3. Although an officer of the old school, Byng used the Cambrai Offensive to test and demonstrate the effectiveness of the latest technology—the tank. However, like the majority of campaigns throughout World War I, the Cambrai operation won no significant breakthrough. Byng enjoyed greater success in resisting the German "Operation Michael" offensive along the Somme during March 21–April 5, 1918, and he was in the forefront of the final Allied offensives in Flanders during August–November.

Following the Armistice, Byng was created a baron (1919) and appointed governor general of Canada, where he served to great popular acclaim from 1921 to 1926. Made a viscount in 1926, he was appointed commissioner of London's Metropolitan Police (1928–31). He brought to this assignment the same keen administrative and logistical skills that he had employed during his military career and was responsible for introducing many reforms that helped make the London police one of the world's premier law enforcement agencies. On his retirement from public life in 1932, Byng was promoted to field marshal.

C

CAESAR, Gaius Julius

BORN: July 13, 100 B.C.

DIED: March 15, 44 B.C.

SERVICE: Third Mithridatic War (75–65 B.C.); Gallic War (59–51 B.C.); War with Pompey (49–45 B.C.)

Greatest military commander of his era; most famous of all Roman emperors

Bearer of the most famous name in Roman history, Julius Caesar was one of the greatest military commanders and most skillful politicians the world has ever known. He was born to a venerable patrician family and claimed direct descent from Venus through Aeneas's son Iulus (Ascanius). Despite their lofty pedigree, Caesar's family was excluded from the inner circles of power. His father was the brother-in-law of Gaius Marius and married to Aurelia, a member of the prominent Aurelii family, yet he died, about 85 B.C., before attaining a consulship. In 84 B.C., his son Caesar entered the priesthood of Jupiter and married Cornelia, daughter of Marius's former partner, Lucius Cornelius Cinna. When **Lucius Cornelius Sulla,** the enemy of Marius, ordered Caesar to divorce her, the young man refused and was compelled to endure a brief period of exile.

Caesar served during this time in the legions posted to Asia, making himself conspicuous for bravery at the siege of Mytilene in 80 B.C. Returning to Rome after the death of Sulla in 78 B.C., he unsuccessfully attempted during 77–76 B.C. to prosecute two of his family's enemies, Gnaeus Cornelius Dolabella and Gaius Antonius Hibrida, both partisans of Sulla. Departing Rome for study in Rhodes, Caesar was captured enroute by pirates, from whom he was ransomed. After assembling a small private army, he succeeded in capturing the pirates in turn and brought about their execution in 75–74 B.C.

War with Mithradates VI of Pontus in 74 B.C. drew Caesar away from Rhodes as he served under Lucullus against this ruler during 74–73 B.C. For his services in battle, Caesar was made a pontiff at Rome in 73 B.C. and elected military tribune. He then saw further action, possibly against the rebellious slave Spartacus in 72 or 71 B.C., and he supported **Pompey,** chief architect of the downfall of the Sullan political system.

Caesar was elected quaestor in 69 B.C., then earned popularity among the Transpadane Gauls by supporting their bid for Roman citizenship in 68 B.C. Following the death in 69 B.C. of his wife Cornelia, Caesar the following year married Pompeia, granddaughter of Sulla and relative of Pompey. He apparently carried out high-level military assignments for Pompey in 67 and 66 B.C., then became aedile in 65 B.C., achieving great popularity by personally financing elaborate public games. During this year, he seems to have participated with Marcus Licinius Crassus in a scheme to annex Egypt. He also promoted the popular land-distribution bill of Publius Servilius Rullus. In 64 B.C. he presided over trials of persons who had committed murder during Sulla's proscriptions against the partisans of Marius, and in 63 B.C., Caesar employed outright bribery on a vast scale in order to become pontifex maximus, head of Roman state religion.

Although Caesar did not participate in the infamous conspiracy to seize power fomented by the demogogue Catiline in 63 B.C., he did oppose the execution of Catiline's accomplices, which gained him much public favor, leading to his election as as praetor in 62 B.C. Shortly after this, he divorced Pompeia on suspicion of infidelity, and married Calpurnia in 58 B.C. In the meantime, he became governor of Further Spain in 61 B.C. and was elected in 60 B.C. to the consulate.

That year, he formed with Pompey and Crassus the First Triumvirate in order to concentrate power and influence. Caesar became proconsul of Illyricum, Cisalpine Gaul, and Transalpine Gaul.

From 58 to 51 B.C., Caesar used the large army that was put at his disposal to fight the Gallic Wars, which resulted in the Roman subjugation of the rebellious Gauls. He emerged from the campaign with tremendous political prestige and leverage. In the meantime, Caesar's daughter, Julia, had married Pompey in 59 B.C. Despite this highly political marriage, friction developed between Caesar and Pompey, which Crassus, jockeying for additional power, deliberately aggravated. The Triumvirate was renegotiated in 56 B.C., but the death of Julia in 54 B.C. and of Crassus the next year combined with Caesar's Gallic triumphs to destroy the relationship with Pompey once and for all. In 50 B.C., Pompey opposed Caesar's bid for a second consulate. On January 10, 49 B.C., Caesar initiated civil war when he committed the illegal act of leading his army across the Rubicon to oppose Pompey in Italy. Pompey's forces collapsed before Caesar's advance, and Pompey retreated to Greece.

In August 49 B.C., Caesar defeated Pompeiian forces in Spain and was created dictator of Rome. He pursued Pompey into Greece, where he suffered a loss at Dyrrhachium, but quickly recovered by destroying Pompey's larger army at Pharsalus on August 9, 48 B.C. Pompey fled to Egypt, with Caesar in pursuit. In this manner, Caesar became involved in the civil war between Cleopatra and her brother Ptolemy XIII. Caesar took Cleopatra as his mistress and made her queen of Egypt.

The next year, Caesar went to Anatolia, where he crushed Pompey's ally Pharnaces, king of Bosporus, in a brief campaign at Zela. So swift was this victory that Caesar reported it in a single memorable phrase: *Veni, vidi, vici* ("I came, I saw, I conquered").

After returning briefly to Rome later in the year, Caesar had to travel to North Africa in December to head off another threat from forces loyal to Pompey. Following victory at Thapsus in April 46 B.C., he was made dictator for ten more years. The sons of Pompey—who had been murdered in Egypt—now mounted a fresh resistance in Spain. Caesar met and defeated them at Munda on March 17, 45 B.C. The following year Julius Caesar was appointed dictator for life and was heaped with additional honors.

Few great soldiers make great political administrators, but Caesar was one of those few. He introduced many reforms, including the enlargement of the Senate, the revision of the system of taxation, and the extension of Roman citizenship to all subjects of the empire. He sought to balance these popular measures with gestures meant to placate those who opposed him, granting unheard of clemency to his enemies. Yet he made the mistake of comparing himself to **Alexander the Great** and proposed to do as he had done: conquer Parthia. Among the Roman aristocracy, such boldness triggered fear of the dictator's overweening ambition. Even those to whom Caesar had extended pardons came to distrust him.

A band of conspirators led by Marcus Junius Brutus and Gaius Cassius Longinus, approached Caesar at a meeting of the Senate in Pompey's theater on March 15, 44 B.C.—the Ides of March. Each assassin stabbed him in turn, and Caesar, as he collapsed at the feet of Pompey's statue, spoke to Brutus not in Shakespeare's Latin—*Et tu, Bruté*—but in Greek: "Even you, lad?"

CAMPBELL, Sir Colin, First Baron Clyde

BORN: October 20, 1792

DIED: August 14, 1863

SERVICE: Napoleonic Wars (1800–15); War of 1812 (1812–15); First Opium War (1839–42); Second Sikh War (1848–49); Crimean War (1853–56); Sepoy Rebellion (1857–58).

Dubbed "Old Khabardar" (Old Careful) because of his conservative approach to battle, Campbell was a British "solder's soldier," who did not believe in squandering lives to win battles

A Scotsman through and through, Campbell was a native of Glasgow, born of modest circumstances to a carpenter. He changed his surname from Macliver to his mother's maiden name after her brother, Colonel John Campbell, purchased for him an ensign's place in the Ninth Regiment of Foot in 1807. His maiden campaign was not a happy one. The Walcheren expedition of 1809 proved disastrous, but his next service, directly under the Duke of **Wellington** in the Peninsula, began to build for him a solid reputation. He was decorated for gallantry at Barrosa, Spain, on March 5, 1811 and served with distinction at Vitoria on June 21, 1813, the siege of San Sebastian (July 25–August 31), and the crossing of the Bidassoa (October 7–8). Fearless in battle, Campbell endured a number of serious woundings.

During the War of 1812, Campbell was dispatched to the United States, where he participated in Sir Edward Pakenham's disastrous assault on New Orleans during December 1814–January 1815, an expedition that was defeated by **Andrew Jackson.** His next war service did not come until 1823, when he took part in the suppression of the Demerara insurrection in British Guiana. He then fought in the First Opium War, but went on to earn significant recognition for his gallantry during the Second Sikh War.

During the Crimean War, Campbell wa given command of the Highland Brigade, which he led at the battle of the Alma (September 20,1854). Most famously, he commanded the Ninthird Highlanders—the celebrated "Thin Red Line"—which halted the Russian advance at Balaklava on October 25, 1854.

With the outbreak of the Sepoy Rebellion, Campbell was appointed commander in chief of colonial forces. He led the second relief of the besieged city of Lucknow on November 16, 1857, and scored signal victories at Tantia Topi and at Cawnpore in December. He invaded and occupied Farrukhabad on January 6, 1858, then took the offensive against Lucknow during March 2–16. The city fell to him on March 21.

On May 5, 1858, Campbell defeated the important rebel leader Khan Bahadur near Bareilly, then conducted final mop-up operations in Oudh during May and June. A grateful Queen Victoria created the commoner Campbell Baron Clyde and, perhaps more significantly, awarded him an annual pension of £2,000. He was promoted to field marshal in 1862, a year before his death.

CANBY, Edward Richard Spring

BORN: 1817

DIED: April 11, 1873

SERVICE: Second Seminole War (1835–42); U.S.–Mexican War (1846–48); Civil War (1861–65); Modoc War (1872–73)

American general; commander of Union forces in the Southwest during the Civil War and later commander of the U.S. Army's Pacific Division; the only regular army general officer killed in the Indian Wars

Edward Richard Spring Canby became well known to Americans as the only regular army general to die in an Indian War when he was murdered by a Modoc Indian named Captain Jack. The public outrage at his death was equalled only by that following the death of **George Armstrong Custer,** who, though holding the brevet rank of general, was a mere lieutenant colonel when he fell in battle at the hands of the Sioux near the Little Big Horn River. Called "The Prudent General" by a biographer, Canby had—unlike the flamboyant Custer—earned in his day a reputation for caution and dependability. Certainly no one had ever accused him of military brilliance, but he was a steady fellow, a deliberate man. Yet it had been an impudent act—perhaps the only impudent act in his long army career—that cost him his life in the Modoc War.

Born in Boone County, Kentucky, Canby had graduated at the bottom of his class at West Point in 1839 (next-to-last, in fact). Nevertheless, he performed well under fire in the Second Seminole War and he did what he was told to do in the tragic removal of the five Civilized Tribes from the Southeast. He held his own in the war between the United States and Mexico, garnering brevets to major and to lieutenant colonel while serving as chief of staff of a brigade. When the Union's defenses collapsed in the Southwest at the onset of the Civil War, Canby rallied the troops sufficiently to continue an effective enough resistance even after his defeat at the Battle of Valverde by Confederate General H. H. Sibley. When the Union's California Column and Colorado

Volunteers arrived to reinforce him, Canby even managed to chase the rebels back into Texas. It was a good, solid performance, and when New York's Draft Riots broke out in 1863, Canby was relieved of the staff duties he had been recalled to the East to perform and was given command of the city's armed forces to quell the disturbance. He picked up the pieces after General Nathaniel Banks botched the Red River campaign, and he was there to cooperate with the Union Navy in the capture of Mobile, for which **Abraham Lincoln** gave Canby his personal thanks. Not surprisingly, then, after such a respectable showing in the Civil War, Canby was promoted in 1866 to brigadier general in the regular army and in 1870 was given command of the Department of the Columbia. Scarcely three years later, he found himself in command of the entire Division of the Pacific.

But it was as head of a peace commission to end the Modoc War, that Canby slipped up, and it cost him his life. Partly, it was arrogance. He knew these Indians, the Modocs, yet he stubbornly ignored warnings of their treachery. He was certain he could depend on his "carrot-and-stick" notions about Indian diplomacy. The last message he sent to Captain Jack before meeting him was, "If you will come in and surrender to us, we will . . . allow you a voice in selecting your future home. If you do not comply . . . we will resign and leave the matter with the military, and you will have no choice in the selection of your future home." At a peace tent pitched in California's desolate Lava Beds, Captain Jack shot Edward Richard Spring Canby in the face.

It was April 11, 1873, and the United States' always troublesome relationship to the Native Americans had taken a turn for the worse. The sympathy that the plight of the plundered Indians had won from the American press and public vanished. It was not enough that Captain Jack should be hanged for his deed—and he would be—but the head of the U.S. Army, **William T. Sherman,** was emboldened to unleash the kind of total war on the Indians that he had perfected against the South in the Civil War. Capturing the mood of the country precisely, Sherman called for the "utter extermination" of the Modocs.

CARLETON, Sir Guy

BORN: September 3, 1724

DIED: November 10, 1808

SERVICE: French and Indian War (1754–63); Seven Years' War (1755–63); American Revolution (1775–83)

Perhaps the ablest British general to fight in the American Revolution

Irish-born Carleton received his ensign's commission on May 21, 1742 and was promoted to lieutenant on July 22, 1751, upon his transfer to the First Foot Guards. During the French and Indian War, he served under Lord Jeffrey Amherst at the siege of the key Newfoundland port town of Louisbourg during June 8–July 26, 1758. After transferring to the Seventy-Eighth Regiment of Foot in August, he was promoted to acting colonel and served under General **James Wolfe,** a close friend. Carleton took part in the long campaign against Quebec (June 26–September 13, 1759), and was wounded while leading his grenadiers across the Plains of Abraham against Quebec City on the final day of the battle.

He sailed back to Europe and, elevated to the rank of acting brigadier general, he participated in the siege of Belle Île, off the coast of France during in 1761. Crossing the Atlantic once again, he distinguished himself at the siege of Havana (June 20–July 30, 1762), where he was severely wounded on July 22. In September 1766, he was appointed royal lieutenant governor of Quebec and then became governor, serving from 1767 to 1770, when he returned to England as colonel of Forty-Seventh Regiment of Foot (promoted April 2, 1772). Shortly after this, on May 25, Carleton was promoted to major general and returned to Canada in 1774. Greeted as a hero, he was appointed governor of Quebec on January 10, 1775.

At the outbreak of the American Revolution, Carleton was subordinate to General **Thomas Gage** until Gage was recalled to England in October 1775. Secure in the confidence that French Canada was now loyal to the British crown and to himself, Carleton dispatched the bulk of the forces under his command to Boston in the wake of the Battle of Concord. This resulted in an uprising in Quebec, and Carleton's declaration of martial law in June.

Carleton responded vigorously to early American offensives in Canada. He withdrew troops from Montreal after General Richard Montgomery advanced on it in November, but successfully repulsed Montgomery's and Benedict Arnold's assault on Quebec (December 31). By May 1776, Carleton had driven the Americans out of Quebec and then pursued them cautiously. When he saw that Benedict Arnold was massing a naval squadron on Lake Champlain, Carleton built one of his own, bided his time, and then penetrated southward, where he cut up Arnold's squadron at Valcour Island on October 11. Once again, he resumed the offensive and demolished the remainder of the flotilla at Split Rock on October 13.

Carleton was less successful in his support of General **John Burgoyne**'s expedition through New York in pursuit of a plan to cut off New England from the other colonies (1777). Ironically, Carleton strongly disliked "Gentleman Johnny" Burgoyne, but was the only British commander who made any attempt to coordinate action with him.

On August 29, 1777, Carleton was promoted to lieutenant general and left Canada to become governor of Charlemont (Armagh), Ireland, in July 1778. The British might well have made better use of his military skills in America; for when he returned to the colonies as commander in chief on May 5, 1782, the war was well on its way to a conclusion.

Carleton left New York on November 25, 1783, and returned to England, where he was created Baron Dorchester in 1786, then returned to Canada in October of that year for a third term as governor. He served until 1798, when he retired to Ireland.

Carleton was an excellent commander serving at a time when England had great need of excellent commanders. Equally valuable to the crown was his even-handed judgment as a civil administrator. Unlike many British governors of Canada, he was content to give Catholic French Canadians wide religious latitude, thereby avoiding much—though not all—potential dissension and preventing the American rebellion from spreading northward.

CARLSON, Evans Fordyce

BORN: February 26, 1896

DIED: May 27, 1947

SERVICE: World War I (1917–18); World War II (1941–45)

Controversial, innovative U.S. general whose elite "Carlson's Raiders" pioneered the commando concept in World War II

A native of upstate Sidney, New York, Carlson enlisted in the army in 1912 and saw service in the Philippines and Hawaii from 1912 to 1915, when he was discharged. He was recalled the following year to serve under General **John J. Pershing** during the "Punitive Expedition" against Pancho Villa along the Mexican–American border. He remained in the army through World War I, gaining promotion to captain and served as assistant adjutant general to Pershing from 1917 to 1919.

Carlson resigned in 1919 and became a salesman, only to reeneter the military in 1922—not as an army officer, but as a private in the U.S. Marine Corps. He was commissioned second lieutenant the following year and was assigned as an intelligence and operations officer in Shanghai during the turbulent closing years of the 1920s. In 1930, he was dispatched to revolution-wracked Nicaragua, then returned to China during 1933–35.

Carlson returned to the United States during 1935–37, when he was assigned to President Franklin D. Roosevelt's military guard at the "little White House" in Warm Springs, Georgia. During this period, he became a close personal friend of the president.

As the Sino–Japanese War heated up in the late 1930s, Carlson was again dispatched to China as an observer (1937–39) and extensively accompanied troops of the Communist Eighth Route Army. He came away from this experience thoroughly impressed with the training, ability, and passion of the troops. His unstinting praise of a Communist military organization, however, brought criticism in the United States and prompted his resignation from the Marines in April 1939.

Carlson became a popular lecturer throughout the United States and authored two books on the Chinese army. As it became clear to him that the United States would inevitably enter World War II, he rejoined the Marines as a reserve major in April 1941, and was transferred to active duty the next month.

When it became apparent early in the Pacific war that unconventional and innovative techniques were required for fighting the Japanese in jungle-island environments, Carlson was called on to command the Second Marine Raider Battalion (1942). Using what he had learned during his Chinese sojourn, Carlson transformed the unit into what was soon called Carlson's Raiders, an elite ranger/commando force that first proved itself in a lightning raid on Japanese positions on Makin Island (August 17–18, 1942).

Carlson's Raider's were instrumental in the "island hopping" strategy of the Pacific war. During the invasion of Guadalcanal (November–December, 1942) Carlson led his unit in operations behind enemy lines. The Raiders served as advance units during the critical landings at Tarawa (November 20–24, 1943) and Saipan (June 15–July 13, 1944). At Saipan, Carlson was gravely wounded. Although promoted to brigadier general, his wounds forced his retirement in 1946.

Carlson never fully recovered. He remained a public figure after the war, vigorously advocating the then-unpopular idea of improved relations with China, but his influence was cut short by his death in the spring of 1947.

CARNOT, Lazare Nicolas Marguerite

BORN: May 13, 1753

DIED: August 2, 1823

SERVICE: French Revolutionary Wars (1782–99); Napoleonic Wars (1800–15)

"The Organizer of Victory" in the French Revolutionary Wars

The son of a lawyer in Nolay, Burgundy, Carnot had the advantages of a thorough education in engineering—traditional preparation for young military officers. He was commissioned in an engineer unit in 1773 and when he was posted at Arras, in 1780, met and befriended the fiery revolutionary Maximillian Robespierre. Carnot soon was elevated to captain and was elected to the Legislative Assembly as deputy for Pas-de-Calais in 1791. He was appointed to the military committee charged with responsibility for inspecting the Army of the Rhine in August 1792, and when he returned from this assignment he was elected to the revolutionary National Convention. He took part in the trial of King Louis XVI and was among those who voted for the monarch's immediate execution in January 1793. With the revolution in full swing, Carnot was named to the Committee of Public Safety (August 14, 1793), with responsibility for military affairs. He well understood that revolutionary France was vulnerable to attack from the outside and quickly reorganized and built up the French military. This accomplished, he served wherever he was needed: with the Army of the North, to relieve a besieged Dunkirk on September 8, 1793, and to Wattignies, to fight against Austro-Allied forces during October 15–16. Through this, Carnot remained politically savvy during the Jacobin regime (1793–94), so that he avoided close "contamination" by his association with Robespierre and thereby escaped the wrath of the anti-Jacobin reaction in August 1794.

Despite his political acumen, Carnot was distrusted by the republic's moderates. He stepped down from the Committee of Public Safety, and although he was promoted to major of engineers (March 21, 1795), he narrowly escaped a vote of proscription in the National Convention. He was saved when a member interjected into the debate that "Carnot has organized victory." The moniker "Organizer of Victory" stuck. Politically rehabilitated, Carnot was elected to the Directory (the principal legislative body of the pre-Napoleonic republic) on November 2, but was soon disappointed and frustrated by the actions of his fellow Directors and by the military, which was defiantly unresponsive to the civil government. At last, Carnot was swept up in the coup of Fructidor (September 1797) and forced to flee to Switzerland. He did not return until **Napoleon Bonaparte** was elected consul.

Under Napoleon's consulship, Carnot worked in the War Ministry beginning in 1800, but voted in 1802 against bestowing the consulship of Bonaparte for life.

He was even less pleased with Napoleon as emperor and retired from public life in 1807. He wrote two important works on geometry and a seminal tactical text on the art of fortification (1810). Despite his dislike of Napoleon, he returned to the colors in January 1814, as a major general when France was threatened with invasion, and directed the defense of Antwerp during January 14–May 4. During the Hundred Days (April–June 1815), Carnot served as minister of the interior and then briefly assumed leadership of the provisional government. With the restoration of Louis XVIII, Carnot was proscribed and exiled as a regicide (he had voted to execute Louis XVI). Carnot died in exile at Magdeburg, Austria.

CASSIUS LONGINUS, Gaius

BORN: 85 B.C.?

DIED: 42 B.C.

SERVICE: Roman Civil War (50–44 B.C.); War against the Assassins (43–42 B.C.)

Mastermind of the assassination of* Julius Caesar *and a powerful, contradictory leader—bold and skilled as a military commander with a passion for discipline verging on that of a martinet; a committed republican, he was nevertheless unpopular with the people

Little is known about the birth or early life of Gaius Cassius Longinus. He was a *quaestor* under Marcus Licinius Crassus (member of the First Triumvirate) during the invasion of Parthia in 53 B.C. When the Romans were routed at Carrhae, Cassius Longinus escaped to Syria and the following year was sent to quell a rebellion in Judea. In 51 B.C., he was instrumental in repulsing a Parthian invasion of Syria and was made a tribune at Rome in 49 B.C. As such, he supported **Pompey the Great** against **Julius Caesar** and served as Pompey's naval commander off Sicily in 48 B.C.

When Pompey was defeated at Pharsalus, Caesar pardoned Cassius, who went on to became a *legate* in 47 B.C. and a *praetor* in 44 B.C. Cassius, of course, proved treacherous and was the principal plotter of Caesar's assassination on March 15, 44 B.C. Following the assassination, Cassius fled Rome, securing a Senatorial appointment as governor of Cyrene. At this point, he quarreled with Mark Antony, broke with him, and took control of Roman forces in the East. The Senate belatedly awarded command of these forces to Cassius and Marcus Junius Brutus, but then proscribed Cassius in August 43 B.C. at the behest of Octavian (later Emperor Augustus).

Cassius and Brutus crushed Rhodes after it declined to support them, then crossed into Thrace. In mid-October of 42 B.C. their forces were attacked by the combined armies of Antony and Octavian at the titanic Battle of Philippi. The battle was a draw, but Antony captured Cassius' camp. Cassius was unaware that Brutus had defeated Octavian's forces. Giving up all for lost, Cassius committed suicide.

CATO, Marcus Porcius Cato ("Cato the Elder")

BORN: 234 B.C.

DIED: 149 B.C.

SERVICE: Revolt in Spain (195 B.C.); Third Punic War (149 B.C.–146 B.C.)

Conservative Roman foe of Carthage

A native of Tusculum, Cato was most likely born into the local aristocracy; however, tradition has painted him as being of peasant origins. He first made a name for himself during the Second Punic War, when he served as a military tribune and distinguished himself at the Battle of Metaurus River (207 B.C.). With the help of the influential L. Valerius Flaccus, Cato was named *quaestor* in Sicily under **Scipio Africanus** in 204 B.C. His rise was steady following this, as he became plebeian *aedile* in 199 B.C. and *praetor* of Sardinia—in effect, governor of the province—in 198 B.C.

By 195 B.C., Cato was consul with Flaccus and fought in Spain, suppressing a revolt (195 B.C.) there and reorganizing the provincial government. On his return to Rome, Cato was granted a triumph—an official hero's welcome. In 191 B.C., he was dispatched to Greece under Manius Acilius Glabrio to fight the Seleucian forces of King Antiochus III the Great. Here he played a key role in the momentous Battle of Thermopylae, in which Aniochus was routed.

With his reputation firmly established through military exploits, Cato became unremitting in his eloquent condemnation of many prominent Romans, especially Scipio Africanus Major. Cato's eloquence and reformer's zeal gained him election to the office of censor along with Flaccus in 184 B.C. Cato pursued an ultraconservative course and was viewed as a stern moralist. He unsuccessfully counseled against unbridled expansion of Rome; however, he developed an implacable hatred and distrust of Carthage and obsessively advocated its utter annihilation. He habitually concluded each of his Senatorial speeches with the sentence, *Ceterum censeo Carthaginem esse delendam* ("Furthermore, I believe that Carthage must be destroyed"). Ultimately, his agitation helped provoke the Third Punic War, the beginning of which Cato—with satisfaction—saw, shortly before his death.

Although opposed to Greek influences, Cato contributed much to Roman culture. He patronized the poet Quintus Ennius and stimulated Roman rhetoric by publishing his own speeches. *Cato's Origins*, the first history of Rome in Latin, set the pattern for Latin prose. Unfortunately only his *On Farming* survives.

CETSHWAYO

BORN: ca. 1826

DIED: February 8, 1884

SERVICE: Civil War (1856); Zulu War (1879)

The last great king of Zululand; an acute and charismatic military leader who united his nation against the British and Afrikaners

Cetshwayo was one of the many sons of the Zulu king Mpande and secured his succession to the throne by killing his half-brother Mbulazi in a civil war during 1856. Having disposed of his principal rival, Cetshwayo devoted the next sixteen years to purging lesser rivals, so that by the time Mpande died in 1872, Cetshwayo remained as the undisputed king.

The next threat, the new king recognized, came not from within, but from British and Afrikaner colonists in South Africa. Accordingly, Cetshwayo set about creating an army of forty thousand warriors. This action did not escape British notice. Early in 1879, British colonial forces made a preemptive strike. It failed—at first. However, reinforcements soon arrived, and the Zulu army was worn down. Cetshwayo was taken prisoner and confined at Cape Town. His kingdom was partitioned among thirteen chiefs designated by the British.

The Zulu emperor did not give up, and in 1882 managed to persuade British authorities to release him. He sailed to London, where he was granted an audience with Queen Victoria, who, overwhelmingly impressed with Cetshwayo, reinstated him as Zulu king. He returned to Africa to try to reunited his disintegrated realm. Civil War resumed among the many factions created and empowered by the British partition, and before he could organize a military force, Ceshwayo mysteriously died, perhaps the victim of poisoning.

As a military leader, Cetshwayo showed a natural ability to organize a large force and use it effectively. Only the superior numbers and weaponry of the British colonial army defeated him.

CHAFFEE, Adna Romanza, Jr.

BORN: September 23, 1884

DIED: August 22, 1941

SERVICE: Cuban Pacification (1906–1907); World War I (1917–1918)

"Father" of the U.S. armored force

Chaffee was the only son of the remarkable Adna Chaffee, Sr., who had joined the army as a private during the Civil War and, by the end of his career, would become chief of staff of the army, with the rank of lieutenant general. The junior Chaffee, steeped in the military, graduated from West Point in the upper half of his

1906 class. Assigned to the First Cavalry, he served in the so-called Army of Cuban Pacification (1906–07), then returned to the States to attend the Mounted Services School at Fort Riley, Kansas, from 1907 to 1909. Chafee excelled at horsemanship, not only creating and commanding a special mounted detachment for the Army War College (1910–11), but competing in the prestigious International Horse Show during the festivities for the coronation of King George V of Great Britain in 1911. Chaffee was given the honor of attending Saumur, the French cavalry school, widely recognized as the finest military equestrian school in the world (1911–12). He returned to Fort Riley as an instructor at the Mounted Services School (1912–13), and then shipped out with the Seventh Cavalry in the Philippines (1914–15), returning in 1916 to West Point as senior cavalry instructor in the Tactical Department.

Promoted to captain, Chaffee was assigned in August 1917, as adjutant to the Eighty-First Division at Camp Jackson, South Carolina, and briefly served as acting chief of staff for the division from December 1917, to February 1918, when it sailed for France and the Western Front of World War I. At Langres, France, Chaffee attended the General Staff College (March–May, 1918) and afterward served there as an instructor. In August 1918, he was appointed assistant G-3 (operations officer) for IV Corps and, soon afterward, G-3 of the Eighty-First Division and of VII Corps (August–October) during offensives at St. Michel (September 12–17, 1918) and Meuse-Argonne (September 26–November 11). During the last days of the war, Chaffee served with the temporary rank of colonel and was G-3 of III Corps. He was decorated with the Distinguished Service Medal, then remained in Europe with III Corps as part of the army of occupation in the Rhineland.

Chaffee returned to the United States as an instructor at the Line and Staff School (later called the Command and General Staff School) at Fort Leavenworth, Kansas, from July, 1919, to May 1920. He had reverted to his permanent rank of captain in August, but was promoted to major in 1920. After brief service with the Third Cavalry during May–August 1920, he was reassigned as G-2 (intelligence officer) of the IV Corps area in Atlanta from August 1920–21), and then as G-3 of the First Cavalry Division at Fort Bliss, Texas, serving there until July 1924.

After attending the Army War College from July 1924, to May 1925, Chaffee was assigned command of a squadron of the Third Cavalry at Fort Myer, Virginia, a plum posting just outside Washington, D.C. He served here until June, 1927, when he was posted to the Operations and Training Division of the War Department General Staff. In this post, Chaffee, now a lieutenant colonel, was instrumental in developing the army's mechanized forces. In 1931, he was assigned to command the newly formed First Cavalry (Mechanized) Regiment at Fort Knox. Serving through February 1934, as regimental commander and then as overall commander of the First Cavalry (February–June 1934), Chaffee did much to establish, train, and organize the army's first armored unit. He was also instrumental in developing basic doctrine of mechanized warfare.

From June 1934, to June 1938, Chaffee headed up the Budget and Legislative Planning Branch of the War Department General Staff, and he used this position to fund increased mechanization. Returning to Fort Knox, he again served as commander of First Cavalry (Mechanized) (June–November, 1938) and, promoted to brigadier general on November 3, was made commander of the Seventh Mechanized Brigade (First and Thirteenth Mechanized Cavalry Regiments).

With the United States perilously close to entering World War II, Chaffee directed the Seventh Brigade in large-scale maneuvers at Plattsburgh, New York (1939) and in Louisiana (summer 1940). This practical experience allowed him to refine and further develop American doctrines of armor. The key concept Chaffee formulated was the combination of armored, infantry, and artillery operations, the integration of which would prove crucial in the coming war.

Chaffee was made commander of the Armored Force on June 10, 1940. This gave him responsibility for all infantry tank and mechanized cavalry units, with supporting artillery, motorized infantry, and engineer units. He created the First and Second Armored

Divisions and was promoted to major general on October 2, 1940. Unfortunately, Chaffee was felled by cancer at the height of his career. He died the summer before Pearl Harbor and thus the U.S. Army was deprived of one of its most valuable officers on the eve of the greatest conflict the nation had ever faced.

CHAMPLAIN, Samuel de

BORN: ca. 1567

DIED: December 25, 1635

SERVICE: French Wars of Religion (1562–98); Indian Wars in Canada (1609–35); Anglo–French War (1626–30)

An able soldier and an energetic explorer; perhaps the single most important early figure in the establishment of New France in North America

Born into a family of sailors in Brouage, a seafaring town on the Bay of Biscay in western France, Champlain became a soldier rather than a seaman and joined the French army to fight against Spanish occupation (1593–97) during the late phase of the French Wars of Religion. He subsequently sailed under Spanish sponsorship to Puerto Rico, Cuba, Mexico, and Central America during 1599 to 1601, then, sponsored this time by the governor of Dieppe, France, he joined a fur-trading expedition to Canada in 1603. Champlain sailed southwest up the St. Lawrence River, reaching the present location of Montreal (he would found the city in 1611). The following year, 1604, Champlain returned to North America and, through 1607, explored the east coast of Canada and Nova Scotia, traveling as far south as Cape Cod and Martha's Vineyard in present-day New England. Champlain carefully charted and mapped his explorations.

Unlike many Europeans who probed the continent, Champlain had a genuine regard for the Native Americans. In 1608, he founded the settlement of Quebec and established Quebec City, which, thanks to him, quickly became a peaceful center of Indian fur trade. To solidify profitable trading relations with the Hurons, Champlain commanded several French military expeditions against the Huron's traditional enemies, the powerful tribes of the Iroquois League. While fighting southward into the Iroquois lands of present-day New York state, Champlain discovered the vast lake that bears his name in July 1609. The culminating battle in which he participated as an ally of the Hurons against the Iroquois came at the present site of Crown Point, New York, in 1609. It was the Hurons' and Iroquois' introduction to the power of European firearms, and, from this point on, firearms became a major item of white–Indian trade.

After establishing Montreal in 1611, Champlain returned to France, where he secured a royal grant giving him a potentially tremendously profitable monopoly on the fur trade. During 1615–16, he returned to North America and penetrated farther west to explore Georgian Bay on Lake Huron. Turning into the area of New York state, he fought and lost a fierce battle (as an ally of the Hurons) against the Onondagas, an Iroquois tribe centered on Lake Oneida, near present-day Syracuse. Champlain was severely wounded and was therefore compelled to winter in a Huron village. He dispatched his companion, Etienne Brule, to explore southward, all the way to the Susquehanna River and Chesapeake Bay.

Champlain served as commandant or lieutenant governor of New France (1612) and vigorously sponsored exploration of the region, most notably the voyage of Jean Nicolet to Wisconsin. He participated in the Anglo–French War of 1626–30, which resulted in his surrender of Quebec to the English on July 20, 1629. However, the English acquisitions in French Canada were restored to the French by the Treaty of Saint-Germain-en-Laye in 1632, and Champlain returned to Quebec the following year as governor of the settlement. He died in Quebec two years later.

Champlain was a remarkable explorer and a highly competent soldier. He was deeply interested in the Indians, about whom he published a book (*Des Savages*) in 1603, describing in detail their customs and politics. The alliances he established with the Hurons were crucial in establishing the French in North America.

CHANDRAGUPTA MAURYA

REIGNED: ca. 321–ca. 297 B.C.
SERVICE: Diadochian Wars (323–281 B.C.); Seleucus' Invasion of India (305 B.C.)

Indian ruler; founder of the Maurya dynasty

Chandragupta Maurya seized power from the last king of the Nanda dynasty, establishing a reign that witnessed the vast expansion of the Indian Empire. He is believed to have been the illegitimate son of a Nanda king and a maid-servant. Early in life, his restless spirit prompted him to seek adventure, and he traveled widely throughout India. One account has him meeting no less than **Alexander the Great** at a camp in the Punjab, during the conqueror's invasion of the region, and offending him "through boldness of speech." For this he was forced to flee to save his life. While hiding, Chandragupta Maurya met a Taxilian Brahman named Chanakya, or Kautilya, who helped him raise an army of mercenaries in order to overthrow the Nanda king. (Another account holds that Chandragupta Maurya was sold as a servant to Chanakya, who then educated the boy and subsequently directed him to overthrow the king.)

Chandragupta Maurya won a decisive battle against the king and took control of the empire in Magadha. At this time, Alexander the Great died, and Chandragupta Maurya took advantage of the resulting disorder to attack the outpost Alexander had established in the Punjab. Some time between 324 and 321 B.C., with a mighty army reputed to have comprised 30,000 cavalry, 9,000 elephants, and 600,000 infantry, he overran all of northern India.

For the next two decades, Chandragupta Maurya seems to have ruled a vast region without challenge. Then, about 304 B.C., the Greek general Seleucos, who had inherited the eastern portion of Alexander the Great's empire, attacked. Chandragupta Maurya soundly defeated him and compelled him to accept humiliating peace terms in 303 B.C., including a forced marriage into the Maurya royal family. In return, Seleucos received the niggardly payment of five hundred elephants. The treaty brought to the Maurya dynasty all lands in northwest India as far as the Hindu Kush mountains.

During the closing years of his life, Chandragupta Maurya renounced both military conquest and the throne itself to become an ascetic devotee of the Jain sage Bhadrabahu. Prolonged drought had devastated his vast empire, and he became convinced that his own failings as king had brought about the disaster. Tradition records that Chandragupta Maurya carried his self-mortification to the point of death by starvation. Whether or not this is true, he was succeeded upon his death by his son Bindusara, and the Maurya dynasty continued to rule India until 186 B.C.

CHARLEMAGNE
(Karl Der Grosse, Carolus Magnus)

BORN: April 2, 742 or 743
DIED: January 28, 814
SERVICE: Saxon Wars (772–99); Lombard War (773–75); Spanish War (778–801); Conquest of Bavaria (787–88); Conquest of the Avars (791–801); Byzantine War (802–12)

Frankish ruler; often regarded as the founder of the Holy Roman Empire

Charlemagne is one of the shining lights of the Dark Ages. He briefly but gloriously united much of western Europe, creating a short-lived period of order and enlightenment during an epoch of prevailing chaos and cultural disaster. He was born to the Frankish king Pepin the Short and his wife Berta. Pope Stephen III anointed him king in 754, along with his brother Carloman. Following the death of Pepin on September 24, 768, Charlemagne took possession of the western part—his portion—of the kingdom on October 9. Carloman died in December 771, and Charlemagne promptly annexed the eastern portion of the kingdom as well. Operating from this base, he embarked the following year on the first of the approximately forty military campaigns of conquest that would occupy him through most of his forty-three-year reign.

The first campaign was a punitive action against a Saxon tribe that had engaged in raiding. Next, invited by the pope to invade Italy and fight the Lombards under Desiderius, he prevailed in 773–74, then quashed a Lombard revolt the next year.

Charlemagne's next major campaign, against the Moors in Spain, met with defeat when the emperor failed to breach the fortress of Saragossa in 778. Hroudland—better known to us as Roland—margrave of the Breton marchi, fought a rearguard action covering Charlemagne's withdrawal and was ambushed by Basques in the Pyrenees valley of Roncesvalles. The story of this defeat spread (and, with it, the further fame of Charlemagne), becoming a favorite subject of the troubadours. Some 300 years after the emperor's death, the story assumed its final form as the 12th-century *chanson* known as *The Song of Roland.*

Returning to the north, Charlemagne was again confronted by fierce resistance from barbarian Saxon tribes. He set into motion a long series of brutal campaigns in Saxony, establishing the bishopric of Bremen in 781, then defeating the Saxons at Detmold and the Hase in 773 and triumphing over the Saxon leader Widukind in 785. Additional campaigns against the Saxons were fought through the 790s, including the suppression of the revolt of 793. During one of his Saxon campaigns, Charlemagne—this man traditionally hailed for having introduced a measure of enlightenment to a Europe immersed in a cultural and moral gloom—beheaded four thousand Saxons in a single day.

Simultaneously with his campaigning against the Saxons, Charlemagne expelled Duke Tossila III from Bavaria during 787–88 and annexed the region. Almost immediately, he next turned to the Avars, who were raiding the Danube region. This resulted in expansion of Charlemagne's realm to Lake Balaton and northern Croatia during 791 to 803. He also invaded Spain for a second time during 796–801, succeeding, on this occasion, in the capture of Barcelona in 801.

In 800, Pope Leo III recognized Charlemagne as the mightiest of European monarchs by crowning him emperor of the Holy Roman Empire. Some of the more enlightened and optimistic inhabitants of Europe hoped that this would be the beginning of the restoration of the Roman Empire as it had existed in classical times. Indeed, Charlemagne had brought much of western Europe under the wing of the Holy Roman Empire. However, his campaigns against the Byzantines for control of Venezia and the Dalmatian coast, during 802–12, were far less conclusive than what he had accomplished in the West. They were effective enough, though, to secure Byzantine recognition of his title—something short of the reunification of the eastern and western empires, but a closer approach to the classical Roman Empire than the world had seen in centuries.

Charlemagne was more than a conqueror. Despite his ruthlessness as a military man, he introduced a degree of humanity into much of Europe, including an extensive system of justice administered by the *missi dominici*—a monk or bishop teamed with a count or military commander who traveled under royal commission through the provinces like circuit judges, hearing criminal and other cases. In this way, the petty tyrannies and corruption of local lords could be effectively counteracted. Although he was barely literate himself, Charlemagne recognized the value of learning and brought scholars from all over Europe to his court at Aachen (Aix-la-Chappelle), including Alcuin and Theodulf, the foremost men of learning of their day.

Toward the end of his life, day-to-day existence in his vast realm was more stable and serene than it had been in perhaps five hundred years. Yet the emperor's great reforms barely outlasted his own passing. After Charlemagne fell ill and died at Aachen on January 28, 814, his empire was divided among his heirs and certainly failed to coalesce into anything like the Rome of classical times.

CHARLES XII

BORN: June 17, 1682

DIED: November 30, 1718

SERVICE: Great Northern War (1700–21)

King of Sweden who was perhaps the most audacious military leader of his age

Charles was born in Stockholm to Charles XI and Ulrika Eleonora. When his father died in 1697, Charles XII, aged fifteen, assumed the throne of the greatest power of northern Europe, with territories encompassing all of present-day Sweden, Finland, the Gulf of Finland, Riga, Pomerania, and the vicinity of Bremen. Charles's father had provided for a regency to guide his heir. However, the regents were quick to conclude that the teenaged king's abilities were so exceptional that they voluntarily terminated the regency in November 1697, after only a few months. The rest of Europe was not similarly convinced of Charles's aptitude for power, and, in 1700, an alliance of Russian, Polish, and Danish forces was formed to take advantage of the boy-king's inexperience. The allies attacked Holstein, a duchy ruled by a duke who had married Charles's sister. Simultaneously, they launched attacks on other of Sweden's continental possessions.

Charles personally took part in initial operations against the Danes, driving the invaders back, then seizing the initiative by leading his troops in the conquest of Zealand (Sjaelland), in the heart of Danish territory, during April 1700. The a greatly abashed Danes capitulated by the Treaty of Travendal on August 28, and Charles turned against his eastern enemies.

He advanced into Livonia (modern Latvia) at the head of an army of 20,000 on October 16, marching to the relief of the Swedish outpost of Narva, which was besieged by the forces of Czar **Peter I.** Charles's forces were greatly outnumbered, but the young king skillfully exploited the element of total surprise and commanded the absolute loyalty of his well-trained troops. In a fierce snowstorm, Charles attacked on November 30, 1700, routing the Russians in the space of a quarter of an hour. He then set out for the Swedish port city of Riga, under siege by a combined Russian-Polish-Saxon army. In a battle of June 27, 1701, he defeated these forces, lifting the siege.

Flushed with these extraordinary victories, Charles went on a full-scale offensive, crossing the Düna River, where he met, fought, and defeated another Russian-Polish-Saxon army at Dünamünde on June 18, 1701. As a result, he came to occupy Kurland and, from there, launched an invasion of Lithuania during August–December, 1701. He advanced next on the Polish capital of Warsaw, occupying it on May 14, 1702, then pressed westward. On July 19, 1702, Charles engaged Polish-Saxon forces at Kliszow, defeating them. Shortly afterward, he captured Krakow.

Augustus II, king of Poland and elector of Saxony, was crushed under the weight of these victories. When Charles defeated yet another large Polish–Saxon force at Pultusk on May 1, 1703, Augustus II was forced to step down from the throne, and the Swedish king put a puppet, Stanislas Leszczynski, in his place in 1704. By 1705, after additional victories, at Punitz and Wszawa, Charles had wholly subdued Poland.

Returning to the Baltic, the tireless monarch relentlessly drove the Russians out of Lithuania, pushing Peter's army as far as Pinsk during the late summer and early autumn of 1705. By the Treaty of Altranstädt, signed on October 4, 1706, Augustus was compelled formally to abdicate the Polish throne and to renounce his nation's alliance with Russia. With the collapse of his Polish ally, Peter had no choice but to offer Charles highly favorable terms for peace. But the Swedish king, who had begun to fight because he was forced to, was now driven by visions of ever greater glory. He declined the czar's terms, and, as a demonstration of his contempt for Peter as well as his offer, he ordered the arrest of the Russian ambassador at Dresden. Citing the fact that the ambassador had been born in Livonia, a Swedish possession, he accused the diplomat of being a traitor and ordered his public execution. Charles then set about raising more troops to augment his army and, during 1707, stockpiled supplies. On New Year's Day, 1708, he commenced an invasion of Russia itself.

At first, the operation rolled ahead, seemingly unstoppable. Charles took Grodno on February 5 and very nearly captured the czar himself. He next advanced on Minsk, where he awaited the spring thaw. On July 12, 1708, he crossed the River Berezina and defeated a large Russian army at Golovchin. He reached the Dnieper River at Mogilev on July 18, meeting with parties of skirmishers. The far more serious problem was famine, due to the desperate Russians' scorched-earth policy. Still, Charles pushed his all-but-exhausted troops toward Moscow, defeating a small force at Dobroje on September 11 and striking up a crucial alliance with **Ivan Mazeppa** and his Ukranian Cossacks, then in rebellion against Peter. To Charles's dismay, however, Mazeppa was ousted in late October 1708, and as the winter closed in, it was apparent that the season was going to be an exceptionally harsh one.

For the first time, Charles faced the prospect of mutiny and desertion among his troops. Through sheer force of personality, he managed to keep his hungry army intact during the terrible winter of November 1708 to April 1709. Impatient to be on the move, he advanced on Voronezh, pausing to lay siege against Poltava. This proved disastrous. Charles's army suffered from an acute shortage of everything needful—provisions, ammunition, and artillery—from May 2 to July 7, 1709. The king himself suffered a severe and crippling wound in the foot during one skirmish against relief forces. At last, on July 8, the Russians bore down with an army of 80,000 on the 23,000 starving Swedes remaining to Charles. The wounded king directed the desperate battle from a litter, then, painfully, on horseback. Although several mounts were shot from under him, his life was spared. He saw his army virtually destroyed by July 9.

With the now-deposed Mazeppa and a mere 1,500 cavalry troops, he slipped across the frontier to Turkish Moldavia, settling into a camp at Bendery. The Turks, regarding any enemy of the Russians as a friend of theirs, treated Charles royally from 1709 to 1714. At length, however, even the Turks wearied of Charles's intrigues, which were aimed at involving Turkey in an all-out war against Peter. Turkish officials arrested their guest, who allowed himself to be taken only after a pitched battle between Turkish Janissaries and his own corps of guards. In November 1714, Charles slipped out of Moldavia and made his way to Swedish Pomerania, reaching Stralsund on November 21. Almost immediately, allied forces laid a year-long siege against this town and Wismar. Charles fled to Sweden in December 1715.

The Swedish army, once the very best in Europe, was a shambles. Undaunted, Charles set about rebuilding it, and this time it was he who offered favorable peace terms to Peter, agreeing to cede to Russia all of his Baltic provinces. In the summer of 1716, he staved off an invasion of Scania, a province in southern Sweden, then raised additional troops in preparation for an invasion of Norway, from which he planned, perhaps, to launch an invasion in an entirely new direction: Scotland. At the very least, he hoped to strengthen the position from which he could negotiate the most advantageous terms with the many enemies still arrayed against his country. He attacked Norway and, by November 30, 1718, had penetrated as far as the fortified town of Fredrikshald (present-day Halden), near Oslo. During the siege of the town, he was struck in the head by a musket ball and killed. His invasion collapsed with his death.

CHIANG KAI-SHEK

BORN: October 31, 1887

DIED: April 5, 1975

SERVICE: Chinese Revolution (1911); Northern Expedition (1926–28); Bandit (Communist) Suppresion Campaigns (1930–34); Second Sino–Japanese War (1937–45); World War II (1939–45); Chinese Civil War (1945–49)

Nationalist Chinese general, political leader, and president of Nationalist China

Chiang Kai-shek was born to a prosperous family in rural Chekiang. His father died when Chiang was young, and he was raised by his mother, who sent him to good schools, where he obtained the classical Confucian education reserved for the children of the moderately well-to-do. Chiang seemed destined for a comfortable berth in the civil government, but when the civil service examination system was suspended as part of the reforms of 1905, he enrolled instead in the Paoting Military Academy in northern China in 1906. The following year, he moved to Japan and the more prestigious Japanese Military Academy.

The corrupt and degenerate Ch'ing (Manchu) dynasty was crumbling, and revolution was in the air. Chiang first joined a revolutionary organization in 1908, but he did not return to China from Japan until he deserted from the Japanese army in 1911.

Chiang Kai-shek (left), with Franklin D. Roosevelt and Winston Churchill. Madam Chiang Kai-shek (Soong Mei-ling) is at the far right. Courtesy National Archives.

He reached Wuhan on October 11, the day after the revolution had begun that would finally topple the Ch'ing dynasty. On November 5, Chiang led an uprising in Chekiang, but the revolutionary movement was badly fragmented, and Chiang found himself on the outside after another revolutionary leader, Sun Yat-Sen, joined forces with General Yüan Shi-k'ai, who had seized nominal control of the government. During July 12–September 1913, Chiang was part of a movement to overthrow the general, whose regime had instantly become dictatorial. The movement failed, and Chiang narrowly escaped arrest by fleeing to Japan. He did not return until late in 1915, when he took part in the so-called Third Revolution, which blocked Yüan's bid to make himself emperor.

From this point until 1918, Chiang lived quietly, even obscurely in Shanghai. He was active during 1916–17 in the scheming of the Green Gang, a secret society that manipulated the volatile Chinese currency during this turbulent period, hoping to make a substantial profit. In 1918, Chiang Kai-shek made amends with Sun Yat-sen, who had joined forces with Ch'en Chiung-ming before building up his own Kuomintang or KMT (the Chinese Nationalist party) with aid from Soviet Russia. By 1923, Chiang had been promoted to the rank of major general in the KMT and the next year was made commandant of the party's Whampoa Military Academy.

Chiang was now well positioned in the most powerful of the Chinese political parties. When Sun died in 1925, his prestige and power in the KMT rose rapidly. During 1923–25, he had successfully consolidated the party's influence in southern China. From July 1926, through May 1927, he led the Northern Expedition, bringing province after province into the KMT fold. Chiang's genius was his ability to build on his military gains by securing backing and support from moneyed interests. He worked with the Shanghai business community to suppress the labor movement and, most importantly, embraced capitalism by breaking with the U.S.S.R. and purging the KMT of Communist influence. In place of Soviet advisors in his army, Chiang recruited many of the contentious warlords, who had hitherto made post-revolutionary China a chaos. During August and September 1927, he defeated Communist forces at Nanching and Hunan.

Despite Chiang's successes, the KMT was continually torn by factionalism, and in 1927 he briefly stepped down from the party leadership, hoping to quell dissent. At the end of the year, in December, he married Soong Mei-ling, who would herself prove a popular figure and a leader in her own right. As Madam Chiang Kai-shek she would be especially popular with Americans and was instrumental in garnering U.S. support for Chiang's Nationalist regime.

Chiang returned to the KMT as military commander in chief and as chairman of the Central Executive Council on January 6, 1928. He resumed the Northern Campaign, advancing on and capturing Peking (Beijing) on June 4. It seemed a momentous occasion: for, in theory at least, the fall of Peking meant that China was now unified under the Nationalist Party.

But China was a vast and diverse nation. It was one thing to proclaim a single government, but quite another actually to govern. Warlords and Communists continued to control many areas, and from December 1930, to September 1934, Chiang led the KMT army in five so-called Bandit Suppression Campaigns in southern China. This was essentially a civil war against the Communists, and all but the last of these campaigns were failures. The final push, which was aided by an increasingly bellicose Germany, made some inroads into Communist control, but in 1931, Chiang's problems with the Communists were overshadowed by the Japanese, who had invaded Manchuria, claiming that it was by rights a Japanese province.

Despite dissension, fragmentation, and virtually continual internal warfare, Chiang, as chairman of the KMT executive council from 1935 to 1945, was the closest thing China had to a single ruler during the post-revolutionary period. His position was highly unenviable. Always an impoverished nation, China was languishing in the great depression of the 1930s, an economic circumstance that was brewing revolution all over the world. He was compelled to tread a razor-thin line between two forces steadily gathering power over China: the Japanese on the one hand, and the Communists on the other. Prior to 1937, Chiang favored the Japanese, believing the Communists to be the more serious threat. But in December 1936, he was kidnaped by Chang Hsüeh-liang—in the so-called "Sian Incident"—and was forced by him to declare a united front with the Communists against the Japanese. This may have moved Japan, on July 7, 1937, to the full-scale invasion of China that commenced the Second Sino–Japanese War and raised the curtain on the Pacific theater of the approaching world war.

Chiang, as commander in chief of KMT forces, fought a doomed battle with his outmoded army against invaders from a country that had transformed itself into a twentieth-century war machine. He was forced repeatedly to retreat to the southwest with his command post. All the while, morale among his forces deteriorated, and, with this, corruption became rampant. Unable to rely on his own people, Chiang turned increasingly to the United States for aid. Yet, even here, he threaded the needle of compromise. After the Japanese attack on Pearl Harbor thrust the United States into World War II, Chiang refused to commit his forces wholly to a military alliance with America, believing that it was more important to husband his resources to fight the Communists, whom, he assumed, would be the only enemy after the war.

The exigencies of desperate warfare did bring a fleeting semblance of unity to China, and, in October 1943, the Executive Council appointed Chiang Kai-shek president of the nation. But, as he had predicted, on the heels of the Japanese surrender in 1945, all pretense of unity vanished. The end of the war had left two major contenders for dominance in China: the KMT and the CCP (Chinese Communist Party). Civil war erupted, and Chiang found himself leader of a party and an army that were rotten with corruption and whose leaders had failed to gain the support of the people. Accordingly, the KMT steadily lost ground, until, on December 7, 1949, Chiang was forced to flee to the island of Taiwan with the tattered remnant of his party. Stubbornly refusing to acknowledge the Communist government on the mainland, he set up a government in exile on the island, over which he presided until his death on April 5, 1975.

Chiang Kai-shek was a controversial leader in large measure because his temperament and philosophy united autocratic Confucianism with ideological pragmatism and a genuine desire for at least a degree of democracy. He was seen by foes and friends alike as crafty and willing to make alliance with whomever could be of aid at the moment. Such moral nimbleness produced short-term gains and was often necessary to immediate survival, yet it fostered the climate of cynicism and corruption that dogged the KMT throughout its existence. Whatever else the Communists had to offer—or failed to offer—they presented a front of idealism and moral commitment.

CHU TEH (Zhu De)

BORN: 1886

DIED: July 6, 1976

SERVICE: Chinese Revolution (1911); Northern Expedition (1926–28); Bandit Suppression Campaigns (1930–34); Second Sino–Japanese War/World War II (1937–45); Chinese Civil War (1945–49)

Communist general during the revolution and civil war in China

Chu was born into extreme poverty in Yunnan; however, his early promise was recognized by his clan, which funded a classical Confucian education for the boy. After graduation, Chu entered the civil service system, but soon turned to the military instead, gaining admission to the Yunnan Military Academy. Upon graduation, he was commissioned in the Yunnan provincial army about 1906. In October 1911, with the outbreak of revolution, Chu seized control of several military local units and led them in an invasion of the neighboring Szechwan province.

In 1912, General Yüan Shih-k'ai took control of the government, and the era of the warlords commenced. Chu eagerly seized the opportunity to e stablish himself as warlord, amassing wealth and becoming addicted to opium habit. Remarkably, however, he was subsequently caught up in the tide of patriotism and nationalism, overcame his addiction, and traveled to Germany in 1919 with the object of studying European military and political systems. In Europe, he met the charismatic Chou En-lai, who persuaded him to the Communist cause. Chu joined the Party, returned to China, and was commissioned a general and division commander in the KMT (Nationalist) army in 1926. In charge of the Nanchang garrison in Kiangsi, he laid the groundwork for Chou's insurrection there, which began on August 1, 1927. Breaking with the KMT, he organized and led Communist forces into the countryside to evade the approach of a large KMT army.

He joined forces with **Mao Tse-tung** in southwestern Kiangsi during the spring of 1928. Chu and Mao made a formidable team, with Mao formulating persuasive political doctrine and Chu providing the military leadership to transform doctrine into action. Chu developed and exploited guerrilla tactics to defend the Kiangsi Soviet territory against the first of **Chiang Kai-shek**'s Bandit Suppression Campaigns during December 1930–June 1933. By the fifth campaign, Chu was directed to meet Chiang Kai-shek's forces head-on in a conventional manner. The result was disaster, as the Communists were outnumbered and outgunned (December 1933–September 1934). With his army on the verge of disintegration, Chu managed to break out from Kiangsi during October and November and commenced the so-called Long March, the 5,000-mile epic journey across China undertaken by the Chinese Communists during 1934–36—a feat some have called the most incredible march in history. At the start of 1935, Chu was instrumental in securing power for Mao from his opponents at the Tsuni conference, and during the march, it was Chu who established and maintained military operations. He became the chief architect of the reconstituted Red Army after the march ended at Yenan, Shensi province in December 1936.

With the outbreak of the Sino–Japanese War in the summer of 1937, Chu assumed command of the Eighth Route Army, which U.S. commando pioneer **Evans Fordyce Carlson** observed and admired. However, the bold Hundred Regiments campaign against the Japanese in northern China (August 20–November 30, 1940) proved disastrous, although Communist propaganda painted it as a victory. Fortunately for the Chinese cause, Chu did not swallow his own party's line and learned from the Hundred Regiments fiasco the folly of going head to head with superior Japanese forces. From this point on, he developed guerrilla and commando tactics aimed at harassing the Japanese and forcing them to commit large numbers of troops who might otherwise be used in battle.

As the war with Japan drew to a close in 1945, the Red Army had been transformed into a highly effective guerrilla force of 800,000 men, which Chu now employed to take control of the northern Chinese countryside and occupy the major cities directly after the war years. He positioned the Red Army as an effective instrument of political power during the post-World War II civil war (August 1945–December 1949). When the People's Republic was declared in 1949, Chu was elevated by Mao to the post of Defense Minister. He moved from that post in 1955 to become one of ten marshals of the People's Republic, then retired to a series of largely ceremonial positions. The Cultural Revolution of 1967–70 nearly brought Chu down, but he was rehabilitated during the backlash against the Maoist extremism that followed the Cultural Revolution. He was named head of the Standing Committee of the National People's Congress and served in that body until his death.

CHURCHILL, Sir Winston Spencer

BORN: November 30, 1874

DIED: January 24, 1965

SERVICE: Malakand Expedition (1897); Mahdist Wars (1883–98); Second (Great) Anglo–Boer War (1899–1902); World War I (1914–18); World War II (1939–45)

Britain's great and inspiring World War II leader—a principal architect of the Allied war strategy

Winston Churchill was the son of Lord Randolph Churchill, descended from the First Duke of Marlborough, and Jennie Jerome, an American. Young Churchill did not take well to schooling (at Harrow) and rejected the university education that befitted his aristocratic station. Instead, he enrolled at Sandhurst (the "British West Point"), graduating in 1894 and receiving a commission in the Fourth Hussars. He took a two-month leave in 1895 to cover unrest in Cuba as a war correspondent, then returned to his regiment and was dispatched to India, where he served in the Malakand expedition to the Northwest Frontier during 1897. He also continued to write as a correspondent and published the first of his many distinguished books, *The Malakand Field Force*.

In 1898, serving in Lord **Horatio Kitchener**'s expedition into the Sudan, he rode in the charge of the Twenty-First Lancers at the Battle of Omdurman and, afterward, published a two-volume account of the Sudanese experience in *The River War* (1899). Early in 1899, Churchill resigned his commission to enter politics. After losing his first bid to enter Parliament, he sailed to South Africa on assignment from *The Morning Post* to cover the Second (Great) Boer War. He was taken captive by the Boers, but pulled off a daring escape that catapulted him to world celebrity and greatly improved his popular political standing.

In 1900, Churchill was elected to Parliament as a Conservative and quickly made a mark as an eloquent debater; however, he soon abandoned the Conservative cause and, by 1904, had declared himself a Liberal. His new party came to power the following year, and Churchill was named under-secretary of state for the colonies. He was made a member of the Cabinet in 1908—as president of the Board of Trade—and two years later was appointed home secretary. In 1911, he was appointed first lord of the Admiralty. It was in this post that Churchill first truly excelled, collaborating with Admiral Lord Fisher of Kilverstone to modernize the Royal Navy and prepare it for what Churchill correctly saw as the coming world war.

At the commencement of World War I, Churchill's confidence in the navy he had helped to create was unbounded. He planned an audacious amphibious assault on the Dardanelles to aid the Russian war effort in 1915. The attack failed disastrously, as did the subsequent associated land campaign at Gallipoli, bringing about Churchill's removal as first lord of the Admiralty and, ultimately, from the government altogether.

Churchill served briefly in the field in France, then rejoined the government under Prime Minister David Lloyd George as minister of munitions (July 1917–December 1918) and was responsible not only for generally increasing the production of guns, but for promoting the development and manufacture of the tank—at the time a most ungainly instrument of war, but one for which Churchill saw an important future. After leaving Munitions, Churchill became secretary of state for war and air during 1918–21, then secretary of state for the colonies (1921–22). In this latter capacity he negotiated important treaties in the Middle East and hammered out the 1921 agreement creating the Irish Free State. Despite these triumphs, Churchill lost his office as well as his seat in Parliament after Lloyd George's government fell in 1922.

Following the collapse of Lloyd George's government, Churchill realigned himself with the Conservatives and was returned to Parliament in 1924, joining the cabinet of Prime Minister Stanley Baldwin as chancellor of the exchequer from 1924 to 1929. His tenure proved disastrous, as his return of the nation to the gold standard in 1925 deepened the depression that had been brought on by the war. When economic hardship triggered a "General Strike" in 1926, Churchill vigorously condemned the strikers, creating a breach with labor that was never healed.

For fully a decade—1929–39—after he stepped down as exchequer Churchill held no government office and became a vehement critic of Baldwin's pro-independence policy for India. He was equally adamant during this period about the developing situation in Nazi Germany, which, he warned, would once again bring the world to war. He advocated putting Britain on a war footing and, in particular, matching Germany's growing air power; for he rightly predicted that the Germans would attack Britain by air in an attempt to bring it to its knees and make it ripe for invasion. Churchill's outrage at the folly of Prime Minister Neville Chamberlain's policy of "appeasing" **Adolf Hitler**'s expansionist ambitions by allowing him to annex the Sudetenland in Czechoslovakia was unbounded. When Chamberlain returned from the Munich Conference (September 29–30, 1938) having reached this agreement and claiming to have achieved "peace in our time," Churchill branded the action "total and unmitigated defeat." Events, of course, proved him right. After Hitler's invasion of Poland brought general war in September 1939, Chamberlain offered Churchill his former post of first lord of the Admiralty. In his characteristically aggressive manner, Churchill advocated an assault on Norway in an effort to dislodge the Germans there. The invasion proved abortive and disastrous, recalling the ill-fated Gallipoli campaign of World War I. Chamberlain resigned, and Churchill replaced him as prime minister during the darkest period of the war.

The darkest period proved to be Winston Churchill's finest hour. As prime minister he assumed all the responsibilities of a war leader. With Europe rapidly falling to the Germans, he turned to the United States, establishing an intensely personal relationship with President Franklin D. Roosevelt, which led to the ostensibly neutral United States supplying Britain with arms and, through a policy of "lend-lease" ships. In June 1940, the British army was beaten back—and nearly annihilated—at Dunkirk, on the North Sea coast of France. Later in the summer, the Battle of Britain commenced as the German *Luftwaffe* (air force) unleashed a reign of of terror over England's skies. Britons prepared to be invaded.

Through it all, Winston Churchill maintained his courage and fired the courage and determination of his people. The Battle of Britain was won by the Royal Air Force (RAF), who defeated the *Luftwaffe* and staved off invasion. In the meantime, Churchill refused to allow Hitler to tie down his forces defensively. He diverted an entire armored division—one of two in Britain—to fight Hitler's armies in the Middle East. At the same time, he struck an alliance with the Soviet Union, aiding it after it was invaded by Germany, even though he was an implacable foe of Communism. With the entry of the United States into the war following the Japanese attack on Pearl Harbor (December 7, 1941), Churchill was instrumental in forging a three-way alliance among the U.S., U.S.S.R., and Britain.

Churchill's controversial war strategy—often debated among the three Allies—was to avoid invading the European mainland until what he called the "soft underbelly of Europe" had been breached by clearing North Africa and the Mediterranean of the enemy. It was not until the summer of 1943, then, that the Allies invaded Sicily and Italy, having fought the first part of the "European" war in northern Africa. A year later, Churchill was instrumental in planning and supporting the Normandy invasion ("D-Day," beginning June 6, 1944).

While Churchill's overall "soft underbelly" strategy governed much of the war, his influence was somewhat diminished following the Normandy campaign. With victory in sight, he advocated a drive by the Western Allies—Britain and the United States—into Berlin in order to prevent Soviet occupation. Already, Churchill perceived the shape of a postwar "Cold War" world. But his strategy was overruled by Allied commander in chief **Dwight D. Eisenhower,** who believed it necessary first to crush the last German resistance in the West. Later, in a speech delivered at a small college in Missouri, Churchill would coin the phrase "iron curtain" to describe the pall of Soviet influence and tyranny that had descended across eastern Europe.

In some ways disappointed by the final conditions of the Allied victory, Churchill received a crushing blow in July 1945, when a general election failed to return him to office following the unconditional surrender of Germany but just before the capitulation of Japan. Yet Churchill refused to be crushed. He received the first dismal election returns while taking a bath: "There may well be a landslide and they have a perfect right to kick us out," he remarked. "That is democracy. That is what we have been fighting for. Hand me my towel."

Churchill managed to return to office in 1951, was knighted, but in July 1953, suffered a stroke. Though ailing, he recovered and continued in office until April 1955, when he was succeeded by his handpicked successor, Anthony Eden. Churchill spent his final decade pursuing his favorite hobby, painting, and seeing to the publication of the last of his great literary works, the four-volume *History of the English Speaking Peoples* (1956–58). Indeed, the prime minister's substantial body of biographical and historical writings (including a six-volume history of World War II, published during 1948–54) earned him the Nobel Prize for literature in 1953.

CIMON

BORN: ca. 510 B.C.

DIED: ca. 451 B.C.

SERVICE: Greco–Persian Wars (499–448 B.C.)

Athenian military commander and statesman

Cimon was a brilliant military tactician, whose victories on the field expanded the Athenian empire. He was, however, an old-line aristocrat, whose conservative allegiances put him out of step with the sweep of democracy in Athens. He was the son of Miltiades, an Athenian aristocrat, and a Thracian princess. Miltiades had defeated the Persians at the Battle of Marathon in 490 B.C., but was disgraced for mishandling a subsequent operation. He died, broken in spirit and bereft of wealth, in 489 B.C. In an effort to recoup some of the family's fortune and standing, Cimon engineered the marriage of his sister to the richest man in Athens. After discharging the massive debt left to him by his father, Cimon went to sea, distinguishing himself in battle against the Persians at Salamis in 480 B.C. On account of this, the grateful citizens of Athens annually elected him *strategos*, a post that combined the functions of war minister and general, from 480 to 461 B.C.

During his period as *strategos*, Cimon negotiated the transfer of the maritime states—newly won from Persia—from Spartan to Athenian control, thereby forming the Delian League, of which he became principal commander. In this capacity, he drove the Persians from much of the Thracian coast and defeated the pirates of Scyros, whom he replaced with solid Athenian settlers. Cimon also exhumed and transported back to Athens what were deemed the remains of Theseus, Athen's ancient and revered king.

But Cimon's greatest triumph was yet to come. About 466 B.C., he led a fleet of two hundred ships against the far superior Phoenician fleet near the mouth of the River Eurymedon. After defeating the Phoenicians on the water, he attacked on land, scoring a stunning victory that greatly undermined Persia's hold on the eastern Mediterranean. Next, he capped this victory with another on the Aegean, driving the Persians out of the Thracian Chersonese (modern Gallipoli). Shortly afterward, he was called on to put down a rebellion on the island of Thasos, which had seceded from the Delian League. Cimon handily defeated the Thasian fleet, then instituted a blockade of the island, maintaining it for about two years, finally forcing the Thasians to surrender in 463 B.C.

Ironically, while Cimon was off obtaining military glory, Athenian democrats were working back home to discredit him. **Pericles** and his followers charged Cimon with having accepted a bribe to refrain from attacking the King of Macedonia, who may have aided the rebellious Thasians. Cimon was summoned back to Athens and stood trial. He was ultimately acquitted, but it was clear to everyone that the aristocracy was on the wane in Athens. The ramifications of this trend were deeply felt in the armed forces. The hoplites, heavy infantry made up of wealthy Athenians, supported Cimon, but the sailors in the fleets—who were of far humbler station—pledged themselves to Pericles and Ephialtes, democratic leaders.

Cimon's position was further eroded by his advocacy of military cooperation with Sparta. In 462 B.C., when Sparta faced a rebellion of the helots—the serf class—Cimon urged joint military operations, arguing that, whatever their differences, Athens and Sparta were as oxen yoked, pulling together for the common good of Greece. Ephialtes countered this position, arguing that Sparta and Athens were rivals, always had been rivals, and that nothing could be gained by lending aid.

In the end, it was Cimon who prevailed. He mounted an assault against the rebels, using 4,000 hoplites. The rebels repelled his forces, however, and Spartan authorities summarily dismissed both Cimon and his troops. It was a gesture of great humiliation, and the fickle citizens of Athens responded to it by sentencing Cimon to a decade-long exile, beginning in 461 B.C.

The judgment of his fellow citizens broke Cimon's heart. From exile, in 457 B.C., he offered his services to Athens when Sparta attacked at Tanagra in Boeotia. Rebuffed by authorities, he appealed directly to the generals to allow him to enlist in the ranks as a common soldier. Even this entreaty was rejected. Despite this treatment, Cimon urged the followers who remained to him to fight bravely for their homeland. This they did, perishing in battle to a man.

After the debacle at Tanagra, even Pericles supported an early end to his adversary's exile. Cimon returned to Athens and served as a key negotiator in peace talks with Sparta. In 451 B.C., after peace between Athens and Sparta was achieved, Cimon was given command of a naval expedition against Persia. During this campaign, he died, though it is unknown whether his death was due to illness or battle wounds.

CINCINNATUS, Lucius Quinctius

BORN: 519 B.C. ?

DIED: 430 B.C. ?

SERVICE: Suppression of the Aequi (458 B.C.)

Roman general and statesman

As a farmer who answered his country's call to military leadership, only to return to the obscurity of his farm when his services were no longer required, Cincinnatus became the archetype of the selfless and dutiful despot. His story is compounded of history and mythology. According to tradition, Cincinnatus was a farmer who was appointed dictator of Rome in 458 B.C. in order to rescue a consular army that was surrounded by the Aequi on Mount Algidus. Thus summoned to leadership, he vanquished the invaders in a single day. Having done his duty and saved his country, he resigned the dictatorship immediately following the crisis and returned to his farm.

Although Cincinnatus resides in the shadows of history, he is perhaps most significant in his revival as an exemplum during the late eighteenth century to describe **George Washington**—hailed as the "Cincinnatus of the West"—who left his plantation at Mount Vernon to answer the call of duty and, victorious, declined to follow the road taken by so many other revolutionary leaders: self-aggrandizement and dictatorship. Indeed, popular American myth paints Washington as only grudgingly having agreed to accept the office of president.

CLARK, George Rogers

BORN: November 19, 1752

DIED: February 13, 1818

SERVICE: Lord Dunmore's War (1774); American Revolution (1775–83)

American frontier general who used meager military resources to secure the lower Ohio valley and Illinois country for the United States

A native of Virginia (born near Charlottesville), elder brother of the William Clark who would co-captain the Lewis and Clark expedition at the beginning of the nineteenth century, George Rogers Clark received scant formal education, but, like **George Washington,** learned the surveyor's art, a trade as valuable to wilderness warfare as it was to the business of frontier real estate. When, on the eve of the American Revolution, conflict between the Shawnee and Virginia's Ohio country settlers broke out (Lord Dunmore's War), Clark put his surveying skills to valuable use as a scout during May–October 1774. Following the war, he settled in Kentucky, where he soon achieved prominence as a military leader, organizing defenses against British-allied Indian raids during the early years of the American Revolution (1776–77).

Clark was a passionate, eloquent, and highly persuasive advocate for the frontier during the Revolution. He traveled to Virginia's capital at Williamsburg and convinced Governor Patrick Henry to provide and fund backing for the defense of the western regions. Clark himself received a militia commission as lieutenant colonel and was given command of the Kentucky militia. He detached an expedition of 175 men down the Ohio to the Illinois country, where he campaigned against Britain's Indian allies beginning in May 1778. On July 4, he took the trading settlement of Kaskaskia—one of the rallying points for pro-British Indians—and then captured Cahokia and Vincennes by late August. **Henry Hamilton,** Britain's infamous lieutenant governor headquartered in Detroit and known to the Indians as Hair Buyer for the bounty he paid on Patriot scalps, recaptured Vincennes during the fall. Undaunted, Clark led his men on an agonizingly difficult winter march—in an age and in a place where armies did not fight during the winter—for a surprise attack on the fort. Hamilton surrendered on February 25, 1779.

Promoted to brigadier general, Clark successfully fought the Shawnee, then repulsed a British–Indian attack on the Spanish settlement at St. Louis in 1780. Although his reputation was made on the colonies' western fringes, he happened to be in eastern Virginia, once again garnering support for the Detroit frontier's cause, when American turncoat Benedict Arnold attacked during January 1781. Clark participated in the defense against the assault.

Following the Revolution, Clark remained in the West as Indian commissioner (1783–86) appointed by

President George Washington. During 1786, Clark led a successful expedition against hostile tribes along the Wabash River, but he lost his office through the political manipulations of James Wilkinson, the most thoroughly crafty and exuberantly corrupt officer ever to serve in the American military. Embittered, Clark carelessly embarked on a number of French and Spanish colonization schemes in the West during 1786–99. Turning his back on the nation that had spurned him, he accepted from France a commission as general in 1793 and, later, one from Spain. He withdrew from all military activity in 1799 and lived out the remainder of his life in retirement in Louisville, Kentucky.

CLARK, Mark Wayne

BORN: May 1, 1896
DIED: April 17, 1984
SERVICE: World War I (1917–18); World War II (1941–45); Korean War (1950–53)

One the finest U.S. generals of World War II, whom Winston Churchill nicknamed the "American Eagle"

Clark was a son of the army, born at Madison Barracks in Sackets Harbor, New York, into the family of a career army officer. As a young man he graduated from West Point and entered the infantry as a second lieutenant in 1917 and in April of the following year was sent to France with the Fifth Infantry Division. He was thrust into the thick of World War I during the Aisne-Marne offensive. Wounded in June, he was assigned to First Army staff during the Saint-Mihiel offensive of September 12–16 and the Meuse-Argonne campaign that closed the war (September 26–November 11). Clark remained on active duty in Germany, serving on Third Army staff during the occupation. On his return to the States in November 1919, he was promoted to captain and served at a variety of posts in the Midwest until he was transferred to the General Staff in Washington, D.C. during 1921–24. He graduated from the Infantry School at Fort Benning, Georgia in 1925 and, in January 1933, was promoted to major. Two years later, he graduated from the Command and General Staff School at Fort Leavenworth, Kansas, and, in the depths of the Great Depression, was assigned to direct a contingent of the Civilian Conservation Corps (CCC) in Omaha, Nebraska (1935–36).

In 1937, Clark graduated from the Army War College and was posted to a staff position with the Third Infantry Division until 1940, when he became an instructor at the Army War College. There he was instrumental in efforts to expand and prepare the army for war. Promoted to brigadier general in August 1941, and major general in April 1942, he was made chief of staff of Army Ground Forces in May. In July, Clark was named commander of U.S. ground forces in Britain and immediately set about organizing II Corps there. Not content to serve in an office, Clark planned and participated in an extremely hazardous espionage operation to obtain intelligence on Vichy French forces in North

Mark Clark. Courtesy National Archives.

Africa in preparation for Operation Torch, the Allied North African landings.

In November 1942, Clark was promoted to lieutenant general and given command of Allied forces in North Africa under **Dwight D. Eisenhower.** Clark was among the chief architects of the invasion of Sicily and Italy launched from North Africa. Leading the Fifth Army, he landed at Salerno on September 9, where his forces held out against heavy defenses until Allied reinforcements and naval action checked German counterattacks during September 10–18. Clark commanded the Fifth Army in its agonizing advance up the Italian peninsula during October 1943–June 1944. On January 22, 1944, elements of his forces landed at Anzio and fought through ultimately to Rome on June 4. Despite the failure of the initial phase of the advance on Rome, the Anzio-Rapido Campaign (January–March 1944), Clark had acted with great enterprise, deliberately defying the directives of the Fifteenth Army Group (combined U.S. Fifth and British Eighth armies) commander, British general Sir Harold Alexander. This ensured that the American army would conquer Rome, but it also allowed a strategic German withdrawal, which meant that Allied forces would continue to encounter resistance in Italy throughout the war.

Clark commanded the Allied advance across the Arno River and north to the Gothic Line during July–December and in December 1944 was named to replace Alexander as commander of the Fifteenth Army Group. As group commander, Clark directed the Allied offensive through the Gothic Line, into the Po Valley, and finally into Austria during April 9–May 2, 1945.

Following the German surrender and the deactivation of Fifteenth Army Group, Clark was named Allied high commissioner for Austria, serving in this capacity from June 1945 to May 1947, when he was made commander of Sixth Army (1947–49), then elevated to chief of Army Field Forces during 1949–52.

In May 1952, Mark Clark became the third overall U.S. commander during the Korean War, succeeding **Matthew B. Ridgway,** who had replaced General **Douglas MacArthur** after the latter was relieved by President Harry S. Truman. Clark remained in command in Korea until after the armistice of July 27, 1953. During his later years in the army, Clark wrote two popular memoirs, *Calculated Risk* (1950) and *From the Danube to the Yalu* (1954). He retired from the army in 1954, when he accepted the post of commandant of the Citadel, South Carolina's prestigious military academy. He served at the Citadel until 1960 and lived the rest of his long life in quiet retirement outside Washington, D.C.

CLAUSEWITZ, Karl Maria von

BORN: June 1, 1780

DIED: November 16, 1831

SERVICE: French Revolutionary Wars (1792–99); Napoleonic Wars (1800–15)

The single most influential military theorist in modern history

Clausewitz was born near Magdeburg to a civil servant who had been an army officer. As a young man, Clausewitz joined the Prussian army in 1792 and was commissioned the following year. He fought in the Rhineland during 1793–94, then endured the boredom of garrison duty for a number of years. During this period—1795–1801—he occupied himself with extensive reading in history and the art of warfare. His knowledge impressed Count **Gerhard von Scharnhorst** sufficiently to admit Clausewitz into his new military school in Berlin in 1801. Clausewitz proved a brillaint student and, following his graduation in 1804, was named aide to Prince August and a member of the Prussian General Staff.

Clausewitz saw action at the Battle of Auerstädt on October 14, 1806 and became a prisoner of war during the Battle of Prenzlau on October 28, gaining his release after nearly a year's captivity. During 1807–11, Clausewitz aided Scharnhorst and Count **Neihardt von Gneisenau** in restructuring the Prussian army. During this period, he also served as military instructor to Crown Prince **Frederick William.** He wrote a treatise for the Crown Prince's instruction, which was subsequently published as *Principles of War* in 1811.

In 1811, with a group of brother officers, Clausewitz indignantly resigned his commission just prior to the French invasion of Russia, protesting Prussia's having been co-opted as a puppet of **Napoleon.** He offered his services to the Russian army to defend against Napoleon, and he persuaded Prussian general Yorck von Wartenburg to accept the Convention of Tauroggen and abandon the French service, taking with him his entire corps (December 30, 1812). William, king of Prussia, spurned Clausewitz until after Napoleon's first abdication. However, even without official military status, Clausewitz traveled with the army as a private observer during the summer of 1813, then was made chief of staff in a small multinational Baltic coast force during the fall. When Clausewitz finally rejoined the Prussian army, he became chief of staff to Thielmann's III Corps in **Blücher's** army, fighting at Ligny (June 16, 1815) and Wavre (June 18). He remained in the capacity of chief of staff during the Prussian occupation of the Rhineland during 1815–17, then was promoted to major general and director of the *Allgemeine Kriegsschule*, or War College, in Berlin in 1818. The following year, he began work on his masterpiece, *Vom Kriege* (*On War*), an effort to create a systematic approach to the art and conduct of war. Clausewitz completed a draft version of the book and was busy on revisions when, serving as chief of staff to the Prussian Corps of Observation during the Polish Revolution (1830), he contracted cholera. He died the following year.

While Clausewitz is best known for *On War*, he wrote many other articles and essays, as well as analyses of particular campaigns. Clausewitz identified and analyzed many concepts essential to modern warfare, including that of "total war": war waged upon the civilian population as well as military targets with the object of reducing a people's will to fight. He also formulated the concept of "friction" in the conduct of war, by which he described such unplanned (but inevitable) factors as chance, fatigue, personality conflicts, and minor errors that interfere with even the best battle plans. Finally, Clausewitz's view of war integrated warfare seamlessly with politics, and, in what is undoubtedly his most-quoted dictum, he defined war as "the continuation of politics by other means."

COCHISE

BORN: ca. 1810

DIED: June 9, 1874

SERVICE: Apache Indian War (1861–90)

Apache chief and war leader

Born, perhaps, in southeastern Arizona, Cochise was most likely the son of chief Pisago Cabezon. Though the Apaches happened to be at peace with Mexico at the time of his birth, the peace was, as usual, temporary. By the time he had become a young man they had returned to fighting the Spanish and Mexicans, and such hostilities would last—punctuated by short truces—the rest of his life. Early on, then, Cochise became a prominent warrior. His first real fight may well have been in May 1832, along the Gila River. Certainly by 1835, Cochise was raiding deep into Sonora. Stretches of unstable cease-fire vanished into renewed outright hostilities, which, especially after 1847, grew incessant. By 1856, Cochise was leading his own band of Chiricahuas and frequently joining the great Apache chief Mangas Coloradas in raids against the Mexicans.

Cochise's Chiricahuas maintained a steady peace, however, with the Anglo-Americans pouring into the American Southwest until 1861. That was the year the Americans accused him, falsely, of kidnaping a young white man and of cattle rustling. The U.S. army officer in charge locally, Lieutenant George Bascom, tried to arrest Cochise and hold him until the Apaches returned their captive. The boy, however, had not been captured by Cochise's Chiracahuas, but by another band of Apaches, and fighting erupted. Both sides now took captives for real—and executed them. Within sixty days of the so-called Bascom affair, the Apaches had killed 150 Americans, destroyed five stage stations, and engaged numerous raids on and small-scale battles with U.S. soldiers and settlers. Before long, however, the United States recalled its loyal troops from the Southwest to fight elsewhere in the American Civil War, and the Indians interpreted the withdrawal as a response to their ferociousness and, hence, not only continued to fight, but stepped up their raids. Shortly after the Civil War had begun, Cochise and Mangas Coloradas combined forces to ambush an advance element of the California Column bound for the Rio Grande to fight Confederates. The Battle of Apache Pass launched twenty-five years of sporadic conflict—some would call it interracial warfare—between Apaches and the Anglo–Americans.

Though Cochise personally would remain at war with the Anglo–Americans for years after the Civil War, now and then he met briefly with military personnel for inconclusive talks. In 1872, General **Oliver O. Howard,** assigned to negotiate an end to the "Cochise War" if he could, persuaded the one "white" man Cochise trusted, Thomas J. Jeffords, to take him to the Apaches' hidden stronghold in the Dragoon Mountains. During a ten-day conference, Cochise agreed to accept a life on a southeastern Arizona reservation, maintain peace with the United States, if not with Mexico, and refrain from depredations against American settlers, provided that Jeffords was named reservation agent. The great chief kept his word until his death, apparently from cancer, two years later, by which time Cochise had become a nearly legendary figure to whites and Native Americans alike, destined to be celebrated in histories, novels, films, and television westerns.

COENUS

DIED: 326 B.C.

SERVICE: Consolidation War (336 B.C.); Illyrian War (335 B.C.); Alexander's Conquest of Persia (334–330 B.C.); Alexander's Invasion of Central Asia (330–327 B.C.); Alexander's Invasion of India (327–326 B.C.)

***Macdedonian general on whose skill and absolute loyalty Alexander the Great** counted*

Nothing is known of the circumstances of Coenus' birth, childhood, or early career. Son-in-law of Parmenion—probably the most capable Macedonian commander under Philip II and, later, **Alexander**—Coenus first gained distinction as a *chiliarch* of *pezetaeri* (brigadier general of pike-armed hoplites) at the Battle of Pelium during Alexander's campaign in Illyria during 335 B.C.

Coenus next fought with valor at the Granicus (334 B.C.) and commanded the right flank of the phalanx at Issus during October 333 B.C. He was with Alexander during the final triumph at Tyre in August 332 B.C., and he led his *chiliarchia* (brigade) against the Persian center and right wing at Arbela/Gaugamela in October 331. This successful assault cost Coenus dearly; he suffered a grave wound. Recovering at length, he accompanied Alexander in his breaching of the Persian Gates and breakthrough to Persepolis in December. It was once again Coenus who was in the vanguard at Bactria during July 330 B.C.

Politically, Coenus was as reliable as he was militarily. He supported Alexander against the conspiracy of his own family members, brother-in-law Philotas, and father-in-law Parmenion, during the fall of 330 B.C.

From 330 to 326 B.C., Coenus was in command of a squadron of Alexander's Companions, campaigning throughout Central Asia. He held independent command during the Sogdian campaign of 329 B.C., and was in charge of the occupation force in Sogdiana (Bukhara) during 329–328 B.C. Early in 328 B.C., he defeated the Sogdian chief Spitamenes at the Battle of Bagae, thereby solidfying Macedonian control over the territory. Following this, he was assigned independent command of military operations west of the Khyber Pass during 327 B.C., then was incorporated by Alexander in his attack on Aornos (modern Kholm, Afghanistan) early in 326 B.C. A commander of great ingenuity and invention, Coenus quickly constructed a pontoon bridge across the Hydaspes River in March 326 B.C., then led the cavalry operation that encircled the right flank and rear of the Indian army (led by Porus) at the Hydaspes. This action brought Alexander victory against the armies of Porus in May.

Following the Hydaspes triumph, Coenus, speaking on behalf of the army, pleaded with Alexander to end, at least temporarily, his expeditions of conquest. His soldiers, weary, ached to return home. The plea was unheeded, and Coenus died—it is not known of what cause—while still far from home with Alexander's armies, near the mouth of the Indus River.

COLIGNY, Gaspard II de

BORN: February 16, 1519

DIED: August 24, 1572

SERVICE: Fourth Hapsburg–Valois War (1542–44); Fifth Hapsburg–Valois War (1547–59); First War of Religion (1560–62); Second War of Religion (1567–68); Third War of Religion (1568–70)

Dubbed "The Hero of Misfortune," Coligny was a courageous and resourceful Huguenot general and admiral

Coligny was born into the aristocracy at Châtillon-sur-Loing (modern Châtillon-Coligny) to Gaspard I de Coligny and Louise de Montmorency, sister of the Constable Anne de Montmorency. He was sent to the court in Paris in 1541, where he was befriended by the influential Duke of Guise, Fançois de Lorraine. He first saw combat on April 14, 1544, at Ceresole, fighting with such valor that he was knighted on the battlefield. During June 19–August 18 that year, he fought in the defense of Saint-Dizier and the following year led an infantry regiment at the siege of Boulogne.

In 1547, following the death of King Francis I and the subsequent return to royal favor of his uncle **Montmorency,** Coligny was named colonel general of the French infantry. In 1551, he was also appointed governor of Paris and Île-de-France, and the following year, the office of admiral of France was added to his posts.

Coligny fought against the Imperials at the battle of Renty on August 12, 1554, and was named governor of Picardy in 1555. After negotiating on behalf of France the Truce of Vaucelles in 1556, he went on to take Lens early the next year. Put on the defensive at Saint-Quentin, he was forced to surrender to the Spanish on August 27 and was confined at Sluys. While a prisoner, Coligny underwent a remarkable coversion to the Huguenot religion and cause. Liberated in 1559, he became a spokesman for and defender of religious toleration. This led to his uniting with Louis de Bourbon, Prince of **Condé,** in the First War of Religion. With Condé, he captured Orléans in the early spring of 1562 and fought in the indecisive battle of Dreux (December 19). He fought at Saint-Denis on November 10, 1567, but was soundly defeated at Jarnac by the Catholic army of Gaspard de Tavannes on March 13, 1569. Coligny rallied, however, and in turn defeated the Catholics at La Roche-l'Abeille in June.

During July and September, he laid siege against Poitiers, where he suffered a massive defeat. During the retreat, Coligny was badly wounded on October 3, but nevertheless rallied his troops in south central France and advanced northward with them to Paris during June 1570. This forced the Catholics to come to favorable terms at the Peace of Saint-Germain, concluded on August 8. The treaty signed, Coligny was restored to favor in the court of King Charles IX by the following year, 1571.

Coligny proposed to the king that French forces invade Flanders, a plan that was opposed by Queen Mother Catherine de Medicis. In the treacherous court of Charles, Coligny soon found himself the target of an assassin; he was wounded on August 22, 1572. Subsequently, Catherine de Medicis persuaded Charles to put to death all Huguenot leaders in the so-called St. Bartholomew's Day Massacre of August 24, 1572. Coligny was among those slain. Attacked by henchmen of Henri de Guise, he was stabbed several times, then thrown from a window.

CONDÉ, Louis II de Bourbon, Prince of

BORN: September 8, 1621

DIED: December 11, 1686

SERVICE: Thirty Years' War (1618–48); Wars of the Fronde (1648–53); Franco–Spanish War (1648–59); War of Devolution (1667–68); Dutch War (1672–78)

French general, veteran of many battles, hailed as "the Grand Condé"

Born in Paris to Henry II de Condé and Charlotte de Montmorency, Condé inherited the title of Duke of Enghien. His early education was at the hands of Jesuit priests, followed by a stint at the Royal Academy in Paris. His abilities were recognized early, as he was appointed governor of Burgundy in 1638, when he was only seventeen. Two years later, he volunteered his services to the Army of Picardy and participated in the siege of Arras (1640), as well as the sieges of Aire (1641), Perpignan, Collioure, and Salces (all in 1642).

Condé made a wise choice in a wife, marrying the niece of Cardinal **Richelieu** in February 1641. Richelieu was deeply impressed by Condé and entrusted him with command of the Army of Flanders in April 1643. The cardinal's confidence was not misplaced. Condé was able to lift the siege of the frontier fortress of Rocroi, routing the Spanish Netherlands Army commanded by Francisco de Melo at the Battle of Rocroi on May 19, 1643. The battle was a triumph that is much studied in military history because of the brilliance of Condé's tactics. Politically, the victory was crucial, bringing to an end many years of Spanish military preeminence.

Following the victory at Rocroi, Condé captured Thionville in August 1623, and in 1624 raised a force in Champagne, which he united with that of the Viscount of Turenne on the upper Rhine. During August 3–9, 1644, the joint forces drove Baron Franz von Mercy's Imperial Bavarian army from its entrenched positions surrounding Freiburg. Following this, in September, Condé detached from Turenne's army and marched to the north to take Philippsburg and the bulk of the Rhine valley, reaching Mainz by November.

Turenne, on May 2, 1645, suffered a defeat at Bad Mergentheim, and Condé rushed to relieve his forces. United once again, they defeated Mercy at Allerheim (August 3, 1645), one of the most brutal battles of the Thirty Years' War. In Flanders, Condé took over the army of Gaston d'Orleans, using it to capture Dunkirk during a siege spanning September 4–October 11, 1646.

In 1647, Condé succeeded his father as prince of Condé and assumed command of French forces in Catalonia, with which he laid siege to Lérida on May 14, 1647. By June 17, he was forced to pull back before the onslaught of an augmented Spanish army. He resumed command in Flanders the following year, and, at Lens, brilliantly defeated the superior forces of Archduke Leopold (August 20, 1648), a victory that sent Spanish emperor Ferdinand II to the peace table.

In 1649, with the outbreak of the first War of the Fronde, Condé took command of Royalist forces sent to blockade Paris during January–March, 1649, and succeeded in pushing the rebels from their entrenched positions at Charenton, near Paris, by February 8, 1649. However, Condé fell into a dispute with the emperor's all-powerful principal minister, Cardinal Mazarin, who imprisoned him from January 18, 1650 to February 13, 1651. Embittered by this experience, Condé offered his services to Spain in rebellion against Mazarin and soon found himself at the head of the rebel army. During April 6–7, 1652, he fought his former comrade-at-arms Turenne to a bloody draw at Bléneau, but broke through to Paris on April 11. However, Turenne defeated him at Saint-Antoine on July 7, 1652, and only by good fortune did Condé escape capture.

Condé was forced to flee to the Spanish Netherlands in October 1652. There he formally joined the army of Spain (in the Franco–Spanish War of 1648–59) and once again engaged Turenne, in the Champagne, during 1653, suffering defeat at Arras on August 24, 1654, but winning the day at Valenciennes on July 16, 1656, which brought relief to the besieged town of Cambrai.

Following the Peace of the Pyrenees, which ended the Franco–Spanish War on November 7, 1659, Condé sought and obtained the pardon of Louis XIV and was repatriated to France in January 1660. Eight years would pass before he was given command of another French army. At last, during the War of Devolution, he led the invasion of Franche-Canté (February 1668), and at the outset of the Dutch War in May 1672, he took his army down the right bank of the Rhine in an effort to capture Amsterdam. The brutal winter of 1672–73 defeated this effort.

From May to August 1674, Condé was on the defensive in the familiar battlefield of Flanders, holding off the advance of William III along the Meuse. Condé scored a victory against William's larger forces at Seneffe on August 11, 1674.

The following year, the Dutch War reached a crisis point following the death of Turenne's at the Battle of Sasbach on July 27, 1675. Leaving a subordinate to command his forces in Flanders, Condé rushed to the Rhine to take over Turenne's army, as well as that of the Marquis of Créqui, who had been made prisoner. In a brilliant relief action, he drove the forces of Prince Raimundo Montecuccoli back across the Rhine during September–October 1675.

The events of 1675 left the aging Condé exhausted and ill, painfully crippled with gout. At the end of the year, he retired to Chantilly, where he quietly lived out the remaining decade of his life, few commanders having seen such wide-ranging and mostly successful action.

CONSTANTINE I THE GREAT

BORN: February 17, ca. 282

DIED: May 22, 337

SERVICE: Roman Civil War (306–307); Civil War with Maxentius (311–12); Civil War with Licinius (314–23); Gothic–Sarmatian War (332–34)

Emperor of Rome in its decline, Constantine made Christianity the state religion of the empire, moved the imperial capital from Rome to Constantinople (the city he founded and built), and greatly reformed and strengthened the army

Constantine (full name, Flavius Valerius Constantinus) was born in Illyria (modern-day Balkans) to Constantius I, who became caesar in 293 after **Diocletian** and his fellow augustus, Maximian, stepped down. Constantine was appointed military tribune in the East in 302, then joined his father, who was fighting in the empire's always rebellious British realm. When Constantius died on July 25, 306, at York, the British legions elected Constantine *Caesar et Imperator*. While it is not clear whether this meant that he had been elevated to emperor of the Roman Empire or just caesar of Britain and Gaul, Constantine took the opportunity to consolidate his power and became involved in a dispute with Maximian over the ascension of the latter's son Maxentius to the Roman throne. Following Maximian's death in 310, the dispute swelled into civil war as Constantine, likewise laying claim to the throne, invaded Italy and defeated Maxentius's forces at Turin and Verona early in 312. In the summer of that year, he defeated and killed Maxentius himself at the momentous Battle of Milvian Bridge and was universally recognized as emperor of the western empire.

The historian Eusebius recorded that, at Milvian Bridge, a vision of a flaming cross appeared to Constantine bearing the legend, *In hoc signo vinces* ("By this sign conquer"), a vision that instantly converted the emperor to Christianity. The fact is that Constantine did not convert until he was on his deathbed in 337; nevertheless, he became a champion of Christianity, ultimately making it the state religion of Rome. This was, undoubtedly, his most momentous contribution to Roman civilization. (His attitude toward religion tended toward contradiction. On June 15, 313, he issued an edict at Nicomedia proclaiming toleration of all religions, including Christianity, yet he also took a role in expunging so-called heresies, expelling the Donatists—a sect of African puritans—in 314 as well as squelching the Arian heresy at the Council of Nicaea in 325, the council in which Christianity became Rome's state faith.)

Once his western holdings were secure, Constantine turned eastward to fight Licinius, the eastern augustus, at Cibalis and Mardia in southern Panonia (modern-day Hungary) during 314. He invaded Thrace in 323, defeating Licinius at Adrianople on July 3 and at Chrysopolis on September 18. Licinius surrendered, was made a prisoner, and, two years later, was executed on Constantine's orders.

Following the civil war with Licinius, Constantine reached the conclusion that the empire could no longer be effectively defended from Rome and therefore decided to build a more strategically favorable eastern capital. In another momentous action, Constantine moved the ancient capital from Rome to Byzantium, which he renamed Constantinople, after himself, in 330. He lavished the empire's treasures on the new capital, bent on making it a focus of learning, the seat not only of the

Roman empire, but of the Christian church. Throughout this effort, the emperor continually fought the barbarian tribes in a dual strategy of repelling them while struggling to assimilate them within the borders of the empire. He allied himself with the Sarmatians against the Goths during 332–34, but when the Sarmatians took this as an invitation to raid and loot throughout the empire, Constantine supported the Gothic king Gelimer in an attack on his erstwhile allies. With the Sarmatians crushed, Constantine now granted the survivors asylum within the empire.

Constantine was not only a highly capable and energetic military commander, he was a ruler of great intelligence, who also had a profound talent for administration. He was able to contain within himself vast contradictions, encouraging learning and enlightenment within the empire, introducing to it a justice system of unprecendented fairness and equality, and decreeing religious toleration, yet ruling as an absolute monarch in the Eastern manner. Generous as well as ruthless, he ordered the murder of his own son Crispius when he became too popular with the people and the legions. With Rome disintegrating and continually threatened by civil dissension from within and by barbarian hordes from without, it is little wonder that Constantine became increasingly wary, even paranoiac, during his later years. He arranged for the murder of a nephew, whom he suspected of plotting against him, and he probably ordered the death of his second wife, Fausta, mother of the son who succeeded him. By the end of Constantine's reign, his court had become a nest of fear and treachery.

CORNWALLIS, Charles, First Marquess

BORN: December 31, 1738

DIED: October 5, 1805

SERVICE: Seven Years' War (1756–63); American Revolution (1775–83); Third Mysore War (1789–92)

A competent British general of wide experience who is best remembered for his surrender to George Washington at Yorktown during the American Revolution

London-born Cornwallis was the eldest son of the First Earl Cornwallis. He was given an aristocrat's education at Eton, then joined the First Foot Guards as an ensign at the outbreak of the Seven Years' War in 1756. He served at Hannover and fought in the Battle of Minden on August 1, 1759, where he was promoted to captain of the Eighty-Fifth Foot.

In 1760, Cornwallis was elected to Parliament, then gained an appointment as lieutenant colonel of the Twelfth Regiment of Foot in May of that year. He returned to Germany, where he fought at Vellinghausen on July 15, 1761, Wilhelmsthal on June 24, 1762, and Lutterberg on July 23. He returned to England to assume his place in the House of Lords after the death of father in June 1762, and three years later was made aide-de-camp to King George III. In March 1766, he was promoted to colonel of the Thirty-Third Regiment of Foot and, four years later, was named by George constable of the Tower of London.

At the outbreak of hostilities in the North American colonies in 1775, Cornwallis was promoted to major general and dispatched to America. He landed in the Carolinas during February 1776 and took part in the failed assault on Charleston during June 16–July 25. Fighting under General Sir Henry Clinton, he performed well at the Battle of the Long Island (off Charleston) during August 22–27, then distinguished himself in fighting on Manhattan Island, New York, where he defeated the American forces at Kip's Bay (September 15). He pursued Washington's forces across the Hudson River, engaging them at Fort Lee in New Jersey on November 18. However, Washington outmaneuvered him at Princeton, where he suffered defeat on January 3, 1777. For this, Cornwallis was severely censured by General Clinton.

Frustrated in New Jersey, Cornwallis favored General **Sir William Howe**'s plan to take Philadelphia, the chief city of the colonies. Cornwallis directed the main attack at the Brandywine Creek on September 11, then led reinforcements to Germantown (in present-day Philadelphia) to check Washington's counterattack there on October 4.

Following his victories in Philadelphia, Cornwallis returned to Britain, where he received promotion to lieutenant general in January 1778. When he sailed back to America in May it was as second in command to General Clinton, and he was key commander at the Battle of Monmouth (New Jersey) on June 28. Cornwallis was compelled to return to England again to attend to his dying wife (December 1778–August 1779), then rejoined Clinton in a new assault on Charleston during February 11–May 12, 1780. When Clinton had to return to New York, Cornwallis was left to command operations in Charleston and also invaded North Carolina, hoping to break the back of the rebellion in the southern theater during the summer of 1780. At the Battle of Camden (North Carolina), Cornwallis bested the army of **Horatio Gates** on August 16, 1780, but was spread thinly through the wilderness region. His subordinates failed at King's Mountain (on the North Carolina–South Carolina border, October 7) and at the Cowpens (South Carolina, January 17, 1781). Despite this, Cornwallis stubbornly pursued his southern strategy, defeating General **Nathanael Greene** at Guilford Courthouse on March 15 in a pyrrhic victory that cost him a third of his men were killed or wounded. In defiance of Clinton's directives, Cornwallis then sped to Virginia to fight the **Marquis de Lafayette.** The Frenchman and his forces narrowly escaped Cornwallis at Green Spring on July 6, and Cornwallis went on to occupy Yorktown in August. The combined French and American forces of Washington and the Count of **Rochambeau** made a surprise attack on the Yorktown base early in September, cutting Cornwallis off from all supply. The stunned commander had no choice but to surrender on October 19, 1781, in what turned out to be the decisive battle of the Revolution.

Made a prisoner of war, Cornwallis was paroled in May 1782 and returned to England. Far from blaming him for final defeat in the Revolution, the crown offered to make him governor-general of India. Initially, he declined, but then accepted on February 23, 1786, on condition that he be granted considerably broad authority. He earned a reputation as an excellent and even-handed administrator. With the outbreak of the Third Mysore War, Cornwallis put himself at the head of an invading army, taking Mysore, assaulting Bangalore, and laying siege to the defiant leader Tipu Sahib in Seringapatam, who surrendered to Cornwallis on March 16, 1792. The following year, Cornwallis was created First Marquess Cornwallis and returned to Britain in February 1794. Three years later, on February 11, 1797, he was appointed commander in chief and governor-general of Ireland and the following year crushed the Irish rebellion and French invasion. As in India, Cornwallis was a just and tolerant administrator, who stepped down from his office in 1801 because King George III refused to grant Catholic emancipation. Cornwallis returned to India as governor-general in 1805 and died that year at Ghazipur.

There is much to admire in Cornwallis, who was a courageous and vigorous campaigner, skilled in tactics, but also headstrong and possessing an imperfect grasp of strategy. His greatest ability may well have been as an enlightened and efficient civil administrator.

CORTÉS, Hérnan

BORN: 1485

DIED: December 2, 1547

SERVICE: Conquest of Mexico (1519–21); Algerian Expedition (1541)

Spanish soldier and conquistador; conqueror of Mexico and destroyer of the Aztec empire

Cortés was born into the lower nobility of Spain in Medellin, Extremadura. He was sent at age fourteen to the University of Salamanca, where he proved himself as brilliant as he was wild: "much given to women," it was said, and all too capable of dispensing gratuitous cruelty. After two years, the restless young man left the university and spent the next several years drifting. At last, his imagination was sparked by accounts of the early transoceanic voyages of exploration and colonial expansion. Accordingly, he sailed in 1504 to Santo Domingo to seek his fame and fortune in the New World.

Cortés' first service was with the conquistador Diego de Velázquez de Cuellar in his campaign to conquer Cuba. This having succeeded, Cortés was named *alcalde* (mayor) of the colonial capital of Santiago. By 1519, the ambitious Cortés had become a figure to be reckoned with in the New World, and he managed to wrest from Velázquez command of an expedition to seize Mexico from the Aztecs. It was a momentous coup, and the conquistador promptly set off to explore the Mexican coastal region as far as modern Veracruz, where he established a city. With great skill, Cortés gathered as much intelligence as he could concerning the political situation in the Mexican highlands, learning about the alliances and enmities that existed among the Native peoples. He marched steadily inland and engaged in battle against the Indians of Tlaxcala, only to end by striking an alliance with them; for they were the traditional enemies of the Aztecs. He led his army onward to the Aztec capital of Tenochtitlan (modern Mexico City), where the Aztec ruler Moctezuma II surprisingly offered no resistance. Later historians have speculated that he believed Cortés to be an incarnation of the mighty deity Quetzalcoatl; others believe that the king simply thought that appeasement was the best way to deal with these invaders. Whatever the reason, the conquistadors entered the capital, unopposed, in November 1519.

Over the next several months, Cortés and his men plundered vast treasures of gold and other valuables in this richest of New World capitals. Cortés was obliged to break off this activity temporarily in order to lead an expedition to the coast, which was being threatened by a rival Spanish army under Panfilo de Narvaez. The inept Narvaez was quickly defeated, but overland travel was slow and, by the time Cortés returned to Tenochtitlan, he found the Aztecs in full revolt. It seems that, in his absence, they had been sadistically brutalized by his second-in-command, Pedro de Alvarado, who had launched a massacre during a religious ceremony. After sustaining heavy losses, Cortés and his men fled the capital during what they called the *Noche Triste* ("sad night") of June 30, 1520. During the combat, however, Moctezuma II had been slain—though it is unclear whether he had been killed by the Spanish (according to Aztec chroniclers) or by the outraged Aztecs themselves (according to the Spanish).

Although he and his men had been driven from the city, the conquistador did not concede defeat. He regrouped his forces and relentlessly besieged Tenochtitlan in 1521. His men cut off all supplies to the city and were careful to destroy the great aqueducts that supplied it with water. Despite this, the defenders held out for three months, until thirst, starvation, and disease had badly cut into their ranks. On August 13, 1521, the Aztecs of Mexico surrendered to Cortés, who thereby acquired vast holdings throughout Mexico and became a fabulously wealthy and powerful man. The Indians of the region were enslaved, a condition ostensibly forbidden by Spanish law, yet institutionalized by Spain in the so-called *encomienda* system, whereby the Native peoples became the "wards" of the conquistador who owned the land they inhabited.

Cortés was undoubtedly cruel, and he was the progenitor of a system of institutionalized colonial cruelty that gave rise to the so-called "Black Legend" of Spain in America. Yet much of the subsequent cruelty—the enslavement and impoverishment of the Native peoples—was the result of the failure of later conquistadors and colonizers to duplicate the success of Cortés, who was perhaps the only Spanish adventurer fully to realize the Old World dream of reaping boundless riches in the New World. Later Spaniards invested their entire fortunes in conquest, only to find poverty and an inhospitable land that could be made to yield barely enough for survival—let alone profit—only if the Indians were driven harder and harder.

As for Cortés, he returned to Spain several times and subsequently led expeditions to Honduras in 1524 and to Baja California in 1536. He participated in the unsuccessful Spanish attack on Algiers in 1541 and died near Seville on December 2, 1547.

CROESUS

DIED: ca. 546 B.C.
SERVICE: Conquest of Ionia (ca. 560 B.C.); war with Persia (550–546 B.C.)

The last king of Lydia, Croesus is better remembered for his mythically proportioned wealth than for his military struggle against Persia

Born into the Mermnad dynasty founded by Gyges in 685 B.C., Croesus succeeded to the Lydian throne after the death of his father, Alyattes, and a power struggle with his stepbrother. After gaining power, he set out to complete the conquest of Ionia begun by his father, taking Ephesus and other cities in western Anatolia (modern-day Turkey).

History and legend best remember Croesus for his enormous wealth, and, indeed, his very name became a byword for great fortune. He used a portion of his riches to improve the conquered city of Ephesus, to rebuild the Artemisium, and to enhance the great Greek shrines, including that of the Oracle at Delphi. It was written by the great historian Herodotus that Croesus was visited by the Athenian lawgiver Solon. Proudly displaying his treasures to Solon, Croesus asked him who was the happiest man the the world. He expected, of course, that the lawgiver would reply that Croesus, himself, was the happiest. Who could not be, possessing such a fortune? Instead, Solon replied with the names of two others and cautioned Croesus to "Account no man happy before his death."

If this exchange actually took place, it was prophetic. For Croesus, following the counsel of the Oracle at Delphi, suffered greatly in his struggle against Persia, led by **Cyrus II the Great.** Cyrus overthrew the Median empire in 550 B.C., whereupon Croesus consulted the Oracle, who foretold the fall of a great empire as a result of war between Persia and Lydia. Assuming that the Oracle meant that he, Croesus, would destroy an empire, and seeking to check the ascension of Persia. Croesus allied his realm with Babylon and, in 546 B.C., invaded Cappadocia, in eastern Anatolia, fighting inconclusively at Pteria before returning to his capital at Sardis. On his way to Sardis, Croesus was captured by Cyrus II, and the Persian army sacked the Lydian capital city.

Later, Croesus sent messengers to the Oracle at Delphi, complaining that he had been misled into war with the Persians. The Oracle replied that he had been punished for the sins of his ancestor Gyges, who, five generations earlier, had murdered the Lydian king and usurped the throne. Nor, the Oracle continued, had Croesus been misled. For a great empire *had* fallen in the Persian–Lydian conflict: that of Croesus himself.

There are differing accounts of the death of Croesus. The poet Bacchylides wrote that he attempted to burn himself on a funeral pyre. Herodotus recorded that Cyrus II condemned Croesus to be burned at the stake, but that he was saved by intercession of no less than Apollo. The Persian physician Ctesias said that the deposed ruler was deliberately spared by Cyrus, who, recognizing his administrative abilities, called upon him as an adviser and made him governor of Barene in Media.

CROMWELL, Oliver

BORN: April 25, 1599
DIED: September 3, 1658
SERVICE: First English Civil War (1642–46); Second English Civil War (1648–49); Irish Expedition (1649–50); Third English Civil War (1650–51); Franco–Spanish War (1635–59)

Great general in the English civil wars; Lord Protector of the Commonwealth

Oliver Cromwell, who combined political genius, great military ability, and all the ruthlessness of the self-righteous, was the chief architect of the English civil wars of the seventeenth century that overthrew the monarchy and instituted republican rule under the Commonwealth. He was born at Huntingdon, the second son in a family whose members had traditionally served in the Parliament. Educated at Sidney Sussex College, Cambridge, during 1615–17, he married in 1620 and soon converted to the Puritan faith. In 1628, he was elected to Parliament as the member for Huntingdon and became famous—or infamous—for unrelenting attacks on the bishops of the Church of England. In 1640, as the nation drifted closer to civil war, Cromwell was chosen to represent Cambridge in the so-called Long Parliament. Again directing his fire against the bishops, this time calling for their total abolition and a general purification of the Church, Cromwell aligned himself with members of Parliament who advocated the overthrow of Charles I. By this point, civil war had become inevitable, and Cromwell boldly acted to block the transfer of silver from Cambridge University to the king's coffers. He next secured Cambridgeshire for Parliament and raised a troop of cavalry from his native Huntingdon. Late in October 1642, Parliament commissioned him commander of a military association of six counties, a double regiment of fourteen troops later dubbed "the Ironsides."

Cromwell soon proved himself a great general, handily defeating Loyalist forces at Grantham on May 13, 1643, and at Gainsborough on July 28. Appointed governor of Ely, he led his troops to victory at Winceby on October 11, 1643, then at Marston Moor on July 2, 1644. Unfavorable terrain prevented Cromwell from effectively employing his superior numbers in pursuit of King Charles at the second Battle of Newbury on October 27, 1644, and a brief truce was struck early in 1645 (the Treaty of Uxbridge). During this period, from January to March, Cromwell successfully argued in Parliament for profound military reforms. The result was the "New Model Army," a standing military force raised by conscription and supported by tax levy. Compared to the old militia system, it was an exceptionally stable and professional fighting force and was, in effect, the first modern army England had.

On June 14, 1645, with the truce dissolved, Cromwell played a key role in the decisive victory at the Battle of Naseby, and, as second in command to Sir **Thomas Fairfax,** he assisted in the taking of Oxford in 1646, which resulted in Charles's surrender and concluded the First Civil War.

Cromwell next attempted—without success—to reach an accord with Charles I. He was also ensnared in a bitter schismatic dispute between Presbyterians in Parliament and Puritans in the New Model Army. This reached a crisis early in 1647, when Parliament voted to disband the army. In response, the army resisted, and Cromwell sided with his troops, taking command of the forces, arresting Charles I, and occupying London. In the meantime, the king opened secret negotiations with Scots Presbyterians during August and through October of 1647, and in November fled to the Isle of Wight. He struck a deal with the Scots, vowing to impose Presbyterianism as the English state religion for a period of three years in return for military support of his effort to regain his kingdom. This provoked Parliament's formal renunciation of allegiance to the king on January 15, 1648, and the Second Civil War commenced.

Oliver Cromwell. From J. N. Larned, A History of England for the Use of Schools and Academies, *1900.*

Cromwell was once again thrust to the forefront of the military action. He besieged and captured Pembroke in south Wales during June and early July

1648, then met and defeated an invading Scots army at the Battle of Preston on August 17–19, 1648. With northern England secured, he returned to London on December 7, where, with his New Model Army backing him, he urged a Parliamentary trial of King Charles. Late in January 1649, as one of 135 commissioners appointed to sit in judgment of the king, he successfully pushed for the monarch's execution, which was carried out on January 30.

England was now without a monarch, and Cromwell was chosen chairman of the brand-new Commonwealth's Council of State. He moved swiftly to put down a mutiny within the ranks of the army during the spring of 1649, then led his forces into Ireland, which, by virtue of a Royalist–Catholic alliance, defied the Commonwealth's authority. The Irish campaign revealed the most sinister side of Oliver Cromwell. Following the victory of a Puritan force led by Michael Jones at the Battle of Rathmines in September, Cromwell wrought upon the Irish countryside a reign of terror during September 1649–May 1650. He laid siege to one Royalist–Catholic stronghold after another, including Drogheda, Wexford, and Clonmel. Following the fall of each of these fortresses, Cromwell summarily slaughtered the defending garrisons, including women and children. It was the most heinous internal military atrocity in British history.

In 1650, Cromwell returned to London, leaving Henry Ireton, Edmond Ludlow, and Charles Fleetwood to complete the reduction of Irish resistance, which, despite the severity of the battles waged against the Irish, endured until the surrender of Galway in May 1652.

Cromwell's return virtually coincided with the commencement of a general uprising in Scotland, sometimes called the Third Civil War. Scottish Presbyterians proclaimed Charles II, son of the executed monarch, the new and rightful king of Great Britain and, on January 1, 1651, he was crowned in Scotland. In the meantime, during July–September 1650, Cromwell launched an invasion of Scotland. At first, the campaign went badly for the Puritans, who were hampered by disease and the Scottish people's scorched-earth policy, which resulted in the miserable deaths—from sickness and starvation—of some 5,000 of Cromwell's 16,000-man force. At Dunbar, a Scottish army of about 18,000 surrounded Cromwell's 11,000 remaining troops, but on September 3, 1650, the Puritan leader staged a surprise cavalry counterattack and scored a decisive victory that resulted in a rout of the Scots. Cromwell himself, however, was felled by malaria. This, combined with a dispute among the Scots, stalled the war for nearly a year.

In the early summer of 1651, the leader, having recovered his health, attempted to negotiate peace with the Scots. This failing, he led a lightning campaign in Scotland, taking Perth on August 2, 1651, then marching headlong back to England, where, at Worcester on September 3, he trapped and destroyed Royalist forces led by Charles II. The Scottish uprising—or Third Civil War—was ended.

Yet Cromwell had another war to fight, during 1652–54, this time with a foreign power, the Dutch, from whom he wished to seize the lucrative East India trade. Yet more pressing, however, was the agitation of the "Rump Parliament"—what remained of the Long Parliament after the Royalists and Presbyterians had been purged from it—which continued to oppose the army and which was dragging its feet in the enactment of the sweeping reforms of church and state called for by the Puritan program. In April 1653, Cromwell forcibly dissolved the Rump Parliament, declaring to them, "You have sat long enough; let us have done with you. In the name of God, go!" He then established a new council, the Little, or Barebones, Parliament, which was charged with nothing less than the restructuring of English government and society. The Barebones Parliament set about its task with a zeal that appalled even Cromwell, who dissolved the body in December 1653. Instead, reform was initiated by a group of New Model army officers, who composed the Instrument of Government, in effect a new constitution, which named Cromwell "Lord Protector" and empowered him to govern with the aid of a council of state and a single-chamber Parliament. In practice, Cromwell and his council—not the new Parliament—enacted the most significant governmental and legal reforms of the Protectorate and ruled the nation. On January 22, 1655, Cromwell dissolved the first

Parliament and created a regionally based system of government led by his most trusted generals.

With a new government in place, Cromwell launched a war against Spain, during 1656–59, in an attempt to wrest from that power control of trade in the West Indies. To prosecute the war, Cromwell needed funds, and to raise funds, he needed yet another Parliament to vote the necessary monies. When the newly constituted body offered to make Cromwell king in 1657, the Lord Protector reacted with scorn and rage, opening up a dispute with Parliament that resulted in its dissolution in February 1658. Later that same year, Cromwell's malarial infection flared, and on September 3, 1658, he died in London.

Oliver Cromwell's fiercely held religious and political beliefs took him to the point of regicide and, in the case of the Irish, even genocide. Yet he also demonstrated a high degree of religious tolerance, allowing the Jews to settle in England for the first time since 1290. It was he who forged Great Britain into a formidable military power, creating a great army and navy, and acquiring the first of many imperial possessions in winning Jamaica and Dunkirk as a result of the war he fought in alliance with France against Spain. Yet the government he created failed to endure very much beyond his own death. His son, Richard Cromwell, became Lord Protector on September 3, 1658, but lacked his father's ability to control Parliament or the army. On May 7, 1659, Richard Cromwell resigned as Lord Protector, and in May 1660, Charles II was restored to the throne.

CROOK, George

BORN: September 23, 1829

DIED: March 21, 1890

SERVICE: American Civil War (1861–65); Indian Wars (1875–77, 1882–86)

The "Gray Fox"; U.S. commander during the Indian Wars and perhaps the most skillful of the army's Indian fighters

Ohio-born George Crook was a West Point graduate (1852), commissioned in the infantry and sent west before the Civil War for service against the Indians. A captain in the regular army, he was made a colonel of volunteers in command of the Thirty-Sixth Ohio Infantry at the outbreak of the Civil War in 1861. He served first in West Virginia during 1862, was breveted major of regulars in May of that year, and brigadier of volunteers in August. He saw important action at South Mountain (September 14) and at bloody Antietam (September 17), where he was breveted lieutenant colonel of regulars.

In 1863, Crook was reassigned from infantry to cavalry command and assigned to lead a division in the Army of the Cumberland. He fought at Chickamauga (September 18–20, 1863), and faced down at Farmington (October 7). Breveted full colonel of regulars, he was briefly assigned command of the Army of West Virginia early in 1864, then, with the rank of major general of volunteers, he fought under General **Philip Sheridan** in the Shenandoah Valley during August 1864–March 1865. During this period, Crook saw action at Winchester (September 19, 1864), Fisher's Hill (September 22), and Cedar Creek (October 19). He was given divisional command in Sheridan's cavalry during the climactic siege of Petersburg during March–April of 1865, and fought at Dinwiddie Courthouse (March 29–31) and Five Forks (April 11).

Like so many officers breveted upward from rank to rank, Crook reverted to lower rank after the war. As a lieutenant colonel, he commanded the Thirty-Second Infantry, which was dispatched to Idaho in 1866 and thence to Arizona, where he did battle with the Apache followers of **Cochise** over an extended period, from 1871 to 1873. Promoted to brigadier general in 1873, he was assigned to command the Department of the Platte in 1875 and served under General Alfred H. Terry against the Sioux and Cheyenne led principally by Crazy Horse during 1876–77. Crazy Horse scored a victory at Rosebud Creek on June 17, 1876, when he attacked Crook's column with superior numbers. However, Crook proved highly effective against the Sioux at Slim Buttes on September 9.

Shortly after the victory against the Sioux, Crook was dispatched to Arizona to pursue **Geronimo.** He was temporarily successful during 1882–83—though Geronimo proved impossible to suppress for long—but had to fight the Apache leader again and was ultimately replaced in Arizona by General **Nelson Miles** in 1885.

Crook was appointed commander of the Department of the Pacific in 1886 and promoted to major general in 1888 and assigned to command the Department of the Missouri. He died of natural causes in his Chicago headquarters in 1890.

A courageous and skilled commander in the Civil War, Crook was even better suited to campaigning against the Indians, which characteristically required persistence, a high degree of wit, and an ability to manage difficult logistical situations—all of this in a context of a chronically inadequate military budget.

CUMBERLAND, William Augustus, Duke of

BORN: April 15, 1721

DIED: October 31, 1765

SERVICE: War of the Austrian Succession (1740–48); Jacobite Rebellion (1745–46); Seven Years' War (1756–63)

Infamous to the Scots as the "Butcher of Culloden"; ruthlessly suppressed the Jacobite uprising in Scotland

Cumberland was the son of George Augustus, Prince of Wales (subsequently King George II), and Caroline of Ansbach. He received his title in July 1726, and was destined by his father to a career in the Royal Navy. However, he soon discovered that the army was more to his liking, and he became colonel of the prestigious Coldstream Guards in April 1740. Two years later, he was promoted to major general. During the War of the Austrian Succession, he acquitted himself with great gallantry at the Battle of Dettingen on June 27, 1743, as his father's subordinate. Following this battle, he was named captain general of all British land forces, as well as commander of the combined British, Hannoverian, Dutch, and Austrian armies. He massed his forces in an attempt to relieve the besieged town of Tournai, but was bested by the French army of Count de Saxe at Fontenoy on May 11, 1745.

At this point, however, Cumberland was urgently recalled to England to counter the invasion of Bonnie Prince Charlie (Charles Edward Stuart) of Scotland during November–December 1745. He pushed and pursued the Jacobite rebels from Derby to Carlisle, then returned to London in January 1746. When the town of Falkirk lay under Scots siege, Cumberland went to Aberdeen to prepare an army to fight the Jacobites. After a month and a half of preparation, Cumberland marched to Culloden Moor near Inverness, where his forces ruthlessly smashed the Jacobites, bringing the Scots rebellion to a very bloody end on April 16, 1746. His military objective achieved, Cumberland sought to ensure an absolute end to rebellion and conducted a brutal campaign of extermination against all Jacobite sympathizers, decimating many of the Scottish Highland clans in the process during April–July, 1746. Highlanders dubbed him "Bloody Butcher" and the "Butcher of Culloden."

Following the suppression of the Jacobite Rebellion, Cumberland returned to Flanders to continue fighting in the War of the Austrian Succession. Again, however, he met his match in Marshal **Saxe,** who defeated him at the Battle of Lauffeld on July 2.

Cumberland was called upon at the outbreak of the Seven Years' War in June 1757 to command an army in Hannover, where he was defeated the French Marshal d'Estrées at Hastenbeck near Hameln on (July 26). The defeat turned into catastrophe, as Cumberland allowed himself to be trapped between the North Sea and the Elbe River. Effectively checkmated and seeing no way to fight his way out, he signed the ruinous Convention of Klosterzeven (September 8), by which he disbanded his army, in effect laying open all of Hannover to French attack. Cumberland's father, now King George II, furiously repudiated the convention, whereupon Cumberland resigned not only from the military, but from all of his offices. Disgraced, he retired to Windsor Castle.

After the death of his father and the ascension to the throne of George III—his nephew—Cumberland once again involved himself in state affairs, skillfully maneuvering to gain influence in court by undermining the king's top ministers, the Earl of Bute (resigned in 1763) and George Grenville (resigned in 1765). By the time of his death in 1765, Cumberland had largely overcome his ignominy and died honored by all—except the embittered Scots.

At best, Cumberland was a brave and efficient commander. His work at Culloden was a brutal lesson in total warfare. However, his failure in Hannover was utter and complete.

CUNNINGHAM, Sir Andrew Browne, First Viscount Cunningham of Hyndhope

BORN: 1883

DIED: June 12, 1963

SERVICE: World War I (1914–18); World War II (1939–45)

Great British naval commander in two world wars

Cunningham joined the Royal Navy in 1898 and saw a wide variety of service prior to World War I, when he was skipper of the destroyer *Scorpion* during the Dardanelles campaign (February 19–November 23, 1915). He showed great skill and initiative in the raid on the German U-boat base at Zeebrugge on April 23, 1918.

Cunningham remained in the navy between the wars and was tapped by King George V as his aide-de-camp during 1932–33. In 1934, he was promoted to rear admiral and given command of the destroyer flotilla attached to the Royal Navy's Mediterranean Fleet. He served in this capacity until 1936, when he was named commander of the battle cruiser squadron and made overall second-in-command of the Mediterranean Fleet. He left this post in September 1938, to become deputy naval chief of staff, serving until the outbreak of World War II in June 1939, when he was again dispatched to the Mediterranean, this time as commander of the fleet.

After the fall of France, Cunningham acted quickly against the French Mediterranean naval force, effectively immobilizing the French fleet at Alexandria by June 1940. Cunningham also triumphed in the first major Mediterranean battle of World War II, decisively defeating Admiral Angelo Campioni off Calabria on July 9, 1940, severely damaging an Italian battleship and cruiser without suffering serious casualties himself. Cunningham launched a daring nighttime torpedo bomber strike from his fleet's single aircraft carrier, H.M.S. *Illustrious*, which devastated the Italian navy. The November 11, 1940, attack on the Taranto naval base sank or badly damaged three battleships and two cruisers, as well as a pair of auxiliary vessels. Then, on March 28, 1941, when Admiral Angelo Iachino dispatched a fleet sortie, Cunningham intercepted it off Cape Matapan, sinking three cruisers and two destroyers. By these two actions, the Italian navy was neutralized for the balance of the war.

Cunningham's fleet covered the evacuations of Greece (April 26–30) and Crete (May 21–June 1), sustaining severe losses from unremitting German air attacks. Cunningham was also responsible for the survival of Malta, supplying the isolated British outpost by sea throughout June 1941–June 1942, again despite continual German air assaults.

Cunningham was posted to Washington, D. C., during June–October 1942, to serve as chief British Admiralty representative to the Combined Chiefs of Staff. After receiving promotion to fleet admiral, he returned to the Mediterranean as Allied naval commander in chief and, on November 8, 1942, provided naval cover for the North African Operation Torch landings. Beginning on July 9, 1943, and through August 27, he provided naval support for Operation Husky, the invasion of Sicily.

On September 8, 1943, Cunningham received the surrender of the Italian fleet, then was recalled to London as first sea lord, following the resignation of Admiral Sir Dudley Pound. Cunningham served in this office throughout the balance of the war and until 1946, when he was created First Viscount Cunningham of Hyndhope. In 1951, he published his memoirs, *A Sailor's Odyssey*, and lived the rest of his life in London, recognized as one of England's greatest sailors and a key presence in the Mediterranean during World War II.

CUSTER, George Armstrong

BORN: December 5, 1839
DIED: June 25, 1876
SERVICE: Civil War (1861–65); Indian Wars (1867–69, 1876–77)

Civil War "Boy General" who went on to lead the Seventh Cavalry to defeat at the Battle of Little Bighorn during the Sioux Wars

Born in New Rumley, Ohio, George Armstrong Custer spent part of his childhood with a half-sister in Monroe, Michigan. Like another casualty of the Indian Wars, U.S. General **Edward Canby,** Custer attended West Point Military Academy and, like Canby, graduated at the bottom of his class on the eve of the Civil War. He may have been a poor student, but Custer excelled in the arts of combat and—during the Civil War—he proved himself a superb field soldier. A staff officer for General **George B. McClellan** and later General Alfred Pleasonton, Captain Custer so demonstrated his potential that he was promoted to brigadier general and given command of the Michigan cavalry brigade. The flamboyant twenty-three year old, with his long yellow hair and a gaudy uniform of his own design, won instant fame. From Gettysburg to Appomattox, he was known for slashing if occasionally ill-considered cavalry charges that often proved decisive, and for a personal fearlessness that earned the respect, even the devotion of his men. By war's end, he had become a major general commanding a full division.

Custer returned to the postwar regular army as lieutenant colonel in the newly authorized Seventh Cavalry Regiment. He made a new name for himself in the West, garbed in fringed buckskin instead of black velvet and gold lace, the very embodiment of the dashing Indian fighter, skilled plainsman, and hunter. The Seventh's commanding colonel was frequently absent, and Custer usually commanded the Seventh in his stead. For all intents and purposes—and in the popular perception—the Seventh was Custer's regiment.

Custer's first experience with Indians, in Kansas in 1867, ended in ignominious failure. Not only did he fail to defeat any Indians, he was court martialed and sentenced to a year's suspension of rank and pay. In 1868, however, he surprised and attacked Chief Black Kettle's Cheyenne village on the Washita River in present-day Oklahoma and, given the rabid and racist tenor of the times, used the brutal and strategically questionable attack on what was in effect a "civilian" target to lay the groundwork for his reputation as an Indian fighter. Guarding railroad surveyors on the Yellowstone River in 1873, he fought the Sioux in two battles that reinforced his Indian-fighting image.

Easterners certainly looked on Custer as the army's foremost Indian fighter. In fact, he was no more successful than some of his peers, less so than a few. His

George Armstrong Custer. Courtesy National Cowboy Hall of Fame and Western Heritage Center.

regiment, moreover, was badly factionalized. Some of his followers worshiped him; others loathed him. But he wrote popular magazine articles and a best-selling book, *My Life on the Plains*, and always, he made good newspaper copy.

In 1874, Custer led the Seventh Cavalry out of his base at Fort Abraham Lincoln to explore the Black Hills of the Dakota Territory. Part of the Great Sioux Reservation, guaranteed to the Sioux by the Treaty of 1868, the Black Hills region had long been coveted by whites who thought its dark recesses held gold. And, indeed, miners traveling with the expedition found gold. To no one's surprise, the news set off a gold rush.

The U.S. government attempted to buy the Black Hills and legalize the mining settlements. When the effort failed, the aggression of the Sioux bands led by **Sitting Bull** and Crazy Horse against friendly Indians gave officials a pretext to wage a war that would solve the Black Hills problem by depriving the Sioux of their independence and, of course, of their power to obstruct the sale. The result was the Great Sioux War of 1876. Custer's Seventh rode with one of the armies that converged on the Indian country, and on June 25, he attacked the village of Sitting Bull and Crazy Horse on Montana's Little Bighorn River. In a battle that will forever remain controversial, the Sioux wiped out Custer and five companies of U.S. troops.

"Custer's Last Stand" stunned and angered Americans and, ultimately, led to the conquest of the Sioux and the acquisition of the Black Hills. But it also awarded Custer himself an immortality that perhaps he did not deserve but that fit his dashing persona. His adoring wife Elizabeth, or Libbie, devoted the rest of her quite long life to defending and glorifying his name. She wrote three books that stirred her contemporaries and that are still minor classics today. The controversy over the battle keeps it alive in American histories, and the image of "Long Hair," erect on his hilltop with soldiers falling around him and shouting Sioux closing in for the kill, remains a shining icon in American folklore.

CYRUS II THE GREAT

BORN: 590–80 B.C.

DIED: ca. 529 B.C.

SERVICE: Conquest of Media (559–550 B.C.); Invasion of Lydia (547–546 B.C.); Conquest of Ionia (545–539 B.C.); Conquest of Chaldea and Fall of Babylon (539–538 B.C.); Campaign against the Masagetae (537–529 B.C.)

Persian military commander and ruler who founded the Achaemenid Empire

As founder of the Achaemenid (Persian) Empire—the largest empire known up to its time—Cyrus the Great looms large in Persian history much as **Alexander the Great** figures in Greek culture or Moses in the Judeo–Christian tradition. He was born in Media or Persis (modern-day Iran) and was most likely one in a long line of ruling chiefs. The kingdom Cyrus inherited, Anshan, was, however, quite small. It is a measure of his accomplishments that, by the end of his reign, he had extended his inheritance so that it encompassed much of the Near East, from the Aegean Sea to the Indus River. Most significantly, he conquered the Babylonian empire and became master of the greatest city of his time, Babylon. As a conqueror in the ancient world, Cyrus was second only to Alexander the Great.

The childhood of Cyrus, known to us through accounts in Herodotus and Xenophon, is wrapped in myth, and Cyrus occupies a quasi-religious position in Persian culture as a founding figure. Astyages, king of the Medes, married his daughter to his vassal, a noble of Persis called Cambyses. The union produced Cyrus. Shortly after the birth, Astyages dreamed that the infant would grow up to overthrow him and therefore immediately ordered the child's death. Astyages's chief advisor, however, secretly gave the baby to a shepherd to raise. Ten years later, Cyrus's extraordinary abilities had already made him famous, and he became known to Astyages, who was persuaded to spare the boy. Grown to manhood, Cyrus did, in fact, revolt against and overthrow his grandfather. Astyages responded in 550 B.C. with an attempt to suppress the rebellion, but so powerful was Cyrus's personality that Astyages' troops deserted to join Cyrus.

As a conqueror, Cyrus was aggressive and relentlessly driven. He began by subduing the Iranian tribes in the region of the Medes, before turning westward. **Croesus,** ruler of Lydia in Asia Minor, long an antagonist of the Medes, sought to capitalize on the overthrow of Astyages by carving out from the Medes a vaster empire for himself. Cyrus responded by attacking Lydia and capturing its capital city of Sardis in 547 or 546. Croesus was either killed in battle, committed suicide by self-immolation, or was taken prisoner by Cyrus. Accounts that claim the latter also note that Cyrus treated Croesus well, a characteristic typical of the ruler, who was celebrated for the tolerance and generosity with which he treated conquered people and their leaders.

As a result of the defeat of Croesus, the Greek cities of the Aegean fell to Cyrus, who, however, was frequently required to suppress revolts among them. The Ionian Greeks, however, submitted to Cyrus peacefully.

Following the defeat of Croesus, Cyrus advanced on Babylonia, which had been destabilized by popular dissatisfaction with its ruler, Nabonidus. Most of the populace seems welcomed Cyrus as a deliverer, and the conquest was quickly accomplished. The fall of Babylon in October 539 B.C., occasioned the most celebrated demonstration of Cyrus' tolerance. For the Old Testament credits him with delivering the Jews from their long Babylonian captivity and permitting their return to their homeland. With equal tolerance, Cyrus permitted the Babylonians to retain their traditional gods and customs of worship, thereby avoiding much of the resentment usually forthcoming from a conquered people.

Having taken control of Babylonia proper, Cyrus acquired dominion over that nation's conquests, Syria and Palestine. This resulted in the submission of other kingdoms as well, most notably Cilicia, which yielded to Cyrus out of a combination of diplomatic necessity and the manifest evidence that Cyrus treated his allies well.

According to the great historian Herodotus, Cyrus met his end at the hands of a woman. One of the nomadic tribes he had conquered shortly after his overthrow of Astyages was the Massagetai, whose chieftain was female. Her son, whom Cyrus had taken prisoner, years later committed suicide while languishing in captivity. Vowing vengeance, his mother finally met Cyrus in battle and killed him.

Cyrus' empire was left to his son Cambyses. Soon after ascending the throne, Cambyses probably murdered his brother Bardiya, a potential rival, then married his sister Atossa, the greater to strengthen his grip on the realm. Under Cambyses and the rulers of his line during the next two centuries, the empire Cyrus had created continued to expand and prosper.

D

DAVID

DIED: ca. 962 B.C.

SERVICE: Judean Civil War (before 1000 B.C.); Conquest of Jerusalem from the Jebusites (after 1000 B.C.); War against the Philistines (after 1000 B.C.)

Jewish military leader and statesman, second king of Israel

Second and greatest king of Israel, David united the very loosely constituted Jewish tribes into a cohesive empire and simultaneously transformed himself into the central and enduring symbol of the bond between God and the Jewish people.

Born in the city of Bethlehem, Judah, David was the youngest son of Jesse and the grandson of Boaz and Ruth. Biblical tradition portrays him as a simple shepherd boy who became his people's champion in single combat against the "giant" Philistine warrior Goliath, slaying him with a well-placed slingshot volley. Whatever the historical basis of this story, young David began his political career proper as an aide in the court of Saul, who regarded the youth as a second son. Moreover, David developed a fraternal relationship with Saul's son and heir, Jonathan, and he married Saul's daughter Michal.

The congenial life of Saul's court might have continued had not David's prowess as a warrior gained him increasing popularity, which triggered the aging king's paranoiac jealousy. Saul hatched a plot to kill David, and the young warrior fled into southern Judah and Philistia, where he gathered about him a popular following by assuming the role of an Old Testament Robin Hood on the desert frontier of Judah. David built a guerrilla army of fellow bandits and refugees, who defended the frontier populace against a variety of marauders and raiders. Thus David won the loyalty of increasing numbers of people.

Yet, unlike most guerrilla leaders later in history, David was not a revolutionary and never designed the overthrow of King Saul. Instead, his actions and growing popularity won over influential Judaean elders, who *invited* him first to become king of Judah in Hebron and then to succeed Saul himself, after the king and Jonathan were both killed by Philistine warriors at the Battle of Mount Gilboa about 1000 B.C. David did not hesitate in prosecuting a civil war with factions loyal to Saul's surviving son, Ishbaal, a struggle that ended when Ishbaal was slain by his own courtiers, whereupon David was accepted as king of Israel, including the far-flung tribes beyond Judah.

David next turned his attention toward Jerusalem, besieging and taking it, expelling the Jebusites who had held it, and establishing it as the capital of the empire. The new king imbued this victory with great spiritual significance by transporting the ark of the covenant, central symbol of the presence of Yahweh (God), into the new capital, thereby anointing Jerusalem a holy city.

Having secured and sanctified his capital, King David campaigned vigorously against the Philistines, defeating them soundly. He went on to annex surrounding tribal kingdoms, including Edom, Moab, and Ammon, creating a large empire over which he ruled for about forty years.

By any measure, David was a successful military commander and political as well as spiritual leader. He neutralized the greatest threat against Israel, the Philistines; he united, as Saul had been unable to do, the fragmented tribes of Israel, forging them into a nation; he expanded his territorial holdings; and he established his empire in both a temporal and a spiritual sense. Nevertheless, his reign was plagued by family dissension and rebellion. In order to unite the various groups from which he had created his nation, David took wives from diverse tribes, cobbling together a family of strangers with nothing in common except mutual hostility. The worst rupture came when David's third son, Absalom, killed his first son, Amnon, because he had raped Tamar, sister of Absalom and half-sister of Amnon. David exiled Absalom, then recalled him. However, during his period of exile, Absalom began to organize a rebellion against his father. When Absalom returned to Jerusalem, he staged a revolt that sent David fleeing from the throne until the king could organize a force of his own to retake his empire. Absalom was killed in battle by Joab, one of David's loyal commanders.

David survived the strife to pass down his empire intact to Solomon, his son by Bathsheba. After Solomon's death, however, the kingdom was divided. While the empire itself ultimately dissolved, the religious traditions of the empire David created have endured through centuries of Judaism.

DAYAN, Moshe

BORN: May 20, 1915

DIED: October 16, 1981

SERVICE: Arab Revolt (1936–39); World War II (1939–45); Israeli War of Independence; (1948–49); Sinai War (1956); Six-Day War (1967)

Flamboyant, courageous Israeli front-line general, who, as Israeli defense minister, gained international recognition during the Arab–Israeli Six-Day War of 1967

Dayan was born in the first *kibbutz*—collective agricultural settlement—established by Zionists in Palestine. As a teenager, he joined the Haganah, the underground army working toward establishing a nation independent from British rule. The *kibbutz* was vulnerable to Arab attack, and the Haganah also served as a perimeter patrol.

Dayan first saw significant action during the Arab revolt of 1936–39, serving with Yizhak Sadeh's guerrilla units. The Jewish settlers sporadically cooperated with the British during the revolt, and Dayan trained under the British general Orde Wingate during this period; however, Dayan's continued activity in the independence movement provoked his arrest at the outbreak of World War II. He was imprisoned at Acre in October 1939 and not released until February 16, 1941. The exigencies of war in the Middle East again brought cooperation between the Haganah and the British, and Dayan became a scout working in advance of the British invasion of Vichy French-held Syria and Lebanon during June 1941. In combat on June 8, Dayan lost his left eye and from that point on wore a trademark black eye patch.

Dayan served on the Haganah general staff prior to the United Nations partition of Palestine in November 1947. In the Israeli War of Independence that followed, he fought against the Syrians in Galilee, successfully defending the Deganya settlements during May 19–21, 1948. Drawing on his Haganah training, as well as his experience with Wingate, he next raised the Eighty-Ninth Commando Battalion, a highly mobile mechanized strike force, which he led on a series of lightning raids against Arab-held positions at Lod and Ramallah between July 9 and and July 19. Following these successes, he was put in charge of the Jerusalem sector (1948–49) and was instrumental in early settlement negotiations with King Abdullah of Jordan in 1949.

After Israeli independence had been established, Dayan went to England in 1953 for study at the Camberley Staff College (1953), returning to Israel later that year as chief of staff of Israeli Defense Forces—the Israeli army. When the Sinai War erupted in 1956), Dayan took general charge of planning and directing the Israeli campaign (October 29–November 5, 1956. Two years after the successful conclusion of the war, he resigned his commission to enter civil politics, gaining election to the Knesset (Israeli parliament) on the Mapai (Labor) ticket in 1959 and serving as Minister of Agriculture from December 1959 to November 1964. In 1964, he broke with his party to join David Ben-Gurion in founding the Rafi (Labor List) party and once again won election to the Knesset.

In June 1967, Dayan was appointed defense minister under prime minister Levi Eshkol. As defense minister, he directed operations during the Arab–Israeli "Six-Day War" of 1967, from which the Israelis emerged having won a decisive victory. Dayan stepped down from the Defense Ministry after **Yitzhak Rabin** succeeded Golda Meir as prime minister in May 1974, but returned to government in 1977, serving as foreign minister in the government of Menachem Begin through 1979. He was instrumental in creating the Camp David peace settlement U.S. president Jimmy Carter mediated between Israel and Egypt. However, growing differences with Begin over policy toward the Palestinian Arabs prompted Dayan's resignation as foreign minister in 1979, two years prior to his death.

DECATUR, Stephen

BORN: January 5, 1779

DIED: March 22, 1820

SERVICE: Quasi-War with France (1798–1800); Tripolitan War (1801–1805); War of 1812 (1812–15); Algerine War (1815)

American naval hero in action against the Barbary pirates and in the War of 1812

Decatur was born on Maryland's Eastern Shore, but was raised in Philadelphia, at the time the young nation's principal city. There he attended the University of Pennsylvania, leaving it to join the navy as a midshipman in April 1798. The following year he was promoted to lieutenant and saw action (as first lieutenant aboard the U.S.S. *Essex)* during the undeclared naval war with France known as the Quasi-War.

In 1801, at the outbreak of the Tripolitan War, he served in Commodore Richard Dale's squadron, then transferred to the U.S.S. *New York*, on which he served from 1802 to 1803, before being given command of the small—sixteen-gun—*Argus* and, afterward, the twelve-gun *Enterprise*.

Commanding the captured ketch *Mastico* (which he renamed *Intrepid*), Decatur led a small band of sailors into Tripoli harbor on February 16, 1804, to set fire to the captured American thirty-six-gun frigate *Philadelphia*, which had earlier been captured as a prize. No less than the great British admiral **Horatio Nelson** called this exploit the "most bold and daring act of the age." The feat earned Decatur promotion to captain in May—at age twenty-five, he was (and remains) the youngest man ever promoted to this naval rank—and he was presented with a sword of honor by Congress. Decatur continued to fight with signal ability and gallanty throughout the war and was a principal in successful negotiations with the Bey of Tunis, which ended the war and did much to establish the integrity of United States sovereignty and enhance the international prestige of the young nation.

Decatur served in the 1808 court-martial that suspended Captain James Barron for negligence in the *Chesapeake–Leopard* incident of June 22, 1807, in which the U.S. warship *Chesapeake* was stopped by the British frigate *Leopard* off Norfolk, Virginia, and four seamen were "impressed" (forcibly removed) into the British service and others killed in an exchange of gunfire. The British policy of impressment on the high seas was one of the precipitating factors in the War of 1812, at the outbreak of which Decatur was given command of the forty-four-gun frigate *United States*, with which he captured the thirty-eight-gun British frigate *Macedonian* in a battle of October 25, 1812. As commodore, Decatur was charged with the defense of New York Harbor (1813), and during June 13–15, 1813, he attempted to run the British blockade of the harbor in the fourty-four-gun U.S.S. *President*. He damaged the twenty-four-gun H.M.S. *Endymion*, but was forced to surrender to superior British forces. Taken prisoner, he was subsequently parolled.

With the outbreak of the Algerine War in May 1815, Decatur was dispatched to the Mediterranean to fight the Barbary pirates, and on June 17 captured the forty-six-gun Algerian flagship *Mashouda*, after which four vessels of his squadron ran the twenty-two-gun *Estedio* aground off Cabo de Gata. It was Decatur who dictated ("at the mouths of cannon") highly favorable peace terms, not only securing release of all U.S. prisoners and an end to U.S. tribute payments (protection money to buy off the Barbary pirates) to Algiers, but also an indemnity payment.

Decatur returned triumphantly to the United States, where he was appointed to the newly created Board of Naval Commissioners in November 1815. Five years later, James Barron resurfaced, embittered by the disgrace he had suffered in the *Cheasapeake–Leopard* court-martial. He challenged Decatur to a duel, mortally wounding the naval hero in that exchange on March 22, 1820, at Bladensburg, Maryland.

DE GAULLE, Charles (André-Marie-Joseph)

BORN: November 22, 1890
DIED: November 9, 1970
SERVICE: World War I (1914–18); World War II (1939–45)

De Gaulle was the military and political leader of the Free French forces and government-in-exile during World War II; after the war, he was the moving force behind France's Fifth Republic

The son of an intensely nationalistic family, Charles de Gaulle was educated at the Military Academy of Saint-Cyr and joined an infantry regiment (under Col. **Philippe Pétain**), as a second lieutenant, in 1913. He impressed Pétain and others with his intelligence, initiative, and, once World War I erupted, with his courage as well. He fought at the do-or-die defense of Verdun, was wounded in combat three times, and was a P.O.W. for two years and eight months. During his captivity, he made five valiant, albeit unsuccessful, attempts to escape.

Following the war, de Gaulle served briefly as a member of a military mission to Poland, then taught at Saint-Cyr. He was chosen for two years of special training in strategy and tactics at the école Supérieure de Guerre (the French War College). When he graduated from the course in 1925, he was promoted by Marshal Pétain to the Staff of the Conseil Supérieur de la Guerre, the Supreme War Council.

Major De Gaulle served during 1927–29 in the army occupying the Rhineland. During this period he became alarmed by the continued danger he saw posed by German aggression. After his Rhineland assignment, he served for two years in the Middle East, then, as a lieutenant colonel, served for four years as a member of the secretariat of the Conseil Supérieur de la Défense Nationale, the National Defense Council.

In addition to his work as a field and staff officer, de Gaulle was a powerful military theorist, who, in 1924, wrote a study of the relation of the civil and military powers in Germany, *Discord Among the Enemy*. He also lectured on the subject of leadership (the lectures were published in 1932 as *The Edge of the Sword)*. Two years later, he published a study of military theory, *The Army of the Future*. In this work, he developed and defended

Charles de Gaulle. Courtesy National Archives.

the idea of a small professional army that was based on a high degree of mechanization for maximum flexibility and mobility. This was in direct opposition to the defensive, static strategy represented most dramatically in the Maginot Line. Not content to allow his ideas to be taken as merely academic, he appealed to political leaders in an attempt to persuade them to his point of view. The result was discord with de Gaulle's commanders and senior officers, including Marshal Pétain himself, who protested de Gaulle's right to publish a historical study titled *France and Her Army*. (The work did appear, in 1938.)

With the outbreak of World War II, de Gaulle was put in command of a tank brigade of the French Fifth Army. He was quickly promoted to the temporary rank of brigadier general in the Fourth Armored Division (this was the highest military rank he was to hold) and proved himself a very able tank commander. He was named undersecretary of state for defense and war on June 6 by French premier Paul Reynaud, who sent him on several missions to England to explore ways in which France might continue the war against Germany.

De Gaulle remained in England when the Reynaud government fell and was replaced by the collaborationist government of Marshal Pétain, who immediately set about seeking an armistice with Germany. On June 18, 1940, de Gaulle broadcast from London his first appeal to those French people who would resist Germany to continue, under his leadership, the war against the Nazis. As a result of this and repeated broadcasts, a French military court tried de Gaulle in absentia, found him guilty of treason, and sentenced him to death, loss of military rank, and confiscation of property (August 2, 1940).

The die now cast, de Gaulle threw himself with great skill, energy, and determination into organizing the Free French Forces and a shadow Free French government. It was an extraordinary task, for he was all but unknown outside of French military circles; even the people of France did not recognize him as a political figure. All that sustained him was his self-confidence, his strength of character, and his ability to lead. Throughout the war, until the liberation of France, de Gaulle continued to broadcast from London (until he moved his headquarters to Algiers in 1943) and to direct the action of the Free French Forces and other resistance groups ("the underground") in France. He worked closely—though not always smoothly—with the British secret services in this effort.

Because his relations with the British government and military were often difficult, de Gaulle moved his headquarters to Algiers in 1943. There he became president of the French Committee of National Liberation. He served at first with General Henri Giraud, but skillfully engineered his ouster and emerged as sole leader of the committee. It was de Gaulle, at the head of the government-in-exile, who returned to Paris on September 9, 1944, after its liberation.

De Gaulle led two successive provisional governments but, on January 20, 1946, resigned over a dispute with the political parties forming the coalition government. He opposed the Fourth French Republic as too likely to repeat the errors of the Third Republic, and, in 1947, formed the Rally of the French People (Rassemblement du Peuple Français), which won 120 seats in the National Assembly in the 1951 elections. Becoming dissatisfied with the RPF, he severed his connection with it in 1953, and it disbanded in 1955. De Gaulle retired for a time, during 1955–56, and wrote three volumes of memoirs.

When insurrection broke out in Algiers in 1958 and threatened to bring civil war to France, de Gaulle was brought back to the national limelight as prime minister designate and, on December 21, 1958, was elected president of the Republic. He served for the next ten years, amid much turbulence, controversy, and opposition from the nation's leftwing political leaders. After his retirement, he continued writing his memoirs, but died of a heart attack the year after he left office.

DEWEY, George

BORN: December 26, 1837

DIED: January 16, 1917

SERVICE: Civil War (1861–65); Spanish–American War (1898)

American admiral whose triumph at Manila Bay during the Spanish–American War was absolute

Dewey was a native of Montpelier, Vermont, who attended Norwich University before enrolling in the U.S. Naval Academy, from which he graduated in 1858. He saw extensive action during the Civil War, serving under **David Farragut** aboard the steam sloop *Mississippi*, at New Orleans (April 24–25, 1862) and Port Hudson (March 14, 1863). He also sailed with the North Atlantic Blockading Squadron, part of an effort to blockade the Confederacy, and bombarded Fort Fisher from the U.S.S. *Colorado* (December 23–27, 1864, and January 13–15, 1865).

Following the war, Dewey rose steadily through the ranks, gaining promotion to commander in April 1872, and to captain in September 1884. He was appointed chief of the Bureau of Equipment in 1889 and president of the Board of Inspection and Survey in 1895, posts that gave him firsthand knowledge of the potential of the modern battleship. He was one of the prime architects of a technologically advanced U.S. Navy.

Promoted to commodore in February 1896, he was reassigned—at his request—to sea duty as commander of the Asiatic Squadron (November 1897). He was in Hong Kong when the Spanish–American War broke out on April 25, 1898, and responded to telegraphed orders from the Navy Department to take his squadron to the Spanish-held Philippines. There he entered Manila Bay by night and, discovering at anchor the Spanish squadron of Admiral Patricio Montojo y Parasón off Cavite, he turned to his flag captain Charles V. Gridley at 5:40 A.M. on May 1, and issued words that entered into American history: "You may fire when ready, Gridley." With that, the one-sided battle commenced, in which all of the Spanish vessels were either sunk or abandoned by noon. Seven U.S. sailors were wounded. Only the lack of ground troops kept Dewey from capturing Manila then and there. But, promoted to rear admiral on May 10, he provided naval support to the U.S. Army, which took the city on August 13.

Dewey was promoted to admiral of the navy on March 3, 1899; it was a rank created especially for him. Upon his return to the United States in September, he was given a spectacular hero's welcome and, exempted from mandatory retirement regulations, was named president of the Navy General Board, in which office he served until his death.

Dewey was an aggressive and enterprising officer, who also believed in careful training, planning, and the acquisition of top-level equipment. All of these factors figured in his extraordinary achievement at Manila Bay.

DIOCLETIAN
(Gaius Aurelius Vaerius Diocletianus)

BORN: ca. 250

DIED: ca. 313

SERVICE: Roman Civil War (284–85); Egyptian Revolt (246–97)

Roman commander, military reformer, and emperor

Diocletian was a common soldier who rose to become perhaps the greatest of the late Roman emperors. Diocletian was merciless with rebels and in his systematic suppression of Christians, but he introduced administrative and military reforms that staved off the ultimate dissolution of the Roman Empire for at least a century. He was born in Dalmatia to humble parents and enlisted in the Roman army, rising quickly through the ranks to become commander of the personal bodyguard of Emperor Numerian in 283. When Numerian was murdered on November 20, 284, Diocletian acted swiftly, arresting and summarily executing the supposed assassin, the praetorian prefect Aper—who also happened to be Diocletian's arch rival. With both Numerian and Aper dead, Diocletian was elected emperor.

His first action was to lead an army against Carinus, the brother of Numerian, who ruled the Western Empire. Diocletian engaged Carinus in battle on the Margus River in Illyricum in the spring of 285, not only defeating Carinus's forces, but killing the Western Emperor himself. Yet Diocletian did not rule long as sole emperor of East and West. He understood that empire was crumbling beneath him, and he reasoned that a single ruler could not successfully defend the nation against the numerous threats from within and without. Accordingly, he elevated his most trusted friend Maximian to the office of caesar—in effect, assistant emperor—in the summer of 285. Shortly thereafter, he named Maximian to the station of Augustus, giving him full co-emperor status. Next, in a program of comprehensive administrative reforms designed to rescue the empire from disintegration, Diocletian appointed two additional caesars, Gaius Valerius Galerius and Flavius Valerius Constantius. Together, these rulers became the Tetrarchy.

Diocletian's menu of reforms included attempts at centralized wage and price controls to curb rampant inflation and a general restructuring of the imperial administration. Diocletian decreed the separation of military from civil authority, and he apportioned the empire into smaller, more manageable provinces, which were further organized into groups of twelve to form larger administrative units called *dioceses*. By far, his most profound reforms came in the military. Diocletian swiftly doubled the size of the Roman army and totally reorganized it. He understood that Rome faced increasingly sophisticated military threats along its far flung frontiers, and that, therefore, the army needed a combination of strong standing forces in place on the frontier in addition to substantial, highly mobile forces available for rapid reinforcement of hot spots. Even as he expanded his army, Diocletian reduced the size of each legion in a bid to facilitate strategic and tactical flexibility.

The new army was put to the test in 294–96, when a Roman usurper named Domitius Domitianus—or Achilleus—established himself as emperor in Alexandria, Egypt. Diocletian personally led an army to Alexandria, which he besieged brutally for eight months before the city fell. He ordered the execution of Achilleus in 296. Diocletian next provided aid to the Upper Egyptians and Nubians in repelling incursions by the Blemmyes, a barbarian tribe. In the meantime, Diocletian looked within the empire proper and unleashed the last mass persecution of Roman Christians during 303–304. He was convinced that the cults were eroding the empire from the inside.

In 305, shortly after this paroxysm against the Christians, a weary Diocletian abdicated, as did Maximian Augustus, leaving the empire to the two caesars. Diocletian retired to his palace in Spalatum—the modern city of Split—on the Dalamtian coast and died about seven years later.

DIONYSIUS I (Dionysius the Elder)

BORN: ca. 430 B.C.

DIED: 367 B.C.

SERVICE: First War with Carthage (397–396 B.C.); Second War with Carthage (392 B.C.); Third War with Carthage (383–378 B.C.); Fourth War with Carthage (368–367 B.C.); Conquest of Southern Italy (390–379 B.C.)

As tyrant of Syracuse, Dionysius I led three wars against Carthage and made Syracuse the strongest Greek city west of the mainland

Dionysius was born of obscure and humble origins and was essentially a public office clerk when conflict broke out in 409 B.C. between Syracuse and Carthage. He managed to maneuver himself into a position of power and influence during this period, becoming *strategos autokrator*—military dictator—in 405 B.C. From this time on, he worked to consolidate his power, also constructing an immense private fortress on the Syracusan island of Ortygia. Systematically, he overthrew the governments of the major city-states of eastern Sicily, sometimes transferring entire populations, sometimes enslaving them in place.

With his power base established, he took on the Carthaginians in 397 B.C., hoping to push the them out of Sicily. While he fell short of this goal, he was able to confine them to the western part of the island by 392 B.C. This accomplished, he turned next to the Italian mainland, taking the cities of Croton in 388 B.C. and Rhegium in 386 B.C., then sacking and looting Pyrgi in about 384 B.C. His wide-ranging ambitions took him next to the Adriatic and Greece, where he unleashed such brutality that a series of popular revolts were provoked against him.

In 383 B.C., he began yet another war with Carthage, suffering a reversal at Cronium about 375 B.C. This setback only strengthened his resolve, and he reopened hostilities again in 368 B.C. but died the following year. His demise came not on the field of battle, but as a result of overindulgence during a victory celebration at the Lenaean tragic festival in Athens.

DONIPHAN, Alexander William

BORN: April 9, 1808

DIED: August 8, 1887

SERVICE: Mormon Crisis (1838); U.S.–Mexican War (1846–48)

Unconventional hero of the U.S.–Mexican War

Doniphan was born near Maysville, Kentucky, and graduated from Augusta College in Augusta, Kentucky, in 1826. He settled in Missouri, where he set up as a lawyer and politician, gaining election to the state legislature in 1836 and 1840. During this period, he also became a prominent figure in the Missouri state militia and was its brigadier general in 1838, when he intervened in the Mormon Crisis. Joseph Smith (founder of the Church of Jesus Christ of Latter-Day Saints, the "Mormons") and other Mormon leaders had been sentenced to death by a court-martial. Doniphan refused to carry out the executions, which he considered judicial murder. He swore that he would arrest and bring to justice anybody who attempted to act against Smith and the others.

In 1846, at the outbreak of war with Mexico, Doniphan organized the First Missouri Mounted Volunteers and, after mustering them at Fort Leavenworth, Kansas, in June, marched with them to Santa Fe, New Mexico, as part of a force under the overall command of Colonel **Stephen Watts Kearny.** At Santa Fe, Kearny left Doniphan in command of New Mexico. A numerically superior Mexican force attacked him near El Paso, Texas, but Donipahn readily routed the attackers on Christmas Day, 1846. He then crossed the border and marched into Chihuahua during February 1847, where he engaged another superior Mexican force, which held fortified positions at the Sacramento River (February 28). Following this triumph, he advanced on Saltillo and then to Matamoros, returning to Saltillo on May 21, having defeated every Mexican force he encountered.

Doniphan proved not only the most effective commander of volunteers during the U.S.-Mexican War, but one of the most effective military leaders of any unit, voluneeer or regular army. Resourceful and cool under fire, he led his men through long, difficult marches and battled against superior numbers while suffering few casualties himself.

Following the war, he returned to Missouri, where he resumed his law practice and was elected again to the legislature in 1854. An opponent of secession as the clouds of civil war gathered in 1861, Doniphan was named as a delegate in the Peace Conference, held in Washington, D.C. during February 1861 as a last-ditch effort to avert war. With the outbreak of war, Doniphan advocated a neutral position for his state, but he briefly commanded the militia as major general. Resigning for personal reasons early in the war, Doniphan moved to St. Louis and later settled in Richmond, Missouri, where he died in 1887.

DÖNITZ, Karl

BORN: September 16, 1891

DIED: December 24, 1980

SERVICE: World War I (1914–18); World War II (1939–45)

*German admiral who replaced **Erich Raeder** as chief of the Nazi navy during World War II; briefly served as head of the Third Reich after **Adolf Hitler**'s suicide*

Dönitz was born in Grünau. He joined the German navy on April 1, 1910, and served on U-boats during World War I. This experience persuaded Dönitz that submarines would play an increasingly important role in naval strategy. With the conclusion of World War I, Dönitz remained in the *Reichsmarine,* the small navy Germany was permitted under the punitive terms of the Versailles treaty. In fact, Dönitz operated clandestinely to build a German submarine force, despite the fact that the treaty specifically prohibited the vessels.

Appointed chief of the Submarine Force in 1935, Dönitz played a key role in expanding this aspect of the navy and was promoted to rear admiral shortly after the

Admiral Karl Dönitz. Courtesy National Archives.

start of World War II. He held a simultaneous post as flag officer in charge of Germany's submarines.

Dönitz took a relatively unprepared U-boat fleet and created a devastatingly effective campaign against Allied shipping in the North Atlantic. With each success, he made louder noises for increased funding of the submarine program. He was soon highly unpopular with other service chiefs and, in particular, with navy commander in chief, **Erich Raeder,** who did not favor submarine warfare. Nevertheless, Dönitz prevailed and won promotion to vice admiral in 1940 and admiral in 1942. As his star rose, Raeder's sank, and Dönitz replaced him as commander in chief of the navy on January 30, 1943. He also continued to serve directly as commander of the U-boat force, which by this time constituted the bulk of the German navy.

Dönitz's triumph did not endure through 1943. By the middle of the year, the Allies had begun to achieve great success against the U-boats. Dönitz introduced and advocated advanced submarine technologies, most notably the snorkel, which permitted shallow-depth operation of the diesel engines, thereby saving battery power and greatly extending the time submarines could operate under water. Yet this innovation had little tangible effect on regaining the initiative from the Allies.

Hitler named Dönitz his successor as chancellor in the will he composed on April 30, 1945, the day he committed suicide. For just over a week, the admiral conducted what little was left of the Third Reich. It was he who negotiated surrender to the Allies on May 7.

Dönitz was tried and convicted of war crimes at the Nuremberg tribunal later in 1945. Sentenced to ten years in Spandau Prison, he served his full term and was released in 1956. He lived out the remainder of his life quietly in a suburb of Hamburg.

Dönitz was an extremely capable commander, whose advocacy of submarine warfare had devastating impact on Allied shipping. He is credited with developing the "wolf pack" tactic, whereby submarines hunted in groups, and he created a support system for the vessels—including seaborne tankers and submarine tenders—which greatly extended their range. He mastered the art of coordinating aerial reconnaissance with submarine attacks on convoys.

DOOLITTLE, James ("Jimmy") Harold

BORN: December 14, 1896

DIED: September 27, 1993

SERVICE: World War I (1917–18); World War II (1941–45)

American Army Air Corps general, best known for his daring and unconventional aircraft carrier-launched bombing raid on Tokyo early in World War II

A native of Alameda, California, Doolittle was educated at Los Angeles Junior College and at the University of California. He joined the Army Reserve Corps in October 1917, after the United States entered World War I and was assigned to the Signal Corps, in which he served as a flight instructor through 1919. The following year, Doolittle was commissioned a first lieutenant in the Army Air Service and gained national attention by making the first transcontinental flight in less than fourteen hours, on September 4, 1922.

Under auspices of the Army Air Corps, Doolittle enrolled in the aeronautical science program at Massachusetts Institute of Technology, from which he earned an Sc.D. degree in 1925. Following this, he worked in several military aviation testing stations and became involved in air racing as well as in demonstrating aircraft during 1925, with the object of promoting aviation generally and military aviation in particular. In September 1929, he successfully demonstrated the potential of instrument flying by making the first-ever blind instrument landing.

Doolittle resigned his commission in February 1930 to become aviation manager for Shell Oil, where he worked on the development of new high-efficiency aviation fuels. He also continued to race, claiming victories in a number of prestigious competitions, including those for the Harmon (1930) and Bendix (1931) trophies. In 1932, he set a world speed record.

With war clouds gathering in July 1940, Doolittle returned to active duty as a major in the Army Air Corps. In the months following the Japanese attack on Pearl Harbor (December 7, 1941), the American and Allied forces were in a desperate defensive position in the Pacific theater. In an effort to raise U.S. morale—and to force the Japanese to divert a portion of their air forces to defense—Doolittle planned and executed an extraordinary and extraordinarily hazardous bombing raid on the Japanese capital city of Tokyo. He led sixteen B-25 bombers from the aircraft carrier *Hornet* on April 18, 1942. The aircraft were not designed to take off from an aircraft carrier, and doing so demanded great skill. Moreover, fuel limitations meant that no return trip was possible; Doolittle and his men would have to land in China, hope to evade capture, and find their way back to Allied lines. The mission was as close to a suicide run as any ever undertaken by American fighting men. Its success was slight in strictly military terms—damage to Tokyo was not extensive; however, the raid on the Japanese homeland was a morale boost of incalculable proportions and did much to spur the American war effort. As planned, it also served to tie down a portion of the Japanese air force to home defense.

Doolittle and most of his raiders survived the action, and Doolittle was promoted to brigadier general. He was sent to England to organize the Twelfth U.S. Air Force in September 1942, and, with the temporary rank of major general, commanded the Twelfth in Operation Torch—the Allied invasion of French North Africa. During March 1943–January 1944, he commanded strategic air operations in the Mediterranean theater and was promoted to the temporary rank of lieutenant general in March 1944. He was given command of the British-based Eighth Air Force's bombing operations against Germany during January 1944–May 1945.

After the war ended in Europe, Doolittle returned to the Pacific, where, with the Eighth Air Force, he provided support in the battle for Okinawa (April–July, 1945) and the massive bombardment of the Japanese home islands.

Following the war, in May 1946, Doolittle left active duty (remaining in the reserves) and took a senior executive position with Shell Oil. He was frequently tapped by the government to serve as an adviser on scientific, technological, and aeronautical commissions during 1948–57. Following his retirement from Shell and the Air Force Reserve in 1959, he continued to work as a consultant, not only in matters of science and aeronautics, but in national security policy issues as well. Doolittle was not only a gallant commander and an able military air tactician, he was an aviation pioneer, who did much to promote the status of the U.S. military's air arm.

DORIA, Andrea

BORN: November 30, 1466

DIED: November 25, 1560

SERVICE: Hapsburg–Valois Wars (1521–25, 1526–30, 1536–38, 1542–44, 1547–59); Turkish War (1532–46)

A Genoese Renaissance admiral and statesman; served under several rulers, including Pope Innocent VIII and Kings Ferdinand I and Alfonso II of Naples

Doria was born of an ancient and noble Genoese family, but was orphaned early and drifted into a life as a *condottiere* (mercenary) in service variously to the pope and other Italian princes. About 1510, he was appointed captain general of the Genoese galleys, which he led on raids against the Barbary corsairs—pirates who preyed upon Genoese shipping in the Mediterranean.

Following the conquest of Genoa by the Holy Roman emperor Charles V in 1522, Doria entered the service of King Francis I of France, who named him admiral of the French Mediterranean fleet. In this capacity, he operated against Imperial supply lines during their invasion of Provence and the abortive siege of Marseilles in 1524. Next, after Francis was defeated and captured at the Battle of Pavia on February 24, 1525, Andrea Doria offered his services to Pope Clement VII, but rejoined Francis after his release in 1527.

His renewed association with the French king did not last long. Feeling that Francis was not grateful for his services, and therefore distrusting him, Doria presented himself to Charles V in 1528. Moreover, he recalled the Genoese fleet from the blockade of Naples, and recaptured Genoa, transferring the Genoese republic to Imperial protection in September.

Doria held no government post, but nevertheless exercised great influence in Genoese political affairs. Now serving as Imperial admiral, he raided throughout the central Mediterranean, particularly along the Greek coast, where he took the towns of Coron and Patras in 1532. He participated in the Holy Roman Empire's campaign against Tunis, which he helped capture during June–July of 1535. At the head of the combined Imperial-Genoese-Venetian fleet, he engaged the corsair leader **Barbarossa** (Khair-el Din) at Préveza on September 27, 1538, but lost. It is a measure of Doria's reputation for mercenary duplicity that many of his contemporaries—and some historians—believe he deliberately lost the battle to take vengeance on his Venetian rivals. Andrea Doria participated in Charles V's unsuccessful Algerian expedition—though he had advised against undertaking it—and his skillful defense during the retreat saved the force form annihilation (September–October 1541).

Doria, one of the great Renaissance schemers, was himself the repeated target of intrigue. He headed off a plot by his political opponent, Giovanni Luigi Fiesco, to overthrow him in 1547 and, the following year, extinguished another conspiracy with violent zeal. In 1550, he led yet another fleet against the Barbary corsairs, but enjoyed little success.

During the Fifth Hapsburg–Valois War, he led an expedition to recover the captured island of Corsica and ended up devoting two years—1553–55—fighting the French there. The results were mixed, and the aging mercenary ultimately withdrew to Genoa, where he lived out the remaining five years of his very long life in quiet retirement as a very wealthy man.

Andrea Doria is an example of the Renaissance mercenary *par excellence*. Courageous, resourceful, and tireless, he was also utterly unprincipled, shifting allegiance for gain as well as out of spite. He is considered one of the greatest tacticians of Renaissance naval warfare.

DU PONT, Samuel Francis

BORN: September 27, 1803

DIED: June 23, 1865

SERVICE: U.S.–Mexican War (846–48); Civil War (1861–65)

U.S. Civil War admiral; responsible for modernizing the United States Navy

Du Pont was born at Bergen Point, New Jersey, near New York City, one of the grandsons of economist Pierre Samuel du Pont de Nemours and, of course, a member of the family that made its fortune in the munitions and chemical industry. In 1815, du Pont was appointed a midshipman in the U.S. Navy and dispatched to duty in the Mediterranean during 1817. By 1826, having served in Europe and South America, he was promoted to lieutenant and, in the glacially slow system of promotions that characaterized the American military of the era, commander in 1843.

Du Pont held command of Commodore Robert Stockton's flagship, U.S.S. *Congress* in 1845 as Mexico and the United States hovered near war. After cruising the Hawaiian Islands in the *Congress*, du Pont took command of the sloop *Cyane* in San Francisco during July 1846. He transported Major **John C. Frémont**'s troops to San Diego in September and conducted coastal operations and landings along the coast of Baja California. His most important operations during the U.S.–Mexican War were the bombardment of Guaymas (October 7, 1846) and support of land operations at Mazatlán (November 11, 1847) and at La Paz (November 17). He was called to relieve the siege at San José del Cabo, but land forces had broken it by the time he arrived on February 14, 1848.

Following the war, du Pont served in a variety of shore assignments, most significantly as an advocate of steam power and as one of the organizers of the U.S. Naval Academy. In 1855, he worked with the Lighthouse Board. Finally, in 1857, he secured another sea assignment, as skipper of the U.S.S. *Minnesota* on an 1857–59 cruise to China. On his return, he was appointed commandant of the Philadelphia Navy Yard (1860).

With the outbreak of the Civil War, du Pont was appointed senior member of the Navy Commission of Conference, which was charged with creating a naval strategy for the war. Du Pont played a key role in executing the Union's strategy of naval blockade of the Confederate states, serving as flag officer of the South Atlantic Blockading Squadron. He prosecuted this command with great vigor, capturing the Confederate forts at Port Royal Sound, South Carolina on November 7, 1861, an action for which Congress officially thanked him. Promoted to rear admiral in July 1862, he next mounted an assault against the defenses at Charleston. These proved far more formidable, and, on April 7, 1863, his ships were driven out to sea by resistance from the Charleston batteries.

Following the Charleston humiliation, du Pont resigned his commission and retired to his home near Wilmington, Delaware, from which he carried on an acrimonious dispute with Secretary of the Navy Gideon Welles over who was responsible for the failure of the assault on Charleston.

Du Pont was one of the few significant naval warriors during the Civil War, but he is also significant as one of the nineteenth-century navy's important innovators, who began the transformation of the force from sail to steam.

E

EDWARD I LONGSHANKS

BORN: June 17, 1239

DIED: July 7, 1307

SERVICE: Civil War (1263–65); First Welsh War (1276–77); Second Welsh War (1282–83); Third Welsh War (1294–95); Anglo-Scottish Wars of 1296, 1297–1305, 1306–1307

"The Hammer of Scots" not only fought the rebellious Scots, but also the Welsh

One of a dozen children born to King Henry III, Edward was granted huge tracts that included Ireland, Wales, the duchy of Gascony, and the earldom of Chester. After 1255, Edward joined a cabal of barons, led by Simon de Montfort, who attempted to overthrow Henry and rule England themselves. Edward first supported Montfort, then backed his father, only to return to Monfort's cause yet again. At last, he renounced the barons (late 1260) and secured his father's forgiveness.

With the outbreak of civil war in 1263, Edward stood by his father; however, he conducted himself with great recklessness on the field at Lewes (May 14, 1264) and suffered a decisive defeat. Montfort took his ersthwhile ally prisoner, but Edward managed to escape in 1265 and immediately took command of Henry's forces—this time with greater skill. He outmaneuvered Montfort on July 8, 1265, trapping the baron behind the Severn River. On August 1, he cut up the reserves at Kenilworth, and, four days later, cornered and killed Montfort himself at Evesham. At this battle, he rescued his father's army, which had been pinned down by the baronial forces.

The next task was generally to subdue pockets of rebellion throughout the English countryside. With that task accomplished, Edward next turned to the Crusades, hoping to join Louis IX of France (1271). On his way to join the French king, however, he learned that the monarch had died. Then, on his way back to England, news reached Edward that his father had also died (November 16, 1272). Fearful that the Barons' Revolt would be renewed, Edward dispatched his ministers to assess the barons' loyalty. He delayed his return to England until he was convinced that the situation with the barons was secure, and, on August 19, 1274, Edward I was crowned at Westminster Abbey.

Edward had learned from his mistakes on the battlefield. He set about creating a methodical, well-organized, efficient, and just government. In effect, he created Parliament by institutionalizing the practice of calling borough and shire leaders together at prescribed intervals. Having created Parliament, he made use of it as an effective tool of policy, enacting wide-ranging reforms in land and trade policy.

Despite his new, measured approach to power, Edward never fully overcame his bellicose nature. When the Prince of Wales attempted to evade his feudal responsibilities to the king, Edward attacked him, annihilating the prince's army and assuming control of all his lands. This provoked a popular rebellion against Edward, who responded with a ruthless march through Wales. Edward punished the rebels and, ultimately, killed both the Prince of Wales and his brother, thereby eliminating the two direct heirs to the throne. Edward then installed his own first son—the heir apparent—as prince of Wales. Ever since this act, the British Crown Prince has assumed the title Prince of Wales.

Edward's reforms began to disintegrate even before he himself died. As he grew older, Edward again had recourse to war in order to subdue rebellion at home. He deliberately provoked war with Scotland, touching off a conflict that would last, sporadically, for 250 years. The aged Edward was victorious on the field, but his military triumphs failed to extinguish rebellion. After eight years of warfare, Edward, worn down, simply gave up. He renewed his efforts against the Scots in 1305, achieving victories and executing rebel leaders. Yet, once again, the rebellion had died down only to revive. While he was returning to Scotland to meet the latest threat, Edward contracted dysentery and died.

EDWARD III OF WINDSOR

BORN: November 13, 1312

DIED: June 21, 1377

SERVICE: Hundred Years' War (1337–1453)

Edward III commenced the Hundred Years' War with France

Edward III was the son of King Edward II of England and Isabella of France. At fifteen, Edward III ascended the throne after his mother and her lover, Mortimer, deposed Edward II in 1327. Edward III came to power as the puppet of Isabella and Mortimer. Then, in 1330, he suddenly came into his own, arranging for Mortimer's execution. That done, Edward III began to rule in his own right.

By 1339, relations between England and France had become warlike. During 1339–40, Edward invaded France, enjoying a modicum of success, but badly straining the treasury. From economic rather than military necessity, Edward withdrew from the continent, though not before declaring himself king of France in 1340, thereby initiating a tradition (which endured until 1801) of English Kings blithely claiming rule over France.

Ruinous to the English treasury, the invasions also sparked the long and debilitating series of conflicts between France and England known as the Hundred Years' War.

When he wasn't fighting, Edward III created the Order of the Garter, Britain's highest order of knighthood, and grandly expanded Windsor Castle. Married to Philippia of York in 1328, he fathered a family of suitably grand proportions, seven sons and five daughters, who, in turn, would produce the generation that created the dynastic conflict called the Wars of the Roses (1455–85). After the death of Queen Philippa, Edward became slavishly devoted to Alice Perrers, his politically ambitious mistress. Through her the unpopular John of Gaunt gained an increasing degree of royal influence, alarming and alienating Parliment as well as Edward's subjects. The king died at a low point in his reign.

EDWARD, The Black Prince

BORN: June 15, 1330

DIED: June 8, 1376

SERVICE: Hundred Years' War (1337–1453); Castilian Civil War (1350–69)

Intolerant but valiant and skilled at arms, Edward was one of the most significant English tacticians of the Hundred Years' War

Born at Woodstock, Edward was the eldest son of King **Edward III** and Queen Philippa. His father named him Prince of Wales in May 1343, and young Edward first saw battle under his father's command in Northern France during 1346–47. At the battle of Crécy, Edward performed with such gallantry that he was knighted on the field (August 26, 1346).

In 1355, Edward was entrusted with his first command independent from his father when he was sent to France. On September 19, 1356, he defeated the French at Poitiers and captured King John II, whom he brought back to England.

As prince of Aquitaine, Edward spent a full decade, beginning in 1362, attending to his realm, which he taxed to the point of rebellion. In 1367, he absented himself from Aquitaine to lead an expedition to Castile to restore King Pedro the Cruel to the throne, defeating the forces of Henry of Trastamara (in Galicia) and the French Constable Bertrand du Guesclin at Ndjera in the process. However, while he was thus engaged, his subjects in Aquitaine rebelled, receiving support from du Guesclin's. The rebellion, thus supported, widened and the Hundred Years' War. As to Edward, he was only partly successful in suppressing the rebellion. In doing so, he ruthlessly sacked Limoges in October 1370, an action that only increased his subjects' hatred of him.

Exhausted by battle, his health failing, Edward returned to England in January 1371 and, the following year, renounced his principality in Aquitaine. There was, however, no rest for Edward. He became embroiled in a political struggle with his brother, John of Gaunt, against whom he sided with the so-called Good Parliament of 1376. He died at the height of the dispute, leaving his son Richard II as heir to the British throne.

EISENHOWER, Dwight David

BORN: October 14, 1890

DIED: March 28, 1969

SERVICE: World War I (1917–18); World War II (1941–45)

American general who served as Supreme Allied Commander in the European theater of World War II

"Ike" Eisenhower was born in Denison, Texas, and was raised in Abilene, Kansas. He graduated from West Point in 1915, but was not sent overseas during World War I. Instead, he was assigned to a variety of stateside training missions. His acute administrative skills were quickly recognized, and in 1920 he was promoted to major. Two years later, he was posted to Panama, returning to the United States in 1924 to attend Command and General Staff School, from which he graduated at the top his class in 1926. He graduated two years later from the Army War College.

From 1933 to 1935, Eisenhower served under General **Douglas MacArthur** in the office of the chief of staff, then accompanied MacArthur to the Philippines, serving there until 1939. Following large-scale maneuvers in the summer of 1941, Eisenhower was promoted to temporary brigadier general (September 1941) in acknowledgment of his performance during the exercises.

Upon America's entry into World War II, Eisenhower was assistant chief of the Army War Plans Division (December 1941–June 1942). Jumped in rank to major general in April 1942, he was named to command the European Theater of Operations on June 25. He served as Allied commander for Operation Torch—the invasion of French North Africa—in November, then directed the invasion and conquest of Tunisia during November 17, 1942–May 13, 1943. The next phase of Allied operations in Europe was the conquest of Sicily, which Eisenhower directed during July 9–August 17, 1944, and the invasion of the Italian mainland, which got underway during September 3–October 8.

Following the landings in Italy, Eisenhower went to London, where he was put in charge of directing plans for the Normandy invasion ("D-Day"). Appointed Supreme Commander of the Allied Expeditionary Force in December, he directed Operation Overlord, the Allied assault on Normandy (June 6–July 24, 1944) that breached **Adolf Hitler**'s "Fortress Europa," then commanded the advance across northern France, which spanned July 25–September 14.

Dwight David Eisenhower. Courtesy National Archives.

In December 1944, Eisenhower was promoted to general of the army ("five-star general"). The massive and desperate German offensive in the Ardennes, known as the Battle of the Bulge (December 16, 1944–January 19, 1945) posed a last-minute threat to the Allied advance across Europe. Eisenhower directed the response to the Bulge offensive, then resumed the Allied advance on February 8, crossing the Rhine in March. He pushed his forces into Germany during March 28–May 8, although, in a controversial political and strategic decision, essentially relinquished occupation of eastern Germany and Berlin to the Soviet troops of the Red Army.

Following the unconditional surrender of Germany on May 7–8, 1945, Eisenhower commanded Allied occupation forces until November, when he returned to the United States, received a hero's welcome, and took up a post as army chief of staff (November 1945–February 1948). After retiring from the army, Eisenhower assumed the unlikely role of president of Columbia University, in which he served admirably from February 1948 to December 1950. He also published a popular memoir, *Crusade in Europe*, in 1948.

In December 1950, faced with a critical Cold War situation, President Harry S. Truman recalled Eisenhower to active duty as Supreme Allied Commander Europe (SACEUR) and commander of NATO forces. Two years later, Eisenhower again retired to run for president on the Republican ticket. He served two terms (1953–61), presiding over a period of international turbulence, yet one in which the United States enjoyed economic prosperity and an unprecedentedly powerful role as a world superpower. As president, Eisenhower was both admired and criticized for his unflappable, easygoing approach to government. He helped bring an end to the Korean War, survived and recovered from a serious heart attack, and continued President Truman's policy of "containing" Communisim through intervention in Lebanon (June 15–August 21, 1958) and by introducing American "military advisors" into war-torn Vietnam.

Following the inauguration of John F. Kennedy in January 1961, Eisenhower entered quiet retirement.

As a military commander, Eisenhower was often criticized for his so-called broad-front strategy, which many felt diluted the war effort. Critics also argue that his policies at the end of the war allowed Soviet Communism to take root in large areas of Eastern Europe. However, few men could have held together the Allied coalition as successfully as Eisenhower. Moreover, his was a level-headed presence that combined political savvy with formidable strategic skill—as well as an innate cheerfulness and forthrightness to which world leaders, military commanders, ordinary soldiers, and the American people responded readily and positively.

EUGENE, Prince Of Savoy-Carignan

BORN: October 18, 1663

DIED: April 20/21, 1736

SERVICE: Austro–Turkish War (1682–99); War of the Grand Alliance (1688–97); War of the Spanish Succession (1701–14); Austro–Turkish War (1715–19); War of the Polish Succession (1733–38)

Austrian field marshal renowned for his tactical skill and speed in an age characterized by unwieldy armies and stately maneuvers

Eugene was born in Paris to Eugene Maurice of Savoy-Carignan and Olympia Mancini. He was grandnephew to Louis XIV's chief minister, Cardinal Mazarin; however, his mother having been banished from France, his request to join the French army was denied by the king, whereupon young Eugene offered his services to the Austrian army.

Fighting under Charles V of Lorraine at the battle of Vienna (September 12, 1683), he achieved distinction and was given command of the Kufstein Dragoon Regiment in December 1683. Commanding this force, he gained further distinction during the reconquest of Hungary for the Holy Roman Empire (1684–88). After persuading Savoy's Victor Amadeus to the Imperial cause in July 1690, he accepted a cavalry command in Italy under Antonio Caraffa and accomplished the relief of besieged Coni (Cuneo), captured Carmagnola, and defeated the French army of Marshal Nicholas Catinat. The timid conservatism of Caraffa and the Duke of Savoy prevented him from capitalizing on his victory.

During the failed invasion of Dauphiné, Eugene's capture of Gap and Embrun were the only bright spots, and these victories prompted his elevation to field marshal in 1693. The following year, he was given overall command in Italy, but enjoyed only mixed success, due in large measure to the poor quality of Savoy's troops.

In 1697, he reorganized the Austrian army to fight effectively along the Hungarian front, winning a significant victory over the Turks at Zenta on September 11, 1697. This represented a personal triumph as well, as Eugene was hailed as the greatest general of his time. He continued to confirm this judgment by captured Sarajevo in October 1697, then, at the outbreak of the War of the Spanish Succession, defeating the superior army of Catinat at Carpi on July 9, 1701. This victory was made the more extraordinary by the speed with which Eugene had marched his small force over the Alps to surprise his adversary in Italy.

On September 1, Eugene handily repulsed the French Duke of Villeroi's assault at Chiari, and while his subsequent offensive against Cremona failed on February 1, he did take Villeroi prisoner. But Eugene's ambition outstripped his logistics. Short of troops and supplies, he was repulsed by Vendôme at Luzzara on August 15 and returned to Vienna. There he was elevated to the presidency of the War Council. Immediately he set about modernizing the cavalry, increasing its size and, having learned from the Luzzara fight, instituting a more flexible system of supply. In Hungary, he reorganized his forces in 1703 and joined the Duke of Marlborough in an attack on Bavaria during May–July of the following year. He united with Marlborough again on August 13 to fight the combined French and Bavarian armies at Blenheim, where he was instrumental in achieving a great—and subsequently much-studied—victory.

On August 16, 1705, in Italy, he once again fought Vendôme, this time at Cassano d'Adda. Although the battle was essentially a draw, Eugene claimed victory. He was destined to face the same commander yet again the next year outside of Turin. After outflanking him, Eugene marched on Turin (July–August 1706), then took Parma on August 15. He joined forces with Victor Amadeus, leader of the Piedmontese in a September 7 attack on French forces under the Duke of Orleans and Marsin at Turin. By December, Eugene and the Piedmontese had driven the French out of Italy. This accomplished, Eugene invaded the south of France during July–August 1707. His ally Victor Amadeus did not wish to press the campaign, however, and little came of the invasion foray.

Eugene sought satisfaction elsewhere. He rushed in advance of his army to join his most trusted comrade-at-arms, the Duke of **Malborough,** in an attack against Vendôme's army at Oudenarde on July 11, 1708. This time, Eugene tasted victory over his formidable adversary. Emboldened, he coordinated campaigns with Marlborough. From August 14 through December 11, he held Lille under siege while Marlborough commanded the covering force. The duo cooperated in the siege and capture of Tournai (June 28–July 29, 1709) and Mons (September 4–October 26). After this, Eugene played a key role in the bloody Battle of Malplaquet (September 11, 1709) before rejoining Marlborough in a campaign through Flanders during the balance of 1709 and into 1710. Together, Eugene and Marlborough took Douai (June 10, 1710) and Béthune (August 30).

Eugene hastened back to Austria after the death of Emperor Joseph I to ensure that Charles VI would be elected Holy Roman Emperor. This accomplished, he returned to Flanders at a critical juncture. The French forces had routed a portion of the Allied army at Denain on July 24, 1712. Eugene was now compelled to take the defensive through August–October 1712, and did so successfully; however, he urged Charles VI that the time had come to make peace. Eugene negotiated favorable terms from Duke Claude Villars at Rastatt on March 6, 1714.

Just two years after peace with France had been concluded, Eugene was sent to Hungary to fight a new war with the Turks (spring 1716). He scored a massive victory against superior forces commanded by Damad Ali Pasha at Peterwardein on August 5, 1716, then marched on Temesvar, which fell to him on October 14. Beginning on June 29 of the following year, he laid siege against Belgrade, also delivering a decisive blow against Shalil Pasha's relief forces on August 16. The city fell to Eugene on the 18th. By July 21, 1718, the highly favorable Treaty of Passarowitz ended this latest war with Turkey.

After the war, Eugene counseled Charles VI to practice moderation in his policies and to avoid dictating harsh terms to those he had conquered, lest war be renewed. Eugene himself returned to combat with the outbreak of the War of the Polish Succession in 1733, rushing to the Rhine to defend against the invasion of Germany during April–September 1733. It was to be his final campaign. Aging and exhausted, he subsequently returned to Vienna and died there three years later.

Few military commanders have enjoyed the degree of success Eugene attained. A believer in training and preparation, he emphasized quick and efficient movement and maneuver, which always gave him an advantage in an age characterized by massive, but unwieldy armies. He was also gifted with an ability to inspire his troops, who adored him. Finally, he was a diplomat of considerable skill; fierce in combat, he was moderate in diplomacy.

EWELL, Richard Stoddert

BORN: February 8, 1817

DIED: January 25, 1872

SERVICE: U.S.–Mexican War (1846–48); Civil War (1861–65)

"Old Baldy" was a Confederate Civil War general of considerable ability

Ewell was born in Georgetown—now part of Washington, D.C.—but moved at an early age to Prince William County, Virginia. He enrolled in West Point and, after graduating in the upper half of his class in 1840, was commissioned in the Dragoons, with which he served in the West, Kansas, Oklahoma, and along the Oregon and Santa Fe trails. With the outbreak of the U.S.–Mexican War, Ewell served under General **Winfield Scott** as he pressed into central Mexico during early 1847. Ewell fought at Veracruz (March 9–28) and Contreras-Churubusco (August 18–20). For his gallantry at Contreras-Churubusco, he was breveted captain.

After the war, Ewell served as a recruiter and then returned to the southwestern frontier—now with the permanent rank of captain—in 1850. Here he fought against Apache raiders, sustaining a serious wound during an 1859 engagement.

Ewell was one of many officers who, while essentially opposed to secession, felt allegiance to his native state. When civil war erupted, he resigned his commission in April 1861, and offered his services to Virginia. The following month he was commissioned a colonel in the Confederate provisional army and was wounded on June 1 in a sharp exchange at Fairfax Courthouse. He was promoted to brigade command on June 17 and led Second Brigade at First Bull Run on July 21, though his unit was not in the thick of the action. On January 24, 1862, Ewell was promoted to major general and assigned under **Thomas J. "Stonewall" Jackson** in the Shenandoah Valley, where he was instrumental in the defeat of General Nathaniel Banks at the Battle of Winchester on May 24–25. Following this, he played roles in the Confederate victories at Cross Keys on June 8 and Port Republic on June 9 (both near Grottoes, Virginia), and then marched to Richmond, where he fought at Gaines' Mill (June 27) and Malvern Hill (July 1) during the bloody series of fights known as the Seven Days.

Never one to shrink from the front lines, Ewell was gravely wounded at Groveton on August 28. His leg was amputated, and he languished for nine months during a painfully slow convalescence. When Stonewall Jackson was killed—accidentally, by one of his own men—at Chancellorsville, Ewell was promoted to lieutenant general and given command of II Corps on May 29, 1863. Despite his disability, he handled himself and his troops well at Gettysburg, capturing the town on the battle's first day, July 1, but on July 2–3 he repeatedly failed to break through the Union lines. He remained at the head of II Corps at the Battle of the

Wilderness during May 5–6, 1864, and Spotsylvania on May 8–18. However, having never fully recovered from his wound, his health deteriorated, and on May 31, he relinquished II Corps to Jubal Early.

Ewell accepted reassignment to the less demanding post of commander of the Department of Richmond. When **Robert E. Lee** withdrew the Army of Northern Virginia from the Richmond-Petersburg defenses during April 2–3, 1865, Ewell attempted to join him with his own forces, but was surrounded and captured at Sayler's Creek during April 6–7. The exhausted and infirm commander was imprisoned for several months at Fort Warren, Massachusetts. In August, he was released and sought the solace of Spring Hill, his Maury County, Tennessee, estate.

Ewell was popular with his men and a tactician of great skill. He was dauntless as well as humane and was one of a handful of Confederate generals who advocated the recruitment of black soldiers into the Confederate army.

F

FABIUS MAXIMUS VERRUCOSUS, QUINTUS

BORN: 266 B.C.
DIED: 203 B.C.
SERVICE: Second Punic War (219–202 B.C.)

The great strategist who bested Hannibal through a patient campaign of attrition

Fabius began public life as consul in Liguria (233 B.C.), then served as censor (230 B.C.), and, once again, consul (228 B.C.). He probably first confronted Hannibal in 218 B.C., when he may have been dispatched to Carthage to demand indemnification for Hannibal's attack on Saguntum (modern Sagunto, Spain). Following the destruction of the army of Gaiss Flaminius at the hands of Hannibal, Fabius was elected dictator in 217. Recognizing the folly of confronting Hannibal's formidable army in the open, Fabius adopted a strategy of guerrilla warfare and attrition, staging raids in hilly terrain where Hannibal's vaunted cavalry was of little utility. When one of Fabius' subordinates, M. Minucius Rufus, having grown impatient with his commander's strategy, foolishly offered battle to Hannibal at Gerunium, he was soon trapped. Fabius marched to his timely relief.

Rufus, now fully persuaded of the wisdom of Fabiuis's mode of battle, supported him against a growing Roman restlessness. At last, however, Fabius's orders were overridden, and the Roman legions attacked Hannibal at Cannae in August 216 B.C. The result was catastrophe for the Romans, who once again turned to Fabius, electing him dictator. From this office, he evolved an even more cautious combat strategy designed to give the legions time to regroup and rebuild.

During 215, 214, and 209 B.C., Fabius was again elected consul and set out on a series of campaigns in Campania, managing to retake Tarentum in 209 B.C. When Scipio Africannus proposed to invade, Fabius, with characteristic reserve, opposed the expedition.

Among the several cognomens of Fabius was "Cunctator" ("the delayer"), which aptly described his favorite strategy employed against the superior forces and superior generalship of Hannibal. The strategy was effective and proved the best choice in a difficult situation.

FAIRFAX OF CAMERON, Sir Thomas Fairfax, 3d Baron

BORN: January 17, 1612
DIED: November 12, 1671
SERVICE: First Bishops' War (1638–39); Second Bishops' War (1640); First Civil War (1642–46); Second Civil War (1648)

Next to* Oliver Cromwell*, the most brilliant commander during the English Civil Wars

A Yorkshireman, born at Denton, Fairfax attended St. John's College, Cambridge from 1626 to 1629, when he joined the army to fight in the Netherlands. He returned to England in 1631 and raised a dragoon unit, the Yorkshire Redcaps, for service in the First Bishops' War (fall 1638–June 18, 1639). For this, Fairfax was knighted.

Fairfax took up arms again in the Second Bishops' War (April–November 1640), but soon found himself persuaded to the point of view of those who opposed the repressive religious measures of King Charles I. Fairfax attempted to present a petition to Charles at Heyworth Moor on June 3, 1642, protesting the levying of royal troops. Not only did the king pay him no heed, Fairfax narrowly avoided being ridden down by the king's escort. Following this, Fairfax put himself in the service of Parliament and was appointed on September 17, 1642, to command the cavalry in his father's Yorkshire army. After briefly blockading York in November, his cavalry occupied Leeds during January 1643, but he suffered a defeat at the Battle of Seacroft Moor in March. By spring, the Parliamentary forces had recovered, and Fairfax took Wakefield on May 20. Fairfax scored a major victory over the royal army at Adwalton Moor ten days later, then rode to the relief of Nantwich, which was held under siege. He drove off the royal army here on January 25, 1644.

On April 11, Fairfax captured Selby, then returned to York as part of a major siege during April–June. He commanded the cavalry in the combined Scottish–Parliamentary army at the Battle of Marston Moor, but suffered a severe wound and had to withdraw under assault of the royal cavalry led by Sir George Goring. The desperate situation was saved when **Oliver Cromwell**'s cavalry rode to the rescue, summoned at the personal request of Fairfax.

In February 1645, Fairfax was commissioned captain general of New Model Army, which he was instrumental in organizing with Cromwell himself. With the new force, he and Cromwell defeated the troops of King Charles and **Prince Rupert** at Naseby on June 14. On July 10, he crushed an army led by Sir George Goring at the Battle of Langport, thereby also relieving besieged Taunton. After taking Bristol on September 10, he withdrew briefly to winter quarters in the West, then, on February 10, 1646, overwhelmed Sir Ralph Hopton's army at Torrington and advanced on, laid siege to, and finally captured Exeter (April 9). Oxford fell to Fairfax's army on June 20.

When the army and Parliament fell into bitter dispute, Fairfax was torn, but ultimately sided with the army and, a political moderate, opposed the arrest of King Charles in 1647. However, he continued to further the cause of the Commonwealth, squashing a Royalist revolt in the southeast during the summer of 1648 and taking the town of Colchester on August 16, following a prolonged siege.

Appointed a judge in the trial of King Charles I, he was not among those who signed the monarch's death warrant. Popular with the army, he remained its commander in chief until just before the Commonwealth went to war with Scotland. Citing his Scottish peerage, he resigned his commission on June 22, 1650. This was only the ostensible reason for his resignation. The establishment of a republic in England did not sit well with his moderate views, and, exhausted by years of bitter campaigning, his health waned. He retired to his Yorkshire estate, returning to public life briefly in 1659 to support General **George Monck**, who, following the death of Cromwell, sided with Parliament after the overthrow of the Rump Parliament in October. Fairfax was chosen to Yorkshire in the Convention Parliament convened on April 25, 1660, and, favoring the Restoration, led a diplomatic embassy to King Charles II in the Hague, seeking his return to England.

FARRAGUT, David Glasgow

BORN: July 5, 1801

DIED: August 14, 1870

SERVICE: War of 1812 (1812–15); Civil War (1861–65)

The most important Union naval commander during the Civil War

Farragut was born a Tennessean (near Knoxville), as James G. Farragut, the son of a naval officer. In 1810, three years after the father moved the family to New Orleans, Farragut's mother died and the boy was adopted by U.S. naval commander **David Dixon Porter.** Farragut joined the navy as a midshipman in December 1810 and served under his foster father aboard the thirty-two-gun frigate U.S.S. *Essex.* He immediately proved himself a fine seaman. When *Essex* captured a prize in the Pacific in 1813, young Farragut was put in comand of her. The midshipman also acquitted himself with distinction in the celebrated engagement between *Essex* and H.M.S. *Phoebe* off Valparaiso, Chile (February 28, 1814), during the War of 1812.

At war's end, Farragut changed his first name from James to David to honor his foster father. After returning to the States, Farragut alternated further schooling with service aboard a number of ships during 1814–22. A brilliant student, he quickly mastered Arabic, French, and Italian under the tutelage of the United States consul in Tunis.

Promoted to lieutenant in 1823, he served in campaign against Caribbean pirates, again under his foster father Porter (February 1823–December 1824). Commanding a shore party at Cape Cruz, Cuba on July 22, 1823, he had a sharp exchange with pirates, which gained him notice as an able scrapper. There followed a period of relative quiet and routine, during which Farragut studied the emerging technology of steamships and achieved promotion to commander in September 1841. He was assigned command of the sloop *Decatur* in 1842 and stationed in Brazil. From 1854 to 1858, he served in California, where he established the Mare Island Navy Yard, the U.S. Navy's first permanent Pacific facility.

With the start of the Civil War, Farragut was called on to command the West Gulf Blockade Squadron. He used his new assignment to stage an assault on New Orleans. Steaming up the Mississippi by night, slipping under the Confederate guns of Forts St. Philip and Jackson (below New Orleans in the Delta). After annihilating a small Confederate river flotilla during April 23–24, 1862, he captured the city on April 25 and forced the surrender of Forts St. Philip and Jackson on the Twenty-Eighth. It was an extraordinary triumph that came at a time when the Union could lay claim to precious few victories.

Farragut conducted additional operations upriver, successfully running past the formidable Vicksburg batteries on June 28. His promotion to rear admiral on July 16 made him the first officer to hold admiral's rank in the U.S. Navy. Farragut continued to operate on the Mississippi, withdrawing past Vicksburg to Baton Rouge during July 22–25, and past Port Hudson on March 14, 1863 to blockade the Red River. The run past the batteries cost him heavily, however. Only two of seven vessels got through.

Returning to New Orleans after Porter's arrival there on May 4, he provided naval support to General Nathaniel Banks at the successful siege of Port Hudson during May 27–July 8, 1863. He was next ordered to capture the defenses of Mobile, Alabama. He took a squadron of four monitors (ironclad steamships) and fourteen conventional wooden ships, including his flagship, *Hartford*, to Mobile on August 5, 1864. The monitor *Tecumseh* struck a mine (in those days called a "torpedo" because of its elongated shape) and sank, causing confusion and panic in the federal squadron. Farragut, who had lashed himself to *Hartford*'s rigging in order to observe and direct the battle, shouted to his flag captain, "Damn the torpedoes! Full speed ahead, Drayton!" It was one of those naval phrases that continue to ring through history.

Farragut successfully sailed through the minefield and defeated the Confederate ironclad *Tennessee*, bringing about the surrender of Forts Gaines and Morgan at the mouth of Mobile Bay. For this victory, Farragut was promoted to vice admiral on December 23, 1864 (again, he was the first U.S. naval officer to hold this rank). By this time, however, ill health forced Farragut's withdrawal from active service. On July 25, 1866, after war's end, he was promoted full admiral—yet another rank created to honor him.

In the slow system of promotion that characterized America's armed forces of the nineteenth century, Farragut did not achieve high rank until he was in his sixties. Nevertheless, he proved himself a vigorous, courageous, and extraordinarily skilled tactician. The losses in men and ships he sustained during operations on the Mississippi River were the result of a rash mission ordered by his superiors. Farragut himself was characteristically more prudent.

FENG, Yü-Hsiang

BORN: September 26, 1882

DIED: September 1948

SERVICE: Warlord power struggles (1920–26); Northern Expedition (1926–28)

A Christian-influenced general of the period of warlord rule that followed the overthrow of the last Chinese emperor

Born at Hsing-chi-chen in the Hopei province, Feng was the son of an impecunious noncommissioned army officer. In 1894, he joined his father's regiment as a private and, taking well to military life, studied in order to prepare himself for entrance into the Paoting Academy. He graduated with an officer's commission and commenced a rapid rise to colonel and regimental command by 1913.

With the fall of imperial China, a welter of warlords rushed to fill the power vacuum. By 1918, Feng was not only commander of the Sixteenth Mixed Brigade, but a member of the Peiyang warlord faction. He fought successfully in the Chihli warlord campaign against rival Manchurian warlord Chang Tso-lin during 1921–22, then was instrumental in the ouster of President Li Yüan-hung in 1923. Early the following year, he was designated a field marshal and placed in command of the Third Army of Wu P'ei-fu's Chihli faction. However, when he was ordered to invade Manchuria, Feng turned against Wu and occupied Beijing, bringing about the collapse of the Chihli faction during October 23–November 2.

After the fall of Wu, Feng invited Sun Yat-sen to preside in Beijing, then reorganized his forces and designated his troops as the Kuominchün, or People's Army, by early 1925. The new army failed against a combined attack by Chang Tso-lin and Wu P'ei-fu in 1926, and Feng was forced to withdraw with his troops to the northwest. At this point, he traveled to the Soviet Union, where he received a modicum of support from the Communist regime, which supplied him with equipment. Returning to China, and with his army at least partially resupplied, Feng made an alliance with **Chiang Kai-shek** toward the end of Chiang's Northern Expedition (1927–28), the military campaign that was begun in 1926 by Kuomintang (Nationalist) Chinese forces to eliminate the various warlord factions and thereby to unite the country. At the time, the Kuomintang included Communist factions, though Chiang sought to suppress the most radical of these. Feng favored this position, but soon mutinied against Chiang, as he had against Wu. His first rebellion was defeated in 1929, whereupon Feng allied himself with Wang Ching-wei and Yen Hsi-hsan and staged a second rebellion, with the object of establishing a new government in Beijing. After a half-year's combat during 1930, Kuomintang forces defeated the new alliance.

Three years after his latest defeat, Feng tried to rally popular support by creating a volunteer army to fight the Japanese in 1933. The effort failed, and Feng traveled internationally in a vain attempt to raise support for the opposition viewpoint within the Kuomintang. His political campaigning was cut short by his accidental death at sea in 1948.

Feng was one of the most significant military leaders during China's warlord period. Different from many of the warlords, he was genuinely concerned for the welfare of the people, a fact reflected in his armies, in which troops were well-fed and humanely treated. Feng enforced a high degree of professional military discipline on his troops. Whereas other warlord factions visited havoc and misery upon the people, Feng restrained his, outlawing the all-too-common raping and looting.

FOCH, Ferdinand

BORN: October 2, 1851

DIED: March 20, 1929

SERVICE: Franco–Prussian War (1870–71); World War I (1914–18)

France's single most able and inspiring commander during World War I

Foch was a native of the Pyrenees region (born at Tarbes), the son of a civil servant. After education in a Jesuit school, he joined the infantry as a private at the start of the Franco–Prussian War. The conflict ended before he saw any action, and he was sent to the École Polytechnique at Nancy in 1871. After graduation in 1873, he was commissioned a second lieutenant. The following year, he joined the Twenty-Fourth Artillery Regiment (1874)

During the years prior to World War I, Foch served in a wide variety of garrison and staff posts. He enrolled in the École Superiere de la Guerre in 1885, and a decade later taught there as a professor, after having served on the General Staff. Foch is one of a very few practical military commanders who made a dramatic impact as a theoretician. He lectured eloquently on such issues as flexibility, on the massing of firepower, and on role of will and morale. In connection with the latter, he developed the concept of the "mystique of the attack," a doctrine that fitted well with such concepts as *élan vital*—the vital force that some French military thinkers believed would drive victory.

Foch published two important and highly influential collections of lectures, *De la Conduite de le guerre* (1897) and *Des Principes de la guerre* (1899). Unfortunately, many of Foch's eager readers simplified the teachings of the master, downplaying much of the detailed strategic and tactical doctrine and underscoring spirit, will, morale, and, *élan*.

Georges Clemenceau, at the time a commissioner in charge of military affairs, made Foch director of the École Superiere, and he was subsequently promoted to general of brigade (1907). In 1913, he assumed command of XX Corps at Nancy, and occupied this post at the outbreak of World War I. His sector fell under heavy attack, and he counterattacked with great vigor during August 14–18, 1914, but was surprised by the counterattack of Crown Prince Rupprecht's Sixth Army at Morhange. Foch fell back with heavy losses during August 20–21. Where lesser commanders would have allowed defeat to turn into a rout, Foch kept his command intact and organized a orderly withdrawal accompanied by damaging counterthrusts that were costly to the Germans.

On August 28, Foch was assigned command of the three corps designated the "Foch Detachment" and subsequently renamed the Ninth Army, which he led at the Battle of the Marne during September 5–10, 1914. The situation rapidly deteriorated. Foch reportedly sent a message to General **Joseph Joffre**: "My center is giving way, my right is falling back, situation excellent, I attack." While this message may well be apocryphal, it describes Foch's aggressive approach to adversity.

After the Battle of the Marne, Joffre appointed Foch his deputy during the so-called Race to the Sea (October–November 1914). Foch was charged with coordinating Allied operations among the British, French, and Belgians. He was in overall charge at the First Battle of Ypres (October 19–November 22) and remained in command of the combined forces through 1915. At the Somme (July 1–November 13, 1916), he fiercely struggled to punch through the German lines, but, short of men and supplies, failed.

When Joffre was replaced by General Robert Georges Nivelle as commander in chief on December 12, 1916, Foch was also temporarily sidelined, since he was closely identified with the discredited Joffre. However, after General **Philippe Pétain** replaced Nivelle on May 11, 1917, he appointed Foch chief of the general staff. In this new post, Foch moved with great speed on the Italian front and enjoyed significant success. He was appointed to the provisional Allied supreme war council, charged with coordinating Allied offensive efforts in November 1917, and then became Allied generalissimo in the West on March 26, 1918, where his first task was to counter the new German Somme offensive of March 21–April 4. Through the final spring of the war, Foch coordinated the Allied response to General **Erich Ludendorff**'s offensive pushes.

For the most part, Foch managed his forces with great skill, taking care to avoid squandering his resources. However, he was taken by surprise at Aisne/Chemin des Dames during May 27–June 4. Foch had anticipated this offensive to come farther north, at Amiens. Again, characteristically of Foch, he was able to recover quickly, moving his forces where they were needed to parry the blow.

On August 6, 1918, Foch was promoted to marshal of France and two days later launched the major Allied counteroffensive that would bring the war to its conclusion on November 11. It was Foch who dictated the terms of the armistice, and it was he who served as president of the Allied military committee at the Versailles treaty conference in January 1920. Foch was also charged with overseeing and enforcing the terms of the armistice and the subsequent peace.

Following Foch's death in 1929, he was laid to rest in a place of supreme honor, under the dome of Les Invalides, with the **Viscount of Turenne** and **Napoleon**. Certainly France's most consistently competent commander during World War I, Foch was among the most significant soldiers and military thinkers of the twentieth century.

FOOTE, Andrew Hull

BORN: September 12, 1806
DIED: June 26, 1863
SERVICE: Civil War (1861–65)

American naval officer instrumental in gaining control of the Mississippi River for the Union during the Civil War

Foote was born in New Haven, Connecticut, to Senator Samuel Augustus Foote. He readily secured an appointment to the U.S. Army Military Academy at West Point, but quickly decided that a seafaring career was more to his liking and, six months after enrolling at the Point, resigned to accept an appointment as a U.S. Navy midshipman in December 1822.

Between 1822 and 1843, Foote saw service in the Caribbean, Pacific, and Mediterranean, and at the Philadelphia Navy Yard. A reformer, he organized a temperance society aboard the U.S.S. *Cumberland*, which developed into a movement that resulted in ending the policy of supplying grog to U.S. naval personnel. During 1849–51, Foote skippered the U.S.S. *Perry*, cruising the waters off the African coast. He was active in suppressing the slave trade there. This experience persuaded him fully to the cause of abolition, and in 1854, he published an antislavery tract, *Africa and the American Flag*. He also became a frequent speaker on the abolitionist circuit.

During 1851–56, Foote served in various shore posts, including on the Efficiency Board created by Commodore **Samuel F. du Pont.** Foote was promoted to commander in 1856 and was assigned command of the sloop *Portsmouth*, with a duty station in the Far East. Assigned to observe British operations against Canton during the Opium War, he found himself under attack from Chinese shore batteries. Foote returned fire, destroying four barrier forts near Canton during November 20–22, 1856.

In 1858, Foote was in the United States again, this time as commandant of the Brooklyn Navy Yard. At the outbreak of the Civil War, he was assigned to command of naval forces on the upper Mississippi (August 1861) and supervised the emergency construction of a fleet of gunboats and mortar craft. He used some of these vessels to capture Fort Henry, but, unable to coordinate with **Ulysses Grant**'s land forces, could not stop the fort's garrison from escaping to Fort Donelson on February 6, 1862. A few days later, on February 14, Foote was severely wounded when his attack on Fort Donelson was repulsed. After his return to duty, Foote supported General **John Pope** in taking the key Confederate position on Island No.10 during March 1–April 7. With the fall of this position, Memphis and the entire lower Mississippi were laid open to the Union forces.

The lingering effects of his wounds forced Foote to retire from his command in June. The following month, however, he was promoted from flag officer to rear admiral and given an appointment as chief of the Bureau of Equipment and Recruiting. Feeling apparently well enough to resume sea command, he was assigned to the North Atlantic Blockading Squadron, but died before he could begin his new assignment.

FORREST, Nathan Bedford

BORN: July 13, 1821
DIED: October 29, 1877
SERVICE: Civil War (1861–65)

General in the army of the Confederate States of America; untutored in the military arts, he was a born commander

Forrest was a native Tennessean, born near Chapel Hill. He had no formal education and spent his early life as a farmhand and cotton farmer, as well as a trader in livestock and slaves. At the outbreak of the Civil War, in April 1861, Forrest enlisted as a private in the Seventh Tennessee Cavalry, then raised a cavalry regiment on his own, leading it by August with a commission as lieutenant colonel. When Union forces assaulted Fort

Donelson, Tennessee, demanding that Forrest surrender the garrison there, he refused and managed to escape with his command intact on February 16, 1862.

Promoted to full colonel in March 1862, Forrest distinguished himself for gallantry at the bloody Battle of Shiloh (April 6–7), where he was gravely wounded. Recovering, he was promoted to brigadier general in July, and given brigade comand under General **Braxton Bragg**.

Forrest was skilled in cavalry hit-and-run tactics and, under Bragg, staged numerous raids against Union supply and communication lines. The most famous of these was a five-day exchange at Rome, Georgia, in April 1863, during where he captured an entire Union cavalry brigade.

At Chickamauga, during September 19–20, Forrest commanded the Confederate right flank. Promoted major general in December 1863, he was transferred west to Mississippi, from which he staged various raids into Union-held Tennessee, capturing Fort Pillow on April 12, 1864. At this battle, one of the most brutal and disgraceful episodes of the war took place. Confederate troops murdered African American Union troops who had surrendered. (Confederates denied this, but most subsequent historians have concluded that the "Fort Pillow Massacre" did occur.) Forrest—in truth, often brutal, even with his own troops—was blamed for the atrocity.

Throughout 1864, Forrest was hailed in the Confederacy and feared in the North for his devastating lightning raids in Tennessee and on Union positions in Alabama, Georgia, and Mississippi. At Brice's Cross Roads, Mississippi, on June 10, 1864, he defeated the superior force of Gen. Samuel D. Sturgis, and in August led a notable raid on Memphis, followed by a raid on Athens, Georgia, in September.

Forrest served under General John Bell Hood as a cavalry commander in the Army of Tennessee and fought a brilliant rearguard action during December 15–16, following the Confederate defeat at the Battle of Nashville. He was promoted early the next year to lieutenant general, but was defated on April 2 at the Battle of Selma (Alabama). A month later, on May 9, he surrendered his command at Gainesville, Georgia.

Forrest's postwar activites included prosperity as a plantation owner and president of the Selma, Marion, & Memphis Railroad. Most significantly, he founded the Ku Klux Klan, serving as its Grand Wizard. In its original incarnation, the Klan was primarily a vigilante group intended to combat the lawlessness of early Reconstruction South. However, even early on, the secret organization was also motivated by racism.

Forrest is associated with the maxim, "Get there the first with the most"—often misquoted as "firstest with the mostest." His strength was rapid, flexible cavalry deployment. His weakness was a tendency toward brutality, and even, as in the Fort Pillow incident, atrocity.

FRANCO (BAHAMONDE), Francisco (Paulino Hermenegildo Teódulo)

BORN: December 4, 1892

DIED: November 20, 1975

SERVICE: Riff Rebellion (1920–26); Spanish Civil War (1936–39)

Spanish general, generalissimo, and fascist dictator

Franco was born in El Ferrol and enrolled in the Toledo Academia de Infantería in 1907. After graduation in 1910, he was commissioned second lieutenant. His service in a 1912 war against Morocco brought him recognition as a highly capable and courageous officer, paving the way to a rapid series of promotions. By 1920 he was deputy commander of the Spanish Foreign Legion in Morocco, leading these forces against Abd-el-Krim during the Riff Rebellion of 1921–26. In 1923, he was promoted to full commander of the Foreign Legion, and in 1925 led a brilliant assault on Alhucemas Bay, which ultimately brought Spanish victory in the lengthy Riff conflict.

Franco's spectacular performance catapulted him to the rank of brigadier general in 1926—Spain's youngest ever. Two years later, he was given the politically influential post of director of the Academia General Militar at Saragossa during the dictatorship of General Primo de Rivera. Franco served at the academy until 1931, when Republican forces, having overthrown the monarchy, accused Franco of monarchist sympathies.

What followed was a confused and violent period on the eve of the Spanish Civil War. Franco, however, having been relieved of his academy post, was transferred to duty in the Balearic Islands from 1931 to 1934. In this way, he managed to steer clear of the military's many conspiracies against the new republic. Recalled in 1934 to suppress a miners' revolt in Asturias, Franco earned the respect of the conservative right wing and the hatred of the left. Fortunately for Franco, it was the conservatives who were on the rise, and he was named chief of the general staff in 1935. But the following year, the left-wing Popular Front gained a majority in the elections, and Franco was once again effectively exiled, this time to a command in the Canary Islands.

From the Canaries, he participated in the military and conservative conspiracy that erupted, on July 18, 1936, into the Spanish Civil War. At this point, Franco flew to Morocco, where he took over the Spanish Foreign Legion garrison, and airlifted a large contingent of legionnaires to Spain later in the month. During July and August, he directed a motorized advance on Madrid, where government forces met him and pushed him back during September and October. By this time, however, the country was divided between government and Nationalist territories. On September 29, 1936, the Nationalists established their own government, with Franco as head of state. In April 1937, Franco also became head of the Falange party and forged a cautious alliance with Fascist Italy and Nazi Germany. With the aid of these two powers, the Nationalists inexorably defeated the Loyalists. After the fall of Madrid on March 28, 1939, the Spanish Civil War ended, and Franco became de facto dictator of Spain.

Both as a politician and a military strategist, Franco was cautious and methodical, but ruthless. After achieving victory in the civil war, he outlawed all political parties and ordered the execution or imprisonment of many thousands of Loyalists. Yet he was careful about Spain's relationship to the Axis powers at the outbreak of World War II. Italy and Germany assumed that Franco would repay the support he had received during the Civil War by throwing in with them in the present war. Instead, however, Franco declared Spain neutral, although it was unmistakable where his sympathies lay, and he not only sent workers to Germany, but created a volunteer Blue Division to fight for the Germans on the Russian front. Nevertheless, as the tide of the war turned against Germany, the pragmatic dictator actually enforced conditions of neutrality, while simultaneously backing down somewhat from absolute dictatorship. In July 1945, he promulgated the *Fuero de los Españoles*, a kind of bill of rights. Next, by referendum in 1947, he agreed to reorganize the government as a monarchy, with himself as regent endowed with the power to choose the next king.

The postwar years saw a general moderation of Franco's brutal fascist stance into the image of a staunch anti-Communist, which was far more palatable to Western Europe and the United States. In 1953, at the height of the Cold War, he agreed to allow the United States to establish and maintain certain military bases in Spain. The moderate trend also increased Spanish economic and political ties with the countries of the West, which, in turn, fostered even more liberalization. By 1955, Spain joined the United Nations, and the following year renounced imperialism by pulling out of northern Morocco.

The 1960s, which brought worldwide unrest, saw an increase in agitation among Spanish students, workers, clergymen, and Basque separatists. Franco reacted by cracking down somewhat, but the trend toward liberalism was too well established to allow for reversal. In July 1969, Franco chose Juan Carlos de Bourbon, grandson of King Alfonso XIII, as heir to the throne. After a protracted illness, Franco died on November 20, 1975.

FREDERICK I
(Frederick Barbarossa, "Red Beard")

BORN: ca. 1123
DIED: June 10, 1190
SERVICE: Second Crusade (1147–49); Six Italian Invasions (1154–83); Third Crusade (1189–92)

The most important German emperor after Charlemagne; through his military prowess, did much to unite medieval Germany

Frederick was the eldest son of the duke of Swabia, **Frederick II** of Hohenstaufen. At the time, Germany was by no means a nation, but rather a loose collection of often violently competitive realms ruled by dukes and princes who met as necessary to elect in common an emperor, but invested in him little authority and even less power. After his father's death in 1147, Frederick became duke of Swabia and followed his uncle, Emperor Conrad III, on the Second Crusade, 1148–49. The abortive Crusade failed, and Frederick returned to Germany. On the death of Conrad in 1152, Frederick was elected emperor at Frankfurt-am-Main on March 4, 1152. It proved a wise choice, for a number of the German dukes and princes were now looking beyond their individual agendas and looking toward the creation of a unified Germany. As separate, small states, they were continually at the mercy of Danish, Polish, and Hungarian aggression. Yet it was difficult to unite Germany when the region's two leading families, the Hohenstaufens (also known as the Waiblingen) and the Welfs, were constantly warring with one another. It was hoped that Frederick, a Hohenstaufen related on his mother's side to the Welfs, might heal the breach by uniting the two hostile bloodlines.

Frederick's own ambitions extended far beyond bringing two rival families together. He saw that Germany's problems did not result solely from internal dissension, but that the region was weakened by chronic conflict between the papacy and the German thrones. Since the time of **Charlemagne,** whom the pope had created Holy Roman Emperor, a succession of popes asserted their domination over the German monarchs. Each German emperor had to be anointed by the pope, and the emperor served essentially at the pope's pleasure. Frederick found in the doctrine of rule by divine right a way to break the chains that bound Germany's emperors. Frederick asserted that it was not the pope who conferred imperial power, but God himself directly. Emperors ruled by divine right. The pope merely confirmed God's will in the matter.

The doctrine was powerful enough to motivate Frederick to press the German sphere of influence into Italy and Burgundy. He invaded Italy in 1144, the first of six such incursions by which he subjugated many of the northern Italian cities by assigning his royal governors (*podestas*) to preside over them. In 1155, he rescued Pope Eugenius from a mob stirred to riot by the religious fanatic Arnold of Brescia. Frederick had Arnold executed, but then aroused the pope's ire by declining to perform the traditional act of submission to papal authority—holding the pope's stirrup as he dismounted his horse. Tension between emperor and papacy intensified further after Hadrian (Adrian) IV succeeded to the papacy and challenged the docrine of divine right by insisting that it was, indeed, the pontiff who had conferred the crown upon Frederick. To this, Frederick responded by publishing a circular in which he formally claimed rule by divine right and election by the princes. The papal coronation was an act of confirmation, not one of creation.

Before undertaking the second of his six invasions of Italy in 1158, Frederick bolstered his position at home in Germany by restoring his powerful Welf kinsman, Duke Henry the Lion, to the throne of Bavaria, which his uncle Conrad had seized. With great political aplomb, however, he also applied a check upon Henry's power by augmenting the authority of Henry's principal rival, Albert the Bear, margrave of Brandenberg. With the major holders of power assuaged, Frederick turned less diplomatically to the lesser rival lords, leading a military campaign of ruthless suppression of feuds. After determining which of the warring nobles was the more offensive, Frederick showed no compunction in putting the offender to death.

Conquest and military action were not the only means by which Frederick consolidated his power. On

June 9, 1156, he married Beatrice of Burgundy, which put him in a position to claim the Burgundian throne.

In June 1158, Frederick captured Milan and other northern Italian cities, then, in November, convened the diet of Roncaglia, by which he asserted his authority over Italy. He appointed a roster of royal officers to administer the country.

In 1159, the quarrelsome Pope Hadrian IV died and was succeeded by Alexander III. Frederick no more trusted the new pope than he had the old, and, to oppose him, engineered the election of a cooperative anti-pope, Victor IV. In response, Alexander excommunicated Frederick in March 1160, triggering an anti-German rebellion in Milan. Frederick responded by sacking Milan, practically wiping it off the map. His move against Frederick having failed, Pope Alexander III fled to France. Here he encouraged the Lombards to rise up against the German administration as well as the anti-pope. Frederick put down this rebellion.

In 1164, the anti-pope Victor IV died, whereupon Frederick secured the election of another anti-pope, Paschel III. Two years later, Frederick invaded Italy again, in October 1166, and attacked Rome, to which Alexander had returned. The city fell to Frederick's assault, driving Alexander into Sicilian exile. Frederick then boldly enthroned his anti-pope.

Having scored a military triumph in Italy, Frederick's army was now menaced by something more terrifying than a human foe. Plague swept through the emperor's ranks, and the princes of Lombardy, exploiting the debilitated state of Frederick's army, joined forces as the Lombard League. Menaced by this coalition, Frederick was forced to retreat out of Italy and back to Germany in the spring of 1168.

During the next six years, Frederick concentrated his military operations in Germany and on the eastern frontiers, installing by force German governments in Bohemia, Hungary, and part of Poland, then putting down the rebellion of Polish Duke Boleslav, known as Curly-Hair. In the meantime, Frederick was also able to conclude cordial relations with the Byzantine Empire, France, and England. He felt ready to stage another Italian invasion by 1174, hoping to defeat the Lombard League. He attempted to form an alliance with Henry the Lion, who declined. Undaunted, Frederick mounted the expedition anyway, and engaged the armies of the League at Legnano on May 29, 1176, where he was decisively defeated, his cavalry cut down by Italian spearmen. Forced to the peace table, Frederick concluded a treaty with Pope Alexander in June 1177, and a six-year truce with the Lombard League followed.

Disappointed in Italy, Frederick was, however, free to return to Germany, where he took vengeance on the uncooperative Henry the Lion, stripping him of all of his fiefdoms in 1179 and giving all of Bavaria to Otto of Wittelsbach. Four years later, Frederick concluded a permanent peace with the Lombard League, agreeing to renounce some of his Italian gains. This done, he betrothed his eldest son, Henry, to Constance, heiress to the Norman kingdom of Sicily, in 1184. This created an alliance that alarmed the papacy, and a growing dispute over the Tuscan lands of Countess Matilda created additional friction. However, Frederick now ruled over a Germany united to a degree previously unattainable. No figurehead or puppet of princes, Frederick was a powerful monarch. Instead of acting against him, Pope Clement III, who succeeded Alexander, appealed to Frederick's sense of Christian responsibility. He persuaded the aging monarch to lead a new Crusade to Palestine.

In May 1189, Frederick mustered his army at Regensburg and marched into Byzantine territory. Officially, Byzantine emperor Isaac II Angelus had granted Frederick's army safe conduct. Nevertheless, his troops were repeatedly attacked by guerillas. Worse, the people of village and town fled before his advance, taking all food and provisions with them and bringing great hardship on Frederick's forces. With difficulty, the emperor was able to negotiate terms for the remainder of his passage through the Byzantine realm and reached the empire of the Seljuk Turks in May 1190. There the sultan, Kilidsh Arslan, pledged his support, only to renege and attack. This contest and the ordeal of passage through Byzantium took a terrible toll on Frederick's army.

A mere six hundred semi-starved knights advanced into the Seljuk capital of Iconium. Incredibly, the now pitiable force took the capital in June, forced the sultan's surrender, and secured much-needed supplies from him. Then came the final catastrophe. On June 10, 1190, either while he was attempting to cross the Saleph River or as he was bathing in it to secure relief from the heat, Frederick Barbarossa drowned. This broke the knights' spirit, and they abandoned the Crusade to return to Germany.

FREDERICK WILLIAM I

BORN: August 15, 1688

DIED: May 31, 1740

SERVICE: Great Northern War (1700–18); War of the Spanish Succession (1701–13)

Prussian monarch and military reformer; father of the modern Prussian Army

As king of Prussia from 1713 to 1740, Frederick solidified Hohenzollern rule by centralizing ruling authority and strengthening the Prussian military. Born in Berlin, Frederick William I was the son of Frederick I, king of Prussia, from 1688 to 1713. His education took place in the strict Prussian military schools, an experience that shaped a career based on hard discipline and rigid structure. Yet Frederick William inherited the throne in 1713, during a period when the Prussian government was weak and its vaunted military in disarray. The new monarch worked immediately and methodically to consolidate power in the crown and to rebuild the military. He instituted a regime of Spartan authoritarianism, demanding from civilians and military alike nothing less than "unconditional obedience" to the state. Suspicious of the nobility, he imposed mandatory military service on the children of noblemen and required state service from the parents themselves. Next, in 1733, Frederick William introduced universal conscription and created an elite officer corps, which was soon unmatched in Europe.

Single-mindedly focused on creating efficient government, Frederick William dissolved the hereditary court and, by a 1719 decree, freed all serfs. In 1723, he established efficient ministries war, finance, and domains, which did away with the traditionally corrupt methods of obtaining and distributing government funds. Having established a central administration and lean government, Frederick William tripled the size of his military to 80,000 men while he amassed a treasury surplus of 8,000,000 taler by 1735. By 1740, Prussia ranked third in Europe—behind only Russia and France—in military strength.

Frederick William was unmistakably a military leader who created a quasi-military state, yet he brought Prussia into no external conflict during his reign. Rather, he laid the foundation for Prussian economic prosperity, political influence, and military strength on which his son Frederick II the Great would build an even more formidable nation. Frederick William died May 31, 1740 in Potsdam.

FREDERICK II

BORN: December 26, 1194

DIED: December 13, 1250

SERVICE: Wars with the Papacy (1227–30, 1240–41, 1244–47); Fifth Crusade (1228–29); war with the Lombard League (1236–39); Civil War in Germany (1247–56)

Medieval German emperor who managed his military affairs with great skill in the context of the extraordinarily complex European politics of the time

Frederick was the son of Emperor Henry VI and Constance, daughter of Norman King Roger II of Sicily; his grandfather was Frederick I, or **Frederick Barbarossa**. Frederick's father caused his election as king of Germany in 1196, and the following year, when Henry VI died, Frederick and his mother went to Sicily. In 1198, his mother also died, and Frederick became King of Sicily. Pope Innocent III acted as the orphan's regent as well as guardian, then relinquished control of Sicily to Frederick when he reached his majority in 1208. A year later, Frederick wed Constance, daughter of King Peter II of Aragon, and that same year, after the death of Emperor Otto IV (son of Henry the Lion, Frederick Barbarossa's great rival) invaded Italy, Pope

Innocent persuaded the German princes to oust Otto and elect Frederick in his stead. He was thus made emperor in September 1211. Frederick hastened to Germany, arriving in September 1212 and quickly set about consolidating support amid the welter of German medieval alliances. He concluded an alliance with France against Otto and King John of England in November, even before his official coronation at Mainz on December 9, 1212.

In July 1214, Frederick's new French allies triumphed over Otto at the Battle of Bouvines, and Frederick's sway over Germany firmed. He traveled to the deposed Otto's headquarters at Aachen, where he was crowned a second time (1215). Frederick now made his son Henry (later Henry VII) king of Sicily (1216), and arranged for the boy's election as king of Italy in April 1220. That same year, Pope Honorius III crowned Frederick emperor in Rome, making him Holy Roman Emperor (November 22). At this coronation, he promised, as he had in 1215, to undertake a crusade to the Holy Land. The crush of affairs in Italy kept him from mounting the effort, however. When the Pope became insistent, Frederick promised, yet again, in 1225. He pledged to undertake the campaign no later than April 1227. However, his summons to the Imperial Diet at Cremona in 1226 prompted the Lombard League to oppose imperial authority in northern Italy. Once again, Frederick had local affairs to attend to, and he postponed departure for the Holy Land to September 1227. When September came, a plague further delayed his departure from the Italian port city of Brindisi. Infuriated, Pope Gregory IX summarily excommunicated Frederick on September 29 and made alliance against him with the Lombard League. Nevertheless, Frederick left Brindisi on June 28, 1228, and set off on the Fifth Crusade.

Frederick enjoyed considerable success in the Holy Land, concluding a 1229 treaty with Egyptian Sultan al-Kamil, by which Jerusalem and certain other towns were returned to Christian rule. Frederick returned to Italy on June 10, 1229, and discovered that the Pope's armies had invaded much of his realm. He mounted a lightning campaign that pushed them out of Italy, and he negotiated the Treaty of San Germano with the Pope on July 23, 1230. He was granted absolution from excommunication in August.

Thus restored to papal favor, Frederick worked fitfully to tighten the imperial grip on Lombardy. However, in September 1234, Frederick's son Henry VII rebelled in Germany. The uprising was quickly squelched by Frederick, who crossed the Alps with an army. Taken prisoner, Henry was confined in a castle in Apulia for the balance of his life (he died in 1244). With the rebellion dismantled, Frederick returned to Italy in August 1236 to fight the Lombard League, scoring a triumph at Cortenuova on November 27, 1237. In the aftermath of this victory, Frederick demanded unconditional surrender of the Lombard League. As a result, the war ground on, and Frederick was denied total victory.

While Frederick was engaged with the Lombard League, Pope Gregory again turned against him and formed an alliance with Genoa and Venice. Disputing with Frederick over control of Sardinia, Gregory excommunicated the emperor a second time on March 20, 1239. In response, Frederick did not seek forgiveness and absolution, but instead revised his entire administration of Italy. This accomplished, he seized Spoleto and the march of Ancona, both of which belonged to the Papal States. The contest thus put into motion continued even after Pope Gregory's death in August 1241. Following the brief pontificate of Celestine IV (October–November 1241), the election of Innocent IV in June 1243 boded a settlement. Negotiations broke down, however, and Innocent fled the forces of Frederick. From Lyons, the Pope denounced Frederick's refusal to compromise over the Lombard League, and officially deposed him on July 17, 1245.

In the wake of the Pope's action, Frederick's Italian kingdom was menaced by revolt in Parma (1246) and Verona (February 1248). His closest advisers also schemed against him. Finally, two opposition leaders, Henry Raspe, landgrave of Thuringia, and William, count of Holland, rose in opposition to Frederick's continued reign in Germany. In Italy, however, Frederick's position improved. Then, suddenly, all of Frederick's earthly struggles ceased. He died unexpectedly at Castel Fiorentino in Apulia.

FREDERICK II THE GREAT

BORN: January 24, 1712

DIED: August 17, 1786

SERVICE: First Silesian War (1740–41); War of the Austrian Succession (1740–48); Seven Years' War (1756–63); War of the Bavarian Succession (1778–79)

Prussian monarch, founder of modern Germany

Frederick the Great, king of Prussia from 1740 until 1786, forged the modern state of Germany, transformed it into a formidable military power, and forever changed the face of European politics and history. He was born in Berlin to the Prussian King **Frederick William I** and Sophia Dorothea of Hanover, daughter of England's King George I. The youthful Frederick loved art and music and was a passionate Francophile in matters of culture and aesthetics. It was his iron-spined father who imposed on the boy the vaunted Prussian military discipline. Young Frederick chafed under this, and, at age eighteen, attempted to escape his father's domination by going to England. His father responded by ordering his son's arrest and imprisonment, then threatened to put him to death. Yielding, Frederick returned to Rheinsberg and even managed to pursue some of his literary and artistic interests.

Upon the death of his father, Frederick was crowned on May 21, 1740. No one in Europe had faith in Frederick's ability to lead, and the Austrian ambassador even wrote that the young man had confessed to him that he was "a poet and can write a hundred lines in two hours. He could also be a musician, a philosopher, a physicist, or a mechanician. What he never will be is a general or warrior." This self-assessment, like the assessment that was universal throughout Europe, could not have been less accurate. Within seven months of his accession to the throne, Frederick marched on Silesia, firing the opening volley in the War of the Austrian Succession. Although his army was badly defeated, he acquired Silesia by treaty in 1742, and the kingdom of Prussia grew in size. Later military actions were more directly successful, and Frederick was destined not only to prove himself an excellent military commander, but the very model of the "benevolent despot." Among Frederick's first acts was the lifting of censorship of the press and the abolition of penal torture. He also decreed general religious toleration.

Frederick was also a great administrator and an economic reformer, who quickly balanced the Prussian budget, then produced a surplus, even while he financed improvements in agriculture and manufacturing. But it was the modern army he created that had the most profound effect on Germany and on Europe. He forged it into a fighting force second to none, and when Russia, France, and Austria became allies in 1756, Frederick invaded Saxony and started the Seven Years' War, the European phase of the French and Indian War in North America. The death of Russia's czarina, Elizabeth, took that country out of the war and thus allowed Frederick to make peace with the other powers.

The result of the long and expensive war—in effect, the first true *world* war—was a reversion to the *status quo ante bellum*, but the fight had not been entirely fruitless; for Frederick had demonstrated to the world that he was a military genius and that Prussia was a military power to be reckoned with.

In 1772, part of Poland was divided among Russia, Austria, and Prussia, thereby further enlarging Frederick's holdings. He immersed himself in the arts and literature once again, becoming a patron of Johann Sebastian Bach, and even composing music of considerable merit himself.

During the early days of August 1786, Frederick reviewed his troops in Potsdam during a torrential downpour. He caught a chill, sickened, and died several days later on August 17. It was an inglorious end, perhaps, but, in later centuries, the likes of **Napoleon I** and **Adolf Hitler** would pay homage at his gravesite.

FRENCH, Sir John Denton Pinkstone, First Earl of Ypres

BORN: September 28, 1852

DIED: May 22, 1925

SERVICE: Mahdist War (1883–98); Second ("Great") Anglo-Boer War (1899–1902); World War I (1914–18)

Distinguished but stodgy British general who performed well in the Second Boer War, but led the British Expeditionary Force to its early disasters in World War I

French was born at Ripple, Kent, to a naval officer, and, indeed, he served first in the Royal Navy as a cadet and midshipman during 1866-70. After joining the militia, he transferred to the army and was immediately gazetted to the Nineteenth Hussars in 1874.

French first saw action as part of the failed Nile expedition led by Sir Garnet Joseph Wolseley's to relieve General **Charles George "Chinese" Gordon** at Khartoum (October 1884–April 1885). Within a short time, by 1889, French was promoted to colonel and given command of the Nineteenth Hussars, which he led until 1893, when he was posted to the War Office staff through 1895.

Promoted to brigadier general, French commanded a cavalry brigade from 1897 to 1899, when he was promoted to major general and dispatched to South Africa to take command of the cavalry division in the expeditionary army of Sir Redvers Henry "Reverse" Buller at the commencement of the Second ("Great") Boer War in October. On the twenty-first, he fought his first engagement during the war, the Battle of Elandslaagte, then defended Ladysmith, from which he and his command narrowly escaped under heavy Boer siege on November 2.

French rapidly reorganized his cavalry division and set about recruiting local mounted troops during November 1899–early February 1900. Just prior to the British invasion of the Orange Free State, French was promoted to lieutenant general, then marched to the relief of Kimberley on February 17. The effort was brilliant but brutal. French exhausted his division, and when he attempted to outflank the Boers at Poplar Grove on March 7, the Boers retreated in great haste, and French's horses were in no condition to give chase. With the Boers in retreat, however, French advanced into Bloemfontein, capital of the Orange Free State (March 13), then went on to take and occupy Pretoria, capital of the Transvaal and the Boer's principal city (June 5).

French was part of the British advance to Komatipoort, which spanned July 21–September 25. He was assigned to sweep the eastern Transvaal, clearing it of resistance from the beginning of 1901 through most of the spring. Establishing a headquarters in the Cape Colony, he directed a lengthy campaign against Boer guerrillas during May 1901–May 1902.

With the conclusion of the war in Africa, French returned to Britain where he became commander of Aldershot (1902–1907) and, promoted to general in 1907, was appointed inspector general of the army. He served in this capacity until 1912, when he was named chief of the Imperial General Staff and, the following year, promoted to field marshal.

The Curragh Mutiny of April 1914 prompted French to step down as chief of the Imperial General Staff, but when World War I began in August 1914, it was French who was chosen to lead the British Expeditionary Force (BEF). Once in France, he concentrated his relatively small army on the left flank of General Charles Louis Marie Lanrezac's Fifth (French) Army in southern Belgium during August 18–19. General **Alexander von Kluck**'s First (German) Army pushed through French's position, driving the BEF back from Mons (August 23) and then Le Cateau (August 25–27) in battles that cost the BEF dearly.

In the face of the German advance, French lost what faith he may have had in Lanrezac and pulled the BEF to a position southeast of Paris. At this point, Lord **Horatio Herbert Kitchener** intervened to persuade French to cooperate with General **Joseph Joffre**'s counterattack at the Marne. French's mission was to attack Kluck's flank, which he did in an action spanning September 5–9.

French directed all BEF operations in France and Belgium during the disastrous opening months of the war. The dismal Allied performance at Ypres I

(October 19–November 22), Ypres II (April 22–May 25, 1915), and the Loos offensive (September 25–November 4) was in significant measure due to French's utter refusal to coordinate his actions with those of his allies. French also proved himself inept in the management of his precious reserves at Loos. At last, on December 15, 1915, French was relieved as BEF commander and replaced by General **Sir Douglas Haig.**

French was removed from combat command. Created a viscount, he was named commander in chief in the United Kingdom and given the office of Lord Lieutenant of Ireland in May 1918. He conducted himself admirably in this office, during a period of acute crisis in Anglo-Irish relations. Retiring early in 1921, he received an earldom and died four years later, at Deal Castle, Kent, near his boyhood home.

French was a highly respected soldier, but his achievements were all won in his youth. By the outbreak of World War I, he had become inflexible and mistrustful of other commanders. His overly cautious refusal to cooperate with the French during the opening months of World War I greatly harmed the Allied cause.

FRUNZE, Mikhail Vasilyevich

BORN: February 2, 1885
DIED: October 31, 1925
SERVICE: World War I (1914–18); Russian Civil War (1917–20)

Russian revolutionary general called the father of the Red Army

Frunze was born to a military family in Pishpek, a Central Asian city that was subsequently renamed in his honor. The young man enrolled in the Vernyi Academy, which he attended on scholarship and from which he graduated in 1904 with a gold medal. He went on to the the St. Petersburg Polytechnic Institute, where he was exposed to radical political philosophy and joined the Bolshevik Party. By 1905, he was a dedicated, full-time revolutionary. The Moscow Russian Social Democratic Workers Party sent him to Ivanovo-Voznesensk (present-day Ivanovo) and Shuya as an agent provocateur. Between 1905 and 1909, he was frequently arrested and sentenced to internal exile. Finally, in 1914, he was condemned to death, but his sentence was commuted to permanent exile in Manzurka, Siberia. This hardly hindered his revolutionary activity. Operating under assumed identities in the Siberian towns of Irkutsk and Chita during 1914-16, he continued to agitate.

Frunze illegally returned to Russia to serve as a statistician in the All-Russian Zemstvo Union on the Western Front in World War I. He assumed leadership of the Bolshevik underground in Minsk, and in May 1917 was elected as a delegate to the First Congress of the Soviet of Peasant Deputies in Petrograd. There he met Vladimir Illych Lenin and was subsequently named chairman of the Soviet of Workers', Peasants', and Soldiers' Deputies in Shuya.

During the Moscow Uprising of October 30, 1917, Frunze commanded the Ivanovo-Voznesensk and Shuya Red Guards, an action that prompted his appointment as Military Commissar of Ivanovo-Voznesensk and Vladimir. In this capacity, during August 1918, he put down an anti-Bolshevik uprising in Yaroslavl. With the outbreak of civil war, he rapidly rose through several promotions, ultimately commanding the Fourth Army of the Eastern Front by December 26, 1918, and taking command of the Southern Group on March 5, 1919.

Frunze enjoyed success against White Army units commanded by Admiral **Aleksandr V. Kolchak.** He was then given command of the entire Eastern Front during July 1919, and succeeded in liberating the Northern and Central Ural Mountains. On August 15, 1919, he was dispatched to the Turkestan Front, which he commanded through September 10, 1920. As commander of the Aktyubinsk campaign, he cleared the Southern Urals of the White Army, then destroyed anti-Bolshevik forces throughout Central Asia, capturing the key cities of Bokhara and Khiva in 1920.

Placed in command on the Southern Front, he organized the Perekop-Chongar campaign on November 7–17, 1920, which defeated the White Army forces of General P. N. Wrangel. Following this, he was named to a number of political offices in the Ukraine during December 1920–March 1924, including membership in the Central Committee—the most important Bolshevik decision-making body—in 1921 and the office of candidate member of the Politburo in 1924. On March 14, 1924, Frunze was appointed deputy director for military affairs, and, supported by **Josef Stalin**, replaced Leon Trotsky as commissar for military and naval affairs on January 26, 1925. That same year, he became a member of the Council of Labor and Defense and served as chairman of the draft committee of the military reform of 1924–25. In this capacity, he was instrumental in creating the "unitary military doctrine," by which offensive military action was linked to political indoctrination and world political action. It became the basis of Western fears of Soviet military world domination. He also became one of the principal architects of the Red Army, writing books on military theory and strategy, and creating an extensive system of military training academies in the Soviet Union.

G

GAGE, Thomas

BORN: ca. 1720

DIED: April 2, 1787

SERVICE: War of the Austrian Succession (1740–48); Jacobite Rebellion (1745–46); French and Indian War (1754–63); American Revolution (1775–83)

British general during the French and Indian War and at the outbreak of the American Revolution

Gage was born to the aristocracy, created an Irish viscount either at birth or shortly thereafter. After graduating from Westminster School in 1736, he received an ensign's commission in the army. By January 1743, he held the rank of captain and served as aide-de-camp to William Keppel, Earl of Albemarle, whom he had first met at Westminster School. In this service, he fought at Fontenoy on May 10, 1745, during the War of the Austrian Succession, then returned to the British Isles, where he served under the Duke of Cumberland at the Battle of Culloden (April 16, 1746), which ended the Jacobite Rebellion in Scotland.

Following Culloden, Gage returned to the Continent to continue fighting in the War of the Austrian Succession. He served in Holland and Flanders (1747–48), then, transferring to the Fifty-Fifth Regiment of Foot (subsequently renumbered Forty-Fourth), he purchased a majority in 1748, gaining promotion to lieutenant colonel on March 2, 1751.

With the Forty-Fourth, Gage sailed to Virginia as part of General **Edward Braddock**'s ill-fated campaign during the French and Indian War. Gage was in command of an advance unit at the Battle of the Wilderness, along Monongahela River (near present-day Pittsburgh), when French and Indian forces out of Fort Duquesne attacked with devastating effect on July 9, 1755. A combination of Braddock's incompetence and his troops' panic resulted in a British rout. Gage, slightly wounded, performed bravely in a hopeless situation. During the long retreat, he met and befriended **George Washington**, who commanded a provincial militia force at the Battle of the Wilderness.

Following the disaster at Monongahela, Gage was given full command of the Forty-Fourth, but did not receive promotion to colonel. He next served as second-in-command to Massachusetts governor William Shirley in the failed Mohawk Valley expedition of August–September 1756, then accompanied the Earl of Loudoun's expedition against French-held Louisbourg, Nova Scotia, which, after interminable preparation, never got off the ground. In 1757, Gage was authorized to raise a local regiment of light troops, the Fiftieth Foot, and finally was promoted to colonel. At the head of this unit, he led the advance guard during James Abercrombie's failed assault on Fort Ticonderoga in July 1758. A year later, in the summer of 1759, Baron Amherst, commander in chief of British forces in America, dispatched him at the head of a column against Fort La Galette (present-day Ogdensburg, New York). Gage determined that his forces were insufficient to take the fort, and, reaching Niagara, he broke off the offensive. He narrowly avoided dismissal by a furious Amherst.

During the winter of 1759–60, Gage commanded the garrison at Albany, New York, then directed Amherst's rearguard actions during the Montreal campaign (May-September 1760). After the fall of Quebec, Gage was appointed governor of Lower Canada, a post in which he served with distinction from 1761–63.

Receiving promotion to major general in 1761, he succeeded Amherst as commander in chief of American forces in November 1763. There he remained, headquartered in New York, for a decade. After a brief sojourn in Britain during June 1773–May 1774, he returned to the now-rebellious colonies as commander in chief of Britain's North American forces and as governor of Massachusetts, the hotbed of the developing revolution.

The shocking British setback at the Battle of Concord (April 19, 1775), which opened the American Revolution, and the subsequent pyrrhic victory at Bunker Hill (Breed's Hill) on June 17 resulted in Gage's recall to Britain. He relinquished command to General **William Howe** on October 10. Once in England, he was formally relieved of command of the North American forces on April 18, 1776.

Gage continued to serve in the British military after his failure in America. He joined Amherst's staff on April 1781 and organized the Kent militia to defend against a feared French invasion. He was promoted to general on November 20, 1782.

While Gage's valor was beyond question, he repeatedly proved himself an overcautious commander. Unjustly, he was made to shoulder the blame for early British reversals during the American Revolution. The fact was that the British government grossly underestimated the resolve and resources of the colonists, leaving Gage with insufficient forces to deal with the rebellion.

GALTIERI, Leopoldo Fortunato

BORN: July 15, 1926
SERVICE: Falklands Islands (Malvinas) War (1982)

Argentine general and president

Leopoldo Fortunato Galtieri, a member of the Argentine military junta that seized power in 1976, served as president of Argentina from 1981 to 1982, when he was arrested on charges of gross negligence in fomenting—and, more importantly, losing—the Falkland Islands (Malvinas) War. He was the son of Italian immigrants to Argentina, born in Caseros, Buenos Aires. Galtieri entered the national military academy in 1943, rising steadily through the officer corps following graduation. By 1980, he had become commander-in-chief of Argentina's armed forces.

In 1979 Galtieri became a member of the military junta that had ousted President Isabel Martinez de Perön in March 1976. Galtieri himself became head of the government on December 22, 1981, when President Roberto Viola resigned due to ill health and unrelenting opposition by the army. Throughout the 1970s and early 1980s, Argentina's economy had badly deteriorated, with inflation soaring to devastating heights. Unrest erupted into open rebellion by March 1982, when Buenos Aires and other large cities were swept by massive anti-government demonstrations. Fearing the ouster of his own regime, Galtieri announced on April 2, 1982, the invasion of the British-held Falkland Islands, which the Argentines called the Malvinas.

It was an immensely popular move, stirring patriotism among the Argentines, who had long considered the islands to be part of their country and resented Great Britain's control of them. Galtieri, it seemed, had found just the issue he needed to quell civil unrest and draw public attention from the economic disaster wrought by his regime. While the Argentine people generally rallied behind the war effort, Galtieri and his military were not equal to the task of successfully invading the islands. Under the leadership of the Margaret Thatcher government, the British resolved to protect the Falklands and dispatched a large and formidable fleet that far outclassed the small and obsolescent Argentine navy. Galtieri was compelled to surrender on June 15, 1982, and, with surrender, public support for his regime instantly collapsed. Within two days, he resigned as president and commander in chief, to be replaced by retired general Reynaldo Bignone.

Nor were Galtieri's troubles over. He was arrested and court-martialled in 1983 for his actions during the Falklands War. That year, he was also tried for human rights violations, but was acquitted in December 1985. In 1986, however, he was convicted of negligence for having provoked the Falklands War—then having lost it. Galtieri was sentenced to imprisonment for twelve years, but was pardoned (along with nearly 280 other military men and accused "leftist subversives") in October 1989 by President Carlos Saul Menem.

GATES, Horatio

BORN: 1728
DIED: April 10, 1806
SERVICE: French and Indian War (1754–63); American Revolution (1775–83)

American general in the Revolution; victor at the key Battle of Saratoga; one of very few professional military men in the Continental Army

Gates was born in Maldon, England, son of the Duke of Leeds's housekeeper and godson of Horace Walpole, the Fourth Earl of Orford and author of the famous Gothic novel *The Castle of Otranto*. After obtaining a commission in the British Army in 1749, Gates saw service in North America. As captain during the French and Indian War, he participated in the disastrous expedition led by General **Edward Braddock** to take Fort Duquesne on the Monongahela (near present-day Pittsburgh) during April–July 1755. Gates was among the many wounded in the engagement (July 9), but he also met **George Washington**, officer in charge of the Virginia militia.

The other major action Gates saw during this period was with a successful British campaign against French-held Martinique in 1761. Following this action, he was promoted to major in 1762, but was nevertheless discouraged by his prospects for further advancement in the army. He resigned his commission in 1772 and immigrated to America, where he called on Washington to help him set up as a planter in Virginia.

At the outbreak of the Revolution in 1775, Washington, as commander in chief of the Continental Army, recommended to Congress that Gates be appointed adjutant general of the army with the rank of brigadier general. This was done on June 17, 1775, and Gates immediately set about organizing the Continental Army around Boston, transforming a loose rabble of volunteers into an effective military force. On May 16, 1776, Gates was promoted to major general and assigned command of the army sent to invade Canada. However, by the time of this appointment, that army was already in full retreat from Canada and, once back in the United States, was under the command of General Philip Schuyler. Gates was therefore quickly reassigned command of troops in New York. In December, he commanded a unit sent to reinforce Washington's forces, then was put in charge at Philadelphia for a brief period before returning to New York in March 1777 to replace Schuyler. Schulyer, however, did not step down willingly, and Gates's first task was to wrest command from him. This accomplished, he directed the defense of New York against an invasion led by British general **John Burgoyne**.

The principal battles in this action, the Saratoga Campaign, were at Freeman's Farm on September 19, 1777, and at Bemis Heights on October 7. No other engagement—save the victory at Yorktown, which effectively ended the war—was more significant than Saratoga. Gates's victory (brilliantly aided by his subordinates Benedict Arnold and Daniel Morgan) crushed British hopes of dividing the colonies along the Hudson River and was the turning point of the war, giving the American cause a credibility that, most importantly, attracted French military aid and alliance.

Gates hoped that his signal success at Saratoga would catapult him to loftier command. He thought he might even replace Washington himself. Indeed, the Conway Cabal, a group of Continental Army officers led by Thomas Conway, an Irish-born officer previously in the French army, maneuvered (with the aid of a faction in Congress and others) to replace Washington with Gates. When Washington discovered their plans, in November 1777, however, the group backed down. It was never clear that Gates, although he craved advancement, approved of the Conway group's activities. In any case, he was able to reestablish a working relationship with Washington, but was never restored to the commander-in-chief's full confidence.

Gates turned down command of an expedition against the Indians in the spring of 1779, but, when Congress asked him—over Washington's objections—to take command in the south on June 13, 1780, he did so. By July, Gates was in the Carolinas rushing to organize a viable mixed force of militia and Continentals. In their first engagement, at Camden, South Carolina on August 16, Gates suffered a severe defeat—perhaps the worst reversal for the American forces during the entire Revolution.

Gates retreated to Hillsboro and regrouped, but, by October, was replaced by General **Nathanael Greene**. Washington graciously accepted the disgraced Gates as his deputy and aide at Newburgh, New York (1782–83), in which post he ended his war service. Following the war, Gates served in the New York legislature from 1800 to 1801.

That George Washington would take Gates back as his deputy, even after the Conway incident and his utter failure at Camden, attests not only to the commander-in-chief's generous spirit, but also to the fact that he recognized Gates's strengths, even as he was well aware of his weaknesses. Gates was an able organizer and administrator, one of precious few professional military officers in the Continental forces. As Washington also recognized, however, he was not an accomplished tactician or strategist. He performed at his best when in subordinate command.

GEIGER, Roy Stanley

BORN: January 25, 1885
DIED: January 23, 1947
SERVICE: World War I (1917–18); World War II (1941–45)

American Marine general who commanded the amphibious assaults on key Pacific islands during World War II

A native of Middleburg, Florida, Geiger graduated from John B. Stetson University in 1907 and practiced law for less than a year before enlisting in the Marine Corps in November 1907. In less than two years, he was commissioned a Second lieutenant, then was promoted to first lieutenant in 1915 after having served at sea and in the Caribbean, the Philippines, and China. In 1917 he was promoted to captain and become the fifth Marine officer to complete pilot training. During World War I, serving with the rank of major, he commanded a squadron in the First Marine Aviation Force in France.

After the armistice, Geiger was sent to Haiti, where he commanded the First Aviation Group, Third Marine Brigade, from 1919 to 1921. He was transferred to Quantico, Virginia in 1921 and graduated from the Army's Command and General Staff School in 1925. In 1929, he graduated from the Army War College. During 1929–31, he commanded Aircraft Squadrons, East Coast Expeditionary Force, stationed at Quantico, then became officer in charge of aviation at Marine Corps headquarters in Washington, D.C., serving there from 1931 to 1935.

Promoted to lieutenant colonel in 1934, he commanded Marine Air Group One, First Marine Brigade from 1935 to 1939, when he was sent to the Navy War College. Following graduation in 1941, he was promoted to brigadier general and given command of the First Marine Air Wing, Fleet Marine Force in September. After the outbreak of World War II, Geiger took command of the air wing on Guadalcanal as soon as the island was captured from the Japanese (September 1942–February 1943). After receiving promotion to major general, Geiger returned to Washington as director of the Marine Division of Aviation. He served in this post from May to November 1943, when he succeeded General **Alexander A. Vandegrift** as commander of I Amphibious Corps (later redesignated III Amphibious Corps). He led the corps in the invasion of Guam during July 21–August 10, 1944. He was in charge of the bloody invasion of Peleliu Island during September 15–November 25, then went on to participate in the landing on Okinawa during April 1–June 18, 1945. During this campaign, his corps was part of the Tenth Army, commanded by General Simon B. Buckner. After Buckner was killed in battle, Geiger assumed command of the Tenth Army until the arrival of General **Joseph W. Stilwell** on June 23. This brief command put Geiger in the record books as the only Marine officer ever to command a field army.

In July, Geiger was named commander of Fleet Marine Force, Pacific, then, following the war, in November 1946, he was assigned to a post in Washington. Unfortunately, Geiger was stricken with a fatal illness a few months after his arrival in Washington. Congress posthumously awarded him the honorary rank of general in July 1947.

Geiger personified the Marine commander: unyielding in the achievement of his objective, a fierce fighter, and a commander who expected—and got—top performance from his men. Nowhere in the Pacific theater was fighting more intense, or more costly, than on the islands of Peleliu and Okinawa.

GENGHIS KHAN (Temujin)

BORN: 1155-67?

DIED: August 18, 1227

SERVICE: Unification of the Mongol Tribes (1190–1206); War against the Western Hsia (1205–1209); War against the Chin (1211–15); War against Khwarism (1218–24); Invasion of North China (1225–27)

Mongol conqueror and ruler; unifier of the Mongols, who became the dominant power in Central Asia

Genghis Khan was born into the Borjigin clan midway through the twelfth century (the dates assigned to his birth vary from 1155 to 1167) and was given the name Temujin. At the age of eight or nine, Temujin's father, Yesukai the Strong, a ranking member of the royal clan, was poisoned by the rival Tartars, another nomadic band, with whom he was feuding. Following the death of Yesukai, a rival family seized control of the clan and cast Temujin and his mother out. The woman responded with a resolution to raise her son to be a Mongol chief, and the first lesson she inculcated in him was to surround himself with absolutely loyal men. Indeed, his talent for doing just this was key to the warrior's rise.

Temujin exhibited great prowess in hunting and warfare at an early age and thereby quickly won a large following. When his new bride was abducted and raped by the Merkit clan, Temujin made a crafty alliance with an acquaintance of his father's, quite literally borrowing a fully-equipped army to lead against the Merkit in 1180. Not only did he regain his bride, but he totally destroyed the Merkit and acquired a larger following, from which he created his own army of 20,000 men.

While Temujin was away fighting the Merkit, the Jurkin clan took advantage of his absence to plunder his property. Upon his return, Temujin exterminated the Jurkin nobility. During operations against both the Merkit and the Jurkin, Temujin formulated his basic military policy: always keep your army's rear free. What this translated into was a policy of destroying enemies one at a time, completely eliminating one threat before moving on to the next.

After destroying the Merkit and the Jurkin, Temujin moved against the far more formidable Tartars, who had become allied with the eastern Mongols. In an act of vengeance for the death of his father, Temujin routed the Tartars in battle in 1201 and then systematically slaughtered every person taller than the height of a cart axle. He reasoned that, by killing the adults, he would prevent their hatreds and prejudices from being passed on to the children, thereby assuring a second generation loyal to himself. Conquests of the Naiman and Karait tribes followed in 1203, and by 1204, after establishing his capital at Karakorum, Temujin was master of Mongolia.

The only other Mongolian clan leader who challenged Temujin was Jamuka, who had been a childhood friend and was now a tenuous ally (he had helped defeat both the Merkit and Tartars). But, in the face of Temujin's conquests, Jamuka's constituents deserted their leader and flocked to Temujin, who convened them, as well as his other followers, at a great assembly by the River Onon. Here Temujin proclaimed his new identity as Genghis Khan—"Universal Ruler."

At this point, all the Mongol clans were now unified under Genghis Khan. They could now look beyond the steppe regions, and, in entering upon an even vaster stage of conquest, Genghis Khan's true military genius came to the fore. In prosecuting tribal warfare against highly mobile nomads, he had employed cavalry exclusively, but in assaulting established cities, he rapidly mastered the art of siege works, catapults, ladders, burning oil, and even accomplished such engineering feats as diverting rivers. Genghis Khan used such tactics in his invasion of the Western Hsia Empire in 1205, 1207, and 1209, enlisting Chinese engineers to help him breach city walls. In April 1211 he crossed the Great Wall itself to begin his conquest of the Chin Dynasty of northern China. By 1215, Beijing (Peking) had fallen, and within two more years the last Chin resistance had been squashed.

Genghis Khan turned next to the south, to Khwarezm, where a local official had made the fatal miscalculation of killing a Mongol trading envoy. Upon this pretext for battle, Genghis Khan hurled 200,000 troops against the Khwarezm, devastating the army, the people, and the land itself. In succession, Transoxiana (Bukhara and Samarkand), Khorasan, Afghanistan, and northwestern India fell, with the culminating battle fought against Shah Mohammed's son Jellaluddin on the Indus River on November 24, 1221.

Genghis Khan next sent his generals into southern Russia during 1221–23 and invaded northwestern China in 1225 to put down a rebellion of the Chin and Hsi-Hsia, winning a great battle on the Yellow River in December 1226. He was about to confront more rebellious Chins in 1227, but fell ill and died on August 18. He was secretly buried on Mount Burkan-Kaldan, the place to which he and his mother had been exiled.

Genghis Khan was not the uncouth barbarian popular legend depicts him as, but was an intelligent strategist, a great organizer, and a ruler of tremendous charisma. His military triumphs were partly the result of his considerable skill as a field commander, but, even more, they were due to the skill of his subordinates, whose absolute loyalty Genghis Khan enjoyed. As his mother had taught, he surrounded himself with men in whom he could place absolute faith and trust. In consequence, his corps of commanders was made up of his own sons and his closest long-time allies. Finally, it is undeniable that Genghis Khan was insatiable in his lust for conquest. Yet he was never a mere plunderer. Part of his success as a conqueror of peoples came from his resolve to provide an efficient and just government for those over whom he had triumphed.

GERONIMO

BORN: ca. 1823

DIED: 1909

SERVICE: Apache Wars (Geronimo's Resistance phase) (1881–86)

Chiricahua Apache war leader who brilliantly eluded vastly superior American and Mexican forces during the Apache Wars

Geronimo was born Goyahkla (He Who Yawns) on the upper Gila River in present-day Arizona or New Mexico. Giving his allegiance to Chief Juh and, later, Chief Naiche (last chief of the Chiricahuas), he gained early and enduring renown as a warrior and (as one Chiricahua admiringly described him) a "wild man." He was married seven or nine times and lived with several Chiricahua bands, depending on which band his wife was a member of.

Geronimo was essentially a guerrilla leader. His skill was so great that he became the most famous Apache among whites during the 1880s.

Following Chief Mangas Coloradas, young Geronimo (as the Mexicans called him) and his family settled in Chihuahua, Mexico, where, on March 5, 1851, Mexican troops killed twenty-one Apaches, including Geronimo's mother, wife, and three children, in a surprise attack. From this point, Geronimo swore vengeance on the Mexicans and engaged in intensive raiding along the U.S.-Mexican border region. His periods of raiding alternated with relatively quiet periods on the San Carlos (Arizona) reservation.

Except for intervals on the reservation, Geronimo was essentially a guerrilla and a fugitive from 1865 until his surrender in 1886, but it was in the late phase of the Apache Wars (a phase of the conflict often called Geronimo's Resistance), from about 1881 to 1886, that Geronimo was of greatest military consequence. Typically leading fewer than a hundred warriors, Geronimo freely raided in the American Southwest and in Mexico, repeatedly eluding army task forces numbering as many as 5,000 men under the command of General **George Crook** and, later, General **Nelson A. Miles**—both of whom were among the most skilled Indian fighters in the army.

Geronimo was finally compelled to surrender on September 4, 1886, after leading his latest pursuers on a 2,000-mile, four-month chase. His surrender to Nelson A. Miles marked the end of the Apache Wars. Geronimo and the other Chiricahuas captured were sent to prisons in the East. Geronimo was first incarcerated in Florida, then Alabama, and, finally, confined to a reservation attached to Fort Sill, Oklahoma.

Geronimo was celebrated by the American public as a warrior hero of legendary proportions. Theodore Roosevelt invited him to appear in his 1905 inaugural parade, with five other Indian leaders, and he also made numerous public appearances, including at the St. Louis World's Fair of 1904 (where he made a significant profit selling autographed pictures of himself). He died on the reservation, in 1909, of pneumonia.

Charismatic and brilliant as a guerrilla leader, Geronimo accomplished little of enduring strategic significance. Certainly, nothing he did improved the lot of the Apaches, though he did gain legendary status for himself and, arguably, garnered respect for the fighting prowess and endurance of the Indian warrior. In World War II, U.S. Army paratroopers used his name—"Geronimo!"—as a jumping cry.

GIAP, Vo Nguyen

BORN: 1912

SERVICE: Indochinese War (1946–54); Vietnam War (1964–75)

The chief architect of North Vietnamese military strategy during the Vietnam War

Giap was born in rural An Xe village, in Quangbinh province. He was educated in the ancient capital city of Hue, at the French-run lycée, then went on to study law at the University of Hanoi. In 1931, he began teaching history at Thang Long School in Hanoi, and simultaneously joined the Communist Party. During the Japanese invasion in 1939, Giap fled to China, where he met **Ho Chi Minh**, with whom he organized the Vietminh Front, a grassroots "people's army," in 1941.

The first mission of the Vietminh was to fight the Japanese invaders. Despite his own lack of military experience, Giap trained his troops in guerrilla tactics, then led them in hit-and-run attacks against the Japanese.

At war's end, Giap was appointed Minister of the Interior in the Democratic Republic of Vietnam (December 1945). Ho Chi Minh, premier of the Democratic Republic, named Giap Defense Minister on December 19, 1946, when war commenced against the French, who returned, after World War II, to reassert authority over their former colony. For eight years, Giap formulated anti-French strategy and generally directed the guerrilla campaign against them. In October 1950, he was victorious at the Battle of Lang Son, which resulted in Vietnamese takeover of outposts along the Chinese frontier. However, his subsequent offensives in the Red River delta failed during November 1950–February 1951. Learning from this experience that it was best to avoid open battle with French forces, Giap resumed guerilla operations and deliberately declined the full-scale battles the French attempted to provoke. Finally, during November 1951–February 1952, the French launched a major offensive against Hoa Binh. That it proved a failure validated Giap's guerrilla tactics.

While frustrating the French colonial forces in Vietnam, Giap also harried French interests and installations elsewhere throughout Indochina. The culmination of his campaign came during November 20, 1953–May 7, 1954, at Dien Bien Phu, a major French stronghold, which Giap held under punishing siege. Giap moved his forces nimbly, and mounted highly effective antiaircraft defenses. The fall of the garrison at Dien Bien Phu brought the French to the peace table and, eventually, resulted in their withdrawal from Vietnam.

After the departure of the French, Giap, as Defense Minister, directed campaigns against the South Vietnamese and then against the Americans, whose presence in Vietnam grew from a relatively small contingent of military advisors in the 1950s, to an army of half-a-million men by the end of the 1960s.

Giap used the same tactics against the Americans that he had employed against the French. He realized that the Americans intended to fight a war of attrition, hoping through unremitting attacks to wear down the North Vietnamese will to fight. Giap countered this strategy by a willingness to sacrifice large numbers of troops. However, on January 30, 1968, at the start of Tet, a Vietnamese lunar holiday, he directed a series of massive offensives, first along the border, assaulting the U.S. base at Khe Sanh, and then attacking South Vietnamese provincial capitals and principal cities. The offensive, which included an assault on the U.S. embassy in Saigon, was costly to U.S. and the South Vietnamese Army of the Republic of Vietnam (ARVN) forces, but it was even more costly to the Viet Cong. In strictly military terms, the three-week offensive was a failure. But, in psychological terms, it was North Vietnamese victory. The offensive convinced many Americans, including politicians and policy makers, that the Vietnam War was unwinnable.

Tet became the high-water mark of American involvement in Vietnam. Gradually, as support for the war waned in the United States, the Americans withdrew, paving the way for Giap's single most successful campaign, the final offensive against the South Vietnamese, culminating in the fall of Saigon to Giap's forces (February–April 30, 1975).

Following the end of the Vietnam War, Giap stepped down as army commander in chief and became deputy prime minister in July 1976. Falling out of favor, he was removed from the Defense Ministry in 1980, then from both the post of deputy prime minister and from the Politburo as well in 1982.

A brilliant, albeit untutored, commander in a unique combat situation, Giap wrote two works of military-political doctrine/history, *People's War, People's Army* (1962) and *Big Victory, Great Task* (1968).

GIBBON, John

BORN: April 20, 1827

DIED: February 6, 1896

SERVICE: Civil War (1861–65); Northern Plains (Indian) War (1875–77); Nez Percé War (1877)

A master artillerist in the Civil War, Gibbon was one of U.S. Army's leading commanders during the Indian Wars

Gibbon was a native of the Philadelphia area, but he was raised in Charleston, South Carolina. He graduated from West Point in 1847 and was commissioned in the artillery. Dispatched to Mexico to fight in the U.S.-Mexican War, he arrived too late to see action. His first taste of combat came in sporadic action against the Seminole Indians in Florida during 1849.

Promoted to First lieutenant in 1850, he was assigned to West Point as an artillery instructor in 1854

John Gibbon. Courtesy Library of Congress.

and served there until 1859. During this period, he wrote the *Artillerist's Manual*, published in 1859, which became a standard reference work.

After receiving a promotion to captain in November 1859, Gibbon was sent to Utah, but was recalled to the East at the outbreak of the Civil War. He was assigned to the division of General **Irvin McDowell** as chief of artillery in October 1861, then was made brigadier general of volunteers in May 1862. In this capacity, he led the First Brigade, First Division, I Corps at the Second Battle of Bull Run (August 29–30, 1862), South Mountain (near Frederick, Maryland) (September 14), and at Antietam (September 17), where a journalist dubbed it the "Iron Brigade." Gibbon's command lost one-third of its strength at Second Bull Run.

On December 13, while commanding Second Division of I Corps at Fredericksburg, Gibbon was gravely wounded, but was able to return to duty to command Second Division of General **Winfield Scott Hancock**'s II Corps at the Battle of Gettysburg during July 1–3, 1863. It was Gibbon's division that took the full force of "Pickett's Charge". On the last day of battle, Gibbon was seriously wounded, returning to duty months later to lead the Second Division at the Battle of the Wilderness (May 5–6, 1864), Spotsylvania (May 8–19), and Cold Harbor (June 1–3).

After receiving promotion to major general of volunteers following Cold Harbor, Gibbon fought at Petersburg (June 15–18), then was appointed commander of XXIV Corps in the Army of the James in January 1865. He led this unit until the end of the war.

Following the war, Gibbon was mustered out of volunteer service in January 1866 and reverted to the rank of colonel in the regular army. He was put in command of the Thirty-Sixth Infantry in July, then transferred to the Seventh Infantry in 1869. He was stationed with his command in Montana, where he participated in campaigns against the Sioux during 1876. It was Gibbon's regiment who were the first to arrive at the Little Bighorn battlefield on June 27, and it was they who relieved the survivors of **George Armstrong Custer**'s command there.

In 1877, Gibbon took part in the long pursuit of the Nez Percé and their great leader, Chief Joseph the Younger. On August 9, Gibbon attacked the Indians on the Big Hole River, but was repulsed. For the third time in his career, Gibbon was seriously wounded.

Promoted to brigadier general in 1885, Gibbon was given command of the army's Department of the Pacific the following year. He served until his retirement, in 1891, at headquarters in San Francisco.

Gibbon was a hands-on commander possessed of tenacity and skill. His multiple woundings attest to his insistence on leading from the front.

GNEISENAU, August Wilhelm Anton, Count Neihardt von

BORN: October 27, 1760
DIED: August 23, 1831
SERVICE: American Revolution (1775–83); Napoleonic Wars (1800–15)

German patriot, crusader for military reform, Prussian field marshal, the strong right hand of Count Gerhard von Scharnhorst

Gneisenau was the son of a Saxon artillery officer. He was educated at Erfurt University, then joined the Austrian cavalry, serving from 1778 to 1780. In 1782, he was sent to Canada as a lieutenant in a mercenary Ansbach regiment fighting in the service of the British crown. He returned to Europe in 1783 and was commissioned a captain in the Prussian army in 1786. For the next two decades, he was assigned garrison duty. At last, in 1806, he fought valiantly at Jena (October 14) and distinguished himself in particular at the defense of Colberg during May 20–July 2, 1807.

During 1807–13, Gneisenau was assigned as staff officer to Gerhard von Scharnhorst. Working with Scharnhorst, Gneisenau brought sweeping reforms to the Prussian military. He was instrumental in creating the general staff and developing the *Krümper* (reservist) system, by which the Prussian army was secretly rebuilt after defeat at the hands of **Napoleon**.

During the 1813 campaign against Napoleon, Gneisenau was appointed to the staff of **General Gebhard von Blücher**. He fought at the Battle of Leipzig during October 16-19, and for his service was created a count. He continued to serve with Blücher (as chief of staff) through the Waterloo campaign of 1815, which resulted in the final defeat of Napoleon. At Ligny, on June 16, he assumed command after Blücher was wounded. His critical decision to retreat toward Wavre, enabled Prussian forces to reinforce the **Duke of Wellington** at the Battle of Waterloo on June 18.

Despite his military accomplishments before and at the culminating campaign of the Napoleonic Wars, Gneisenau's liberal politics brought censure from the conservative Prussian government. In 1816, yielding to pressure, Gneisenau resigned from the army and served briefly as governor of Berlin. In 1825, however, his achievements were recognized by his promotion to field marshal, and he returned to active duty as commander of the Observation Army sent to protect Prussia's eastern borders during the Polish insurrection of 1831. Along with his chief of staff—no less a figure than **Karl von Clausewitz**—Gneisenau fell ill and succumbed to epidemic cholera while in his headquarters as Posen.

GONZALO DE CORDOBA, Hernandez

BORN: 1453

DIED: December 1, 1515

SERVICE: Civil War in Castile (1474–79); Granada War (1481–92); Caroline War (1494–98); Neapolitan War (1501–1504); War of the Holy League (1511–14)

Spanish general known as "El Gran Capitán"—one of the most significant figures in European warfare of the Renaissance century

Gonzalo de Cordoba was the son of Pedro Fernández de Cordoba, intendant (governor) of Andalusia. The boy was sent to court at an early age and served in 1466 as a page to Alfonso, stepbrother of King Henry IV of Castile. After this, he became a page to Alfonso's sister Isabella. After Isabella ascended to the throne of Castile, Gonzalo de Cordoba fought in her service during 1474–79 against the Portuguese and supporters of Alfonso, who contested the throne. From 1481 to 1492, he served in the war against the Moorish kingdom of Granada, where he showed special skill in capturing forts. He was appointed one of two commissioners to negotiate with Boabdil for the surrender of Granada late in December 1491.

Following his successes in Granada, Gonzalo de Cordoba was secure in the favor of Queen Isabella, who sent him to Naples with a small expeditionary force to aid Naples' King Alfonso, a relative of Isabella's husband, King Ferdinand of Aragon. Gonzalo de Cordoba landed in Calabria on May 26, 1495, and augmented his band of 2,100 with local militiamen. Nevertheless, he was quickly overrun by French and Swiss troops at the First Battle of Seminara on June 28, 1495. Instead of panicking, however, he fought an agile delaying action to evade the brunt of French heavy cavalry and the Swiss pikemen. Using the time he gained, Gonzalo de Cordoba organized a stronger infantry unit, with which he was able, at length, to drive the French and Swiss out of the Kingdom of Naples during the course of 1495–98.

After the last of the French had left Naples, Gonzalo de Cordoba returned to Spain and was dispatched as part of a Franco-Spanish expedition into the Ionian Islands in 1500. He captured Cephalonia. However, the alliance soon collapsed, and Gonzalo de Cordoba found himself in southern Italy once again, fighting the French. He was also, once again, outnumbered, and had to withdraw in the face of an army led by the Duc de Nemours. Holing up in Barletta, Apulia, he was blockaded from August 1502 to April 1503, when he managed to receive reinforcements. Thus augmented, Gonzalo de Cordoba decided to break the blockade. Assuming a position at Cerignola, his forces repelled a direct attack by the French and by Swiss pikemen. The French commander, Duc de Nemours, was killed by a Spanish harquebus shot, and the enemy forces withdrew in disarray on April 26, 1503.

Gonzalo de Cordoba retook Naples on May 13, then laid siege against nearby Gaeta. This proved unsuccessful; Gonzalo de Cordoba retreated at the approach of a vastly superior Franco-Italian army in October. Wisely, the Spaniard resorted to guerrilla tactics, raiding the countryside from October through December. Gonzalo's break came when the French commander, Louis de la Tremoille, fell ill and was replaced, in turn, by two mediocre leaders, the Marquis Gonzaga of Mantua and the Marquis Ludovico of Saluzzo. Moreover, harsh winter weather had set in. Gonzalo took advantage of the weather *and* the poor command. He made a surprise attack across the Garigliano River, using hastily constructed pontoon bridges, and quickly routed the superior French army during December 28–29. Gaeta, which had resisted siege, now fell to the Spanish on January 1, 1504.

In March 1504, King Ferdinand rewarded El Gran Capitán with an appointment as viceroy of Naples. However, by 1507, Ferdinand, always distrustful of others with power, recalled Gonzalo to Spain. In 1512, after the northern Italian city of Ravenna fell to the Papal States during the War of the Holy League, Gonzalo was once again placed at the head of a Spanish army. In 1515, he was stricken with a fever and left Italy to return to Spain. There, in Granada, he died.

The Spanish army Gonzalo de Cordoba forged held sway over many European battlefields for nearly a century after the commander's death. In addition to brilliance as a strategist and tactician, Gonzalo did much to promote the use of firearms, which he used most effectively at Cerignola, a battle that transformed European warfare.

GORDON, Charles George ("Chinese Gordon")

BORN: January 28, 1833

DIED: January 26, 1885

SERVICE: Crimean War (1853–56); Second Opium War (1859–60); Taiping Rebellion (1850–64); First Mahdist War (1883–85)

British general who won posthumous fame for his remarkable but failed defense of Khartoum (1884–85) against Sudanese rebels led by the Mahdi

Gordon was born at Woolwich, the fourth son of General H. W. Gordon. By 1848, he was enrolled as an officer-cadet at Woolwich and, after graduating in June 1852, was commissioned a second lieutenant in the Royal Engineers. He was dispatched to the Crimea, arriving at Balakielava on January 1, 1855. He fought through to the collapse of Sevastopol on September 9, 1855. His superiors took note of his exceptional intelligence and coolness under fire. He received an appointment to the international commission responsible for surveying the Russo-Turkish frontier during April 1856–November 1858, then volunteered for service in China during the Second Opium War (1859–60).

Gordon participated in the major actions of the Second Opium War, including the landing at Beitang on August 1, 1860, and the attack on the forts of Dagu on the 21st. He was present at the taking of Peking (Beijing) on October 6, then remained in China after the conclusion of the war to accept appointment, during April 1863, by the Chinese government to command of the Ever Victorious Army against the Taipings. Gordon transformed a small, rag-tag rabble into a well-disciplined army of modest proportions. The force's several victories over the numerically superior Taipings became legendary, as did the figure "Chinese Gordon" cut, marching at the head of his army, armed only with a walking stick.

After participating in the capture of Suzhou in November 1863, Gordon returned to England, where he was appointed to command the Royal Engineer post at Gravesend in September 1865. In September 1871, he served on the Danubian commission in Galati, Romania, then accepted appointment as governor of Equatoria (southern Sudan) in late 1873.

Gordon landed in Cairo on February 6, 1874 and made his way to Gondokoro, Sudan, where he established his capital by mid-March. He was tireless in traveling his territory, building a series of military outposts and forts and acting vigorously against the slave trade. He also found time to make the first accurate map of the Nile and the lakes. He briefly returned to England on December 24, 1876, then returned to Sudan in late March 1877 as governor-general, in the service of the khedive of Egypt. Gordon was faced with having to suppress uprising after uprising, twice subduing Walad-el-Michael, brother of Ethiopian emperor John, in April 1877 and fall 1878, then suppressing a rebellion in Daffur province (western Sudan) in May 1877. The latter was accomplished more through the force of his powerful personality than by force of arms. He also did much to interdict the slave trade in Sudan from January-August 1879.

At the end of 1879, the restless Gordon resigned his post in Sudan and returned to England, where he joined the staff of the Viceroy of India designate, the Earl of Ripon in May 1880. Almost immediately, he gave up this post to return to China, where he succeeded in dissuading the political leader Li Hung-chang from rebellion while also convincing the Chinese government not to declare war on Russia over the Ili River dispute during June and July 1880. Once again, Gordon returned to England, spending a year there before assuming command of the engineer post on Mauritius (April 1881–April 1882). From this assignment, he went to the Cape Colony to reform and reorganize local militia forces during the summer of 1882. After returning to England for a brief stay, he sailed for Palestine, where he began an intensive study of the Bible and biblical history during January–December 1883.

The beginning of 1884 found Gordon in Brussels, where he accepted an appointment from King Leopold I to govern the Congo. However, British prime minister Gladstone requested that Gordon travel to Khartoum, Sudan, to restore order in the midst of the so-called Mahdist Revolt, which had begun in 1881. Failing the restoration of order, Gordon was directed to evacuate the province.

Gordon arrived in Khartoum on February 18, 1884, and made his first priority the evacuation of 2,000 women and children. On March 13, 1884, forces led by the **Mahdi,** a Sudanese rebel leader, blockaded Khartoum. Although British authorities organized a relief expedition under Sir Garnet Wolseley in August, it was unable to set out from Cairo until November. In the meantime, Gordon did all that he could to defend Khartoum. But it was hopeless from the beginning. On January 26, 1885, the Mahdi's army stormed the city. Gordon was slain on the steps of the Governor's Palace.

"Chinese Gordon," already legendary in his lifetime, assumed near mythic proportions upon his death. He symbolized high Victorian notions of British imperial valor—calm bravery and limitless resourcefulness in the name of civilization standing against the unthinking force of "primitive" peoples. While the values that drove the British campaign of conquest and colonization are distasteful to modern sensibilities, it is undeniable that Gordon was, in every respect, an extraordinary soldier, whose like would not be seen again until the emergence of **T. E. Lawrence**—Lawrence of Arabia—during World War I.

GÖRING, Hermann William

BORN: January 12, 1893
DIED: October 15, 1946
SERVICE: World War I (1914–18); World War II (1939–45)

German Reichsmarschall (Imperial Marshal) who was architect of the Nazi Luftwaffe (air force) in World War II

Göring was born at Rosenheim, Bavaria, to a former cavalry officer, who had also served as German consul general in Haiti. The young Göring was destined for a military career and was sent to Karlsruhe Military Academy in 1905, followed by the main cadet school at Lichterfelde in 1909. Graduating in 1912, he was commissioned a lieutenant in the 112th Infantry, but soon transferred to the air service. At the outbreak of World War I in 1914, Göring served with distinction as an observer, then trained and qualified as a pilot in October 1915. Shot down before the end of the year, he was badly wounded and did not return to duty until 1916.

After racking up an excellent record, he was promoted to squadron commander in May 1917 and, following the death of Germany's most celebrated air ace, Baron Manfred von Richthofen, he succeeded to command of Richthofen's squadron in July 1918 and led the squadron in a manner worthy of the "Red Baron."

He was demobilized following the November 11, 1918, armistice with the rank of captain and was

Hermann Göring. Courtesy National Archives.

hired as a test pilot for the Dutch Fokker aircraft manufacturing firm and the Swedish Svenska Luftraflik (1919–20). During 1921, he attended Munich University and met **Adolf Hitler**. Impressed, he joined the fledgling Nazi Party, in which he served as commander of its paramilitary SA (*Sturmabteilung*, Storm Troops or Brownshirts). He participated in the abortive Munich (Beer Hall) Putsch of November 9, 1923, and was badly wounded in the resulting melee. He was arrested, but escaped and found refuge in Austria.

Göring returned to Germany in 1927 and won election to the Reichstag (German parliament) the following year. In 1932, when the Nazi party was firmly entrenched on the German political scene, he became Reichstag president. The following year, after Hitler's elevation to chancellor of Germany, Göring became Reichsminister, Minister of the Interior, Prussian prime minister, and air commissioner. It was Göring who masterminded the creation of the feared secret police force known as the Gestapo, and it was he who ordered the construction of the first concentration camps, which were originally used to detain political dissidents, but soon were greatly expanded to confine Jews and others the Reich deemed "undesirable." By April 1934, Göring turned over management of the camps to Heinrich Himmler and, later that year, was appointed master of the Reich Hunt and Forest Office. He proved quite effective in this office, creating wildlife preserves and introduced game laws and forest-management reforms that are still in use in Germany today.

If Göring was an able steward of his nation's natural resources, he did not hesitate to eliminate political enemies. At Hitler's behest, he played a major role in the purge of the SA during the so-called Night of the Long Knives (June 30, 1934), in which potential rivals for power were murdered *en masse*. The next year, as Reichsminister for Air and commander of the Luftwaffe, Göring secretly began to organize and build what would become the world's most advanced and, for a time, most powerful air force. While this was under way, Göring assumed power second only to Hitler himself, as director of the Four-Year Plan (1936). Göring had total authority in matters of the German economy, and he reorganized state-owned industries as the Hermann Göring Works during 1937–41. In 1939, Göring was officially appointed to succeed Adolf Hitler and, the following year, was given the title of Reichsmarschall.

The opening months of World War II proved the mettle of Göring's Luftwaffe, which was chiefly responsible for the rapid defeat of Poland during September 1–28, 1939. The air arm was integral to the leading Nazi strategy of Blitzkrieg—lighting war: quick, massive attack aimed not just against military targets, but also against civilian populations.

Göring's reign as Hitler's favorite was not destined to last long. During the successful invasion of France (May–June 1940), his Luftwaffe proved unable to interdict the evacuation of Anglo-French forces from Dunkirk (May 28–June 4). Had the air force destroyed the Allied armies there, invasion of England would have been a foregone conclusion. As great as this missed opportunity was, the Luftwaffe experienced a failure of even greater significance in the Battle of Britain (August 1940–May 1941). Göring originally targeted Royal Air Force (RAF) bases, with the object of destroying aircraft on the ground, effectively neutralizing the RAF as a threat. However, Göring acquiesced in Hitler's decision to bomb major cities. Not only did this fail to break the British will to fight (quite the contrary), it allowed the RAF a chance to regroup and meet the Luftwaffe in the skies above England. Defending their homeland, the RAF pilots proved more than a match for those of the Luftwaffe, which ultimately lost the Battle of Britain. This defeat spelled the end of a German invasion threat.

Toward the end of 1942, Göring made another serious strategic error. When the German Sixth Army was being battered by the Russian winter and the Red Army at Stalingrad, Göring pledged to resupply the troops. Hitler counted on this, but the Luftwaffe failed to come through during November–December 1942. As a result, the decimated Sixth Army surrendered. The war on the Russian front now turned against Germany.

Following the Russian debacle, Hitler no longer trusted Göring, who indeed became thoroughly corrupt. He embezzled from his own government and looted the art treasures of conquered nations. He built a palace and decorated it with the spoils of war. Here he lived a life of dissipation, which disgusted Hitler. As the war turned inexorably against Germany, Göring took up the use of morphine once again; he had originally become addicted to it when the drug had been used to treat him after his injury in the 1923 Putsch.

As Göring lost the confidence of Hitler, so he soon lost the faith of the German populace. He had pledged to Germany that Allied bombs would never rain upon the Fatherland. By 1944, German cities were bombed day and night. Göring had joked that he would change his name to "Meier" (a common German name) if a single bomb ever fell on Germany. By 1944, the people regularly referred to him by that epithet.

The end of Göring's power came in April 1945, when he volunteered to succeed Hitler, who was holed up in his underground bunker in Berlin. Enraged, Hitler summarily removed Göring from all of his offices and charged him with high treason. He was placed under house arrest at Berchtesgaden, Hitler's mountain retreat, on April 23. When Berchtesgaden was overrun by U.S. troops, Göring surrendered to them. Later, he was charged with war crimes at the Nuremberg tribunal. Found guilty, he was sentenced on October 1, 1946 to be hanged, having declined the privilege of death by firing squad. Before sentence could be carried out, he committed suicide by swallowing a capsule of cyanide he had secreted in his anus.

As a young flier in World War I, Göring had shown rare courage. As the father of the Luftwaffe, he had demonstrated a talent for tireless organization. However, he was not a great strategist, and, in the end, this failing was disastrous for the German air force and the German war effort. Personally, too, Göring lacked character and was all too eager to participate in the Nazi orgy of death.

GRANT, Ulysses Simpson

BORN: April 27, 1822

DIED: July 23, 1885

SERVICE: U.S.-Mexican War (1846–48); Civil War (1861–65)

As general-in-chief of the Union army, Grant led the North to victory in the Civil War

Grant was born at Point Pleasant, Ohio, as Hiram Ulysses Grant, the son of farmer Jesse B. Grant. Raised on his father's farm, Grant received an appointment to West Point in 1839. When he learned that he was listed on the academy's rolls as "Ulysses," a name he often went by (Simpson was his mother's maiden name), he accepted it as his name henceforth.

Grant was a mediocre cadet, graduating in 1843, twenty-first of out of a class of thirty-nine. Commissioned a second lieutenant, he was assigned to the Fourth Infantry and, two years later, during September 1845, was sent as part of **Zachary Taylor**'s command to Texas, as Mexico and the United States stood at the brink of war.

Grant fought with distinction in major battles of the war, including Palo Alto (May 8, 1846), Resaca de

Ulysses S. Grant. Courtesy Library of Congress.

la Palma (May 9), and Monterrey, Mexico (September 20–24). When General **Winfield Scott** replaced Taylor during March 1847, Grant was transferred to his command and participated in the capture of Veracruz (March 9–29, 1847). He fought at Cerro Gordo during April 17–18, at Churubusco on August 20, and at Molino del Rey on September 8. In the latter battle, Grant earned a brevet promotion to first lieutenant for gallantry. He distinguished himself further during the September 13 assault on Chapultepec, for which he was breveted captain. On September 16, he was formally commissioned first lieutenant.

At the conclusion of the U.S.-Mexican War, Grant returned to the United States to marry Julia T. Dent (in August 1848) and was posted variously in New York, Michigan, California, and Oregon during 1848–54. At length promoted to captain (August 1853), he grew impatient with the army's slow system of advancement. Resigning his commission, he returned to his family in Missouri in July 1854.

Grant discovered that he had little talent for anything other than soldiering. He was plagued by chronic financial problems, failed business ventures, and a tendency to drunkenness. In 1860, he moved to Galena, Illinois, where he joined his father and brothers in the family leather business and was working in the business as a clerk when the Civil War began in April 1861.

Because of his military experience, Grant was chosen to train the Galena militia company, after which he went to the Illinois state capital, Springfield, where he worked in the state adjutant general's office until June 1861, when he was appointed colonel of the Twenty-First Illinois Volunteer Infantry regiment. In August, he was promoted to brigadier general of volunteers and given command of the District of Southeast Missouri, headquartered at Cairo, the southernmost tip of Illinois.

Acting on his own initiative, Grant seized Paducah, Kentucky, on September 6, 1861, but was forced to retreat from Belmont, Missouri, on November 7. The aggressive Grant found himself at odds with his superior, General **Henry Wager Halleck**, a mediocre and overly cautious commander. At length, Grant persuaded Halleck to allow him to move against Fort Henry on the Tennessee River, which he took on February 6, 1862. By this action, Union forces were able to begin to seize the initiative in the West.

Grant's next objective was Fort Donelson on the Cumberland River, which he invested on February 14. The next day, the Confederate garrison mounted a massive breakout attempt, which Grant was barely able to check. By the sixteenth, however, the garrison surrendered. Grant had presented his terms, stating that nothing less than "unconditional surrender" would be acceptable. From that point on, Grant acquired a new name—not Ulysses Simpson, but "Unconditional Surrender" Grant.

Henry Halleck was conditionally delighted with Grant's victories. He felt that Grant verged on insubordination and temporarily relieved him from command, but restored him to his position late in March. Confederate general **Albert Sidney Johnston** surprised

Grant at Shiloh, but Grant was able to save the day, aided by the timely arrival of reinforcements. Although the Union army incurred very heavy losses during Shiloh (April 6–7, 1862), the Confederates had been driven back.

Halleck took direct command of western forces after Shiloh, and Grant was cast into the background. However, when Halleck was made general in chief of the Union army in July, Grant took command not only over his own army but that of William Starke Rosecrans as well. Although Rosecrans was not a brilliant commander, he achieved victories at the Mississippi towns of Iuka (September 19) and Corinth (October 3–4), which were important because they allowed Grant to launch his Vicksburg campaign in November 1862. Heavily-fortified Vicksburg was the key to the Mississippi River.

From December 1862–March 1863, Grant tried various tactics to lay effective siege against the fortress town. After all of them failed, he took his army south of Vicksburg and, under covering fire furnished by **David Dixon Porter**'s gunboats, was able to cross back to the east bank of the Mississippi during April 30–May 1. He took Grand Gulf, just below Vicksburg, on May 3, then captured Jackson on the fourteenth. This split the armies of Confederate generals John C. Pemberton and **Joseph E. Johnston**. After beating Pemberton's army at Champion's Hill (May 16, 1863) and driving it back upon Vicksburg, Grant laid siege to the city. It did not fall to him until July 4, 1863. Together with the Union victory at Gettysburg, which came on the same day, Vicksburg was the definitive turning point of the Civil War. The Confederate cause was doomed.

Grant was promoted to major general in the regular army and was assigned command of the Military Division of the Mississippi on October 4. His first task was to break the Rebel siege of Union-held Chattanooga, Tennessee. Arriving outside of the city on October 23, he and his subordinates, General George Thomas and General **William Tecumseh Sherman**, broke the siege during October 25–28, then went on to defeat the army of **Braxton Bragg** in the so-called "Battle Above the Clouds" at Lookout Mountain and Missionary Ridge during November 24–25.

Hitherto relatively obscure, Grant impressed President **Abraham Lincoln,** who had appointed one general after another to lead the Union armies. Each of his appointments had fallen short. In Grant, however, Lincoln saw a general who was willing to fight, to endure casualties, and to fight some more. Early in 1864, Grant was promoted to lieutenant general and named the new general in chief of all the federal armies. He did not disappoint the president. Grant not only took in the overall picture of the war, he remained a fine tactical commander. In northern Virginia, he attempted an outflanking maneuver against **Robert E. Lee** in the Battle of the Wilderness during May 4–7, 1864. Although Grant was able to push farther south, Union casualties were so heavy at the Wilderness that some began calling the general "Butcher Grant." The fact was that Grant's greatness as a general lay precisely in his willingness to trade casualties for strategic objectives. He understood the mathematics of the Civil War. The North had far more men to lose than the South. Constant, unremitting aggression would win the war—at a cost, to be sure—but it would win the war.

Grant fought Lee again at Spotsylvania on May 8–17, driving him back to the North Anna River during May 23-26. His advance southward was checked by Lee at Cold Harbor on June 3, prompting Grant to withdraw from in front of Cold Harbor and move south across the James River to attack Petersburg during June 12–17. Although his subordinate **Ambrose Burnside** delayed moving into position, thereby spoiling the Battle of the Crater (July 30), Grant doggedly invested Petersburg from August 1864 through March of the next year. He emerged victorious at Five Forks (March 29–31), and, from here, invaded Richmond (the Confederate capital, from which President Jefferson Davis and his cabinet had fled) and Petersburg.

The war was all but ended. Grant pressed his pursuit of Lee's Army of Northern Virginia through early April, finally accepting surrender of the army at Appomattox Court House on April 9, 1865.

Following the war, Grant returned to Washington, where he was put in charge of the massive demobilization and the military role in postwar Reconstruction. In recognition of services to the nation, he was promoted to the newly created rank of general of the army in July 1866.

Grant served briefly as interim secretary of war under President Andrew Johnson during 1867–68, but his insistence on measures to protect the army of occupation in the South caused a permanent rift with Johnson, who was a Tennessean. Grant then embraced the strong (and often punitive) Reconstruction policies of the radical wing of the Republicans, and he easily achieved the Republican nomination for president in 1868. Defeating Democrat Horatio Seymour, he assumed office and was almost immediately engulfed in scandal. He was, however, popular enough to gain election (over Democrat Horace Greeley) to a second term.

All eight years of the Grant presidency were riddled with political chicanery, ineptitude, and scandal. No one impugned Grant's integrity, but it was clear that he had little political sense and was often the victim of his cabinet and his advisors. He retired to private life after the close of his second term, settling in New York in 1881. As he had failed in every civilian enterprise he undertook before the Civil War, so he failed after the war. Not only was his presidency corrupt, his personal finances were disastrous, and he was bankrupt by 1884. At the urging of the humorist Mark Twain, who owned a successful publishing company, Grant, impoverished and afflicted with cancer of the throat, wrote his *Memoirs*. Completed only four days before his death, the work is a literary masterpiece and an extraordinary living history of the Civil War. Published by Twain's firm, its success saved Grant's widow from penury and rescued her husband's reputation from the taint of a bad presidency, presenting him instead as the great commanding general that he was.

GRASSE, Count François Joseph Paul de

BORN: 1722

DIED: January 14, 1788

SERVICE: War of the Austrian Succession (1740–48); American Revolution (1775–83)

French admiral best known for his invaluable aid during the American Revolution

Grasse was born at Bar, near the village of Grasse in southern France. When he was only ten years old, his family sent him to the naval school at Toulon, and he joined the navy of the Knights of Malta in 1734, eventually serving as page to the Grand Master of the Knights.

When the War of the Austrian Succession began in 1740, Grasse left the service of the Knights of Malta to join the French navy. He served nearly to the end of the war, until he was taken prisoner at the battle of Cape Finisterre on May 3, 1747. His release came the following year, when the war ended, and he served off India and in the Caribbean and Mediterranean. Grasse's first major command was given him in 1773, when he took charge of the Marine Brigade at Saint-Malo. Two years later, he was made captain of the twenty-six-gun frigate *Amphitrite* on a cruise to the troubled French colony of Haiti during June 1775–76. He was promoted to comand of the much larger *Intrépide*, a seventy-four-gun vessel, in 1777, and then was named *chef d'escadre* (the equivalent of commodore) on June 1, 1778. On July 27, he was sent to command a division at Ushant, then sailed, on July 6, 1779, to Grenada, to take command of a squadron under the Comte d'Estaing, who had been dispatched to aid the American revolutionary cause. Late in 1779, after d'Estaing returned to France—he had been wounded in an unsuccessful attempt to wrest Savannah, Georgia, from British control—Grasse briefly commanded the French fleet in the West Indies, then led a squadron at Martinique in 1780.

Late in 1780, Grasse fell ill and had to return to France, arriving in Brest on January 3, 1781. On March 22, he was promoted to rear admiral and, that very day, sailed with twenty ships-of-the-line, three frigates, and 150 transports for the West Indies. He attacked the fleet of British admiral Alexander Hood, successfully disrupting his Caribean operations during April-August. On August 13, he received **George Washington**'s request for naval support at Yorktown, Virginia, and arrived off Yorktown on August 30. He opened up on the fleet of Admiral Thomas Graves at the battle of the Virginia Capes (September 5–9), which blocked the relief of the British forces led by **Charles Cornwallis.** This led directly to Conwallis's surrender of his army, an act that assured American victory in the Revolution.

Following the action at Virginia Capes, Grasse sailed for the West Indies on November 4 and captured St. Kitt's on February 12, 1782; however, where Admiral Hood had failed, Admiral **George Rodney** triumphed. He defeated Grasse at the Saints on April 12 and captured him as well as his flagship, *Ville de Paris*. In all respects, the battle was a disaster for Grasse. Rodney penetrated the center of the French line and managed to capture seven of the twenty-nine ships under the French admiral's command.

During his genteel imprisonment in England, Grasse acted as an intermediary between the French and British governments throughout the peace negotiations of August. On his release and return to France, however, he was blamed for the debacle at the Saints. Grasse refused to accept blame, in turn criticizing the action of his subordinates. However, a court of inquiry cleared most of them, with the result that Grasse fell out of favor with the court of Louis XVI in May 1784. Two years later, he was officially returned to favor.

The admiral's service to the United States was not forgotten. When his five children fled the bloodbath of revolutionary France, they all found asylum and new homes in the United States.

GREENE, Nathanael

BORN: August 7, 1742

DIED: June 19, 1786

SERVICE: American Revolution (1775–83)

Usually deemed the best general in the Continental Army next to George Washington; he was largely responsible for U.S. victories in the South during the American Revolution

A native of Potowomut (modern Warwick), Rhode Island, Greene was the son of an iron founder, to whom he apprenticed. He became prominent in Rhode Island political life, gaining election to the colony's General Assembly in 1770 and serving from 1770 to 1772 and in 1775. Although Greene was brought up as a Quaker, his revolutionary fervor caused him to break with that pacific faith. He zealously set about raising a militia company during October 1774. Ironically, because Green was partially lame, the militia company he had helped raise declined to elect him their captain. Instead, he volunteered for service as a private. In May 1775, however, he was appointed brigadier general of militia and, on June 22, brigadier general in the Continental Army.

Greene's natural abilities as an officer were first manifested in the siege of Boston, where he showed a mastery of logistics and a keen ability to coordinate his efforts with others'. He marched his brigade to Long Island, New York, on April 1, 1776, was promoted to major general on August 9, but missed the August 27 Battle of Long Island because he had been stricken with a fever.

Greene attacked Staten Island on September 12, then counseled George Washington to hold Fort Washington (in upper Manhattan). Unfortunately, this advice contributed to the loss of the garrison there when the British overran the fort on November 16. Greene amply redeemed himself at the momentous Battle of Trenton on December 26, however, which was a signal triumph for the American cause.

Greene's next major engagement came at the Brandywine in Pennsylvania. With great brilliance and dash, he force-marched his brigade from Chadds Ford, Pennsylvania, to the town of Brandywine to check a British flanking movement on September 11, 1777, thereby protecting Washington's main column. On October 4, Greene himself led the main column at Germantown (now part of modern Philadelphia). When American positions along the Delaware River became untenable in the fall of 1777, Greene skillfully evacuated them.

Because of his natural flair for logistics, Washington asked Greene to accept the post of quartermaster general to the Continental Army. It was not a position the fighting general much wanted, but he took it on (March 2, 1778), having secured a promise that he would also retain field command troops in the field. Accordingly, he led the right wing at the Battle of Monmouth on June 28, 1778, distinguishing himself there and in subsequent engagements at Newport and Quaker Hill in August.

Nor did Greene neglect his duties as quartermaster general. He struggled to supply the troops during the winter of 1778–79 and in the wide-ranging campaigns of 1780. His antagonist in this struggle was less the British than the Continental Congress, which was both parsimonious and inefficient in funding supply. After many bitter quarrels with Congress over supply, Greene resigned as quartermaster general in 1780 and was appointed to succeed Benedict Arnold as commander in the Hudson Highlands. However, he soon found himself dispatched to the South after General **Horatio Gates** ignominiously surrendered Camden, South Carolina, on August 16, 1780.

Greene assumed command in the South on December 2, joining forces with Daniel Morgan after Morgan triumphed at the Battle of Cowpens, South Carolina on January 17, 1781. In the wake of this victory, however, the superior army of British general **Charles Cornwallis** gave chase, and Greene and Morgan quickly withdrew into southern Virginia during January and February, 1781. Greene handled the maneuver with great skill and dispatch, so that the British pursuers were exhausted. At Guilford Courthouse on March 15, he turned on Cornwallis, letting the British commander make a costly attack. The result was a tactical defeat for the colonials and a pyrrhic victory for Cornwallis, whose forces were badly cut up. Cornwallis withdrew from the Carolinas into Virginia. As to Greene, he moved back into the Carolinas, where he did battle against an army led by Lord Rawdon. Suffering a defeat at Hobkirk's Hill on April 25, Greene recovered to lay siege against Augusta, Georgia, during May 22–June 5. He also besieged the settlement of Ninety Six, South Carolina, during May 22–June 19, but had to withdraw before Rawdon's forces.

Greene allowed his exhausted forces to rest and recover, then went on the offensive again, this time against Rawdon's successor, Lt. Col. Duncan Stuart, at Eutaw Springs. It was a close battle, but, on September 8, the British finally drove Greene back. It was another pyrrhic victory for the British, however, and British strength in the South was now greatly reduced. By this point, they controlled only Savannah and Charleston—coastal towns—having relinquished interior positions. After Washington's victory at Yorktown on October 17–19, 1781, southern engagements became minor. Greene established headquarters in Charleston, South Carolina, after the British evacuated on December 14, 1882. He remained there until news reached him in August 1783 that peace had been formally concluded. Like many other heroes of the Revolution, Greene was now faced with repairing his beleaguered personal finances. A shameless government bickered over reimbursing him for expenses he incurred personally supporting his army when funding from Congress was unavailable. After sorting out his finances, Greene retired from public life, taking up residence on an estate near Savannah, Georgia.

Like Washington, Greene was an archetypal commander of the American Revolution. He depended on skill at rapid maneuver and, with characteristically strained resources at his disposal, learned how to make every British victory excessively costly.

GRIERSON, Benjamin Henry

BORN: July 8, 1826

DIED: August 31, 1911

SERVICE: Civil War (1861–65); Indian Wars (1865–91)

Perhaps the Union army's finest cavalry officer in the Civil War; did much to mold the U.S. Cavalry after the war

Grierson was born in Pittsburgh, Pennsylvania, but moved west as a young man and settled in Illinois, where he set up as a music teacher before entering business. Financially wiped out by the Panic of 1857, he turned to Republican politics and became moderately influential in Illinois. Like many relatively small-time politicians, Grierson obtained a military appointment when the Civil War began. Unlike most such political appointees, however, he soon proved himself extremely able.

He was made a staff assistant to Brigadier General Benjamin M. Prentiss in May 1861, then commissioned a major in the Sixth Illinois Cavalry in October. Promoted to colonel in April 1862, he first distinguished himself in his pursuit of General Earl Van Dorn's Confederate raiders after their December 20 assault on Holly Springs, Mississippi. Following this action, during spring 1863, General **Ulysses S. Grant**, at the time commanding western forces, chose Grierson to make a diversionary raid across Mississippi from Tennessee to Louisiana in order to allow Grant to position his army against the key Mississippi River stronghold of Vicksburg. Grierson left La Grange, Tennessee with 1,700 troopers and six small cannon on April 17 and conducted no mere diversion, but a spectacularly successful raid through Mississippi. With dash and brilliance, he evaded his pursuers at every turn, while harrying them and destroying precious railroad track, rolling stock, telegraph lines, and supplies. Equally important, his raids panicked the locals and tied up Confederate military resources. Remarkably, when he arrived in Baton Rouge, Louisiana, on May 2, his force was virtually unscathed.

In recognition of his remarkable exploit, Grierson was commissioned a brigadier general of volunteers June 3 and then participated in the siege of Port Hudson, near Baton Rouge, during May 21–July 2. Afterward, he returned to Tennessee to command the cavalry division of XIV Corps, with which he pursued the Confederate raider **Nathan Bedford Forrest** during the spring and summer of 1864. In command of Union cavalry at Brice's Cross Roads, he was defeated by Forrest on June 10, but then was a key commander (Under General A. J. Smith) among the forces that defeated Forrest at Tupelo, Mississippi on July 14.

Following Tupelo, Grierson was put in command of the District of West Tennessee's Cavalry Corps, then, on November 6, was given command of the Fourth Division, Cavalry Corps, Military Division of the Mississippi. However, he was relieved by General James H. Wilson the next month, and General Napoleon J. T. Dana retained him in the Division of the Mississippi. Serving in this division, he raided Confederate communications and supply lines in Mississippi, part of the pursuit of General John Bell Hood's Army of Tennessee, which was in full retreat from Nashville during December 21, 1864–January 5, 1865.

In February, Grierson was breveted major general of volunteers and ended the war in command of four-thousand cavalry troopers as part of General **Edward R. S. Canby's** Military Division of West Mississippi. His brevet rank was made permanent.

Grierson's cavalry expertise was called on to assist **George Armstrong Custer** and General Wesley Merritt in organizing cavalry forces in Texas. After a brief stint with the army of occupation in Alabama, Grierson was mustered out of the army on January 15, 1866. However, he quickly rejoined as colonel of the Tenth (Negro) Cavalry on June 28, 1866 and was breveted to regular brigadier and major general by March 1867.

Grierson led the Tenth Cavalry for some twenty-five years, during which the unit performed admirably in Kansas and Oklahoma (1866–75) and New Mexico and Arizona (1875–88). The African-American troopers earned from the Indians the nickname of Buffalo Soldiers, a term of respect. Grierson led his unit in operations against the Apaches under Victorio during the summer of 1880 and succeeded General **Nelson A. Miles** as commander of the Department of Arizona on December 1, 1888. During 1873–74, while he served with the Tenth, Grierson was also superintendent of the General Mounted Recruiting Service, headquartered at St. Louis.

Grierson retired from the army on July 8, 1890 and lived the remainder of his life in Jacksonville, Illinois. His accomplishments as a cavalry officer were the more extraordinary in that he lacked formal military training. His Tenth Cavalry was arguable the most professional and effective single cavalry unit serving in the West.

GUDERIAN, Heinz

BORN: January 17, 1888

DIED: 1953

SERVICE: World War I (1914–18); World War II (1939–45)

German general who played a key role in the development of German mechanized forces

Guderian was born at Kulm (Chelmno) into the family of a Prussian army officer. He attended the cadet school in Karlsruhe from September 1900 to April 1903, then enrolled in the main cadet school at Gross Lichteffelde, from which he graduated in December 1907. He entered the Tenth Hannoverian Jäger Battalion on January 27, 1908, as a second lieutenant, and, on the eve of World War I, attended the *Kriegsakademie* (War College) during 1913–14.

At the commencement of war in August 1914, Guderian was in command of a wireless station and by April 1915 had become assistant signals officer in Fourth Army. He served in this post through April 1917, when he moved through a variety of staff appointments, culminating in an appointment to the Great General Staff in February 1918. He held this post until the armistice.

After the war, Guderian participated in *Freikorps* operations in Latvia during March–July 1919 as chief of staff of the Iron Division. He was chosen as one of four-thousand officers of the 100,000 *Reichswehr*—the greatly reduced German army permitted by the Treaty of Versailles—late in 1919. Subsequently, in January 1922, he was assigned to the Inspectorate of Transport Troops in the *Truppenamt*, which was the cover designation of the General Staff, officially banned by the Versailles treaty. During 1922–24, he served in a transport battalion in Munich, then became an instructor in tactics and military history on the staff of Second Division during 1924–27.

Guderian returned to General Staff duty during October 1927–February 1930. During this period, he was briefly attached to a Swedish tank battalion. Afterward, given command of a motor transport battalion, he reorganized it as a provisional armored reconnaissance battalion (1930–31). When he was appointed chief of staff to the inspector of Transport Troops in October 1921, he began work on plans for the creation of tank forces.

In October 1935, Guderian left the staff post to assume command of the Second Panzer Division at Würzburg. The following year, he was promoted from colonel to major general and, in 1937, gained wide attention with a small book entitled *Achtung! Panzer* (*Attention! Armor*). In it, he digested his theories of mechanized warfare.

During 1937–38, Guderian commanded XVI Corps, comprising three panzer divisions, then, during the *Anschluss* (the annexation of Austria to Germany), Guderian led the Second Panzer Division through Linz to Vienna (March 12–13, 1938). At the commencement of World War II, he led XIX Panzer Corps through the Polish campaign (September 1–October 5, 1939), demonstrating the role of armor in the Biltzkrieg by advancing with great speed from Pomerania and across the Polish Corridor, on September 4, to capture Brest during September 16–17. Reinforced, the XIX Corps was next placed under Panzer Group Kleist for the campaign in France. Guderian and his unit were in the vanguard of the invasion of France on May 10, 1940, crossing the Meuse River at Sedan and reaching the English Channel coast on May 19. Guderian was rushing toward Dunkirk when his unit was halted by order of **Adolf Hitler** on May 24.

When France capitulated, XIX Corps was on the Swiss frontier near Basel, and, by November 1940, was expanded into Second Panzer Group. It was part of Field Marshal Fedor von Bock's Army Group Center during Operation Barbarossa, the German invasion of the Soviet Union, which commenced on June 22, 1941.

Guderian's Second Panzers, along with the Third Panzer Group, encircled Soviet forces at Minsk on July 10 and went on to surround Smolensk and to capture Roslavl' during July 12–August 8, 1941. From here, Guderian was dispatched south to coordinate operations with the Fourth Panzer Group of Field Marshal **Gerd von Rundstedt**'s Army Group South in a massive maneuver to encircle 600,000 Red Army troops in the "Kiev pocket" during August 21–September 6.

Guderian's offensive push toward the Soviet capital of Moscow was stalled by a combination of brutally cold weather and increasingly fierce Soviet resistance during October 23–November 7, 1941. In a desperate situation, Guderian made repeated requests for permission to withdraw from exposed positions around Tula during December 5–26. In response, he was finally relieved of command and replaced by General by Gunther von Kluge on December 26, 1941. Following this, Guderian fell ill and did not recover until 1943, when he was recalled to duty as inspector general of panzer troops in February. By this time, the panzer forces had been badly mauled at Stalingrad, and Guderian set about rebuilding German armor, working closely with Armaments Minister Albert Speer to step up and improve tank production.

On July 21, 1943, Guderian replaced General Kurt Zeitzler as army chief of staff. As the war drew to a close, a desperate Adolf Hitler dismissed Guderian from his position on March 28, 1945. Guderian was held under arrest by the Allies for several months after the German surrender, but, unlike many top generals, he was not charged with war crimes.

Guderian was one of the chief architects of German armor in World War II. He was an excellent theoretician, an efficient staff officer, and a battlefield commander of unquestioned courage and tactical ability. His relationship with Hitler was a stormy one, since he refused to submit meekly to the Führer as a "yes man."

GUSTAV II ADOLF (Gustavus Adolphus)

BORN: December 9, 1594

DIED: November 16, 1632

SERVICE: War of Kalmar (1611–13); Polish Wars (1617–29); Thirty Years' War (1618–48)

Swedish king and champion of the Protestant cause

King Gustav II Adolf of Sweden was hailed as the "Lion of the North," concluding three wars begun by his father and committing Sweden to the Thirty Years' War on the side of the German Protestant princes. Born in Stockholm, he was the son of Charles IX and Christina of Holstein. In 1611, following his father's death, he succeeded to the throne, aided by the clever political maneuvering of Axel Oxenstierna, who, as a member of the ruling council, led the aristocracy to accept Gustav as king in exchange for concessions from the crown. Yet the throne he had inherited was far from secure. In 1599, Charles IX had seized power from Sigismund III Vasa, who was also king of Poland, touching off a dynastic struggle that resulted in intermittent warfare between Sweden and Poland over the next sixty years. Moreover, Gustav inherited ongoing wars with Denmark and Russia. The war with Denmark ended with the Peace of Knared (1613), by which Sweden relinquished control of its only port on the North Sea, Alvsborg, to Denmark as security for the payment of a war indemnity. The Russian conflict had begun when Charles IX attempted to fill the vacant Russian throne with a Swedish nobleman. Gustav II Adolf managed to conclude that war In 1617, signing the Peace of Stolbova, which secured Ingria and Kexholm for Sweden and cut Russia off from the Baltic.

Even as Gustav prosecuted the conflict with Poland, he had to act to secure the throne domestically. Oxenstierna had asked the council to draw up a list

Gustav II Adolf (Gustavus Adolphus). From M. Guizot, A Popular History of France from the Earliest Times, *n.d.*

of grievances and then advised Gustav to sign a charter of guarantees, in effect granting the council the authority to strip the monarchy of all power if its grievances went unanswered. Gustav's personal charisma, however, continued to command the loyalty of the nobility as he enacted broad government reforms. With the domestic situation stabilized, Gustav took advantage of a Turkish attack on Poland in 1621 to renew his war with Sigismund. He moved vigorously against Riga and Livonia, capturing them, then moving his war headquarters to Prussia. The conflict dragged on until 1629, when the Truce of Altmark was concluded, whereby Sigismund renounced his claim to the Swedish throne.

Freed of his inherited battles, Gustav plunged headlong into the Thirty Years' War (1618–48), sending an army into Germany on the side of the Protestant princes and against the Habsburgs. The forces of the Catholic Habsburgs were ably led by Generals **Johan Tilly** and **Albrecht von Wallenstein,** and Gustav feared that a Catholic victory in Germany would lead to the restoration of Catholicism in Sweden. He sailed for Germany in May 1630 with 13,000 troops, a number soon augmented by an additional 27,000. Nevertheless, he met with defeat in his first engagement at Mecklenburg, but then went on to a pair of victories that swept him through Germany: at Breitenfeld on September 7, 1631, and at the Lech River in Bavaria in April 1632.

With much of central and northern Germany under his control, Gustav made overture to create a Protestant League as a permanent bulwark against the forces of Catholicism. He was unable to win the loyalty of all the Protestant princes, however, who saw Gustav as ambitions and overly eager to impose Swedish administration in conquered territories. While Gustav engaged in fruitless debate, Wallenstein put his troops on the move once again. On November 6, 1632, Gustav's army attacked Wallenstein at Lutzen, Saxony, and the Lion of the North was killed in battle.

Although Gustav II Adolf failed to create the Protestant League he so desired, his entry into the Thirty Years' War secured the survival of Germany's Protestant princes during the Counter-Reformation. The corollary effect of this was to delay the emergence of a unified Germany until the nineteenth century.

H

HADRIAN (Publius Aelius Hadrianus)

BORN: 76

DIED: 138

SERVICE: Dacian Wars (101–102, 105–106); Jewish Revolt (132–135)

Unpopular but effective Roman ruler who did much to consolidate and centralize power

Hadrian was born in the Iberian town of Italica, near modern Seville, Spain. He was orphaned in 85 and raised by his cousin **Marcus Ulpius Trajanus,** who became the emperor Trajan. Under Trajan, Hadrian served with distinction in the Dacian Wars and was rewarded for his service by an appointment as governor of Lower Panonia (modern-day Hungary) in 107. Seven years later, he was also made governor of Syria.

Trajan ascended the imperial throne in 98, and Hadrian had long been his favorite and a favorite of his wife, Plotina. Nevertheless, Trajan deferred formally adopting the young man he had raised until he lay on his deathbed on August 8, 117. Not only did he acknowledge him as his adopted son, he also designated him his successor. On August 11, the day of Trajan's death, Hadrian became emperor—in spite of much internal opposition.

Hadrian was not nearly as expansionist as Trajan had been. He was more interested in consolidating the empire than in expanding it. Although he retained the province of Dacia (modern Romania), he withdrew from Parthia, into which Trajan had made inroads. Dacia was troublesome enough. During 117–18, Hadrian had to move swiftly in the suppression of a rebellion led by four of Trajan's generals. He captured and executed the conspirators.

With internal strife at least temporarily quelled, Hadrian traveled through the provinces during 121–32, inspecting the prospects for imperial growth and maintenance. The tour confirmed his original impulse to consolidate rather than expand, and Hadrian developed a strategy of containment and defense. He ordered the construction of great defensive works—fortified walls—on the German frontiers and in northern England, where large stretches of "Hadrian's Wall," begun in 121 or 122, now attract tourists.

Elsewhere, particularly in Greece and Palestine, Hadrian was determined to set the imprint of empire by means of massive building projects. In Palestine, he decreed the construction of a new city, Aelia Capitolina, on the site of Jerusalem. Its central shrine, a temple dedicated to Jupiter Capitolinus, was to replace the Temple of Jerusalem. As if that weren't a sufficient cultural blow, the emperor promulgated a ban against ritual circumcision. These actions provoked the Jewish leader Bar Kochba to lead a bloody revolt that spanned 132–35. The war, devastating to the Jews, proved very costly to Hadrian as well. The emperor ultimately laid waste much of Palestine.

After the Palestinian wars, Hadrian withdrew to his villa at Tivoli, outside of Rome and concentrated on grooming his aging brother-in-law, L. Julius Ursus Servianus, as his successor. However, he suddenly adopted L. Ceionius Commodus, bestowed on him the name L. Aelius Caesar, and summarily nominated him as heir in 136. When Aelius died soon after this, Hadrian adopted T. Aurelius Antoninus in January 138. It was Aurelius who assumed the throne, after Hadrian's death on July 10, 138, and ruled as Antoninus Pius.

HAIG, Alexander Meigs, Jr.

BORN: December 2, 1924
SERVICE: Vietnam War (1965–73)

American general who served in Vietnam and became most famous as an advisor to President Richard M. Nixon and as secretary of state under President Ronald Reagan

Philadelphia-born Haig attended the University of Notre Dame during 1943 before entering the U.S. Military Academy at West Point, from which he graduated in 1947. Commissioned a second lieutenant that year, he later enrolled in the Naval War College, from which he graduated in 1960. The following year, he received a Master's degree from Georgetown University.

Haig served as staff officer in the Office of the Deputy Chief of Staff of the Army for Operations during 1962–64 and was military assistant to the Secretary of the Army in 1964. Appointed Deputy Secretary of Defense in 1964, he served through 1965, leaving that post to take up a field command as battalion and brigade commander with the First Infantry Division in Vietnam during 1966–67.

Haig returned to the United States as regimental commander and deputy commandant of the Military Academy at West Point, serving in this post from 1967 to 1969, when he was appointed military assistant to Presidential Assistant for National Security Affairs Henry Kissinger (1969–70). In 1970, Haig was elevated by President Nixon to the post of deputy assistant to the President for National Security Affairs. He served until 1973, when he became vice chief of staff of the army. During the last months of Nixon's embattled presidency, Haig also served as chief of the White House staff (May 1973–August 1974).

From 1974 to 1979, Haig was commander-in-chief, U.S. European Command and Supreme Allied Commander, Europe (SACEUR)—in effect, the commander of NATO forces. After retiring from active duty in 1979, he became president, CEO, and director of United Technologies, Inc., until President Ronald Reagan appointed him secretary of state in 1981. Haig played an important role in creating strongly anti-Soviet policy; however, he was often at odds with the president and resigned in 1982.

Haig returned to the private sector as an executive with a number of firms, then campaigned unsuccessfully for the Republican presidential nomination in 1988. His memoir *How America Changed the World* was published in 1992.

HAIG, Douglas, First Earl

BORN: June 19, 1861
DIED: January 29, 1926
SERVICE: Mahdist War (1883–98); Second ("Great") Anglo-Boer War (1899–1902); World War I (1914–18)

Underrated commander of the British Expeditionary Force (BEF) in World War I

Haig was born in Edinburgh to the wealthy family of a distiller. He received a superb education at Clifton and Brasenose Colleges, Oxford, then went on to Sandhurst, the British army's military academy, from which he graduated at the top of his class in 1885. He was commissioned in the Seventh Hussars and served with them in India and Britain. During the Nile campaign of the war against the **Mahdi,** he fought at the battles of the Atbara River (April 8, 1898) and Omdurman (September 2), distinguishing himself in both.

With the outbreak of the Second Anglo-Boer War, Haig was appointed chief of staff to General **Sir John French** and, promoted to colonel, was given command of the Seventeenth Lancers, known as the "Death or Glory Boys." He led the unit in operations against Boer general **Jan Smuts**'s guerrillas in Cape Province during 1901–02.

Following the Second Anglo-Boer War, Haig was posted to India, where he served as inspector general of cavalry from 1903 to 1906. Promoted to major general in 1905, he was appointed director of military training at the War Office and served in that capacity through 1909. During this period, he published a book, *Cavalry Studies* (1907). In 1909, he was sent back to India as chief of the general staff there, was promoted to lieutenant general in 1910, and continued serving in India until 1912, when he took over command at Aldershot.

At the outbreak of World War I, Haig commanded I Corps of the BEF under his former commander, Sir **John French.** He fought at Mons (August 23, 1914) and the Marne (September 5–9), as well as in Picardy and Artois during the Race to the Sea (October-November). In February 1915, Haig was appointed commander of First Army, and in this post he directed a major attack at Neuve-Chapelle during March 10–13. The operation was moderately successful, but Haig's gains were readily contained by the Germans. During May 9–26, he launched another offensive, this time in Artois, but was stopped at Festubert. Hitting German positions at Loos during September 26–October 14 resulted in negligible gains and severe losses.

Blame for the tragically disappointing performance of the BEF during the opening months of the war fell not on Haig, but on General French, who was recalled and dismissed as BEF commander. Haig replaced him on December 17, 1915, and immediately set about planning a massive British offensive at the Somme in an effort to relieve pressure on the French at Verdun. So desperate was the French situation, Haig was forced to begin the Somme offensive before he was fully ready. Initial gains were more substantial than in previous attacks, but, in an effort to relieve the French, Haig continued to apply pressure against the Germans, even after the British advance had stalled. The result was punishing both to the Germans and to the BEF, which incurred very heavy losses during June 24–November 13, 1916. Nevertheless, Haig was promoted to field marshal at the end of 1916.

Placed under command of French general Robert Georges Nivelle, he launched a moderately successful attack at Arras during April 9–15, 1917. Unfortunately, under Nivelle, the French armies were racked by mutiny and approached collapse. Nivelle's own offensives had failed, and Haig embarked on the desperate Passchendaele offensive of July 31–November 10, which proved costly and, for the most part, futile.

Haig braced for what he predicted would be major German offensives in Flanders early in 1918 and called for 600,000 reinforcements. He received a mere 100,000 troops, and the German Somme offensive, when it came, during March 21–April 5, nearly destroyed the British Fifth Army led by Hubert del Poer Gough.

Following this ordeal, the Allies finally agreed to Haig's proposal that an Allied high command be established, with French general **Ferdinand Foch** as its chief, the better to coordinate allied operations. When German general **Erich Ludendorff** pressed his second offensive around the Lys River during April 9–17, the British lines nearly dissolved again. Haig, however, rallied his troops, who ultimately halted the German advance. This accomplished, he directed a counterattack at Amiens during August 8–11, which was a breakthrough. Ludendorff called August 8, 1918, "the black day of the German Army." Working to execute Foch's final grand offensive, Haig directed the last Allied attacks in Flanders from September until the armistice on November 11, 1918.

Haig will forever be identified as the chief advocate of the strategy of attrition in World War I—a policy that resulted in few gains and heavy losses. However, he did not so much advocate the strategy, as he recognized that it was the only possible alternative in a stalemated contest between entrenched armies. For confronting and recognizing the terrible realities of World War I, Haig is often unjustly reviled as slow-witted and heartless. An able soldier, he was neither. After the war he worked vigorously for the welfare and relief of veterans, and, as president of the British Legion, he instituted the "Poppy Day" fund-raising program for disabled soldiers. Created First Earl Haig in 1919, he died seven years later in London.

HALLECK, Henry Wager

BORN: January 16, 1815

DIED: January 9, 1872

SERVICE: U.S.-Mexican War (1846–48); Civil War (1861–65)

Prominent Union general in the Civil War; poor as a field commander, but an able administrator

Born in Westernville, New York, Halleck graduated from West Point in 1839 and was commissioned in the Engineers. He was sent on a European tour to inspect fortifications in 1844 and, of a scholarly turn of mind, digested what he had learned in a series of lectures on military science as well as in an 1846 book, *Elements of Military Art and Science*.

Halleck served in the U.S.-Mexican War from 1846 to 1848, but saw little combat. His principal assignment was building fortifications. Breveted captain in May 1847, he was appointed secretary of state in the military government of California and was a central figure in drafting the state's constitution, completed in September 1849. Halleck resigned from the military in August 1854 to take up the practice of law, in which he achieved prominence. He subsequently served as president of the Pacific & Atlantic Railroad.

With the outbreak of the Civil War, Halleck was appointed major general of militia, becoming a major general of the regular army by August 1861. In November, he was assigned to command the Department of the Missouri (later called the Department of the Mississippi), and he did a creditable job organizing rag-tag volunteers. However, when he took command of the Army of the Ohio during the Corinth campaign (May–June 1862), he revealed himself as a poor field officer. He was soon replaced by his subordinate, **Ulysses S. Grant**, and was "kicked upstairs," to Washington, D.C., in July, where he served as general in chief of the Army. Grant replaced him here as well in March 1864, and Halleck was once again elevated, to chief of staff.

In the closing months of the war, Halleck served as commander of the Military District of the James (March–April 1865), and, after the war, became commander of the Division of the Pacific (August 1865–March 1869) and then Division of the South (1869–72).

Well-meaning and learned, Halleck was an office general rather than an effective field commander. He was far too cautious to use men effectively in battle. However, he performed such duties as recruiting, organizing, and training with great efficiency and, in this way, contributed to the Union war effort.

HALSEY, William Frederick, Jr.

BORN: October 30, 1882

DIED: August 16, 1959

SERVICE: World War I (1917–18); World War II (1941–45)

U.S. admiral in World War II; advocate of using aircraft carriers to "hit hard, hit fast, and hit often"

Born in the gritty working-class town of Elizabeth, New Jersey, "Bull" Halsey was the son of a naval officer. He graduated from the naval academy at Annapolis in 1904 and was commissioned ensign in 1906. His first assignment at sea was under Admiral **George Dewey**—the around-the-world cruise of the Great White Fleet (August 1907–February 1909).

Halsey attended torpedo school at Charleston, South Carolina, and was assigned duty aboard several destroyers and torpedo boats. He was given command of U.S.S. *Flusser* (DD-20), serving aboard her from August 1912 to February 1913, and of U.S.S. *Jervis* (DD-38), serving from 1913 to 1915. He commanded the vessel during the occupation of Veracruz (April–October 1914).

During 1915–17, he was attached to the Executive Department at the Naval Academy and was promoted to lieutenant commander during August 1916. During World War I, he was assigned command of two destroyers, U.S.S. *Duncan* (DD-46) and U.S.S. *Benham* (DD-49), performing hazardous convoy escort duty from a base in Queenstown, Ireland. After the war, Halsey skippered destroyers in the Atlantic as well as the Pacific (1918-21).

In 1921, Halsey was transferred from sea duty to the Office of Naval Intelligence. The next year, he was assigned the prestigious post of naval attaché in Berlin, as well as Norway, Denmark, and Sweden (1922–24). He then returned to sea duty aboard destroyers in the Atlantic, then transferred to the battleship U.S.S. *Wyoming* (BB-32) as executive officer during 1926–27. Promoted to captain in February 1927, he was given command of the U.S.S. *Reina Mercedes* (IX-25), the post ship at Annapolis, which had been captured from the Spaniards in 1898. In 1930, he was made

William "Bull" Halsey (far right). Courtesy National Archives.

commander of Destroyer Squadron 14, serving until 1932, when he attended the Naval War College (graduated 1933) and the Army War College (graduated 1934).

Seeing the future of naval warfare in carrier-based aviation, Halsey, age fifty-two, completed flight training at Pensacola, Florida, in May 1935 and assumed command of the aircraft carrier *Saratoga* (CV-3) in July. Two years later, he was assigned command of the Pensacola Naval Air Station (1937–38). After securing promotion to rear admiral in March 1938 he was given command of Carrier Division 2 (1938–39), followed by Carrier Division 1 (1939–40).

Halsey next moved up to vice admiral in June 1940 and was assigned to command Aircraft Battle Force as well as Carrier Division 2. Halsey was at sea with the carriers U.S.S. *Enterprise* (CV-6) and *Yorktown* (CV-5) during the Japanese attack on Pearl Harbor (December 7, 1941). He used the carriers in the months that followed to raid outlying Japanese islands in the Central Pacific (January–May 1942). He collaborated with Army Air Corps colonel **James H. Doolittle** in executing Doolittle's daring bombing raid on Tokyo, using B-25 bombers launched from the carrier *Hornet*.

Late in May 1942, Halsey was stricken with a serious illness and turned over command to Raymond Ames Spruance. Because of his illness, Halsey missed the critical Battle of Midway (June 4, 1942), often cited as the turning point in the Pacific war.

By October, Halsey was returned to active duty and replaced Adm. Robert L. Ghormley as commander of South Pacific Force and Area. Narrowly defeated at Santa Cruz during October 26–28, he nevertheless gained an ultimate strategic victory by maintaining station off Guadalcanal, which land forces were in the process of invading. During November 12–15, 1943, Halsey defeated the Japanese at sea off the island, then commanded naval support efforts for the capture of the rest of the Solomon Islands during June–October, 1943.

The fall of Bougainville, toward the end of 1943 and beginning of 1944, isolated the key Japanese base at Rabaul, rendering it vulnerable.

Halsey was named commander of Third Fleet in June 1944 and took the battleship U.S.S. *New Jersey* (BB-62) as his flagship in August. That autumn, he directed landings at Leyte in the Philippines (October 17–20). Here, however, he faltered. Always motivated by his hit hard, fast, and often strategy, he pursued the remnant of the Japanese carrier force off Luzon on October 25, 1944. Although he sunk four Japanese vessels, he left San Bernardino Strait covered only by a weak force of escort carriers and destroyers, which were set upon by Admiral Takeo Kurita's superior Central Force. The outnumbered and outclassed Americans managed to repulse the attack in the Battle of Samar (October 25), and Halsey dashed to reinforce the beleaguered detachment in an operation that became known as "Bull's Run."

The incident stained Halsey's otherwise spotless reputation. Halsey suffered a further reverse when his Third Fleet, supporting amphibious operations in the Philippines, was struck by a typhoon that sank three destroyer escorts in December. Halsey nevertheless went on to sweep through the South China Sea, destroying massive amounts of Japanese tonnage during January 10–20, 1945.

Halsey turned over command to Spruance, then returned to sea-going command during the last stages of the Okinawa campaign (May–June 22) and the raids against the Japanese home islands during July and August. Japan's official surrender took place aboard his new flagship, U.S.S. *Missouri* (BB-64) in Tokyo Bay on September 2. Halsey turned over command of Third Fleet to Admiral Howard Kingman in November and was promoted to fleet admiral the following month, assigned to special duty to the office of Secretary of the Navy until he retired in April 1947.

Following his retirement from the navy, Halsey held a number of executive and advisory positions in private business and found himself vigorously defending his sometimes reckless actions during the Philippines campaign. The critics aside, however, the admiral was greatly loved by the public, who saw Bull Halsey as a hero bigger than life, who put the navy on the offensive when other commanders might well have adopted a far more conservative policy.

HAMILCAR BARCA

BORN: ca. 285 B.C.

DIED: ca. 229 B.C.

SERVICE: First Punic War (264–241 B.C.); war against mercenary troops (241 B.C.); Invasion of Spain (237 B.C.)

Carthaginian general, father of Hannibal Barca and Hasdrubal

Hamilcar Barca is most famous as the father of **Hannibal Barca,** but he was also a great general whose military operations established Carthaginian rule in Spain. He was appointed general late in the First Punic War and was given command of the Carthaginian forces in Sicily in 247 B.C., during a time of grave crisis, when Carthage had already lost almost all of its Sicilian possessions. His response was to seize the initiative and undertake a bold offensive against the coastline of Lucania and Bruttium, seizing Mt. Pellegrino near Palermo. He fortified this strategic high ground and built a harbor for his fleet, which served as a base from which he could raid and generally harass the Italian coast.

In 250 B.C., Hamilcar left Mt. Pellegrino to stage a surprise assault on Roman forces at Mt. Eryx, splitting the Roman legions in two and gaining a great strategic advantage, which he exercised to the utmost, mercilessly hammering at the coastal settlements of Italy and Sicily for the next nine years, until 241 B.C., when massed Roman forces finally bested the Carthaginian fleet, cutting Hamilcar off from his lines of supply and communication. Having been authorized to treat with the Romans, he negotiated a favorable peace and began the march back to Carthage. His mercenary troops, who had long endured without pay, finally mutinied, and Hamilcar waged war against these elements of his own forces. To the prisoners he took he offered clemency in return for service in his regular army. This policy came to an end after the penultimate battle of the rebellion, during which the rebel leaders murdered and mutilated their Carthaginian prisoners.

Hamilcar's loyal forces killed more than ten thousand mercenaries, yet the hardened rebels regrouped and besieged Carthage itself. In countering the siege, Hamilcar demonstrated masterful generalship, driving the besiegers into a gorge and there annihilating them. Following this, Hamilcar was put at the head of an expedition to Spain in 237 B.C. intended to recover territory lost in the First Punic War. Moving swiftly, he retook the whole of southern Spain, establishing the city of Akra Leuke on the hills of Alicante to defend the newly conquered area. While withdrawing from siege operations against the town of Helice, however, Hamilcar drowned in the River Alebos.

HAMILTON, Henry

BORN: unknown

DIED: 1796

SERVICE: French and Indian War (1754–63); American Revolution (1775–83)

Infamous British colonel and colonial lieutenant governor of Canada who put a bounty on Patriot scalps during the American Revolution

Nothing is known about the birth and early life of Henry Hamilton. During the French and Indian War, he served in the Louisbourg campaign (May 3-July 27, 1758) and in Quebec from June to September 13, 1759. He was appointed lieutenant governor of Canada and assigned to command the British garrison at Detroit from 1775 to 1779 during the American Revolution.

During the Revolution, most Indian tribes believed that the British were the lesser of two evils—that their main interest was trade with the Indians, not usurpation of Indian land, whereas the Americans meant to keep moving west and plant settlements. Many Indians were, therefore, receptive to fighting on the side of the British, and Hamilton was both skilled and ruthless at recruiting and organizing Native American guerrilla action. The Indians came to call him "Hair Buyer" because he paid them a bounty on the Patriot scalps they brought to him as a result of the Ohio Valley raids he sanctioned.

On February 25, 1779, "Hair Buyer" was taken prisoner by **George Rogers Clark** during the Battle of Vincennes (in present-day Indiana). Hamilton was transported back east to Williamsburg, Virginia, where he was held for several months until he was paroled to New York. He was inactive throughout the balance of the war.

After the war, Hamilton served as lieutenant-governor of Quebec from 1784 to 1785 and then as governor of Bermuda (1790–94) and Dominica (1794–95).

HAMILTON, Sir Ian Standish Monteith

BORN: January 16, 1853

DIED: October 12, 1947

SERVICE: Second Afghan War (1878–80); First Anglo-Boer War (1881); Mahdist War (1883–98); Third Burmese War (1885–86); Tirab Campaign (1897–98); Second ("Great") Anglo-Boer War (1899–1902); World War I (1914–18)

British general best known for action in the Second (Great) Anglo-Boer War and at Gallipoli in World War I

Born on the island of Corfu, Hamilton joined the army in 1872, and subsequently transferred to the Ninety-Second Highlanders, with which he served in the Second Afghan War during November 1878–September 1880. This won him early attention from General Frederick Roberts, who noted his gallantry and calmness under fire. Sent next to South Africa in the First Anglo-Boer War, he served under General Sir George P. Colley at the Battle of Majuba Hill, where he was wounded and taken prisoner on February 27, 1881. He was soon released, but his wound permanently crippled his left wrist.

During the Mahdist War, Hamilton participated in the abortive Nile expedition to relieve **Charles George "Chinese" Gordon**, whose tiny garrison was under siege at Khartoum (September 1884–April 1885). Following this failed mission, Hamilton was posted to

India, where he saw action in the Third Burmese War during November 1885–January 1886. Some ten years later, now a colonel, Hamilton commanded a brigade in the Tirah campaign southwest of Peshawar on the Northwest Frontier during October 1897–April 1898. The next year, he was transferred to South Africa as chief of staff to General Sir George White, who was commanding troops in Natal.

With the outbreak of the Second Anglo-Boer War, Hamilton was put in command of the infantry brigade at Elandslaagte on October 21, 1900. The Boer forces laid siege to him at Ladysmith during November 2–February 28, 1900. During the siege, Hamilton incurred very heavy losses at the Battle of Wagon Hill (January 6), due to his failure to fortify his position. This disaster notwithstanding, Hamilton not only received a promotion to local lieutenant general, but Lord Roberts entrusted him with command of the division assigned to take Doornkop in the last major set-piece battle of the war (May 16). Following this battle, Hamilton was given command of another division, which was assigned to put down guerilla action during June-November. After Lord Roberts left Africa for Britain on November 29, Hamilton replaced him as chief of staff to Lord **Kitchener**.

During the remainder of the war, Hamilton continued to serve in the field as well as on Kitchener's staff. He won a signal victory at Rooiwal on April 11, 1902 and was among the dignitaries present at the May 31 peace treaty ceremony in Vereeniging.

With the conclusion of the war, Hamilton returned to Britain, where he was appointed quartermaster general of the British army, serving in this post from 1903 to 1904. With the outbreak of the Russo-Japanese War in 1904, Hamilton was made head of a military observer mission that accompanied the Japanese army in Manchuria through September 1905. He wrote eloquently about his experiences in China, publishing his memoir as *A Staff Officer's Scrap-Book* in 1907.

Hamilton was appointed commander of the Southern Command and Adjutant General during (1909–10) and used this position to oppose Lord Roberts's proposals for conscription. Thus, when Britain would enter World War I, it was with nothing more than a relatively small—albeit professional—army.

Appointed Commander in Chief in the Mediterranean in 1910, Hamilton was promoted to general at the outbreak of World War I, in 1914, and served first in Britain as Commander in Chief of the Home Defense army. However, in March 1915, he was assigned as commander of the Mediterranean expeditionary force and, on April 25, directed the landings at Gallipoli. The operation soon turned into a disaster—for which Hamilton subsequently shouldered much blame, but which was not entirely his fault. He received little support, lacked proper equipment, and had poor compliance from his subordinates. When reinforcements arrived early in August, Hamilton renewed efforts to push inland, but the campaign ended in a costly failure, which Hamilton steadfastly refused to acknowledge. He opposed withdrawal so vehemently that he was dismissed from command and recalled to Britain in October.

After Gallipoli, Hamilton held no further commands. When the war ended, he took up the cause of providing for the welfare of British veterans and wrote several memoirs as well as a book on the Gallipoli campaign.

A controversial figure, Hamilton was an officer of great intelligence and courage, but utterly inflexible and unwilling to cut his losses.

HANCOCK, Winfield Scott

BORN: February 14, 1824

DIED: February 9, 1886

SERVICE: U.S.-Mexican War (1846–48); Third Seminole War (1855–58); Civil War (1861–65); Northern Plains Indians War (1866–69)

Among the most able of the Union's Civil War generals; less effective during the Indian Wars

Hancock was born in rural Montgomery County, Pennsylvania, and graduated from West Point in 1844 with a second lieutenant's commission in the Sixth Infantry. He was sent to service in Mexico during the U.S.-Mexican War under his namesake, General

Winfield Scott, who had replaced **Zachary Taylor** as principal commander in the war. Hancock distinguished himself in the advance on Mexico City and was breveted to first lieutenant following the battles of Contreras and Churubusco (August 19–20, 1847).

After the U.S.-Mexican War Hancock served in a number of posts before being promoted to captain in November 1855. In December, he was dispatched to Florida to fight in the Third Seminole War (December 1855–May 1858), which consisted mostly of frustrating guerrilla exchanges.

During the turbulent period immediately preceding the Civil War, Hancock served in "Bloody Kansas" with units attempting to keep the peace between pro- and anti-slavery factions (1859–60). Hancock saw brief service in California during 1860–61, and returned to the East at the outbreak of the Civil War in 1861. He was appointed brigadier general of volunteers in September 1861 and served gallantly under General **George B. McClellan** during the Peninsula campaign, winning particular distinction at Williamsburg (May 5, 1862) and Seven Pines-Fair Oaks (May 31–June 1). Hancock was a divisional commander (Sumner's II Corps) at the pivotal Battle of Antietam on September 17.

In November 1862, Hancock was promoted to major general of volunteers and again distinguished himself, first at Fredericksburg on December 13 and then at Chancellorsville during May 2–4, 1863. At Gettysburg (July 1–3, 1863), he was in command of II Corps, and his defenses on Cemetery Ridge were key in repulsing the celbrated Confederate charge named for General **George Pickett** (July 3). Hancock was wounded in this engagement.

It was not until May 4–5, 1864, that Hancock again saw action, during the Wilderness campaign, at Spotsylvania (May 8–18), Cold Harbor (June 3–12), and at Petersburg (June 15–18). In August 1864, Hancock was promoted to brigadier general in the regular army and was assigned to command Washington's defenses.

After the war, in July 1866, Hancock was promoted to major general and given command of the Department of the Missouri (encompassing Missouri, Kansas, Colorado, and New Mexico) the following month. In this capacity, he directed operations against the Indians—mostly Cheyenne—in Kansas (1866–67). These actions were variously called Hancock's Campaign and Hancock's War and resulted in futile cavalry pursuits and inconclusive spasms of violence and reprisal. Following this, Hancock was replaced as commander of the department by **Philip Sheridan**.

Hancock was sent next to command the Department of Louisiana and Texas during 1867–68, the Division of the Atlantic in 1868–69, and the Department of Dakota from 1869 to 1872. At last, he was transferred back east as commander of the Division of the Atlantic and the Department of the East, a post he held from 1872 to 1886. During this period, he was responsible for executing many of the harsh and punitive Reconstruction policies of the Radical Republican. He found these repugnant and frequently disputed with politicians over them. At last, in 1880, he accepted the Democratic nomination for president, narrowly losing to Republican James A. Garfield. Following this defeat, he resumed his duties as commander of the Department of the East and, six years later, died at his desk in his Governor's Island, New York, headquarters.

Hancock was a fine commander and especially valuable in the early months of the Civil War, when valiant, skilled, and aggressive Union officers were a rarity. He was also a man of great moral fiber, who balked at using military might to enforce Reconstruction policies he considered harsh and unjust—although his critics accused him of excessive zeal in conciliating former slave owners at the expense of the rights and welfare of former slaves. His reputation in history is marred by his poor performance as commander of the Department of the Missouri. Ordered to create a strong military presence in Kansas, he provoked a futile Indian war.

HANNIBAL

BORN: unknown

DIED: 406 B.C.

SERVICE: Carthaginian resurgence in Sicily (409–406 B.C.)

First of the line of celebrated Carthaginian generals

Born in Carthage, Hannibal was the son of Gisco and the grandson of Hamilcar, prominent Carthaginian leaders and warriors. He was *suffete* (chief magistrate) of Carthage. When the Segestans requested Carthage to aid them in their struggle against neighboring Selinus in 410 B.C., Hannibal responded by organizing a large force during the winter and launching an amphibious assault at Lilybaeum (present-day Marsala) in Sicily during the spring of 409 B.C.. He laid siege against Selinus, capturing the city-state after nine days. His retribution against Selinus was severe and final: he killed everyone within the city and then tore down its walls.

Hannibal used the adventure into Selinus to begin an expansion of Carthaginian influence throughout Sicily. He besieged Himera (near Termini Imerese), which yielded only after a prolonged siege. He took more than 3,000 prisoners, whom he subsequently executed. After looting the city, he left a garrison there and returned to Carthage in the late summer.

Hannibal led a second expedition to Sicily, in partnership with his nephew Himilco, in 407 B.C. and laid siege against Acragas (modern Agrigento) the following year with a very large force. Hannibal's military career was cut short by his death from plague.

Hannibal was ancestor to the more illustrious **Hannibal Barca**, better known simply as Hannibal, who was born in 247 B.C.

HANNIBAL BARCA

BORN: 247 B.C.

DIED: 183 B.C.

SERVICE: Conquest of Spain I (237–220 B.C.); Second Punic War (219–202 B.C.)

Most famous of the three Carthaginian military leaders called Hannibal; the "Father of Strategy" who terrified Rome during the Second Punic War

Son of Hamilcar Barca, the Carthaginian general who had fought the Romans ably during the First Punic War, Hannibal traveled to Spain during his father's campaign there in 237 B.C., but returned to Carthage to finish his education after his father's death in 228 B.C. He then returned to Spain in command of cavalry under his elder brother Hasdrubal in 224 B.C. When Hasdrubal was assassinated in 221, Hannibal became overall commander of the army and immediately set out across northwest Iberia (modern Spain), which he pacified during two lightning campaigns in 221 and 220 B.C.

Following his victories in Iberia, Hannibal determined to exact vengeance on Rome for its victory in the First Punic War. His plan was to attack Italy itself, but he needed a strategy that would avoid the Roman-controlled Mediterranean; therefore, he staged a brilliant, epic overland campaign. He first took Saguntum, an Iberian ally city-state of Rome, which fell after an excruciating eight-month siege late in 219 B.C.

In 218 B.C., Hannibal left Iberia and invaded Gaul (July), evading and outmaneuvering the Roman legion led by Publius Cornelius Scipio the Elder at Massilia (modern Marseille). In August, he crossed the Rhone River and, by the fall, was confronted with the Alps. Undaunted, he organized and executed a spectacular Alpine crossing—elephants and all—entering Italy during September–October 218 B.C. He engaged the Roman cavalry and *velites* (lightly armed troops) at the Battle of the Ticinus in November, then trounced the main Roman force commanded by T. Sempronius Longus at the River Trebbia in December. Next came the army of G. Flaminius, against which Hannibal assumed the defensive, making a devastating surprise attack at Lake Trasimene in April 217 B.C.

At the spectacular battle of Cannae, on August 2, 216 B.C., Hannibal decisively defeated the superior forces of G. Terentius Varro and L. Aemilius Paulus, inflicting some 55,000 Roman casualties in one of the Legion's greatest military disasters. Hannibal's campaign against M. Claudius Marcellus in southern Italy, during November 216–June 214 B.C., ended inconclusively, and his assault against the citadel fortress of Tarentum (modern Taranto) failed in 213 B.C.

Hannibal marched next on Rome itself during the summer of 211 B.C., defeating two legions under General Fulvius Centumalus at Herdonea during the following summer. However, even Hannibal could not long sustain what he had gained. He gradually lost support among Roman colonial possessions in southern Italy and found himself the victim of the patient strategy of attrition employed by **Fabius Maximus Verrucosus**, known as "The Delayer." His situation worsened with the death of his younger brother Hasdrubal, who was coming to him with reinforcements, in 207 B.C. Finally, Carthage itself withdrew its support for his extended campaigning, and Hannibal at last withdrew to Africa to defend against the invasion of **Scipio Africanus** during the fall of 203 B.C.

Hannibal met defeat at the hands of Scipio and Masinissa at the Battle of Zama in the spring of 202 B.C. and negotiated peace with Rome. Elected *suffete* (magistrate) in 196 B.C., Hannibal instituted a series of reforms, for which he earned many enemies. Denounced to Rome by a group of them, he fled to Antiochus III the Great of Syria, offering to him his military service. He raised and commanded a Phoenician fleet against Rhodes in 190 B.C., but was defeated there by Eudamus of Rhodes and L. Aemilius Regilus at the Eurymedon. Once again, Hannibal was forced to flee, landing first at Crete and then at Bithynia (in modern Turkey). His situation desperate, he took his own life by swallowing poison.

For all his fame, Hannibal remains a shadowy figure. No Carthaginian chronicler recorded his accomplishments, which are known only from the records of his Roman enemies. What is clear is that he was one of history's greatest generals, a master of grand strategy and innovative tactics. He was at his best in adversity, when confronting a superior force.

HANNIBAL GISCO

ACTIVE: 264–260 B.C.

SERVICE: First Punic War (264-241 B.C.)

Least illustrious of the Carthaginian military leaders sharing the name of Hannibal; he was defeated at Mylae

Hannibal Gisco was the son of the Carthaginian general Gisco. He commanded a flotilla off Lipara (modern Lipari) when the Mamertine, rebels against Carthage, were besieged by Hiero II of Syracuse in 264 B.C. Hannibal placed a garrison at Messina, but it nevertheless fell to the Roman legion summoned by the Mamertines. Hannibal broke through Roman siege lines Agrigentum (modern Agrigento, Italy), leaving the city to fall; it was razed in the spring of 262 B.C.

Hannibal was made an admiral, and from his base at Panormus (modern Palermo, Sicily), he raided the southern Italian coast during 262–261 B.C. At Mylae, in 260 B.C., on the Sicilian coast, he was attacked by a Roman fleet under Gaius Duilius. The Romans infiltrated the Carthaginian formation and boarded the vessels. The fleet was dispersed, and Hannibal's flagship was captured, as were about half of his other ships. Hannibal himself managed to escape and fled to Carthage, from which he was again dispatched, this time in command of a fleet out of Sardinia. Roman forces bottled up the fleet in harbor, sinking most of the vessels. Enraged, Hannibal's own subordinate officers assassinated him.

HARDEE, William Joseph

BORN: October 12, 1815

DIED: November 6, 1873

SERVICE: Second Seminole War (1835–1843); U.S.-Mexican War (1846–48); Civil War (1861–65)

Confederate general in the Civil War who distinguished himself as a corps commander at Shiloh (1862), Murfreesboro (1862), and Missionary Ridge (1863), earning the sobriquet "Old Reliable"

Hardee was a native of Savannah, Georgia, and graduated from West Point in 1838 with a second lieutenant's commission. Posted to the Second Dragoons, he fought briefly in the long, sporadic Second Seminole War in 1840. He was sent in 1840 to France to study cavalry tactics.

Hardee was promoted to captain of dragoons on September 18, 1844, and, two years later, served with General **Zachary Taylor** in the U.S.-Mexican War. Captured early in the conflict, he was liberated in a prisoner exchange and fought at Monterrey during

September 20–24, 1846. When Taylor was replaced as overall commander by General **Winfield Scott**, Hardee fought at Veracruz (March 27, 1847) and at Mexico City (September 13-14). He acquitted himself with such distinction at these battles that he was breveted to major and then to lieutenant colonel.

After the war, Hardee reverted to the regular-army rank of major in the Second Cavalry and wrote *Rifle and Light Infantry Tactics*, which was published in 1855. The manual, universally called *Hardee's Tactics*, became a U.S. Army standard. In 1856, Hardee was promoted to lieutenant colonel and appointed commandant of West Point, in which post he served until the formation of the Confederacy.

In January 1861, Hardee resigned his commission in order to accept a colonelcy in the provisional Confederate army. By June, he had been promoted to brigadier general and was dispatched to Arkansas to form Hardee's Brigade. In September, he was sent to Kentucky and was promoted to major general before the end of the year. Commanding III Corps, Army of Mississippi, he fought at Shiloh (April 6–7, 1862) and Perryville (October 8). Following the latter battle, he was promoted to lieutenant general and led III Corps at Stones River (December 31, 1862–January 2, 1863).

In November, Hardee was in charge of the right flank of **Braxton Bragg**'s Army of Tennessee at the Battle of Chattanooga (November 24–25) and temporarily replaced Bragg as commander of the Army of Tennessee in December.

Hardee served as a corps commander during the Atlanta campaign (May–August 1864), leading a gallant surprise attack on James B. McPherson's Army of the Tennessee at Peachtree Creek (in present-day Atlanta) on July 20. After a fierce battle, he was repulsed. Friction developed between Hardee and the fiercely stubborn John Bell Hood, overall commander during the Atlanta campaign. Hardee sought and was granted transfer to command of the Department of South Carolina, Georgia, and Florida in September. He failed to stop **William Tecumseh Sherman**'s juggernaut advance through southern Georgia and withdrew from Savannah, Georgia on December 20, and Charleston, South Carolina in February 1865. After rejoining **J. E. Johnston**'s Army of Tennessee, he surrendered with that army at Durham Station, North Carolina in April. After the war, he retired to private life in Selma, Alabama.

Hardee was an excellent tactician and a brave commander. *Hardee's Manual* remained a standard in both the armies of the North and South and after the war as well.

HARRISON, William Henry

BORN: February 22, 1773

DIED: April 4, 1841

SERVICE: Indian Wars (1790–95, 1811–13); War of 1812 (1812–15)

American military commander, territorial governor, and short-lived U.S. president; called "Old Tippecanoe" because of his victory against Tecumseh's followers along the Tippecanoe River in Indiana Territory

Harrison was born at Berkeley plantation in Charles City county, Virginia, a son of Virginia politician Benjamin Harrison. He was educated at Hampden-Sidney College during 1787–90, then enrolled in the College of Physicians and Surgeons at Philadelphia. He did not receive his degree, however, because he left to accept an ensign's commission in the First Regiment at Fort Washington, Cincinnati in 1791. Promoted to lieutenant the following year, he was appointed aide-de-camp to **General Anthony ("Mad Anthony") Wayne** and participated in the victory against the Indians of the Old Northwest at Fallen Timbers on August 20, 1794. Promoted to captain in May 1797, Harrison was given command of Fort Washington until he resigned his commision in June 1798 to accept appointment as secretary of the Northwest Territory. The territory voted him their representative to the U.S. Congress in 1799, and the following year he was appointed governor of Indiana Territory.

During his long tenure as territorial governor, Harrison negotiated a series of treaties with Indians, securing the cession of vast tracts of land. As was

William Henry Harrison. Courtesy Library of Congress.

common practice in concluding treaties between the government and various Indian groups, Harrison identified pliable tribal "representatives" who agreed to the treaties—and then held the treaties as binding upon the entire tribe. Often, the majority actually disapproved of the agreements and repudiated them—a situation that inevitably led to violent conflict. During this period, the great Shawnee leader **Tecumseh** and his brother Tenskwatawa (known as "The Prophet") siezed upon a Harrison treaty to organize pan-tribal resistance against the incursions of white settlement.

On November 7, 1811, Harrison attacked Tecumseh's encampment along the Tippecanoe River while Tecumseh was absent on a recruiting expedition. Harrison's victory, although narrow, resulted in Tenskwatawa being discredited among Tecumseh's followers.

Harrison planned to capitalize on this victory, but had to focus instead on the War of 1812. Harrison angled for high command in the regular army, but was compelled to content himself with a commission as major general of the Kentucky militia after British forces captured Detroit in August 1812.

In September, Harrison marched to the relief of beleaguered Fort Wayne (in the present state of Indiana), then received promotion to brigadier general of regulars and an appointment as commander in chief in the Old Northwest region. In January 1813, Harrison moved against Fort Malden on Lake Erie. One of his generals, James Winchester, impatient for action, made a premature attack on January 22 along the Raisin River just south of Detroit. The result was a terrible rout, in which only thirty-three out of a force of 960 American troops evaded death or capture.

Following this disaster, Harrison had to take a defensive posture and set about fortifying the frontier with Fort Meigs (at Maumee, Ohio) and Fort Stephenson (at Fremont, Ohio) while he awaited reinforcements. Promoted major general in March 1813, he carefully rebuilt his army and, by fall, launched a major offensive, which resulted in the retaking of Detroit on September 29. Next, supported by the improvised Lake Erie fleet of **Oliver H. Perry**, Harrison engaged and defeated British and Indian forces at the climactic battle of the Thames on October 5. In this engagement, Tecumseh was slain.

Harrison resigned from the army in May 1814, settled in Ohio, and was elected to Congress, serving during 1816–19, when he was elected to the Ohio state senate. In 1824, he was elected to the U.S. Senate, serving until 1828, when he was appointed first U.S. minister (ambassador) to Colombia. In 1829, he resigned from that post to return to Ohio, where he lived in quiet semi-retirement until 1836, when he campaigned unsuccessfully for president against Martin Van Buren. He ran again in 1840, with John Tyler as his running mate, and won election by a substantial margin, having campaigned under the most famous slogan in American political history: "Tippecanoe and Tyler too!"

Unfortunately, Harrison caught a chill during his inauguration on March 4, 1841, fell ill with pneumonia, and died a scant month after taking office.

HASEGAWA, Yoshimichi

BORN: 1850
DIED: 1924
SERVICE: Restoration War (1868); Satsuma Rebellion (1877); First Sino-Japanese War (1894–95); Russo-Japanese War (1904–1905)

Japanese field marshal who gained the attention of the Western world during the Russo-Japanese War

Hasegawa was born in what is now the Yamaguchi prefecture of Japan and, during the Meiji Restoration War, he fought on behalf of the Choshu clan from January to March 1868. In 1871, shortly after the new government army was formed, he joined it with the rank of captain. By 1877, he was a major in command of a regiment during the Satsuma Rebellion (February 17–September 24, 1877). He participated in the relief of Kumamoto Castle on April 14.

Hasegawa traveled to France during 1885–86 to observe French military practices. The year of his return to Japan he was promoted to major general and given command of a brigade in the First Sino-Japanese War. At the Battle of Pyongyang, Korea, he won distinction for his valor and for the performance of the brigade (September 15, 1894). He and his unit acquitted themselves similarly at Haicheng during December 1894–January 1895.

With the outbreak of the Russo-Japanese War in the spring of 1904, Hasegawa was put in command of the Guards Division in General Kuroki's First Army. He fought with great distinction at the Battle of the Yalu on April 30–May 1 and was promoted to general in June. He was assigned command of the Korea Garrison Army during September 1904–December 1908 and served as chief of staff of the army from 1912 to 1915. A stern military traditionalist, he protested personally to the emperor in 1913 when the government revised regulations to allow service ministers to be chosen from reserve officers rather than exclusively from regular army officers. Hasegawa was promoted to field marshal in 1915.

Hasegawa was an excellent officer, who favored expansion of military control of the civilian government. In this, he anticipated the trend that would dominate Japanese politics throughout the first half of the century, culminating in World War II.

HASTINGS, Francis Rawdon-Hastings, Earl of Moira and First Marquess of (Lord Rawdon)

BORN: December 9, 1754
DIED: November 28, 1826
SERVICE: American Revolution (1775–83); French Revolutionary Wars (1792–99); Gurkha War (1814–16); Third Maratha War (1817–18)

One of the more capable commanders Britain sent to North America to fight the Revolution

Hastings was born at Moira, County Down, Ireland and enrolled at Oxford. While he was still a student there, in 1771, he was commissioned an ensign in the Fifteenth Regiment of Foot. After completing his education, he joined the regiment en route to Boston, Massachusetts, in July 1774, to put down the incipient rebellion there. Hastings's first engagement was at the Battle of Bunker (Breed's) Hill, where he led his company after its captain was shot. Hastings was also wounded in the June 17, 1775 engagement.

Hastings rose rapidly in the North American service, securing an appointment as aide-de-camp to Sir Henry Clinton and, later, as a staff officer to General William Cornwallis. As a field officer, Hastings fought throughout the central colonies and was singled out in dispatches for his actions at Long Island (August 27, 1776), White Plains (October 28), and Fort Washington, in present-day New York City, (November 16–20). By 1778, he had been promoted to lieutenant colonel and was chosen by Clinton to form a provincial regiment in June 1778. The Volunteers of Ireland, as the unit was called, fought at the battle of

Monmouth on June 28, then joined Clinton at Charleston, South Carolina, in April 1780.

Hastings and his regiment distinguished themselves at the Battle of Camden, on August 16, a major British victory in South Carolina. After Cornwallis stubbornly resumed the offensive in the wake of the defeat of his subordinate, Banestre Tarleton, at the Battle of Cowpens on January 17, 1781, Hastings was assigned to hold the British gains in South Carolina and Georgia. Hastings used a small field force to attack General **Nathanael Greene** at Hobkirk's Hill, winning a victory there on April 25. He also successfully relieved and evacuated the settlement of Ninety-Six, South Carolina, on June 19–20, but was stricken with illness and had to relinquish command to Colonel Paston Gould.

Hastings set sail for England on July 20, 1781, only to be captured by a French privateer and imprisoned in Brest. He was soon released and returned to England, where he was promoted to full colonel and made aide-de-camp to King George III in 1782. The following year, he was created Baron Rawdon, then added the surname "Hastings" to his own when his mother succeeded to the barony of Hastings in 1789. Hastings succeeded his father as Earl of Moira in 1793 and was promoted to major general in October of that year.

In 1794, Hastings was dispatched to lead troops in support of the Vendean insurgents during the French Revolutionary Wars. However, after the French victory at Fleurus on June 26, 1794, he and his force were rushed to Ostend, Holland, to reinforce the main British army. With great cleverness, he was able to elude the main French army and reach the Allies at Amsterdam.

Hastings was promoted to lieutenant general in 1798 and then general in 1803. The following year, he was named commander in chief in Scotland. In 1806, Hastings was appointed constable of the Tower of London and became a close friend and adviser to the Prince of Wales (subsequently George IV).

In 1812, Hastings was sent to India as governor general of Bengal and commander in chief on the subcontinent. He arrived at his new post in 1813 and was immediately confronted by Gurkha raids. Finally, on November 14, 1814, he declared war on Nepal and directed a successful campaign against the Gurkhas that resulted in a favorable and enduring peace by 1816. For this triumph, he was created Marquess of Hastings in 1817.

During 1816–18, Hastings next waged war against the piratical Pindari—mercenaries in the employ of various local rulers—and the Marathas. Hastings presented the Marathas with an ultimatum in November 1817: either lend assistance in the suppression of the Pindaris or face war. When the Maratha leaders refused aid, Hastings dispatched a force under Sir Thomas Hyslop to fight them. He was victorious in a key December 21 battle. Following this, Hastings turned his attention to the Pindaris and the Maratha holdouts. On June 2, 1818, the last of the hostiles surrendered.

Hastings was an excellent colonial administrator, who negotiated the purchase of Singapore in 1819 and who did much for the people of Bengal, building public works, rendering the ancient Mogul canal system operational again, and, most significantly, fostering education as well as a high degree of native Indian self-government. Nevertheless, he became involved in shady business activities and was recalled to England on January 1, 1823. The following year, he was appointed governor and commander in chief at Malta, where he died two years later.

HENRY II

BORN: March 5, 1133

DIED: July 6, 1189

SERVICE: Civil War with King Stephen (1135–54); Wars with King Louis VII of France (1157–80); Subjugation of Wales (1158–65); Invasion of Ireland (1169–75); War with Scottish Rebels and France (1173–74); War with King Philip II of France (1180–89)

Perhaps the greatest king of England; came to rule an Anglo-Norman domain of vast extent

Born at Le Mans, Henry was the son of Geoffrey Plantagenet, Count of Anjou, and Matilda, the daughter of King Henry I of England. Henry I had taken the first steps in forging for his kingdom an English, as opposed to Anglo-Saxon or Norman, identity. By granting royal charters to many English towns, he rescued the just-emerging merchant classes from domination by the barons, yet he ultimately fell short of establishing a full measure of reform and good government. Worse, his only son, William the Atheling, died before his father, and, while the nobles pledged to accept Henry's daughter, Matilda, as their queen, Stephen, Count of Blois, claimed the throne when Henry I died in 1135.

The usurper was by no means universally welcomed, and Matilda invaded England to claim her right of rule, igniting a civil war during which young Henry was taken to England to be educated. He went next to Normandy in 1144, which Geoffrey Plantagenet had just captured from Stephen, and, in 1147, Henry returned to England to campaign against Stephen. His first essay in generalship ended in disaster. Two years later, he allied himself with King David of Scotland and attacked Stephen again, once more meeting with defeat and only narrowly escaping with his own life.

In 1150, Henry became Duke of Normandy and succeeded his father as Count of Anjou the next year. He married Eleanor of Aquitaine, former wife of France's Louis VII, in 1152, and thereby came to control large territories of southern France. This gave him the power base he needed to attempt a third invasion of England in January 1153. Militarily, the Battle of Wallingford, fought in July 1153, was inconclusive, but it did garner for Henry a large measure of popular support. Putting him in the position to make a favorable peace, albeit an uneasy one. Through the mediation of Theobald, Archbishop of Canterbury, it was agreed that Stephen would continue to rule until his death, at which time Henry would succeed him.

The young warrior did not have long to wait. Stephen died in October 1154, and Henry II was crowned on December 19. During this period, England was only a part of Henry's realm—called the Angevin Empire—which included much of present-day France and, in size, was second only to the Holy Roman Empire among European states. But Henry is deemed an English king because it was upon England that his rule left its most enduring mark. He mounted an offensive in the north, recovering the northern regions of Northumberland, Cumberland, and Westmoreland from Scottish domination during 1157. From 1159 to 1165, he campaigned in Wales, temporarily subjugating this most rebellious realm, and in 1171 he annexed Ireland to the English crown. During this tumultuous period, from 1159 through 1174, he also prosecuted a campaign on the continent against his French rival, Louis VII.

As vigorous as Henry II was in military operations, he instituted perhaps even more significant changes on the civil front, beginning with the replacement of feudal service with a system of scutage, "shield money" paid to the crown in lieu of actual military service. Next, he reformed the system of English justice, concentrating all authority in the hands of royal circuit judges, thereby making the administration of law solely the responsibility of the crown. Moreover, he established the primacy of English common law traditions, enriched by Norman refinements, which ensured that the administration of justice would be governed by set principles rather than by the whim of a monarch. Among the most significant rights Henry introduced were trial by jury and the right of appeal. Such judicial reforms brought him into sharp conflict with his childhood friend Thomas à Becket, now Archbishop of Canterbury. Becket defended the supremacy of authority of church courts over that of royal justice, and the conflict escalated, with brief periods of reconciliation, during 1162–70. Henry II had sought to resolve the conflict once and for all by the Council of Clarendon in 1164, at which the bishops were called upon to pledge their obedience to the king. Rather than sign, Becket went into exile for six years, during which time he attracted a following among clergymen as well as the general populace, who saw in the Church a reassuring alternative to royal tyranny. When Becket returned to England in 1170, he inflamed a Christmas audience with a sermon in which he excommunicated certain knights loyal to Henry. He also predicted his own martyrdom. Henry, fighting in France at this time, declared in exasperation, "Will no one rid me of this turbulent priest?" Four of the king's knights took this as a royal commission to assassinate Becket.

Henry was devastated by the deed and took upon himself full responsibility for it, making a public act of contrition and undertaking acts of penance prescribed by the Pope. He retained his power following the death of Becket, but the assassination triggered rebellion among the English barons in 1173, which coincided with warfare against France and the Scots. Henry hastened to France, where he successfully fought off attacks on Normandy and Anjou during the summer and fall, returning without pause to England, where he countered a Scots invasion, capturing the Scots King William the Lion at the Battle of Alnwick on July 13, 1174.

With seemingly inexhaustible vigor, Henry quelled internal rebellion and external threats, but nevertheless neglected the revolt simmering within his own family. He endowed his sons Richard (later, as **Richard I,** called the Lion-Hearted), Geoffrey, and John (later King John) with lands and titles, yet sternly withheld funds from them. To avoid the kind of strife that had followed the death of Henry I, he anointed his eldest son, Henry, king. Young Henry repaid this act by rebelling against his father in 1183, forcing Henry II to put down the revolt. His rebelling son died of a fever soon afterward. Painful as this was, Henry II made matters yet worse by formally recognizing the right of his next son, Richard, to succeed him, even while he openly demonstrated favoritism to Geoffrey and John. Henry II's wife, Eleanor of Aquitaine, encouraged Richard and Geoffrey to act against their father, and, toward this end, Richard allied himself with the new king of France, **Philip II Augustus.** Together, Richard and Philip began a war of rebellion against Henry II, a rebellion secretly joined by John.

In 1189, the forces of Henry II were beaten back at the Battle of Le Mans. The king fell ill and called for a truce, ultimately agreeing to generally unfavorable peace terms, whereupon he retired to his castle in Chinon. There his condition worsened, and it was on his deathbed that he was told of the treachery of John, his favorite son. He died, on July 6, 1189, in the arms of his youngest son Geoffrey, muttering final words worthy of a Shakespeare history play: "Shame, shame on a conquered king."

HENRY V

BORN: September 16?, 1387

DIED: August 31, 1422

SERVICE: Hundred Years' War (1337–1453); Glendower's Revolt (1402–1409); Revolt of the Percies (1403–1408)

Probably England's most beloved king; his inspired military leadership brought stability to the nation and expanded England's French domains

Henry was born at Monmouth, the first son of Henry, Duke of Lancaster (later King Henry IV). When his father was forced into exile by King Richard II in 1399, Henry V was taken into Richard's household. His father returned later in the year, and Henry V fought in his father's Irish campaign. When Henry IV was crowned king of England on October 15, 1399, Henry V was made Prince of Wales, and from 1402 to 1409, Henry V fought against Welsh rebels led by Owen Glendowner.

Henry V often disagreed with his father's policies. After 1408, he frequently found himself opposing Henry IV and his ministers. Nevertheless, the two maintained good relations, and, on March 20, 1413, following the death of his father, Henry V assumed the throne. The first crisis confronting him was a rebellion instigated by a radical Christian sect called the Lollards, who believed in the supremacy of individual conscience and who collaborated with others in a political revolt that boded anarchy. They had been moderately active during the reign of Henry IV, and now Henry V waged full-scale war on them, ruthlessly crushing resistance during December 1413–January 1414 and finally executing their leader, Sir John Oldcastle.

With the Lollards suppressed, Henry was now confronted with a new revolt, this one among a group of nobles who wanted to place Edmund Mortimer, Earl of Monmouth, on the throne. Henry launched another vigorous military campaign, which crushed the rebellion by July 1415.

While attending to domestic matters, Henry remained acutely aware of foreign affairs. The political situation in France during this period was even more chaotic than what prevailed in turbulent England. Taking advantage of the disarray, Henry declared war on France, claiming that Normandy was, by absolute right, an English possession. He sailed for the continent on August 10, 1415, and besieged the fortress town of Harfleur during August 13–September 22. Henry was at the head of a handpicked professional army drawn from men of mixed background and loyalties, including Welshmen, Irishmen, Gascons, and others. While the king trusted their fighting ability absolutely, he mainlined their political loyalty by enforcing upon them an iron discipline. The result was victory, when Harfleur surrendered on September 22. Henry left there an English garrison. While he permitted most of the town's inhabitants to remain, he demanded ransom payments from the well-to-do, and he exiled the aged, infirm, and very young, whom he deemed useless mouths to feed. For this action, he was respected by some and reviled by others. It was, however, in keeping with the military practice of the time.

Leaving Harfleur, Henry marched with 900 men-at-arms and some 5,000 archers to Calais, reaching an area just south of the Somme by October 1515. There French constable Charles d'Albret assembled a vastly superior army of 35,000 against Henry's 5,900. The Battle of Agincourt, on October 25, seemed as if it would be hopelessly one-sided. However, Henry commanded his forces with inspiring calmness and skill, while the French managed theirs ineptly. Most important were Henry's long-bowmen, who exacted a heavy toll on the lumbering French artillery and on mounted knights arrayed in ranks so tightly packed that maneuvering was impossible. The French lost between 3,000 and 8,000 men at Agincourt, while the English suffered a mere 250 casualties. This key battle, one of the most famous in European history, marked the end of the heavily armored medieval knight as an effective agent of warfare. It also elevated the "common" soldier to paramount importance in modern warfare.

Agincourt ended when Henry made another harsh military decision. The French had launched—and lost—two of the three cavalry waves at their disposal when a band of raiders suddenly attacked the English baggage train at Henry's rear. The king realized that, distracted by the raiders, his troops were now quite vulnerable to the third wave of French cavalry. Worse, Henry held so many French prisoners, that the attackers might well act in concert with them. Victory could still belong to the French. Acting swiftly, Henry ordered that all prisoners be put to death. At this, the last wave of knights withdrew from the field.

Following Agincourt, Henry conquered all of Normandy during 1417–19. Rouen fell to him after a siege that lasted from September 1418 to January 1419. He marched on Paris in May 1420 and made an alliance with Burgundy that resulted in the Treaty of Troyes, by which the weak French king, Charles VI, acknowledged Henry V heir to the throne of France. For the present, Henry was declared regent. With this power in his belt, the king marched on to subdue northern France, laying siege to and capturing Meaux during October 1421 to May 1422. This was the high-water mark of Henry's reign, as England and France stood on the verge of union.

Unfortunately, years of unremitting warfare took their toll on the English monarch. He soon succumbed to a bout of dysentery and died at Bois de Vincennes. His successors were unable to hold what he had gained in France.

HERACLIUS

BORN: ca. 575

DIED: February 11, 641

SERVICE: Persian War (603–28); Avar War (617–21); Arab War (634–42)

Byzantine emperor who crushed the Persians in a series of brilliant campaigns during 622–28, temporarily forestalling the disintegration of the empire

Heraclius was born of Armenian lineage in eastern Anatolia. While his father (also called Heraclius) was serving as governor of the Roman province of Africa during the reign of Emperor Phocas, important factions in Constantinople prevailed upon the elder Heraclius to rescue the crumbling empire from misrule. The father sent the son at the head of an army to overthrow Phocas. Heraclius landed in Constantinople in October 610, deposed the emperor, and was himself crowned.

It was a dubious office. For Byzantium was torn by civil strife and pressured from without by Slavic, Persian, and Turkish raiders. The empire's frontiers were fluid and violent, while the empire's economy was strained by payment of exorbitant tributes demanded by the invaders and raiders. Recognizing that this situation could not long stand, Heraclius began his reign by reorganizing the imperial bureaucracy, strengthening the armies, and generally reforming the administration of government.

In 614, the Persians took Syria and Palestine, capturing, among other prizes, Jerusalem and appropriating the holiest of holy Christian relics, a fragment of the True Cross. In 619, Egypt and Libya fell to a Persian army of occupation. However, Heraclius could not counter these moves until he had protected the central empire against invasion by the Avars. Some time during 617–619, he negotiated with them at Thracian Heraclea, but they soon broke the truce, and Avar tribesmen attempted to take Heraclius prisoner. The emperor outrode his pursuers all the way back to Constantinople, and it was not until 622 that Heraclius finally brought about peace with the Avars.

Long in coming, peace with this tribe allowed Heraclius to direct his resources against the Persians. He organized a massive campaign aimed at recovering Jerusalem and the Cross. It was, de facto, the first Crusade, and the emperor fought brilliantly, driving the Persians out of Anatolia. At this point, Heraclius offered a truce to the emperor Khosrow II, who rejected the terms and, stupidly, confirmed the Byzantine resolve by insulting Christ and Christianity. Heraclius had little trouble raising and motivating an army for a two-year holy campaign that began in Armenia as the staging area for a massive invasion of the Persian empire.

Heraclius personally commanded his army. In 625, while he and his troops were camped on the west bank of the Sarus River in Anatolia, his men spied the Persian forces on the east bank. They made an unauthorized charge across the bridge, and the Persians emerged from ambush. They were on the verge of annihilating Heraclius's forces, when the emperor took up his sword, advanced to the bridge, personally struck down the leader of the Persian troops, and rallied his men for a devastating assault. The Persians counterattacked at the Bosporus the following year, hoping to unite with the Avars in an assault on Constantinople, but Heraclius sunk the Persian fleet, which left Persian land forces with neither transport nor supplies. With Persians stranded, the Avar assault was unsupported and failed.

Late in 627, Heraclius took the offensive and invaded Persia, meeting the Persian forces in a great battle near Nineveh. There he killed three generals in one-on-one combat and led his troops deep into the Persian lines. He next personally killed the chief commander of the Persian forces. Leaderless, they disintegrated. Early the next year, Heraclius entered the capital city of Dastagird. At his approach, Khosrow II was overthrown by his son, who willingly concluded peace terms favorable to the Byzantines. These included redemption of all captives, the return of captured Roman lands, and the return of the Holy Cross. The relic Heraclius triumphantly bore back to Jerusalem in 630 and reinstalled in the Church of the Holy Sepulcher.

Heraclius recognized that conquest was one thing and sustaining his gains was quite another. He believed that the key to maintaining the empire intact was to unite the fractious Christian world by conciliating the diverse theologies of Egypt, Syria, and Armenia, all of which had traditionally been subject to persecution at the hands of Christianized Roman emperors. The effort to reconcile all these faiths proved more exhausting than combat, and Heraclius failed. He was in ill health and near collapse in 634 when Arab forces invaded Syria. In his reduced state, he was unable to assume personal command of his army and had no choice but to entrust leadership to other commanders. The Arabs dealt the Byzantines a severe defeat at the Battle of Yarmuk in 636, and, in that single stroke, Heraclius lost all he had won in Syria and Egypt. The emperor hastily ordered the removal of the True Cross fragment from the Church of the Holy Sepulcher and evacuated it.

In his withdrawal from the Middle East, the emperor, fearless in battle, fell victim to a water phobia. He delayed an entire year on the Asian side of the Bosporus, fearful of crossing the strait. At last, his troops constructed a pontoon bridge screened with leaves that hid the water from view. Heraclius crossed this into Constantinople, but there grew increasingly debilitated with a painful illness apparently related to enlargement of the prostate. He was a brilliant administrator and one of the world's great generals. Unfortunately, the precipitous decline of the Byzantine Empire was beyond even his managing.

HILL, Ambrose Powell

BORN: November 9, 1825

DIED: April 2, 1865

SERVICE: U. S.-Mexican War (1846–48); Operations against the Seminoles (1855–58); Civil War (1861–65)

Confederate general called by Robert E. Lee "the best soldier of his grade with me"

A native of Culpepper, Virginia, Hill graduated from West Point in 1847 and was commissioned a second lieutenant in the artillery in time to see service in the U.S.-Mexican War. After the war, he was stationed in Texas, then fought in policing action against the Seminoles during 1855. From 1855 to 1860, as captain, he participated in the U.S. Coastal Survey, Washington, D.C.

Like many other Southern-born officers, Hill resigned his commission during secession, in March 1861. He was made a colonel in the provisional Confederate army and given command of the Thirteenth Virginia Regiment at First Bull Run on July 21. In February 1862, he was promoted to brigadier general and distinguished himself at Williamsburg on May 5. By the end of the month, on the 26th, Hill was promoted to major general and fought valiantly at Fair Oaks five days later.

During the so-called Seven Days, Hill led his division in very heavy fighting from June 26 to July 2, and he supported **Thomas "Stonewall" Jackson** at Cedar Mountain on August 8. At the end of the month, he fought at Second Bull Run (August 29–30). Hill was the spoiler at Antietam (September 17), leading a forced march to reinforce Confederate lines and thereby deprive the Union army of what would have been its first decisive victory. At Fredericksburg, on December 13, Hill had charge of the right flank at Fredericksburg.

Wounded at Chancellorsville (May 2–4, 1863), Hill was promoted to lieutenant general and commander of the newly formed III Corps on May 24. He led this unit into battle at Gettysburg during July 1–3 and at Bristoe Station on October 14. He was also a key commander at the Battle of the Wilderness on May 5–6, 1864 and at Spotsylvania Court House during May 8–18.

Hill was a frontline commander, who believed in leading from the front. He was cut down while rallying his troops during the defense against the Union assault on Petersburg, April 2, 1865. He was the last major Confederate commander killed in battle.

A. P. Hill was a thoroughly remarkable soldier. Tireless, he fought in many of the Civil War's major battles. He meted out stern discipline, but treated his men with great fairness and, because of his example of courage under fire, was universally respected by all in his command. Like the best Confederate commanders, he specialized in rapid movement and flexible deployment.

HILL, Daniel Harvey

BORN: July 12, 1821

DIED: September 24, 1889

SERVICE: U.S.-Mexican War (1846–48); Civil War (1861–65)

Courageous but contentious and gloomy Confederate general who nevertheless earned the adoration of his command

Born in the York District of South Carolina, Hill graduated from West Point in 1842 and was commissioned in the artillery. He fought in the U.S.-Mexican War, in which he distinguished himself at Contreras and Churubusco (August 19–20, 1847), earning a brevet to captain. After Chapultepec (September 13, 1847), he was breveted again, to major.

Hill was a man of complex nature, who craved the excitement of battle, yet was also intensely introspective and intellectual. He resigned his commission in 1849 to become a mathematics professor at Washington College (now Washington and Lee University), then moved on to Davidson College, North Carolina, in 1854. In 1859, he was given an opportunity to combine his military and academic experience as superintendent of the North Carolina Military Institute.

Hill rejoined the military at the outbreak of Civil War, becoming the colonel of the First North Carolina Regiment at Big Bethel on June 10, 1861. By September, he was promoted to brigadier general, and in March 1862 was made major general. Hill fought at Fair Oaks on May 31, 1862, and, at South Mountain, fought a holding action that gave **Robert E. Lee** time to concentrate his army behind Antietam Creek. In the ensuing Battle of Antietam, Hill led his division in very heavy fighting on September 17. On December 13, he fought at Fredericksburg, and the following year was in charge of the Richmond defenses.

In July 1863, Hill was promoted to lieutenant general commanding a corps in the Army of Tennessee under **Braxton Bragg.** After serving with Bragg at Chickamauga during September 19–20, Hill accused him of incompetence and was largely responsible for his removal from command. However, Hill was himself temporarily relieved as well. When he returned to duty, it was as divisional commander at Bentonville, North Carolina during March 19–21, 1865. It was his final command of the war.

After the conclusion of the Civil War, Hill retired to Charlotte, North Carolina, where he became involved in publishing and in writing religious tracts. From 1877 to 1884, he served as president of the University of Arkansas, then moved to the presidency of Middle Georgia Military and Agricultural College (now Georgia Military College), a position he held until his death.

HILL, Sir Rowland, First Viscount

BORN: August 11, 1772

DIED: December 10, 1842

SERVICE: French Revolutionary Wars (1792–99); Napoleonic Wars (1800–15)

"Daddy Hill" was a British general best known for his service during the Napoleonic Wars; Wellington deemed him one of his best subordinates

Hill was a Shropshireman, born at Hawkstone, the second son of Sir John Hill. He joined the Thirty-Eighth Regiment as an ensign in 1790, but transferred to the Fifty-Third Regiment the following year. During 1791–93, he studied at the Strasbourg Military School, then went on to an appointment as aide-de-camp during the siege of Toulon (August 27–December 19, 1793). Following this action, he was promoted to major in the Ninetieth Regiment, which he led in the Egyptian campaign against **Napoleon** in 1801. He was wounded at Aboukir (March 20–21).

Following his return from Egypt, Hill was dispatched to Ireland with the rank of brigadier general in 1803, then served as major general in the Hanover expedition of 1805. He commanded a brigade in the Peninsula at Rolica on August 17, 1808, Vimeiro (August 21), and Corunna (January 16, 1809). During the Peninsula campaign he fought at Oporto on May 12, 1809 and commanded the Second Division at Talavera on July 28. In 1810, he commanded a corps defending the Portuguese border.

Illness sidelined Hill for a time in 1810–11, but he returned to active duty in Spain, commanding the Second and Fourth Divisions in Extremadura, in order to cover **Wellington's** right flank at the siege of Badajoz (May 7, 1811–April 6, 1812). He scored a victory at Arroyomolinos on October 27, then took the offensive against the French positions at Almarax on May 19, 1812. Hill was in command of the right wing at Vitoria on June 21, 1813 and blockaded Pamplona from June 22 to October 31. He took the offensive again when he broke through the French lines at the Nivelle on November 10 and then along the Nive River (November 12).

Hill defeated **Nicolas Soult** at Saint-Pierre on December 13, then went on to additional victories at Orthez (February 27, 1814) and Toulouse (April 10). He was created Baron Hill of Almaraz and Hawkstone Salop.

In 1815, Hill was dispatched to Brussels to supervise the mobilization of the Prince of Orange's troops, and Hill commanded a corps at the climactic Battle of Waterloo on June 18. It was he who led a critical counterattack against the French Imperial Guard.

Following the final exile of Napoleon, Hill was put in charge of the Army of Occupation in Paris (1815–18), then retired. He returned to duty, however, in 1828, when he was named commander in chief of the army, and he served in this most senior post until his death in 1842.

HINDENBURG, Paul Ludwig Hans von

BORN: October 2, 1847

DIED: August 2, 1934

SERVICE: Seven Weeks' War (1866); Franco-Prussian War (1870–71); World War I (1914–18)

German field marshal, hero of World War I, and president of Germany's Weimar Republic

Born in Posen, Prussia, Paul von Hindenburg was the son of a Prussian military officer and was raised in an atmosphere steeped in his nation's military traditions. His career formally began when he was only eleven years old and was enrolled in the famed Wahlstatt military school. Later, as a young officer, he participated in numerous conflicts, including the Austro-Prussian war of 1866 and the Franco-German War of 1870–71. It was during the latter war that Hindenburg earned his reputation as a superior strategist, securing a post on the staffs of German field marshals **Helmuth von Moltke** and **Alfred von Schlieffen** in 1878. Having served his country with distinction, Paul von Hindenburg retired from the army in 1911, with the rank of general.

It was destined to be a brief hiatus. The events of August 1914—the outbreak of World War I—brought Hindenburg command of the Eighth Army division in the East. His skills as a strategist proved brilliantly effective, as he decisively defeated the advancing Russian armies at the Battle of Tannenberg in August 1914 and then at Masurian Lakes the following month. Victory in Poland at the battle of Lodz led to Hindenburg's appointment as chief of the general staff in 1916. He was now in command of all of Germany's armed forces.

The cornerstone of Hindenburg's strategy was to defeat France and England before the Americans could join the conflict. But, while Russia was clearly losing the war, the conflict on the Western Front had reached a terrible stalemate. To break it, Hindenburg authorized a campaign of submarine warfare against Great Britain, intended to blockade and starve that nation into surrender. Unfortunately for Hindenburg's strategy, this policy, helped propel the United States into the war, especially after the torpedo attacks on the British passenger vessels *Sussex* and *Lusitania* in 1916 (with the loss of American lives) and the formal declaration of unrestricted submarine warfare in 1917. In essence, submarine warfare *did* break the Western Front stalemate—by bringing the Americans to bear, and the result, of course, was not a German victory, but a defeat and the nation's acceptance of the humiliating terms of the Treaty of Versailles.

Such was the nature of World War I that the German military was not popularly blamed for the defeat. Rather, it was a series of politicians that took the blame, and when Hindenburg again retired from military service in 1919, he was revered as a hero. In 1925, he was elected president of the second Weimar republic and served two terms. The aging Hindenburg could not, however, reconcile the factions of a country plagued by punitive peace terms, torn by political unrest, and worn down by the economic effects of a worldwide depression. In desperation, a faltering Hindenburg looked to the charismatic Adolf Hitler for support. Personally, he detested the bellicose Nazi leader, but he conceded to him ever greater measures of authority, at last appointing him chancellor of his second cabinet in 1933. With Hindenburg by this time senescent, Hitler was *de facto* leader of Germany, and after Hindenburg's death the following year on August 2, he became the country's unquestioned dictator and Paul von Hindenburg's most devastating legacy to Germany and the world.

HITLER, Adolf

BORN: April 20, 1889

DIED: April 30, 1945

SERVICE: World War I (1914–18); World War II (1939–45)

German founder of the Nazi Party; dictator and conqueror of Europe, infamous for brutality and genocidal slaughter

The man who would dominate not only Germany but much of Europe, and who would bring an unprecedented degree of darkness and terror to the world came from a background of banal squalor. He was born in the Austrian town of Braunau am Inn, but was raised mainly in Linz, the son of a minor customs official. Alois Hitler, an illegitimate child, had used his mother's maiden name, Schickelgruber, until 1876, when he took the name Hitler. He was a brutal father, who was particularly put out by what he regarded as Adolf's penchant for day dreaming. Indeed, the boy was unfocused and proved a poor student, leaving secondary school in 1905 without a diploma. He had a restless but desultory ambition to become an artist, but his uninspired drawings and watercolors twice failed to gain him admission to the Academy of Fine Arts in Vienna. After the death of his mother, whom he idolized, Hitler went to Vienna anyway, hoping to make a living as an artist. From 1907 to 1913, he eked out a marginal existence by painting advertisements, postcards, and the like. He became reclusive and moody, his depression animated solely by a developing set of racial hatreds, focused most intensely on the Jews, whom Hitler decided were a threat to the Germanic—or "Aryan"—race.

Adolf Hitler moved to Munich in 1913, apparently to escape conscription into the Austrian army. He was nevertheless recalled to Austria in February 1914 for examination for military service, only to be rejected as medically unfit. Yet, in August 1914, with the outbreak of World War I, Hitler rushed to enlist in the Sixteenth Bavarian Reserve Infantry (List) Regiment. War service transformed the lackluster youth into a passionately militaristic nationalist. Short and slight, he served in the front lines as a runner, achieving promotion to corporal, and earning four decorations, including the Iron Cross First Class on August 4, 1918. He was seriously wounded in October 1916 and was gassed at the end of the war.

Hitler remained with his regiment until April 1920, serving as an army political agent and joining the German Worker's Party in Munich in September 1919. In April 1920, he left the army to go to work fulltime for the party's propaganda section. This was a period of intense crisis in Germany. The Treaty of Versailles, which ended World War I, was heavily punitive, the German economy was in shambles, and the nation was rocked by an abortive Communist revolution. Seizing on the unrest, Hitler agitated to transform

Adolf Hitler. Courtesy National Archives.

the German Worker's Party by August 1920 into the *Nazionalsozialistische Deutsche Arbeiterpartei*, commonly shortened to NSDAP or Nazi party. The emerging party leader forged an alliance with Ernst Röhm, an army staff officer, and, with his aid, was elected president of the party in July 1921. Hitler proved a riveting street corner orator, assaulting all those he identified as Germany's enemies, especially Communists and Jews, as well as the nations that had forced Versailles upon the German people. Perhaps swayed by the power of his own rhetoric, Hitler was emboldened during November 8-9, 1923, to lead the Munich Beer Hall Putsch, an attempt to seize control of the Bavarian government. The rebellion was quickly suppressed, however, and Hitler was arrested, tried, and convicted of treason. Not wishing to create a martyr, the tribunal gave Hitler a light sentence: five years in the relative comfort of Landesberg prison, near Munich. There he wrote his political autobiography, *Mein Kampf* ("My Struggle"), the crystallization of the Nazi philosophy and a manifesto of hatred directed against Jews, Communists, effete liberals, and exploitive capitalists the world over. *Mein Kampf* also presented a digest of popular doctrines of German racial superiority and purity, positing the nation's unstoppable will. Hitler wrote of a Germany that would rise to become the dominant power in the world, a nation that would rightly claim its *Lebensraum*—living space—in central Europe and in Russia.

Hitler was released from prison after serving only nine months of his sentence, and he immediately set about strengthening his party, especially in the industrial German north. He recruited the men who would lead the country into mass atrocity and all-consuming war: **Hermann Göring**, popular World War I air ace; Josef Goebbels, master propagandist; Heinrich Himmler, skilled in strongarm, terror, and police tactics; and Julius Streicher, a popular anti-Semitic journalist. With a theory and personnel in place, the party was ready to seize the moment. And that came in 1929, with the worldwide economic collapse and the depression that followed. Germany, already groaning under the harsh sanctions of Versailles, now reeled on the edge of revolt. Forging an alliance with the Nationalist party headed by industrialist Alfred Hugenberg, the Nazis increased the number of Reichstag seats they held from 12 to 107, becoming the second largest party in Germany. Nor did Hitler confine his party's activities to the Reichstag. He created the SA (*Sturmabeteilung*, or Brownshirts) as his party's paramilitary arm—an organized mob of the unemployed and the discontent, who literally beat down the opposition in the streets of Germany.

Adolf Hitler ran for president of the German republic in 1932, narrowly losing to the incumbent **Paul von Hindenberg**, ancient hero of World War I. The July elections gained the Nazis 230 Reichstag seats—37 percent of the vote—making it the largest party represented, and Hindenberg, weary and given to fits of outright senility, had no choice but to appoint Hitler *Reichskanzler* (Reich Chancellor, or prime minister) on January 30, 1933.

Hitler moved to consolidate his power. When fire destroyed the Reichstag on February 27, 1933, he found a pretext for legally abolishing the Communist party and imprisoning its leaders. Next, on March 23, 1933, he engineered passage of the Enabling Act, which granted him four years of dictatorial powers. These he used to systematically dismantle all German parties, except for his own NSDAP, to purge Jews from all government institutions, and to bring all government offices under the direct control of the party. He then turned on his own ranks, purging them of potential rivals during the so-called Night of the Long Knives, June 30, 1934, murdering his ally and mentor Ernst Röhm and hundreds of other Nazis whose radicalism posed a threat to Hitler's absolute domination. Shortly after this, in August 1934, Hindenberg died and Hitler assumed the functions of the presidency, but he took the title of *Führer*—Supreme Leader—of the Third Reich.

The Führer disbanded the SA—Brownshirts—and replaced them with the more strictly military SS—*Schutzstaffel*, or Blackshirts—under the command of the trusted Heinrich Himmler. Together with a secret police organization called the Gestapo, the SS rapidly built a network of concentration camps to which political enemies, Jews, and other "undesirables" were "deported." In 1935, Hitler enacted the Nuremberg Racial Laws, which deprived Jews of citizenship. These outrageous policies of terror and persecution were carefully orchestrated by propaganda minister Josef Goebbels as aspects of programs of economic recovery. It was a recovery financed by a headlong program of rearmament on a massive scale and in complete defiance of Versailles. Hitler created the *Luftwaffe* (air force) under Hermann Göring, remilitarized the Rhineland (in 1936), and multiplied his land forces in size.

In October 1936, Hitler struck an alliance with **Benito Mussolini**, Fascist dictator of Italy. In March 1938, Hitler made his first major military move, invading and annexing Austria in the *Anschluss*, then menacing Czechoslovakia into relinquishing the Sudetenland, a border region Germany long coveted. In the face of all this aggression, the two major western European military powers failed to act, seemingly paralyzed into a policy of craven and cowering "appeasement." In 1935, England agreed to an Anglo-German Naval Pact, then, at the Munich Conference of September 29–30, 1938, France and England agreed to the dismemberment of Czechoslovakia, feeling that this would appease the Führer. Hitler quickly annexed not only the Sudentenland, but the remainder of western Czechoslovakia as well, then went on to claim the "Memel strip" from Lithuania in March 1939. After concluding a non-aggression pact with **Josef Stalin** of the Soviet Union on August 23, 1939, Hitler invaded Poland on September 1.

Thus World War II began, and history came to know the tactic of *Blitzkrieg*—lightning war. Hitler overran Poland, Denmark, and Norway during April 9–June 9, 1940. France and England belatedly responded, but France fell quickly, during May 25–June 25, 1940. Only Great Britain held out, preparing to repel an anticipated invasion. But preparatory to mounting a proposed mass invasion of England, Hitler sought air supremacy, and during July–October 1940, the Luftwaffe and the Royal Air Force fought the Battle of Britain, in which the German military was dealt its first reversal. Failing to conquer the British skies, the Germans were forced to abandon their British invasion plans.

Nevertheless, during the war's early years, such Allied victories were rare. Hitler's armies controlled territory from North Africa to the Arctic and from France to central Europe. In April 1941, the German army invaded the Balkans, occupying Yugoslavia and Greece. Then, on June 22, abrogating the Nazi-Soviet non-aggression pact, Hitler executed "Operation Barbarossa": the invasion of the Soviet Union. Victories were quickly forthcoming in this vast country until, like **Napoleon** before him, the Russian winter, combined with the dogged resistance of the Russian people and their army, stalled Hitler's forces, first outside Moscow in December 1941, then, during the winter of 1942–43, at Stalingrad. The Russians began to exact a tremendous toll on the German army, draining Hitler's resources.

Even though Germany's Russian offensive was showing signs of its eventual collapse, Hitler earned a reputation as a military genius, the invincible leader of invincible forces. In fact, he was not a great military strategist, and he was often at odds with his advisors. Worse, following the Japanese attack on Pearl Harbor on December 7, 1941, the United States declared war on Japan and Germany—a contingency for which Hitler had not planned. The Führer also turned much of his attention to consummating the "Final Solution" to the "Jewish Question" and instituted the Holocaust: the genocide of some six million Jews, mostly in concentration camps now transformed into death camps expressly designed for mass murder. Hitler was no longer a political genius or a great military leader. He was an obsessed and desperate mass murderer.

By 1943, the tide of the war was turning hopelessly against Germany. The ruinous retreat from Russia was well under way, North Africa was lost, and Mussolini's regime had fallen to the Allied invasion of Italy. American and British bombers pummeled German cities day and night, and in June 1944, the Allied D-Day operation commenced, troops were landed on the coast of France, and the invasion of western Europe had begun. In the face of these defeats, the Führer made increasingly desperate, reckless, and irrational military decisions. On July 20, 1944, a cabal of his top officers attempted to assassinate him with a bomb hidden in a briefcase. The plan miscarried, and, although the bomb went off, Hitler survived—seriously wounded and emotionally devastated. In a series of show trials, he condemned the plotters to grisly execution (they were hanged from piano wire).

From December 16, 1944, to January 1945, Hitler committed his last reserves to a final offensive in the Ardennes, hoping to arrest the Allied advance and retake Antwerp. The result was the Battle of the Bulge, costly to Germany as well as to the Americans, but it was the Fürher's last gasp, and the offensive was crushed.

Hitler retreated to his *Fürherbunker*, a hardened underground command shelter in Berlin. From this headquarters, he ordered combat to the death of the last man, civilian as well as military. Finally, on April 29, 1945, as American, British, and Free French forces closed in from the west and the Russian army approached from the east, Hitler hastily married his long-time mistress, Eva Braun. The next day, the couple committed suicide. Admiral **Karl Dönitz**, whom Hitler had appointed as his successor, sued for peace, and the Third Reich dissolved in ashes and blood.

HO CHI MINH

BORN: May 19, 1890

DIED: September 3, 1969

SERVICE: World War II (1940–45); Indochinese War (1946–54); Vietnam War (1964–75)

First president of the Democratic Republic of Vietnam (North Vietnam); a skilled guerrilla strategist

Ho was born Nguyen That Thanh (and also called Nguyen Al Quoc). His father was an impecunious scholar in the village of Kim Lien, and the boy was raised in severe want. He was educated at the grammar school in Hue and became a schoolmaster for a time, then was apprenticed at a technical institute in Saigon. He left Vietnam (then called French Indochina) in 1911 to work as a cook, first on a French ocean liner and then at a London hotel. With the end of World War I, he moved to France, where he became an enthusiastic Socialist. During the 1919 Paris Peace Conference ending World War I, he unsuccessfully petitioned on behalf of civil rights in Indochina. Rebuffed here, he became increasingly radicalized and founded the French Communist party. He traveled to the Soviet Union to study revolutionary methods and was inducted into the Comintern, the international Communist organization controlled from Moscow, which assigned him the formidable task of spreading Communism throughout East Asia. Ho did not balk. He founded the Indochinese Communist party in 1930, and lived for the rest of that decade in the Soviet Union and China.

With the commencement of World War II, Ho Chi Minh returned to Vietnam, where, in 1941, he organized the Communist-controlled League for the Independence of Vietnam, or Viet Minh, which became the focus of the resistance movement against Japanese occupation. During the war, despite a period of imprisonment by the anti-Communist Nationalist Chinese in 1942–43 (during which he adopted "Ho Chi Minh"—He Who Enlightens—as his name), Ho formed a relationship with the American OSS, precursor of the CIA, which worked with him to develop a Vietnamese underground and guerilla movement to fight the Japanese. Ironically, after the war, this network would become the core of Communist resistance, first to colonial French domination and then to U.S. efforts to overthrow the North Vietnamese regime during the Vietnam War.

At war's end, on September 2, 1945, Ho Chi Minh proclaimed the independence of the Democratic Republic of Vietnam and became its first president. For the next quarter-century, he served as president of a divided and embattled people. He led the Viet Minh in eight years of guerrilla warfare against French colonial forces from 1946 to 1954, creating a remarkably efficient guerrilla government. Although Ho was active in directing military policy, he left tactical and strategic matters to General **Vo Nguyen Giap**.

Ho and Giap decisively defeated the French at Dien Bien Phu in 1954, then devoted themselves to another fifteen years of battle against the anti-Communist South Vietnamese regime, the capital of which was established in Saigon pursuant to a 1954 conference in Geneva. Beginning about 1959, the United States became involved in this struggle as part of its Cold War policy of "containing" Communism wherever it threatened to take over a nation. The U.S. began by lending advisory and material support to the South Vietnamese, and, gradually, becoming more deeply involved in a direct military sense. By 1969, 500,000 American G.I.s were fighting in Southeast Asia.

Throughout the Vietnam War, Ho Chi Minh advocated—and symbolized—unity under Communism for the two Vietnams, no matter what the cost. While his active role in the war steadily decreased beginning in 1959—when he began to suffer from the ill health that would plague his last ten years—Ho's message motivated his followers to endure incredible hardships and losses. While the Viet Cong, successor to Ho's Viet Minh, was highly skilled in guerrilla tactics, it was their willingness to die in great numbers that ultimately won the war for Communism. The United States had determined to wear down the Viet Cong through attrition, but the Viet Cong were prepared to suffer attrition, even to accept repeated military defeat in order to achieve political victory. Ho Chi Minh did not live to see that victory, which came in 1975, when U.S. force totally withdrew from South Vietnam. That hardly mattered. A hands-on political leader, he was a military leader as well by virtue of spirit and the example of sacrifice to an idea.

HOOKER, Joseph

BORN: November 13, 1814

DIED: October 31, 1879

SERVICE: Second Seminole War (1835–43); U.S.-Mexican War (1846–48); Civil War (1861–65)

"Fighting Joe" Hooker was a capable Union commander during the Civil War, but also irascible and contentious

A native of Hadley, Massachusetts, Hooker graduated from West Point in 1837 and was commissioned a second lieutenant of artillery. His first combat assignment was in the Second Seminole War, and he made first lieutenant in November 1838. During the U.S.-Mexican War, he was assigned as a staff officer, first under General **Zachary Taylor** and then **Winfield Scott**, who replaced Taylor. In combat during the U.S.-Mexican conflict, he demonstrated conspicuous gallantry, for which he received three brevet promotions, ending the war as a regular lieutenant colonel.

Following the war, from 1849 to 1851, Hooker served in the Division of the Pacific, taking a leave of absence at the end of this period, then resigning in 1853 to take up farming in Sonoma, California.

Hooker enjoyed little success as a farmer and left agriculture to become superintendent of military roads in Oregon from 1858 to 1859, when he was appointed colonel of the California militia. In May 1861, at the outbreak of the Civil War, he was made brigadier general of volunteers, then moved to the East Coast to become divisional commander in III Corps under **George McClellan** during the Peninsula campaign. He fought at the siege of Yorktown during April 5–May 4, 1862 and at Williamsburg on May 5.

Promoted to major general of volunteers, he transferred to the Army of Virginia and led a division at Second Bull Run during August 29–30, 1862. In September, he became commander of I Corps, Army of the Potomac and fought at South Mountain on September 14. Three days later, he was wounded at Antietam, but recovered to command the center Grand Division (a combination of III and V Corps) at Fredericksburg on December 13.

After General **Ambrose Burnside** proved a disastrous failure as commander of the Army of the Potomac, Hooker was called upon to replace him. He set about effectively reorganizing the force, yet was nevertheless soundly defeated by **Robert E. Lee** at Chancellorsville during May 2–4, 1863 and asked to be relieved.

Hooker was reassigned as commander of XI and XII Corps in September, serving under William S. Rosecrans and **Ulysses S. Grant** at the Battle of Chattanooga, where he redeemed himself as an excellent field commander at Lookout Mountain on November 24. For his action in this engagement, he was breveted major general of regulars. Hooker next marched with **William Tecumseh Sherman** in the Atlanta campaign during May–August 1864. However, after he was passed over for command of the Army of the Tennessee, he asked to be relieved. Instead, he was given command of the Northern Department (1864–65). At the end of the war, he was in charge of the entire Department of the East, taking command of the Department of the Lakes in 1866. Two years later, Hooker retired with the permanent rank of major general.

Hooker's reputation has always been somewhat controversial. Perhaps this is because, while he was a highly capable tactician and field commander, a man of unquestionable bravery and aggressiveness, who also knew how to administer and organize an army, he was not a great strategist and did not fare particularly well in independent command. He was most effective as a high-level subordinate—but his desire for higher command led to fits of temper and threats of resignation.

Sam Houston. Authors' collection.

HOUSTON, Sam(uel)

BORN: March 2, 1793

DIED: July 26, 1863

SERVICE: War of 1812 (1812–15); Texas War of Independence (1835–36)

Architect of Texas independence from Mexico; he was a statesman and a highly effective general

Houston was born near Lexington, Virginia and was raised there and, later, in Tennessee. He had scant formal schooling and, as a boy, could have been a model for Mark Twain's Huckleberry Finn. He resisted "civilization," running off to live with the Cherokee Indians rather than submitting to a clerkship at a local store (1808-11). However, after his sojourn among the Cherokee, Houston set up briefly as a schoolteacher, then, in 1813, was commissioned an ensign in the army. He served in under **Andrew Jackson** during the Creek War phase of the War of 1812 and participated in Jackson's victory at Horseshoe Bend (March 28, 1814), which effectively ended the Creek threat.

Houston took to army life and served as a military subagent during the first phase of the removal of the Cherokees to Indian Territory (modern Oklahoma) in 1817. The following year, he was promoted to first lieutenant, but shortly after this his career was cut short after he was scolded by Secretary of War John C. Calhoun for appearing before him in Indian dress and for making an inquiry that called into question Calhoun's official integrity. Angered by Calhoun's affront, Houston abruptly resigned his commission in May 1818 and turned to the study of law.

Sam Houston was admitted to the bar in Nashville, Tennessee late in 1818. He became intensely interested in Tennessee politics and was easily chosen major general of the state militia in 1821. Two years later, he was elected to Congress, in which he served until 1827, when he was elected governor of Tennessee. Reelected in 1829, he resigned in April of that year after his bride of three months deserted him. Deeply dejected, he moved west, where, once again, he took up residence among the Cherokees, who formally adopted him into their tribe in October 1829. Houston volunteered to act as their representative and traveled to Washington, D.C., several times to appeal for better treatment for them and other Indians. President Jackson commissioned Houston to negotiate with several tribes in Texas during 1832.

While Houston was in Texas, he became increasingly involved in local politics and participated in the San Felipe Convention in April 1833, which drew up a constitution as well as a petition to the Mexican government for Mexican statehood. Houston decided to settle in Texas and was soon appointed commander of the small Texas army (November 1835).

When Texans received no satisfaction from the Mexican government, they met in another convention, which declared independence on March 2, 1836. Houston was a participant in the convention, and he mobilized his army—740 strong—after the Alamo fell to Mexican forces under **Antonio Lopez de Santa Anna** and Colonel Fannin's force at Goliad were captured and massacred.

In one of the most extraordinary feats in American military history, Houston led his tiny army to victory over Santa Anna's vastly superior force of 1,600 professional soldiers at the Battle of San Jacinto on April 21. Not only was the Mexican force routed, but the Texans suffered few casualties—Houston was wounded in the leg—and Santa Anna himself was taken prisoner the next day. A condition of Santa Anna's release was that he recognize Texas independence. The war was won.

Houston was twice elected president of the Republic of Texas, from September 1836 to December 1838, and from 1841 to 1844. When Texas was finally admitted to the Union, Houston was elected senator in 1846 and was reappointed to the Senate in 1852. Houston was a Democrat, but his unwavering support of the Union cost him support in the Texas legislature and ensured that he would not be returned to the Senate in 1858, on the eve of secession and civil war. He was, however, elected governor in 1859 and tried without success to block his state's secession from the Union in 1861. When he then refused to swear allegiance to the Confederacy in March 1861, he was removed from office and retired to Huntsville, Texas, where he died two years later.

Sam Houston is one of the most colorful and thoroughly admirable men in American history. He combined tolerance, integrity, honesty, and the spirit of public service with courage and natural ability as a military commander.

HOWARD, Oliver Otis

BORN: November 8, 1830

DIED: October 26, 1909

SERVICE: Civil War (1861–65); Nez Percé War (1877); Bannock War (1878)

American general who lost his arm in the Civil War and who fought vigorously during the Indian Wars

Howard was born and raised in rural Leeds, Maine and graduated from Bowdoin College in 1850, then attended West Point, from which he graduated in 1854. Commissioned a second lieutenant, he was assigned as an artillerist, then returned to West Point as a mathematics instructor. With the outbreak of the Civil War, he was appointed colonel of the Third Maine Volunteers in June of 1861. By the time of the first Battle of Bull Run, July 21, 1861, he was a brigade commander, and, in September, brigadier general of volunteers.

Howard served under **George B. McClellan** in the Peninsula Campaign, fighting at Fair Oaks on May 31, 1862, where he lost his right arm. Returning to combat, Howard fought at South Mountain on September 14 and at Antietam on September 17. In November 1862, he was promoted to major general of volunteers and divisional commander. He led II Corps, at Fredericksburg on December 13, then was assigned command of XI Corps on April 2, 1863.

At Chancellorsville, during May 2–4, 1863, XI Corps was routed by **Thomas "Stonewall" Jackson**, but Howard redeemed himself amply at Gettsburg (July 1–3), when his unit performed with sufficient distinction to merit the special thanks of Congress.

In September 1863, Howard transferred to the Army of the Cumberland, where he won distinction at Chattanooga during November 24–25. In command of IV Corps (April 2, 1864), he served under **William Tecumseh Sherman** in the Atlanta campaign (May–August), then was elevated to command of the Army of the Tennessee in July. Serving under Sherman again, he accompanied the general on his march of destruction through Georgia and the Carolinas during September–December 1864. In December, Howard was promoted to brigadier general of regulars and in March 1865 breveted to major general.

Following the war, Howard was very active in the Reconstruction government and was particularly concerned for the welfare of the liberated slaves. He was appointed commissioner of the Freedman's Bureau, the agency charged with assisting African Americans, serving in this capacity from May 1865 through June 1872. In 1867, he founded (and later became president of) Howard University in Washington, D.C. The institution remains the foremost center of higher education for African Americans in the nation.

During the 1870s, Howard was also closely involved in Indian affairs. He negotiated with **Cochise** for the return of the Chiricahua Apaches to their reservation in 1872, and, two years later, returned to active military duty as commander of the Department of the Columbia. In this capacity, he unsuccessfully negotiated with a faction of the Nez Percé Indians for their removal from lands desired by the government. After the collapse of negotiations, Howard directed an exhausting military campaign against the band and its leader, Chief **Joseph** the Younger. The pursuit of the Nez Percé consumed five months, beginning in June and ending on October 5, 1877, at the Battle of Bear Paw Mountain, where Howard's subordinate **Nelson A. Miles** forced a surrender. (It is a measure of Howard's essential integrity that, following the defeat of Chief Joseph, he petitioned the president and Congress for permission to resettle the Nez Percé on the land they had wished to occupy. The petition proved unsuccessful.)

During 1878, Howard campaigned against the Bannock Indians, who were raiding in the Northwest. Following the intensive and exhausting campaign over some of the most rugged and unforgiving terrain on the continent, Howard returned to an assignment back east, becoming superintendent of West Point in January 1881. He served in this post until September 1882, when he was appointed commander of the Department of the Platte (1882–86) and then the Division of the East (March 1886–November 1894).

Belatedly, in 1893, Howard was honored with the Medal of Honor for action at Fair Oaks during the Civil War. He retired the following year and founded Lincoln Memorial University in Tennessee in 1895. He then returned to New England to write military history and an autobiography.

Howard was a thoroughly remarkable man, who combined philanthropy, profound religious belief, and a passion for education with valor and skill as a military commander.

HOWE, Richard, Earl

BORN: March 8, 1726

DIED: August 5, 1799

SERVICE: War of the Austrian Succession (1740–48); Seven Years' War (1756–63); American Revolution (1775–83); French Revolutionary Wars (1792–99)

Extraordinary British admiral, who fought in the American Revolution and in the French Revolutionary Wars; brother of General Sir William Howe

London-born Howe was the elder brother of William Howe, British general in the French and Indian War and the American Revolution, and younger brother of another officer, George Augustus Howe. Richard entered the Royal Navy in 1740, and first served aboard H.M.S. *Severn* in **George Anson**'s circumnavigation of the globe (1740–44); his ship did not complete the voyage.

Howe's skill was recognized early, and he enjoyed rapid promotion, becoming captain of the sixty-gun H.M.S. *Dunkirk*, aboard which he distinguished himself in the first naval action of the Seven Years' War (1755) when he captured the French *Alcide*, a vessel of sixty-four guns. In the Atlantic service during the balance of the war (1756–63), he against distinguished himself, under Admiral Edward Hawke, at the battle of Quiberon Bay on November 20, 1759.

Upon the death of his elder brother George Augustus near Fort Ticonderoga, New York (July 6, 1758) during the French and Indian War, William became Viscount Howe and, in 1762, was elected to Parliament for Dartmouth. He was elevated to membership on the Admiralty Board, serving in 1763 and 1765, and became treasurer of the Navy (1765–70). Promoted to rear admiral (1770) and vice admiral (1775), he was appointed commander in chief of the North American station. With his brother Sir William Howe, he was appointed peace commissioner (February 1776), charged with negotiating terms to end the revolt of the American colonies. When negotiations collapsed, he provided key naval support for the capture of New York by Sir William Howe (July 2–September 12, 1776).

A dispute involving another peace commission prompted his resignation from his North American post. However, he remained on station long enough to protect New York and to checkmate a combined French and American assault against Newport, Rhode Island in August 1778.

Howe returned to active duty when Lord Sandwich left office as First Lord of the Admiralty in March 1782, then took up command of the Channel Fleet. His ships undermanned and outnumbered (thirty-three to forty-six), he nevertheless relieved the siege of Gibraltar in October 1782.

Howe was made First Lord of the Admiralty, serving during January 28–April 16, 1783 and again, from December 1783 to August 1788. Created Baron and Earl Howe in the Irish peerage in 1788, he resumed sea duty again as commander of the Channel Fleet when war commenced with revolutionary France in 1793. He attacked the French convoy escort fleet of Admiral Louis T. Villaret de Joyeuse in a battle that became known as the Glorious First of June (May 29–June 1, 1794). The French escort was defeated—although the food convoy made it to France.

Howe retired after the Glorious First, but returned to play a role in ending the Spithead mutiny (April 16–May 15, 1797); he was one of the few high officers the sailors trusted.

HOWE, Sir William, Second Earl

BORN: August 10, 1729

DIED: July 12, 1814

SERVICE: French and Indian War (1754–63); American Revolution (1775–83)

Commander in chief of the British forces during the American Revolution; brother of Admiral Richard Howe

Born in London and educated at Eton, Howe was commissioned a cornet in the Duke of Cumberland's Light Dragoons on September 18, 1746. He was promoted to lieutenant the following year and transferred to the Twentieth Regiment of Foot in January 1750. He rose rapidly, becoming lieutenant colonel of the Fifty-Eighth Foot under Sir Jeffrey Amherst at the siege of Louisbourg (Nova Scotia) during June 2–July 27, 1758.

On the death of his elder brother George Augustus at Fort Ticonderoga in July 1758, William succeeded him in Parliament as member for Nottingham. He remained in North America, however, commanding a light infantry battalion under General **James Wolfe** in operations against Quebec during June–September, 1759. It was Howe who led the ascent to the Plains of Abraham during the night of September 12–13, which positioned the British army for the final assault on Quebec City and the defeat of French general the **Marquis de Montcalm**, which spelled the end of French dominance in North America.

Howe commanded a brigade during General James Murray's advance on Montreal in September 1760 and, the following year, led a brigade in the siege of Belle Isle. During June–August, 1762, Howe was adjutant general at the siege of Havana, then, in 1764, became colonel of the Forty-Sixth Foot in Ireland.

In 1768, Howe was appointed lieutenant governor of the Isle of Wight and was promoted to major general in 1772. Two years later, he perfected a new system of light-infantry training, which was universally adopted by the army, and in 1775, he was dispatched to America as second in command under General **Thomas Gage.** Howe he arrived in Boston on May 25, 1775, and directed the attack on Bunker (Breed's) Hill. Although he personally led the June 17 assault valiantly, the British victory was very costly. Nevertheless, it was Gage who was relieved, and Howe succeeded him to command of the army in Boston on October 10.

On March 17, 1776, Howe made a strategic withdrawal from Boston to Halifax, Nova Scotia, from which he organized and launched the army component of a joint army-navy invasion of New York, supported by his brother, Richard Howe. Landing on Staten Island on July 3, William Howe crossed over to Long Island during August 22–25, where he outflanked and defeated the American army under General Israel Putnam. When General **George Washington** took over from Putnam on August 27, Howe defeated him as well and went on to capture New York City on September 12. At this point, however, Howe's pursuit of Washington became lumbering, and Washington temporarily checked it at Harlem Heights, near the northern tip of Manhattan on September 16. Howe defeated Washington at White Plains on October 28, capturing Fort Washington (in modern New York City) and Fort Lee (New Jersey) during November 12–16). It was for the capture of New York that Howe was later knighted.

Following the collapse of New York, William and Richard Howe attempted to negotiate peace terms with the colonists. This failing, William Howe sailed from New York on July 23, 1777, to capture Philadelphia. He landed at Elkton, Maryland, on August 25 and marched against Philadelphia, destroying resistance at Cooch's Bridge (Newark, Delaware) on September 2, and defeating Washington at the Battle of Brandywine Creek on September 11. Philadelphia fell on September 26, and Howe easily turned back an ill-conceived attack at Germantown (in modern Philadelphia) on October 4. Supported by his brother's fleet, Howe took Forts Mifflin and Mercer (both near Philadelphia, on either side of the Delaware) during October 22–November 20.

Howe wintered in Philadelphia, awaiting acceptance of his April 14, 1778, letter of resignation. Both he and his brother had officially resigned over a dispute concerning a peace commission. On May 25, Howe turned over command to Sir Henry Clinton and returned to England.

Praised for having captured New York and Philadelphia, Howe was also criticized for having failed to coordinate with and support General **John Burgoyne**—although he had received no orders to do so. Nevertheless, Howe was promoted to lieutenant general of ordnance in 1782 and, in 1793, to general. During the French Revolutionary Wars, he held several important commands within England, and in 1799 succeeded to Richard's earldom upon his brother's death.

By the beginning of the 1800s, Howe was suffering from a painful and debilitating disease, which caused him to resign his ordinance post in 1803 and generally retire from military service.

Howe was a skilled commander and a good man, who was greatly admired by his troops. He was in many ways typical of liberal British attitudes toward colonial independence. To be sure, he was dedicated to his duty as a soldier, but he sympathized with the colonists. The fact that his heart was not in the cause for which he was fighting may have contributed to his lack of overall success in opposing the rebellion.

HULL, Isaac

BORN: March 9, 1773

DIED: February 13, 1843

SERVICE: Quasi-War with France (1798–1800); Tripolitan War (1801–1805); War of 1812 (1812–15)

U.S. naval commander of the frigate *Constitution,* ***which he led to victory in the War of 1812***

Born in Derby, Connecticut, Hull was the son of a militia captain and nephew of Revolutionary War general William Hull. He first saw service at sea as a cabin boy in 1787, and by 1792 was master of a merchant vessel.

After being twice captured by French privateers in the West Indies during 1796–97, he joined the U.S. Navy with a commission as fourth lieutenant aboard the forty-four-gun frigate *Constitution* on March 9, 1798. Hull quickly distinguished himself during the Quasi-War with France during 1798–1800. He captured the French privateer *Sandwich* near Puerto Plata harbor.

In 1803, Hull was given command of the sixteen-gun *Argus* and provided naval support for land forces taking Derna during the Tripolitan War (April 27, 1805). Hull was made captain of the *Constitution* on June 17, 1810 and was dispatched in that vessel to Europe to deliver loan payments to Holland and to escort U.S. minister Joel Barlow to France in 1811. Hull returned to the United States shortly before the outbreak of the War of 1812.

The commander's first encounter during the war was against a far superior British squadron off Egg Harbor, New Jersey. Pursued for sixty-six-hours during July 17–20, 1812, he evaded capture. The next month, off Barnegat Bay, New Jersey, he sank the British thirty-eight-gun frigate *Guerrière* in a battle that was over within a half hour on August 19. Because the British vessel's cannonballs seemed to bounce off the hull of *Constitution*, the vessel was nicknamed "Old Ironsides," and it is by that name that she—the oldest continuously commissioned vessel in the U.S. Navy—is still known.

Later in the war, Hull was given command of harbor defenses in New York City, and in 1815 was appointed one of the first three members of the new Board of Naval Commissioners. He was also appointed commandant of the navy yards at Boston and Portsmouth. From 1824 to 1827, he commanded the forty-four-gun frigate *United States* as commodore of the Pacific Squadron. In 1829, he served ashore once again, as commandant of the navy yard at Washington, D.C. He left this post in 1835 and, three years later, became commander of the Mediterranean Squadron, serving until July 1841. He died two years later in Philadelphia.

Hull was consistently a first-rate sailor and proved an excellent administrator as well. His most celebrated triumph, over *Guerrière*, was especially significant because, in a stroke, it shattered the myth of Britain's invincibility at sea, and at lifted the morale of the nation at a particularly dark period of the War of 1812.

HUNG HSIU-CH'ÜAN

BORN: January 11, 1811

DIED: June 1, 1864

SERVICE: Taiping Rebellion (1850–64)

Chinese rebel leader

Hung was born into a farm family in Kwangtung (Guangdong) province, and, his abilities recognized early, he was educated at the expense of the Hakka clan so he could enter the civil service. However, Hung failed four times to pass the civil service examination—in 1828, 1836, 1837, and 1843—and, after his third failure, fell dangerously ill. During his sickness, he had visions in which God told him to create a Heavenly Kingdom of Great Peace. His fourth failure prompted him to take to the road as an itinerant preacher. Soon his religious message was blended with advocacy of a socialist utopia. By 1844, his preaching became very popular among the poor, and within three years, he had established a utopian community of 3,000 in the mountains of Kwangtung. By 1850, the community had grown to 10,000. Members of the community observed the community's law, and its religious basis was a blend of Christianity and Confucianism.

By the time the movement had reached significant proportions, the government began to crack down with a program of harassment and outright persecution. Hung responded by proclaiming a full-scale rebellion on January 11, 1851. His small army was intelligent and highly dedicated. They took the city of Yungan on September 25, 1851. When government troops besieged the city, Hung's forces broke through on April 3, 1852 and moved into the Yangtze valley, where they spread their doctrine throughout the countryside, attracting thousands of new adherents.

Now known as the Taipings, they moved against Wuhan, which fell in January 1853 to an army numbering nearly half a million. After the fall of Wuhan, the Taipings captured Nanking during March 19–21. Hung set up his headquarters-cum-capital in Nanking and sent out from it expeditions north (1853–55) and west (1856–63). These failed to win new adherents. As to Hung's utopian ambitions, they were very imperfectly realized. He crushed an uprising among erstwhile supporters in the fall of 1856, then retired from the administration of his "Heavenly Kingdom," leaving it in the hands of his two brothers, who were men of middling ability. Hung's leading general, Li Hsiu-ch'eng, used military means to preserve the state Hung had founded, but Hung grew increasingly distant and disconnected from events. When Imperial forces besieged Nanking, Hung failed to rally his people or lead resistance. He remained passive until the city's food ran out, and then he committed suicide on June 1, 1864.

The Taiping Rebellion extended far beyond Hung's direct control. Not only did it come close to toppling the Ch'ing Imperial government, it grew into an engulfing conflict involving sixteen of China's eighteen provinces. Perhaps as many as ten million persons were killed during one of the worst civil wars in world history.

HUNYADI, Jänos

BORN: ca. 1387

DIED: August 11, 1456

SERVICE: Hussite Wars (1419–23); sporadic border fighting with Turks (various dates); Civil War (1439–40); Turkish War (1441–43); Varna Crusade (1443–44); Turkish War (1448–49 and 1455–56)

Hungarian general and national leader who introduced modern warfare—strategy and weaponry—into Hungary

Hunyadi was the son of a magyarized Vlach clansman named Vojk and a Magyar woman, Elizabeth Morsina. As a teenager, Hunyadi entered the service of King Sigismund of Hungary (last of the Luxemburg dynasty of Holy Roman emperors) and fought in at least one of Sigismund's three unsuccessful expeditions against the Hussites, followers of the Czech religious reformer John Huss, who were striving for Bohemian nationalism. Hunyadi gained his early fame, however, for capturing the Turkish-held fortress of Semendria in 1437.

As a reward for his services, King Albert (who succeeded Sigismund on his death in 1437) granted Hunyadi several estates and a seat on the royal council. He was also appointed *ban* (governor) of Szörény in western Wallachia, and he distinguished himself in border fighting against the Turks in the Szörény region.

When Albert died in 1439, Hunyadi supported the royal claims of Ladislas I (King Wladyslaw III of Poland) against those of Ladislas V (Ladislas Posthumus, infant son of Albert). Hunyadi's actions brought victory to Ladislas I's partisans during the civil war of 1439–40, and Hunyadi was rewarded with the captaincy of the fortress of Belgrade and made *voivode*—lord—of Transylvania. He then secured Ladislas' blessing for a number of campaigns against the Turks, defeating them at Semendria in 1441 and at Hermannstadt and the Iron Gates the following year.

In 1443, Hunyadi pressed deeper into Ottoman territory, capturing Nish and Sofia, then uniting his forces with those of King Ladislas I to win a signal victory over Sultan Murad II at Snaim (1443). Following this, Hunyadi's army drove the Turks out of the modern regions of Albania and the southern Yugoslav territory during the fall of 1443 to the early spring of 1444. Hunyadi formulated a grand plan to eject the Turks from all of Europe, but Murad sued for peace, and Hunyadi concluded a ten-year truce with him in 1443.

Hunyadi next joined Ladislas in creating an alliance with Venice. This concluded, he violated his truce with Murad and, in 1444, led the Hungarian army and fleet down the Danube. However, the Venetian fleet failed to block Murad's return to Europe from Asia in July 1444, and George Brankovic of Serbia supported the Sultan, thereby preventing the Albanian national hero, Skanderbeg (George Castriota) from joining the Hungarians at Varna. Thus deprived of allies, Hunyadi's forces were badly beaten by the Turks at Varna (November 10), and Hunyadi himself barely escaped death or capture. Ladislas I was not so fortunate and was killed in the battle.

Upon Ladislas' death, Hunyadi was elected governor of Hungary and regent for young King Ladislas V (the son of Albert). The boy was, however, held by the Austrians, and, in 1446, Hunyadi invaded Austria to force Frederick III to release him. In the process, his forces ravaged Styria, Carinthia, and Carniola, finally threatening Vienna itself. At this point, Frederick sued for peace and struck a two-year truce, though Ladislas V remained captive.

In 1448, Hunyadi once again resumed war against the Ottomans. His army of 25,000 was severely defeated at Kossovo by a vastly superior force of Turks—some 100,000 men—led by Sultan Murad II on October 17. The following year, Hunyadi did conduct a successful punitive expedition against George Brankovic.

In 1450, Hunyadi finally succeeded in negotiating the release of Ladislas V, and he peacefully surrendered the Hungarian regency to him two years later. In return Ladislas V made Hunyadi Count of Bestercze and captain general of the kingdom. Yet Ladislas would not fund military campaigns against the Ottomans, and Hunyadi renewed his war against the Turks in 1455, using his personal fortune to provision and man the fortress of Belgrade, which he put under the command of his eldest son Láslö and his brother-in-law Mihàly Szilàgyi. When the Turks laid siege to the fortress, Hunyadi raised a force of mercenaries during May-June 1456 and advanced with it in two-hundred small river galleys to the relief of the defenders. Such was the popularity of Hunyadi that his mercenaries were eagerly joined by peasant volunteers who had been rallied by his ally, the Franciscan friar Giovanni da Capistrano. Hunyadi's fleet succeeded in destroying the riverborne flotilla of the Ottomans on July 14, and Hunyadi scored the greatest triumph of his career when he defeated the army of Sultan **Mohammed II the Conqueror** outside Belgrade during July 21–22.

Flushed with victory, Hunyadi laid plans to continue his campaigns against the Turks. He was, however, felled by plague and died in his camp outside Belgrade.

One of Hungary's most celebrated national heros, a force for national unity in a time of great chaos and fragmentation, Hunyadi was tireless and valiant, as well as a many-faceted military leader. He was an excellent strategist and tactician, who took steps to create a permanent—and modern—Hungarian army. Personally inspiring, he was adored by his troops.

HUSSEIN, Saddam (Takriti)

BORN: 1937

SERVICE: Iran-Iraq War (1980–88); Persian Gulf (Kuwait) War (1990–91)

Oppressive president of Iraq; invaded and attempted to annex Kuwait, thereby triggering the brief, destructive Persian Gulf War

Orphaned at nine months, Saddam (Takriti) Hussein was raised by an uncle, Khairallah Talfah, an anticolonialist who led an unsuccessful coup and bid for independence in 1941. Lacking the required family connections, Saddam was refused enrollment in the Baghdad Military Academy and turned instead to membership in the radical Ba'ath (Arab Socialist Renaissance) Party, founded by Michel Aflaq. The party first supported Abdul Karim Kassim's overthrow of the Iraqi monarchy in 1958, then turned against Kassim. Saddam, having already killed a Communist politician who ran against his uncle in a parliamentary election, eagerly volunteered to assassinate Kassim. However, his attempt failed, and Saddam, wounded in the leg, fled to Syria. From there he went to Cairo, where he studied law.

Saddam remained in exile for three years. As a ploy to confuse the police, he dropped the name Takriti (by which he was known at the time), taking as his last name his father's first, Hussein. He returned to Baghdad, organized a secret Ba'ath militia, which, in February 1963, at last succeeded in deposing and executing Kassim. One of Saddam's relatives, Ahmed Hassan Bakr, became premier; five years later (July 17, 1968), Bakr overthrew President Aref (who had installed him), and a decade after that, in July 1979, stepped down himself in favor of Saddam Hussein.

Domestically, Saddam's rule was always characterized by police-state terror. He struggled particularly with the rebellious Kurds, against whom he employed lethal nerve gas on March 16, 1988. Lacking military training, he nevertheless has shown himself to be bellicose and (many have suggested) unbalanced. Commander in chief of the world's fifth-largest army, he began, in September 1980, an eight-year war against Iran that proved so costly it nearly led to a military coup in Iraq. Ultimately, however, the operation succeeded in wearing Iran down, and Saddam concluded a favorable peace.

In 1990, Saddam invaded and attempted to annex the small, oil-rich nation of Kuwait. While he had an easy time defeating the Kuwaiti forces, his own army, air force, and navy were quickly defeated by a United Nationssanctioned coalition of twenty-eight countries led by the United States (January 17, 1991–April 10, 1991), and Saddam was compelled to withdraw from Kuwait. During the brief "Persian Gulf War," Iraqi armed forces inflicted minimal damage against military targets. Against civilian targets in noncombatant Israel, however, Saddam launched numerous small-missile ("Scud") attacks. In Kuwait, Iraqi forces terrorized citizens and set ablaze some three hundred oil fields. Through 1991 and into 1992, oil-field fires and massive, deliberate oil spills in the Persian Gulf posed a grave ecological threat.

As a military leader, Saddam Hussein was very poor. Iraqi military losses in the Persian Gulf War were estimated in excess of 80,000 men, with mass desertion and overwhelming loss of materiel. The civilian infrastructure also suffered severe damage. Although a degree of civil unrest followed the ruinous war, especially in outlying provinces (particularly the always rebellious Kurdistan), Saddam Hussein maintained power and, throughout 1991 and 1992, even exhibited bouts of defiance against sanctions levied by the United Nations.

I

IVAN III THE GREAT
(Ivan Vasilievich)

BORN: January 21, 1440
DIED: October 27, 1505
SERVICE: Campaign against the Tartars (1458); Campaigns against the Golden Horde (1467–69); Conquest of Novogorod (1478)

Russian ruler and founder of Russian nation

Ivan III combined military proficiency with the skills of a diplomat to unite the feudal Russian lands into one kingdom. He was born on January 22, 1440, in Moscow during a bitter power struggle between his father, Vasily II, and his uncles. From the moment of his birth, Ivan's early life was constantly in danger. In 1446, Vasily II was abducted and blinded, and Ivan was secreted in a monastery, only to be betrayed to his father's enemies. Strangely, at this point, his father's captors suffered a change of heart, repented of their crimes, and reaffirmed their allegiance to Vasily. He was released, blinded as he was, to rule again, and Ivan returned, safe and sound, from his hiding place.

As heir, young Ivan was given nominal command of a military unit and was always present when his father made policy decisions. Thus he was apprentice to power. At the age of eighteen, in 1458, he led a successful campaign against the Tatars, and when his father died in 1462, Ivan succeeded Vasily as grand prince of Moscow. The experiences of his childhood made Ivan determined to free his throne of all feudal obligations and to unite the Russian lands under his sole control.

He began with brilliant campaigns against the Golden Horde during 1467–69, freeing Russia of subservience to the Tatars in the east. Next, he turned to the powerful city-state of Novgorod. Only after a number of costly sieges did Ivan take Novgorod in 1478. He stripped it of its political autonomy, annexed its possessions, and colonized it with a populace loyal to him. He moved next to the lands of Yaroslavl and Rostov, which he acquired through diplomatic negotiation by 1474. The city of Tver made overtures of union with the Lithuanians, but ultimately bent to Ivan's will in 1485. By this year, only two major Russian city-state retained any degree of independence.

Ivan was a master in combining military force with clever psychology, dividing the tribes that were allied against him by playing them against one another. For example, when Khan Ahmed of the Golden Horde allied himself with Poland-Lithuania, Ivan made an alliance with Khan Girei of the Crimea, then led troops against Ahmed and crushed him, thereby freeing Moscow from all Tatar influence.

As was often the case with nationalist leaders, Ivan's campaign of unification and consolidation produced its own sets of conflict. Following his territorial conquests, Ivan failed to distribute any land to his brothers. Worse, when his brother Yurii died, Ivan promptly seized all his possessions without sharing any among his surviving brothers. Finally, another brother voluntarily ceded his lands to Ivan, which left two brothers, Boris and Andrei, aligned against Ivan. The pair approached King Casimir IV of Poland with a proposal for alliance, but were rebuffed, and plans for opposition collapsed. Andrei nevertheless continued sporadically to dispute with Ivan and was arrested in 1491, Ivan seizing all his land. After the death of Boris in 1494, half of his land was ceded to Ivan, the remainder going to one of Boris's sons.

By the turn of the century, Ivan had successfully acquired and united almost all of the Russian lands, but he was not as successful in designating an heir to his newly constituted kingdom. The son of his first marriage died, but not before siring a son, Dmitry. Ivan's second marriage did produce a son, Vasilii, but because this second wife was from Byzantium, her Orthodoxy—and, in consequence, her son's—was open to question. Moreover, Ivan's second wife was an unattractive and unpleasant woman, universally disliked in the Kremlin court. In view of this, Ivan designated Dmitry as heir, which prompted Vasilii and his mother to plot a revolt in 1497. Ivan acted with bloody efficiency to put down the rebellion, casting Vasilii as well as his mother into prison. With Dimitry set to succeed him, the heir-designate now became associated with a heretical religious party, which distressed Ivan and alienated the metropolitan (bishop) of Moscow. Turning rapidly about face, Ivan retrieved Vasilii from prison, named him heir, and incarcerated both Dmitry and *his* mother. Thus it was that, upon Ivan's death on October 27, 1505, Vasilii succeeded him unopposed, assuming the throne of a Russia that would dominate Eastern European politics into the twentieth century.

J

JACKSON, Andrew

BORN: March 15, 1767

DIED: June 8, 1845

SERVICE: American Revolution (1775–83); War of 1812 (1812–15); First Seminole War (1817–18)

American frontier general and Indian fighter; seventh president of the United States; hailed as the champion of democracy and of the "common man"

It has never been clear whether Jackson was born in South or North Carolina, though it is more likely that he was born on the South Carolina side of the Waxhaws settlement, the son of humble Scotch-Irish immigrants. His father died before he was born, and Jackson was given scant formal education. His young life was torn apart in 1780 by the British invasion of the Carolinas during the American Revolution. His brothers, fighting in the Revolution, were captured by the British at the Battle of Hanging Rock on August 1, 1780; one of them, Robert, died. Young Andrew, who had himself become a patriot partisan, was captured and severely wounded when an officer struck him with his sword after he refused an order to black the man's boots. In the meantime, Jackson's mother had volunteered to minister to American prisoners of war confined on a ship anchored in Charleston Harbor. She contracted prison fever and died, leaving Andrew Jackson an orphan at age fourteen. These traumas planted within Jackson an abiding hatred of the British.

After the war, Jackson quickly dissipated his modest inheritance, then undertook the study of law, gaining admission to the North Carolina bar. He moved to Nashville, Tennessee, where, in 1791, he became attorney general for the Southwest Territory and, subsequently, circuit riding solicitor in the area surrounding Nashville. During this period, many of Jackson's private clients were merchants who had lost money as a result of the United States' extending federal authority over the territory. From this early point in his career, therefore, Jackson acquired a reputation as a fighter—not only against military and imperial tyranny, but against federal tyranny as well.

In 1791, Jackson married Rachel Donelson Robards; both he and Rachel believed that she and her first husband had been legally divorced. When this proved not to be the case, they remarried in 1794, but the incident would haunt Jackson's private and political life for many years, ultimately becoming the issue of a duel. In the meantime, Jackson's political star rose. He served as a delegate to the Tennessee constitutional convention in 1796 and was elected to Congress, serving from 1796 to 1797. He earned a fearsome reputation as an opponent of the **Washington** administration's conciliatory stance toward Great Britain and the Indian tribes that had sided with the British during the Revolution. In 1797, Jackson was appointed to serve out the senatorial term of his political mentor, William Blount, who had been expelled from the Senate as a result of his involvement in a British plan to seize Florida and Louisiana from Spain.

Jackson had problems of his own, most of them involving his precarious finances. In 1798, on the verge of bankruptcy, he resigned from the Senate and returned to Tennessee. There, from 1798 to 1804, he served as a Tennessee superior court judge, then stepped down to devote himself full-time to rebuilding his fortune. It was during this period that he built his famous plantation, the Hermitage, outside of Nashville. To all appearances, Jackson's political career seemed over. Worse, in 1805, he expressed support for Aaron Burr's illegal—even traitorous—scheme to create an empire stretching from the Ohio River to Mexico. He even became involved in the scheme to a very limited extent. Jackson subsequently repudiated Burr, but he was rebuffed by President Thomas Jefferson when he sought office as governor of the newly purchased Louisiana Territory.

Disappointed, Jackson was in no mood for the actions of one Charles Dickinson, a young resident of Nashville, who, in 1806, insulted Rachel Jackson and published personal attacks on Jackson in the local newspaper. Jackson responded by challenging him to a duel, killing the young man on May 30—Jackson was himself seriously wounded—and the public heard little more of Jackson until William Blount, now governor of Tennessee, commissioned him major general of volunteers in the War of 1812.

Jackson's most distinguished service came toward the end of the war, in 1814, when he marched against the pro-British "Red Stick" Creek Indians, defeating them in the decisive Battle of Horseshoe Bend on March 27. He then capitalized on his victory by compelling all of the Indians of the region—hostiles, friendlies, neutrals, and even allies—to cede enormous tracts of land throughout Alabama and Georgia. In a single stroke, Jackson had shown himself to be a natural military genius and a high-handed manipulator of the Indians.

Jackson was quickly appointed to command of the defense of New Orleans, which was imperiled by the British. He mounted a brilliant defense and, on January 8, 1815, soundly defeated the attackers in one of the few set-piece battles of the War of 1812 which Americans could claim victory. Ironically, by the time the Battle of New Orleans was fought, the warring nations had signed a peace treaty (in Ghent, Belgium, on December 24), but the news failed to reach the commanders in the field. To Americans this hardly mattered. Jackson's triumph was hailed as a vindication of national honor in war that had consisted of one American military disaster after another. Andrew Jackson was transformed into an overnight legend and a veritable icon of American courage, skill, and righteousness.

Jackson continued his military career in 1817, when he led the nation's first war against the Seminole Indians, mercilessly evicting them from their tribal homelands and pursuing them deep into Spanish Florida through the spring of 1818. Jackson did not scruple to interpret his mission as anything less than a mandate to conquer Spanish Florida. He campaigned not only against Indians, but audaciously deposed Spanish colonial authorities and others. He captured one Lieutenant Robert Armbrister of the British Royal Marines and an elderly Scottish trader named Alexander Arbuthnot. Charging both with aiding the Indians, he organized a drumhead court, convicted both men, and hanged them. He then seized Spanish Pensacola on May 26, 1818.

Jackson's peremptory actions led to the Adams-Onis treaty of 1819, by which Spain formally ceded Florida to the United States. The United States had gained a vast territory, but Jackson's lawless conduct provoked conservative elements in Congress to seek his censure. They failed. The advocate of the "common man" had become far too popular.

Jackson resigned his army commission in 1821 to become provisional territorial governor of Florida, and the following year, the Tennessee legislature nominated him for the presidency, then, in 1823, elected him to the U.S. Senate. In 1824, Jackson did run for president, but was narrowly defeated by John Quincy Adams in a bitterly contested election that had to be decided by the House of Representatives. Jackson ran again in 1828 and won by a comfortable margin. He served two terms as president, bringing about such profound changes in American government that the era became known as the Age of Jackson.

The military implications of Jackson's presidency were also profound. He endorsed and enforced the 1830 Indian Removal Act, by which most of the eastern Indian tribes were ordered removed west of the Mississippi River to Indian Territory—an area encompassing modern Oklahoma and parts of other states. Although the Indian Removal Act provided for compensation and land exchanges, also specifying that the Indians removed had to agree to removal by treaty, in practice the legislation provided the basis for a cruel program of military force. Many Indians were marched westward quite literally at the point of a bayonet.

Jackson also responded vigorously and with implied military threat to the Nullification Crisis of 1832. The president's supporters were sharply divided on question of tariffs, with Jackson's Southern constituents opposing a protective tariff (and believing Jackson would lower all duties), and supporters in other regions—where manufacturing was struggling to establish itself—favoring protectionist tariffs. Southerners protested that the tariffs established by John Q. Adams in 1828 were high, and they were disappointed when Jackson lowered them only slightly in 1832. On November 24, 1832, South Carolina acted in accordance with the doctrine of nullification espoused by John C. Calhoun. Nullification held that an individual state had the right to nullify within its borders any federal act the state judged to be unconstitutional. South Carolina therefore declared both the tariffs of 1828 and 1832 null and void and prohibited the collection of tariffs in South Carolina after February 1, 1833. Jackson replied with his Nullification Proclamation of December 10, 1832, announcing his intention to enforce the law—though also pledging a compromise tariff. South Carolina backed down, and civil war was postponed for almost thirty years.

Those who expected unquestioning bellicosity from Jackson were disappointed when he declined to take a decisive official stand in favor of Texas independence from Mexico. Indeed, Jackson delayed recognizing the Lone Star Republic until the day before he left office in 1837. He did not relish war with Mexico.

Jackson's later years, including his later years as president, were plagued with ill health. He retired to the Hermitage after his second term ended, but remained a powerful force in the Democratic party. His handpicked candidate, Martin Van Buren, easily captured the Democratic nomination and became the nation's eighth president in 1844.

JACKSON, Thomas Jonathan ("Stonewall")

BORN: January 21, 1824

DIED: May 10, 1863

SERVICE: U.S.-Mexican War (1846–48); Civil War (1861–65)

*Second only to **Robert E. Lee** in the pantheon of Confederate generals*

A native of Clarksburg, Virginia (now West Virginia), Jackson graduated from West Point in 1846, was commissioned a second lieutenant of artillery, and was sent to Mexico with the army of **Winfield Scott**. He quickly distinguished himself for his steely calm gallantry at Veracruz (March 27, 1847), Cerro Gordo (April 18), and Chapultepec (September 13). At the last battle, he was breveted major.

Thomas J. "Stonewall" Jackson as a young officer. From Harper's Pictorial History of the Civil War, *1866.*

Jackson resigned from the army in February 1852, but served as professor of artillery tactics and natural philosophy at Virginia Military Institute from 1851 to April 1861, when he was commissioned colonel of Confederate volunteers. Promoted to brigadier general on June 17, he fought brilliantly at First Bull Run on July 21. His unshakable defense of Henry Hill against Union attack won him the nickname "Stonewall," and in October he was elevated to major general. The next month, he was appointed commander of all forces in the Shenandoah Valley.

Jackson executed an extraordinary campaign against far superior numbers that blocked reinforcement of **George B. McClellan**'s advance on Richmond during 1862. Although he was repulsed at Kernstown (near Winchester, Virginia) on March 23, with characteristic aplomb he outmaneuvered and then defeated Union forces at Front Royal on May 23. This was followed by victory at Winchester (May 24–25), Cross Keys (June 8), and Port Republic (June 9). Ultimately, Jackson drove the Union forces out of the Shenandoah. This accomplished, Jackson was then ordered to Richmond, where he supported Lee's efforts to drive McClellan from the Peninsula during the series of battles known as the Seven Days (June 26–July 2).

On August 27, Jackson engaged General John Pope's army in northern Virginia, first destroying its depot at Manassas Junction, then beating back its attack at Groveton (August 28). During the next two days, Jackson was instrumental in the Confederate victory at Second Bull Run. Jackson also played a central role in the invasion of Maryland, winning special distinction at Antietam (September 17). By October, he received a promotion to lieutenant general and was made commander of II Corps.

Jackson commanded the right flank at Fredericksburg on December 13, then led a devastating flanking maneuver at the expense of General **Joseph Hooker** at Chancellorsville on May 2–4, 1863. During this battle, Jackson frequently exposed himself to grave danger and was accidentally shot by his own troops. His left arm, badly mangled, was amputated. Lee remarked to him: "You have lost your left [arm]. I have lost my right arm."

That might well have served as the general's epitaph. Jackson developed pneumonia and died within a week of his wounding, at Guinea Station, Virginia.

Called "Old Blue Light" for the other-worldly blue of his eyes, Jackson was a truly inspired commander. He was not only an outstanding tactician, he was personally fearless and always aggressive. Deeply religious and a stern disciplinarian, he was nevertheless idolized by his command. He had a gift for inspiring his troops to great and heroic deeds.

JAMES II

BORN: October 14, 1633

DIED: September 5, 1701

SERVICE: Franco-Spanish War (1635–59); Second Anglo-Dutch War (1665–67); Third Anglo-Dutch War (1672–74); War of the League of Augsburg (1685–97)

British monarch overthrown in the Glorious Revolution; also a brave soldier and a skilled admiral

James II was the younger son of Charles I and Henrietta Maria. Created duke of York in January 1634, he lived in Oxford during the First Civil War (October 1642–June 1646), which deposed his father (who was ultimately executed). James was held by Parliament under house arrest at St. James's Palace, but he managed to escape on April 20, 1648 and joined his mother, already in exile in France.

James entered the French army in April 1652 and served with distinction under the **Viscount of Turenne** in four separate campaigns. When **Oliver Cromwell** formed an alliance with France, James, with personal regret and great reluctance, left the French service and joined the other exiled Royalists in an alliance with Spain. He fought in the Spanish army at the Battle of the Dunes (June 14, 1658).

James returned to England after the restoration of Charles II to the throne in 1660. He was immediately appointed Lord High Admiral (May 1660) and, aided and advised by Samuel Pepys (a naval reformer who was twice secretary of the admiralty, but who is best remembered by history for his detailed diary of London life during the period) and Matthew Wren, he built up and modernized the Royal Navy. James also supported a number of colonial enterprises, including the seizure of New Amsterdam from the Dutch in North America (September 7, 1664); the city was renamed New York.

James assumed command of the British fleet during the early phase of the Second Anglo-Dutch War and defeated the Dutch fleet, under Admiral Opdam, at the Battle of Lowestoft on June 13, 1665. While the British people hailed this victory, the battle had been more than usually bloody, and James was pressured into relinquishing command to Prince **Rupert** and **George Monck** by a public who feared risking the life of the heir to the throne.

James's conversion to Roman Catholicism—between 1668 and 1671—prompted the House of Commons to attempt, unsuccessfully, to bar his ascension to the throne. On June 7, 1672, he once again led the fleet, at the Battle of Sole Bay against the Dutch admiral Michel de Ruyter. James's fleet was caught off guard at anchor and was badly mauled before Prince Rupert arrived to drive Ruyter off.

James resigned command rather than swear the anti-Catholic oath required of military officers by the Test Act of 1673; however, he met with surprisingly little opposition when he finally came to the throne on February 7, 1685. But he moved swiftly and harshly to suppress the rebellions of Argyll and Monmouth during the summer of 1685. He also enlarged the army—a move that was seen by opponents as preparation for an alliance with the expansionist (and Catholic) Louis XIV of France. James issued two declarations of indulgence to protect Catholics, an action that alienated the Church of England, and he circumvented the Test Act by promoting Catholics to high office and military commissions. In 1688 he put seven Anglican bishops on trial for refusing to order his declarations to be read in all the churches; the bishops were acquitted.

The final straw came in June 1688 with the birth of his son. With a Catholic succession in the offing, James's Protestant opposition invited the king's Dutch Protestant nephew and son-in-law, William of Orange, to come to England. William and his wife, James's older daughter, Mary, landed at Torbay on November 5, 1688 and advanced on London, whereupon James's army deserted him. James fled on December 11, 1688, was taken prisoner in Kent, but allowed to escape to France on December 23. The "Glorious Revolution" that brought William and Mary to the British throne had been brief and bloodless.

James, however, was not quite through. He raised an army in France and augmented it when he landed in Ireland in March 1689. He was declared king by a parliament summoned in Dublin, and then was handily defeated by William's army at the Battle of the Boyne on July 1, 1690. James once again fled to France, where he retreated increasingly into religiosity. He died in exile, at St. Germain, in 1701.

James's early career showed him to be a capable and courageous soldier and a highly skilled admiral. However, as king, he was impolitic, heavy handed, and dense. It was these qualities, rather than any nefarious schemes he may have harbored, that caused his downfall.

JELLICOE, John Rushworth, First Earl

BORN: December 15, 1859

DIED: November 20, 1935

SERVICE: Boxer Rebellion (1900); World War I (1914–18)

Journeyman admiral of the British Royal Navy, skilled as a tactician but lacking as a strategist; his single great battle was Jutland, during World War I

Jellicoe, son of a merchant ship's master, entered the navy as a cadet in 1872 and was commissioned a sublieutenant in 1880. He graduated from the gunnery theory course at the Royal Naval College with honors—an £80 prize—in 1883 and the following year was assigned to H.M.S. *Excellent* as fully qualified gunnery officer. From 1886 to 1888, he sailed on H.M.S. *Monarch* as gunnery lieutenant, and in 1890 was appointed to the Admiralty Board as assistant to director of naval ordnance.

Jellicoe served on Admiral Sir George Tryon's flagship H.M.S. *Victoria* in the Mediterranean. He was nearly killed when H.M.S. *Camperdown* rammed and sank *Victoria* in maneuvers off Beirut on June 22, 1893. Transferred to H.M.S. *Ramillies* (1893–96), still in the Mediterranean service, he was promoted to captain on January 1, 1897. That same year he served as a member of an important ordnance committee, then was made commander of H.M.S. *Centurion* the following year, serving as flag captain to Admiral Edward Seymour on the China Station. During the Boxer Rebellion, Jellicoe served in the First Peking Relief Expedition (June 10–26, 1900) and was severely wounded in one exchange.

Jellicoe was transferred to shore duty for a time, as naval assistant to the controller (third sea lord) and returned to sea in August 1903 as captain of the armored cruiser *Drake*.

In 1905, Jellicoe was appointed director of naval ordnance at the Admiralty, then was promoted to rear admiral (August 1907) in the Atlantic Fleet. In 1908, he returned to the Admiralty as third sea lord, then was made acting vice admiral in command of the Atlantic Fleet in December 1910. Transferring to the Home Fleet as commander of the Second Division, he was confirmed as vice admiral in November 1911. The following year, he supervised gunnery experiments on H.M.S. *Thunderer* and *Orion* and was instrumental in the adoption of Sir Percy Scott's modern director firing system in 1912.

As second sea lord—1913—he participated in naval maneuvers, commanding the "Red" fleet. He served as second in command of the Home Fleet under Admiral Sir George Callahan, whom he replaced as Commander in Chief of the Home Fleet at the outbreak of World War I on August 4, 1914. Jellicoe's first concern was to improve the fleet's combat readiness. In March 1915, he led the entire Home Fleet on a mission to intercept German admiral Reinhard Scheer's sortie to break the British blockade in the North Sea. The Battle of Jutland (May 31–June 1, 1916) was the only great sea battle of World War I. The battle was a punishing one. Strategically indecisive, it was clear that the Germans had displayed greater tactical skill and that their ships were generally sturdier than the British vessels. (Jellicoe, a calm and understated man, observing that British vessels exploded and sank after taking relatively few hits, remarked to an aid, "Something seems to be wrong with our ships today.") British losses were heavier than those of the Germans. On the other hand, the German High Seas Fleet not only failed to break the British blockade, it was bottled up in port again for the duration of the war. While the battle was hardly a glorious triumph for Jellicoe, he did avoid potential disaster—as one commentator put it, he could have "lost the war in an afternoon" —and the German High Seas Fleet was effectively neutralized.

Jellicoe was elevated to first sea lord (November 28, 1916–December 1917) and was created Viscount Jellicoe of Scapa in December 1919. That same year he was made admiral of the fleet, and from 1920 to 1924 served as governor-general of New Zealand. He was created earl in 1925.

JOAN OF ARC (Jeanne d'Arc)

BORN: ca. 1412

DIED: May 30, 1431

SERVICE: Hundred Years' War (1338–1453)

French peasant girl who led the French army against the English during the Hundred Years' War; French national heroine and patron saint

Born in the village of Domrémy, between Champagne and Lorraine, Joan was a quiet, pious child, who, beginning when she was about thirteen years old, began to hear "voices." She subsequently identified these as the voices of Saint Catherine, Saint Margaret, and Saint Michael. The voices told her to undertake the mission of liberating France from English domination. After keeping her voices secret for some five years, she left Domrémy in 1429 and, under escort, traveled to the court of the dauphin (later King Charles VII), who had been deprived of his rights as heir to the French throne by the 1420 Treaty of Troyes.

Joan attempted to convince the dauphin to organize resistance to liberate France. He subjected Joan to tests by a group of clerics, whose report to him persuaded him to reassemble his troops and place them under Joan's command in an expedition to relieve Orléans, which had been under English siege for eight months. During eight days in May 1429, Joan (with the aid of Etienne de Vignolles, called La Hire) lifted the siege. She persuaded the dauphin to mount an offensive with the objective of breaking through to Reims, where the dauphin could be crowned. She defeated the English army of Sir John Fastolf and John Talbot, Earl of Shrewsbury, at Patay on June 15, not only taking Shrewsbury prisoner, but largely destroying the English army. The dauphin's army reached Reims on July 16, and the coronation took place.

Following the coronation, Joan gained the submission of several towns to the newly crowned King Charles VII, among them Senlis, Beauvais, and Compiègne during July–August. She marched on Paris, but failed to take the city and was wounded in battle on September 8. The army was disbanded at Gien on September 22, and Joan wintered with Charles from October until April 1430. Early in April, she set out to defend Compiègne from the Burgundians. Arriving on May 14, she began the defense, but John of Luxembourg's Burgundian army laid heavy siege on May 22, and Joan was unhorsed during a sortie out of the town. She surrendered to John on May 23 and was imprisoned at Beaulieu Castle.

The Burgundians tried and convicted her of heresy at Rouen during March 25–April 24, 1431. She recanted her visions and was sentenced to be abandoned to secular authority. In turn, an English court sentenced her to "perpetual imprisonment"; however, the judges visited her in her cell shortly after the trial, and they determined on May 29 that she had abjured her recantation. For this, she was burned at the stake on May 30.

Joan was officially rehabilitated by the French in 1450, but was not canonized until May 16, 1920. Whether or not one perceives Joan as having been divinely inspired, she was an extraordinarily inspiring leader and a natural soldier, who possessed a sound intuitive grasp of tactics.

JODL, Alfred

BORN: May 10, 1890

DIED: October 16, 1946

SERVICE: World War I (1914–18); World War II (1939–45)

One of Adolf Hitler's top generals; directed all campaigns of World War II, except for Operation Barbarossa

Jodl was born in Würzburg, the son of a distinguished military family. There was little question that he, too, would pursue a military career, and he served as an officer during World War I. After the war, in 1919, he was appointed to the General Staff with the rank of captain. In 1923, he met Hitler and was immediately captivated, utterly convinced of Hitler's genius. Jodl rose steadily through the General Staff, becoming a member of the key the war plans division in 1935. In April 1938, he was appointed head of the Land Operations Department of *Oberkommando der Wehrmacht* (OKW). At this time, he was promoted to general major, and the following year became chief of the Wehrmacht Operations Staff. He held this post throughout the war, until 1945, and reported directly to **Wilhelm Keitel** and Hitler. Effectively, Jodl directed all of the Wehrmacht's major campaigns, except for the invasion of the Soviet Union, Operation Barbarossa.

Increasingly, Hitler, fancying himself a strategist, assumed a guiding role in the OKW, and Jodl was put in the difficult position of creating a shadow General Staff, assuming independent control over certain aspects and theaters of the war, issuing secret orders without Hitler's knowledge or approval. Hitler had absolute faith in Jodl's loyalty and ability. He did not look over the general's shoulder, as he did with most of his other commanders. Doubtless, Jodl's shadow command postponed disaster for Germany; however, as the war ground on, Jodl's judgment faltered. Late in the war, he managed the final German offensive—in the Ardennes, the so-called Battle of the Bulge—ineffectively. As a result, the western front collapsed by January 1945.

After Hitler's suicide in Berlin on April 30, 1945, Admiral **Karl Dönitz** appointed Jodl as his representative at the signing of the surrender to the Allies at

Rheims on May 7. A prisoner of war, Jodl was tried by the International Military Court at Nuremberg and was found guilty on four counts of war crimes. He was sentenced to death and was hanged at Nuremberg on October 16, 1946. Ironically, on February 28, 1953, a German denazification court reviewed the evidence against Jodl and determined that he had acted strictly in a military sphere and had not, in fact, violated international law. He was posthumously exonerated.

Jodl was a talented officer, who managed for many months of the war the nearly impossible task of operating effectively behind Hitler's back. By the end of the war, Jodl's judgment deteriorated, even as the Nazi war machine broke down.

JOFFRE, Joseph Jacques Césaire

BORN: January 12, 1852

DIED: January 3, 1931

SERVICE: Franco-Prussian War (1870–71); Sino-French War (1883–85); War in Indochina (1885–95); conquest of Madagascar (1894–96); World War I (1914–18)

Marshal of France who, as "Papa Joffre," led the French army in World War I until he was relieved in 1916

Joffre was born in Rivesaltes, *département* of Pyrénées Orientales and graduated from the école Polytechnique with a degree in military engineering. During the Franco-Prussian War, he served in the defense of the siege of Paris as a junior engineering officer (September 19, 1870–January 26, 1871). By the time of the Sino-French War, he had been promoted to captain and fought in Formosa (March–June 1885). He went on to see action in Indochina and West Africa.

Joffre taught as an instructor at the artillery school, then served under Joseph Simon Galliéni as fortifications officer on the island of Madagascar from 1900 to 1905. By 1905, he was promoted to general of division. Three years later, he was a corps commander, and in 1910 had become a member of the Supreme War Council and director of the rear. Galliéni recommended Joffre's appointment as chief of the General Staff (Grand Quartier General) in 1911. In this capacity, Joffre and the General Staff developed Plan 17, which served as the battle plan of the French conduct of World War I during its opening weeks. The plan created unmitigated disaster that allowed the German army to penetrate deep into French territory.

During World War I, Joffre took overall charge of the French war effort for the first thirty months of the conflict. The qualities of calm and resolute refusal to panic, which had impressed his superiors in earlier days and helped gain him rapid promotion, were valuable during the darkest days of the war. Yet Joffre was also perceived by many as slow-witted and unwilling to take the offensive. If his even-handed skill during the disastrous Battle of the Frontiers (August 14–25, 1914) and the headlong French retreat that followed (August 26–September 4) saved the French army from total annihilation, it was the misconceived Plan 17 that created the circumstances of defeat in the first place. However, once Joffre became convinced of the threat presented by the German right wing, he came through with a vigorous counterattack at the Marne during September 5–9. Unfortunately, the outflanking attempts that followed this—the so-called Race to the Sea of September–November 1914—failed, as did offensives in Champagne (December 20, 1914–March 1915) and Artois (January 1–March 30, 1915). Papa Joffre could not stand to see so many of his men cut down in battle, and he aborted the offensives. When he returned with additional attacks in Artois (May 16–June 30, September 25–October 30) and Champagne (September 25–November 6), casualties again were terribly high, but results were poor.

Then came the German offensive at Verdun during February 21–December 18, 1916. Joffre was thoroughly surprised. Worse, he had earlier removed the guns from the city's fortifications in order to reinforce the besieged town of Liège. Joffre rushed General **Henri Phillippe Pétain** to command the defense of Verdun. Vowing "They shall not pass," Pétain held the city, but incurred terrible losses. These losses, as well as those incurred at the Somme, prompted Joffre's relief from command on December 13, 1916. Despite this, he was created Marshal of France on the day after Christmas.

During 1917, Joffre served as head of the French military mission to the United States, then retired from military and public life after the war.

Joffre was much reviled for Plan 17 and the early disaster it brought. On the other hand, his courage and methodical fortitude saved the French army from complete annihilation in the opening weeks of the war. The Marne was his signal triumph—a success that eluded him in subsequent engagements.

JOHN I ZIMISCES

BORN: 924

DIED: 976

SERVICE: Campaigns in the East (956–969); War with Prince Sviatoslav (969--971); War against the Muslims (973–976)

A Byzantine emperor of supreme ruthlessness, political adroitness, and military skill

John I Zimisces was descended from Armenian nobility and was related to the Kurkuas and Phocas families, both extremely influential in Byzantine government. He quickly acquired a reputation for military prowess, fighting in the service of Romanus II and Nicephorus II Phocas, and emerging as the hero of the Battle of Samosata, in northern Mesopotamia, in 958.

Zimisces acquired Theophano, the wife of Nicephorus, as his mistress, and, with her, schemed to assassinate Emperor Nicephorus II, who was murdered on December 10 or 11, 969. Zimisces rushed to assume the throne and also wasted no time in securing the sanction and support of the patriarch of the Orthodox Church, Polyeuktos of Constantinople. Unfortunately, the patriarch deemed the relationship between Zimisces and Theophano sinful. He made his support of the new emperor conditional on the exile of Nicephorus faithless widow. Zimisces accordingly banished her.

During 969–972, Zimisces had to mount resistance against the Russian prince Sviatoslav and his Bulgarian allies, who were nibbling at Byzantine territory. In 970, Zimisces achieved a major land victory at Arcadiopolis and then emerged victorious as well from a sea battle that repulsed the Russian fleet and allowed the establishment of a blockade across the Danube.

In April 971, Zimisces attacked the Bulgarian capital city of Great Preslav. After the city fell, he chased Sviatoslav's forces, pushing them northward to the Danube River fortress of Dorostorum. Here Sviatsoslav took his stand, resisting a two-month siege before he finally surrendered in July 971, relinquishing Bulgaria to the Byzantines. Zimisces quickly transferred his attention to the Muslims, whom he pushed back beyond the middle Euphrates River. This action extended the Byzantine empire deeply into Syria.

In May 975, Zimisces conquered Damascus and advanced on Jerusalem. However, as he approached the city, his army met strong resistance from the Fatimids, and, while his forces fought desperate, Zimisces fell ill with typhoid. Gravely afflicted, he aborted the assault against Jerusalem and returned to, where he died early the next year.

JOHN II COMNENUS

BORN: 1088

DIED: April 8, 1143

SERVICE: Seljuk War (1120–21); Petcheneg Invasion (1121–22); naval war with Venice (1122–26); Hungarian Dynastic War (1124–26); First War against the Danishmends (1132–35); Reconquest of Cilicia (1134–37); War with Antioch (1137–38); Second War against the Danishmends (1138–41)

Byzantine emperor whose energy and ruthlessness were rare in the empire of the Dark Ages

John was the son of **Alexius I Comnenus** and Irene Ducas, and he succeeded his father as emperor on August 16, 1118. Alexius had established a practice of settling prisoners taken in battle as soldier-farmers, who were granted estate leases in return for military service. With this hired army, John defeated the Seljuks and reconquered most of Anatolia during 1120–21. When the marauding Petchenegs invaded the empire in 1121, John responded vigorously, not only pushing them out of Byzantine territory, but defeating them so thoroughly that they ceased to be a threat.

Beginning in 1122, John waged a limited naval war against Venice, competitor in trade, but the results were inconclusive. By 1126, John withdrew from the conflict and affirmed Venetian trading concessions. His marriage to a Hungarian princess embroiled him in that kingdom's dynastic disputes. John ultimately compelled the Hungarians to acknowledge Byzantine suzerainty through a series of sharp armed conflicts during 1124–26.

With the conquest of Hungary behind him, John turned to the East, He suppressed the emirate of the Danishmends of Melitene in Anatolia (present-day Turkey) during 1132–35, then led an overlapping campaign into Cilicia (southern Turkey) during 1134–37. He reconquered this kingdom, which had been ruled by Armenians since the battle of Manzikert in August 1071.

After regaining Cilicia, John laid siege to Antioch, the city-state his father had tried in vain to retake for the empire. The city fell in 1138, and its ruler, Raymond of Poitiers, swore fealty to John. Flushed with victory, the emperor left Antioch to fight a second war against the Danishmends. The conflict spanned 1139–41 and ended in a Byzantine triumph. However, no sooner was the war ended than Raymond, supported by the Latin clergy, repudiated his oath of allegiance to John (1142). Accordingly, John prepared to undertake an expedition against Antioch, but, while hunting, he was wounded by a poisoned arrow. Presumably, this was no accident, but an assassination attempt. John never recovered from the wound.

Among a roster of weak and incompetent Byzantine emperors, John II Comnenus stands tall. An able statesman, he was also a tireless military commander who enjoyed a high rate of success.

JOHN III (John Sobieski)

BORN: August 17, 1629

DIED: June 17, 1696

SERVICE: Chmielnicki's Cossack Revolt (1648–54); First Northern War (1655–60); Russo-Polish War (1658–66); Polish-Turkish War (1671–77); Austro-Turkish War (1682–99)

Polish ruler whose greatest military accomplishment was the liberation of Vienna from Turkish domination

Sobieski was born at Olesko near Lvov, son of the castellan of Cracow and his wife, who was the granddaughter of the Cossack hetman Stanislaw Zolkiewski. Sobieski's early life was one of privilege. Educated at Cracow's famed university, he made a grand tour of Western Europe during 1646–48, then returned to Poland to serve in battle against the Cossacks of Bogdan Chmielnicki. After war broke out with Sweden in 1655. Sobieski joined the Swedes in the absence of the Polish king John II Casimir, but soon returned to the Polish fold—in 1656—and fought alongside Stephan Czarniecki to defeat the Swedish forces.

Sobieski went on to distinguish himself in a series of campaigns against the Tartars and Cossacks in the Ukraine. As a reward for this service, he was made grand marshal and field commander of the Polish army in 1665, and, three years later, was named commander in chief of all Polish forces (February 5, 1668).

After John II Casimir died in 1669, a diet was convened to elect a successor. Sobieski accepted bribes from King Louis XIV of France to support a French candidate, who failed to gain election. Sobieski then plotted against the newly elected king, Michael Wisnowiecki, whom he personally despised. The Polish gentry sided with Michael in opposition to Sobieski's plan to expand the army. When Sobieski led the army against the Turks in 1671, the king and diet concluded the humiliating treaty of Buczacz, on October 18, 1672, ceding territory to the Turks. Despite this, Sobieski fought on, defeating the Turks four times in ten days during early 1673, then routing a massive Turkish army outside Chocim fortress on November 11, 1673.

Sobieski went on to take the city of Chocim, but broke off operations to return to Warsaw when he received word of King Michael's death. Marching into the Polish capital with 6,000 victorious veterans, he brushed the diet opposition aside and easily gained election as King John III on May 21, 1674.

He did not enjoy a peaceful reign. The Turks mounted a renewed invasion in 1675, and John III rushed to meet them. He smashed their army at Lvov, then returned to Cracow on February 2, 1676 to be crowned. Following this, he hastened to the Ukraine, where his camp at Zorawno was repeatedly attacked by the massive Turkish army under Ibrahim Pasha during September–October. Ibrahim Pasha failing to breach the camp, he concluded the Treaty of Zorawno on October 16–17, by which all of Ukraine was returned to Poland except for Podolia.

John III next set his sights on taking East Prussia from the Turks. He concluded an alliance with the French to no avail and turned next to the Austrians, with whom he concluded the Treaty of Warsaw on March 31, 1683. In a brilliant battle, John personally led a mass cavalry charge that broke Turkish resistance outside of Vienna on September 12, 1683. Routed, the Turks fled, with John in pursuit. In October, the Polish leader liberated Gran from Turkish domination.

Despite John's magnificent victories against the Turks, Poland remained torn by internal dissension, which crippled the balance of the king's reign.

JOHNSTON, Albert Sidney

BORN: February 2, 1803

DIED: April 6, 1862

SERVICE: Black Hawk War (1832); War of Texan Independence (1835–36); U.S.-Mexican War (1846–48); Civil War (1861–65)

Brilliant Confederate general whose death early in the conflict Jefferson Davis called "the turning point of our fate"

A native of the village of Washington, Kentucky, Johnston graduated from West Point in 1826 and was commissioned a second lieutenant in the infantry. He was appointed regimental adjutant during the Black Hawk War of 1832, but, like so many promising officers in the U.S. Army, was soon discouraged by prospects for advancement. He resigned his commission in April 1834 to become a farmer. Two years later, however, he volunteered as a private in the Texas army for service against Mexico during the War of Texan Independence (1836). By January 1837, Johnston was commander in chief of the Texas army, and the following year was appointed secretary of war in the cabinet of **Sam Houston,** president of the Texas Republic.

Johnston resigned from Houston's cabinet in March 1840 to take up farming again, but at the outbreak of the U.S.-Mexican War in 1846, he served under General **Zachary Taylor** as colonel of the First Texas Rifle Volunteers and fought at Monterrey (September 20–24, 1846).

After the war, Johnston rejoined the U.S. Army as a major and, by December 1849, was U.S. Army paymaster. In 1855, he was given command of Second Cavalry, then became commander of the Department of Texas in April 1856. During the Mormon Expedition (1857–58), Johnston was given a brevet promotion to brigadier general and put in charge of the operation against Mormon insurgents. Virtually on the eve of the Civil War, late in 1860, he was serving as commander of the Department of the Pacific, but resigned in April 1861 to serve the Confederacy.

Johnston was named commander of the Western Department in September 1861 and recruited and organized the Army of Mississippi, which he rushed to the defense of a long front extending from the Mississippi River to Kentucky and the Alleghenies. His troops, thinly stretched along this front, were overwhelmed by **Ulysses S. Grant** and had to yield Fort Henry on the Tennessee River (February 6, 1862) and Fort Donelson on the Cumberland (February 14). These were terrible blows to the Confederacy, since the forts served effectively as gateways to the Mississippi River.

After Nashville fell to Union forces on February 28, Johnston hastily organized the defense of Corinth, Mississippi. He then took the offensive with a devastating surprise attack against Grant at Shiloh, Tennessee. It was at this bloody battle that Johnston sustained a fatal wound April 6, 1862. His death reversed the course of the battle, and Johnston's leaderless army was defeated on April 7.

Johnston was a colorful figure, a true warrior, and a commander of great dash, whose full measure as a military leader can never be taken, however, because of his death early in the war.

JOHNSTON, Joseph Eggleston

BORN: February 3, 1807

DIED: March 21, 1891

SERVICE: Second Seminole War (1835–42); U.S.-Mexican War (1846–48); Mormon Expedition (1857–58); Civil War (1861–65)

Confederate general who was a favorite of Robert E. Lee

Born near Farmville, Virginia, Johnston graduated from West Point in 1829 and was commissioned a second lieutenant in the artillery. During service in the Second Seminole War, he was promoted to first lieutenant (July 1836). Even with his promotion, Johnston was discouraged by the paucity of prospects for advancement in the military, and he resigned in May 1837 to enter civil engineering. However, a year later, he returned to the army, once again fighting the Seminoles. Johnston joined Corps of Topographical Engineers rather than the artillery.

During the U.S.-Mexican War, Johnston served under General **Winfield Scott** and distinguished himself at Cerro Gordo, on April 18, 1847, after which he was breveted to colonel. After serving in the Mormon Expedition of 1857–58, Johnston was promoted to brigadier general and appointed quartermaster general of the army in June 1860. Less than a year later, in April 1861, he resigned his commission and joined the provision army of the Confederacy as a brigadier general in command of the Army of the Shenandoah. At First Bull Run, on July 21, 1861, Johnston was the ranking Confederate officer, helping to give the South its first major victory.

In August 1861, Johnston was promoted to general and made commander of the Confederate Army of the Potomac. He faced **George B. McClellan** in the Peninsular Campaign (1862), during which he was wounded at Fair Oaks on May 31 and was subsequently replaced by **Robert E. Lee** as commander of the Confederate Army of the Potomac. After Johnston's return to duty, he was named commander of the Department of the West in November 1862 and directed a heroic defense of Vicksburg, Mississippi, against the onslaught of **Ulysses S. Grant**. In command of dwindling forces, he was driven from his base of operations at Jackson, Mississippi, by General **William T. Sherman** on May 14 and could do nothing to prevent the fall of Vicksburg on July 4, 1863.

Following Vicksburg, Johnston became commander of the Army of Tennessee (December), then exacted a measure of revenge against Sherman when he capitalized on Sherman's ill-judged offensive maneuver and defeated the Yankee commander at Kennesaw Mountain, near Atlanta, Georgia, on June 27, 1864. Johnston next defended Atlanta itself, but Confederate president Jefferson Davis was appalled by Johnston's inability to halt Sherman and relieved him of command, replacing him with the heroically reckless John Bell Hood.

Robert E. Lee personally restored Johnston to command in February 1865, and Johnston made a valiant effort to check Sherman's March to the Sea at Bentonville, North Carolina, on March 19–21. Vastly outnumbered in North Carolina, Johnston surrendered to Sherman at Durham Station on April 26. At that point, the Civil War was effectively over.

Following the war, Johnston left military life and entered the insurance business in Georgia. He wrote his memoirs during 1865–78) and was elected to Congress (1879–81). He served as commissioner of railroads in 1885.

Johnston was one of the Confederacy's best commanders, aggressive but not foolish. He was a fine tactician as well as strategist, who was deeply admired by his men.

JOMINI, Antoine Henri, Baron de

BORN: March 6, 1779

DIED: March 22, 1869

SERVICE: Napoleonic Wars (1800–15); Russo-Turkish War (1828–29)

Controversial Swiss-French general, historian, and military theorist

Swiss born (at Payerne in the canton of Vaud), Jomini began his professional life as a banker in Basel (1796). He moved to Paris in 1798, where his friends and associates included expatriate Swiss radicals. Late in 1798, he returned to Switzerland to work in the war ministry of the newly formed Swiss Republic. By 1800, he held the rank of major, and the following year returned to his business career in Paris.

At this time, Jomini began work on *Traité des grandes opérations militaires*, an analytical study of strategy based on the campaigns of **Frederick the Great**. Among the readers of the *Traité* in manuscript was no less a figure than Marshal **Michel Ney**, **Napoleon**'s most celebrated subordinate, who invited Jomini to join his staff. Ney also helped Jomini publish his work in 1805.

Jomini served as aide-de-camp to Ney at Ulm during October 1805 and at the Battle of Austerlitz on December 2. After reading his book, Napoleon personally promoted Jomini to colonel on December 27. At the beginning of the Prussian campaign, Jomini was serving on Ney's staff, but soon transferred to the staff of the emperor and served with Napoleon at Jena (October 14, 1806) and Eylau (February 7–8, 1807). He was returned to service under Ney as chief of staff in Ney's corps. Although Ney did not always get along well with his subordinate, Jomini served with the marshal in Spain during 1808–09.

In November 1809, Jomini returned to France to serve under **Louis Alexandre Berthier**, who did not take well to Jomini's arrogance. Nevertheless, Jomini was promoted to *général de brigade* on December 7, 1810 and was put in command of the general staff's historical section (January 1811). During the Russian campaign, he was appointed military governor of Vilnius and, later, Smolensk. He also fought during the campaign, at the crossings of the Berezina, where he suffered great hardship and subsequently fell ill. He took a leave of absence early in 1813 to recover his health, then returned to the army as Ney's chief of staff on the eve of Lützen (May 1–2). He also fought at Bautzen on May 21, after which Ney recommended him for promotion to general of division. Instead of promotion, however, Jomini suffered arrest on orders of Berthier, who charged him with being late with the corps returns. Angered by this action, Jomini summarily deserted Napoleon and joined the Allies on August 14, 1813.

Jomini was made a lieutenant general in the Russian army and was appointed aide-de-camp to Czar **Alexander I**, whom he served throughout the balance of the war (1813–14). Declining to enter Paris with the Allies in 1814, he visited Vienna and Switzerland in 1815, then returned to Russia in 1816, where he was appointed military tutor to grand dukes Nicholas and Michael. Later, Jomini returned to Paris and lived there for some time.

During 1817–26, Jomini published a series of works on military history and theory. He left Paris for Russia, once again, in 1826 and became aide-de-camp to Czar **Nicholas I**, holding the exalted rank of general-in-chief. He saw further combat during the Turkish War of 1828–29, at the siege of Varna (August 5–October 11, 1828).

Following the war with Turkey, Jomini served as military adviser to the Nicholas. In 1838, while visiting Paris for an extended period, he published his masterwork, *Précis de l'art de la guerre*. Following service—on the Russian side—in the Crimean War, Jomini retired to Passy, where he died in 1869).

Jomini never lived down his desertion of the French, even though Napoleon forgave him for it and declared that his action was fully justified. Jomini's theoretical works have not endured like those of **Karl von Clausewitz**, but, in his own time, Jomini was as influential as Clausewitz. Jomini worked variations on the theme of simplifying strategy, massing force against the enemy's "decisive point" in order to win a quick and unambiguous victory.

JONES, John Paul

BORN: July 6, 1747

DIED: May 30, 1792

SERVICE: American Revolution (1775–83); Second Russo-Turkish War (1787–92)

American naval hero of the American Revolution

Jones was born John Paul near Kirkbean, Kirkcudbrightshire, on Solway Firth, Scotland. He was the son of a gardener and, in youth, went to sea a cabin boy aboard a merchant vessel. His first voyage, in 1759, was to Virginia, where he visited an elder brother living in Fredericksburg. After the bankruptcy of his employer in 1766, Jones's cabin-boy apprenticeship came to an abrupt end, and he became chief mate on a Jamaican slaver. Soon disgusted by the slave trade, he quit in 1768 and sailed for Scotland. This proved a fateful voyage. The captain and first mate died en route, and Jones assumed command to take the ship safely home. The owners of the vessel rewarded him with captaincy of the merchantman *John*, out of Dumfries, Scotland. However, during a voyage to Tobago, West Indies, Jones disciplined a sailor by flogging—a common practice of the era. The sailor then jumped ship and died, presumably of his wounds, aboard another vessel in April 1770. On his return to Scotland in November, Jones was arrested and jailed for murder, but released on bail to search for evidence that would exonerate him. Incredibly, while on a voyage to Tobago in search of that evience, he unintentionally killed a mutinous sailor in December 1773. Fearing that the worst would result from a trial, John Paul disguised his name by adding Jones to the end of it and fled to America, where he participated in the settlement of his late brother's Fredericksburg estate in 1774.

With the outbreak of the American Revolution, Jones was hired by Congress to fit out the twenty-gun *Alfred*, the first ship purchased by the new government. After raising the new nation's banner over the ship, on December 3, 1775, Jones was commissioned senior lieutenant in the Continental Navy four days later. He then sailed with the fleet to the Bahamas, where he participated in the capture of New Providence during March 1776. On April 6, commanding *Alfred*'s main battery, he distinguished himself against H. M.S. *Glasgow* and two days later was promoted to captain, with command of the twelve-gun *Providence*.

Jones was ordered to conduct raiding operations until he ran out of provisions. With this mandate, he set out on a cruise that resulted in the sinking of eight ships and the capture of eight others from the Bahamas to Nova Scotia during August 21–October 7. Returning to take command of the *Alfred*, he sailed in company with the *Providence* for Nova Scotia in November and captured a merchant ship as well as a military transport off Louisbourg. He also raided the coast of Nova Scotia, burning an oil warehouse and a ship, then capturing three vessels off Cape Breton.

Jones returned in triumph to Boston with his prize ships in December. He was commissioned captain of the *Ranger* on June 14, 1777 and sailed in it to France on November 1 to acquire the new frigate, *Indien*. Jones had the honor of receiving the first salute of the new United States flag from French ships in Quiberon Bay on February 14, 1778.

While in Europe, Jones raided two forts at Whitehaven, England, though he failed in an attempt to kidnap Lord Selkirk at St. Mary's Island in Galway Firth on April 23. Nevertheless, he captured the twenty-gun *Drake* off Carrickiergus, Ireland, on April 24, then put in at Brest on May 8. From here, he was assigned command of a U.S.-French expedition out of L'orient on August 14, 1779. Jones sailed in the *Bon Homme Richard* (of forty-two guns), and the flotilla also included the U.S. vessel *Alliance* (thirty-six guns), and two French ships, the *Pallas* (thirty-two guns) and the *Vengeance* (twelve guns). When they encountered a British merchant fleet off Flamborough Head, Jones attacked and defeated the escort, *Serapis*, in a spectacular night battle on September 23.

After this, Jones took command of the *Alliance*, but enjoyed no success on the subsequent cruise. Worse, he was abandoned ashore when the vessel was seized by its former captain on June 1780. The following year, Jones returned to the United States aboard the *Ariel* and on April 14 received the thanks of Congress. He was assigned command of the seventy-four-gun *America*, which was subsequently given to France. Eager to remain in command of the vessel, Jones volunteered to serve in the French fleet to the West Indies in December 1782.

Following the conclusion of the revolution, Jones went to France as a prize agent in November 1783, then returned to the United States, where Congress presented him with a gold medal on October 16, 1787. The following year, he accepted an appointment in the Russian navy as a rear admiral and, in this capacity, defeated the Turks at the battle of Liman in the Black Sea during June 17-27. However, Jones did not take pleasure in his Russian assignment and, with his health failing, he retired to Paris in 1789. He lived there in relative obscurity for the rest of his life. His grave, in Paris, was not identified until 1905. In 1913, his remains were returned to the United States for entombment at the Naval Academy in Annapolis.

Jones was a great naval commander, a man of courage and inexhaustible resourcefulness. As much as his victories aided the American cause in the Revolution, his greatest contribution may well be the example of gallantry he set for the U.S. Navy during its earliest years.

CHIEF JOSEPH

BORN: ca. 1840
DIED: September 21, 1904
SERVICE: Nez Percé War (1877)

American Indian leader during the so-called Nez Percé War of 1877

Chief Joseph, often called Young Joseph to distinguish him from his father, Joseph the Elder or Old Joseph, was born in the Wallowa Valley of Oregon. His Nez Percé name, variously transliterated as Heinmot Tooyalaket, In-mut-too-yah-lat-lat, Hin-mah-too-yah-lat-kekt, and Hinmaton-yalatkit, means "A thunder coming from water over land." (Old Joseph, who had adopted the Christian faith, baptized his son Ephraim.)

Old Joseph was among the Nez Percé leaders who, in 1855, signed a treaty with **Washington's** territorial governor Isaac Stevens, ceding much of their land to the federal government in return for the guarantee of a large reservation in Oregon and Idaho. Stevens almost immediately violated the treaty by pushing settlement into the reservation, provoking the Yakima War of 1855–56. Old Joseph managed to keep the Nez Percé out of the conflict until 1861, when gold prospectors encroached upon the Wallowa Valley. In 1863, government negotiators at the Lapwai Council demanded a revision of the 1855 treaty, calling for a great reduction in the size of reservation. For the most part, those Nez Percés whose homes remained undisturbed by the proposed revision, signed the new treaty and agreed to sell the old lands, while those—like Chief Joseph the Elder—whose homes lay outside the new boundary resisted and refused to sign. These so-called Non-Treaty Nez Percés nevertheless lived in relative peace because white settlement was slow to invade the Wallowa Valley, and there was no attempt to move the Non-Treaty bands by force.

In 1871, Chief Joseph the Elder died, and Young Joseph carried on his father's policy of passive refusal to move. Young Joseph had earned universal admiration and respect in part because of his great skill as a buffalo hunter, exhibited in hunting expeditions with his younger brother Olikut (1845–77) in Montana. Shortly after his father's death, homesteaders finally began to push into the Wallowa Valley, and Young Joseph gained even greater prestige by successfully protesting the incursion to the Indian Bureau. The result, in 1873, was a proclamation by President **Ulysses S. Grant** establishing the Wallowa Valley as a reservation.

Joseph's triumph was illlusory. Settlers ignored the reservation boundaries, established homesteads at will, and soon became a political bloc powerful enough to prompt Grant to reverse his decision in 1875. Although a number of other Nez Percé chiefs, older and more established, effectively outranked Joseph, he was recognized by white authorities as the most influential of the Indian leaders and it was with him that General **Oliver O. Howard** decided to negotiate. He met with Joseph and another leader, Old Toohoolhoolzote, at Fort

Lapwai on November 13 and 15, 1876. Joseph succeeded in winning Howard's sympathy as well as his acknowledgment that Old Joseph had never sold the Wallowa Valley, yet the general persisted in his mission and gave the Indians one month to move to a reservation or be driven off by force.

Characteristically of Joseph, he demonstrated great restraint, resisting tribal pressure (especially from his own brother, Olikut) to fight, because he realized that war against the far more numerous white settlers would be fruitless. However, on June 13 and 14, 1876, a number of young warriors led by Wahlitits, disheartened and drunk, murdered four whites who were implicated in the death of Wahlitits's father. Despite Joseph's protests to authorities that the killings had not been sanctioned by the tribal council, Howard dispatched one hundred cavalrymen to pursue the Non-Treaty Nez Percés, who had begun to move south, toward the Salmon River. The killing of fifteen more settlers increased the urgency of the mission, yet the Indians consistently eluded the pursuers. It was at Joseph's insistence that a delegation of Nez Percés met with Captain David Perry on June 17 for a parlay. The first to encounter the Indian party were undisciplined civilian volunteers, who opened fire, thereby igniting general warfare between the Nez Percés and the U.S. Army.

That first battle went to the Indians, who quickly sent the volunteers running and then readily flanked Perry's main column, killing one-third of his command. What followed was a series of Nez Percé victories, always against superior numbers, as Chief Joseph led his band of eight hundred for some three months in a trek over 1,700 miles of the most forbidding terrain the West had to offer. At each turn, Joseph eluded the pursuing army, and when an engagement took place, it was the army that was took the greater punishment.

Yet the pursuit steadily took its toll on the Nez Percés. Joseph sought haven for his people among the Crows, but, when he discovered that Crow scouts had been working for Howard, he determined to press on to Canada, hoping that the great Hunkpapa leader **Sitting Bull**, self-exiled there, would welcome them as his brothers. On September 30, 1877, Joseph and his followers were encamped a mere forty miles south of the Canadian border, on the northern edge of the Bear Paw Mountains. Three-hundred-and-fifty to four-hundred troopers commanded by **Nelson A. Miles** attacked in a bitter, snow-whipped battle that developed into a six-day siege. Realizing that the situation was hopeless, Joseph counseled surrender, but was resisted by two other leaders, Looking Glass and White Bird, who wanted to fight to the end. When Looking Glass was struck in the head and killed by a stray bullet on October 5, Joseph at last spoke to Miles in speech of extraordinary sorrow and dignity, concluding "Hear me, my chiefs! I am tired; my heart is sick and sad. From where the sun now stands I will fight no more forever."

Consigned with his followers to a reservation, Chief Joseph spent many years petitioning the government for permission to return to the Wallowa Valley. In these efforts he was aided by the military adversaries whose respect and admiration he had won: Oliver O. Howard and Nelson A. Miles. The petitions nevertheless proved fruitless, and Joseph died on the Colville Reservation, in the state of Washington, in 1904.

JOUBERT, Petrus Jacobus

BORN: January 20, 1831

DIED: March 27, 1900

SERVICE: First Anglo-Boer War (1880–81); Second ("Great") Anglo-Boer War (1899–1902)

Distinctly unmilitary in behavior, attitude, and bearing, Joubert was a Boer general who consistently defeated the British

Joubert was born into an impoverished farm family in the Oudtshoorn district of Cape Colony. He was taken by his family to settle in Natal and subsequently moved on to Transvaal, where his family settled in the Wakkerstroom district near Natal. The untutored Joubert taught himself English, Dutch, and law, then applied himself to farming, the law, and business, soon amassing a considerable fortune. He also became active in politics, gaining election as a member of the *Volksraad* for Wakkerstroom in 1860 and serving until 1876. A supporter of South African President T. F. Burgers, he served as vice president during Burgers's trip to Europe (1875–76). Joubert broke with Burgers over the president's expedition against the native chief Sekukuni in 1876. His reluctance to take up arms was further demonstrated in 1877 when he offered only the mildest verbal resistance to the British annexation of Transvaal (1877). Nevertheless, Joubert joined Paul Kruger in the independence movement and became one of the triumvirate of independence leaders, including Kruger and Andries Pretorius, that proclaimed the Transvaal independent on December 30, 1880.

Joubert was made commandant general of the Boer forces and immediately took the offensive. With 2,000 troops, he invaded Natal and handily defeated General Sir George Colley at Laing's Nek on January 28, 1881. Colley then took up the high ground at Majuba Hill, but Joubert skillfully counterattacked before Colley could outflank him. His Boers annihilated Colley's forces, thereby securing the Transvaal's independence on February 27.

Despite his military accomplishemnts, Joubert failed to best Kruger in the 1883 presidential election to Kruger. He ran unsuccessfully against him three more times, in 1888, 1893, and 1898, losing very narrowly in the last election. As he pursued his frustrating political career, Joubert also built up and trained the *staatsartillerie*, the artillery force that was the sum total of Transvaal's standing army. By the late 1890s, he had transformed the grassroots force into a small modern army, equipped with the latest ordnance from European makers. When the Second Boer War broke out in October 1899, Joubert was able to defeat General George White at the second Battle of Laing's Nek (October 12), at Talana (October 15), at Elandslaagte (October 21), and at Nicholson's Nek (October 30). Then he pushed White and his force into the town of Ladysmith on November 2 and contained them there. However, a bad fall from his horse forced him to resign command abruptly on November 25, 1899, and he retired to Pretoria, where he died some months later.

Joubert was of indecisve and mildly effeminate manner. As a businessman, he had the reputation of being a sharp dealer, and was called "Slim" Joubert—the word *slim* meaning crafty. He was not someone likely to achieve military success, and yet he did just that. No one seemed more amazed by this than Joubert himself, who attributed his victories to God rather than his own skill.

JUEL, Niels

BORN: May 8, 1629

DIED: April 8, 1697

SERVICE: First Anglo-Dutch War (1652–54); First Northern War (1655–60); Scanian War (1675–79)

Bold Danish admiral who developed extraordinarily successful tactics of naval warfare and who built the Danish fleet into a formidable weapon

Juel was born in Christiana (modern Oslo) to a noble family; as the younger son, however, he was not in line for the primary inheritance and decided to stake out a career at sea. He enrolled at the Soroe Academy, then studied navigation and seamanship in France and Holland. He apprenticed himself to Maarten van Tromp and Michel De Ruyter of the Dutch navy, and he served in this force during First Anglo-Dutch War. He fought at the battles of Dungeness (December 10, 1652), Portland (February 18–20, 1653), North Foreland (June 11), and Scheveningen (August 10).

During 1655-56, Juel fell seriously ill. He used his long recuperation to study the art of shipbuilding, then, in 1656–58, he sailed with De Ruyter against the Barbary corsairs. He next fought against the Swedes at the siege of Copenhagen during July 1658–February 1659, and is believed to have participated in the relief of Admiral Opdam (November 8, 1658) and De Ruyter (February 11, 1659). In recognition of his services, Juel was awarded the Order of the Dannebrog and promoted to admiral. He was also made director of the royal shipyard at Holde (near Copenhagen), a post he held from 1659 to 1660. However, he failed to achieve supreme command of the navy, having been passed over in favor of his rival Admiral Adelaer in 1663. Despite this disappointment and, as Juel saw it, affront, he continued to serve loyally, especially during the Scanian War against Sweden.

In 1676, Adelaer died suddenly, and Juel was given supreme command. He continued to press against the Swedes, launching a bold amphibious assault on Gotland, which fell to him in May 1676. A much larger Swedish fleet attacked him repeatedly off the Jasmund peninsula on Rügen Island during June 5, 1676, and Juel was able to hold this force at bay until Cornelis van Tromp arrived to reinforce him. Juel then participated in the Danish-Dutch victory off Oland on June 1, 1676. Following this triumph, the Danes were able to invade Scania. On June 1, 1677, Juel defeated another Swedish fleet (led by Erik Sjöblad), then went on to a decisive victory pover the superior fleet of Admiral Evert Horn at Kjöge Bight on July 1.

Juel was appointed vice president of the Admiralty Board (1678–83), then served as president until his death. He was an extraordinary naval commander, who specialized in such innovative tactics as cutting off part of the enemy line, isolating it, then hitting it with overwhelming force. He was adept at defeating an enemy in detail. Unlike many combat commanders, he was also a talented administrator, who played a central role in creating a formidable Danish fleet.

JUNOT, Jean Andoche, Duke of Abrantes

BORN: October 23, 1771

DIED: July 29, 1813

SERVICE: French Revolutionary Wars (1792–99); Napoleonic Wars (1800–15)

One of Napoleon's ablest—and most fanatical—generals

Junot was born into a well-off agricultural family at Bussy-le-Grand, in Burgundy, and joined the Côte d'Or Volunteers in 1792, fighting at the Battle of Longwy on August 23, during which he was wounded. After recovering, he saw service at the siege of Toulon from September 7 through December 19, 1793 and became secretary to **Napoleon** Bonaparte, whom he subsequently served as aide-de-camp in the Army of Italy during 1794–97.

In August 1796, Junot was seriously wounded in the head during a cavalry action and did not return to service until 1798, when he was part of the Army of the Orient. He fought at Malta on June 10 and at the Pyramids on July 21. In January 1799, he was promoted to provisional general of brigade and served with the Army of Syria, fighting at Nazareth and Mount Tabor in April. Returning to Egypt, he fought at Aboukir on July 25 and was taken prisoner by the British in October while he was on his way back to France. Ultimately released, he was promoted to general of division and, in 1803, appointed commandant of Paris.

Junot, who had by now earned the sobriquet of "The Tempest," set about training an army in Arras in preparation for an invasion of England. When this did not materialize, he was appointed by Napoleon ambassador to Lisbon in 1805, but soon left to serve on Napoleon's staff at Austerlitz. After serving in Italy, Junot was appointed governor of Paris in July 1806, then, as commander of the Corps of Observation of the Gironde, he was directed to invade and occupy Portugal. He captured Lisbon, whereupon Napoleon appointed him governor general of Portugal (1807).

Junot was defeated by Arthur Wellesley (later the **Duke of Wellington**) at Vimeiro on August 21, 1807 and was compelled to yield and evacuate Portugal by the terms of the Convention of Cintra, which was concluded the day after the battle. Junot then went to Spain, where he was given several corps-level commands. Leading III Corps, he besieged Saragossa from December 20, 1810 to February 20, 1809, then, in command of VIII Corps, he returned to Portugal as part of the Army of Portugal under Marshal **André Masséna**. He fought at the siege of Ciudad Rodrigo during June 16–July 19, 1810, then at the battles of Bussaco (September 27) and Sobral (October 11). Shortly after this, he received a bad wound to the face at Rio Maior.

On March 5, 1811, he fought rearguard action to cover the retreat of the army from the Lines of Torres Vedras and subsequently fought at Fuentes de Onoro during May 3–5. Junot then accompanied Napoleon's campaign into Russia during 1812, holding numerous commands in that epic invasion disaster. Perhaps in part due to the hardship endured in Russia, as well as to the head and face wounds he had received during his career, Junot became insane after his return to France.

He went to live with his parents in Montbard and, in their home, committed suicide.

Junot had great natural skill as a soldier, but was untutored in the military art. His advancement in Napoleon's army was due in no small measure to his intense devotion to the emperor. He was more capable as a tactician and a leader of men than as a strategist.

JUSTINIAN I THE GREAT

BORN: ca. 482

DIED: November 14, 565

SERVICE: First Persian War (524–32); Vandal War (533–34); Italian (Gothic) War (534–54); Second Persian War (539–62); Spanish War (554); Bulgar Invasion of Thrace (559)

Emperor of the Roman Empire from 527 to 565; his ambition was to restore Rome to its classical glory, but he gave his leading general insufficient support

Justinian I was born Flavius Petrus Sabbatius in the Macedonian Balkans to parents of Latin-speaking peasant background. As a child, he accompanied his uncle Justin, a prominent army officer, to Constantinople where, despite his humble origins, he received a thorough education. It was in honor of this uncle that he took the name Justinian.

Justin became emperor on the death of Anastasius I in 518, and Justinian served as his uncle's adviser. The childless Justin legally adopted the young man and conferred upon him the title of caesar in 525. On April 4, 527, Justinian became co-emperor, and, later that year, on August 1, he became sole emperor after Justin died.

The years of war with Persia had depleted the empire's treasuries, and the early years of Justinian's rule were harsh, condemned by the Senate as autocratic and oppressive. At last, on January 13–18, 532, the Nika Riots threatened Justinian holds on the throne. Rioting among the circus faction of the Hippodrome escalated into violent cries for governmental reform. The riot ended in an attempt by the aristocracy to overthrow Justinian, whereupon the emperor released masses of troops upon the rioters. The bloodbath ended the uprising.

Also in 532, however, Justinian seized on a chance for peace in the East, an opportunity to reduce military spending and improve conditions at home. He concluded with the new Persian king, Chosroes I, a "Perpetual Peace," which freed Roman military resources to be used in the liberation of territories in the West taken by the barbarians. The emperor directed his armies first at North Africa, where his ablest general, **Belisarius,** defeated the Vandals during 533–34. Next, Justinian turned to Italy, and, by 540, negotiated through Belisarius a treaty with the Ostrogoths. This peace endured for more than a decade before the Ostrogoths rebelled again and had to be suppressed, this time by Narses, a eunuch who was one of Justinian's most trusted generals.

Even in peace, Italy had been ravaged by war for so long that its economy refused to recover, and Justinian could not afford to keep up the massive defenses needed to repel future invasions—notably by the Lombards a few years after the emperor's death.

In 540, Chosroes violated the "Perpetual Peace" by invading Syria-Palestine. Once again, the emperor was bogged down in wars fought on far-flung battlefields. Peace was not secured until 562, when Justinian agreed to pay greater tribute sums to Persia.

The most enduring act of Justinian's reign was the codification of Roman law. Begun in 528 by the lawyer Tribonian, the work included all valid laws, opinions by Roman jurists, a textbook for students, and new laws. While Justinian's codification and strict enforcement of laws—especially those against tax evasion and illegal use of power—angered the aristocracy, the work had enormous influence on the future development of European law. Religious affairs also occupied much of the emperor's attention. Dissident groups within the Christian church fought over the divinity of Christ, the separation of religious and secular authority, and other matters. Justinian sought to reconcile these differences and convened the fifth ecumenical council. The council failed, and the eastern dissidents—the Monophysites—eventually transformed themselves into a separate church, dominant in Syria and Egypt.

Thus the last years of Justinian's reign were burdened by deepening economic crisis, insoluble religious differences, and, finally, a pandemic of bubonic plague, which swept the empire during 542–43. During these later years, opposition to the emperor steadily mounted, and repeated barbarian attacks from the Balkans came dangerously close to the capital. At last, when Justinian died on November 14, 565, the people of Constantinople rejoiced.

While Justinian enjoyed considerable success in fending off and even suppressing the barbarians, his empire was set upon on so many fronts that victory and prosperity were ultimately impossible to achieve.

K

KALB, Johann (Baron de Kalb)

BORN: 1721

DIED: August 19, 1780

SERVICE: War of the Austrian Succession (1740–48); Seven Years' War (1756–63); American Revolution (1775–83)

German general who served the Patriot cause in the American Revolution

Kalb was born in Bavaria of humble peasant stock. He left home in 1737 and, by 1743, was a lieutenant in a French infantry regiment. In the French service, he fought under Marshal **Maurice de Saxe** during the War of the Austrian Succession, and was promoted to major by 1756. Kalb next fought with distinction during the Seven Years' War, but retired from the army shortly after marrying a wealthy heiress in 1764.

In 1768, at the behest of the French foreign minister, he traveled to North America to assess the attitude of the British colonies toward the mother country. Within a few years, Kalb returned to Europe and reentered the French army under Count de Broglie. Promoted to brigadier general by November 1776, he was recruited by Silas Deane, the American diplomat who did much to persuade France to join the American cause, and sailed with the **Marquis de Lafayette** to revolutionary America on April 20, 1777. Spurned by a Congress unwilling to accept his services, he threatened to sue on breach of a contract signed with Deane. However, before this became an issue, he was commissioned a major general in the Continental Army on September 15, 1777. After wintering at Valley Forge (1777–78), he was appointed second in command under Lafayette in what turned out to be an abortive invasion of Canada.

On April 3, 1780, Kalb was ordered to march from Morristown, New Jersey, to relieve forces at beleaguered Charleston, South Carolina. Leading the Maryland and Delaware Continentals, he made the epic march, but was stunned on his arrival to discover that Charleston had already surrendered. He then joined forces with **Horatio Gates** at Deep River, North Carolina. Relinquishing command to Gates on July 25, he implored the general to attack. The overcautious Gates hesitated, thereby losing the vital element of surprise. With that, the Battle of Camden, at which Kalb served, turned into a rout. Yet even after the left and center had been routed by the British on August 16, Kalb fought on. Wounded no fewer than eleven times, Kalb was taken prisoner and died of his injuries on August 19, 1780.

Kalb is remembered as one of the handful of European professional soldiers who performed valuable service to the cause of American independence. He was an able solder, but was always in subordinate command positions. Had Gates taken his advice at Camden, the tragic battle might have ended differently for the Americans.

KARAGEORGE (George Petrovich)

BORN: November 14, 1762

DIED: July 25, 1817

SERVICE: Austro-Russo-Turkish War (1787–91); Serbian Revolt (1804–13)

Fierce Serbian nationalist who led a peasant army to repeated victories

George Petrovich was born in Visevac, Serbia, to a peasant family and was early on nicknamed Karageorge—Black George—because of his swarthy complexion and his equally dark moodiness. He herded swine and worked as a stable groom until he killed some Turks in a fight in 1786 and was forced to flee to Urbica. Here he married and then moved on to Austria in 1787, finding work as a forest ranger. The following year, he enlisted in the Austrian *Freikorps and* fought in Italy and during the Austro-Russo-Turkish War. After Austria concluded an armistice with Turkey in 1790, Karageorge remained in Serbia, where he continued to conduct a guerrilla campaign against the Turks.

Karageorge gave up battling the Turks after the Treaty of Sistova was concluded on August 4, 1791. He moved to the village of Topola, where he started to trade in cattle. By the turn of the century, Karageorge had become a wealthy merchant and the father of seven children. (The third of these later became prince of Serbia.)

Karageorge was elevated to prominence when the Serb peasantry resolved to rise against the Turks. They elected Karageorge *vozhd*—leader—of the independence movement. The target of the movement was the Janissaries—rebellious soldiers of the Ottoman Sultan—who exercised tyrannical control of Serbia. Karageorge waged war against this group successfully, but after the Janissary leader had been killed, Karageorge refused to stop fighting. He wanted to compel the Turks to grant Serbia autonomy in exchange for their assistance in suppressing the rebellious Janissaries. The Turks rebuffed the demand, whereupon Karageorge led his peasant army to a remarkable victory over highly trained Turkish regulars at the Battle of Ivankovac in 1805.

Following his victory over the Turks at Ivankovac, Karageorge created the Serbian State Council—the first step toward establishing independence. He then met the Turks in battle at Misar (1806) and, after triumphing there, went on to seize Belgrade. With this, Serbian independence had been achieved.

Karageorge strengthened Serbia's position in 1807 when he concluded an alliance with Russia in its war against Turkey. However, when Russia excluded Serbia from the armistice of Slobodzen later in 1807, Karageorge was filled with rage and whipped his followers into a similar state. At his urging, the Serbian State Council promulgated the first Serbian Constitution. Karageorge was named the "first and supreme hereditary leader" in 1808.

In 1809, war broke out again between Turkey and Russia. Serbia and Russia made common cause against the Turks, and Karageorge led a mixed Serbo-Russian army to victory over the Turks at Varvarin in 1810. This led to peaceful Serbo-Russian relations for a time, but the Treaty of Bucharest, which ended the Russo-Turkish War, provided a vague amnesty and autonomy for Serbia. The treaty left the question of independence wide open, stipulating that it be settled between the Turks and Serbs. This was yet another betrayal by Russia. The Turks threw all of their forces against the Serbs, who were defeated in the course of 1812–13. Karageorge was compelled to flee to Austria, where authorities interned him for a year before he was permitted to go on to Bessarabia in 1814. However, the Russians refused to allow him to return to Serbia during the revolt that took place the following year there. In response, Karageorge went to St. Petersburg to argue on behalf of his cause. The Russians ordered him interned in the Ukraine, and it wasn't until 1817 that he managed to travel secretly through Bessarabia back to Serbia. There, he was assassinated while he slept. His assailants were agents of the pro-Turkish Serbian leader Milosh Obrenovich and the Turkish vizier of Belgrade.

Karageorge was a thoroughly remarkable military leader, who proved capable of organizing a peasant army that could defeat a better-equipped and better-trained force. Morose and moody, Karageorge was also a terrifying presence, who was subject to fits of sudden violence. He personally killed some 125 people, including his father. An iron disciplinarian, he punished transgressions among his troops with summary execution.

KEARNY, Philip

BORN: June 1, 1814

DIED: September 1, 1862

SERVICE: U.S.-Mexican War (1846–48); Italian War (1859); Civil War (1861–65)

Fearless American general, whom Winfield Scott called "a perfect soldier"; nephew of the more famous Stephen Watts Kearny

Nephew of "The Pathfinder," **Stephen Watts Kearny,** Philip Kearny became a distinguished soldier in his own right. He was born in New York City and was educated at Columbia College, graduating in 1833. He joined the First U.S. Dragoons that year as a second lieutenant and served in 1834 under his uncle on a disease-plagued but otherwise peaceful expedition into Pawnee and Comanche country. From Stephen Watts Kearny he learned the art of soldiering, benefiting greatly from the training program his uncle had instituted at Fort Leavenworth, Kansas, which molded the First Dragoons into a first-class unit. He completed his military education in France, to which he was sent in 1839 to study cavalry tactics. In the French service the following year—on station in Algeria—he learned enough to write a manual on cavalry tactics, which was adopted by the U.S. Army.

On his return to the United States, Kearny served Generals Alexander Macomb and **Winfield Scott** as aide de camp, then resigned his commission in 1846. However, almost immediately, with the start of the U.S.-Mexican War, Kearny raised and equipped his own cavalry company, secured a captain's commission in December 1846, and went to Mexico with his company to serve under Scott.

Kearny performed with characteristic gallantry at the Battle of Churubusco on August 20, 1847, leading a magnificent charge into enemy's center. He lost an arm in the engagement, but gained a brevet promotion to major and earned the admiration of the usually dour Scott, who called him "the bravest man I ever knew."

Kearny once again retired from service in October 1851, but was enticed to return to France as an officer in the French Imperial Guard under Napoleon III. He won distinction at two major battles of the Italian War—Magenta, on June 4, 1859, and Solferino, on June 24—and was the first American to receive the Cross of the Legion of Honor.

Kearny returned to the United States at the outbreak of the Civil War in 1861, gaining an appointment as brigadier general of the New Jersey Volunteers and serving in General Samuel P. Heintzelman's III Corps, Army of the Potomac, as a divisional commander. He fought in the disheartening Peninsular campaign during 1862, performing with the vigor and utter disregard for personal safety that had become his trademark. Kearny fought at Williamsburg (May 5) and at Fair Oaks (May 31). At the Second Battle of Bull Run, during August 29-30, Kearny played a key role in stopping **Thomas "Stonewall" Jackson**'s pursuit at Chantilly. However, he was felled by a sniper's bullet while reconnoitering.

Kearny represented the ideal of the gallant commander of the nineteenth century. Fearless and chivalrous, he was a skilled and polished officer, who earned the adoration of his command.

KEARNY, Stephen Watts

BORN: August 30, 1794

DIED: October 31, 1848

SERVICE: War of 1812 (1812–15); U.S.-Mexican War (1846–48)

Called "The Pathfinder"; Kearny played a major role in the U.S.-Mexican War in the American West and was an admirable frontier officer

A native of Newark, New Jersey, Kearny studied at Columbia College in nearby Manhattan for two years, from 1808 to 1810, joining the army on the eve of the War of 1812. Commissioned a first lieutenant in March 1812, he fought at Queenston Heights with the Thirteenth Infantry, suffering a wound and capture on October 13. Upon his parole, he was promoted to captain on April 1, 1813, and served at Sacket's Harbor, Long Island, during the balance of 1812 and part of 1814, before being transferred to duty at Plattsburgh, New York later in the year.

Kearny remained with the army after the war, transferring to the Second Infantry, with which he was dispatched west in 1819 and, the following year, pioneered a route from Council Bluffs, Iowa, to the St. Peter's River in Minnesota. In 1821, he was assigned to duty at Fort Smith, Arkansas, as paymaster and inspector. In June of 1821, he joined the Third Infantry in Detroit and, two years later, was breveted to major.

Joining the First Infantry at Baton Rouge, Louisiana, he was assigned to the Second Yellowstone

Expedition up the Missouri River during 1824–25. He next established and built Jefferson Barracks in St. Louis as the headquarters for the army's Great Plains region in 1826 and two years later was sent to Prairie du Chien, Wisconsin, as commander of Fort Crawford. While serving there, he played a key role in negotiating the 1830 treaty of Prairie du Chien with the Indians.

During 1831–32, Kearny reestablished Fort Towson in Indian Territory (present-day Oklahoma), then returned to the East to serve as superintendent of recruiting in New York City (1832–33). Appointed lieutenant colonel of the newly created First Dragoons on March 4, 1833, he returned to Jefferson Barracks and set out under General Leavenworth on an expedition into Pawnee and Comanche country during 1834. Although peaceful, the expedition was plagued by disease. Accompanying Kearny on this assignment was his nephew, **Philip Kearny,** who became his uncle's protégé and was destined to serve as a soldier of great distinction in the U.S.-Mexican War and the Civil War.

Stephen Watts Kearny was promoted to colonel of his regiment on June 4, 1836, and took the unit to Fort Leavenworth, Kansas, where he crafted it into an elite frontier regiment. During this period, he wrote *Carbine Manual, or Rules for the Exercise and Maneuvers for the U.S. Dragoons* (published in 1837), which served as a textbook for frontier warfare. In 1837, Kearny laid out a military road from Fort Leavenworth to the Arkansas River and, two years later, successfully intervened to prevent a war between the Potowatomies and the Otoes. In 1840, he was instrumental in heading off civil disturbances among the Cherokee, and in 1842 prevented the Indians from participating in a Texas-Mexico border clash. For these accomplishments, Kearny earned a reputation as an able administrator committed to peaceful resolution of conflict wherever possible. He was honored on all sides as a man of integrity, who was committed to justice and fair play.

In July 1842, Kearny returned to Jefferson Barracks to command the newly created Third Military District, which encompassed all of the Great Plains. He led an expedition over the Oregon Trail to Wyoming and back by way of Colorado and the Santa Fe Trail in 1845. The following year, with the outbreak of the U.S.-Mexican War, he was called on to mount an expedition to capture New Mexico and then to occupy California. Organizing his dragoons as well as a force of Missouri mounted riflemen (under **Alexander Doniphan**), he led his 1,700-man army to the conquest of Santa Fe on August 18. Capturing the town was one thing, Kearny realized, and holding it was another. His skill as an administrator and his reputation as a just man won over the native Mexican population.

Stephen Watts Kearny. Courtesy Library of Congress.

Receiving intelligence that Commodore R. E. Stockton and Lieutenant Colonel John Charles Frémont had taken California, Kearny set out for the Pacific Coast with a detachment of 120 dragoons. When he reached the vicinity of San Diego on December 2, he found a rebellion in progress. Encountering a force of Mexicans at San Pasqual, he fought a sharp but inconclusive engagement on December 5, then pressed onward to join Stockton just outside San Diego. Kearny had been given authority as

U.S. commander in California, but Stockton, who had proclaimed himself governor, challenged him. Setting aside ego and recognizing that Stockton's men constituted the bulk of available U.S. forces, Kearny acknowledged Stockton's authority and secured his cooperation in a combined army and navy force assault on Los Angeles. Kearny defeated the Mexicans at San Gabriel on January 8, 1847 and at Mesa on January 9. With this, U.S. authority in California was reestablished and the rebellion suppressed. However, a dispute with Stockton (and his ally Frémont) erupted again over command authority of land forces. Kearny secured confirmation of his authority from Washington, but Stockton remained uncooperative and unyielding. At last, Stockton was relieved by Commodore W. Brandford Shubrick, but, as a parting gesture, appointed Frémont governor. Prudently, Kearny waited until he had command of reinforcements before he—peacefully—replaced Frémont as governor.

Kearny quickly set about creating a stable government for the nation's new territory, then returned to Fort Leavenworth, with Frémont in tow. On arrival at Fort Leavenworth, he arrested Frémont and initiated court-martial proceedings against him for insubordination. Frémont was convicted, despite the efforts of his father-in-law, the powerful Missouri senator Thomas Hart Benton. Benton also tried unsuccessfully to block Kearny's assignment in Mexico as commander of Veracruz in April 1848.

Stricken with yellow fever, Kearny was sent to Mexico City to recuperate. Recovering, he took command of Second Division and was appointed military governor of Mexico City. He returned to Veracruz, then sailed for the United States on July 11, broken in health, reaching Jefferson Barracks on July 30. Just before his death three months later, Kearny was breveted major general—despite a two-week filibuster from Benton in opposition to the promotion.

Kearny was one of the most admirable military figures of the western frontier. A capable and courageous soldier, and a fine leader, he was also a believer in solid, peaceful civil government. His essentially peaceful capture and occupation of Santa Fe was a prime example of his modus operandi.

KEITEL, Wilhelm

BORN: September 22, 1882

DIED: October 16, 1946

SERVICE: World War I (1914–18); World War II (1939–45)

Nazi German general and war criminal fanatically devoted to Adolf Hitler

Keitel was born in Helmscherode, Braunschweig, and early in life determined on a professional military career. He served as an artillery officer during World War I, later gaining an appointment to the General Staff. Keitel was seriously wounded in action.

After the war, Keitel was active in the *Freikorps*, serving in a number of regimental commands before he was appointed head of the Army Organization Department in 1929. In 1934, he was promoted to *generalmajor* and, the following year, was assigned to the War Ministry as head of the Armed Forces Office. He served in this office through 1938, gaining promotion to *generalleutnant* in 1936 and *general der artillerie* in 1937. On February 4, 1938, Keitel replaced General Werner von Blomberg as chief of staff of the Armed Forces High Command and in November became *generaloberst*. With the fall of France in July 1940, he received his final promotion, to *feldmarschall*, and was charged with concluding armistice negotiations with the French at Compiègne, in the very railway wagon in which the Germans had signed the armistice ending World War I.

Throughout World War II, Keitel served as **Hitler's** senior military adviser. His loyalty to Hitler was absolute and unquestioning, even when the Führer's military judgment faltered and then disintegrated. Hitler adored Keitel, calling him "the greatest commander of all time," whereas the German general staff secretly reviled him with the nickname *Lackeitel* (from *Lakai*, "lackey").

After Germany's surrender on May 8, 1945, Keitel reaped the reward of his loyalty to Hitler. He was convicted by the International Military Court at Nuremberg of war crimes and was sentenced to death.

Specifically cited in the indictment against him was his approval of mass murders in Poland, and his support of the SS *Einsatzgrüppen*, the "Special Action Units" whose mission was the mass murder of "undesirable" civilian populations in Russia. Most infamous was his *Nacht und Nebel* (Night and Fog) orders, which authorized the secret and summary arrest of any persons deemed "endangering German security." Keitel was hanged in the Nuremberg Prison.

Keitel's rise through the Nazi military was meteoric, and he doubtless possessed considerable ability as a military organizer. His unthinking obedience to Hitler, however, was no service to the German war effort and ultimately led to the crimes against humanity for which he was tried and executed.

KESSELRING, Albert von

BORN: November 30, 1885

DIED: July 16, 1960

SERVICE: World War I (1914–18); World War II (1939–45)

One of Nazi Germany's best commanders; Kesselring excelled at defensive strategies in the later phases of World War II

Kesselring was born at Marktsteft, Bavaria, and joined the Bavarian Army as an artillery lieutenant in 1906. By the time of World War I, he was a staff officer, who served in various assignments. He remained in the much-reduced *Reichswehr* after the disarmament decreed by the Treaty of Versailles, always serving in staff posts.

In October 1933, Kesselring transferred to the *Luftwaffe* (air force), at the time a secret and illegal military branch, having been proscribed by the Versailles treaty. Kesselring was promoted to chief of the Luftwaffe general staff in June 1936, with rank of lieutenant general. The following year, he was promoted to general and assigned command of *Luftflotte* I (Air Fleet 1). In this capacity, from 1938 through January 1940, he led the devastating air operations during the invasion of Poland. He transferred to the command of *Luftflotte 2* in January 1940 and directed air operations during the invasion of France (May–June 1940). Following this triumph, Kesselring was promoted to field marshal on July 19, 1940, and commanded direct his Luftflotte during the Battle of Britain. From August 8 to September 30, 1940. Kesselring's strategy was to attack airfields in southern England. It was a highly successful policy that threatened to annihilate the RAF on the ground. However, **Adolf Hitler** ordered Kesselring to redirect his efforts against London on September 7, and the Blitz (after "*Blitzkrieg*," or lightning war) began.

Although the Blitz was devastating to the citizens of London and other civilian targets, it allowed the RAF to operate and ultimately resulted in the defeat of the Luftwaffe in the skies over England. By deviating from Kesselring's original strategy, Germany had lost its bid for air supremacy in England. Despite the failure of the Luftwaffe, Kesselring was appointed *Oberbefehlshaber (OB) Sud*—commander in chief, south—in December 1941. His command responsibilities included the Mediterranean Basin, and he shared with General **Erwin Rommel** direction of the campaign in North Africa.

Kesselring's mission became one of defense by 1943. He supervised the largely successful evacuation of Tunisia in May, then directed the defense of Sicily from July 9–August 17, which cost the Allies dearly. Appointed commander of Army Group C in November 1943 after **Benito Mussolini** had been overthrown (on July 24) and the Allies had invaded the Italian mainland (during September 8–9), Kesselring was faced with a desperate command. He managed his assignment with great skill, making the Italian campaign slow and costly for the Allies, especially the Americans. The defense spanned 1943-45.

Kesselring was dispatched to the German Western Front in March 1945 in a desperate effort to halt the Allied advance. It was a forlorn hope. In May 1945, Kesselring was taken prisoner by Allied units and was tried in Nuremberg for war crimes, accused of having authorized the massacre of 320 Italian prisoners in the Ardeatine Caves. He was sentenced to death in May 1947, but the tribunal commuted his sentence to life imprisonment in October 1947. He was released in October 1952 because of ill health and lived quietly until his death at Bad Nauheim in 1960.

Kesselring was a formidable commander. Had he been permitted to carry out his original air strategy against Britain in 1940, that nation might well have suffered invasion. His subsequent military accomplishments were almost all defensive operations performed under increasingly desperate circumstances.

KIMMEL, Husband Edward

BORN: February 26, 1882

DIED: May 14, 1968

SERVICE: World War I (1917–18); World War II (1941–45)

American admiral, who with General Walter Campbell Short, bore the brunt of responsibility for U.S. unpreparedness at Pearl Harbor

Born in Henderson, Kentucky, Kimmel graduated from the Naval Academy at Annapolis in 1904 and was assigned to service aboard several battleships in the Caribbean during 1906–07. He was assigned to the U.S.S. *Georgia* (BE-15) during the around-the-world cruise of the Great White Fleet from December 16, 1907 to February 22, 1909. Kimmel was wounded during the U.S. occupation of Veracruz, Mexico, in April 1914.

In 1915, Kimmel was appointed aide to assistant secretary of the navy Franklin D. Roosevelt, then was detached to an adviser post with the British Grand Fleet, where he taught British officers new gunnery techniques during World War I. Recalled to United States service when the nation entered the war, he served as squadron gunnery officer with the U.S. Sixth Battle Squadron from 1917 to 1918. He was executive officer aboard U. S. S. *Arkansas* (BB-33) from 1918 to 1920, then served ashore as production officer at the Naval Gun Factory in Washington, D.C. from 1920 to 1923. Dispatched to the Philippines, he headed the Cavite navy yard and was subsequently given command of Destroyer Divisions 45 and 38 during 1923–25.

After completing the senior course at the Naval War College in 1926, Kimmel was promoted to captain in July and assigned to the office of the chief of naval operations from 1926 to 1928. Assigned command of Destroyer Squadron 12 in the Battle Fleet from 1928 to 1930, he moved onto an assignment as director of ships' movements in the office of the Chief of Naval Operations in 1930 and served there until 1933, when he was given command of U.S.S. *New York* (BB-34). In 1935, he was given another shore assignment, this time in the Navy Budget Office, was promoted to rear admiral in November 1937, and was appointed commander of Cruiser Division 7 in July 1938. The following June, Kimmel became commander of Battle Force Cruisers, and of Cruiser Division 9.

Kimmel was jumped over forty-six more senior admirals to the post of CINCPAC (commander in charge of the Pacific), with his flag aboard U.S.S. *Pennsylvania* (BB-38) in Pearl Harbor, territory of Hawaii. By February 1941, he held the rank of admiral and commenced thorough preparations for anticipated war.

Tragically, the one eventuality he did not prepare sufficiently for was a preemptive surprise attack on Pearl Harbor itself. He was, of course, not alone in being surprised by the devastating December 7, 1941, Japanese air assault. However, with his army counterpart, General **Walter Short,** he bore much of the direct blame for the disaster. On December 17, he was relieved of his post and sent to Washington to testify in the initial inquiries.

Kimmel was never officially blamed for Pearl Harbor, but his career was destroyed. He retired from the Navy with the rank of rear admiral on March 1, 1942, and was periodically summoned to additional Pearl Harbor inquiries through 1946. After working for an engineering firm from 1946 to 1947, he retired from public life completely, later publishing *Admiral Kimmel's Story* as a defense of his actions.

Kimmel was generally an excellent officer, who was highly thought of in professional military circles. His culpability for Pearl Harbor remains a topic of debate among historians. Part of the problem was poor communications and cooperation between army and navy contingents on Oahu, which failed to share intelligence adequately. While Pearl Harbor destroyed Kimmel's career, it also deprived the U.S. Navy of the services of a highly capable admiral during the balance of World War II.

KING, Ernest Joseph

BORN: November 23, 1878

DIED: June 25, 1956

SERVICE: Spanish-American War (1898); World War I (1917–18); World War II (1941–45)

Veteran of the Spanish-American War, King was a key, if irascible, U.S. admiral during World War II; was an important naval strategist

A native of Lorain, Ohio, King was a midshipman during the Spanish-American War and sailed aboard U.S.S. *San Francisco* patrolling off the East Coast during April-December 1898. After the war, he returned to the Naval Academy, from which he graduated near the top of the class of 1901. As an ensign in 1903, he served aboard U.S.S. *Cincinnati*, from which he observed naval action during the Russo-Japanese War (February 1904–September 1905).

King was promoted to lieutenant in June 1906 and, from then until 1909, he was an ordnance instructor at the Naval Academy. From 1909 to 1913, he pulled sea duty aboard battleships of the Atlantic Fleet. After promotion to lieutenant commander in July 1913, he served ashore with the Engineering Experimental Station at Annapolis, then, in 1914, was given command of the destroyer U.S.S. *Terry* (DD-25) off Veracruz during the April–November Mexican crisis.

King was promoted to commander in 1917 and to the temporary rank of captain in September 1918, serving in the Atlantic Fleet during America's involvement in World War I. Following the Armistice, he was chosen to head the postgraduate department at the Naval Academy, then, in 1921, chose sea duty again, as commander of a refrigerator ship off the East Coast. In 1922, he undertook submarine training at New London, Connecticut, and afterward assumed command of Submarine Division II through 1923, then became commandant of the Submarine Base at New London, until he was appointed senior aide to Captain H. E. Yarnell, commander of Aircraft Squadrons Scouting Fleet (1926–27). While working under Yarnell, King enrolled in aviator training and received his pilot's wings at the age of forty-eight in May 1927.

Proficient in naval aviation, King was appointed assistant chief of the Bureau of Aeronautics (1928–29, then became commander of the naval air base at Hampton Roads, Virginia. From 1930 to 1932, he was captain of the aircraft carrier U.S. S. *Lexington* (CV-3). King graduated from the Naval War College senior course in 1933 and was promoted to rear admiral in April of that year, securing an appointment as chief of the Bureau of Aeronautics, a post he occupied until 1936.

In 1936, King was given command of the Aircraft Scouting Force, and, in 1938, following promotion to vice admiral, became commander of the five-carrier Aircraft Battle Force. He left this post to join the General Board in August 1939, then assumed command of the Fleet Patrol Force in the Atlantic in December 1940. Promoted on February 1, 1941 to admiral and commander in chief of the Atlantic Fleet, he directed the undeclared antisubmarine war with Germany off the U.S. East Coast.

After Pearl Harbor, King was named chief of naval operations (December 1941) and then commander in chief of the United States Fleet (March 13, 1942). In this post, he played a critical role in formulating Allied strategy and was a participant in all of the major Allied conferences. His principal focus was the Pacific theater.

Promoted to fleet admiral on December 17, 1944, he retired after the war, in December 1945, but continued to serve in an advisory capacity to secretaries of the Navy and of Defense, as well as to President Harry S. Truman. He published a memoir in 1952.

King had extensive sea service, although he never personally saw combat. Nevertheless, he was a sound strategic thinker and a man with a fierce reputation for getting things done. He gained notoriety during World War II by remarking on his appointment as Chief of Naval Operations: "When they get into trouble, they always call for the sons-of-bitches."

KINKAID, Thomas Cassin

BORN: April 3, 1888

DIED: November 17, 1972

SERVICE: World War I (1917–18); World War II (1941–45)

World War II American admiral who excelled at coordinating cooperative effort among the services

A native of Hanover, New Hampshire, Kinkaid graduated from the Naval Academy in 1908 and sailed with the Great White Fleet aboard U.S.S. *Nebraska* (BB-14) and U.S.S. *Minnesota* (BB-22) during 1908–11. He made himself an expert in naval gunnery and attended the ordnance course at the Naval Postgraduate School in Annapolis in 1913. Promoted to lieutenant (jg) in June 1916, Kinkaid served patrol duty off the East Coast as the United States prepared to enter World War I. Promoted to lieutenant in November 1917, he was sent overseas as gunnery officer of U.S.S. *Arizona* (BB-39) in April 1918.

Following the Armistice, Kinkaid returned to the States as an officer in the Bureau of Ordnance, where he served until 1922. That year, he was promoted to lieutenant commander and became an aide to Admiral Mark Bristol, sailing near Turkey during 1922–24. Kinkaid received his first sea command, of the destroyer U.S.S. *Isherwood* (DD-284) in 1924, then returned to shore duty in 1925 as an officer at the Naval Gun Factory in Washington, D.C. He returned to sea as commander and gunnery officer with the U.S. Fleet from 1927 to 1929, when he enrolled in the Naval War College (graduated 1930).

From 1933 to 1934, Kinkaid was executive officer on the battleship U.S.S. *Colorado* (BB-45) in the Battle Force. He then directed the Bureau of Navigation's Officer Detail Section from 1934 to 1937 and, promoted to captain, was given command of the cruiser U.S.S. *Indianapolis* (CA-35). He left this command in 1938 to become naval attaché and naval air attaché in Rome (November 1938) and naval attaché in Belgrade, Yugoslavia. In March 1941, Kinkaid returned to the United States and received promotion to rear admiral.

Given command of Cruiser Division 6 the month before Pearl Harbor, he was dispatched after the Japanese surprise attack to support raids against Rabaul and New Guinea (March 1942). Kinkaid fought in the battle of Coral Sea (May 4–8) and at Midway, during June 2–5, the turning point of the naval war in the Pacific. He assumed command of Task Force 16, built around the aircraft carrier U.S.S. *Enterprise* (CV-6). In this role, he supported the crucial landings on Guadalcanal (August 7) and fought in the carrier battles of the Eastern Solomons during August 22–25, and off the Santa Cruz Islands (October 25–28). During November 12–15, he fought Japanese surface ships in the vicinity of Guadalcanal.

Named commander of the North Pacific Task Force, Kinkaid directed the recapture of the Aleutian Islands. He retook Amchitka on February 12, 1943, and then Attu on May 11–30. He landed troops unopposed at Kiska on August 15, the final action of the Aleutian campaign.

After promotion to vice admiral in June, Kinkaid was transferred to command of Allied Naval Forces in the Southwest Pacific Area and, on November 26, the U.S. Seventh Fleet as well. In this capacity, he supported General **Douglas MacArthur**'s amphibious advance along the New Guinea coast toward the Philippines. In coordination with Admiral **William F. "Bull" Halsey**'s Third Fleet, Kinkaid's Seventh Fleet covered the American landings on Leyte (October 20, 1944). A crucial point in this campaign came when Kinkaid responded to intelligence of a massive Japanese counterattack by deploying his superannuated battleships under Admiral **Jesse B. Oldendorf** to blockade the southern entrance to Leyte Gulf at Surigao Strait. Olendorf checkmated the advance of the Japanese southern force and, during a spectacular night battle on October 25, destroyed it. However, the main Japanese attack force, which arrived off the east coast of Samar, was met only by a small group of escort carriers under Admiral Clifton E. Sprague. When Sprague was joined by a detachment of Oldendorf's battleships on October 25, the Japanese withdrew.

Kinkaid next directed additional Philippine amphibious operations against Mindoro (December 15) and at Lingayen Gulf on Luzon (January 9, 1945). After promotion to admiral in April 1945, Kinkaid directed the landing of American occupation forces in China and Korea during September, then left the Seventh Fleet to take command of the Eastern Sea Frontier at New York from January to June 1946. He was named commander of the Atlantic Reserve Fleet in January 1947, in which post he served until he retired on May 1, 1950.

Kinkaid was a personally warm and likable commander, the diametric opposite of **Ernest Joseph King.** In large part due to his charm and geniality, he was able to secure effective cooperation from army, navy, and Army Air Force units. This made him especially adept at directing amphibious efforts, both in the Aleutians and in the Philippines. His subordinates responded well to the large degree of autonomy he invested in them.

KIRBY SMITH, Edmund

BORN: May 16, 1824

DIED: March 28, 1893

SERVICE: U.S.-Mexican War (1846–48); Civil War (1861–65)

Confederate general in charge of the Trans-Mississippi Department during the Civil War

Kirby Smith was a born in St. Augustine, Florida, and was a graduate of West Point, Class of 1845. Commissioned in the infantry, he served in the U.S.-Mexican War, fighting in most of the major battles from Palo Alto (May 8, 1846) to Mexico City (September 13–14, 1847). At Cerro Gordo (April 18, 1847), he was breveted to first lieutenant, and at Contreras (August 19–20), he made brevet captain. From 1849 to 1852, he taught mathematics at West Point. Promoted to regular captain in 1855, he served in the Second Cavalry under **Albert Sidney Johnston** against the Indians in the Southwest.

Kirby Smith resigned early in the secession period prior to the Civil War and was appointed colonel in the provisional Confederate cavalry in March 1861. In June, he was upgraded to brigadier general and given command of a brigade at First Bull Run on July 21, where he sustained a severe wound. During 1862, he was a major general and divisional commander under P. G. T. Beauregard, and he led **Braxton Bragg**'s invasion of Kentucky during July–October. Taking Richmond, Kentucky on August 30, he cleared all Union forces out of the Cumberland Gap before he was forced to withdraw because of lack of reinforcement and support. Rejoining Bragg at the Battle of Perryville on October 8, he also fought at Stones River (December 31, 1862–January 3, 1863).

Kirby Smith made lieutenant general in October 1862, and in February of the next year was given command of the Trans-Mississippi Department. After July 4, 1863, however, the fall of Vicksburg cut off the Confederate West from the East, and Kirby Smith was isolated. Nevertheless, he made his department self-sufficient and enforced iron discipline over his ragtag troops, who referred to their region as Kirby Smithdom.

Promoted to general in February 1864, Kirby Smith was highly effective during the Red River campaign of March–May, when he checked the Union advance at Sabine Cross Roads on April 8 and forced the Federals to retreat down the Red River all the way to the Mississippi. Kirby Smith remained in command for the duration of the war and was, in fact, commander of the last major Confederate force in the field when he finally surrendered at Galveston, Texas on May 26, 1865.

Following the war, Kirby Smith traveled in Mexico and Cuba, returning to the States at the end of 1865. He enjoyed a successful civilian life, as president of the Atlantic and Pacific Telegraph Company (1866–68), president of the Western Military Academy (1868–70), and president of the University of Nashville (1870–75). From 1875 until his death, he was mathematics professor at the University of the South.

As Kirby Smith had been the last principal Confederate officer to relinquish his command, so he was the last surviving full general of the Civil War. He was a highly resourceful and independent commander, who excelled at what the best Confederate generals did well: making do with very little.

KITCHENER, Horatio Herbert, First Earl Kitchener of Khartoum

BORN: June 24, 1850

DIED: June 5, 1916

SERVICE: Franco-Prussian War (1870–71); Mahdist War (1883–98); Second Anglo-Boer War (1899–1902); World War I (1914–18)

Britain's foremost general at the beginning of the twentieth century

Irish-born (near Listowel in County Kerry) son of a British army officer, Kitchener attended the Royal Military Academy, Woolwich, and graduated with a commission in the Royal Engineers in January 1871. He saw his first combat service not with the British army, however, but with the French, as a volunteer during the Franco-Prussian War (1870–71). After this, during 1874–82, he served an extensive tour in Palestine, Anatolia (Turkey), and Cyprus, performing military survey and intelligence work. Late in 1882, he was posted to Cairo as second in command of an Egyptian cavalry regiment, then served with signal distinction as intelligence officer in Viscount Wolseley's failed Nile expedition (October 1884–March 1885) to relieve General **Charles ("Chinese") Gordon**, who was besieged by the Mahdi's forces at Khartoum.

After service on Zanzibar, Kitchener returned to Sudan as governor general of the Red Sea coast. He received a severe facial wound in a battle with Mahdist general Osman Dinga on January 17, 1888. Kitchener then served as adjutant general in Cairo, after which he secured appointment in 1892 as commander in chief (*sirdar*) of the Egyptian army—this despite the fact that he held the rank of colonel in the British army. Kitchener worked wonders for the disorganized and backward Egyptian army, remolding it into an efficient, formidable, and modern force.

In 1896, Kitchener persuaded his superiors to allow him to invade Sudan in order to punish the **Mahdi's** followers and avenge the death of Gordon. Kitchener advanced up the Nile, building a railroad as he went. He took Dongola on September 21, 1896, and Abu Hamed on August 7, 1897. He engaged and defeated Mahdist forces under Osman Dinga and Khalifa Abdulla at the Battle of Atbara River on April 7, 1898, then attacked the main Mahdist army outside Omdurman. There, thanks in large part to the skill of a subordinate, General Hector MacDonald, who parried a strong counterattack, he defeated this large force on September 2, 1898. Kitchener continued upriver and confronted a small French unit commanded by Major Jean B. Marchand at Fashoda in southern Sudan. Pressed by Kitchener, the French withdrew from the region on November 3, and Kitchener was appointed governor general of the Sudan, serving from 1898 to 1899.

With the outbreak of the Second ("Great") Boer War, Kitchener was appointed Sir Frederick Roberts's chief of staff in South Africa (December 18, 1899) following initial British defeats. Arriving in Cape Town on January 10, 1900, Kitchener was instrumental in formulating with Roberts a plan to outflank Boer positions in southern Orange Free State during February 11–16. Kitchener then took charge of the Battle of Paardeberg. His attack on Pieter Cronjé's forces failed on February 18, but Cronjé surrendered on February 27. Next, Kitchener spearheaded the British advance through Bloemfontein on March 13 to Johannesburg, on May 31, and finally to Pretoria on June 5.

After Roberts was recalled to Britain to succeed Sir Garnet Wolseley as commander in chief, Kitchener was named commander in South Africa on November 29. He launched a vigorous campaign to annihilate Boer guerrillas, but met with little success until he waged total warfare against the Boers, destroying their farms in an effort to deprive them of supplies, then forcing Boer families into what he called "concentration camps." With the guerrillas' support network disrupted, Kitchener pressed the attack against them, also crisscrossing the countryside with block-houses and barbed-wire to restrict mobility. Ultimately, the guerrillas were forced to surrender, and the Boers accepted British sovereignty by the terms of a May 31, 1902 treaty.

Kitchener returned to England as a national hero in July 1902. He was created viscount, but turned down an appointment in the War Office to accept appointment in India as commander in chief there. He set about radically reorganizing the army during 1902–1909, molding it into an efficient colonial force. In 1909 he was promoted to field marshal, declined appointment as Viceroy of India, but accepted appointment as Viceroy of Egypt and the Sudan, where he served from 1911 to 1914.

Kitchener was back in England, on leave, when World War I broke out. With reluctance, he accepted appointment, in July 1914, as Secretary for War and made the unpopular—but all too realistic—prediction that the war would be long. He called for full-scale mobilization, which included expansion of British industry as well as the creation of a vast army. Accustomed to taking full charge of all efforts, he refused to delegate authority and was soon swamped. In May 1915, he was relieved of his duties overseeing industrial mobilization, then, while he was touring the stalemated beachheads of Gallipoli in November 1915, the British Cabinet ended his strategic authority. Disgusted, Kitchener remained a patriot and stayed on in the Cabinet. He was dispatched on a mission to Russia and perished when the cruiser H.M.S. *Hampshire* struck a mine and sank off the Orkney Islands on June 5, 1916.

Kitchener was an extraordinary soldier. His ambition was unbounded, and he was reviled as ruthless and arrogant. His personality was cold and remote. Yet he was both a courageous and efficient commander, who was ideally suited to colonial assignments. Although he died relatively early in World War I—and had been stripped of strategic authority even earlier—it was largely because of his vigorous efforts that Britain mobilized so extensively on the war's Western Front.

KLÉBER, Jean-Baptiste

BORN: March 9, 1753

DIED: June 14, 1800

SERVICE: French Revolutionary Wars (1792–99)

Brilliant French general of the Revolutionary Wars

An Alsatian, Kléber was born in Strasbourg, the son of a mason. He planned on becoming an architect, but was soon drawn toward a military career and accepted a lieutenancy in the Austrian army in 1776. Soon discouraged by the lack of prospects for advancement, he resigned his commission and returned to Alsace in 1785. He was appointed inspector of public buildings in Belfort, then, in 1789, joined the National Guard during the Revolution. He served along the upper Rhine as a lieutenant colonel in the Alsace Volunteer Battalion in 1792, winning great distinction at the siege of Mainz (April 10–July 23, 1793). He was jumped to the rank of general of brigade on August 17, 1793 and sent to fight the Vendee rebels, defeating them at Cholet on October 17 and at Le Mans on December 13. In this battle, his forces inflicted more than 15,000 casualties. At Savenay, on December 23, he enjoyed another stunning victory and was promoted to general of division in April 1794. Transferring to the Army of the Sambre and Meuse under Count Jean-Baptiste Jourdan, he forced the surrender of the Austrian garrison at Charleroi during June 19–June 25, then fought valiantly at Fleurus on June 26.

Kléber was put at the head of a 100,000-man army and marched through Belgium in a campaign that ended with the triumphant siege of Maestricht during September 22–November 4. Recommended for the post of commander in chief of the army, he declined, pleading ill health. Instead, he was assigned to command a corps under Jourdan in June 1795 and kept it together during the retreat over the Rhine in 1796. Although he brought his corps back to Bamberg, he refused what he considered a suicidal order to counterattack at Würzburg in September 1796 and then resigned his command.

Kléber remained inactive until he was offered the command of a division in **Napoleon** Bonaparte's expeditionary Army of Egypt in 1798. Severely wounded in the assault on Alexandria on July 2, he was appointed military governor of that city while he recuperated. During February 8–19, 1799, he laid siege to El Arish, then fought at Jaffa on March 3. He besieged Acre from March 19 to May 20 and coordinated with Napoleon to triumph over a vastly superior Turkish army at Mount Tabor on April 16.

Kléber was named commander in chief in the Orient on August 22 after Napoleon left for France.

Kléber concluded the Convention of El Arish on January 28, 1800, securing the evacuation of his army from Egypt. However, British and Turkish commanders repudiated the agreement in March, and Kléber found his forces in a critical situation. He was confronted by vastly superior odds and a hostile population. However, rather than surrender, he took the offensive. At Heliopolis, he defeated the Turks on March 20 and recaptured Cairo on April 21. This bought much-needed time for his exhausted army. But, while serving as administrator for Egypt, Kléber was knifed by a fanatic in Cairo on June 14, 1800. His wounds were fatal.

KLUCK, Alexander von

BORN: May 20, 1846

DIED: October 19, 1934

SERVICE: Seven Weeks' War (1866); Franco-Prussian War (1870–71); World War I (1914–18)

German general who played a key role in the opening months of World War I

A native of Münster, Germany, Kluck saw his first action against Austria as a young officer during the Seven Weeks' War in June–August 1866. He served next in the Franco-Prussian War, during July 1870–May 1871, after which he became a staff officer who rose rapidly. In the years prior to World War I, Kluck became *general der infanterie* (1906) and then inspector general of the Seventh Army District (1913).

At the outbreak of World War I, he was chosen to lead the First Army on the extreme right of what was planned as a massive offensive sweep through Belgium and northern France. Kluck succeeded admirably through August, taking Brussels on August 20, and, after suffering a temporary setback against the British at Mons on August 22–23, outflanking the British II Corps at Le Cateau on August 26 and dealing it much devastation. He pressed the advance, but was forced to make a turn that deviated from the German master plan for the war—the **Schlieffen** Plan—when the Second Army's General Karl von Bülow, having suffered a setback at Guise, asked for support. Kluck had planned to approach Paris from the south and west and now found himself north of the capital. This shift, combined with breakdown in coordination among the commanders, bought the Allies an opportunity to counterattack at the Marne during September 4–9. While Kluck gained a tactical success at the Battle of the Ourcq, he now found himself vulnerable to encirclement and had to retreat. In this complex of events, the German army lost the initiative, and World War I became a long, futile war of attrition, a static contest with much death and destruction on both sides, but little movement.

Kluck remained in command of First Army until he was severely wounded in March 1915. He retired in October 1916, and the Germans lost an extremely capable commander, whose actions might have won the war in its first few months—had his colleagues cooperated more efficiently.

KNOX, Henry

BORN: July 25, 1750

DIED: October 25, 1806

SERVICE: American Revolution (1775-83)

American general who served with distinction as an artillery officer in the Revolution and later became secretary of war

Bostonian by birth and bookseller by trade, Knox joined the Massachusetts militia in 1768 and the Continental Army in 1775. He served under General Artemas Ward at Bunker (Breed's) Hill on June 17, 1775 and became colonel of the Continental Regiment of Artillery on November 17. **George Washington** dispatched him to retrieve the artillery captured at Fort Ticonderoga. He managed to transport some sixty pieces (weighing about 120,000 pounds, with shot) by sledge over three-hundred miles through the snow to Cambridge, Massachusetts (December 5, 1775–January 25, 1776). This action, which he humorously dubbed the "noble train of artillery," saved Boston from capture by the British; the artillery allowed Washington to break the siege of Boston (April 19, 1775–May 17, 1776).

Knox's regiment fought at Long Island (August 27, 1776) and was instrumental in covering Washington's long retreat through New York and New Jersey. Knox's artillery played a key role in Patriot victories at Trenton on December 26 and at Princeton on January 3, 1777. Following the Trenton victory, Knox was promoted to brigadier general.

Knox used the "down time" from January through May 1777—with his army in winter quarters—to establish the Springfield Arsenal in Springfield, Massachusetts, and the Academy Artillery School—which was the precursor of the U.S. Military Academy at West Point—at Morristown, New Jersey. When combat recommenced, Knox fought at Brandywine on September 11, but drew criticism for his role in the lost Battle of Germantown on October 4. However, at Monmouth (June 28, 1778) and at Yorktown (May-October 17, 1781), it was clear that he had trained his artillerymen well. Triumphs here were crucial to the outcome of the war.

On March 22, 1782, Knox became the youngest major general in the army and was appointed the first commandant of West Point on August 29, 1782. When Washington left the army to become president of the United States, Knox replaced him as commander in chief of the army. He was appointed first U.S. Secretary of War under Washington and served from September 12, 1789 to December 31, 1794.

Knox retired to his estate at Thomaston, Maine in 1795, but accepted an appointment as major general during the crisis with France in 1798.

Knox was an excellent commander, who developed the Continental Army's artillery branch into an effective weapon of war. He is generally given credit for having founded the U.S. Military Academy, West Point.

KOLCHAK, Aleksandr Vasiliyevich

BORN: November 4, 1874

DIED: February 2, 1920

SERVICE: Russo-Japanese War (1904–05); World War I (1914–18); Russian Civil War (1917–20)

Russian admiral and anti-Bolshevik

Kolchak was born in St. Petersburg, the son of a naval artillery officer. He graduated from the Russian Naval Academy, second in his class, in 1894, then sailed on two Russian Academy of Sciences arctic expeditions. During the Russo-Japanese War, he commanded a destroyer, then was in charge of a battery during the siege of Port Arthur (June 1904–January 2, 1905). Captured by the Japanese, he was briefly held as a prisoner of war in Japan.

In the wake of the Russo-Japanese War, Kolchak joined other officers in agitating for reform of the navy. He was instrumental in creating the Russian Naval General Staff during 1906–09. During 1908–11, Kolchak participated in a third polar expedition to discover a northern sea route to the Far East. During 1911 until the outbreak of World War I in 1914, he served on the Naval General Staff. He was captain of the Baltic Sea Fleet flagship when World War I began in August 1914 and was appointed chief of the Bureau of Operations of the Baltic Fleet the following year. Working with undermanned forces and poor supplies, he nevertheless created an adequate coastal defense system. Given command of the Baltic destroyer forces, Kolchak fought with distinction in the Gulf of Riga during 1916 and in August of that year was promoted to vice admiral, the youngest officer ever to hold that rank in the Russian Navy. He was assigned as commander in chief of the Black Sea Fleet.

After the February (March) Revolution of 1917, Kolchak threw his support behind the Provisional Government of Alexander Kerensky. However, he was continually threatened by mutiny in his command, and he resigned in June. Kerensky dispatched him to the United States to study the American navy and to assess the prospects of invading the Bosporus. On his way to the U.S., he visited Admiral **John Jellicoe,** who befriended him. His visit to the United States was discouraging, and he found little support for his efforts to learn from the U.S. Navy. He sailed from San Francisco to Japan, touching port there after the Bolshevik October (November) Revolution of 1917. Appalled by the Bolshevik's "separate peace" negotiations with the Germans at Brest-Litovsk in January 1918, he offered his services to the British Royal Navy.

Suddenly, the Anti-Bolshevik Socialist government ordered Kolchak to duty in Siberia, and he was subsequently appointed Minister of War and Navy in the anti-Bolshevik Socialist Government established in Omsk. Kolchak assumed control of the Anti-Bolshevik government in a military coup d'etat at Omsk on November 18, 1918. Very soon, the anti-Bolshevik government's liberals turned against Kolchak, who had become a dictator and whose regime was riddled with corruption. The Czechoslovak Legion, strong supporters of the anti-Bolsheviks, also turned against him, and Kolchak suffered a series of military reversals as the Civil War entered the summer of 1919. At last, on January 4, 1920, Kolchak relinquished the supreme command of the anti-Bolshevik forces to General Anton I. Denikin and sought asylum with the Allies. The Czechs, however, delivered him to Bolshevik authorities at Irkutsk on January 15, 1919. He was executed by firing squad and his body thrown into the Angara River.

KOSCIUSKO, Tadeusz Andrezj Bonawentura

BORN: February 4, 1746

DIED: October 15, 1817

SERVICE: American Revolution (1775–83); Russian Invasion of Poland (1792); Polish Insurrection (1793–94)

Polish general who served in the Continental Army during the American Revolution

Kosciusko was born into an impoverished gentry family at Mereczowszczyzna, Byelorussia, and was sent to the Royal Military School at Warsaw in 1765. Entering the army as a lieutenant, he soon achieved promotion to captain and was sent to France in 1769 for further military study. Returning to Poland in 1774, he was rapidly discouraged by his lack of advancement in the Polish army and resigned his commission. He supported himself as a tutor, giving drawing and mathematics lessons to the daughters of Hetman Jozef Sosnowski. He had an affair with the youngest of the Sosnowski daughters and was forced to flee for his life to France after being wounded by one of Sosnowski's retainers. From France, he traveled to America, where he offered his services to the colonists, who were preparing for a war of rebellion.

In Philadelphia, Kosciusko worked as a civilian engineer for the Pennsylvania Committee of Defense, assisting in plans for fortifications along the lower Delaware River during August 1776. On October 19, 1776, the Continental Congress commissioned him a colonel in the Continental Army, and he spent the winter planning and constructing Fort Mercer on the New Jersey side of the Delaware, just below Philadelphia. In March 1777, he marched north with General **Horatio Gates** and assumed command of the northern army. He planned and built the field fortifications at Saratoga, which aided greatly in the American victory there during August–October.

Kosciusko planned and constructed the defenses at West Point, New York during March–June 1780, and when Gates was appointed to command of the Southern Department, he, in turn, appointed Kosciusko chief engineer of the department in July 1780. The Pole did not arrive in the Carolinas until after Gates had been defeated at Camden (August 16, 1780), but he remained to serve under General **Nathanael Greene**, distinguishing himself as Greene's chief of transport during the Race to the Dan (January–February 1781). Although Kosciusko mishandled the siege of Ninety-Six, South Carolina, and thereby allowed the Tory garrison to escape during May 22–June 19, he redeemed himself at Charleston, South Carolina, commanding a cavalry unit and gathering intelligence during the summer of 1782.

At the conclusion of the war, Kosciusko was granted U.S. citizenship and was given the thanks of Congress, along with a generous pension, land grants, and the permanent rank of brigadier general. Nevertheless, he returned to Poland late in 1784, where he was appointed major general in the Polish army on October 1, 1789. He led these forces to defend against the Russian invasion in the spring of 1792, but resigned his commission when King Stanislaw II ordered a halt to Polish resistance. Disheartened, he traveled to France after the Third Partition in 1793, hoping to secure aid from the French Republican government.

In France, Kosciusko was informed of plans for an insurrection in Poland. He hesitated to join it, but ultimate journeyed to Cracow, arriving on March 24, 1794. He immediately set about raising an army, which was very poorly equipped. Despite a shortage of provisions and weapons, his forces defeated a Russian army at Raclawice on April 4, and Kosciusko advanced on Warsaw to drive out the Russian garrison on April 17. But Russian and Prussian forces converged on Warsaw, and Kosciusko was defeated at Szczekociny on June 6. Withdrawing into Warsaw proper, Kosciusko conducted a spirited defense against siege during July 26–September 5. Successful, he was promoted to lieutenant general, then attacked the Russian army at Maciejowice. Unfortunately, anticipated support failed to appear, and he himself was wounded and taken prisoner on October 10. Held for two years, he was released on November 26, 1796 after promising not to take up arms against Russia again.

Kosciusko left Poland for the United States, where he was welcomed as a hero on August 18, 1797 and was awarded almost $20,000 in back salary as well as a land grant in the Ohio country. In a grand humanitarian gesture, he used the money to purchase slaves, whom he then set free. He returned to France in 1798, where he lived in semiretirement. General William R. Davie, of the United States Army, asked him to write a treatise on artillery, and Kosciusko produced *Maneuvers of Horse Artillery*, which was published in 1800 and subsequently adopted as part of the course work of West Point.

When the Congress of Vienna ended hope of independence for Poland, Kosciusko retired to Solothurn, Switzerland in 1816 and, the following year, freed the serfs living on his Polish lands. Upon his death in 1817, his will was read. It provided for the sale of his Ohio lands, the proceeds to be used to free more U.S. slaves. Kosciusko's body was taken to Cracow for burial. Polish citizens built an earthen mound in his honor.

A thoroughly admirable man, Kosciusko was a patriot and soldier, whose zeal was tempered by great humanity. His services to the Continental Army were highly valuable, especially at the crucial Saratoga campaign.

KRUEGER, Walter

BORN: January 26, 1881

DIED: August 20, 1967

SERVICE: Spanish-American War (1898); Philippine insurrection (1899–1903); Mexican expedition (1916–17); World War I (1917–18); World War II (1941–45)

One of the ablest tactician-planners in the U.S. Army during World War II

Krueger was a native of Platow, West Prussia, whose family immigrated to the United States in 1889. Raised in Cincinnati, Ohio, he dropped out of high school to enlist in the army during the Spanish-American War (1898). He participated in the Santiago de Caba campaign during June 22–July 17, and, after returning to the States, joined the regular army in mid-1899. He was dispatched to the Philippines during the insurrection there (1899–1903) and was promoted to second lieutenant in the Thirtieth Infantry in June 1901.

After returning to the United States, Kreuger attended the Infantry and Cavalry School, graduating in 1906, then went on to the Command and General Staff School, from which he graduated in 1907. He was assigned to a second tour in the Philippines during 1908–09, then served on the faculty of the Army Service School from 1909 to 1912. During this period, he translated several German works on tactics, becoming a de facto authority on the German army and its military practices.

Krueger was promoted to captain in 1916 and served under General **John J. Pershing** during the

Walter Krueger. Courtesy National Archives.

Punitive Expedition in Mexico from March 1916 to February 1917. He then joined the American Expeditionary Force in France in February 1918, attending the General Staff College at Langres, France, then becoming assistant chief of operations for the Twenty-Sixth Division. After transferring to the 84th Division, he became chief of staff of the Tank Corps during the Meuse-Argonne offensive in October.

After the war, Krueger remained in France as chief of staff for VI Corps, then became chief of staff for IV Corps in Germany. He held the temporary rank of colonel, but reverted to captain when he returned to the United States in July 1919. Krueger graduated from the Army War College in 1921 and was assigned to the War Plans Division of the General Staff, where he served from 1923 to 1925. In 1926, he graduated from the Naval War College and taught there from 1928 to 1932, when he was promoted colonel. Four years later came promotion to brigadier general and an appointment as assistant chief of staff for War Plans.

Krueger left his staff post in 1938 to assume command of Sixteenth Brigade at Fort Meade. He was promoted to major general in February 1939 and given command of the Second Division, and then VIII Staff Corps in October 1940. Promoted to temporary lieutenant general, he was appointed commander of the Third Army and Southern Defense Command in May 1941. Once World War II began, the Sixth Army was activated under his command, and he took it to Australia in January 1943, commanding it through a series of combat operations in the Southwest Pacific Theater under General **Douglas MacArthur**. Krueger had charge of the landings on Kiriwina and Woodlark Islands on June 30, 1943, then directed the invasion of New Britain from December 15, 1943 to March 1944. This invasion was followed by operations on the Admiralty Islands and along the northern coast of New Guinea during February–August 1944. Krueger was in command during the seizure of Morotai, the last of the islands taken before the landings on the Philippines in October.

Krueger directed the landings on Leyte on October 20, beginning the process of the U.S. return to the Philippines. He was in charge of landings on Mindoro (December 15), and on Luzon (January 9, 1945). Fighting in central Luzon stretched from February to August. In March, Krueger was promoted to general, just after the fall of Manila to U.S. forces on March 14. Krueger's men drove the remnants of Japanese resistance into the mountains of northeastern Luzon and liberated most of the island by late June.

Krueger remained in the Pacific after the surrender of Japan, leading the Sixth Army in occupation duty on Honshu in September 1945, but retired less than a year later, in July 1946.

Krueger was one of the most skillful tacticians of World War II, a great believer in careful planning. Although he led some of the most difficult operations in the Pacific, he was known as a commander who did his utmost to minimize casualties, and his operations were indeed characterized by low casualty rates.

KUROPOTKIN, Aleksei Nikolaevich

BORN: April 10, 1848

DIED: January 25, 1925

SERVICE: Conquest of Samarkand (1868); Conquest of Kokand (1876); Russo-Turkish War (1877–78); Russo-Japanese War (1904–1905); World War I (1914–17)

Pre-revolutionary Russian general

Kuropotkin was born in Shemchurino, Pskov province, and entered the army in 1864. He first saw action in the taking of Samarkand in 1868, then became a diplomat along Russia's eastern borderlands, in Kashgaria (western Xinjiang) during 1874. He fought in the conquest of Kokand in 1876 and, during the Russo-Turkish War of 1877–78, served as chief of staff for an infantry division under General M. D. Skobolev. Under Skobolev, he participated in the ill-conceived attack at Third Plevna (September 11, 1877), but also served in Skobolev's signal victory at Senova during January 8–9, 1878.

Promoted to major general in 1882, Kuropotkin found the leisure to write a personal history of the Balkan operations during the Russo-Turkish War, which was published in 1885. From 1890 to 1898, he was given command of an army in Transcaucasia, then was appointed Minister of War.

An advocate of eastern expansion, he nevertheless warned his government of Japan's preparations for war and, feeling that the Russian military was inadequate to meet the Japanese threat, advised withdrawal from Manchuria and the Liaotung peninsula in 1903. The Tsar accepted his counsel, but was subsequently dissuaded from his agreement by hyper-expansionist factions in his court. The result was war with Japan by February 1904.

Kuropatkin was appointed commander in chief of land forces in the Far East in April. Overcautious, passive, and utterly lacking confidence in his army, he fought in southern Manchuria, but quickly withdrew to Liaoyang during August 1–25. He half-heartedly took the offensive, launching several unsuccessful attacks from favorable positions during August 25–September 3. He concluded that he had been defeated and withdrew farther northward, to the Sha Ho. There he attacked the Japanese again, this time penetrating the Japanese left during October 5–11. However, the Japanese counterattacked in force directly on his center during October 11–13, and, once again, Kuroptkin folded and withdrew by the seventeenth. Desperate, he rushed into an attack at Sand-epu. The operation, conducted in a snowstorm, surprised the Japanese and went well—but, falling victim to his customary caution, Kuropotkin failed to press the attack during January 26–27, 1905. He withdrew yet again, to Mukden, where he decided to dig in, establishing forty miles of trenches in February.

Clearly, Kuropotkin had lost the initiative. The Japanese mounted an all-out offensive against his right, compelling him to commit his last reserves. After two fierce weeks of fighting during February 21–March 10, he withdrew yet again, to Tieling and Harbin and was relieved of command. If Kuroptkin was not a great a general, he was at least an honest one. In 1909, he published a full and frank account of his action in *The Russian Army in the Japanese War*.

Kuroptkin was called upon to command again during World War I, but was not given command of troops at the German front. Instead, he had charge of the northern army group during February–July 1916 and was dispatched with it to Turkestan as governor general and commander in chief. When the 1917 revolutions erupted, he was arrested by revolutionaries and sent to Petrograd (St. Petersburg) in April 1917. The provisional government of Aleksandr Kerensky acknowledged his services to Russia and released him. He dropped out of public and official life, retiring to his home village, where he became a schoolteacher and sat out the remainder of the revolution and the civil war that followed.

Kuropotkin is typical of the Russian army during the late tsarist period. He was able enough as an administrator, but, like his army, lacked the vigor to achieve victory.

KUTUZOV, Mikhail Ilarionovich Golenischev, Prince of Smolensk

BORN: September 16, 1745

DIED: April 28, 1813

SERVICE: Russian intervention in Poland (1764–69); First Russo-Turkish War (1768–74); Second Russo-Turkish War (1787–92); Napoleonic Wars (1800–15)

Superannuated Russian field marshal of the Napoleonic Wars

Kutuzov was born in St. Petersburg, the son of a general. Destined for a military career, he was educated at an engineering and artillery school and graduated in 1757. He received his commission in the artillery in 1761 and saw his first action during Russia's intervention in Poland during 1764–69. In 1770, he left Poland for the Crimea to fight the Turks and lost an eye in battle.

Although Kutuzov showed no great tactical or strategic flair, he was a fearless and tireless commander, and he rose rapidly through the officer corps. By the time of the Second Russo-Turkish War, he was a major general, and served with great distinction at the siege of Ochakov on December 17, 1788, suffering another serious wound.

In 1790, he distinguished himself at the December 22 Battle of Izmail, then, during 1793–1802, was posted to a number of administrative and political assignments, including a stint as ambassador to Constantinople, governor of Finland, ambassador to Berlin, governor of Lithuania, and military governor of St. Petersburg. The aging Kutuzov retired in 1802, but was recalled to active duty and appointed commander of the Russian forces of the Third Coalition against **Napoleon** in 1805. In his first engagement, he succeeded in delaying Napoleon at Dürrenstein on November 11, but then suffered a catastrophic defeat, along with the Austrians, at Austerlitz on December 2.

Kutuzov was relieved of command and given the military governorship of Kiev in October 1806, followed by a similar appointment in Vilnius during June 1809. In 1811, he commanded the Army of Moldavia against the Turks and succeeded in crushing the Turkish army at Ruschuk on July 4. The following year, thanks in large part to this victory, the Treaty of Bucharest was concluded (May 28, 1812), which added Bessarabia to Russia's territorial holdings.

Kutuzov was again called on to confront Napoleon, this time during the emperor's invasion of Russia in 1812. Named commander in chief of all Russian forces, he at first followed the leading strategy of drawing Napoleon deeply into Russia, forcing him to extend his supply lines tenuously. However, popular pressure objected to this strategy of retreat, and Kutuzov at last yielded, taking a stand in front of Moscow at Borodino on September 7. There his forces suffered a severe defeat.

The French withdrew from Moscow on October 19, and Kutuzov fought an indecisive battle at Maloyaroslavets during October 24–25). Promoted to field marshal, he attacked the retreating French at Vyaz'ma (November 3), Krasnoye (November 17), and the Berezina River (November 26–28), finally driving them out of Russia altogether; Kutuzov pressed his pursuit through Poland and into Prussia during January 1813. Ailing, he was replaced by Prince Ludwig Wittgenstein in April. Shortly after this, Kutuzov collapsed and succumbed to old age and exhaustion.

Kutusov was by no means a brilliant commander, but he was tough and well loved by his soldiers. He also knew the value of partisan peasant troops and used them well in harrying Napoleon's long retreat from Russia. That he was able to press his pursuit of Napoleon so vigorously is the more remarkable because of his advanced age—and his advanced addiction to alcohol.

L

LAFAYETTE, Marie Joseph Paul Yves Roch Gilbert du Motier, Marquis de

BORN: September 6,1757

DIED: May 20, 1834

SERVICE: American Revolution (1775–83); French Revolutionary Wars (1792–99)

French general and champion of liberty who rendered invaluable service to the cause of American independence

Lafayette was born at Chavaniac in the Auvergne. His military officer father died on August 1, 1759 at the battle of Minden, and his mother died in 1770, leaving him an inheritance that made him a wealthy young man. Lafayette joined an infantry regiment in 1771, but two years later transferred to the dragoons. Promoted to captain in 1774, he was approached by U.S. minister to France Silas Deane, who, in December 1776, negotiated an agreement by which Lafayette would provide military aid to the colonies in the American Revolution. Lafayette arrived in America in company with **Baron de Kalb** during April 1777. Neither salary nor rank had been agreed upon.

Lafayette first saw action at the Battle of Brandywine, where he fought with distinction and suffered a wound on September 11. But even more than his skill in combat, it was his steadfastnessnes and resolution during the terrible winter at Valley Forge and his unwavering support of **George Washington** when the so-called Conway Cabal maneuvered to remove him as commander in chief that earned him a prominent place in the Continental Army. He was put in charge of an invasion of Canada during March–April 1778, but the plan soon had to be abandoned. However, Lafayette continued to distinguish himself in battle: at Barren Hill (May 18); at Monmouth (June 28), where he led a division; and at Newport (July–August).

Lafayette returned to France on leave. Promoted to colonel while there in April 1779, he created the initial plans for a full-scale French expeditionary force to aid the colonies. When he returned to America, he was given command of the Virginia light troops in April 1780. During September, he was a member of the court martial board that heard the case of Benedict Arnold's confederate, Major André, whom the board sentenced to death as a spy. Lafayette faced Arnold himself in Virginia during March 1781, when he was sent to counter Arnold's raiding activity there.

In Virginia, Lafayette successfully evaded **Charles Cornwallis**'s repeated attempts to draw his outnumbered force into battle during April–June. When he received reinforcements early in June, he rushed to coordinate with General **Anthony Wayne** in a fierce battle against Cornwallis at Green Spring. In the July 6 melee, Lafayete had two horses shot from under him. Finally, he pursued Cornwallis to Yorktown in August and played a key role in the siege there during September 14–October 19.

Returning to France once again, he was promoted to major general in December 1781. Three years later, during July–December 1784, he toured the United States, where he was greeted as a hero. Named a member of the French Assembly of Notables in 1787, he represented Auvergne in the Estates General of 1789 and was then appointed commander of the newly established National Guard on July 26, 1789. In October, he saved the royal family from the Paris mob.

Lafayette was promoted to lieutenant general in 1791 and the following spring was given command of the Army of the Center. However, his moderate views soon drew suspicion from the radical Jacobins. He was relieved of command and, justly fearing the guillotine, fled to Belgium in 1792. He was apprehended by Austrian authorities and imprisoned, then imprisoned a second time by the Prussians, first at Magdeburg and then at Olmutz during 1792–97. Released on September 23, 1797, he returned to France, but had a profound distrust of **Napoleon** Bonaparte. He therefore retired from public life, living quietly on his wife's estate at La Grange Bleneau. It was not until the introduction of Napoleon's liberal constitution of 1815 that Lafayette reemerged, this time as vicepresident of the Chamber of Deputies. In this capacity, he was instrumental in securing Napoleon's second abdication.

Lafayette served during the Bourbon restoration as a deputy in 1818 and quickly became a leader of the liberal opposition. Defeated in his 1824 bid for reelection to the Chamber of Deputies, he toured the United States and was once again accorded an enthusiastic welcome. Returning to France, he was reelected to the Chamber in 1827 and was made commander of the National Guard during the revolution against Charles X in July 1830. At first a supporter of Louis Philippe, he subsequently denounced him in 1832 because of his reneging on his liberal pledges. Lafayete died two years later.

Few men born into Lafayette's privileged position have been so thoroughly dedicated to the principle of liberty. Fewer still have been able to act on principle with so much success. Barely experienced in the military art, Lafayette proved an excellent commander, able tactician, and inspiring leader of men.

LALLY, Thomas Arthur, Count de

BORN: January 13, 1702

DIED: May 6, 1766

SERVICE: War of the Polish Succession (1733–38); War of the Austrian Succession (1740–48); The "Forty-Five" (1745–46); Seven Years' War (1756–63)

Lally was a prominent French general of the War of the Austrian Succession and the Seven Years' War

Born at Romans in Dauphiné, Lally was the son of Sir Gerard Lally, a radical Irish Jacobite in the French service, and a French mother. At birth, he was enrolled in his father's regiment, the Regiment de Dillon of the Irish Brigade, and was commissioned an officer at age seven. In 1728, he was given actual command of a company, and by the time of the War of the Polish Succession in 1733, he was a veteran.

In the War of the Polish Succession, he fought in the Rhineland during the spring and summer on 1734, and, ten years later, during the War of the Austrian Succession, he fought against the British at Dettingen on June 27, 1743. Commissioned colonel of the Irish regiment of Lally on October 1, 1744, he led his unit to great glory under Marshal **Maurice de Saxe** at Fontenoy on May 10, 1745. His regiment was key in defeating the superior English-Hanoverian force that had punched through the French center. Lally was wounded in the engagement.

Lally participated in Jacobite conspiracies to depose George II of Britain and was a partisan of "Bonnie Prince Charlie," Charles Edward Stewart, who invaded Scotland in 1745 and was ultimately defeated at the bloody Battle of Culloden on April 16, 1746. The French created Lally baron de Tolendal in 1746, and he served with Marshal de Saxe during the later phase of the War of the Austrian Succession, campaigning in the Low Countries and fighting at Lawfeld on July 2, 1747, as well as at Maastricht during April 15–May 7, 1748. Following the successful siege there, Lally was named *maréchal de camp*.

After the war, Lally lived in semi-retirement until the commencement of the Seven Years' War on May 17, 1756, when he was recalled to active service and given command of an expedition to India. Arriving in Pondicherry on April 28, 1758, he launched a vigorous campaign against the British, scoring some relatively minor victories, including the capture of Fort St. David, south of Madras during March–June 2. He next laid siege against Madras, beginning on December 13. However, he received poor naval support, and a British relief force arrived to defeat Lally at Masulipatam on January 25, 1759. Breaking off the siege on February 17, Lally enjoyed no further victories. At last on January 22, 1760, his tiny army of fewer than 3,000 men—1,500 Frenchmen and 1,300 Indian troops—was mauled by

General Eyre Coote's superior forces at Wandiwash. Lally withdrew what remained of his army to Pondicherry, where he was besieged on April 7. He held out until January 15, 1761, then finally surrendered and was taken to London. While in captivity, he received word that charges had been brought against him in Paris, and he demanded to return there to face his accusers. The British authorities granted his request. Lally returned and was imprisoned for two years, during 1763–65. Brought to trial, he was convicted and sentenced to death on May 6, 1766, with execution passed three days later.

Trial and execution unjustly ended the career of a courageous and determined soldier, who had been the victim of inadequate manpower, inadequate supply, and utter lack of cooperation and coordination in India. His imprisonment and death outraged many elements in France and led no less a figure than the philosopher Voltaire to write a bold indictment of the affair in *Fragments on India.*

LANNES, Jean, Duke of Montebello, Prince of Sievers

BORN: April 10, 1769

DIED: May 31, 1809

SERVICE: French Revolutionary Wars (1792–99); Napoleonic Wars (1800–15)

Called the "Roland of the French Army"; one of* Napoleon's *greatest marshals

Lannes was a Gascon, born at Lectoure, and was a humble dyer's apprentice until he joined the Revolutionary Army at Gers on June 20, 1792. He fought in the Army of the Pyrenees against Spain, then transferred to the forces of **Napoleon** Bonaparte in Italy during 1795. Here his rise through the ranks was meteoric. By 1796, he was general of brigade, fighting with great distinction at Dego (April 15), Lodi (May 10), and Bossano (September 8). A fierce combatant, he was felled three times by wounds in the course of three days at Arcole during November 15–17. He personally shielded Bonaparte, winning from his commander great admiration.

During Napoleon's campaign into Egypt, Lannes participated in the capture of Alexandria, then was instrumental in putting down the Cairo uprising that followed. During the subsequent Syrian and Egyptian campaigns, he received a gunshot wound to the temple and was left for dead on the field at the siege of Acre on May 8, 1799. He recovered, only to be severely wounded in the thigh at Aboukir on July 25.

For his performance at Acre, Lannes was promoted to provisional general of division, and was subsequently given command of the vanguard division of the Army of Italy. During the coup of Eighteenth Brumaire, his enthusiastic support of Napoleon Bonaparte led to his elevation as inspector general of the Consular Guard and a promotion to the permanent rank of divisional general on May 10, 1800.

Napoleon's confidence was not misplaced. In Italy, his performance at Montebello on June 9, 1800 and his unyielding will to fight, though vastly outnumbered, at Marengo on June 14 were crucial to securing victory. As a reward for his services, Napoleon sent him to Portugal as a diplomat in 1802, but he soon proved unsuited to the role. He eagerly returned to France to take command of a unit preparing for the invasion of England in 1803. Then, on May 19, 1804, Emperor Napoleon I created Lannes, along with seventeen others, a Marshal of the Empire.

In 1805, during the Ulm campaign, Lannes carried off a stunning feat in the capture of the Danube bridge at Spitz. Warning the Austrians that an armistice was in effect and that they could not, therefore destroy the bridge, he sent his grenadiers stealthily forward, taking the key crossing intact on November 12. Lannes performed similarly valuable service at Austerlitz on December 2, 1805, when his division held Santon Hill, becoming the point on which the Grand Armée pivoted to attack and decimate General **Mikhail Kutuzov**'s Austro-Russian army.

During the Prussian campaign, Lannes was commander of V Corps, leading the vanguard to victory at Saalfeld on October 10, 1806 and arriving first on the field at Jena on the fourteenth, where he assumed command of the center during the battle. He was wounded again at Pultusk in December 1806. His injury led to a fever that put him out of action for five months. He returned to the army in time to serve at the siege of Danzig in May 1807, then at the Battle of Heilsberg on June 10, and at Friedland on the fourteenth. At Friedland, he was able to pin down the Russians, thereby allowing Marshal **Michel Ney** to cut off their retreat and force their surrender.

In 1808, Napoleon created Lannes Duke of Montebello and dispatched him to Spain, where he defeated a Spanish army at Tudelo on November 30, 1808, although he seriously injured himself in a fall from his horse. Quickly recovering, he took over the siege of Saragossa, forcing the town to capitulate on February 20, 1809.

Later in 1809, he was in Germany, fighting at Ahensberg (April 20), Landshut (April 21), and Eggmühl (April 22). A frontline commander, he personally led a division against the walls of Ratisbon on April 23. He was seen to seize an assault ladder from a group of hesitant troops and declare, "Well, I will let you see that I was a grenadier before I was a marshal, and still am one!" He stole off through enemy fire until his men, finally shamed, took the ladder and scaled the walls. The town fell.

Lannes was given command of II Corps and was present during the occupation of Vienna. It was his corps that held tits position at Aspern-Essling during a massive Austrian assault of May 21–22. Only an order from the emperor himself compelled its retreat. Following this battle, Lannes was sitting cross-legged, mourning the death of a comrade at arms, when a partially spent cannon ball hit him in the legs. Both were crushed. The right leg had to be amputated, and, in the days before sterile surgical technique, Lannes developed an infection and fever. He died on May 31.

The loss of Jean Lannes—the first of Napoleon's marshals to die in battle—was a heavy blow to the emperor. He was a great advance-guard commander, fearless, inspiring, and a consummate tactician.

LASALLE, Count Antoine Charles Louis

BORN: May 10, 1775

DIED: July 6, 1809

SERVICE: French Revolutionary Wars (1792–99); Napoleonic Wars (1800–15)

Dashing, if reckless, general of Napoleonic light cavalry

Lasalle was born at Metz in Lorraine and joined the Royalist infantry as a second lieutenant in 1786, but transferred to the cavalry during the Revolutionary Wars and led a hussar unit in the Army of Italy. He drew attention to himself at the Battle of Rivoli, on January 14, 1797, by capturing an entire Austrian battalion. The following year, he left the Army of Italy for the Army of Egypt and fought with distinction at the Pyramids on July 21, 1798. Two years later, he was made a prisoner of war when the army surrendered, but he was soon liberated.

Lasalle went on to serve again in Italy and the Gironde. On February 1, 1805, he was promoted to general of brigade and fought valiantly at Austerlitz (December 2) and Schleiz (October 9, 1806). In the aftermath of Jena-Auerstadt on October 14, he tenaciously pursued the fleeing Prussian army, thereby forcing its surrender at Prenzlau on October 28. The next day, he tricked the Prussian garrison at Stettin into surrendering to his small force of eight hundred hussars.

After fighting at Golymin on December 26, he received promotion to general of division (on December 30) and was given command of twelve regiments of hussars and chasseurs under Marshal **Joachim Murat**. Although his division was present at Eylau (February 7–8, 1807), it did not fight. Lasalle did lead it into battle at Heilsberg on June 10, a fierce encounter in which Lasalle saved Murat's life and then Murat rescued him when his division was dispersed by Russian cavalry and artillery fire. Following this harrowing experience, Lasalle was created a Count of the Empire on March 10, 1808.

Transferring to the Army of Spain in 1808, Lasalle commanded a division of light cavalry at Torquemada on June 6 and Medina de Rioseca on July 14. Most spectacular was his action at the Battle of Medellin on March 28, 1809, when he led the Twenty-Sixth Dragoons in a charge through a 6,000-man square. The result of this single stroke was the rout of the Spanish army.

On April 22, Lasalle moved yet again, this time to the Army of Germany, where he assumed command of the light cavalry of Jean-Baptiste Bessières's Reserve Cavalry Corps. Leading the light cavalry, he fought at Aspern-Essling durng May 21–22 and at the siege of Raab during July 15–24. Then, during the second day of the battle of Wagram, on July 6, he was shot through the forehead and instantly killed while leading a charge.

Lasalle was a great exponent of light cavalry tactics. Personally daring, even reckless, he was nevertheless a strict disciplinarian, who, however, earned great loyalty from his men.

LA TRÉMOILE, Louis II de, Viscount of Thouars

BORN: ca. 1460

DIED: February 24, 1525

SERVICE: The Mad War (1488–91); Caroline War (1494–95); Neapolitan War (1500–03); War of the Holy League (1511–14); First Hapsburg-Valois war (1521–26)

Valiant French commander of the late Renaissance period; called "Le Chevalier sans reproche"

Of noble birth—eldest son of Louis I de La Trémoile—La Trémoile joined the French royal army about 1481 and came to prominence seven years later when he played a principal role in defeating the rebel dukes of Orléans and Brittany at Saint-Aubindu-Cormier. Subsequently, during 1491–92, he participated in the suppression of sporadic rebellion throughout Brittany. After this, he was part of King Charles VIII's expedition to Italy, and as cocommander with Charles of the main body at the French victory of Fornovo on July 6, 1495. Four years later, he took part in Louis XII's invasion and occupation of Milan during September–October 1499, occupying Milan in 1500.

During 1501–02, La Trémoile commanded forces in southern Italy before leaving for Burgundy, of which he briefly served as governor. In 1502, he was appointed admiral (not a naval office, but a political post similar to military governor) of Guienne, and the following year he commanded French forces in southern Italy after the death of the Duke of Nemours at Cerignola on April 23, 1503. Unfortunately, La Trémoile became seriously ill and had to return to France. In his absence, his army, under the Marquis of Mantua, was defeated and destroyed at the Battle of River Garigliano (in central Italy) on December 29, 1503.

La Trémoile fought under Louis XII during the War of the League of Cambrai against Venice, and probably was present at the French victory at Agnadello on May 14, 1509. Following French reverses in Italy during 1512, he once again invaded and captured Milan on May 1513, but was defeated by Swiss troop at the Battle of Novara on June 6, 1513. He retreated across the Alps. But was nevertheless named admiral of Brittany the following year, and, during July–August 1515, he returned to Italy with the army of Francis I. He defeated the Swiss at Marignano during September 13–14, then retired from active service for a time.

At the commencement of war with Charles I of Spain and Charles V of Germany in 1521, he returned to duty and was given command in Picardy. During 1522–23, he defended against English naval raids led first by the Earl of Surrey and then by the Duke of Norfolk. During August–September 1524, he united with Francis's army at Lyon and accompanied it into Italy. He was among those slain at the debacle of Pavia on February 24, 1525.

La Trémoile was a brave and tireless fighter, who served his king and country well past middle age. He is more notable as a subordinate than as an independent commander.

LAWRENCE, James

BORN: April 6, 1802
DIED: June 4, 1813
SERVICE: Tripolitan War (1801–05); War of 1812 (1812–15)

American naval officer famed for his last words: "Don't give up the ship!"

Born in Burlington, New Jersey, Lawrence joined the U.S. Navy as a midshipman on September 4, 1798. He was promoted to lieutenant on April 6, 1802 and the following year sailed for the Mediterranean aboard U.S.S. *Enterprise*, to fight in the Tripolitan War. He was second-in-command under Lieutenant **Stephen Decatur** during the bold raid into Tripoli harbor to burn the captured frigate *Philadelphia* on February 16, 1804—a feat praised by no less a figure than **Horatio Nelson**.

Lawrence participated in all five bombardments of Tripoli during August 31–September 3, 1804, and subsequently remained in service in the Mediterranean waters until 1808, when he served aboard a number of vessels, including, most importantly, the U.S.S. *Constitution*. On November 3, 1810, he was promoted to master commandant and given command of the eighteen-gun sloop *Hornet*. Initially, he sailed with Commodore John Rodger's squadron during June-August 1812, but was subsequently assigned to cruise on his own. He sunk the British sloop *Peacock* off British Guiana on February 24, 1813, then returned to the United States to receive his promotion to captain on March 4.

Lawrence was given command of the U.S.S. *Chesapeake*, a thirty-eight-gun frigate, on June 1. He sailed out of Boston and, against orders, attacked H.M.S. *Shannon* on June 4. With his ship manned by a green crew, the attack was foolhardy, and *Shannon* quickly defeated *Chesapeake*. Lawrence himself fell mortally wounded during the exchange and, as he was taken below, he uttered words that have entered history: "Don't give up the ship!"

There is much to admire in the courageous Lawrence, and had his unauthorized attack against *Shannon* succeeded, he would have been hailed as a very great naval hero. As it is, even admirers of the officer's enterprise, criticized Lawrence for disobeying orders and acting with poor judgment.

LAWRENCE, Stringer

BORN: March 6, 1697
DIED: January 10, 1775
SERVICE: War of the Austrian Succession (1740–48); Jacobite Rebellion ("The Forty-Five") (1745–46); Second Carnatic War (1749–54); Seven Years' War (1756–63)

British general often regarded as the founder of the Indian army

Lawrence was born in Hereford and entered the army about 1726 as an ensign. He was posted to Gibraltar, where he rose to captain. In 1745, he fought in Flanders and the following year was recalled to England, from which he was dispatched to Scotland to fight in the bloody Battle of Culloden on April 16, 1746.

Lawrence was attached to the British East India Company as a major and arrived in India to take charge of the company's troops at Madras early in 1748. "Troops" hardly describes the motley assortment of men, Europeans and sepoys, unfit for most employment, Lawrence had to work with. Nevertheless, he molded them into an effective and well-disciplined force. Leading them, Lawrence checkmated the French attack at Cuddalore in June.

Lawrence was taken prisoner during an attack on the French post at Ariancopang in August, but he was released that fall after news of Anglo-French peace reached his captors. In 1749, he was in command of an operation that took Devikottai, and he also led an unsuccessful mission to Nasir Jang in 1750. That same year, Lawrence resigned in protest over a pay dispute and returned to Britain. Two years later, promoted to lieutenant colonel, he returned to the East India Company service. He took charge of successful operations around Trichinopoly, defeating the Frenchman Jacques Law in 1752 and emerging victorious at Bahur on August 26. The French massed a major attack against Trichinopoly during late 1752 and harried the city through part of 1754. Lawrence conducted a consistent and highly effective defense, which ultimately resulted in the recall of the French commander.

Despite his extraordinary service, Lawrence was replaced by Colonel John Adlercorn in 1754. However, Lawrence refused to serve under him until 1757, and during December 1758–February 1759, he successfully defended Madras against Baron **Thomas Lally**'s long siege. At last, in 1761, Lawrence was elevated to local major general and was once again named commander in chief of East India Company forces. He served until 1766, when he retired to London.

Lawrence was the first in a long line of British colonial commanders who transformed an unpromising rabble into an effective military force. He is rightfully regarded as the man who founded the British Indian army.

LAWRENCE, T[homas] E[dward]

BORN: August 15, 1888
DIED: May 13, 1935
SERVICE: World War I (1914–18)

British organizer of Arab irregular forces in World War I; author of one of the twentieth century's great autobiographies; famed as "Lawrence of Arabia"

Born at Tremadoc, Caernarvonshire, Lawrence was the illegitimate son of Sir Robert Chapman by his daughters' governess, Sara Maden. He received an excellent education at Jesus College, Oxford. In 1909, he traveled to the Middle East as a scholar with a special interest in Crusader castles. His study became the basis of his first-class honors thesis, completed in 1910. During 1911–14, he was awarded a traveling endowment from Magdalen College, and he joined the expedition excavating Carchemish (Barak) on the Euphrates. Early in 1914, he explored northern Sinai, ostensibly for historical and archaeological reasons, but actually to reconnoiter the situation in Turkish Palestine as the world hovered near war.

Just before the outbreak of war, Lawrence returned to Britain and was commissioned in the British army. When war commenced in August, he was assigned to the War Office's map department. However, after Turkey declared war on England, Lawrence was dispatched to Cairo in December as an intelligence officer attached to the Arab section. Sent to accompany another officer, Ronald Storrs, on a mission to Sharif Husein in the Hejaz during October 1916, Lawrence remained to visit the army of Husein's son Faisal outside Medina. On his return to Cairo in November, he was assigned to Faisal's army as political and liaison officer. Lawrence worked with Faisal to mold the ongoing Arab revolt into a force that would aid the British cause in the Middle East. He became an organizer of irregulars, whipping up the flagging insurrection and resupplying the army. He also assumed strategic leadership of Faisal's forces. Soon, it became clear to Lawrence that the Turkish garrison in Medina was less critical an objective than the capture of Wejh on the Red Sea coast. After taking it on January 24, 1917, he organized guerilla strikes against the Hejax railway during March–April.

Lawrence soon gathered about himself a private army of devoted followers. He himself adopted Arab dress and became the scourge of Turkish desert garrisons. After he won a significant victory at Tafila during January 21–27, 1918, he was promoted to lieutenant colonel. As 1918 wore on, Lawrence concentrated on the capture of Damascus and was given command of Arab forces functioning as the right wing of **Edmund Allenby**'s final offensive at Megiddo. On September 27, he captured Der'aa, then raced ahead to take Damascus on October 1.

Lawrence was chosen as a member of the British delegation to the Versailles peace conference, at which he struggled in vain to prevent dismemberment of the Levant. During 1921–22, he was adviser on Arab affairs to the Middle Eastern Division of the Colonial Office, but was soon disillusioned with colonial government policy and left government service in 1922. He found himself hounded by the near legendary status his wartime exploits had earned for him, and he enlisted in the Royal Air Force under a pseudonym in August 1922. When a newspaper revealed his secret in January 1923, he resigned and enlisted in the Royal Tank Corps in March, calling himself T. E. Shaw (in 1927, he legally changed his name to T. E. Shaw).

Lawrence transferred back to the RAF in 1925, and the following year published his wartime memoirs, a massive literary masterpiece entitled *The Seven Pillars of Wisdom*. The first edition was strictly limited to a mere 150 copies and was by his own direction the only edition of the book published while he was alive. However, in 1927, he published to a wider audience a shorter version of the memoir, *Revolt in the Desert*.

Lawrence left the RAF in 1935 and was fatally injured in a motorcycle accident near his home, Cloud Hill, in Dorset.

T. E. Lawrence was not only a remarkable military leader—whose exploits recall those of **Charles "Chinese" Gordon** in the previous century—but was a thoroughly remarkable man. His barely controlled manic energy and masochistic drive lured him from one adventure to another. He was also one of the great memorists of the twentieth (or, indeed, any) century.

LAWTON, Henry Ware

BORN: March 17, 1843

DIED: December 19, 1899

SERVICE: Civil War (1861–65); Apache War (1885–86); Spanish-American War (1898); Philippine Insurrection (1899–1901)

American general whose policy of leading from the front ultimately resulted in his death

Lawton was born and raised just outside of Toledo, Ohio, and enrolled in Methodist Episcopal College, leaving it in 1861 to enlist in the Ninth Indiana Volunteers at the outbreak of the Civil War. He entered with the rank of first sergeant and saw combat in Western Virginia before mustering out. Less than a month after he left the Ninth Indiana, he joined the Thirtieth Indiana Volunteers as a first lieutenant. With this unit, he fought at Shiloh during April 6–7, 1862.

Lawton transferred to the Army of the Cumberland in 1863 and fought at Stones River during December 31, 1862–January 3, 1863, and at Chickamauga during September 19–20, 1863. He received promotion to captain, then fought in the Atlanta campaign during July–August 1864. On August 3, he took and held a key Rebel position against heavy fire. For this action, he received the Medal of Honor and a promotion to lieutenant colonel.

Lawton went on to distinguish himself at Franklin Tennessee, on November 30 and at Nashville during December 15–16, 1864, after which he was breveted colonel. Mustered out of the army on November 25, 1865, he enrolled at the Harvard law school, but soon left to rejoin the army. He secured a postwar appointment as Second lieutenant in the Forty-First Infantry on July 28, 1866 and was promoted to first lieutenant the following year.

Lawton served under General **George Crook** in Arizona as regimental quartermaster, but in the glacially slow promotion system of the postwar army, he did not make captain until 1879. In 1886, he participated in operations against the Apache resistance leader **Geronimo**, leading a force on a grueling 300-mile pursuit of the Geronimo over the Sierra Madre mountains and deep into Mexico. After much frustration and hardship, Lawton's command ran the Indian leader to ground.

Promoted to major and inspector general in 1888, Lawton made lieutenant colonel in 1889. Eager for action, he volunteered for "service in any capacity" at the outbreak of the Spanish-American War in April 1898. He was commissioned brigadier general of volunteers and led a unit at the make-or-break Battle of San Juan Hill/El Caney on July 1. Lawton was among the U.S. commissioners who accepted the surrender of Santiago, and he was subsequently appointed military governor of Santiago during July–September. He returned to the States in October and was assigned command of IV Corps, then was dispatched to the Philippines in December in command of elements of Fourth and Seventeenth Infantries. He assumed overall command in the Philippines on his arrival there and immediately set about putting down the ongoing insurrection. He captured the rebel stronghold of Santa Cruz on April 10, then took the rebel headquarters villages of San Rafael-San Isidro on May 15. While leading an attack on San Mateo on December 19, 1899, Lawton was felled by a rebel bullet.

LEE, Charles

BORN: 1731

DIED: October 2, 1782

SERVICE: French and Indian War (1754–63); Russo-Turkish War (1768–74); American Revolution (1775–83)

Controversial U.S. Revolutionary general

Lee was born in Cheshire, England, and showed a flair for languages during his continental education. He joined the British army as an ensign in 1747, and by 1751 was a lieutenant in the Forty-Fourth Regiment under **Edward Braddock**. He served with that ill-fated general in North America during the French and Indian War, purchasing a captaincy while abroad. During service in New York's Mohawk Valley, he married the daughter of a Seneca chief and lived among the Mohawks.

Critically wounded at Fort Ticonderoga on July 7, 1758, he recovered in time to fight at Fort Niagara during June–July 1759 and at Montreal during September 1760. On his return to England, Lee was breveted to major and joined the 103rd Regiment on August 10, 1761. He served with **John Burgoyne** in Portugal against the Spanish invasion during 1761–62, and in 1762, his brevet promotion was confirmed. However, when his regiment disbanded, Lee retired on half-pay.

He did not long remain idle. The colorful commander joined the Polish army in 1765 and within two years was promoted to major general. He took a two-year hiatus in England, then returned to Poland, where he fought the Turks until, broken in health, he was sent back to England in 1770. The Poles made him a lieutenant colonel in 1772.

Lee immigrated to America in 1773, where he purchase a Berkeley County, Virginia (now West Virginia) estate in May 1774. The rumblings of revolution were impossible for Lee to resist, and he appealed to Congress to appoint him a major general on June 17, 1775. He was third in command of the Continental army, behind **George Washington** and Artemus Ward. Lee was instrumental in the successful siege of Boston during April 19, 1775–March 17, 1776, then was assigned command of the Southern Department on February 17, 1776. He mounted an effective defense of Charleston on June 28 before rejoining Washington's up North at White Plains, New York, on October 28. However, when Lee did not join Washington on his long and tactically effective retreat through New Jersey, it began to appear that he was scheming to replace Washington as commander in chief of the Continental Army.

Having failed to join the main body of the army, Lee was captured by the British at Basking Ridge, New Jersey on December 13, 1777 and turned traitor on March 29 by furnishing them with a plan to defeat the Americans. Nevertheless, Lee was exchanged in April of the following year and returned to divisional command at the Battle of Monmouth on June 28. He failed miserably in the battle and was openly insubordinate to Washington, who saw to his court-martial during July 4–August 12, 1778. Lee was found guilty of disrespect and was suspended from command for a year. His behavior also brought a flurry of challenges, and Lee was wounded (December 3) in one of a number of duels he fought.

In July 1779, Lee quietly retired to his estate, but appealed to Congress to restore his command. Instead, that body responded by officially dismissing him from the army on January 10, 1780. Lee died two years later.

Lee was an officer of great dash, flamboyance, and promise. His overweening ambition wrecked his career and—apparently—spoiled his military judgment.

LEE, Fitzhugh

BORN: November 19, 1835

DIED: April 28, 1905

SERVICE: Civil War (1861–65); Spanish-American War (1898)

Dashing Confederate cavalry commander during the Civil War and general in the U.S. Army later

Lee was born on his family's Fairfax County, Virginia, estate and was the grandson of **Henry "Light-Horse Harry" Lee** and a nephew of **Robert E. Lee.** He graduated near the very bottom of his West Point class in 1856 and was commissioned a second lieutenant in the cavalry. Despite his relatively poor showing at West Point, Lee was made an instructor at Carlisle Barracks, Pennsylvania from 1856 through 1858, then served with the Second Cavalry in Texas under Colonel **Albert S. Johnston.** He saw some combat against the Indians and was wounded in one exchange. Sent to teach at West Point in 1860, he resigned his commission on May 21, 1861, and offerred his services to the Confederacy.

Lee was a first lieutenant on the staff of General **Joseph E. Johnston** at First Bull Run on July 21. The next month, he was jumped to lieutenant colonel and given command of the First Virginia Cavalry. By March 1862, he was a full colonel and led his regiment against **George McClellan** during the Peninsular campaign. He participated brilliantly in **J.E.B. Stuart's** flanking ride around McClellan's army during June 12–15, then saw action during the Seven Days (June 25–July 1). Following this, Lee was promoted to brigadier general on July 24.

Lee commanded a cavalry brigade at Second Bull Run (August 29–30), but was slow in getting his forces into position. The delay allowed Union general John Pope to escape and prevented Robert E. Lee from more effectively capitalizing on his victory. On September 14, 1862, Fitzhugh Lee again led a brigade, at South Mountain (Maryland) and then, on September 17, at Antietam. He participated in General Stuart's raids on Occoquan and Dumfries in December, and the following year distinguished himself in a defensive action at Kelly's Ford, where he successfully repulsed an attempt to overwhelm his brigade on March 17, 1863.

Lee next performed a valuable service at Chancellorsville, not only screening General **Thomas J. "Stonewall" Jackson,** as he performed a critical turn, but also discovering the vulnerable position of Union general **Oliver O. Howard** corps during May 1–6. Lee fought well at Gettysburg during June–July and was promoted to major general, with command of a division on August 3. He showed his mettle at Spotsylvania, where he fought a fierce delaying action that bought **James Longstreet**'s corps time enough to gain the crossroads during May 8–12, 1864. However, Lee was again late at Trevilian Station (near Richmond)—though he did arrive in time to check George **Armstrong Custer's** attack on June 11–12. His action ultimately won the day for the Confederates.

Lee was transferred to the Shenandoah Valley under Jubal Early in August 1864 and fought at the Battle of Winchester on September 19. Here he was gravely wounded and saw no action until March 1865, when, as temporary commander of the Cavalry Corps, he led an attack through Federal lines at Farmville on April 9, just before Appomattox. Fitzhugh Lee surrendered his corps on April 11 and retired to a farm in Stafford County, Virginia.

Unlike many of his compatriots. Fitzhugh Lee was a young man at the end of the war. He boldly reentered public life, gaining election as governor of Virginia (1885–90), and although he was defeated in a bid for a U.S. Senate seat in 1893, he secured an appointment as American consul-general in Havana in 1896. On post there at the outbreak of the Spanish-American War in April 1898, he offered his services to the U.S. Army and was appointed major general of volunteers. Commanding VII Corps, he never saw action, but was appointed governor of Havana and Pinar del Rio province in January 1899. Pursuant to the volunteer service act, he was made a brigadier general of regulars in March and given command of the Department of the Missouri. He retired from the army in March 1901.

Fitzhugh Lee was a very good cavalry commander who was full of high spirits and certainly looked and acted a dashing part. He also showed great concern for his troops and horses and was singularly successful at providing for both through resourceful foraging.

LEE, Henry ("Light Horse Harry")

BORN: January 29, 1756
DIED: March 25, 1818
SERVICE: American Revolution (1775–83)

Bold and impetuous American commander of "Lee's Legion" in the Revolutionary War

Born in Prince William County, Virginia, Lee graduated from New Jersey College (present-day Princeton University) in 1773 and three years later was a captain in a regiment of Virginia cavalry. The unit was made part of the First Continental Dragoons on March 31, 1777, and Lee served under General **George Washington.** He distinguished himself at Spread Eagle Tavern on January 20, 1778 and was promoted, by resolution of Congress, to major-commandant on April 7. At this time, he was given his first independent command, a corps consisting of three troops of dragoons and three companies of light infantry—100 horse and 180 foot. Subsequently dubbed "Lee's Legion," the unit raided Paulus Hook on August 19, 1779, winning a significant victory there.

Lee was promoted to lieutenant colonel on November 6, 1780 and received reinforcements to Lee's Legion. With his augmented force, he joined General **Nathanael Greene's** troops in South Carolina on January 13, 1781 and provided support to "Swamp Fox" **Francis Marion** in his raid on Georgetown, South Carolina, on January 24, 1781. He triumphed over Tories at Haw River on February 25, then achieved victory at Guilford Courthouse on March 15.

"Light Horse Harry" Lee proved so effective as a commander of light cavalry that his unit was detached from the main army as an independent strike force. Lee's Legion fought at Fort Watson (near Summerton, South Carolina) during April 15–23; at Fort Motte (near St. Matthews, South Carolina) on May 12; at Fort Granby (near Columbia, South Carolina) on May 15, and at Augusta (Georgia) and Ninety-Six (South Carolina) during May–June. On September 8, he linked up again with the main army at Eutaw Springs and was sent with key dispatches to Washington at Yorktown during May–October. He fought at Gloucester on October 3, then returned to Greene's army and fought at Dorchester (South Carolina) on December 1. He planned the successful attack on Jolus Island (South Carolina) during December 28–29, but left the Continental Army in February 1782 due to exhaustion.

After the war, Lee recovered his strength and reentered public life. He served in Congress (1785–88), the Virginia legislature (1789–91), and was three times elected governor of Virginia during 1791–94. He led a small force again in 1794, this time to put down the Pennsylvania Whiskey Rebellion.

In 1799, it was Lighthorse Harry Lee who eulogized Washington in a congressional resolution as "first in war, first in peace, and first in the hearts of his countrymen." He served in the House of Representatives from 1799 to 1801, then returned to private life, debt-ridden and hounded by creditors. He was twice imprisoned for debt.

Lee devoted 1808–09 to composing his popular memoirs. During the War of 1812, he was caught up in a violent antiwar riot in Baltimore during July 1812. While trying to protect a friend from the mob, he was badly injured and crippled for life. He sailed for the West Indies the following year to convalesce. On his way back to Virginia in 1818, he died.

Lee was an extraordinary cavalry commander and was especially skilled at guerrilla and quasi-guerilla actions. Awarded a Congressional Medal for his service in the Revolutionary War, he also was the father of **Robert E. Lee,** perhaps the most universally respected military commander the nation ever produced.

LEE, Robert Edward

BORN: January 19, 1807
DIED: October 12, 1870
SERVICE: U.S.-Mexican War (1846–48); Civil War (1861–65)

Confederate army general in chief; the most universally respected, admired, and beloved military commander in U.S. history

Robert E. Lee. Courtesy Library of Congress.

Robert E. Lee was born at Strafford, Virginia, the third son of Revolutionary War hero **Henry "Light Horse Harry" Lee** and his second wife, Ann Hill Carter. He graduated from West Point, second in the Class of 1829, and was commissioned in the Corps of Engineers. He was posted to service along the southeast coast, where he met and married another illustrious ancestor of the Revolution, Mary Custis, great-granddaughter of Martha Washington.

Lee was a brilliant military engineer, whose work on the Mississippi at St. Louis and the New York Harbor defenses during 1836–46 gained him a high degree of recognition. He was chief engineer under General John E. Wool during the U.S.-Mexican War. On expedition to Saltillo, Lee built bridges and selected routes between September 26 and December 21, 1846. He then joined General **Winfield Scott** at the Brazos River during January 1847 and served brilliantly as a staff officer at the capture of Veracruz on March 27 and at Cerro Gordo on April 18. It was Lee's reconnaissance efforts that located the route for Scott's outflanking force. Lee's reconnaissance was also critical to the successes at Contreras and Churubusco—both just outside of Mexico City—and at Chapultepec Castle (September 13, 1847).

Following the war, Lee was appointed superintendent of West Point, serving there from 1852 to 1855, and was then promoted to colonel commanding the Second Cavalry, headquartered in St. Louis. From 1855 to 1857, Lee served in Texas and the Southwest before he was recalled to Virginia after the death of his father-in-law in 1857. Lee was still in Virginia when he was ordered to Harper's Ferry to put down John Brown's raid, which he did, capturing Brown, on October 18, 1859. From February 1860 to February 1861, Lee commanded the Department of Texas and was then recalled to Washington on February 4. He resigned his commission when Virginia seceded. It is widely believed that, on April 20, 1861, **Abraham Lincoln** offered Lee command of Federal forces. Unwilling to fight against his home state, Lee refused, accepting instead command of Virginia's military and naval forces that month. He was soon also serving as personal military advisor to Confederate president Jefferson Davis.

Lee did not enjoy success with his first field command in western Virginia. Here, the locals were hostile to Tidewater Virginia and broke with the state to unite with the Union. Lee was also hampered by uncooperative subordinates. He met with defeat at Cheat Mountain (West Virginia) during September 12–13 and was then driven out of the mountains. Davis then withdrew Lee from West Virginia to apply his engineering expertise to strengthening the defenses of the southeast coast at Charleston, Port Royal, and Savannah during October 1861–March 1862. Following this, he was recalled to Richmond once again to advise Davis as Union forces bore down on General **Joseph E. Johnston**'s army, pushing it back toward the Confederate capital in March. It was Lee who urged Major General **Thomas J. ("Stonewall") Jackson** to undertake the highly effective Valley Campaign during May 1–June 9, 1862.

When **J. E. Johnston** was wounded at Seven Pines during May 31–June 1, Lee was called on to replace him. He created the Army of Northern Virginia and led it in a successful defense against **George B. McClellan** in the Seven Days (June 26–July 2). Lee also spectacularly outgeneraled John Pope, humiliating him at Second Bull Run during August 29–30. Lee took the offensive by invading Maryland and was attacked by McClellan at Antietam on September 17. The battle was in most respects a bloody draw. Lee repulsed McClellan, but he himself was forced to retreat to Virginia.

In Virginia, Lee confronted General **Ambrose Burnside**, whom he defeated at Fredericksburg on December 13, 1862. At Chancellorsville, during May 2–4, 1863, Lee scored a brilliant victory over **Joseph Hooker**, then invaded the North again. He was, however, defeated by General **George G. Meade** at the make-or-break Battle of Gettysburg (July 1–3, 1863), which was the turning point of the war. Lee was forced to withdraw into Virginia, and the North was secure from invasion.

Lee conducted no major campaigns until General **Ulysses S. Grant** was made Union general in chief on March 9, 1864. He then employed a brilliantly successful defense against Grant at the Wilderness during May 5–6, 1864. Lee bested Grant again at Spotsylvania (May 8–12), but was then forced out of his entrenchments at the North Anna River on May 23 when Grant enveloped him. However, Lee successfully repulsed Grant's assault at Cold Harbor on June 3, but suffered envelopment that forced him back across the James River, which Grant crossed during June 12–16. Lee mounted a desperate defense of Petersburg, but the long siege there badly weakened the Army of Northern Virginia (June–October).

An increasingly desperate Jefferson Davis named Lee general in chief of the Confederate armies on February 3, 1865, but after the failure of John Brown Gordon's surprise assault on Fort Stedman (near Petersburg) on March 27, 1865, Lee once again found himself outflanked by Grant (at Five Forks, during March 29–31, He withdrew from Richmond and Petersburg during April 2–3, 1865, but Grant applied the pressure and pursued Lee as earlier Union generals had not. With his route of escape through the Carolinas blocked, Lee surrendered the Army of Northen Virginia to Grant at Appomattox Court House on April 9, 1865. For all intents and purposes, this ended the Civil War.

Lee was a prisoner of war for a short period before he was paroled. In September 1865, he accepted the presidency of Washington College (later renamed Washington and Lee University) in Lexington, Virginia. Prematurely aged, he died five years later.

Lee was a great all-around commander. He was a brilliant strategist and tactician and, even more, a penetrating battlefield psychologist, whose ability to anticipate enemy action was uncanny. Beyond this, he was admirable for the strength of his character, which commanded absolute loyalty—and even love—from his troops.

LEFEBVRE, Francis Joseph, Duke of Danzig

BORN: October 25, 1755

DIED: September 14, 1820

SERVICE: French Revolutionary Wars (1792–99); Napoleonic Wars (1800–15)

French Marshal of the Empire under Napoleon I

Born at Rouffach, Alsace, Lefebvre was the son of a miller who had also seen service in the hussars and who commanded the local *garde bourgeoise* (home guard). Although Lefebvre's parents wanted the boy to pursue a career in the church, he was far more interested in the army and went off to Paris to join the French Guards as a private on September 10, 1773. He rose slowly through the noncommissioned ranks, making first sergeant by 1788. The following year, he entered the National Guards as a sublieutenant.

While serving as escort to the king and the royal family, Lefebvre was twice wounded in scrapes. He joined the regular army in 1792 as captain in the Thirteenth Light Infantry and saw action in the armies of the Center and of the Moselle. In contrast to his early years, his rise now was meteoric. After securing appointment as adjutant general, he was promoted to general of brigade on December 2, 1793. As such, he fought at Geisberg (December 26) and at Fleurus (June 26, 1794). In the latter engagement, he performed with extraordinary flair and gallantry, inducing his brigade to repulse no fewer than three Austrian onslaughts. Following this exchange, Lefebvre was promoted to general of division and transferred to the Army of Sambre-et-Meuse, where he served under Louis Lazare Hoche, **Jean-Baptiste Kléber,** and, finally, Count Jean-Baptiste Jourdan. On October 2, 1794, he participated in the victory at Aldenhoven, and his division subsequently furnished the advance guard of the Army of the Rhine, also under Jourdan. Lefebvre was the first to cross the Rhine in September 1795 at Duisburg, after which he scored a string of victories: at Opladen, Hennef, Siegburg, and Altenkirchen, before transferring to the armies on the Belgian coast led by Hoche.

Lefebvre served under Hoche in the Army of the Sambre-et-Meuse and was given provisional command of that army in 1798 after Hoche's death. In 1799, he was transferred to the Army of the Danube, under Jourdan, and was put in command of the advance guard. That year, however, Lefebvre was badly wounded in the arm and had to return to France. Named commander of the Seventeenth Military Division in Paris, he particpated in the coup of 18th Brumaire (November 9–10). He commanded his grenadiers to turn out the Council of Five Hundred and rescue **Napoleon.** In the new government, in 1800, Lefebvre served as a senator, subsequently rising to the presidency of the Senate. Finally, on May 19, 1804, Napoleon created Lefebvre a Marshal of the Empire, chief of the fifth cohort, grand officer, and grand eagle of the Legion of Honor. He was given the honor of carrying the sword of Charlemagne at Napoleon's coronation in 1805.

The emperor chose Lefebvre to command II Corps of Reserve in the Ulm-Austerlitz campaign, and the following year, in 1806, Lefebvre was made commander of V Corps and, late in the year, was also named to command the infantry of the Imperial Guard. He saw action at Jena on October 14 and distinguished himself during the siege of Danzig (March 18–May 27, 1807). A grateful Napoleon created Lefebvre Duke of Danzig on September 10, 1808 and, two months later, was put in command of IV Corps in Spain. After defeating **Blake** at Durango on October 31, he triumphed at Valmaseda on November 5. He pursued Blake's army, running him to ground at Espinosa on November 10–11. With **Claude Victor-Perrin**, Lefebvre delivered a final blow to Blake's army, which was neutralized for the remainder of the war.

By December. Lefebvre was occupying Segovia, but in March 1809 he was summoned to Germany as commander of the Bavarian VII Corps. He distinguished himself at Abensberg (April 20) and Eggmühl (April 22), only to suffer a reversal at Rastatt the following month. During the spring and early fall, he saw action throughout the Tyrol, capturing Innsbruck and suppressing the Tyrolean revolt led by Andreas Hofer.

Exhausted by continuous campaigning, Lefebvre sought and was given a two-year leave. He then returned to active duty as commander of the infantry of the Old Guard for the Russian campaign in 1812. He fought at Borodino on September 7, then received the crushing news of the death of his only son in Russia. Broken-hearted, he continued to serve, fighting in the German campaign of 1813, in command of the Imperial Guard, which he led at Dresden (August 26–27) and Leipzig (October 16–19). In 1814, on French soil, he led the Imperial Guard with great skill at Champaubert (February 10) and Montmirail (February 11).

After Napoleon's first abdication, the Bourbons created Lefebvre a peer of France, and he was again made a peer when Napoleon returned. By this time, however, he was far too old for field command. With the final return of King Louis, his appointment as Marshal of the Empire was confirmed, but he was removed from the chamber of peers and not restored until 1818, two years before his death.

Lefebvre was not the most dashing or brilliant of Napoleon's marshals, but he was among the most reliable. He led by courageous example, and although he was known as a strict disciplinarian, he also had great regard for the welfare of his men, whether they were Poles, Saxons, Bavarians, or Frenchmen.

LEMAY, Curtis Emerson

BORN: November 15, 1906
DIED: October 1, 1990
SERVICE: World War II (1941–45)

American army and air force officer, proponent of strategic bombing doctrine, and one of the architects of the modern air force

LeMay was born in Columbus, Ohio, and failing in his bid to obtain an appointment to West Point, attended Ohio State University. He left the university after completing the ROTC program and joined the army in 1928, becoming a cadet in the Air Corps Flying School in September. He earned his wings on October 12, 1929, and was commissioned a second lieutenant in January 1930.

LeMay was posted to duty with the Twenty-Seventh Pursuit Squadron, headquartered in Michigan. During the next few years, he completed the civil engineering degree he had begun at Ohio State. Awarded the degree in 1932, he did work for the CCC (Civilian Conservation Corps) and flew the air mails when President Franklin Roosevelt assigned Army fliers to air mail operations in 1934. After being promoted to first lieutenant in June 1935, LeMay attended an over-water navigation school in Hawaii.

In 1937, he transferred from pursuit planes to the 305th Bombardment Group at Langley Field, Virginia, and became involved in exercises demonstrating the ability of aircraft to find ships at sea. LeMay became one of the first army pilots to fly the new B-17 bombers. He led a flight of them on a goodwill tour to Latin America during 1937–38, after which he attended the Air Corps Tactical School (1938–39). Promoted to captain in January 1940, he was given command of a squadron in Thirty-Fourth Bomb group and the following year was promoted to major. Once the United States entered World War II, his advancement came even faster. By January 1942, LeMay was a lieutenant colonel, and three months later was promoted to full colonel. He assumed command of the 305th Bombardment Group in California in April, bringing that unit to Britain as part of the Eighth Air Force.

Once in place in Britain, LeMay set about perfecting precision bombing tactics by the risky means of abandoning evasive maneuvering over targets and by also introducing careful target studies prior to missions. Soon, LeMay had doubled the number of bombs placed on target. In June 1943, LeMay was assigned command of the Third Bombardment Division, which he led on the so-called "shuttle raid" on Regensburg in August. The following month, LeMay was promoted to temporary brigadier general, followed by promotion to temporary major general in March 1944. He was then sent to China to lead the Twentieth Bomber Command against the Japanese.

LeMay took command of the Twenty-First Bomber Group on Guam in January 1945 and stunned his air crews by modifying their B-29s to carry more bombs. To do this, he stripped them of defensive guns (and gun crews and ammunition), then ordered the planes to attack targets singly and at low level. Remarkably, crews survived, and bombing effectiveness was dramatically increased. The Twenty-First annihilated four major Japanese cities with incendiary bombs—and LeMay's fire-bombing raids proved far more destructive than the atomic bombing of Hiroshima and Nagasaki.

As the war wound down, LeMay was named commander of the Twentieth Air Force (Twentieth and Twenty-First Bomber Groups) in July 1945 and then deputy chief of staff for research and development, a post he held through 1947. In that year, he was promoted to temporary lieutenant general in an air force made independent from the army in 1947. He was given command of U.S. Air Forces in Europe on October 1, 1947 and was a key planner in the great Berlin airlift of 1948–49, which kept Berlin supplied during the Soviet blockade of the city.

In October 1948, LeMay returned to the United States as head of the newly created Strategic Air Command, which, until the development of ICBMs, functioned as the armed forces' sole delivery system for atomic and thermonuclear weaponry. Under LeMay, the Air Force greatly expanded and entered the jet age with B-47 and B-52 bombers and in-air refueling tankers (KC-135s), which greatly extended the bombers' range. By the 1950s, LeMay oversaw the introduction of missiles into the Air Force's inventory of nuclear-capable weapons.

LeMay was promoted to general in October 1951, becoming the youngest four-star general since **Ulysses S. Grant**. In 1957, he was named vice chief of staff of the Air Force and became chief of staff in 1961. During the 1960s, his hard-nosed conservatism frequently brought him into conflict with the administration's of John F. Kennedy and Lyndon Johnson. His relations with Secretary of Defense Robert S. McNamara were particularly strained and even bitter. LeMay became increasingly irascible, and on February 1, 1965 retired from the Air Force. His political conservatism hardening during the turbulent late 1960s, he became the running mate of Alabama Segregationist governor George Wallace in his failed 1968 bid for the presidency.

LeMay was an uncompromising commander, who demanded "maximum effort" from his air crews and from their aircraft. He was a forward-thinking planner, who shaped the modern air force and established its place as the most strategic of the "triad" of services (army, navy, air force). He developed the doctrine of precision bombing and then brought the U.S. Air Force into the jet age and the nuclear age.

LETTOW-VORBECK, Paul Emil von

BORN: March 20, 1870

DIED: March 9, 1964

SERVICE: Boxer Rebellion (1900–01); Hottentot-Herero Rebellion (1904–08); World War I (1914–18)

German general and guerrilla in Africa during World War I

Born at Saarlouis, Lettow-Vorbeck was the son of a prominent Prussian army officer. Marked from birth for a military career, Lettow-Vorbeck was commissioned as an artillery officer and graduated from the prestigious *Kriegsakademie* in 1899. His promise was early apparent, and he was tapped for service on the General Staff (1899–1900).

In 1900, Lettow-Vorbeck was dispatched to China with the German expeditionary force as part of an international coalition to punish the so-called Boxers, Chinese nationalists who rebelled against European interests in their country. Following the conclusion of the Boxer Rebellion in 1901, Lettow-Vorbeck served in German Southwest Africa (present-day Namibia), fighting the Hottentots and Hereros during 1904–08. This proved an extraordinarily arduous undertaking for a commander trained in the arts of European wars. Lettow-Vorbeck was wounded in an ambush in 1906 and sent to a hospital in South Africa. Upon his return to Germany, he was promoted to lieutenant colonel and, early in 1914, was sent back to Africa as commander of forces in German East Africa (present-day Tanzania). His native troops—*askaris*—were armed with obsolete weapons and, as Lettow-Vorbeck knew, would be cut off once a world war commenced. With limited supplies and very limited manpower, he resolved to strike preemptively. As soon as World War I began in August 1914, he staged raids against the British railway in Kenya and then attempted to capture Mombasa. In this effort, he was driven back by September, but successfully defended against a British amphibious attack on the port town of Tanga in northeastern Tanzania during November 2–3, 1914. He made the attack particularly costly for the British, not only inflicting heavy losses, but acquiring a large cache of badly needed arms and ammunition.

Lettow-Vorbeck continued to strike at the British, who were themselves poorly supplied and ordered to maintain a defensive posture only. When the Royal Navy sank the German cruiser *Koningsberg* in the Rufiji River, the ship on which Lettow-Vorbeck depended heavily for support, he set about salvaging most of her guns, and he commandeered the vessel's crew as soldiers.

In March 1916, Lettow-Vorbeck finally faced a formidable offensive, led by South African general **Jan Christiaan Smuts**. With brilliant patience, Lettow-Vorbeck made the climate and terrain his ally, fighting nothing more than delaying actions to wear down the British troops. In the end, tropical diseases did far more than his men's bullets. Nevertheless, as British numbers increased, Lettow-Vorbeck gradually yielded to invasion, exacting from the invaders what price he could.

Whenever possible, he turned on his pursuers and lashed back, frequently surprising the British. At Mahiwa, during October 15–18, 1917, he cost the British forces—which outnumbered his four to one—1,500 casualties while incurring himself a mere one hundred.

It was nevertheless clear to Lettow-Vorbeck that the British would ultimately drive him out of German East Africa. He therefore invaded the Portuguese colony of Mozambique in December and supplied his four-thousand-man army by raiding Portuguese garrisons. He raided as far south as Quelimane on the coast during July 1–3, 1918, then turned north again and reentered German East Africa during September–October. After this, he launched an invasion of British-held Rhodesia (present-day Zimbabwe) and took the principal city of Kasama (modern Zambia) on November 13—two days after the Armistice. Isolated from the rest of the world, Lettow-Vorbeck heard only vague rumors of the German surrender, but these were sufficient to prompt him to open negotiations with the British. He surrendered his undefeated army to the British at Abercorn (Mbala, Zambia) on November 23.

Following the end of World War I, Lettow-Vorbeck remained in Africa to arrange for the repatriation of German soldiers and POWs. When he returned home in January 1919, it was to a hero's welcome and a promotion to *generalmajor*. Enormously popular, he became a right-wing extremist who organized the occupation of Hamburg by a force of rightist volunteers and the Freikorps, ousting left-wing Spartacists from the city in July. Arrested and briefly jailed for this action, he nevertheless gained election to the Reichstag, serving from May 1929 to July 1930. Although a rightist, he opposed the Nazis, but, finding that opposition a lost cause, he retired from politics and from public life generally. He lived out the remainder of his long life quietly in Hamburg.

As a German officer, Lettow-Vorbeck was among the least likely candidates for greatness as a guerrilla leader, but he had a natural talent for resourcefulness and making the most of very limited resources. Moreover, he won the respect and loyalty of European and native troops alike.

LIGGETT, Hunter

BORN: March 21, 1857

DIED: December 30, 1935

SERVICE: Philippine Insurrection (1899–1903); World War I (1917–18)

World War I American general who earlier directed the War College, making it an effective center for the study of battlefield problems

Liggett was a native of Reading, Pennsylvania, who graduated from West Point in 1879 and was commissioned in the infantry. During service in the frontier West, including minor operations against Indians, he was promoted to first lieutenant (June 1884) and captain (June 1897). With the outbreak of the Spanish-American War in April 1898, Liggett accepted a volunteer commission as major and assistant adjutant general. During the war, he served Stateside as a division adjutant in Florida, Alabama, and Georgia from June 1898 through April 1899. During April–October 1899, after the war with Spain had ended, he was posted in Cuba, then sent to the Philippines in December 1899, where he commanded a subdistrict on Mindanao from 1900 to 1901.

Returning to the United States, Liggett rejoined the Fifth Infantry at his permanent, regular army rank of captain in October 1901, but was promoted to major the following year and transferred to Twenty-First Infantry in Minnesota. He was appointed adjutant general of the Department of the Lakes, serving in this post from 1903 to September 1907, when he was given command of a battalion in the Thirteenth Infantry, headquartered in Kansas. Promoted to lieutenant colonel in June 1909, he graduated from the Army War College the following year, then remained at the school as its director from 1910 to 1913. From 1913 to 1914, he served as president of the War College. During his tenure at the institution, he introduced a program of thorough and logical study of battlefield problems, helping to make the War College one of the military's most prestigious training schools and the focus of the army's strategic planning effort. Liggett played a major role in creating the army's officer education program and was instrumental in preparing plans for intervention in Mexico and the Caribbean, as well as for the defense of the Philippines.

Liggett attained promotion to colonel in March 1912 and to brigadier general in February 1913. In 1914, Liggett assumed command of Fourth Brigade, Second Division in Texas, then was sent to the Philippines in 1915. The following year, he was appointed commander of the Philippines Department. On March 4, 1917, he was promoted to major general and returned to the United States in May as commander of the Western Department.

In September 1917, Liggett was appointed to command the Forty-First Division, which he took to France in October. In Europe, he was named commander of I Corps on January 20, 1918 and had administrative control of the handful of American units in action during the spring. Liggett led I Corps (which included the Second and Twenty-Sixth U.S. and the 167th French divisions) into action near Chateau-Thierry on July 4, 1918. I Corps played a key role in the Aisne-Marne offensive of July 18-August 5.

Liggett and his corps were transferred to Lorraine, where Liggett prepared for the first major American offensive of World War I: reduction of the Saint-Mihiel salient (September 12–16). He also participated in the first phase of the Meuse-Argonne campaign during September 26–October 12, then replaced General **John J. Pershing** as commander of First Army on October 16. Liggett prepared and reorganized First Army for the second phase of Meuse-Argonne (October 16–31), which proved a great success and, in its final phase (November 1–11), the culminating battle of the war.

Liggett stayed on in Europe after the Armistice as commander of First Army until it was disbanded on April 20, 1919. He then commanded Third Army, the occupation force in Germany, from May 2 to July 2, 1919. After returning to the United States, he resumed command of the Western Department until his retirement, with the rank of major general, on March 21, 1921. In June 1930, he was advanced to lieutenant general on the retired list.

Liggett was a solid army officer, who was noted for his tactical and operational skill as well as for his compassion for his troops. Indeed, some criticized him for being overly kind. While his service during World War I was gallant and brilliant, Liggett's most enduring contribution to the American military was his work at the War College, which he molded into a highly effective instrument of training as well as planning.

LINCOLN, Abraham

BORN: February 12, 1809
DIED: April 14, 1865
SERVICE: Black Hawk War (1832); Civil War (1861–65)

U.S. president during the Civil War; worked closely with his generals to plot the Union's strategy

Lincoln was born in very modest circumstances near Hodgenville, Kentucky and moved to Indiana in December 1816. In March 1830, he settled in Illinois. He was largely self-taught, having had virtually no formal education. As a militiaman, he volunteered for service in the Black Hawk War during April–August 1832 and, always popular, gained election as captain of his unit. However, he saw no action in the war. This constituted the sum total of his direct military experience.

Lincoln passed the Illinois state bar in 1836 and set up a law practice in the state capital, Springfield, in 1837, while serving in the state legislature (1834–40). He was sent to the U.S. House of Representatives in 1847 and served one term.

Lincoln's law practice prospered, and he gained a reputation for honesty and fairness as a circuit lawyer. He joined the newly formed Republican Party in 1856, but lost his bid for a Senate seat to Stephen A. Douglas, whom he debated in a series of oratorically brilliant exchanges that have become classics of American politics and that articulated the crisis of nation on the verge of civil war. Despite his loss to Douglas, the debates gained Lincoln national prominence, and he was nominated as the Republican presidential candidate on May 15, 1860. With the Democratic party split by secession, Lincoln was elected on November 6, an event that prompted the secession of South Carolina from the Union on December 20. Six other southern states soon followed suit, and the Confederate States of America was created before Lincoln's inauguration on March 4, 1861.

Lincoln's first excruciatingly difficult decision was whether or not to fight to preserve the Union. He determined to go to war if necessary and ordered the provisioning of the Federal garrison at Fort Sumter in South Carolina's Charleston harbor. The Confederates demanded the surrender of the fort and opened fire on April 12, 1861. The Civil War had begun.

Lincoln acted quickly. Without securing Congressional approval, he called for the raising of 75,000 volunteers and immediately ordered a blockade of Confederate ports. No one was more acutely aware than Lincoln that he lacked military experience. However, he had an instinctive grasp of strategy and understood that the greater population and industrial capacity of the North gave the Union a profound advantage over the South. He also understood, however, that the Confederate army now possessed many of the best officers who had formerly served in the U.S. Army. Lincoln understood that the Confederates were counting on a series of quick, decisive victories to break the will of the North to prosecute the war. His strategy was to prevent these victories and, ultimately, to fight a war of attrition against the relatively underpopulated and under-industrialized South.

Lincoln's strategy was repeatedly frustrated by a series of generals whose overly cautious approach to battle brought defeat after disappointing defeat. In succession, Lincoln appointed **George B. McClellan**, **Henry W. Halleck**, John Pope, **Ambrose Burnside**, **Joseph Hooker**, and **George G. Meade** either as principal field commanders in the East or as overall commanders of the Union forces. Each of these generals brought varying degrees of failure. At last, in March 1864, Lincoln chose **Ulysses S. Grant** to lead the Union armies. Grant embraced the strategy of attrition and threw his resources aggressively into battle. Aided by the likes of **William T. Sherman, Philip H. Sheridan**, and George H. Thomas, Grant made steady progress, even against a military genius like **Robert E. Lee**.

Grant's strategy was a painful one, involving massive loss of life. Despite frequent popular outcries, Lincoln threw his full support behind his general, whom, he believed, would ultimately deliver final victory.

Once he was confident of victory, Lincoln set about formulating generous terms of peace—as he put it in his second inaugural address, "with malice toward none and charity for all." Surrender of the Confederacy would involve yielding only to twin conditions: reunion and freedom for the slaves. This latter issue had grown from a secondary position—Lincoln insisted that his principal aim was to save the Union, not free the slaves—to the central moral focus of the war. On January 1, 1863, Lincoln promulgated the Emancipation Proclamation and, on January 31, 1865, pushed through Congress the Thirteenth Amendment, abolishing slavery. (It was ratified by the states on December 18, 1865.)

Lincoln was reelected in 1864 and focused his energy on creating the conditions that would allow for reunification of the nation without punishing the South. In perhaps the single greatest stroke of tragedy ever to befall the United States, the president was shot on April 14, 1865, by the fanatical Southern sympathizer, John Wilkes Booth, shortly before the war ended. His death the following morning likewise killed the nation's best hope for a healing and conciliatory peace. The result was a long and bitter period of Reconstruction, which created great cultural and racial rifts that have yet to fully heal.

By any measure, Abraham Lincoln was an extraordinary man and a great leader. Intelligent and compassionate, he also proved to be a naturally skilled military strategist, despite his lack of formal training and practical military experience.

LIN PIAO

BORN: 1907

DIED: September 13, 1971

SERVICE: Bandit (Communist) Suppression Campaigns (1930–36); Second Sino–Japanese War (World War II) (1937–45); Chinese Civil War (1946–49); Korean War (1950–53); Sino–Indian War (1962)

Chinese Communist marshal who played a principal role in the Cultural Revolution, but subsequently vied with Mao Tse-tung for power

Born in Hupeh province, Lin Piao was the son of a factory owner, whose business ultimately faltered under the crushing burden of taxes. Nevertheless, Lin was given a privileged education at a preparatory school. He joined the Nationalist Party—Kuomintang (KMT)—and in 1926 graduated from the Whampoa Military Academy. He then saw service with the KMT Fourth Corps during the first half of the Northern Expedition (July 1926–April 1927).

After gaining promotion to major, Lin took over his regiment and induced it to leave the KMT and go over to the Communist side in the Nanchang (Jiangxi) uprising led by Ho Lung and Yeh T'ing during August and September. When that uprising faltered, Lin adroitly withdrew and united with **Chu Teh.** Later, they joined with Mao Tse-tung in creating the Kiangsi Soviet (1927–28).

Lin became prominent and powerful in the Kiangsi Soviet resisting the KMT's efforts during the Bandit Suppression campaigns of 1934. Leading I Corps, he spearheaded the Communist breakout from KMT encirclement in October, then was among the military leaders of Mao's legendary Long March (1934–35).

Following the Long March, Lin Piao ran the Pao-an Military Academy and was then named to command of the new Eighth Route Army's 115th Division in 1937. He served in this capacity through the Sino–Japanese War (World War II), building the division until, by 1945, it had assumed the proportions of a full-scale field army. Lin performed brilliantly against the Japanese, ambushing an entire regiment at P'ing-hsing Pass in September 1937, then serving prominently in **Chiang Kai-shek**'s ill-advised Hundred Regiments offensive during August 20–November 30, 1940.

During most of the war, Lin engaged in a steady effort to raise armies and establish bases behind Japanese lines throughout northern China.

With the end of the war, Lin Piao led a large force of 100,000 into Manchuria, where he was mauled by a KMT army under Ch'en Ch'eng and Tu Yu-ming during March–June 1946. He retreated into the north, recouped, then renewed the attack with half a million troops in May 1947, driving the KMT southward before him. Systematically and relentlessly, Lin cut off the KMT garrisons of the major cities. With the fall of Mukden on November 1, 1948, the KMT was finished in Manchuria. This region secured, Lin pushed on to Beijing, which he was instrumental in taking on January 22, 1949, after a prolonged siege.

Following the fall of Beijing, Lin Piao marched south and crossed the Yangtze in April, taking Canton in October. It was this army group that he led into Korea after the United Nations advanced against the North Koreans to the Yalu River in November 1950. Lin pushed **Douglas MacArthur**'s U.N. troops relentlessly southward, beyond the South Korean capital of Seoul during November 25, 1950–January 15, 1951. Spectacular as this counteroffensive was, Lin could not hold the ground he had taken and was in turn forced back north during January–April 1951. After this, he was replaced by General P'eng The-huai and returned to China, where he was appointed a Vice-Premier of the People's Republic in 1954. The following year, he was named one of ten marshals of the People's Liberation Army.

Lin replaced P'eng as Minister of Defense in 1959 and immediately set about purging the armed forces of Russian influences. He also ratcheted up the army's combat readiness, then directed operations during the conflict with India in the Himalayan border region during October 20–November 21, 1962. It is likely that Lin was also responsible for directing military assistance to North Vietnam during the Vietnam War.

Lin Piao was a supporter of Mao Tse-tung's sweeping Cultural Revolution of 1966–69, and when the

Cultural Revolution brought China near to anarchy, it was Lin who helped Mao restore order in the rural areas. Following the Cultural Revolution, Lin was designated Mao's heir apparent and, indeed, was second only to the chairman in political power. However, Mao soon learned to fear the growing power of the People's Liberation Army, and felt that Lin was too radical in his support of the leftist fringe of the government. As relations between Mao and Lin deteriorated, Lin planned a military coup, but was politically checkmated by Mao. Forced to flee the country, he died near the Mongolian border when his plane crashed.

Lin was a very capable military commander, who combined boldness with thoughtful strategy. Politically adept, he was nevertheless no match for Mao Tse-tung, who outmaneuvered him and brought about his downfall.

LI SHIH-MIN (T'ai Tsung)

BORN: 598

DIED: 649

SERVICE: Revolts against *Yang* Ti (613–18); Wars with the Eastern Turks (629–41); War with the Western Turks (639–48)

Chinese emperor who distinguished himself as a military strategist; largely responsible for the conquest of the eastern capital of Lo-yang and the eastern plain

Li was born into a Wei noble family and was the second son of Li Yüan. The traditional view is that Li was the major force behind his father's uprising against the decaying Sui dynasty in 617. However, more recent evidence suggests that he played a fairly minor role in instigating the revolt. But Li and his elder brother, Li Chien-ch'eng, both served as T'ang commanders in the taking of the Sui capital at Chang'an. With the capture of the capital, Li's father proclaimed the T'ang dynasty in 618 and took the Imperial name of Kao Tsu.

Li then launched a number of military campaigns to bring all China under T'ang rule. Most importantly, he repulsed a massive raid by the Eastern Turks, which pressed to the very walls of Chang'an in 624. Three years after this, Li killed his elder and younger brothers, then compelled his father to abdicate the throne in his favor. He proclaimed himself T'ai Tsung, emperor of China as T'ai Tsung.

Once he assumed the throne, Li personally led fewer campaigns, but he still took an active role in military affairs. He established a high training standard for his infantry crossbowmen, decreeing that they should be able to hit a man-size target two times out of four at 300 paces. Doubtless, he also established other requirements for his military. More importantly, he continued to set strategy for all military campaigns and, in this, exhibited extraordinary brilliance. His armies defeated the Eastern Turks in 630. In two campaigns—639–40 and 647–48—Li wrested the Tarim basin from the Western Turks, thereby gaining control of the Great Silk Road, the principal trading and military artery of the East.

In 641, Li's forces destroyed Tibetan invaders led by Song-tsan Gampo. With consummate diplomacy, Li then forged an alliance with the Tibetans. Ultimately, Li Shih-min was acknowledged the Great Khan (supreme ruler) over most of Central Asia.

Not all of Li's campaigns were successful. His two invasions of Korea, in 644 and 646, gained him little, as did his foray into India in 646.

Li Shih-min was a complex character. He was ruthless, as his murder of his brothers and treatment of his father showed. He was an extraordinary military commander, both when personally leading troops as well as when setting strategy. He was also an able civil ruler, who was guided by the Confucian maxim of governing for the good of the governed. He supported education, literature, and the arts.

LI TSUNG-JEN

BORN: 1891

DIED: 1969

SERVICE: Chinese Civil War (1916–27); Northern Expedition (1926–28); Bandit (Communist) Suppression Campaigns (1930–34); Second Sino-Japanese War (World War II) (1937–45); Chinese Civil War (1946–49)

Chinese warlord who became a general in the service of the Nationalists (KMT)

Li was the son of a poor but educated family and was accorded a classical Chinese elementary education, then sent to the Cotton Weaving Institute. After this, he attended the Kwangsi Military Academy and, in 1916, joined the Kwangsi provincial army. He rose rapidly through the ranks, but in 1921, the Kwangsi army was destroyed in the course of China's anarchical civil warfare. Li led 2,000 followers from the army into hiding in the hills and there joined other warlords, most notably Pai Ch'ung-hsi, in a campaign of guerrilla operations against the Canton government.

By 1925, Li and his allies dominated Kwangsi province, and he joined the Nationalists—Kuomintang (KMT)—with the proviso that he retain control in Kwangsi. Li fought in the first phase of the Northern Expedition as a corps commander and supported **Chiang Kai-shek**'s campaign against leftists and Russian advisers during July–September 1927.

Li was in command of an army during the final advance on Beijing (April 7–June 4, 1928). By the end of 1928, his power rivaled that of Chiang Kai-shek. He now controlled Kwangsi, as well as much of central China. Accordingly, he turned against Chiang, resisting that leader's efforts at consolidating and centralizing power under a single Nationalist government. In 1929, Chiang defeated Li, who agreed to leave the country. However, he returned to Kwangsi in 1930 and created a relatively effective government for the province. By 1936, he launched a new rebellion against Chiang Kai-shek, but then returned to the KMT fold to meet the Japanese threat in 1937.

In the war with Japan, Li was given command of the Fifth War Area and commanded the Chinese counterattack that led to a victory at the Battle of Taierhchwang in April 1938. It was the first KMT victory in ten months. Li continued in high command throughout the balance of the war, and when Japan surrendered in 1945, Li moved quickly to recover the occupied territory for the Nationalists. Chiang Kai-shek, however, distrusted Li and ordered a halt to his actions. Li held no military commands after 1947, but was elected vice president in that year and, after Chiang resigned as president on January 21, 1949, Li briefly became president. He made peaceful gestures of accord toward the Communists in April, but was spurned. That summer, realizing that the Nationalist cause was doomed and Communist victory a certainty, he left China for the United States, living here in exile until he returned to mainland China in 1965. No longer a military or political threat, he was greeted as a hero of the struggle against Japan and lived in quiet retirement in China for the remainder of his life.

Li was highly unusual as a warlord. He was an able administrator, who, though desirous of power, also had the good of his people at heart. As a military commander, Li was both competent and aggressive, giving the KMT some of its few triumphs against the invading Japanese. It is unfortunate for the cause of Chinese Nationalism that he and Chiang conflicted more than they cooperated. The rivalry between them greatly weakened the KMT.

LOGAN, John Alexander

BORN: February 9, 1826

DIED: December 26, 1886

SERVICE: U.S.–Mexican War (1846–48); Civil War (1861–65)

American Civil War general (Union); originator of Memorial Day

A native of rural Murphysboro, Illinois, Logan was largely self-educated before volunteering for service in the U.S. Army as a second lieutenant in the U.S.–Mexican War. After the war, he entered the University of Louisville (Kentucky), graduating with a law degree in 1851, and in 1858, he was elected Democratic congressman from Illinois, but resigned his seat in 1861 to join the Union Army as a private in a Michigan regiment. After fighting in First Bull Run on July 21, 1861, he returned to Illinois to form the Thirty-First Illinois Regiment and was appointed its colonel in September.

Colonel Logan served under General **Ulysses S. Grant** at the Battle of Belmont on November 7 and at the assaults on Forts Henry and Donelson during February 6–15, 1862. In March 1862, he was promoted to brigadier general and fought at Vicksburg during January–July 4, 1863, commanding a division in General James B. McPherson's XVII Corps.

In November 1863, Logan was promoted to major general and given command of XV Corps of the Army of the Tennessee. When McPherson was killed during the Atlanta campaign (May 5–September 1, 1864), Logan assumed temporary command of the army on July 22, but was relieved by General **William T. Sherman**. Sherman lacked confidence in Logan's experience, claiming in particular that he paid little attention to logistics, and returned him to command of a corps. (Sherman assigned command of the Army of the Tennessee instead to General **Oliver O. Howard**.)

Logan was reelected to the House of Representatives—now as a Republican—after the war and served from 1867 to 1871. He was then elected to the Senate and played a key role in the impeachment of President Andrew Johnson.

Immediately after the war, in 1865, Logan became one of thte founders of the Grand Army of the Republic (GAR), an organization of Union Army veterans, and headed it as commander in chief for many years. In 1868, he inaugurated the observance of Memorial Day—also called Decoration, Day—when he asked GAR members to decorate soldiers' graves with flowers on May 30.

LONGSTREET, James

BORN: January 8, 1821

DIED: January 2, 1904

SERVICE: U.S.–Mexican War (1846–48); Civil War (1861–65)

Confederate general of considerable capability, but whose actions at Gettysburg clouded his reputation

Longstreet was born and raised in Edgefield District, South Carolina. He graduated from West Point in 1842 and was commissioned a second lieutenant in the infantry. During the U.S.–Mexican War, he served under General **Zachary Taylor** and, subsequently, under **Winfield Scott**.

Promoted to first lieutenant in January 1847, he was breveted to captain after distinguishing himself at Churubusco on August 20 and was breveted again, to major, at the Battle of Molino del Rey on September 8. A few days later, on September 13, during the assault on Chapultepec just outside Mexico City, Longstreet was severely wounded.

After the war, Longstreet reverted to the permanent rank of first lieutenant but was promoted to captain in December 1852 and then to major in July 1858.

In June 1861, after South Carolina seceded from the Union, Longstreet resigned his commission and accepted a Confederate commission as brigadier general. Commanding the advance guard at Blackburn's Ford (near Centreville, Virginia) on July 18, he repulsed General **Irvin McDowell** in a prelude to First Bull Run (July 21). During the Yorktown campaign, he served under General **J. E. Johnston** and fought well at Williamsburg on May 5, 1862, but was dilatory at Fair Oaks on May 31, which drew criticism. At Second Bull Run, during August 29–30, he commanded five divisions coordinated to support General **Thomas ("Stonewall") Jackson.**

Longstreet did not see eye to eye with **Robert E. Lee** and disputed the wisdom of invading Maryland. Nevertheless, he acquitted himself well at South Mountain on September 14 and at Antietam on September 17. The following month he was promoted to lieutenant general and given command of I Corps in Lee's Army of Northern Virginia. At Fredericksburg, on December 13, he successfully defended Marye's Heights, but because his units were detached from the main army for campaigning in Suffolk, Virginia, during April 1863, he missed Chancellorsville (May 2–4), where his forces were sorely needed.

Once again, he sharply differed with Lee during the invasion of Pennsylvania. In particular, he faulted Lee's tactical plan at Gettysburg (July 1–3). That he was apparently slow in executing Lee's orders on July 2 has been unjustly ascribed to his dispute with his commander. His tardiness dogged the remainder of his career, and he unfairly bore more than his fair share of the burden for the Confederate defeat at Gettysburg.

Following Gettysburg, Longstreet was dispatched to reinforce **Braxton Bragg**'s Army of Tennessee. With Bragg, he fought at Chickamauga during September

19–20, pushing forward in an impressive advance that was ultimately checked by Union general George Thomas. Bragg sent Longstreet against **Ambrose Burnside**'s forces at Knoxville. Longstreet ended up doggedly laying siege to Burnside in the city and was therefore unavailable to Bragg at Chattanooga, during November 24–25, when Bragg was badly in need of support.

In April 1864, Longstreet returned east to rejoin the battered Army of Northern Virginia. He fought at the Wilderness (May 5–6), where he was severely wounded by friendly fire. His right arm permanently paralyzed, he returned to duty in October and fought throughout the balance of the war, distinguishing himself at Petersburg and Richmond. He surrendered with Lee at Appomattox Court House on April 9, 1865.

After the war, Longstreet made himself intensely unpopular among his Southern compatriots for voicing his admiration of President **Ulysses S. Grant** and for joining the Republican party. He was given several government posts, including that of customs surveyor (1869) and postmaster (1873) of New Orleans. In 1880, he was appointed U.S. minister to Turkey and the following year became U.S. marshal in Georgia. In 1898, he was appointed U.S. railroad commissioner. His wartime memoirs, *From Manassas to Appomattox*, appeared in 1896.

Longstreet was able, but also ambitious and disputatious. Despite his friction with Lee, the Confederate general in chief did not hesitate to promote him. Longstreet was a better tactician and administrator (much beloved by his troops) than he was a farsighted strategist.

LUCAS, John Porter

BORN: January 14, 1890

DIED: December 24, 1949

SERVICE: Punitive Expedition into Mexico (1916–17); World War I (1917–18); World War II (1941–45)

American general in World War II; criticized for his direction of the Anzio campaign

John Lucas. Courtesy National Archives.

Lucas was born in rural Kearneysville, West Virginia, and was a graduate of West Point, Class of 1911. Commissioned a second lieutenant in the cavalry, he was posted to the Philippines from December 1911 to August 1914, then returned to the States, where he was assigned to the Thirteenth Cavalry at Columbus, New Mexico, during the depredations of the Mexican political outlaw Pancho Villa. Lucas was a first lieutenant in 1916 and was given command of the Thirteenth Cavalry's machine-gun troop when Villa raided Columbus on March 9, 1916. Lucas's soldiers were instrumental in driving the bandit off. Lucas served throughout the next phase of operations against Villa, the Punitive Expedition led by General **John J. Pershing** (March 15, 1916–February 5, 1917).

Promoted to captain, Lucas was appointed aide-de-camp to General George Bell, headquartered at El Paso, Texas, during February–August 1917. With U.S. entry into World War I, Lucas was attached to the Thirty-Third Infantry Division and promoted to the temporary rank of major, with command of the division's 108th Field Signals Battalion in January 1918. By the time the Thirty-Third Division was shipped out to France in May, Lucas had been promoted to temporary lieutenant colonel. Early in June, he was wounded near Amiens, seriously enough to warrant return to the States.

By the Armistice, Lucas was sufficiently recovered to served home duty in Washington, D.C. After reversion to his permanent peacetime rank of captain, he taught military science at the University of Michigan (1919–20), then transferred to the field artillery and received promotion to major in 1920. In June 1921, Lucas graduated from the Field Artillery School, then stayed on there as an instructor during 1921–23. In June 1924, he graduated from the Command and General Staff School, going on to a long stint as professor of military science and tactics at Colorado Agricultural College from 1924 to 1929, when he was given command of the First Battalion, Eighty-Second Field Artillery, at Fort Bliss, Texas.

Lucas left Fort Bliss in June 1931 to enroll in the Army War College, from which he graduated in June 1932. From 1932 to 1936, he was posted to the Personnel Division, G-1, of the War Department General Staff—and was promoted to lieutenant colonel in 1935.

From 1936 to 1937, Lucas commanded the First Field Artillery Regiment at Fort Bragg, then served on the Field Artillery Board from December 1937 to July 1940. After a short stint as commander of the Fourth Field Artillery in 1940, Lucas was promoted to brigadier general and served briefly as commander of the Second Infantry Division. In July 1941, with the nation of the verge of World War II, Lucas was named commander of the Third Infantry Division and was promoted to temporary major general on August 5. He conducted amphibious maneuvers in Puget Sound, then was assigned command of III Corps, in Georgia, during April 1942–May 1943.

In the spring of 1943, Lucas was sent to England to join the staff of **Dwight D. Eisenhower.** He was named commander of VI Corps, Fifth Army, in September 1943 and led his troops through Campania, Italy. He advanced to the Venafro line, which he reached on January 3, 1944, but was withdrawn after the Allied advance ground to a bloody halt south of Rome. Lucas was assigned to land his corps at Anzio (January 22, 1944) in a bid to go around the German defenses and take Rome. A methodical campaigner, Lucas proceeded with excessive caution—in part because his objectives were never made fully clear—and he managed to do little more than eke out a beachhead before German forces checkmated him.

Lucas's lack of progress brought down a storm of criticism from the British Mediterranean Theater commander, General Sir Harold R. L. G. Alexander. As a result, General Eisenhower relieved Lucas and replaced him with General Lucian K. Truscott. Lucas was sent back to the United States, where he was assigned in March to command the Fourth Army in Texas.

After the war, from June 1946 to January 1948, Lucas was assigned as chief of the U.S. Military advisory group to the Nationalist (Kuomintang—KMT) forces of Generalissimo **Chiang Kai-shek.** Lucas was promoted to the permanent rank of major general (retroactive to August 1944) at the end of this assignment and returned to the United States as deputy commander of the Fifth Army, headquartered in Chicago. He served in this capacity until his death.

Lucas was an extremely capable officer, possessed of great personal courage. A deliberate tactician, he was overly cautious, as the Anzio episode demonstrated.

LUDENDORFF, Erich

BORN: April 9, 1865

DIED: December 20, 1937

SERVICE: World War I (1914–18)

German (Prussian) general largely responsible for military strategy in the late phase of World War I; later, an arch militarist and early Nazi enthusiast

Ludendorff was born at Kruszewnia, near Poznan, in Prussia, the son of a landowner on hard times. With few other prospects, Ludendorff entered the Cadet Corps in

1877 and then the Thirty-Ninth Fusilier Regiment of the German army in 1883. That he was the very model of the Prussian officer did not escape his superiors, who admired his ramrod bearing as much as they did his obvious intelligence. In 1893, the promising young officer was admitted to the prestigious *Kriegsakadamie*, training ground for the General Staff, which Ludendorff joined in 1895.

From 1904 to 1908, Ludendorff was a member of the Mobilization Section of the General Staff, gaining appointment as its chief from 1908 to 1913. He earned the admiration of Count **Alfred von Schlieffen**, author of the plan by which Germany intended to win the next European war, and the younger Count **Helmuth von Moltke**, chief of the General Staff and nephew of the great **Moltke** who had served Bismarck. With Moltke, Ludendorff labored on revision of the Schlieffen plan.

Despite his close ties to men of high influence, Ludendorff fell afoul of the German War Minister and temporarily left the General Staff in 1913 for command of the Thirty-Ninth Fusiliers and, later in the year, an infantry brigade at Strasbourg. Promoted to *generalmajor* (brigadier general), he was assigned as deputy chief of staff for Karl von Bülow's Second Army in 1914, just before the outbreak of World War I.

During the assault on Liège, from August 5 to August 16, 1914, he took command of the Fourteenth Brigade after its general was killed. He rose to the occasion, capturing the Belgian city during August 6–7 and then the fortress there. Late in August, General **Paul von Hindenburg**, duly impressed, chose Ludendorff as his chief of staff to replace Max von Prittwitz and Count Alfred von Waldersee, who had floundered on the Eastern Front. Together, Hindenburg and Ludendorff created and executed a winning strategy in the East and soon turned the tide. The stunning victory against the Russians at Tannenberg during August 25–31, 1914 was only partially planned by Hindenburg and Ludendorff, but the triumphs that followed were their work: the Masurian Lakes (September 10–13); the campaign in central Poland (September 17–early December)—including the capture of Lódz (November 21–December 6, 1914); and the victory of the Second Masurian Lakes (January 31–February 21, 1915). Ludendorff also planned the East Front summer offensive of Gorlice-Tarnów (May 2–June 27, 1915), which resulted in the fall of most of Poland to the Germans—though it failed in its other objective, the destruction of the Russian army.

Back on the Western Front, the German offensive at Verdun faltered, and Erich von Falkenhayn was relieved as chief of the General Staff and replaced by Hindenburg on August 29, 1916. Ludendorff was made deputy chief of staff and soon eclipsed Hindenburg in popularity, not only with the troops, but with the German people. In effect, the Hindenburg-Ludendorff duo became de facto military dictators of wartime Germany.

Ludendorff had free rein to plot out the Austro-German Caporetto offensive of October 24–November 12, 1917, which very nearly took Italy out of the war. In defiance of those who worried about America's coming into the war, Ludendorff advocated and supported unrestricted submarine warfare to break the British blockade.

Once the United States did declare war, Ludendorff, with Hindenburg's concurrence, determined to make an all-out effort was required immediately to win victory on the Western Front *before* U.S. entry became genuinely effective. Accordingly, he planned three major offensives: the Somme (March 21–April 4, 1918), the Lys (April 9–29), and the Aisne (May 27–June 4). All three offensives gained precious ground, but failed to break the British forces. Ludendorff then struck at Montdidier-Soissons (June 9–13) and Champagne-Marne (July 15–17), but accomplished even less. Worse, Germany's own resources flagged, and, now augmented by the Americans, the Allies struck back with massive counteroffensives between July 18 and October 28. Reeling, Ludendorff pronounced July 18, 1918, the "Black Day of the German army."

Amid the new Allied onslaught, the Ludendorff–Hindenburg team began to crack. Playing on his greater popularity, Ludendorff publicly argued with Hindenburg. At last, by the end of September, Ludendorff called for an armistice, pointing out that Germany was in a hopeless position. When he made unauthorized intervention in Prince Max of Baden's protracted peace negotiations on October 24, he was relieved of command two days later.

Following the armistice, Ludendorff temporarily settled in Sweden to write his memoirs. In 1919, he returned to Germany, where he threw himself into extremist right-wing politics. Early on, he became an enthusiastic supporter of the Nazi party and applauded the Beer Hall Putsch in Munich (November 8–9, 1923), which landed **Adolf Hitler** in prison. Ludendorff was likewise tried, but acquitted.

In May 1924, Ludendorff was elected to the Reichstag on the Nazi ticket, then failed in his bid for the German presidency in April 1925. After this, he fell away from Hitler and the other Nazis, feeling that they were inept. He withdrew from politics and public life generally, and began writing treatises on military strategy, in particular elaborating upon **Clausewitz**'s theory of total war. As part of an overall military strategy, he advocated government by military dictatorship in which the entire population would be effectively mobilized for war.

LUXEMBOURG, François Henri de Montmorency-Boutteville, Duke of

BORN: January 8, 1628

DIED: January 4, 1695

SERVICE: Franco–Spanish Wars (1635–59; includes phases of the Thirty Years' War and the Wars of the Fronde [1648–53]; Dutch War (1672–78); War of the League of Augsburg (1688–97)

Protégée of Louis II de Bourbon, prince de ***Condé*** *("the Great Condé"); distinguished himself fighting under Condé; a Marshal of France*

Luxembourg was born in Paris, the posthumous son of François de Montmorency, Count of Boutteville, who had suffered execution for having killed a man in a duel. Young Boutteville was raised in the household of Charlotte de Montmorency, Princess of Condé and the mother of the Great **Condé.** Although Condé was six years older than Boutteville, he treated the younger boy as a brother and, despite Boutteville's deformity (he was a hunchback), the two grew close. As Condé became a soldier, so he prepared Boutteville, who followed him on his campaign into Catalonia in 1647 and Belgium the next year.

It was at Lens, on August 10, 1648, that the young man first distinguished himself, prompting Condé to propose him for regimental command.

Condé was imprisoned on January 18, 1650 after a quarrel with Cardinal Mazarin, whereupon Luxembourg joined **Henri de Turenne** and the Second Fronde. When Turenne suffered defeat at Rethel on October 15, Luxembourg was made a prisoner of war. Subsequently released, he campaigned for Condé along the Saône during March 1652–June 1653, then joined Condé in the Spanish service, fighting by his side in Flanders and building renown as an extraordinary commander of cavalry.

He was at Condé's side when Condé and Don John of Austria were defeated by Turenne at the Dunes on June 14, 1658. After the Treaty of the Pyrenees concluded the Franco–Spanish Wars on November 7, 1659, he and Condé returned to France. Boutteville married the superannuated Duchess of Luxembourg, thereby gaining the title "Duke of Luxembourg" in 1661. Although the new title did not sit well with many in the French court, Condé managed to get Luxembourg reinstated into the French by 1668.

In March 1672, at the commencement of the Dutch War, Luxembourg was dispatched to lead the army of the bishop of Cologne. He pushed into the Low Countries until he was checked at Groningen in July. Then, establishing headquarters at Utrecht, he campaigned against Leiden and the Hague during the winter. An unseasonable thaw during December 1672–February 1673 forced his withdrawal.

With Condé, he fought at Seneffe on August 11, 1674, then, after the death of Turenne on July 27, 1675, he was created Marshal of France along with seven other leading military figures on July 30. Tapped to replace Condé in command in the Rhineland, he failed to best William III of Orange during 1676–77, who took the offensive against him at Saint-Denis, near

Mons, on August 14, 1678. Luxembourg took heavy casualties, but ultimately repulsed William. Ironically, this bloody victory came four days *after* a peace had been concluded in the Dutch War.

Luxembourg no sooner returned to Paris than he was charged with having committed sacrilege and having practiced the "black arts." He was cast into the Bastille on January 24, 1680, and although he was acquitted, he was exiled from Paris in May. Only through the intercession of Condé was he allowed to return to court in 1681. Seven years later, with the outbreak of the War of the League of Augsburg, he was denied a command—until Marshal d'Humière's defeat at Walcourt, whereupon Luxembourg was put at the head of an army in Flanders in (April 1690).

In Flanders, Luxembourg triumphed over George Frederick of Waldeck's inferior force at Fleurus on July 1, 1690. Following this, he laid siege to Mons and stormed the city (March 15–April 8, 1691). He moved on to capture Hal, Belgium, in June, then dealt Waldeck another hard blow at Leuze. This September 19, 1691, cavalry attack came after a stealthy night march.

During May 25–July 7, 1692, Luxembourg supported and covered the siege of Namur, and on August 3 defeated William III's attack at Steenkirk. On July 29 of the following year, 1693, Luxembourg took the offensive against William, pounding him in his entrenchments at Neerwinden. After three failed assaults, Luxembourg's cavalry penetrated the defenses, cut up Williams's army, and captured eighty-four of the allied armies' ninety-one cannon. The next year, he bested William in a campaign of maneuver through Flanders, then returned to Paris to hero's welcome. Luxembourg's vindication and triumph were short-lived, however. He died suddenly a few days after New Year, 1695.

LYSANDER

DIED: 395 B.C.

SERVICE: Peloponnesian War (431–404 B.C.); Corinthian War (395–386 B.C.)

Spartan admiral of the Peloponnesian fleet, who defeated the Athenians and led the Spartans in war against Thebes

Lysander was the son of Aristocritius, about whom nothing is known. Neither is anything known of Lysander's early career, save that he had been raised in poverty and according to a regimen of strict Spartan discipline. He enters history—and, apparently first came to power—in 408 B.C. as *navarch* (admiral) of the Peloponnesian fleet. Setting up a base at Ephesus, he established friendly relations with Cyrus the Younger and the Persian satrap at Sardis. The following year, he defeated an Athenian fleet at Notium. This victory notwithstanding, he was forced to renounce the title of *navarch*, because Spartan law absolutely limited the term to one year. But, in 406, when the new *navarch*, Callicratidas, was defeated, Lysander's partisan demanded his reappointment. Yet Spartan law dictated that no man could be *navarch* twice. Thus Lysander was without offical title or authority. He remained powerful nevertheless. Securing funding from Cyrus the Younger's satrapy, he sailed out of Ephesus to attack the Athenians.

Lysander assaulted Lampsacus on the eastern shore of the Hellespont and, once the city was breached, allowed his men to plunder at will. He engaged the Athenians Aegospotami, taking three or four thousand Athenian prisoners, whom he executed. In the end only a fraction of the Athenian fleet remained afloat. Defeat was devastating and decisive, and it marked the end of the Peloponnesian War.

Lysander sailed on to Athens, which he blockaded until the citizens, near starvation, sued for peace. Lysander set up the Tyranny of Thirty in 404 as the government of Athens and installed Spartan governors over former Athenian cities in Greece and Asia Minor. He then went on to pillage the coasts of Asia, sending treasure to his supporters in Sparta. Utterly ruthless, he summarily put to death his enemies and those of his friends. To those he exiled from Sparta, he pledged amnesty, only to execute them upon their return.

Lysander's relentless brutality moved the Persians to protest to Sparta, and Lysander was summoned to appear before the Spartan *ephors*, who censured him. After this, perhaps recognizing the need to remove himself from Sparta for a time, he obtained permission from the *ephors* to make pilgrimmage to the temple of Zeus Ammon in Libya. When he returned from Libya, he was welcomed by the Spartan masses as a hero. He used his popular influence to help Agesilaus II secure the throne. Once this was accomplished, he persuaded the king to wage war on the Persians. Lysander accompanied Agesilaus to Asia, fully expecting to be named commander. However, the king instead named Lysander official Carver-of-Meats. It was an incredible insult, and Lysander returned to Sparta, his influence greatly diluted.

Lysander returned to military service in 395, when the Spartans attacked Thebes at the start of the Corinthian War. While leading a force against Haliartus in Boeotia, Lysander was slain in a surprise raid. It was a fitting end to an arrogant and brutal commander. Lysander appears not to have been motivated in his conquests by a lust for wealth. He took little or no booty for himself, instead sending all captured prizes to his friends and to the treasury of Sparta. Instead, he was motivated by a desire to enrich Sparta and expand its influence—and perhaps also by a desire to conquer for the sake of nothing more or less than conquest.

The wealth Lysander brought to Sparta had an unexpected effect. It tended to erode the strict discipline and morality for which the Spartans were famed, actually weakening the city-state.

M

MACARTHUR, Douglas

BORN: January 26, 1880
DIED: April 5, 1964
SERVICE: World War I (1917–18); World War II (1941–45); Korean War (1950–53)

U.S. general; Supreme Allied Commander in the Pacific during World War II

The son of General Arthur MacArthur (later the army's senior ranking officer), Douglas MacArthur was born at Little Rock Barracks, Arkansas, and received an appointment to West Point, from which he graduated in 1903, first in his class. He was commissioned a second lieutenant of engineers and sent to the Philippines, then served as aide to his father during a military tour of Asia in 1905–06.

During 1906–07, MacArthur was appointed as aide to President Theodore Roosevelt, then was given command of a company of the Third Engineers at Fort Leavenworth, Kansas during 1908–09. Chosen as an instructor at the General Service and Cavalry schools, he served there from 1909 to 1912 and was subsequently appointed to the general staff, on which he served from September 1913 to 1917. During this period, MacArthur participated the Veracruz expedition, during April–November 1914.

With the United States' entry into World War I, MacArthur was instrumental in creating the Forty-Second "Rainbow" Division and served as the unit's chief of staff when it was sent to France in October 1917. MacArthur served with the unit at Aisne-Marne (July 25–August 2), then commanded a brigade during the assault on the Saint-Mihiel salient from September 12 to September 17. He led a brigade at Meuse Argonne (October 4–November 11, 1918), and commanded the entire division in the "race to Sedan" at the end of the war (November 6–11).

At war's end, MacArthur served with occupation forces in Germany, returning to the United States in April 1919, when he received appointment as superintendent of West Point. He left the Point in 1922 to accept a command, as major general, in the Philippines, remaining there until January 1925. After returning to the States (and serving on the court-martial tribunal that heard the high-profile case of General **Billy Mitchell**), MacArthur was returned to the Philippines as commander of the Department of the Philippines from 1928 to 1930.

In 1930, MacArthur returned to the States again, this time as chief of staff of the U.S. Army. He served in this high post through 1935, but brought down upon himself and the army a storm of controversy when, in the summer of 1932, he personally led a detachment of troops against the so-called Bonus Army (Depression-ridden World War I veterans demanding early payment of promised government benefits) camped in and around Washington, D.C. The action, unnecessarily brutal, brought MacArthur much negative publicity.

In October 1935, MacArthur was sent back to the Philippines to organize its defenses in preparation for granting it independence from the United States. The new Philippine Government appointed MacArthur field marshal in August 1936, and he resigned his commission in the U.S. Army to accept the appointment, because he did not want to be transferred from the Philippines before completing preparations for its defense. However, he accepted recall to American service on the eve of war with Japan (July 26, 1941) and was promoted to lieutenant general, with overall command of U.S. Army Forces in the Far East (USAFFE). He remained headquartered in the Philippines.

Like other senior American officers, MacArthur was surprised by the Japanese bombing of Pearl Harbor on December 7, 1941, and then by air attacks on Clark and Iba airfields in the Philippines on the following day. Although hopelessly undermanned and underequipped, MacArthur mounted a skillful defense of the Philippines, withdrawing to fortified positions on Bataan during a long retreat (December 23, 1941–January 1, 1942) that he made costly to Japanese ground forces. MacArthur personally commanded the defense of Bataan and the Manila Bay forts until President Franklin D. Roosevelt ordered his evacuation to Australia. He left on March 11, reluctantly, promising "I shall return."

Decorated with the Medal of Honor, he was made supreme commander of Allied forces in the Southwest Pacific Area in April 1942. In this post, he directed the reconquest of New Guinea as a first step in the liberation of the Pacific. He successfully directed the repulse of a strong Japanese assault on Port Moresby during July–September 1942, then took the offensive in an advance across the Owen Stanley range during September–November, ultimately assaulting and taking the Buna-Gona fortifications during November 20, 1942–January 22, 1943. Following this hard-won gain, MacArthur directed the island-hopping strategy by which the Allied forces were to retake the Pacific islands and advance inexorably against mainland Japan. After campaigning along the north coast of New Guinea, he invaded western New Britain during December 15–30, 1943, cutting off the major Japanese base at Rabaul. Next came victories at Hollandia Jayapura and Aitape, which isolated the Japanese Eighteenth Army in April 1944. Moving west along the New Guinea coast, MacArthur took Sansapor on July 30, then, in September, coordinated a massive offensive with Admiral **Chester Nimitz** in the central Pacific. MacArthur's forces took Morotai in the Molucca islands while Nimitz pounded then invaded the Palau Islands (Carolines). On October 20, 1944, MacArthur personally commanded ground forces invading Leyte, thereby redeeming his pledge to return to the islands.

Intent on liberating the Philippines, MacArthur concentrated on expansion of Philippine operations to Mindoro on December 1 and Luzon on January 9, 1945. (Critics thought that his obsession with the Philippines compromised other aspects of the Pacific war effort.) Following the Luzon campaign (February 3–August 15, 1945), MacArthur readily liberated the rest of the Philippines, Simultaneously, on Borneo, he took the coastal oilfields that fueled much of the Japanese war effort.

Douglas MacArthur. Courtesy National Archives.

In April 1945, MacArthur was named commander of all U.S. ground forces in the Pacific. It was he who would have overall charge of the anticipated invasion of Japan. However, the dropping of newly developed atomic bombs on Hiroshima and Nagasaki in August prompted the Japanese to surrender before the invasion was launched. Promoted to (five-star) general of the army, MacArthur was accorded the honor of accepting the Japanese surrender, which took place aboard the United States battleship *Missouri* riding at anchor in Tokyo Bay on September 2, 1945.

MacArthur served masterfully as supreme commander of Allied occupation forces in Japan, governing the broken nation with a strong hand tempered by benevolence. MacArthur administered both the rebuilding and the democratization of Japan. In most quarters throughout the conquered nation, MacArthur became a popular figure.

On June 25, 1950, communist North Korean troops invaded South Korea, and MacArthur was named supreme commander of United Nations forces in Korea by a U. N. Security Council resolution on July 8. He set about directing the defense of the Pusan perimeter during August 5–September 15), then staged perhaps the greatest military operation of his career by landing a force at Inchon on September 15, thereby enveloping the North Koreans and ultimately destroying North Korean forces in the South. He next secured U.N. and the U.S. approval to invade North Korea in October, and pushed the communist forces deep into the north, all the way to the Yalu River. Then, during November 25–26, communist Chinese forces entered the war, pushing the United Nations and South Korean armies southward. MacArthur conducted a fighting withdrawal and settled into a defensive front just south of the South Korean capital of Seoul.

Thus embattled, MacArthur publicly advocated bombing targets in China itself, a move that American politicians, chief among them President Harry S. Truman, feared would trigger a nuclear world war. When MacArthur persisted in insubordination and flatly refused to conduct a limited war, Truman relieved him of command on April 11, 1951—despite the fact that he had just recaptured Seoul on March 14.

Replaced in Korea by Lieutenant General **Matthew B. Ridgway,** MacArthur returned to the United States, hailed as a national hero. On April 19, 1951, he delivered a stirring retirement address to Congress—in which declared that "old soldiers never die, they just fade away"—and, although there was much talk of his entering politics at the presidential level, he retired from public life.

By any measure, Douglas MacArthur was an extraordinary soldier, his strategic and tactical brilliance matched only by his boundless ego—which often assumed proportions of a Greek tragic hero. Certainly one of World War II's greatest generals, he was master of the amphibious strategy without which the war in the Pacific would have been a lost cause. Moreover, for all his brashness, egocentricism, and even hunger for glory, MacArthur was deeply concerned for the welfare of his troops. He planned his amphibious operations meticulously and methodically, with the object of keeping casualties at a minimum. In this, he was remarkably successful, especially considering the determined, even fanatical nature of the Japanese adversary.

MACDONALD, Jacques Etienne Joseph Alexandre, Duke of Taranto

BORN: November 17, 1765

DIED: September 7, 1840

SERVICE: French Revolutionary Wars (1792–99); Napoleonic Wars (1800–15)

One of Napoleon's most gallant, if impulsive, marshals

Macdonald was born at Sedan to Nael Stephan Macdonald, a Scots Jacobite who had fled to French exile after Charles ("Bonnie Prince Charlie") Stuart's defeat at the Battle of Culloden (1745). Macdonald joined the Irish Legion in French service, then transferred to the Dillon Regiment, gaining promotion to lieutenant in 1789 and captain in 1792. In that year, he was appointed aide-de-camp to General Charles Dumouriez and for gallantry at Jemappes, also in 1792, he was jumped to the rank of colonel and given command of the Ninety-Fourth Infantry. Just one year later, he made general of brigade (August 27, 1793).

Macdonald saw action at Tourcoing and Hondschoote in 1794, and at the end of the year was promoted to general of division, then, from 1795 to 1798, served in the armies of the Sambre-et-Meuse and of the North. In 1798, he transferred to the Army of Italy, ultimately replacing Jean Championnet as the army's commander in 1799.

On June 7, 1799, Macdonald defeated Hohenzollern's Austrian army at Modena, but, on June 17–19, was repulsed at Trebbia and incurred grievous losses. He was given command of the Army of Reserve at Dijon the following year and, with it, made a spectacular Alpine crossing at Splilgen in November–December.

During 1801–02, Napoleon employed Macdonald as a special envoy to Denmark. He was further honored by appointment as a grand officer of the Legion of Honor in 1804. However, after Macdonald defended his friend **Jean Moreau** against a charge of treason, he tumbled from the emperor's favor. It 1807, he joined the Neapolitan army, fighting under Eugene de Beauharnais at the Piave in May 1809. During July 5–6, when he brilliantly led a corps at Wagram, he won back the confidence of Napoleon, who presented him his marshal's baton on the field for having pierced the Austrian center. The following December, the emperor created him Duke of Taranto and assigned him to replace General Pierre Augereau in Catalonia during 1810–11.

In 1812, he was called on to participate in the invasion of Russia, but was bogged down in a frustrating siege of Riga. He ultimately broke off the siege and withdrew when he received news of Napoleon's retreat. He suffered a further humiliation when Yorck's Prussian Corps, attached to his command, defected to the Russian side on December 30.

Macdonald returned to the field in the 1813 German campaign as commander of XI Corps. He fought at Lützen on May 2 and at Bautzen during May 20–21. But, on August 26, now commanding the newly created Army of the Bober, he deliberately defied Napoleon's orders and gave chase to **Gebhard Blücher**'s Prussian forces, running them to ground at the Katzbach River. Blücher turned on Macdonald and attacked his corps, which was backed up against the river. It was a disaster. Badly cut up, the remains of the XI withdrew toward Dresden. Next came Leipzig, during October 16–19, when he was ordered to take charge of the rear guard and soon found himself desperately swimming the Elster River to evade capture.

Macdonald fought with greater success at Hanau on October 30, then assumed command of the lower Rhine defenses. He remained fiercely loyal to the emperor through the desperate battle within France during 1814, but, while successful in a handful of minor engagements, had to give up Troyes to the enemy in March. Among the last of the marshals to accept Napoleon's first abdication, he was presented with the sword of Murad Bey, which Napoleon had personally taken in Egypt; and was made a peer of France as well as knight of the Order of St. Louis. Louis XVIII appointed Macdonald governor of the Twenty-First Military Division, in which capacity he escorted the Bourbon king to Belgium when Napoleon returned from Elba.

Despite his dedication to Napoleon, Macdonald did not take part in the Waterloo campaign, and, after Napoleon's final exile, rose in the Bourbon court. He was appointed grand chancellor of the Legion of Honor and a member of the King's Privy Council. In 1825, he was made one of the four marshals in command of the Royal Guard.

Macdonald was a dashing, at times erratic commander, whose glory came during the campaigns before 1809. Although he was capable of independent command, his judgment was at best uneven. The fiasco at the Katzbach River is ample evidence of this.

MACDONOUGH, Thomas

BORN: December 31, 1783

DIED: November 10, 1825

SERVICE: Quasi-War with France (1798–1800); Tripolitan War (1801–05); War of 1812 (1812–15)

U.S. naval officer who won one of the most important victories in the War of 1812 at the Battle of Lake Champlain

MacDonough was a native of New Castle County, Delaware, son of a wealthy and prominent physician and judge. Young MacDonough was apprenticed as a clerk in Middletown, Delaware, when his elder brother returned from sea, wounded, having lost a leg in the Quasi-War with France. Thomas MacDonough immediately quit his job and joined the U.S. Navy as a midshipman on February 5, 1800. He shipped out on a corvette bound for the West Indies, where he took part in the capture of three French vessels during May–September.

Following the Quasi-War, MacDonough was assigned to the thirty-eight-gun frigate *Constellation* in 1801 and, from 1801 to 1803, participated in naval action against Tripoli. He then served on the frigate *Philadelphia* in 1803, and after it was captured by Tripolitans on October 31, 1803, he was transferred to the twelve-gun sloop *Enterprise*, skippered by Lieutenant **Stephen Decatur.** With Decatur, he took part in the daring raid into Tripoli harbor to burn the *Philadelphia* on February 16, 1804. For this exploit, MacDonough was promoted to lieutenant and assigned to the sixteen-gun schooner *Syren*.

MacDonough was next posted to Middletown, Connecticut, where **Isaac Hull** was building gunboats. After working with Hull there, MacDonough was given his own command, the eighteen-gun U.S.S. *Wasp*, which he sailed to Britain and the Mediterranean, and which he also sailed on patrol duty to enforce President Jefferson's anti-British embargo during 1807–08.

During 1810–12, MacDonough obtained a leave of absence to sail a merchant vessel to Britain and India. He returned to active duty at the outbreak of the War of 1812 and was first assigned to U.S.S. *Constellation*, when the frigate was being fitted out in Washington. Impatient for action, MacDonough transferred to command the gunboats defending Portland, Maine. He was then dispatched to Burlington, Vermont, where he assumed command of naval forces on Lake Champlain in October. In July 1813, MacDonough was promoted to master commandant and, despite acute shortages of everything—trained personnel, supplies, arms—he managed to ready three sloops and two gunboats for action against the British. In August, one of the sloops was lost, but MacDonough rushed to completion another three, in addition to four new gunboats. With these, he drove the British squadron back to Canadian waters during the fall.

In 1814, the British mounted a major push into New York; however, the British commander did not want to advance below Plattsburgh without first gaining control of Lake Champlain. British Commodore Robert Downie sailed his fleet against MacDonough's small flotilla. MacDonough braced for battle and prepared with careful deliberation. Anchoring his ships near Plattsburgh, he waited. At first, the British vessels, led by the thirty-seven-gun *Confiance*, clearly appeared to have the advantage in the September 14 battle. But MacDonough hammered away at them. Then, in a brilliant stroke of seamanship, MacDonough used cables to swing round his twenty-six-gun flagship, U.S.S. *Saratoga*, in order to expose her undamaged port side to the British. This gave him a firepower advantage over the battered British vessel and led to victory. With the major British craft sunk or captured, the British land forces broke off the New York invasion and withdrew to Canada.

Following his victory on Lake Champlain, MacDonough accepted the thanks of Congress and received promotion to captain. Suffering from advanced tuberculosis, he was assigned shore duty as Isaac Hull's replacement as commander of the Portsmouth Navy Yard on July 1, 1815. In 1818, he returned to sea, however, as captain of the captured U.S.S. *Guerrière*, which he sailed to the Mediterranean. He was then named to command the large—seventy-four-gun—frigate U.S.S. *Ohio* under construction in New York. Subsequently, he commanded the *Constitution* ("Old Ironsides") on a cruise to the Mediterranean in 1824, but was increasingly ill. He set sail for home, put in at Gibraltar en route, and died.

MacDonough was brave and thorough. His careful preparation at the Battle of Lake Champlain, together with his consummate seamanship and grasp of tactics, won a pivotal battle in the War of 1812.

MACHIAVELLI, Niccolò

BORN: May 3, 1469

DIED: June 21, 1527

SERVICE: Pisan War (1495–1509); War of the Holy League (1511–14)

Florentine patriot and political-military theorist, whose principal work,* The Prince *(1513), was an uncompromising statement of the use of power and force in the service of the state

Florentine by birth, Machiavelli was the son of an impoverished lawyer named Bernardo Machiavelli. In

1498, the young man advanced in the government of the Florentine Republic, becoming head of the second chancery after the execution of Cirolamo Savonarola. As secretary to the magistracy of the *Signoria* (grand council) during 1500–02, Machiavelli became intimately involved in the complex workings of diplomacy and power politics. Then, when Piero di Tommaso Soderini gained election as *gonfalonier* (chief executive) for life, Machiavelli was appointed his principal assistant in 1502.

Machiavelli crafted and engineered passage of a militia law that created a 10,000-man citizen army for Florence in December 1505. The following year, he became secretary to the Council of Nine, the body that controlled the militia. Machiavelli commanded the militia in operations against Pisa, and was instrumental in forcing the city to surrender on June 8, 1509. He was a key player in the diplomatic and military maneuvering preparatory to the War of the Holy League during 1510–11, but was ousted from his offices after Florence fell to a Papal–Spanish army and the Medici were restored to power in 1512.

Machiavelli was cast into prison and subjected to torture on suspicion of sedition and treason, but was ultimately released for lack of evidence. He retired from public life in 1513, living on his modest estate, Sant' Angelo, just outside Florence. For the succeeding fourteen years, he studied and wrote, producing *The Prince*, his most famous work, in 1513. It is a brilliant discourse on Renaissance politics, and also a kind of handbook for the use of power. Deeply cynical—some have said amoral—*The Prince*, holds that the state's iron necessity justifies the use of all power. This proposition became known as the "reason of state."

In addition to *The Prince*, Machiavelli wrote *The Discourses on Livy* (1516–19), *The Art of War* (1520), and *The History of Florence* (1521–25), as well as a number of plays, works of fiction, and poems.

By the 1520s, Machiavelli had gained favor with Cardinal Giulio de' Medici, through whom he obtained the post of official historian of Florence in November 1520. During 1521–26, he was sent on a number of relatively minor diplomatic missions, then received appointment as secretary to the inspectors of fortifications in April 1526. He marched with the Papal army in 1526–27, returning to Florence after the sack of Rome in May 1527. He was not able to find a place in the new republican government that had replaced the Medici in Florence. Discouraged, he became ill and died.

While *The Prince* is deservedly Machiavelli's best-known work, his *Art of War*, is also a penetrating discussion of how moral factors bear upon warfare. In addition, it is an attempt to rationalize a system of strategy. While this is not fully successful, the effort makes Machiavelli the first strategic theorist of modern times.

MACKENSEN, August von

BORN: December 6, 1849

DIED: November 8, 1945

SERVICE: Franco–Prussian War (1870–71); World War I (1914–18)

Among Germany's most successful commanders on the Eastern Front during World War I

Born at Haus Leipnitz, Saxony, Mackensen joined the elite Death's Head Hussars as a cadet in 1869. By the time of the Franco–Prussian War (August 1870–February 1871), he was a junior officer in the unit. Following the war, Mackensen worked his way up the ranks as a career officer, attaining appointment to the General Staff in 1882 and given the honor of accompanying Kaiser **Wilhelm II** on a tour of Palestine in 1898.

Mackensen made *general der cavallerie* in 1908, with command of XVII Corps. He led this unit at the outbreak of World War I in operations along the Eastern Front, first at Gumbinnen on August 20, 1914, and then at Tannenberg, on August 27–31, in which the Russian armies were dealt a severe blow. He also fought at the first Masurian Lakes engagement, during September 9–15, and was then attached to the new Ninth Army, becoming its commander on November 4. In this new post, he directed the offensive at Lódz during November 11–21.

When the Austro–Hungarian Carpathian offensive of March–April 1915 collapsed, Mackensen was rushed south with German reinforcements to assume command of the German–Austrian Eleventh Army in April. He also took control of the Austro–Hungarian Fourth Army, and, with this and his Eleventh, he attacked at Gorlice-Tarnów. Smashing through the Russian lines on May 2–4, he captured a staggering number of prisoners—120,000—and effectively annihilated the entire Russian Third Army.

From this triumph, Mackensen proceeded to take Lemberg (Lvov) on June 22 and then Brest-Litovsk on August 29. In recognition of his achievements, he was promoted to field marshal on June 20 and transferred to the Balkans, where he commanded an offensive against Serbia in September. He wielded the combined forces of the German Eleventh, Austrian–German Third, and Bulgarian First Armies against the battered and badly outnumbered Serbs beginning on October 7. Two days later, Belgrade fell to him, and Mackensen had cleared Serbia by December 4.

In the spring of 1916, anticipating the entry of Romania into the war, Mackensen was assigned command of the Bulgarian–Turkish–German Danube Army in northern Bulgaria. Mackensen attacked Romanian forces on September 1, capturing the rail lines from Constanza to the Danube at Cernavoda. He took the town of Turtukai on September 6 and Silistra three days later. After obtaining Turkish reinforcements, he continued northward to the Danube, which he crossed on November 23, driving from here northeast toward Bucharest, which he took on December 6. During December 1916–January 1917, he consolidated German control of all Romania.

The Romanian operations were Mackensen's last significant actions of the war. He was interned by the French at Neusatz after the Armistice and released in December 1919. The following year, he retired from the army and, in 1933, was appointed state councillor under the Nazi government. Remaining loyal to his old army comrades, he attempted—without success—to warn Chancellor **Paul von Hindenburg** about the impending mass murder of **Adolf Hitler**'s rivals (including certain army officers) during the Night of the Long Knives (June 30–July 1). Mackensen was inactive during World War II.

MACKENZIE, Ranald Slidell

BORN: July 27, 1840

DIED: January 19, 1889

SERVICE: Civil War (1861–65); Indian Wars (1874–81)

One of the U.S. Army's most successful Indian fighters during the Indian Wars of the 1870s and 1880s

New York City-born Mackenzie was the son of the prominent naval officer Alexander Slidell Mackenzie. He was educated at Williams College, which he left to accept an appointment to West Point, from which he graduated in 1862, first in his class. He was immediately sent into action at Second Bull Run (August 29–30), where he was wounded and breveted to first lieutenant for gallantry. His next battle was Fredericksburg (December 13), followed by Chancellorsville (May 1–6, 1863), and Gettysburg (July 1–3). Breveted to major, he was promoted to the permanent rank of captain in November, only to be breveted yet again, to lieutenant colonel, for gallantry during the Petersburg campaign of June 13–18, 1864.

Named colonel of volunteers, Mackenzie assumed command of the Second Connecticut Volunteers, which he led in combat in the Shenandoah Valley during July–October 1864. Shot at Cedar Creek on October 19, he lost two fingers from his right hand (later, during the Indian Wars, the Indians would call him Bad Hand). Breveted to colonel in the regular army, he was also promoted to brigadier general of volunteers and given command of a cavalry division in the Army of the James.

Mackenzie went on to distinguish himself in fighting at Five Forks on April 1, 1865, an action for which he was breveted a final time, to major general. Although General **Ulysses S. Grant** praised him as "the most promising young officer in the army," like others who had received multiple brevets in combat, he reverted to his permanent rank—captain—after the war. After serving briefly in the Corps of Engineers, he was appointed colonel of the Forty-First Infantry, an African-American unit, in 1867. Mackenzie molded these proud and determined "Buffalo soldiers" into a crack frontier unit, which served with distinction along the Texas frontier. In 1869, the Forty-First and Thirty-Eighth Infantry were consolidated as the Twenty-Fourth Infantry, and Mackenzie was assigned as colonel of the new regiment.

In 1871, he transferred to command of Fourth Cavalry at Fort Concho in San Angelo, Texas. As he had done with the black infantry units, he transformed this cavalry regiment into the pride of the United States Army. He took the regiment into action against Comanches and Kiowas who had bolted from their reservation and engaged in raiding throughout southern Texas.

After defeating the Comanche war leader Mow-way in the summer of 1872, Mackenzie took his unit west to fight Apaches raiding from bases deep within Mexico. During May 18–21, Mackenzie led a lightning raid into Mexican territory to lay waste three Apache villages near San Remolino (now El Remolino). He then participated in the so-called Red River War against Comanches, Kiowas, and southern Cheyennes, achieving a signal victory at Palo Duro Canyon, Texas on September 28, 1874. The Palo Duro fight was a classic action of the Indian Wars, its object being to destroy shelter, food, clothing, and horses.

In 1876, Mackenzie served under General **Philip Sheridan** against the Sioux and northern Cheyenne and achieved victories over Red Cloud and Red Leaf in Nebraska during October, then defeated Dull Knife at Crazy Woman Creek on November 25–26. The destruction of Dull Knife's camp in the depths of a late fall siege of freezing weather led to the defeat of the Sioux under Crazy Horse.

In 1880, when the Utes domiciled at the White River Agency in northwestern Colorado threatened an uprising, Mackenzie quickly intervened, overseeing the peaceful transfer of 1,400 Utes to a new reservation in Utah in August 1881.

On October 30, 1881, Mackenzie was named to command of the District of New Mexico. Immediately, he set about stemming raids conducted by renegade Apaches and soon brought relative peace to the territory. Promoted to brigadier general, he was given command of the Department of Texas on October 30, 1883, but did not remain in this post for long. He collapsed with a devastating physical and mental breakdown and had to be relieved of command. On March 24, 1884, he retired from the army, mentally unbalanced and apparently suffering from the neurological effects of tertiary syphilis. He returned to the East, where he was cared for by his sister until his death five years later.

Mackenzie was a remarkable officer. Conspicuous for his gallantry during the Civil War, he was just the kind of officer the rigors of the Indian Wars demanded: tireless, relentlessly aggressive, and a disciplinarian who could also inspire his men with pride and loyalty. His illness and premature death were a great loss to the army.

MAGRUDER, John Bankhead ("Prince John")

BORN: August 15, 1810

DIED: February 18, 1871

SERVICE: Second Seminole War (1835–42); U.S.–Mexican War (1846–48); Civil War (1861–65)

Flamboyant Confederate general, who entered the Mexican service after the Civil War

A native of Winchester, Virginia, Magruder graduated from West Point in 1830 and entered the army as a second lieutenant of infantry, subsequently transferring to the artillery. He saw service in the West, as well as on the East Coast, including in Florida during the Second Seminole War.

Promoted to first lieutenant in 1836, he made captain a decade later, just as the United States and Mexico prepared to go to war with one another. He fought at Palo Alto (May 18, 1846) and commanded an artillery battery at Cerro Gordo on April 18, 1847, after which he was breveted major. Another brevet, to lieutenant colonel, was forthcoming after the assault on Chapultepec on September 13.

Magruder was assigned to garrison duty in Maryland, Rhode Island, and California after the war with Mexico. He resigned his commission on the eve of the Civil War and joined the Confederate forces as an infantry colonel on March 16, 1861. In June, he saw his first action, at Big Bethel, Virginia, successfully repulsing General Benjamin Butler's attempt to advance from Fortress Monroe on June 10. Magruder was rewarded with a promotion to brigadier general in July, which was followed by a step up to major general in October.

Magruder had charge of Confederate forces on the Yorktown Peninsula, so went head to head with Union general **George B. McClellan** during the Peninsula campaign. He skillfully maneuvered his mere 12,000 troops, thereby duping McClellan—who always tended to overestimate the enemy's strength—into devoting time and resources to a fruitless siege of Yorktown during April 4–May 4, 1862.

Magruder fought gallantly in the Seven Days (June 25–July 1), but his tardiness in executing **Robert E. Lee**'s orders at Malvern Hill on July 1 drew Lee's anger. Lee exiled Magruder to service in the District of Texas in October. His responsibilities were extended to include Arizona and New Mexico as well, and Magruder excelled in his new assignment, taking Galveston, Texas, seizing the Union gunboat *Harriet Lane*, and driving off the blockading squadron on January 1, 1863. He also participated in General Richard Taylor's operations against General Nathaniel Banks's Red River campaign during March 1864.

Magruder fled the United States for Mexico after the war and was commissioned as major general in the service of doomed Mexican emperor Maximilian I in 1865. He did not return to the States until 1869. He spent his final years in Houston.

MAGSAYSAY, Ramón

BORN: August 31, 1907

DIED: March 17, 1957

SERVICE: World War II (1941–45); Huk Rebellion (1946–54)

Filipino soldier and president, best known for defeating the communist-led Hukbalahap (Huk) movement

A native of Iba in Zambales province, southwestern Luzon, Magsaysay supported himself through course work at the University of the Philippines from 1927 to 1931 and graduated from José Rizal College in 1933. He went to work as an auto mechanic, but worked his way into management as branch manager of a transport company by 1941. With the outbreak of World War II in this year, he joined the Philippine as a captain and fought on Luzon and at Bataan during December 1941–April 1942. After the surrender of American forces on the islands, Magsaysay became a guerrilla operating out of Zambales.

Magsaysay was appointed military governor of Zambales after the province was liberated by American troops in 1945, and he was elected to two terms in the Philippine Congress on the Liberal ticket during 1946–50. President Elpidio Quirino appointed him secretary of defense, with a specific mandate to put down the Hukbalahap (or Huk) Rebellion in 1950. He began the task by first reforming the army and using army personnel in place of local constabulary units to operate against the Huks. Magsaysay kept up vigorous pressure against the Huks, even as he offered those who surrendered very generous terms that included land and farm implements. This combined approach made headway against the Huk movement, but Magsaysay resigned as secretary of defense in 1953 and also left the Liberal Party (which he felt was corrupt), joining the Nationalists instead.

Magsaysay gained election to the presidency and was inaugurated on December 30, 1953. He also served as his own secretary of defense until Louis Taruc, leader of the Huks, surrendered and the movement collapsed in May 1954.

Magsaysay was popular, but his close friendship with the United States provoked frequent criticism and opposition. He was engaged in a struggle to redistribute land and reform land policy when he was killed in an airplane crash.

MAHAN, Dennis Hart

BORN: April 2, 1802

DIED: September 16, 1871

SERVICE: No combat service

Influential U.S. Army instructor and military science author at West Point

Mahan was born in New York City, the son of Irish immigrants. His ambition was to be an artist, and he secured an appointment to West Point mainly to take advantage of the drawing courses required as part of the engineering program. He excelled intellectually, and Major Sylvanus Thayer made him acting assistant professor of mathematics during his second year. He graduated first in the Class of 1824 and was commissioned in the engineers, but remained at West Point as an instructor during 1824–26.

During 1826–30, Mahan studied in Metz, France, at the School for Application for Engineers and Artillery, then returned to West Point as assistant professor of engineering in 1830. After he achieved an appointment as full professor, he resigned his second lieutenant's commission on January 1, 1832, and taught at West Point as a civilian. His most significant contribution to coursework at the academy was his fourth-year course in military science,"Engineering and the Science of War." He wrote the texts for the course, and in 1836 published *Complete Treatise on Field Fortification* and the following year, *Elementary Course on Civil Engineering*. His *Elementary Treatise on Advance-Guard, Out-Post, and Detachment Service of Troops* was published in editions that appeared in 1847, 1853, and 1863. It was nothing less than a comprehensive treatment of tactics and strategy.

Mahan believed that an offensive campaign of maneuver was generally the most effective means of winning a war, and he stressed the necessity of outposts and reconnaissance. His ideas influenced all of the major commanders of the Civil War, on both the Union and Confederate sides.

In addition to the works just mentioned, Mahan was the author of *Industrial Drawing* (1852), *Descriptive Geometry as Applied to the Drawing of Fortification and Stereotomy* (1864), and *An Elementary Course of Military Engineering* (1866–67).

West Point was Mahan's life, and when the academy's Board of Visitors recommended his retirement because of his advanced age in 1871, the profoundly depressed Mahan leaped off a steamboat to his death in the Hudson River near Stony Point, New York.

MAHMUD OF GHAZNA

BORN: 971

DIED: 1030

SERVICE: Conquest of Khorasan (1000); Ilak Invasion of Khorasan (1006–07); Uprisings in Ghor and Khwarezm (1011–16); Persian Expedition (1029)

Sultan of Ghanza who expanded his holdings into a vast empire

As sultan of Ghazna, Mahmud built the modest realm into a vast empire, transforming its capital into the cultural center of Central Asia, and, in the course of seventeen full-scale invasions, carrying Islam deep in to India. He was the son of Amir Sabuktigin, a Turkish slave who became ruler of Ghazna in 977. Mahmud was twenty-seven years old in 998 when he succeeded his father. He vastly expanded the small sultanate until it included Kashmir, Punjab, and most of Iran.

Mahmud began by striking an alliance with the mighty Abasid caliph in Baghdad, whose recognition was necessary to legitimate many of Mahmud's conquests. Early in his reign, Mahmud vowed to invade India annually, and he very nearly did just that, leading no fewer than seventeen expeditions into the subcontinent between 1001 and 1026. On his first invasion, Mahmud rode at the head of 15,000 cavalry troops. At

Peshawar, he was met by 12,000 cavalrymen, 30,000 infantry soldiers, and 300 elephants commanded by Jaipal, ruler of the Punjab. Greatly outnumbered, Mahmud nevertheless routed Jaipal's forces, killing 15,000 of his men and capturing Jaipal himself, along with fifteen of his relatives and commanders. As a demonstration of magnanimity, Mahmud released Jaipal, who, however, immediately abdicated to his son, Anandpal, prepared his own funeral, ascended the pyre, and immolated himself.

Anandpal built an alliance of rajas from all over India, gradually assembling a vast army to counter the repeated assaults against the subcontinent. By 1008, Anandpal fielded a huge force against Mahmud at a place between Und and Peshawar. For forty days and nights the rival armies faced one another in stillness. Finally, Anandpal attacked, making extensive use of 30,000 fierce Khokar tribesmen, who pounded Mahmud's army to the verge of retreat. At what seemed the point of triumph for Anandpal, his elephant took fright and stampeded from the battlefield. Anandpal's troops saw their leader apparently in headlong retreat. They panicked, broke ranks, and fled, and Mahmud found himself victorious once again. He marched deep into India, annexing the Punjab, and looting its riches.

Mahmud reaped magnificent riches from his many invasions, and these he used to transform the capital city of Ghazna into a wealthy metropolis of magnificent buildings. He avidly fostered the development of the arts and learning, and soon his city rivaled Baghdad itself as a cultural center. Nor was it all gaudy show. Vigorous in conquest, Mahmud was a cultured leader who generally treated the people he conquered with respect. Although in 1024, after sacking the city of Somnath, he defiled the Hindu temple there, breaking its sacred lingam, he was generally tolerant of non-Islamic faiths. He saw his mission in part as the dissemination of Islam throughout India, yet he also maintained a large unit of Hindu troops under the command of Indians, and he declined to practice religious persecution. In this way, he was able long to sustain the fruits of his military conquests.

MAJORIANUS, Julius Valerius (Majorian)

DIED: August 7, 461
SERVICE: Hun Invasion (451–453); Vandal Wars (455–460)

The only great Roman emperor of the fifth century; enjoyed military success against the Huns and the vandals

Majorian was the scion of a venerable military family and quickly made a reputation as warrior when he distinguished himself against the Huns at Chalons in 451. Majorian helped Duke Ricimer overthrow Emperor Avitus in 455, for which he was rewarded with an appointment as *magister militum*—supreme military commander—in 457. He led the Roman armies in a successful defense against invading Alamanni at Bellinzona in March 457, then, with the support of Ricimer, was proclaimed emperor on April 1.

In the summer of 457, the new emperor took the field to push back invading Vandals under Gaiseric in Campania. This accomplished, he turned his attention to domestic reforms, working to bring a measure of justice to the provinces and to eliminate ruinous abuse of taxation. In 458, he returned to military matters, supervising construction of a fleet with which he intended to reconquer Africa from the Vandals. While these preparations were under way, he traveled to Gaul to consolidate support for the enterprise there. He made an alliance with King Theodoric II of the Visigoths in the spring of 460, then traversed the Pyrenees into Spain, where he joined his new 300-ship fleet at Cartagena during May–June.

The emperor was betrayed, and the Vandals surprised and destroyed the fleet. Nothing daunted, Majorian made preparations for a new expedition, but had to rush back to Italy when Ricimer turned on him and fomented a general mutiny in Lombardy. Disgusted with the people of Rome, Majorian abdicated on August 2, 461. He died, mysteriously, five days later. It is likely that he was the victim of assassination, probably perpetrated by Ricimer.

MANGIN, Charles Marie Emmanuel

BORN: 1866
DIED: 1925
SERVICE: Colonial posts (1898–1914); World War I (1914–18)

French World War I general renowned and reviled as "The Butcher" and "The Eater of Men"

Mangin was a native of Lorraine and saw his first action as a military man in far-flung colonial outposts. He participated in Jean-Baptiste Marchand's 1898 expedition to the Nile at Fashoda (modern Kodok), Sudan, and then served Morocco from 1908 until the outbreak of World War I in 1914, when he was recalled to France.

Just before the commencement of the war, Mangin was promoted to general of brigade and earned early distinction leading his unit at Charleroi during August 21–22. The following year, Mangin was promoted to general of division and given command of the Fifth Division III Corps, under General Robert Nivelle at Verdun during February–December 1916. Mangin performed with his typical aggressiveness and apparent disregard for the lives of his troops that earned him his sobriquets: "The Butcher" and "The Eater of Men." In an army whose spirit was flagging, such aggressive qualities were highly prized, and Mangin was promoted to command of the Sixth Army early in 1917. His mission was to carry out an offensive on the Aisne, but his troops were repeatedly repulsed and ultimately unable to penetrate the third line of German defenses along the Chemindes-Dames. By mid April, the offensive bogged down, and Nivelle, who was now in overall command of the Western Front, affixed blame for the offensive's failure solely on Mangin, whom he relieved from command in May.

Mangin spent little time on the sidelines, and was soon brought back to command the Tenth Army, which he led in a vigorous counterattack against the Germans at Villers-Cotterets. He checked the German advance there and was therefore instrumental in bringing to a halt what turned out to be the final German offensive of the war.

One of France's boldest senior commanders, Mangin was prodigal with his resources and was steeped in *élan* that gave him boundless self-confidence. Yet he was never gratuitously reckless and always planned his campaigns carefully. His treatment by Nivelle, who was faltering under the burden of chief responsibility for the French war effort, was unjust and unjustifiable.

MANNERHEIM, Carl Gustav Emil von

BORN: June 4, 1867
DIED: January 27, 1951
SERVICE: Russo-Japanese War (1904–05); World War I (1914–17); Finnish Civil War (1918–20); Russo-Finnish War (1939–40); World War II (1941–44)

Russian and Finnish officer who served Finland valiantly in the cause of independence

Born in Villnäs, Finland, Mannerheim attended various military schools and, in 1889, was commissioned a lieutenant of cavalry in the Russian army. He was immensely popular with his troops and other officers and was known as a skilled horseman. In 1895, he was chosen as one of the guard of honor at the coronation of Czar Nicholas II and Tsarina Alexandra. He saw his first combat action during the Russo-Japanese War (February 1904–September 1905), from which he emerged with the rank of colonel. With the outbreak of World War I, he rose swiftly, attaining the rank of lieutenant general, with command of a corps by the middle of 1917.

Mannerheim left the Russian army in 1917, after its collapse and the November Revolution. He returned to Finland, and he answered his country's call after it declared independence from Russia on December 6, 1917. The conservative Mannerheim was not an enthusiastic supporter of the revolutionary government; however, he was intensely opposed to the Communists, and he willingly accepted command of the anti-Communist White forces in Finland on January 18, 1918. From a base at Vasa, western Finland, he led his forces against a communist uprising, engaging the Red Guard on March 16 outside of Tampere. He captured the Karelian Isthmus on April 29, also containing communist breakout attempts. Mannerheim became regent of Finland on December 12, 1918 and served in this capacity until a republic was formally established on June 17, 1919. In the meantime, he quelled additional, minor hostilities along the border until the conclusion of the Treaty of Dorpat on October 14, 1920 ended the Russo–Finnish War.

Once the war was over, Mannerheim retired, only to return to public life as chairman of the defense council. Increasingly concerned over the Soviet threat, he worked toward obtaining increased military funding, and he oversaw construction of border fortifications in the Karelian Isthmus. These defensive forts became known as the Mannerheim Line by their completion in 1939.

During the period of Soviet–Nazi non-aggression at the beginning of World War II, Mannerheim was appointed commander in chief of Finnish forces when the U.S.S.R. invaded Finland on November 30, 1939. At first, Mannerheim enjoyed significant success, but the Soviets' numerical superiority wore down the Finnish defenders. Mannerheim had imposed a costly victory on the Soviets by the time he surrendered to them on March 12, 1940. When war against Russia resumed on June 25, 1941, Mannerheim was again in overall command, directing operations on the Karelian Isthmus and Eastern Karelia.

Mannerheim was promoted to field marshal on June 4, 1942, and, after President Risto Ryti resigned following the successful Russian summer offensive of 1944, Mannerheim stood for election and won, becoming Finland's president on August 4, 1944. He concluded an armistice with the Russians the following month, by which Finland agreed to lend military aid to clear Lapland of German troops during September–December 1944. Mannerheim continued in office until shortly after the war, resigning in 1946 because of illness.

A skilled commander, Mannerheim made the most of very limited forces. He was a conservative by nature, but his patriotism motivated him to fight for his country.

MANSTEIN, Erich von

BORN: November 24, 1887

DIED: June 10, 1973

SERVICE: World War I (1914–18); World War II (1939–45)

German field marshal in World War II; a consummate officer of the old Prussian school, he openly disagreed with* Adolf Hitler *on military policy

Born Erich von Lewinski in Berlin, to General Eduard von Lewinski and his wife, his father died, leaving his mother unable to support her ten children. Erich was adopted by a childless aunt married to General George von Manstein, and thus the boy was given a new name. He enrolled in cadet school, from which he graduated in 1906 with a lieutenant's commission in the Third Foot Guards Regiment. This crack unit was commanded by General **Paul von Hindenburg,** the young man's uncle. Having shown great promise, Manstein was allowed to enroll in the *Kriegsakademie*, but his study was interrupted by the commencement of World War I in 1914.

Manstein was wounded in the first year of the war, in November, and was taken out of the front lines and assigned to staff duty while he convalesced. He remained in staff assignments for the balance of the war. At the end of the war, Manstein remained in the military, serving in the *Reichswehr,* and by 1929 had achieved appointment to the General Staff. In 1936, he was made deputy to the chief of the General Staff, General Ludwig Beck, but was removed from this position two years later when Defense Minister General Werner von Blomberg and army commander in chief General Werner von Fritsch, opponents of Hitler's plans for conquest, were removed from office. Manstein was reassigned to command of an infantry division in Silesia, and then became chief of staff of the German occupation army in Czechoslovakia.

In August 1939, Manstein was appointed chief of staff of the Eastern Army Group under General **Gerd von Rundstedt** and participated in the blitzkrieg invasion of Poland in September 1939. Always an independent thinker, Manstein objected to the General Staff's

Erich von Manstein. Courtesy National Archives.

plan for the invasion of France, insisting on a plan that thrust the German forces through the Ardennes. Ultimately, Manstein's views prevailed, and the result was a German triumph and a quick French collapse. Manstein was given the satisfaction of a field command in the follow-on force during the later stages of the invasion of France (May 1940).

In March 1941, Manstein was assigned to command a Panzer (tank) corps—LVI Panzer Corps—and led it to dramatic success against the Soviets in June 1941. In July, Manstein was promoted to field marshal and given command of the Eleventh Army in the Crimea, then was put in charge of Army Group Don, in November 1942, which was sent to relieve the German Sixth Army at Stalingrad, where it had been encircled and was being destroyed. Manstein openly disputed with Hitler over Hitler's orders that the Sixth not break out and link up with relief forces. Army Group Don was unable to stem the tide at Stalingrad.

Put in command of Army Group South, Manstein set about salvaging the collapsing German front in southern Russia. In March 1943, he recaptured Kharkov in a surprise attack. He also directed the right pincer of the doomed German assault on the Kursk salient in July 1943, then directed the fighting withdrawal of German forces in southern Russia. He was relieved in March 1944 and remained inactive, effectively retired, for the balance of World War II.

Manstein surrendered to British forces in May 1945 and was indicted for war crimes during the later phases of the Nuremberg tribunal, in August 1949. Of the principal Allies, only the Soviet Union was interested in prosecuting Manstein, who was sentenced to eighteen years' imprisonment. His sentence was reduced, however, and he was released in 1953. During 1955–56, Manstein was chairman of the West German Parliament's military subcommittee and was charged with the task of reorganizing West German forces and developing military doctrine for them.

MAO TSE-TUNG

BORN: December 26, 1893

DIED: September 9, 1976

SERVICE: Chinese Revolution (1911–12); Northern expedition (1926–28); Bandit (Communist) Suppression Campaigns (1930–34); Sino–Japanese War (1937–45); Chinese Civil War (1946–49)

One of the founders of the Chinese Communist Party (1921) and the founder of People's Republic of China (1949); a grassroots military leader of great skill and a formidable inspirer of men

Mao Tse-tung was born to a prosperous family of Hunan peasant landowners. He was sent to the local elementary school, where he received a classical Chinese education, which emphasized Confucian thought. Mao broke off his education in October 1911, after forces under Sun Yat-sen overthrew the Ch'ing (or Manchu) dynasty. Mao, caught up in the revolution, served during 1911–12 as an orderly in a militia unit, but was then called home by his authoritarian father, who insisted that the boy learn a trade at a commercial school, which he attended during 1912–13.

From 1913 to 1918, Mao lived in the provincial capital of Changsha, where he enrolled in the normal school, intending on becoming a teacher. However, he moved to Beijing in 1918 and lived there for the year, supporting himself as a clerk in the library of Beijing University. In 1919, he returned to Hunan and secured an appointment as a teacher at the Changsha Normal School. By this time, he had already acquired more than a local reputation as a political intellectual.

Mao married Yang K'ai-hui, daughter of one of his teachers, in 1920, then, in 1921, served as Hunan's chief delegate to the founding congress of the Chinese Communist Party (CCP), held in Shanghai. With the rest of the CCP, he joined the Nationalist Party—the Kuomintang (KMT)—in 1923 and was chosen as an alternate member of the KMT Shanghai Executive Committee in 1924. Illness soon forced his return to Hunan. During this period, he gravitated inexorably to the left. He organized unauthorized unions of laborers and peasants, soon provoking authorities to issue a warrant for his arrest. Mao fled to Canton in 1925 and there worked as a radical journalist. He soon worked his way into the inner circle of KMT leader **Chiang Kai-shek,** ultimately becoming head of the KMT's propaganda section.

Mao and Chiang soon conflicted, and in May 1926 Mao was removed from the propaganda post. He joined the Peasant Movement Training Institute, a radical, far-left CCP cell. By April 1927, the gulf between the KMT and the CCP had become too wide to bridge. Chiang Kai-shek repudiated the KMT alliance with the CCP and launched his Northern Campaign against CCP units. Mao retreated underground and, independently even of the CCP, put together a revolutionary army. With it, he led the Autumn Harvest Uprising in Hunan during September 8–19. When the uprising failed, Mao was ejected from the CCP.

Mao gathered the remnants of his army—his staunchest followers—and retreated with them into the mountains, where he struck an alliance with another CCP outcast, **Chu Teh.** Together, in 1928, they formed a peasant army called the Mass Line, with which they set about creating their own republic, the Kiangsi Soviet. By 1934, the Soviet numbered some 15 million people. The existence of the Kiangsi Soviet defied not only Chiang Kai-shek's KMT, but also the Russian-dominated international Communist party, which ordered would-be Communist revolutionaries to concentrate on capturing cities—in accordance with orthodox Marxist doctrine—rather than concern themselves with the rural peasantry. Mao and Chu Teh resisted, however, and, between 1929 and 1934, using guerrilla tactics, they repulsed four KMT attempts to wipe out the Soviets. In 1930, however, the KMT executed Mao's first wife, Yang K'ai-hui, and, after a fifth assault on the Kiangsi Soviet in 1934, Mao was forced to evacuate with some 86,000 men and women. The evacuation became the "Long March" over a distance of some 6,000 miles to the province of Shensi. It was without doubt the most remarkable military march in modern history.

By October 1935, with a mere 4,000 followers, Mao established a new party headquarters at Yen-an. The KMT was compelled to forebear further attacks on the CCP when the Japanese invasion of China made common cause between the two groups. Mao likewise made peace with Chiang Kai-shek in December 1936 and launched the Hundred Regiments offensive against the Japanese during August 20–November 30, 1940. It was generally ineffective.

Mao did little else to fight the Japanese during the war years. Instead, he consolidated the CCP position in northern China and his own leadership of the party. In defiance of orthodox Marxism, he continued to organize peasants, then conducted a program of purges that secured, by April 1945, his election as permanent chairman of the party's central committee. Mao also used the war period to write and publish a series of essays promulgating the basis for Chinese communism. The CCP had 40,000 members in 1937. Under Mao, it grew to 1.2 million by the end of World War II.

With the end of the Japanese threat came an end to the rapprochement between the CCP and KMT. There was an attempt to establish a workable coalition government, but a civil war soon broke out, wherein Mao's forces dealt the armies of Chiang Kai-shek one defeat after another, from 1946 to 1949. At last, Chaing's Nationalists fled to the island of Taiwan, setting up there a government in exile. Late in 1949, Mao and his fellow Communists declared the creation of the People's Republic of China on the vast mainland.

The United States remained loyal to its wartime ally Chiang Kai-shek and to the legitimacy of Nationalist China. The nation rejected Mao's attempts to establish diplomatic relations, thereby propelling him into a close alliance with **Josef Stalin**'s Soviet Union. Mao was ruthless, devoting much of the years 1949 through 1954 to a purge of opponents within the party. He instituted agricultural collectivization, forcing landowners off their land. Then, from November 1950 through July 1953, he directed a massive and determined intervention in the war between North and South Korea. The armies of Communist China and the United States contended on the battlefield.

With the death of Josef Stalin in 1953, Mao assumed center stage as the preeminent Marxist leader in the world. He professed dissatisfaction with the slowing pace of revolutionary change in the Chinese countryside, pointing out that senior party members often behaved like members of the old upper classes. Mao initiated the Hundred Flowers movement during 1956–57, under the slogan, "Let a hundred flowers bloom, let a thousand schools of thought contend." He began by encouraging intellectuals to criticize the party and its methods of government and administration. Whether by design or out of fear of the hostile tenor of the criticism that was quickly forthcoming, Mao soon turned the Hundred Flowers movement against the dissidents and worked to create a worshipful cult of personality about himself. Simultaneously, he applied renewed pressure to achieve the complete transformation of rural ownership, calling for the total elimination of private property and the formation of people's communes. He promulgated a program called the Great Leap Forward, an attempt to accelerate industrialization on a grand scale.

The result of Mao's oppressive handling of the Hundred Flowers Movement, forced collectivization, and accelerated industrialization was administrative chaos on an unprecedented scale, together with growing popular resistance. This and a siege of adverse weather brought widespread famine.

Late in 1958, Mao voluntarily stepped down as head of state and was replaced by Liu Shao-chi. He returned to public life by the mid 1960s with a brilliantly orchestrated attack on Liu Shao-chi that Mao fashioned into a great proletarian "Cultural Revolution." From about 1966 to 1969, Mao and his third wife, Chiang Ch'ing, engaged the nation in a frenzied national debate on its political future and, after Mao resumed his position as party chairman and head of state, propelled China into perpetual revolution—a whirl of violent and deliberate chaos. His object in this was to transform the nation into a purely Marxist society, from which every vestige of traditional government and culture had been expunged. But the Cultural Revolution produced a mass army of radical Maoist students known as the Red Guards, who wrought havoc on China far beyond Mao's ability to control it. Mao called upon military, led by **Lin Piao** (whose support he secured by arranging to have him named his successor in the 1969 constitution of the Chinese Communist Party) to suppress the Red Guards, who were considerably subdued by 1971—the year Lin Piao died in an airplane crash, apparently after having plotted to assassinate Mao.

By the early 1970s, Mao was back in control and had moderated his views considerably, even making overtures to open diplomatic and economic relations with the United States. He received President Richard M. Nixon in a historic meeting in Beijing in 1972. Four years later, Mao died.

Mao's moderate policies came too late to forestall reaction against the would-be heirs of his power. Chiang Ch'ing and her closest associates, reviled as the "Gang of Four," were arrested and put on trial. Mao's handpicked successor, Hua Kuo-Feng, was ousted from the party's inner circle, and the government came under control of avowed moderates. Today, the memory of Mao occupies an ambiguous place in Chinese history and culture. He was once the focus of an extraordinary cult of personality, whose image was as ubiquitous in China as were copies of the "little red book"—a collection of *Quotations from Chairman Mao*. Now, officially, he is seen as a pioneering revolutionary, whose Cultural Revolution is nevertheless condemned for its excesses.

MARCELLUS, Marcus Claudius

BORN: ca. 268 B.C.

DIED: 208 B.C.

SERVICE: First Punic War (264–241 B.C.); Insubrian War (222 B.C.); Second Punic War (219–202 B.C.)

***Roman consul and general; among the best commanders Rome fielded against* Hannibal**

Nothing is known of Marcellus' early life, before he saw his first military action during the First Punic War. Following this conflict, he served in a variety of political and military positions, then was elected consul in 222. That same year, he fought the Insubres in northern Italy, successfully breaking their siege of Clastidium. When he went on personally to kill their chief, Viridomarus, he earned renown throughout Rome.

Marcellus fought the Second Punic War after the disastrous setback at the Battle of Cannae in 216. Going up against **Hannibal Barca,** he repulsed the great Carthaginian general's attack on the Roman army at Nola late in 216 and was subsequently elected consul in 215, 214, 210, and 208. In 209, he served as proconsul. He captured Casilinum (modern Capua) in 214, then fought Hannibal (to a draw) at Nola again, in 214. The next year, he led an expedition to Sicily to dissuade Syracuse from allying with Hannibal's Carthage or to position himself for an attack on Sicily if such an alliance were concluded. However, he annihilated the Syracusan border post of Leontini at this time, which propelled Syracuse into an alliance with Carthage, whereupon Marcellus laid siege to Syracuse. He was greatly surprised by defensive engines designed by Archimedes, and by 212 had still been unable to breach the city's walls.

Marcellus was now faced with defending against a force from Carthage; however, that army was wiped out by a malaria plague before it reached Marcellus' position. With the Carthaginian threat thus neutralized, Marcellus entered Syracuse through treachery and, as Roman troops ravaged the city, Archimedes was slain in 211.

Between 210 and 208, Marcellus jousted with Hannibal in southern Italy, but fought no major set battles. In a minor exchange near Benusia, Marcellus was killed. His death was a tremendous loss to the Roman military, which possessed few commanders capable of tangling with Hannibal.

MARCUS AURELIUS (Antoninus)

BORN: April 26, 121

DIED: March 17, 180

SERVICE: Parthian War (162–166); Marcomannic War (170–174)

One of a handful of truly admirable Roman emperors; an able commander who employed great deliberation and systematic thought to military campaigning

Born Marcus Annius Verus, the son of Annius Verus and the nephew of Antoninus Pius's wife, he was adopted in 138 by Antoninus, Emperor Hadrian's adoptive heir, who also adopted at the same time L. Ceionius Commodus. Marcus Aurelius was of a scholarly bent, but was sufficiently popular to gain election as consul three times—in 140, 145, and 161—and faithfully served his adoptive father until Antoninus' death on March 7, 161, at which time he declared himself emperor, naming Commodus (now known as L. Aurelius Verus) as co-emperor.

When Parthia invaded Syria late in 161, Marcus Aurelius declared war on the nation the following year, delegating most of the field command responsibilities to Verus and his general, Gaius Avidius Cassidus. The legions returned victorious, but brought with them the plague, and this weakened Rome just as Germanic tribes, including the Quadi and Marcomanni peoples, penetrated the Danubian frontier and wreaked havoc on the empire's border areas during 167–168. Despite the plague, Marcus raised a defensive army and led it in person in 167. He embarked on a vigorous campaign of fortress building and ultimately forced the Marcomannito to accept a truce in 168.

After Verus died in 169, Marcus once again had to confront the Marcomanni, and the emperor found himself in the field for the next three years, during 169–72. The result of this arduous campaign was the pacification of the frontier region. Triumphant against one troublesome tribe, Marcus turned on the Quadi, which he destroyed by 174. No sooner was this accomplished, however, than the Sarmatians invaded Moesia (the northern region of present-day Bulgaria). Furthermore, various Germanic tribes caused unrest along the Rhine. Marcus responded by sending subordinates to fight the Germans while he fought the Sarmatians, defeating them by 175.

Marcus Aurelius was almost immediately faced next with the revolt of Avidius Cassidus in Egypt and was preparing an expeditionary force when he learned that Avidius had been slain by his own troops. Instead of going to Egypt, then, Marcus extensively toured the eastern provinces of the empire and then elevated his son Commodus to coemperor in 177. The following year, the Danube frontier again erupted, and he set out to neutralize the Bohemian tribes once and for all. The campaign proved costly—albeit ultimately successful—and provoked much unrest in Rome. Marcus Aurelius was occupied with the political fallout from his Danubian campaign when he died of an illness in Vindobona (Vienna).

Marcus Aurelius is remembered as one of the noblest of Roman emperors, an image greatly enhanced by his magnificent *Meditations*, which he penned while campaigning in the field. He was a good organizer of highly successful military campaigns, and when he personally commanded an army, he did so with great vigor and relentless determination.

MARION, Francis

BORN: 1732

DIED: February 27, 1795

SERVICE: Cherokee War (1760–61); American Revolution (1775–83)

America's "Swamp Fox" of the Revolution; master of guerilla tactics

Marion was born in Winyah, South Carolina, and, as a young man, shipped out to sea on a merchantman, but was cured of his seafaring ambitions after his ship was wrecked. He returned to dry land as a farmer about 1748. Marion saw his first military action in 1761, when he served as a lieutenant of South Carolina militia in a campaign against the Cherokee. He earned the admiration of his fellow militiamen during this expedition. Gaining in reputation throughout back-country South Carolina, Marion was sent to the South Carolina Provisional Congress as a delegate in 1775.

With he outbreak of the Revolution, Marion was commissioned captain of the Second South Carolina Regiment on June 17, 1775. On February 22 of the following year he was promoted to major and given command of the left face of Fort Sullivan (later Fort Moultrie) during the first defense of Charleston. It was he who had the satisfaction of firing the last shot at the retreating British fleet on June 28.

Marion made lieutenant colonel on November 23 and on September 23, 1778 was put in command of the Second South Carolina. He fought at Savannah on October 9, 1779, then narrowly evaded capture when Charleston was finally captured by the British on May 12, 1780. Fleeing the city, Marion holed up in the low country swamps, where he organized a remarkably effective guerrilla operation throughout South Carolina. Commanding a tiny force of only fifty-two men, he scattered a Loyalist militia force of 250 at Blue Savannah on September 4. Moving on to Black Mingo, he destroyed the Loyalist outpost there on September 29, then squashed a Loyalist uprising at Tearcoat Swamp on October 26.

Marion became a thorn in the side of British general **Charles Cornwallis,** who dispatched a force under Colonel Banastre Tarleton to run him to ground. It was a frustrated Tarleton who remarked, "But as for this damned old fox, the devil himself could not catch him," and the sobriquet "Swamp Fox" stuck.

Marion was promoted to brigadier general of militia in 1781 and, using Snow's Island as a headquarters, he organized Marion's Brigade, an enlarged guerrilla force that continued raiding the British throughout South Carolina. On January 24, 1781, he staged a spectacular raid on Georgetown, and August 13 distinguished himself in his relief of a trapped Patriot unit at Parker's Ferry. Given command of North and South

Carolina militia forces, Marion fought his biggest engagement at Eutaw Springs on September 8. His next major battle did not come until the following year, at Fair Lawn—just outside of Charleston, South Carolina—on August 29.

Marion served as commandant of Fort Johnson in Charleston after the war and then used his considerable renown to help him secure election to South Carolina's Senate in 1782, 1784, and 1786. He served as delegate to the state's constitutional convention in 1790.

Marion was, in the phrase of cliche, a legend in his own time. He commanded absolute loyalty from his men, even though he was sometimes brutal to them. His courage was matched only by his skill as a guerrilla fighter. His greatest ally was the South Carolina swampland, which he knew intimately, always exploiting it to tactical advantage. In the South, where the Loyalist presence was strong and Patriot resources slim, Marion's actions were crucial to winning the Revolution.

MARIUS, Gaius

BORN: 157 B.C.

DIED: 86 B.C.

SERVICE: Jugurthine War (112–105 B.C.); War against the Cimbri and Teutones (105–101 B.C.); Social War (91–88 B.C.); Civil War (88–82 B.C.)

Roman general and statesman who transformed the Roman army from a body of citizen-soldiers into a professional force

Marius was born near Arpinum (present Arpino), the son of a plebeian family of equestrian rank. In 119 B.C., he became a tribune, but fell afoul of other officials when he was accused of political bribery in 116 B.C. Acquitted, he went on the next year to become quaestor, and in 114 B.C. was appointed governor of Farther Spain.

Marius was the equivalent of a staff officer under Q. Caecilius Metellus Numidicus in Rome's war against Jugurtha, king of Numidia (approximately modern Algeria) in 109 B.C. Disgusted with his commander's conduct of the war, Marius returned to Rome, where he accused Metellus of incompetence. Whether or not Marius' accusations had merit, they gained him much political favor and support, and in 107 B.C. he was elected consul and also replaced Metellus at the head of the army in Africa.

Marius launched a successful campaign against Jugurtha during 107–105 B.C., although he was disappointed that it was **L. Cornelius Sulla** and not himself who actually took Jugurtha prisoner. Nevertheless, he was given a triumph when he returned to Rome on January 1, 104 B.C. and that year was elected consul. He was given command of the armies against the Cimbri, and the following year he was elected consul again, serving through 100 B.C.

His office as consul gave Marius authority to reform the army in a sweeping fashion. He abandoned the traditional militia system and transformed the army into an organization of professional soldiers recruited on a voluntary basis. He reorganized the tactical structure of the force, creating the cohort structure that is at the heart of the Roman legion system. The new army was put to the test against the Teutones at Aquae Sextiae in 102 B.C. and against the Cimbri at Vercellae the following year. In both campaigns, Marius was victorious.

Despite his military triumphs, the political policies Marius attempted to introduce into the Roman republic were defeated, and, discouraged, he left Italy to travel in Asia during 99–94 B.C. After his return, the Social War (against the Socii, or Allies) erupted, and Marius led an army against the Marsi in 91–90 B.C. His friend P. Sulpicius Rufus tried to usurp from Sulla and give to Marius command of an army to be sent against Mithridates IV of Pontus, but failed. Sulla then invaded Rome in 88, killed Rufus, and went after Marius, who fled to Africa, returning to Italy in 87 B.C. He allied himself with L. Cornelius Cinna, with whom he raised an army and retook Rome from Sulla. It was now Marius' turn to exact vengeance, and he unleashed a brutal campaign against his enemies—except for Sulla, who was fighting in Greece at the time.

Marius' final triumph was short lived. Less than two weeks after taking Rome, he died.

MARLBOROUGH, John Churchill, First Duke of

BORN: May 26, 1650

DIED: June 16, 1722

SERVICE: Third Dutch War (1672–74); War of the League of Augsburg (1688–97); War of the Spanish Succession (1701–14)

British captain general; one of England's greatest soldiers, who achieved victories over Louis XIV of France, notably at Blenheim (1704), Ramillies (1706), and Oudenaarde (1708)

Churchill was born at Ashe, in Devon, the son of an impecunious royalist gentry family. He attended St. Paul's school during 1664–65, then became page to James, Duke of York, in 1655. Two years later, he was commissioned an ensign in the Foot Guards and was sent to serve in the Tangier garrison during 1668–70. He then returned to court and served under James at the naval battle of Solehay on May 28, 1672, distinguishing himself for valor, which earned him a promotion to captain.

Churchill next served in the Duke of Monmouth's English contingent in the service of French King Louis XIV, and he again distinguished himself, this time at the siege of Maastricht, which spanned June 5 to 30, 1673. After performing valiantly at Sinsheim on June 16, 1674, Marlborough was promoted to colonel. On October 4, 1674, he led a notable charge at Enzheim.

On his return to court, Marlborough met and fell in love with Sarah Jennings, whom he married in the winter of 1677–78. Created Baron Churchill of Aymouth in 1682, he was appointed colonel of the First Dragoons the following year. He was fiercely loyal to his king, James II, who promoted him to major general and second in command during Monmouth's rebellion. He was in overall command at Sedgemoor on July 6, 1685—a royalist triumph. Although Churchill was made colonel of the Life Guards for his services in quashing Monmouth's rebellion, he grew increasingly dissatisfied with James II's intense Catholicism. At last, he broke with James and threw in his lot with William and Mary after their landing at Torbay in November 1688.

William and Mary created Churchill Earl of Marlborough and appointed him privy councillor in 1689. They dispatched him to command an English brigade in the Prince of Waldeck's army in Flanders in 1689 during the War of the League of Augsburg. At the Battle of Waldcourt, on August 25, 1689, he drew Waldeck's enthusiastic praise.

During the summer of 1690, Marlborough returned to England and was instrumental in planning strategy for William III's Irish campaign. In the fall, Marlborough took Cork and Kinsale with a small force. Despite this success, William turned against him, relieved him of command, exile him from court, and even briefly imprisoned him during May 1692. However, after Marlborough intervened to effect the reconciliation of Princess Anne and William III after Mary's death in 1694, he came back into royal favor and served in a series of offices: as governor to Anne's son, the Duke of Gloucester (1698) and as a royal justice (1698–1700). He was also given a new military command in 1700, of the English troops in the Netherlands and the home army.

On May 15, 1702, when Anne succeeded to the throne, Marlborough was made Knight of the Garter and appointed master general of the ordnance as well as captain general of the army. In the early phase of the War of the Spanish Succession, Marlborough captured Venlo (September 15, 1702), Roermond (October 7), and Liège (October 18), despite poor cooperation from the Dutch. Created Duke of Marlborough after these triumphs, he quickly went on to capture Bonn during May 1703, but fell afoul of his Dutch allies at Antwerp in the fall.

With Prince **Eugene,** he moved to block French invasion of Austria via Bavaria, advancing to the Danube with German and British forces during May–June 1704. He was in command at the storming of the Schellenberg on July 2, 1704, and he took Donauwörth. Linking up with Eugene's army on August 12, they fought the titanic Battle of Blenheim the next day, which became Marlborough's most celebrated and consequential victory. It was the first major defeat for the French army in more than half a century, and it saved Vienna from invasion by a Franco–Bavarian army. Moreover, the Blenheim victory preserved the alliance of England, Austria, and the United Provinces against France. As a result of Blenheim, Bavaria withdrew from the war. Following Blenheim, Marlborough took Trier and Trarbach (Traben-Trarbach), and in 1705, Leopold I, the Holy Roman (Austrian) emperor, created Marlborough Prince of Mindelheim.

Marlborough's results were far less conclusive against the armies of François Villeroi in Flanders during 1705. Although he broke through the defensive Lines of Brabant at Elixem in July, once again the Dutch allies faltered, and Marlborough could not capitalize on his gains. However, on May 23, 1706, he defeated Villeroi at Ramillies, then captured in rapid succession Ostend (July 6), Menin (August 18), Dendermonde (September 5), and Ath (October 2).

The alliance was riddled with dissension by 1707, which wrecked Marlborough's strategy and compelled him to work with Eugene to reorganize his campaign. During May–June 1708, he fought a successful defensive action against the Duc de Vendôme, then turned the tables on him, making a forced march that caught him at Oudenarde. Coordinating with Eugene, he defeated the French general on July 11, 1708.

From August 14 to December 11, 1708, he commanded the army that covered Eugene's siege of Lille, then pushed the French out of western Flanders during December 1708–January 1709, capturing Ghent on January 2, 1709 in the process. France's Louis XIV, weary of the war, negotiated with Marlborough to end the conflict during the winter of 1708–09, but the talks broke down, and Marlborough laid sieged to Tournai on June 26, 1709. The town fell to him on September 3, 1709. He then pounded the army of Claude Villars at Malplaquet on September 11, 1709—a costly victory that left him free to besiege and capture Mons during September 24–October 20 and retake Ghent during December 25–30, 1709.

Despite growing political conflict, greatly aggravated by the terrible cost of Malplaquet, he and Eugene captured Douai during April 23–June 27, 1710 and then breached Villars' lines of Ne Plus Ultra during August 4–5, opening the way for the capture of Bouchain, which fell to Marlborough on September 13, 1710.

All of these triumphs notwithstanding, Marlborough was recalled to England for political reasons. The new Tory government wanted peace and saw to the removal of Marlborough from all his offices on December 31, 1710. The duke retired to the continent until the accession of George I in 1714, when he was restored to most of his military office. However, by this time ill health plagued Marlborough, and he suffered a series of strokes. Retiring from public life in 1716, he died six years later.

Marlborough was one of the very greatest soldiers ever to serve England. His record of victories is extraordinary, and his skill as a tactician and strategist was matched only by the degree of affection and loyalty he commanded from his men and officers.

MARMONT, Auguste Frederic Louis Viesse de, Duke of Ragusa

BORN: July 20, 1774

DIED: March 2, 1852

SERVICE: French Revolutionary Wars (1792–99); Napoleonic Wars (1800–15)

French Marshal of the Empire who betrayed Napoleon by surrendering Paris (March 30, 1814) and taking his troops into the Allied lines

Marmont was born at Châtillon-sur-Seine in Burgundy, the son of a retired Royalist officer who owned a prosperous iron works. Marked from childhood for a military career, Marmont attended the artillery school at Châlons and was commissioned in the Army of the Moselle as a lieutenant of artillery in 1792. The next year, he was promoted captain and transferred to the Army of the Alps and then to the Army of the Pyrenees. In this latter assignment, he served directly under **Napoleon** Bonaparte. He distinguished himself at the siege of Toulon in 1793 and the next year served with the Army of Italy and then, under General Louis Desaix de Veygoux at Mainz during October 1795. The following year, Napoleon selected him as his aide-de-camp, promoting him to major. Under Napoleon's watchful eye, he distinguished himself at Lodi on May

10 and at Castiglione on August 5. Marmont returned to a hero's welcome in Paris, bearing the Austrian colors he captured at Rovereto, Bassano (Bassano del Grappa), and San Giorgio.

Promoted to colonel, Marmont next served in the Army of the Orient and at Malta in 1798. It was reported that, in the attack on Malta, he personally seized the colors of the Knights of St. John with his own hands. Whatever the truth of this, he was promoted to general of brigade for his gallantry and moved on with Napoleon to Egypt, where he fought at Alexandria on July 2, then accepted appointment as governor the conquest of the city.

After fighting in the Battle of the Pyramids on July 21, Marmont was selected to return to France with Napoleon, whom he supported in the coup of the Eighteenth Brumaire (November 9–10, 1799). Napoleon, having been named First Consul, rewarded Marmont with an appointment as councillor of state.

Marmont accompanied Napoleon at Marengo on June 14, 1800, performing again with distinction and skill, which earned him promotion to general of division on September 9. Marshal Guillaume Brune chose Marmont to sign the Armistice of Treviso in January 1801, and Napoleon approved his appointment as first inspector general of artillery the following year.

Marmont served as commander of artillery of the encampments at Boulogne from 1803 to 1805, and in 1805 he was made colonel general of the elite *chasseurs á cheval.* At Ulm, during October 7–20, he took charge of II Corps, then transferred to Italy, where, in July 1806, was made governor general of Dalmatia.

Marmont's single most spectacular victory came on September 30, 1806, when, commanding 6,000 men, he triumphed over 16,000 Russians and Montenegrins near Old Ragusa. Two years later, this feat was recognized with his creation as Duke of Ragusa.

Napoleon called Marmont to Austria in 1809 to command XI Corps, and he performed magnificently at Wagram during July 5–6 and at Znaim on July 10. These actions earned him his marshal's baton in a field ceremony. After this, he served in Illyria until 1811, when he was dispatched to Portugal to assume command of the VI Corps under Field Marshal **André Masséna,** whom he replaced as commander of the Army of Portugal in May.

Marmont outgeneraled the Duke of **Wellington** in Spain in 1812, but faltered at Salamanca on July 22, 1812, and fell victim to Wellington's devastating counterattack. Marmont valiantly tried to salvage the situation, but was severely wounded by a shell burst. He was out of action for nearly a year and did not return to the field until May 2, 1813, when he led VI Corps at Lützen and then at Bautzen, during May 20–21. On August 26–27, he fought at Dresden, then Leipzig (October 16–19), and, finally, Hanau (October 30).

Napoleon never relied more on Marmont than in the desperate campaign for France in 1814. However, the marshal failed at Laon during March 9–10, but then held his ground against vastly superior forces at Montmartre—just outside Paris—on March 30. But then, on April 5, Marmont surrendered his corps and gave up Paris to the Allies.

With the restoration of Louis XVIII, Marmont was elevated to the peerage and named captain of the king's bodyguard. When Napoleon returned from his first exile in Elba, Marmont accompanied Louis into exile. Following Napoleon's final defeat at Waterloo in 1815, Marmont was one of the four marshals of France who presided at the court martial of Marshal **Michel Ney** for having betrayed his pledge of loyalty to the Bourbon Louis. Marmont voted for the death sentence to be executed upon his former comrade at arms.

Louis dispatched Marmont to squash a rebellion in Lyons in 1817, then made him a minister of state and member of the Privy Council. He was given command of the First Military Division in 1821 and was briefly ambassador to Russia. With the outbreak of the July Revolution of 1830, Charles X called on him to restore order, but he failed, and Paris fell to the rebels. Marmont fled France, ultimately settling in Venice, where he devoted the last two decades of his life to retirement and the composition of his memoirs.

Marmont was a superb, if sometimes reckless, commander. His many accomplishments as one of Napoleon's most reliable marshals were, in the minds of many Frenchman, eclipsed by his betrayal of Napoleon.

MARSHALL, George Catlett

BORN: December 31, 1880
DIED: October 16, 1959
SERVICE: Philippine Insurrection (1899–1903); World War I (1917–18); World War II (1941–45)

American chief of the joint chiefs of staff during World War II and, after the war, as secretary of state, architect of the Marshall Plan

A native of Uniontown, Pennsylvania, Marshall graduated from Virginia Military Institute in 1901 and was commissioned a second lieutenant of Infantry on February 3, 1902. He was dispatched to the Philippines during the conclusion of the insurrection on Mindoro and served there with the 30th Infantry Regiment during 1902–03. Back in the States, he attended Infantry and Cavalry School at Fort Leavenworth, graduating at the top of the class of 1907 and staying on at the Staff College during 1907–08. Promoted to first lieutenant in 1907, he taught as an instructor at the schools from 1908 to 1910.

After serving in a number of miscellaneous assignments during 1910–13, Marshall returned to the Philippines for three years, serving as aide to General **Hunter Liggett** and gaining promotion to captain in 1916. After returning to the United States, he was assigned as aide to General James F. Bell, in the Western Department and then in the Eastern Department in 1917. In June of 1917, Marshall shipped out to France as a staff officer with the First Division. Assigned as Division operations officer, he was one of the planners of the first U.S. offensive of the war in May 1918.

After promotion to temporary colonel in July, Marshall was attached in August to General **John J. Pershing**'s General Headquarters at Chaumont, where he was the key member of the team that planned the massive Saint-Mihiel offensive of September 12–16. He was also in charge of the transfer of half a million men from the neutralized Saint-Mihiel salient to the Meuse-Argonne front. The movement was accomplished so swiftly that Marshall earned praise as a logistician and was named chief of operations for the First Army in October. The next month, he was made chief of staff of VIII Corps.

George C. Marshall. Courtesy National Archives.

With the end of the war, Marshall was assigned to service with the army of occupation in Germany. In September 1919, he returned to the States and reverted to his prewar rank of captain, but was appointed aide to Pershing, who was now Army Chief of Staff. Marshall served in this capacity through 1924 and collaborated with Pershing on aspects of the National Defense Act. He was also instrumental in writing the general's reports on the American Expeditionary Forces in the war.

Marshall was promoted to major in July 1920 and lieutenant colonel three years later. After leaving Pershing's staff, he served in Tientsin, China, as executive officer of the Fifteenth Infantry, then returned to the United States in 1927 and became assistant commandant of the Infantry School at Fort Benning through 1932. He was promoted to colonel and worked with the Civilian Conservation Corps (CCC) in 1933, then became senior instructor to the Illinois National Guard from 1933 to 1936, when he was promoted to brigadier general and given command of Fifth Infantry Brigade at Vancouver Barracks, Washington.

In 1938, Marshall left his assignment at the Fifth Infantry and came to Washington, D.C., as head of the War Plans Division of the Army General Staff. Promoted to major general in July, he was appointed deputy chief of staff. On September 1, he was made a temporary general and appointed chief of staff. From this position, he launched a rapid expansion of the army preparatory to war. Under his direction, the army would grow from its prewar strength of 200,000 to 8 million during the war.

After Pearl Harbor and U.S. entry into World War II, Marshall reorganized the General Staff and, by March 1942, effected the restructuring of the army into three major commands: Army Ground Forces, Army Service Forces, and Army Air Forces. Serving on the Joint Chiefs of Staff, he was a principal military adviser to President Franklin D. Roosevelt, and he was present at all the great Allied conferences of the war, first with Roosevelt and then with President Harry S. Truman. Marshall was one of the key architects of American and Allied military strategy, as well as U.S. political strategy bearing on the war.

In December 1944, Marshall was elevated to the rank of general of the army—five-star general—and concluded his service as chief of staff, on November 20, 1945. Just five days after he resigned as chief of staff, however, President Truman sent him to China as his special envoy. For the next year, Marshall attempted to mediate a peace between **Chiang Kai-shek** (and his KMT) and **Mao Tse-tung** (and the Chinese Communist Party), but without success. After his return to the United States, he replaced James F. Byrnes as secretary of state in Truman's cabinet on January 1947.

In June 1947, Marshall spoke at a Harvard University commencement ceremony and proposed a sweeping program of economic aid to rebuild war-ravaged Europe and thereby not only aid humanity, but forestall the spread of communism in economically devastated areas. This was the germ of the European Recovery Program, soon known as the Marshall Plan. It was perhaps the single most important "weapon" in the Free World's Cold War arsenal.

Marshall resigned from the Cabinet in January 1949, but returned in September of the next year as secretary of defense and served in that post during the opening of the Korean War. Incredibly, Red-baiting Senator Joseph McCarthy targeted Marshall as being "soft on communism," and the secretary suffered public attack on is reputation. In ill health, Marshall stepped down as secretary of defense and retired from public life in September 1951. Three years later, in December 1953, he was awarded the Nobel Peace Prize, chiefly in recognition of the Marshall Plan. He was the first soldier ever honored with the prize. It was a fitting culmination to an extraordinarily distinguished career.

MARTINET, Jean

DIED: July 1, 1690
SERVICE: Dutch War (1672–79)

French general instrumental in the reorganization and reform of the army; his name is now a synonym for an overly strict and punctilious disciplinarian

Nothing is known of Martinet's early life, other than that he was a commoner. He became one of the prime movers of the reform and reorganization of the French army during the 1660s and 1670s.

Martinet was a lieutenant colonel of the infantry regiment Du Roi from 1662 and worked to model this regiment into a thoroughly disciplined, highly drilled body of soldiers. Du Roi was intended to set the example for the rest of the infantry to follow. It was a very large regiment, with 54 rather than the customary 10 to 16 companies, and was intended as a kind of training ground for officers, who would imbibe Martinet's methods and take them to other regiments throughout the army.

In addition to providing the example of military reform, Martinet invented copper pontoons for use in constructing temporary bridges. He also established an elite corps of grenadiers within the infantry, and by 1670, most French infantry regiments included a company of grenadiers. He also advocated modernizing the army through use of the plug bayonet in place of the pike. This was the only one of his reforms that failed.

Martinet died in command of an assault on the fortress of Duisburg on June 21, 1672. However, the brilliant performance of Louis XIV's armies at the end of the seventeenth century is testimony to the effectiveness of Martinet's work.

MASSÉNA, André, Duke of Rivoli and Prince of Essling

BORN: May 6, 1758

DIED: April 4, 1817

SERVICE: French Revolutionary Wars (1792–99); Napoleonic Wars (1801–1815)

*One of the greatest of **Napoleon's** marshals; a brilliant commander, but one who had little respect for the emperor's regime*

Born in Nice, Masséna was orphaned young and went to sea in 1771 as a ship's boy. In 1775, Masséna, whose ancestry was Italian, enlisted in the Royal Italian regiment in the French service. He had attained the rank of sergeant when he retired after marrying the daughter of a surgeon in 1789. Making his home in Antibes, he got his living primarily by smuggling. With the outbreak of the French Revolution in 1791, he rejoined the army and was chosen captain of guides for France's new Army of Italy. In February 1793, he fell afoul of army officials for looting, but his intimate knowledge of the local geography made him too valuable to dismiss. Not only did he fight at the siege of Toulon during September 7–December 19, 1793, he was promoted to general of division in December.

During the spring of 1794, Masséna repeatedly triumphed in Italy and was given command of one of three divisions of the Army of Italy during November 1795, achieving a remarkable success at Loano on November 25. **Napoleon** entrusted to him command of the center during the opening of the Italian campaign proper in April 1796. He emerged victorious at Montenotte (April 12) and Dego (April 14), notwithstanding a minor reversal at Dego the following day.

Masséna led the vanguard in the advance on Turin, and he participated in the great victory at Lodi Bridge on May 10. He was instrumental in major victories at Castiglione (August 5), Bassano (September 8), and at Rivoli (January 14–15, 1797). In the advance on Vienna, he led his troops with singular dash during March–April 1797.

At the conclusion of the Viennese campaign, Masséna returned to Paris, but was recalled to Italy in February 1798 to fight under Marshal **Louis-Alexandre Berthier.** His troops mutinied in protest of failure to receive pay, and they expelled Masséna from his command. He was subsequently given command of a corps under Barthelmy-Catherine Joubert in Switzerland, in November 1798, and succeeded him as commander of army after Joubert was beaten at Stockach on March 25, 1799. Masséna's first action as overall commander was to repulse an attack at Zürich on June 4, but thereafter prudently withdrew to the Aar River and regrouped. He then took the offensive, routing a Russian army at the Second Battle of Zürich on September 25, 1799, and pursuing it northward across the Rhine.

Following his triumph in Switzerland, Masséna was recalled to Italy to command what remained of the Army of Italy in the wake of the Eighteenth Brumaire coup. He performed brilliantly in executing a defensive campaign in the Ligurian hills during April 3–10, 1800, only to fall under siege at Genoa from April 24 to June 4. He had no choice but to surrender and was subsequently repatriated with his men across the River Var.

On October 18, 1804, Masséna was named Marshal of the Empire and assigned command of forces in Italy just prior to the Ulm campaign during September 1805. He boldly attacked the numerically superior forces of Archduke Charles, whose army he pursued into the Julian Alps. Turning next upon Calabria during July–December 1806, he effected that region's pacification.

Masséna was created Duke of Rivoli in 1808 and given command of IV Corps early the following year, as France prepared to go to war against Austria. He served with mixed success at Abensburg-Eckmühl during April 20–22 to block the escape of Austrian forces through Landshut, then led an offensive at Aspern-Essling during May 21–22, but reinforcements failed to reach him and he had to break off. At Wagram, July 5–6, 1809, despite a severe injury that prevented his riding a horse (he led the battle from a carriage), he fought an effective holding action that allowed Napoleon to flank Archduke Charles and crush his army.

Created Prince of Essling in January 1810, Masséna was given command of the Army of Portugal in April and led it in an advance on Portugal from Spain during June. He took the fortress of Ciudad Rodrigo on July 10, then stabbed at **Wellington**'s army at Bussaco on September 25, 1810, but failed. Moreover, he was unable to penetrate the Lines of Torres Vedras outside Lisbon and had to retire, hampered by shortage of supply, during November 1810. When he moved next to relieve Almeida in April 1811, he was once again bested by Wellington, at Fuentes de Oñoro on May 5, and, later that month, was replaced by Marshal **Auguste Marmont.** Masséna did not command another army, but returned to Paris, where, during the Hundred Days (April–June 1815), he remained neutral.

Masséna was among Napoleon's most brilliant strategists. Wellington, who twice triumphed over him in Spain, remarked that "he was their best." His failures in Spain were due in part to poor lines of supply and to his own declining health.

MATSUI, Iwane

BORN: 1878

DIED: 1948

SERVICE: Russo–Japanese War (1904–05); Siberian Expedition (1918–22); Second Sino–Japanese War (1937–45)

Japanese general convicted of war crimes against the Chinese

Matsui was born in Nagoya and graduated from the Military Academy in 1897. A promising young officer, he attended the Army Staff College, graduating in 1904, in time to serve in the Russo–Japanese War. Following that conflict, he was appointed resident officer in China during 1907–12 and was promoted to major in 1909.

Matsui became resident officer in France during 1914–15, then returned to China as resident officer there from 1915 to 1919. During this period, he was promoted to colonel (1918) and served as a regimental commander from 1919 to 1921, when he was appointed a staff officer in the Siberian Expeditionary Army, in which he served until 1922.

From 1922 to 1924, Matsui was posted to Kwantung (Guangdong) Army Headquarters in Manchuria as chief of the Special Services Agency at Harbin and in 1923 was promoted to major general. During 1924–25, he commanded a brigade, then became chief of the Intelligence Division, Army General Staff, serving in this capacity from 1925 to 1928. Promoted to lieutenant general in 1927, he negotiated with Chinese Nationalist (KMT) leaders, then returned to Japan to take command of the First Division in Tokyo during 1929–31.

Matsui served on the Japanese delegation to the Geneva Disarmament Conference of 1931–33 and, in 1933, was promoted to general. At this point, the apparent culmination of his military career, he began a period of semiretirement, although he served on the Supreme War Council during 1934–35. In July 1937, however, he was recalled to active duty as commander of the Shanghai Expeditionary Army and he directed the battle for Shanghai during August 8–November 8. After this, he fought up the Yangtze River and took Nanking during November–December. He had overall command of the Japanese forces in China and at Nanking during the Rape of Nanking, in which Japanese soldiers killed some 200,000 civilians (children among them) and committed other war crimes and crimes against humanity.

Matsui then served as commander of the Japanese Central China Army during December 1937–38, after which he returned to Japan as a cabinet councillor and retired on the eve of World War II, in 1940.

At the conclusion of World War II, Matsui was tried and convicted for the Rape of Nanking. He was executed in 1948, and while it is true that the actual atrocities were the work of his subordinates, it was judged that, as overall commander, he bore primary responsibility for the crimes. Certainly, there is no evidence that he did anything to avert these actions.

MATTHIAS I CORVINUS

BORN: February 23, 1440

DIED: April 6, 1490

SERVICE: Turkish War (1463); Bohemian War (1468–78); Second Turkish War (1475); Wars with Austria (1477–85)

Hungarian ruler who, at the time of his death, was the most powerful monarch in Central Europe

Matthias was born at Koloszvár and was the son of Hungary's national hero, **Jänos Hunyadi,** and Elizabeth Szilägyi. When he was only twelve, he fought alongside his father and, in 1453, was made Count of Besztercze (Bistrila). He fought in the relief of Belgrade on July 21–22, 1456, for which he was knighted.

His career might have been brief indeed; for, on August 11, 1456, Hunyadi died, and his father's enemies persuaded Matthias to return to Buda (modern Budapest). There he was seized, summarily tried, and condemned to death on a trumped-up charge of conspiracy. His elder brother Ladislav, also arrested and tried, was put to death, but Matthias was ultimately spared on account of his youth. Subsequently, he was held by George of Podebrad—at the time, governor of Bohemia—but was later released. On his return to Hungary, he was elected king by acclamation on January 24, 1456—though against opposition from a group of nobles led by László Garai and Miklos Ujlaki.

The new king saw his nation under chronic attack from a number of enemies. Moreover, some of Hungary's key nobles opposed him. He acted quickly to forge an alliance with George of Poderbrad, who was now his father-in-law and had been elected king of Bohemia in 1458. With this alliance secured, Matthais led an expedition into Turkey—only to find himself compelled to return when the rebellious nobles crowned Holy Roman Emperor Frederick III as king of Hungary on March 4, 1459. Reentering Hungary, Matthias used the military to drive out Frederick, who yielded and recognized Matthias as king of Hungary in April 1462. Two years later, Matthias was officially crowned, on March 29, 1464.

Matthias now joined the Catholic League *against* George of Podebrad, and several years of chaotic warfare ensued, in which allegiances and couterallegiances were traded in murky fashion. Matthias by turns fought George, the Poles, and Emperor Frederick, who bested Matthias when he attacked Vienna in 1477, compelling the Hungarian to accept the terms of the Peace of Olomouc, by which he acknowledged Wladyslaw as king Vladislav II of Bohemia. Matthias, however, retained three Bohemian provinces as security for a payment to Vladislav of 400,000 florins in 1478.

With the confused situation in Europe cleared up, Matthias renewed his campaign against the Turks with two fairly modest expeditions in 1463 and 1475. These expeditions were undertaken not for territorial gain, but to pacify and stabilize the border regions. In the meantime, Matthias exacted an additional 100,000 florins from Frederick in return for granting him recognition as king of Hungary in the event that Matthias died without male issue. That agreement came in 1478; three years later, in 1481, Matthias started another war against Frederick—this time with far more success. Vienna fell to him on June 1, 1485, and then Matthias took Styria, Carinthia, and Carniola—all in northeastern Italy. Quickly, he concluded alliances with the Swiss, the Saxons, the Bavarians, and the prince-archbishop of Salzburg. Now presiding from his capital in Vienna, Matthias had risen to a position of tremendous power and was the most influential monarch in Central Europe.

Ruthless as a politician, Matthias backed his bold maneuvering with a fine army, certainly the most feared and respected in Central Europe. As his father had done before him, Matthias steadily increased the strength and level of training of his army. His elite Black Companies terrorized those he wished to subdue. In contrast to the disciplined ferocity of his army, Matthias's court was cultivated to a degree that the rest of Europe would not equal for at least another century. He amassed a personal library that attracted scholars from far and wide, and he established Hungary's first printing press. He also founded a university at Pozsony (Bratislava).

MAUDE, Sir Frederick Stanley

BORN: 1864

DIED: Late November 1917

SERVICE: World War I (1914–18)

Masterful British general of the Middle Eastern theater in World War I

Little is known of Maude's early career in or out of the military, but at the outbreak of World War I, in August 1914, he was posted to the staff of III Corps and served in France on the Western Front through most of 1915. He was then transferred to the Dardanelles as commander of the Thirteenth Division. He participated in the evacuations of the Suvla beachheads during December 18–19, 1915) and those at Helles during January 8–9, 1916.

After the debacle at the Dardanelles, Maude's division was transferred to Mesopotamia to relieve the beleaguered forces under General Sir Charles Townshend at Kut al Amara. Unfortunately, Maude was unable to reach Townshend before he surrendered his garrison on April 29, 1916.

In September, Maude was named to command all British forces in Mesopotamia and meticulously prepared an offensive in Basra. The operation commenced on December 13, but had to be suspended on December 26 because of heavy rains. Maude resumed on January 6, 1917 and slogged his way through to the defenses surrounding Kut on February 16. His two corps—about 48,000 men—forced General Nur-ud-Din's 20,000-man corps to retreat on February 24, and Maude pressed on past Ctesiphon to Baghdad. Here the Turkish forces made a stand, but Maude swept them aside in a short battle. He took Baghdad on March 11.

After reorganizing, Maude continued his advance. He captured Samarra on April 23, then repulsed a Turkish counterattack. However, once again, the brutal climate intervened. Plagued by 120-degree heat, Maude suspended operations for four months, then advanced up the Euphrates until he reached Ramadi, which he took after a brief battle on September 27–28. He then moved to the Tigris River, capturing Tikrit on November 2. Later that month, however, he succumbed to the cholera endemic to the region. He died in Baghdad.

Maude was an extraordinary commander in a little-discussed theater of World War I. Following Townshend's surrender, Mesopotamia fell out of British control, until, singlehandedly, Maude regained all that had been lost. His operations were characterized not only by their effectiveness, but by their light casualties. Maude must be counted among Great Britain's best generals of World War I.

MAURICE (Flavius Tiberius Mauricius)

BORN: 539

DIED: 602

SERVICE: Persian War (572–591); Avar War (591–595)

Byzantine Emperor, general, and author of* Strategikon, *a highly influential encyclopedia of the science of war

Maurice earned his early fame as a general under Tiberius II Constantine in the war against the Persian emperor Chosroes I. In 575, he drove the Persians out of his native Cappadocia (southeastern Turkey). At this time, too, he wrote the *Strategikon*, which became the blueprint for the Byzantine military for centuries afterward.

On his deathbed, Tiberius named Maurice caesar on August 5, 582, and a few days later, on August 13, he was crowned emperor. Maurice resumed the ongoing war against Persia, personally leading an army that defeated Bahram at Nisibis in 589. With great political aplomb, he exploited the internal chaos then raging in Persia and maneuvered Chosroes II, the grandson of Chosroes I, into line for the throne. Once the new ruler was ensconced, he gained a favorable peace treaty, which ended the Persian War in 591.

In the meantime, at the empire's western frontier, the Avars were raiding and visiting terror. Working closely with his general-in-chief, Priscus. Maurice campaigned against the Avars from the Black Sea to the Theiss (Tisza) River in 595. The culminating battle took place at Viminacium on the south bank of the Danube. There, in 601, Maurice defeated the Avar chieftain Bayan.

Years of unremitting warfare created a widespread economic crisis, and Maurice found himself unable to pay his Danubian army. A revolt arose among them, led by General Phocas, who marched the renegade troops against Constantinople in 602. Maurice responded by abdicating, hoping thereby to restore peace. But Phocas was too ambitious and ordered the death of Maurice as well as his two sons.

MAXIMILIAN I, Duke and Elector of Bavaria

BORN: April 17, 1573

DIED: September 27, 1651

SERVICE: War of the Cleves–Jülich Succession (1617); Thirty Years' War (1618–48)

German ruler who was an effective military champion of the Roman Catholic side during the Thirty Years' War

Munich-born, Maximilian was the son of Duke William V and Renée of Lorraine. He was sent to the Jesuit college of Ingolstadt for his education and then assumed the ducal throne of Bavaria after his father abdicated in 1597.

On July 10, 1609, Maximilian founded the Catholic League in reaction to the Protestant union of Anhausen, which had been established on May 14, 1608. He then placed his brother Ferdinand as archbishop and elector of Cologne, also securing for him a number of additional sees following the death of his uncle Ernst in 1612. Maximilian continued to consolidate his family's power by using the Catholic League's army, led by General **Count Johan Tilly,** to take Jülich, the Duchy of Berg, and the Duchy of Ravenstein, all of which he turned over to his brother-in-law, Wolfgang William of Palatinate-Neuberg in 1614. (This was the principal action of the War of the Cleves-Jülich Succession.)

Maximilian I. From Charlotte M. Yonge, Pictorial History of the World's Great Nations, *1882.*

In 1617, Maximilian founded a new League to shut out the Hapsburgs, whose influence had become too strong in the original Catholic League. Moreover, in 1619, Maximilian essentially blackmailed the Holy Roman emperor, Ferdinand II, by withholding aid to him in his war against the Bohemian rebels until he agreed to withdraw from the Catholic League. In the summer of 1620, Maximilian accompanied Tilly's army on its invasion of Bohemia, and he fought at the battle of White Mountain on November 6 as nominal commander of the League army. After the Bohemians were suppressed, Maximilian acquired control of the Upper and Rhenish Palatinates and, in 1623, was made elector of the Palatine of the Rhine, an office formerly held by Frederick V, who had been king of Bohemia and head of the Protestant opposition to Catholic Austria.

Becoming closely allied with the Holy Roman emperor, Maximilian vied for power with allied closely with the Emperor, his position was threatened by **Albrecht von Wallenstein,** the Bohemian soldier-statesman who acted as commanding general of armies in the service of the Holy Roman emperor. Fearing the independence both of Wallenstein and his troops, Maximilian managed to get him dismissed.

Wallenstein commanded the Bavarian army after the death of Tilly in April 1632. Failing to stop the Swedish invasion of Bavaria, he reluctantly joined forces with Wallenstein at Alte Veste late in August of 1632. Within two years, however, Maximilian engineered Wallenstein's downfall.

Maximilian concluded the Peace of Prague, a partial and mildly favorable settlement of the issues behind the Thirty Years' War, in 1635. He then joined forces with France during September–October 1648, managing to hold on to all he had gained in the war, except for the Rhenish Palatinate.

Maximilian was a skilled and devious statesman, who developed a grand strategy and adhered to it, even at the expense of plunging his people and much of Europe into the costly Thirty Years' War.

MAZEPA (Mazeppa), Ivan Stepanovich

BORN: ca. 1644
DIED: September 8, 1709
SERVICE: Great Northern War (1700–21)

Hetman (leader) of the Cossacks of the Russian Ukraine who turned against the Russians and joined the Swedes during the Great Northern War

Mazepa was born on his family's estate, Mazepintsy (near Belaya Tserkov) and served as a page to Polish King Casimir beginning in 1659. Casimir sent the boy to western Europe for an education and on his return to the Ukraine, in 1663, Mazepa entered the service of Petro Doroshenko, hetman of Ukrainian Cossacks and ally of the Turks. For reasons that are unclear, Mazepa left Doroshenko in 1674 to join the service of another hetman, Ivan Samoilovich, who was the pro-Russian and anti-Turk. In 1682, Mazepa became adjutant general of Ivan's Cossack band and, in 1682, succeeded Ivan as hetman of the Cossacks of the left bank of the Dnieper River.

Mazepa threw his support behind the Russian prince V. V. Golitsyn's second military expedition against the Crimean Tatars (vassals of the Turks) in 1689. Like the first expedition of 1687, this proved a disastrous failure, and Mazepa restored himself to the favor of Czar **Peter the Great** through an adroit diplomatic effort.

But Mazepa was no friend of the Russians, and he increasingly resisted Russian influence in the Ukraine, especially when Peter began requiring Cossack troops to serve in posts distant from their homeland. Moreover, Russian taxes on the Ukraine had become oppressive. Mazepa had been in communication with Sweden's bellicose King **Charles XII** as early as 1700, with the outbreak of the Great Northern War. During the summer of 1707, he openly sided with him and took his troops over to Charles when Peter called on his Cossacks for support *against* the Swedes in October 1708.

Most of Mazepa's army did not share their leader's hatred of the Russians, and they looked at their hetman's allegiance to Sweden as a betrayal. All but 700 of Mazepa's 5,000 troops left him by the time he fought the Battle of Poltava on July 8, 1709. Severely outnumbered, he was beaten, but managed to escape and was granted asylum with Charles XII among the Turks. He died a short time later, at Bendery, Moldavia.

MCCLELLAN, George Brinton

BORN: December 3, 1826
DIED: October 29, 1885
SERVICE: U.S.–Mexican War (1846–48); Civil War (1861–65)

For a time, the Union army's overly cautious general in chief during the Civil War

McClellan was a native of Philadelphia and graduated, second in his class, from West Point in 1846, just in time to serve under **Winfield Scott** in the U.S.–Mexican War. McClellan was commissioned a second lieutenant in the engineers. He distinguished himself at Contreras and Churubusco (August 19–20, 1847), for which he was breveted first lieutenant, then fought with conspicuous gallantry once again at

George B. McClellan. Courtesy Library of Congress.

Chapultepec (September 13) and was promoted to the brevet rank of captain.

After the war, McClellan returned to West Point as an instructor in engineering (1848–51), then was assigned as chief engineer for the erection of Fort Delaware (near Delaware City, Delaware) during 1851–54. Promoted to the regular rank of captain, he was assigned to cavalry in March 1855 and was dispatched to the Crimea as a U.S. observer of the ongoing Crimean War. Even as a cavalryman, McClellan thought like an engineer, and he designed a new saddle, which was adopted by the army in 1856. The McClellan saddle is still widely used.

McClellan resigned his commission in January 1857 to become chief engineer of the expanding Illinois Central Railroad, then left that position to become president of the Ohio & Mississippi Railroad in 1860. However, with the outbreak of the Civil War, McClellan accepted an appointment as major general of Ohio Volunteers in April 1861. The next month he was commissioned a major general of regulars and given command of the Department of the Ohio.

His first engagement, at Rich Mountain on July 11, resulted in a victory that secured for the Union what subsequently became West Virginia. This was great news in the context of the gloomy dispatches that reached the North during the early months of the war, and McClellan was hailed in the popular press as the "Young Napoleon." He was tapped for service as commander of the Department (later called Army) of the Potomac in July, and then was chosen by **Abraham Lincoln** to replace the aging Winfield Scott as general in chief of the army in November.

McClellan worked wonders with the Army of the Potomac, transforming it from an inefficient rabble to a well-trained and highly capable force. However, while McClellan proved an excellent administrator and organizer, he was habitually overcautious as a field commander. He launched the Peninsular campaign with the object of taking Richmond, but repeatedly hesitated, always asking for reinforcements and grossly overestimating Confederate strength. In March 1862, Lincoln, complaining that the "Young Napoleon" had a bad case of "the slows," relieved him as general in chief, though he remained in command of the Army of the Potomac.

During April 5–May 4, he laid siege to Yorktown, which finally fell to him. He successfully repulsed **Joseph E. Johnston** at Fair Oaks on May 31, then fought **Robert E. Lee** to a draw at Mechanicsville on June 26, but refused to press his advantage, choosing instead to withdraw. He then fought the bloody series of battles known as the Seven Days during June 25–July 1, ultimately driving Lee back at Malvern Hill.

McClellan was put in charge of preparing defenses for Washington, D.C., which was continually threatened by the Confederates. He checked Lee's ill-conceived invasion of Maryland, defeating the Confederates at South Mountain on September 14 and earning a terribly costly victory at Antietam on September 17. Yet, once again, he did not capitalize on what he had gained. He did not press his advantage, and in November 1862, McClellan was relieved of command by President Lincoln.

In 1864, McClellan ran against Lincoln as Democratic candidate for president and was soundly defeated. Having resigned his commission when he ran for president, he now took a protracted European tour. In 1870, he accepted a position as chief engineer for New York City's Department of Docks and simultaneously served the Atlantic & Great Western Railroad as trustee, then president (1872). He was elected governor of New Jersey in 1877, serving until 1881.

McClellan was in many respects an admirable military man: a fine organizer, a competent strategist, and a good motivator of his men—who affectionately called him "Little Mac." However, he lacked dash and decisiveness. His caution was crippling, and his adversary, Lee, was quick to comprehend this. Lee readily outgeneraled him.

MCCULLOCH, Ben

BORN: November 11, 1811

DIED: March 7, 1862

SERVICE: Texan Revolution (1835–36); Indian Wars (1840); U.S.–Mexican War (1846–48); Civil War (1861–65)

Hero of the Texan Revolution and U.S.–Mexican War, and Confederate general in the Civil War

McCulloch was born in Rutherford County, Tennessee, and moved with his family to Alabama when he was nine years old. Ten years later, he returned to Tennessee, where he was befriended by Davy Crockett, whom he followed to Texas. Immediately, he was plunged into the Texan Revolution and participated in the victory against General **Santa Anna** at the Battle of San Jacinto on April 21, 1836.

Returning to Tennessee briefly, he settled near Gonzales, Texas, and made his living as a surveyor. In 1839, he gained election to the Texas Congress, and the next year fought against Indians who raided the plains. His most significant battle was not in Texas, but at Plum Creek, Colorado, which put an end to the Great Comanche Raid of 1840. Following this, McCulloch became prominent in the Texas militia movement, becoming by May 1846 a major general in command of militia forces west of the Colorado River. In July, he was also commissioned a major of volunteers and personally organized a company of rangers, which he put at the disposal of General **Zachary Taylor** at the Battle of Monterrey (September 20–24) and the Battle of Buena Vista (February 22–23, 1847) during the U.S.–Mexican War.

In 1849, McCulloch went to California in search of gold and became sheriff of Sacramento. He returned to Texas in 1852 and was appointed U.S. marshal for the eastern district of the state in March 1853. (He resigned in 1859.) In 1858, he served as one of two commissioners empowered to negotiate with the Utah Mormons to end the violence known as the Mormon War.

McCulloch enthusiastically supported the secession of Texas from the Union on the eve of the Civil War, and it was he who accepted the surrender of federal stores and troops at San Antonio from Gen. David E. Twiggs, commander of the Department of Texas in February 1861. In May, he was commissioned a brigadier general in the provisional Confederate army, with authority over a large region that included Arkansas and Indian Territory (modern Oklahoma). Under General Sterling Price, he fought in Missouri at Wilson's Creek and was instrumental in checking General Franz Sigel's maneuver to envelope his forces. He linked up with the main portion of Price's army, fighting through to a pyrrhic victory over Captain Nathaniel Lyon on August 10. He then returned to Arkansas, where he was assigned command of a corps under General Earl van Dorn. At the Battle of Pea Ridge, he was killed by a Union sharpshooter.

McCulloch was a frontiersman first and a soldier second. Nevertheless, he was highly skilled in combat and was a most effective leader of men.

MCDOWELL, Irvin

BORN: October 15, 1818

DIED: May 4, 1885

SERVICE: U.S.–Mexican War (1846–48); Civil War (1861–65)

Union general in the Civil War, best remembered for defeats at the two Bull Run engagements

McDowell was born in Columbus, Ohio and received his education in France before returning to the United States to enroll at West Point, from which he graduated in 1838 with an artillery officer's commission. He was dispatched to duty along the Canadian border with the First Artillery Regiment during 1838–41, then returned to West Point as an instructor, serving from 1841 to 1845.

In October 1845, McDowell (who had been promoted to first lieutenant in 1842) was appointed aide-de-camp to General John E. Wool and served under him during the U.S.–Mexican War. Distinguishing himself at the Battle of Buena Vista during February 22–23, 1847, he was breveted to captain, a promotion that became permanent the following year, when he was assigned to the adjutant general's corps. From 1848 until the outbreak of the Civil War, McDowell served in a variety of staff posts. During this period, in 1856, he was promoted to major, and, in 1858–59, he was sent to Europe to study military administration.

With the commencement of the Civil War in 1861, McDowell was jumped to brigadier general and made field commander of the Washington area in May 1861—largely because he had consistently impressed the army's general-in-chief, **Winfield Scott,** and Treasury Secretary Salmon P. Chase. They had chosen a capable man. With great energy and thoughtfulness, McDowell organized an army from newly recruited men and assembled a force of 37,000 around Washington, D.C. by early July. Although he felt that further preparation was necessary—his men were still green, as were many of his officers—the government wanted decisive action against the "rebels." Accordingly, McDowell developed a plan to outflank the army of Confederate general P. G. T. Beauregard at Centreville in order to severe it from its base of supply at Richmond.

McDowell led five divisions toward Manassas, Virginia, on July 16 and occupied Centreville on the eighteenth. He prepared his troops as best he could during the next three days, then crossed Bull Run on July 21. The deliberation with which McDowell proceeded gave Beauregard ample opportunity to redeploy advantageously. Worse, Confederate general **Joseph E. Johnston** arrived with reinforcements from the Shenandoah Valley. The result was a Union defeat that turned into panic as raw recruits fled back to Washington.

Following the Bull Run debacle, McDowell was replaced in August by General **George B. McClellan** as commander of what was now the Army of the Potomac, and McDowell was assigned command of I Corps with the rank of major general of volunteers in March 1862. President **Abraham Lincoln** had sufficient confidence in McDowell to entrust to him the defense of Washington during McClellan's Peninsular campaign in April–July 1862.

During August 29–30, McDowell, now commanding III Corps in General John Pope's Army of the Rappahannock, fought the Second Battle of Bull Run and, once again, suffered defeat. Relieved of command, he was brought before a court of inquiry, which exonerated him. Nevertheless, he was removed from theaters critical to the war. Assigned to command the Department of the Pacific in July 1864, he was reassigned to the Department of California in July 1865, then, after the war, given command of the Department of the East in July 1868. In November 1872, promoted to major general, he was named commander of the Department of the South. From June 1876 until his retirement in October 1882, he returned to command of the Department of the Pacific.

Few officers in the U.S. Army were as well prepared as McDowell at the outbreak of the Civil War. However, he lacked the character and spark that make a great combat commander. Personally unattractive, he was a glutton and moody, aloof to the point of inattentiveness. He was not capable of motivating the inexperienced troops he commanded at First Bull Run.

MCMORRIS, Charles Horatio

BORN: August 31, 1890

DIED: February 11, 1954

SERVICE: World War I (1917–18); World War II (1941–45).

World War II American admiral; directed the Aleutian campaign

McMorris was born in the little Southern town of Wetumpka, Alabama, and gained appointment to the Naval Academy, graduating at the top of his class in 1912. His intellectual accomplishments were widely recognized by his classmates, who affectionately nicknamed him "Soc," short for Socrates. The sobriquet remained with him for life.

McMorris's first assignments were aboard battleships and an armored cruiser in the Atlantic. He participated in the U.S. landing at Veracruz, Mexico, on April 21, 1914. During World War I, McMorris served aboard destroyers in the North Atlantic from January 1917 through June 1919 and was given command of the destroyer *Walke* (DD-34).

For a brief time after the war, he filled a recruiting assignment in Pittsburgh, then was transferred to the Bureau of Navigation during 1919–22. His next seagoing assignment was as executive officer of the minelayer U.S.S. *Burke* (DM-11), on which he served for two years, until 1924.

From 1925 to 1927, McMorris was assigned as an instructor in seamanship at the Naval Academy, then was given command of the destroyer *Shirk* (DD-318) in the Pacific. After skippering this vessel from 1927 to 1930, he returned to the Naval Academy as an instructor in history and English. He left in 1933 to serve as navigator aboard the battleship *California* (BB-44), again in the Pacific, returning to the Bureau of Navigation in 1935.

McMorris left the Bureau to attend the Naval War College in 1937 and graduated a year later, becoming operations officer of the Scouting Force commanded by Admiral Adolphus Andrews. Promoted to captain in January 1941, he was named director of war plans under Pacific Fleet commander Admiral **Husband E. Kimmel** and served in this capacity through the early months of World War II, until April 1942.

In May 1942, McMorris took command of U.S.S. *San Francisco*, a cruiser, and the following month participated in the Marine landings on Guadalcanal, providing naval bombardment support. On October 12, he also took part in action off Cape Esperance.

McMorris was promoted to rear admiral in November and the following month was named to command Task Force 8 from his flagship, U.S.S. *Richmond*. In February of the next year, he bombarded the Japanese-held Aleutian island of Attu, then, off Komandorski Islands on March 25, 1943, intercepted a Japanese relief force bound for the garrison on Kiska Island. Going up against Admiral Boshiro Hosagoya's two heavy and two light cruisers, and four destroyers with two light cruisers and four destroyers, McMorris was badly outgunned. Nevertheless, he made for the Japanese transports. Although he was driven off—with severe damage to one of his cruisers—he forced the Japanese vessels to break off the engagement without supplying Kiska. In this way, his objective had been achieved.

In May 1944, McMorris was appointed chief of the joint staff under Admiral **Chester Nimitz,** commander in charge of the Pacific (CINCPAC) and was instrumental in planning the Central Pacific island-hopping offensives. Promoted to temporary vice admiral in September 1944, he commanded the Fourth Fleet after the war, during February–September 1946, then accepted appointment to the General Board, serving as its president from December 1947 to August 1948. McMorris was promoted to permanent vice admiral in July 1948 and became commander of Fourteenth Naval District and the Hawaiian Sea Frontier. He served in this post until his retirement in September 1952.

MCNAIR, Lesley James

BORN: May 25, 1883

DIED: July 25, 1944

SERVICE: World War I (1917–18); World War II (1941–45)

American general who was one of the principal architects of the World War II U.S. Army

McNair was born in Verndale, Minnesota, and graduated from West Point near the top of his class in 1904. Commissioned a second lieutenant of artillery, he served in Utah, Massachusetts, New Jersey, and Washington, D.C. through 1909, gaining promotion to first lieutenant in June 1905 and captain in May 1907.

After service with the Fourth Artillery Regiment in the West from 1909 to 1913, he was sent to France to observe artillery training techniques, then returned to the United States in time to participate in the expedition to Veracruz (April 30–November 23, 1914). He also took part in General **John J. Pershing**'s Punitive Expedition into Mexico in pursuit of Pancho Villa, during March 1916–February 1917.

McNair was promoted to major in May 1917, then was put on detached duty with the General Staff. He shipped out to France with First Division during World War I, but in August was transferred to General Headquarters, American Expeditionary Force, with the rank of lieutenant colonel. In June 1918, he was promoted to colonel, and in October became a brigadier general—at the time the youngest general officer in the army.

At war's end, McNair was senior artillery officer in the General Staff's Training Section, but then reverted to his permanent rank of major. Upon his return to the United States, he taught at the General Service School (1919–21), then transferred into a staff post in Hawaii, serving there from 1921 to 1924. He returned to the mainland as a professor of military science at Purdue University from 1924 to 1928, when he was promoted to lieutenant colonel and sent to the Army War College. After graduation the following year, he became assistant commandant of the Field Artillery School and also worked with the Depression-era Civilian Conservation Corps.

Promoted to colonel in May 1935, McNair was given command of the Second Field Artillery Brigade in Texas in March 1937, and was promoted once again, to brigadier general (March 1937). Named commandant of the prestigious Command and General Staff School at Fort Leavenworth in April 1939, he served until October 1940, also functioning in the capacity of chief of the newly organized General Headquarters there. His responsibilities included training, organization, and mobilization.

Promoted to major general in September 1940, he was again promoted, to temporary lieutenant general, in June 1941. Later, when General **George C. Marshall** instituted a sweeping reorganization of the army, McNair was named chief of Army Ground Forces in March 1942. Headquartered at the Army War College, McNair directed the expansion of AGF from 780,000 men to its maximum wartime strength of 2.2 million in July 1943. He was an energetic, hands-on commander, who traveled throughout the country and to the various war theaters to ensure that his troops were combat ready. On one of these trips, in Tunisia, he was severely wounded by a shell fragment in 1943.

McNair was dispatched to England in June 1944, to free up General **George Patton** for invasion operations in France. McNair replaced him as commander of the decoy "First U.S. Army Group," designed to mislead the Germans as to the entry point for the D-Day invasion. The next month, while McNair was in Normandy, observing the Eighth Air Force bombing of German positions, he was killed when some of the bombs fell short. It was a great loss to the United States Army, for McNair had streamlined the U.S. Army's traditional two-brigade, four-regiment "square" division into a three-regiment "triangular" division, which proved much more flexible in World War II. He was a brilliant staff officer, who was instrumental in building the wartime army.

MEADE, George Gordon

BORN: December 31, 1815

DIED: November 6, 1872

SERVICE: Second Seminole War (1835–42); U.S.–Mexican War (1846–48); Civil War (1861–65)

Union general in the Civil War, whose single most significant victory was at Gettysburg

Meade was born in Cadiz, Spain, to the family of an American naval agent. He grew up in Philadelphia and attended West Point, graduating in 1835, with a commission in artillery. He was dispatched to Florida, where he served during the opening of the Second Seminole War, but was soon felled by fever and had to return north. He was so debilitated from this illness that he resigned his commission on October 26, 1836 and became a civil engineer. He rejoined the army on May 10, 1842, and was once again commissioned a Second lieutenant—this time of engineers. Four years later, with the outbreak of the U.S.–Mexican War, he served under General **Zachary Taylor** and fought at Palo Alto (May 8, 1848) and Resaca de la Palma (May 9). He performed with distinction at Monterrey during September 20–24, for which he was breveted to first lieutenant.

With the conclusion of the war, Meade did survey work for the army in the Great Lakes region, in Philadelphia, and in Florida. In August 1851, his promotion to first lieutenant was confirmed, and he made captain in May 1856. With the start of the Civil War, he was jumped to brigadier general of the Pennsylvania volunteers and was rushed to an assignment as one of the defenders of Washington. He then served under **George McClellan** in the Peninsula campaign (April–July 1862), but was badly wounded on June 30 at Frayser's Farm, outside of Richmond, during the Seven Days.

Meade recovered by late summer and was present at Second Bull Run during August 29–30, after which he was promoted to command a division. He fought at South Mountain, Maryland, on September 14, then took command of I Corps at Antietam after General **Joseph Hooker** was disabled on September 17. In November, Meade was promoted to major general of volunteers and given command of Third Division, I Corps at Fredericksburg, where his troops eked out a temporary success on the Union left (December 13).

Following Fredericksburg, Meade was given command of V Corps, then commanded the Center Grand Division (III and VI Corps) of the Army of the Potomac in December, but, with the recovery of Hooker, reverted to command of V Corps.

Meade was present at Chancellorsville during May 1–6, 1863, but was not heavily engaged. However, after Hooker's miserable performance there, Meade was named to replace him as commander of the Army of the Potomac. In the meantime, **Lee** invaded Maryland on June 28, and then Pennsylvania. At Gettysburg, Pennsylvania, Meade commenced an entirely unplanned battle with Lee. Like Meade, Lee had not planned on fighting at Gettysburg either, but he believed that a victory there might turn the tide of the war by once and for all breaking the Northern will to fight. He therefore committed his forces to it in a major effort.

Meade's strategy was defensive, and it was brilliantly executed. With an engineer's eye, he sized up the terrain and used it as an ally during the July 1–3 battle. However, after the disastrous failure of **George Pickett**'s charge—the final Confederate attack—Meade failed to press a counterattack. He then chased Lee's beaten army cautiously, failing to inflict a decisive blow against it. Nevertheless, Meade was promoted to regular army brigadier general as of July 3 and remained in command of the Army of the Potomac. When General **Ulysses S. Grant** became general in chief and field commander in March 1864, Meade's authority was reduced to that of Grant's executive officer.

A man with a greater ego than Meade would have resigned or protested. Instead, Meade served his country and worked very well with Grant. He took Grant's direction at the Wilderness (May 5–6), Spotsylvania (May 8–18), and Cold Harbor (May 31–June 12), but here finally persuaded Grant to refrain from further frontal attacks, which were proving costly and unproductive.

Still commanding the Army of the Potomac at Petersburg during June 15–18, he directed the long and arduous trench warfare campaign around that city from July to April of the next year. He also directed the final operations of the army during the Five Forks–Appomattox campaign of March 30–April 9 and was among those present when Grant accepted Lee's surrender at Appomattox on April 9, 1865.

Meade remained with the army after the war as commander of the Third Military District (encompassing Alabama, Georgia, and Florida), and was one of very few Northern military governors who attempted to discharge their duty with fairness and humanity. In 1869, Meade became commander of the Military Division of the East. Unfortunately, Meade never fully recovered from the wound he received early in the war. As a result of this injury, he developed pneumonia in 1872 and succumbed to it.

Like George McClellan, Meade was highly intelligent and thoroughly capable, but, also like McClellan, he was overly cautious. While he conducted the Gettysburg campaign with brilliance, his failure to capitalize on the victory may well have extended the war unnecessarily.

MEDICI, Giovanni de' ("The Invincible," "Giovanni of the Black Bands")

BORN: April 6, 1498

DIED: November 30, 1526

SERVICE: First Hapsburg–Valois War (1521–25) Second Hapsburg–Valois War (1526–29)

Italian* condottiere *(mercenary leader) who was the valiant and skillful captain of forces in service to Pope Leo X

Son of Giovanni ("il Popolano") de' Medici and Caterina Sforza, Giovanni de' Medici was early schooled in the brutal and devious ways of Italian Renaissance politics. Cesare Borgia imprisoned his mother in 1500, and the young man had ample exposure to the machinations of the Medici in Florence. His first military command was of a cavalry company in Giuliano de' Medici's campaign against Francesco Maria della Rovere during March–June 1516. Pope Leo X commissioned Giovanni to capture the city of Fermo, which he accomplished with 4,000 infantryman and a handful of cavalry troops. During the late summer and autumn of 1521, Giovanni served under Prospero Colonna against the French surrounding Milan. This action culminated in the capture of Milan on November 23, 1521 and thereafter Giovanni was known as "the Invincible." With the death of Pope Leo on December 1, 1522, Giovanni ordered his troops to wear black armbands and carry black banners. After this, they were known as the Black Bands, and Giovanni had earned another sobriquet.

His forces participated in Prospero Colonna's victory of La Bicocca on April 26, 1522, but, after the French withdrew, he suffered the fate that befalls most mercenaries at one time or another: lack of employment. Nevertheless, Giovanni maintained a force of more than 3,000 men and six guns at Reggio nell' Emilia, paying them himself against the day when he would be called to service again. The call came in the spring of 1523, when the Medici pope Clement VII needed him to repel another French invasion. He again served under Colonna, and, following Colonna's death, under Charles de Lannoy. He was successful at repelling the French and fought with distinction at the battle of the Sesia on April 3, 1524. In the fall, however, Pope Clement transferred his allegiance from Charles V to Francis I, and Giovanni likewise switched to the French side.

He served at the siege of Pavia during October 27, 1524–February 24, 1525. Wounded in the foot, he was given a safe-conduct pass to Piacenza, so avoided being caught up in the French defeat at Pavia. When the war heated up again in the spring of 1526, he commanded a body of papal troops under Francesco Maria della Rovere, Duke of Urbino. A cannonball hit his thigh, necessitating the amputation of his leg. It is reported that Giovanni himself held the candle to illuminate the surgery. As was often the case after such a procedure, infection and gangrene supervened, to which Giovanni succumbed.

Giovanni de' Medici never commanded a vast army, but he was one of Renaissance Italy's most skillful and courageous soldiers. His speciality was the hit-and-run raid.

MERRILL, Frank Dow

BORN: December 4, 1903

DIED: December 11, 1955

SERVICE: World War II (1941–45)

World War II American general in the arduous Burma campaign

Merrill was born in Hopkinton, Massachusetts and enlisted in the army in 1922, serving in Panama through 1925, He received an appointment to West Point, from which he graduated in 1929 and was commissioned a second lieutenant of cavalry. During

1931–32, he attended Ordnance School, then, from 1934 to 1935, Cavalry School, in which he became an instructor during 1935–38.

In 1938, he was attached to the U.S. embassy in Tokyo and took the opportunity to study both the Japanese language and the Imperial military organization. Promoted to captain in 1939, he left the embassy assignment the following year and joined the intelligence staff of General **Douglas MacArthur**'s Philippine Command. In 1941, he was promoted to temporary major and was on a mission in Rangoon when the United States entered World War II on December 8, 1941. Remaining there, he joined the command of Lieutenant General **Joseph W. Stilwell** when Stilwell reached Burma with Chinese forces in March 1942.

Merrill served with Stilwell during the first Burma campaign, and accompanied his retreat to India in May. He was promoted to temporary lieutenant colonel at that time and then to full colonel early the following year. Stilwell appointed him to command a provisional U.S. infantry regiment, which he sent into combat in northern Burma as part of the joint American–Chinese offensive to reopen the Burma Road in February 1944. Merrill's troops marched a hundred miles into Burma and spearheaded a broad Chinese–American envelopment action.

During this arduous jungle campaign, Merrill suffered from heart trouble and had to be hospitalized twice. At last, in mid August, he was transferred to lead a liaison group of the Allied Southwest Asia Command in Ceylon and was promoted to major general in September. He was then appointed chief of staff of General Simon B. Buckner's Tenth Army in the Okinawa campaign from April 1 to June 22, 1945, then served under Stilwell again when Stilwell took over after Buckner's death.

After the war, Merrill served as chief of staff of the Sixth Army, headquartered in San Francisco, then, in 1947, was appointed chief of the American Advisory Military Mission to the Philippines. He retired from the army in July 1948 and served the state of New Hampshire as commissioner of roads and public highways.

MIDDLETON, Troy H.

BORN: October 12, 1889

DIED: October 9, 1976

SERVICE: World War I (1917–18); World War II (1941–45)

American general who was praised in World War II for his calm rationality, even as some criticized his lack of drive

Middleton was born in the rural South, near Georgetown, Mississippi, and was appointed principal cadet officer in his graduating class at Mississippi A & M College in 1909. He tried unsuccessfully to secure an appointment to West Point, swallowed his disappointment, and enlisted in the army in March 1910. Two years later, he was commissioned a second lieutenant and, in April–November 1914, participated in the occupation of Veracruz.

Promoted to captain when the United States entered World War I in 1917, he was assigned to training duties before securing promotion to major and shipping out to France in May 1918. There he commanded the First Battalion, Forty-Seventh Infantry, Fourth Division, in the Aisne-Marne counteroffensive during July 18–August 6. He participated only marginally in the offensive against the Saint-Mihiel salient during September 12–16, but won great distinction as lieutenant colonel commanding the Thirty-Ninth Infantry, Fourth Division, during the Meuse-Argonne offensive in October–November.

Following the Armistice, Middleton was promoted to full colonel and served in occupation forces on the Rhine. When he returned to the States in late in 1919, he reverted to captain, then became a member of the first faculty of the new Infantry School at Fort Benning through 1921. During 1921–22, he studied in the advanced infantry course, and in 1924 graduated from the Command and General Staff School, remaining as an instructor from 1924 to 1928.

In 1929, Middleton graduated from the Army War College and was posted to Louisiana State University as commandant of the ROTC program there from 1930 to 1936. He served in the Philippines during 1936–37, then returned to the States and retired from the army.

U.S. Maj. Gen. Troy H. Middleton. Courtesy National Archives.

He was appointed dean of administration at LSU in 1937 and, subsequently, acting vice president and comptroller until he was recalled to active duty, with the rank of lieutenant colonel, in January 1942.

Promoted to brigadier general, Middleton was appointed assistant commander of the Forty-Fifth Infantry Division in June and subsequently received promotion to major general in command of the Forty-Fifth. He led this division in combat in Sicily, receiving accolades for the performance of his troops during July 10–31, 1943.

Middleton's men participated in the Salerno landings during September 10–14, falling under heavy German counterattack on the twelfth and thirteenth. Middleton's unflappable leadership was crucial in averting disaster and ultimately bringing success. After leading his division through a month and a half of difficult campaigning along the Volturno River during October–November, he was shipped back to the States for treatment of a serious knee injury and was next sent to Britain, where he was given command of VIII Corps in March 1944. The corps landed at Normandy on June 12, and Middleton was in command at the Saint-Lô breakthrough during July 25–August 1.

General **George Patton** counted Middleton as one of his best corps commanders and sent him to clear Brittany. He began by laying siege to Brest from August 9 to September 19. In November, however, he was transferred to the main front in eastern France, where his corps was assigned to hold an eighty-eight-mile sector in what was deemed the relatively quiet Ardennes. This turned out to be the focus of the Battle of the Bulge (December 16, 1944–January 15, 1945), the last major, desperate German offensive of World War II. With only 70,000 men, either war weary or green, Middleton's VIII Corps took the main thrust of the initial offensive. Middleton handled the crisis with his customary calm efficiency, holding his sector, and preventing a rout.

During January–March 1945, Middleton was instrumental in the conquest of the Rhineland and the lightning advance across Germany during March–May.

With the end of World War II, Middleton again retired from the army—as lieutenant general—and resumed employment with LSU as that institution's comptroller. In 1951, he was appointed president of the university and served until 1962.

MILES, Nelson Appleton

BORN: August 8, 1839

DIED: May 15, 1925

SERVICE: Civil War (1861–65); Red River War (1874–75); Sioux War (1876–77); Nez Percé War (1877); War with Geronimo (1882–86); Ghost Dance Uprising (1890); Spanish–American War (1898)

American general; perhaps the army's single most effective Indian fighter

Miles was born near Westminster, Massachusetts, and attended school there and in Boston. He was in Boston at the outbreak of the Civil War and, always ambitious and with full confidence in his abilities, borrowed funds to raise a volunteer company. Soon, however, he was commissioned in the regular army as a lieutenant in the Second Massachusetts Regiment in 1861 and was quickly chosen as an aide to General **Oliver O. Howard,** accompanying Howard during **George McClellan**'s Peninsula campaign (April–July 1862).

Miles was a dashing officer of boundless courage, and promotion came swiftly. He was promoted to lieutenant colonel in the field at Fair Oaks on May 31 and was then jumped to colonel when he took command of the Sixty-First New York at Antietam on September 17. Wounded at Fredericksburg on December 13, he was fit for duty again in the spring of 1863 and fought at Chancellorsville during May 2–4. He was again wounded in that battle and was later awarded the Medal of Honor, in 1892.

Promoted to brigadier general of volunteers in May 1864, Miles commanded a brigade in the Battle of the Wilderness during May 5–6; at Spotsylvania, May 8–18; Cold Harbor, May 31–June 12; and at Petersburg, June 15–18. He fought with such distinction at Ream's Station, near Petersburg, on June 22, that he was breveted to major general in August and, during March–April 1865, was given command of a division during the last phase of operations around Petersburg.

In the aftermath of the war, during October 1865, Miles was promoted to major general of volunteers and made commander of II Corps. He also served as commandant of Fort Monroe, Virginia, and was therefore assigned as the jailer of the Confederacy's president, Jefferson Davis. He did not treat Davis kindly, but confined him, shackled, in a dark and miserable cell. This act raised indignation, not only in the South, but in the North as well.

Miles earned a commission as colonel in the regular army and was assigned in July 1866 to command the newly formed Fortieth Infantry, an African American regiment. This assignment displeased Miles, as did the duties of civil government and police work required by Reconstruction policies. Miles lobbied for a change of command, and, during this period, in 1868, he made and advantageous marriage, to Mary Sherman, niece both of Senator John Sherman of Ohio and his brother, General **William T. Sherman.** The following year, in March 1869, Miles was given command of the Fifth Infantry in the West, an assignment very much to his liking.

The colonel took charge of one of four columns sent against the Comanche, Kiowa, and southern Cheyenne during the Red River War on 1874–75. When **George Armstrong Custer** and his command were annihilated at the Little Bighorn on June 25, 1876, Miles was one of few army officers who remained undemoralized and eager to take the offensive. His vigorous campaigning against the Sioux and northern Cheyenne resulted in their withdrawing to assigned reservations or fleeing into Canada. On January 8, 1877, he defeated Crazy Horse—one of those closely associated with the Custer battle—at Wolf Mountain (near the present Miles City, Montana).

Miles's next major victory came when General **Oliver O. Howard** called him to Eagle Creek in the Bear Paw Mountains (near Chinook, Montana) to coordinate with him in the encirclement of a band of Nez Percé led by **Chief Joseph.** The battle, which took place in driving snow during October 1–5, 1877, resulted in the surrender of Joseph and his band, which withdrew to a reservation.

Miles was promoted to regular army brigadier general in 1880 and replaced General **George Crook** as commander of U.S. forces in Arizona. Miles took up the pursuit of **Geronimo** and his band of renegade Apaches with great singleness of purpose during 1885–86. However, Geronimo repeatedly evaded capture. In August 1887, he finally persuaded a disheartened and exhausted Geronimo to surrender, and the Indian leader was taken to confinement in a federal stockade in Florida.

Promoted to major general in 1890, Miles commanded the final operation of the Indian Wars. He directed the suppression of the much-feared Ghost Dance cult uprising in 1890. The massacre of Sioux at Wounded Knee Creek, perpetrated by the Seventh Cavalry under Colonel James W. Forsyth, had not been ordered by Miles, who, in fact, condemned it and relieved Forsyth of command following it. Miles ordered a court of inquiry and was outraged when the court exonerated Forsyth. Over Miles's protests, General John M. Schofield, Miles's superior, reinstated

Forsyth. Nevertheless, Miles's public reputation was badly tarnished by Wounded Knee, although his military career forged ahead. In 1895, he was appointed commanding general of the army, in which post he had a stormy relationship with Secretary of War Russell A. Alger. In bad odor with the civilian administration, he was denied a major role in the Spanish–American War, which commenced in April 1898. He did direct the conquest of Puerto Rico, an operation he planned and executed with great precision during July 25–August 13.

The Spanish–American War occasioned yet another dispute between Miles and the War Department, which he publicly accused of supplying spoiled beef to the troops in Cuba—charges that were subsequently substantiated and led to a major reform movement not only in government, but in federal regulation of food handling and processing.

Miles was promoted to lieutenant general in February 1901, but could not leave controversy behind. He was reprimanded for publicly interfering in navy business by commenting on Admiral **George Dewey**'s report on charges against Admiral Winfield S. Schley. He also insisted on meeting with rebel leaders while touring the Philippines during the insurrection of 1902—an action that provoked official outrage. At last, when he opposed Secretary of War Elihu Root's plans for wholesale army reform in 1903, Miles was forced into retirement.

Miles was an extraordinarily energetic and aggressive field commander, who was driven by boundless confidence and a degree of arrogance that matched it. Talented and highly capable, his abrasive personality outraged subordinates and superiors alike. He was particularly resentful of civilian "interference" in military affairs.

MINAMOTO, Yoshitsune (Ushiwaka)

BORN: 1159

DIED: 1189

SERVICE: Gempei War (1180–85)

Military hero of medieval Japan

The fifth son of Yoshitomo Minamoto, he was an infant when his father was assassinated by one of his own men. The early life of Minamoto is shrouded in legend, including tales of his having mastered swordsmanship under the tutelage of wood goblins. His brother Yoritomo called on him to fight alongside him against the dominant Taira clan in 1180, and, after Yoritomo relinquished field command, Yoshitsune and his cousin Yoshinaka allied themselves against the Taira during 1181–83.

When Yoshinaka succeeded in taking Kyoto, Yoritomo suspected him of greater ambition and dispatched Yoshitsune and his older brother, Nonyori, against Yoshinaka. A number of battles developed around Kyoto during 1184. Ultimately, Yoshitsune emerged victorious and Yoshinaka was slain.

Having finished off a potential rival, Yoshitsune then attacked the Taira in their fortified camp along the coast near present-day Kobe. Yoshitsune led a small force over a mountain and down a treacherous path in order to achieve the element of surprise by attacking from an unexpected direction. The victory at Ichinotani (1184) was devastating, and he succeeded in defeating the Taira again at Yashima on Shikoku early the next year. He then gave chase, pushing the Taira to Danno-ura, off the western tip of Honshu. When they turned to fight, he destroyed their army as well as their fleet in April 1185.

Yoshitsune was showered with treasure and high court rank by the retired Emperor Go-Shirakawa. But his brother Yoritomo grew as jealous and as suspicious of him as he had of Yoshinaka. Yoritomo isolated and undermined his brother's political position. Ultimately, Yoshitsune had to flee the court to preserve his life. By 1188, he found refuge with Hidehira Fujiwara, the last remaining powerful clan outside Minamoto's control, in Mutsu province on northern Honshu. The following year, Hidehira's successor, Yasuhira, betrayed Yoshitsune to his brother, and the warrior was slain at the Koromogawa River. (He may have died by his own hand rather than allow himself to be captured.)

Yoshitsune is celebrated in traditional Japanese literature and lore. Although it is not always possible to separate fact from fantasy, it is clear that he was a commander of great skill and daring.

MITCHELL, William ("Billy")

BORN: December 29, 1879

DIED: February 19, 1936

SERVICE: Spanish–American War (1898); World War I (1917–18)

American Army officer who was an early and controversial champion of U.S. military aviation

Mitchell was born, in Nice, France, to John Lendrum Mitchell, who would become U.S. senator from Wisconsin. The young man was raised in Milwaukee and educated at Racine and Columbian (later George Washington) Universities, then enlisted in the First Wisconsin Infantry at the outbreak of the Spanish–American War in 1898. He fought in Cuba, rising to lieutenant of volunteers and was subsequently commissioned a lieutenant in the regular army. Assigned to the Signal Corps, he attended the Army Staff College at Fort Leavenworth, Kansas during 1907–09, then served for a short time on the Mexican border before securing an assignment to the General Staff in 1912.

In 1915, Mitchell resigned his coveted staff post to transfer to the aviation section of the Signal Corps. (At this time, the army's infant aviation program was the province of the Signal Corps.) He enrolled in flight school at Newport News, Virginia, and got his wings in 1916. In that year, he was sent to Europe as an observer, and, in April 1917, with U.S. entry into World War I, he was appointed air officer of the American Expeditionary Force and promoted to lieutenant colonel in June. In May 1918, he became air officer of I Corps with the rank of colonel and was the first U.S. officer to fly over enemy lines.

Mitchell commanded a successful French–American bombing mission consisting of 1,500 aircraft—the greatest number of planes ever massed to that time—against the Saint-Mihiel salient in September. The mission demonstrated just what air support of ground action could do. Appointed to command the combined air services for the Meuse-Argonne offensive—now with the rank of brigadier general—Mitchell led another massive formation of bombers against targets behind enemy lines on October 9.

With the conclusion of World War I, Mitchell was named assistant chief of the Air Service in 1919 and embarked on a controversial campaign to create a separate air force. He also advocated unified control of military air power. Both proposals were stoutly resisted by the military establishment. Worse, he outraged navy officials when he boasted that the airplane had made the battleship obsolete. In 1921, he demonstrated his point by bombing the captured German dreadnought *Ostfriesland*. Witnesses were impressed by the spectacle of Mitchell's bombers sinking the vessel in 21 1/2 minutes, though, in truth, the ship had been less successfully bombarded from high attitude for a longer period and was already taking on water when it was finally sunk. This spurred the navy to conduct further tests during 1923, the results of which initiated the navy's development of the aircraft carrier as an offensive weapon.

Mitchell was relentless in his campaign for enlargement of the present Air Service and creation of an independent air force. Frustrated superiors saw to his demotion to colonel and reduced responsibility as air officer of the VIII Corps Area in San Antonio, Texas (April 1925). Despite this, he kept the controversy alive, making many statements to the press. When the navy dirigible *Shenandoah* crashed in a thunderstorm on September 3, 1925, Mitchell went to the papers with accusations of War and Navy Department "incompetency, criminal negligence, and almost treasonable administration of the National Defense."

This proved the final straw. Mitchell was court martialed and convicted of insubordination in December 1925 and sentenced to five years' suspension from duty without pay. Instead of accepting this, he resigned his commission on February 1, 1926, continuing to speak out from his Middleburg, Virginia, home until his death a decade later.

Clearly, Billy Mitchell was ahead of his time, and most of his theories about the role of aviation in warfare proved true—including a remarkable (and much scoffed-at) assessment that the navy's fleet at Pearl Harbor in the Hawaiian Islands was vulnerable to carrier-launched air attack and that the attack would be made by Japan. His inpolitic and caustic manner did not help his cause. However, after death, his positions were vindicated, and he is considered the founding father of the U.S. Air Force. His memory was honored when the World War II B-25 bomber was named for him: the Mitchell bomber.

MITSCHER, Marc Andrew

BORN: January 26, 1887

DIED: February 3, 1947

SERVICE: World War I (1917–18); World War II (1941–45)

American admiral and naval aviation pioneer

Mitscher was a Midwesterner—born in Hillsboro, Wisconsin, and raised in Oklahoma City—but he had a passion for the sea and gained admission to the U.S. Naval Academy. After graduating in 1910, he sailed aboard two armored cruisers. He was serving on the U.S.S. *California* during 1913–15 when he took part in the landings at Veracruz in April 1914. The next year, he took flight training at the Pensacola Naval Air Station and earned his wings in June 1916. He stayed in Pensacola for advanced flight training and served aboard the attack cruiser *Huntington*, based at Pensacola, performing balloon and aircraft catapult experiments during April 1917.

Following a stint of convoy escort duty in the Atlantic after the United States entered World War I, Mitscher was posted to Montauk Point Naval Air Station on Long Island, New York, then was appointed to command the Rockaway, Long Island, NAS in February 1918. The following year, he was transferred to command of the Miami NAS.

In May 1919, Mitscher attempted to fly across the Atlantic, but made it only as far as the Azores—a feat for which he received the Navy Cross. In the winter of 1920, he transferred to the Pacific as commander of the Pacific Fleet's air unit based in San Diego. He then took charge of Anacostia NAS in Washington, D.C. and also served with the Plans Division of the Bureau of Aeronautics during 1922–26. During this period, in 1922, he led the Navy team at the international air race at Detroit and, in 1923, at St. Louis.

From July to December 1926, Mitscher served aboard U.S.S. *Langley*, the navy's first aircraft carrier, in the Pacific. He transferred to the U.S.S. *Saratoga* for precommissioning duty and was appointed the ship's air officer when she entered the fleet in November 1927. Promoted to commander in October 1930, Mitscher returned to shore duty in Washington at the Bureau of Aeronautics, serving until 1933, when he was named chief of staff to Base Force commander Admiral Alfred W. Johnson and served aboard the seaplane tender *Wright* for a year before being appointed executive officer of the *Saratoga* in 1934. Once again, Mitscher returned to the Bureau of Aeronautics as leader of the Flight Division from 1935 to 1937.

Mitscher was given command of U.S.S. *Wright* late in 1937 and was promoted to captain the following year. He then took command of Patrol Wing 1, operating out of San Diego. Then, during June 1939, Mitscher was appointed assistant chief of the Bureau of Aeronautics. He served in this post until the eve of World War II, when he was given command of the new aircraft carrier U.S.S. *Hornet* in July 1941; it was Mitscher who brought the ship into commission in October, and it was Mitscher who was in command when Army Air Corps colonel **Jimmy Doolittle** used the carrier as the base from which he launched sixteen B-25 bombers on his celebrated raid against Tokyo in April 1942.

Mitscher was skipper of the *Hornet* at the turning-point battle of Midway during June 3–6, 1942, and was promoted to rear admiral and commander of Patrol Wing 2 in July. In December, he was appointed commandant of Fleet Air, based at Noumea, and when U.S. forces took Guadalcanal, he moved his base there in April 1943. Mitscher next directed combined operations of army, navy, marine, and New Zealand air units during the Solomons campaign before returning to sea duty as commander of the Fast Carrier Task Force, which operated against Japanese positions in the Marshall Islands, Truk, and New Guinea from January to June of 1944.

Promoted to vice admiral in March 1944, Mitscher took charge of carrier operations at the Battle of the

Philippine Sea and succeeded in decimating the Japanese carrier force in the celebrated "Marianas Turkey Shoot" of June 19–21. His next assignment was support of amphibious landings at the Bonins and Palau during August and September, then he was assigned command of air operations to provide cover for the landing at Leyte, Philippines, in October. During the Battle for Leyte Gulf (October 24–26), Mitscher directed carrier operations that resulted in the destruction of most of the remaining Japanese carriers.

Mitscher played a supporting role at Iwo Jima in February 1945 and at Okinawa in April. In the Battle of the East China Sea, on April 7, his carriers sank the battleship *Yamato* and most of her escorts. In July, Mitscher returned to Washington, D.C., as deputy chief of Naval Operations (Air) and, shortly after the war, in March 1946, he was promoted to admiral. He commanded the Eighth Fleet briefly before his death, at age sixty, from illness.

Slight and of diminutive stature, Mitscher did not look the part of daredevil aviator or highly placed commander. However, he was a brilliant tactician and a true pioneer of naval aviation. He was especially noted for his uncanny calm and for his concern about the safety of his aviators.

MODEL, Walther

BORN: January 24, 1891

DIED: April 21, 1945

SERVICE: World War I (1914–18); World War II (1939–45)

Nazi field marshal of World War II whose problem-solving zeal earned him the nickname of "The Fürhrer's Fireman"

Model was born in Genthin near Magdeburg. His father was a music teacher, but young Model resolved on a career as a professional soldier and joined the army in 1909. By the time of World War I, he was serving in a number of staff and adjutant positions and was decorated with the Iron Cross in October 1915. Following the war and the restrictive terms of the Treaty of

Walther Model. Courtesy National Archives.

Versailles, Model was selected as one of the small group of 4,000 officers to serve in the *Reichswehr.*

Model was early on attracted to **Adolf Hitler** and Naziism. He joined the Nazi Party and, in 1935, gained command of the Army General Staff's Technical Department. He was promoted to *generalmajor* on March 1, 1938, and the following year was given command of IV Corps in the blitzkrieg invasion of Poland (September–October 1939). After promotion to *generalleutnant* on April 1, 1940, he took command of the Third Panzer Division, which he led through Flanders and France during May–June 1940.

Model was named *general der panzergruppen* and sent to the Russian front on October 1, 1941. He briefly commanded XXXI Panzer Corps and was then promoted to *generaloberst* on February 28, 1942, with command of the Ninth Army, which he led from January 1942 to January 1944. He was in command of the northern arm of what turned out to be an abortive offensive at Kursk during July 5–12, 1943, and was then assigned command of Army Group North from January

1994 to March, when he was promoted to *generalfeldmarschall* and named to command Army Group North Ukraine on March 31.

In the final stages of the war, on June 28, 1944, Model was given command of what remained of Army Group Center. In the wake of the July 20 assassination attempt on Hitler—in which a cabal of officers attempted to blow him up with a briefcase bomb—Model affirmed his loyalty to the Führer and was named commander of Army Group B and commander in chief (OB) West on August 17, 1944. The latter part of this appointment proved short lived, and, on September 5, Model's authority reverted to command of Army Group B only.

It was Model who checked the Allied attack at Arnhem during September 17–26, and it was he who directed the final, desperate German offensive of World War II, at the Ardennes—the so-called Battle of the Bulge—from December 16, 1944 to January 15, 1945. By April 1945, Model's army was trapped, surrounded in the Ruhr Pocket. Model put up a valiant, albeit doomed resistance and held out against an Allied pounding for eighteen days. He gave up his 300,000 remaining troops in surrender. However, rather than surrender himself, he committed suicide at Lintorf on April 21.

Unlike many of Germany's best military men, Model was a thoroughgoing Nazi, whose dedication to Hitler was absolute. Model's skill as a general was matched by his energy and near fanaticism. He was among Hitler's most consistently aggressive commanders.

MOHAMMED II THE CONQUEROR

BORN: March 30, 1432

DIED: May 3, 1481

SERVICE: Albanian Wars of Independence (1443–68); Conquest of Serbia and Bosnia (1456–64); Conquest of Greece (1458–60); Venetian War (1463–79); Rhodian War (and Italian expedition) (1480–81)

Ottoman ruler and conqueror of Constantinople

Mohammad II the Conqueror. From Stanley Lane-Poole, The Story of Nations: Turkey, *1888.*

Sultan Mohammed II earned his byname after greatly expanding the Ottoman Empire into Europe, finally taking the city of Constantinople, which had long resisted Ottoman domination. Mohammed II was born at Edirne. He was the eldest son of Murad II and served as sultan for a time when his father retired to Magnesia in 1444, and then assumed the throne permanently after his father died in 1451. Mohammed II presided over a vast empire extending to the east and west of Constantinople. Mohammed's father had been content to allow the city, sole remnant of the once-mighty Byzantine empire, to exist independently, surrounded by his realm. Restless and ambitious, Mohammed resolved to commence his reign by conquering Constantinople.

Mohammed set about methodically to accomplish this formidable task, beginning in 1451–52 by building a fortress, Rumeli Hisar, just outside of the city, effectively covering both sides of the Bosphorus. Byzantine emperor Constantine XI protested this outrage, giving Mohammed just what he wanted: a pretext for a declaration of war. Constantine XI commanded a small army of no more than 10,000 against an array of Ottoman forces that totaled 80,000, including an elite corps of Janissaries and, under the command of renegade Hungarian artilleryman Urban, a siege train of seventy heavy cannon. Yet Constantine enjoyed certain advantages. Constantinople was virtually surrounded by water, and its landward defenses were extremely strong. When Mohammed attempted to push his fleet into the Golden Horn, he was prevented by a boom thrown across the entrance by the city's defenders. Thus stymied, Mohammed decided to transport his fleet overland from the Bosphorous into the Golden Horn. And he built a massive one-mile-long plank road between the Bosphorous and the Horn, greased it with vast quantities of animal fat, and slid eighty vessels across it. Beginning on April 2, 1453, these ships, combined with land-based artillery, set up a withering barrage. The city withstood the bombardment for nearly two months, until May 29, 1453, by which time the artillery had made a sufficient breach to allow a Janissary charge. Constantine XI was killed in the battle that ensued, and, contrary to Mohammed's wishes, his troops pillaged and sacked the city for three days.

Mohammed II set about to repair the ruin and to restore as much of the city as possible, encouraging the learned and the cultured to remain. But the city fell into step decline and decay. While Mohammed had conquered Constantinople, he was largely deprived of its true value.

In 1456, the sultan, now known as El Fatih—the Conqueror—invaded Serbia and laid siege to Belgrade. John Hunyadi defeated him in a naval battle there on July 14, 1556, and on land during July 21–22, but while the sultan withdrew from Belgrade, he overran Serbia during 1457–59. His forces also penetrated southern Greece during 1458–60, crushed the small empire of Trebizond in 1461, and successfully invaded Bosnia during 1463–64. Perpetually hostile toward Venice, which had briefly aided Constantine XI in the defense of Constantinople, Mohammed declared a long war against it in 1463. He next raided Dalmatia and Croatia in 1468, then launched a devastating and brilliant amphibious assault on the Venetian fortress of Negroponte in Euboea (Évvoia), which spanned June 14 to July 12, 1470. He captured the city with few losses to himself. Venetian diplomats, however, persuaded the Persians to attack Mohammed at Erzinjan in 1473. The sultan prevailed against this force, then marched against the Crimean city of Kaffa (Feodosiya). Capturing it from the Genoese.

In the meantime, in 1468, Mohammed reconquered Albania, which had been lost in the rebellion of the Janissary leader Skanderbeg (George Castricata) in 1443. Mohammed went on to capture most of the Venetian ports along the Albanian coast, then dispatched raiders from Croatia across the Alps into Venetia, bringing all of northeastern Italy to its knees. The Venetians at last made peace, agreeing to recognize the Turks' conquests. Early in 1480, Mohammed's forces crossed the Adriatic and seized Otranto, going on to besiege the Knights of St. John on Rhodes during 1480–81. Here the sultan suffered a bad defeat. He withdrew to Tekfur Cauiri to plan a second assault on Rhodes, but fell ill and died there on May 3, 1481.

Mohammed's many military triumphs tend to overshadow his cultural accomplishments. He was among the most cultivated of Turkish sultans, an enthusiastic patron of learning and the arts, whose reign also produced a seminal legal work, the *Quanun-nmae*, which codified Ottoman government institutions and the administration of justice.

MOLTKE, Count Helmuth Johannes Ludwig von

BORN: May 25, 1848

DIED: June 18, 1916

SERVICE: Franco–Prussian War (1870–71); World War I (1914–18)

German general who served as supreme commander in the West during the opening months of World War I

Moltke was born at Gersdorff in Mecklenburg. His uncle was **Helmuth Karl Bernhard von Moltke,** the great field marshal of Otto von Bismarck. This connection would certainly aid in the younger Moltke's career advancement, but it figured even more as an insupportable burden. The younger Moltke was inevitably judged against his uncle—and, inevitably, was always found wanting.

He entered the army in 1870, at the outbreak of the Franco–Prussian War, but saw no combat during the conflict. In 1882, he served as adjutant to his uncle, then saw service with a number of guard units. From 1902 to 1904, he commanded First Division in the Guard Corps and was promoted to *General Quartiernseister*—deputy chief—of the General Staff. Two years later, Moltke succeeded **Alfred von Schlieffen** as chief of the General Staff. As Moltke had inherited his uncle's reputation, so he now inherited the Schlieffen Plan. A strategy for fighting the next European war, the plan called for holding off Russia in the East, while defeating the French in the West with a grand outflanking maneuver through Belgium.

As war approached, it fell to Moltke to adapt the plan in to meet changes in the situation since 1906. This was no simple task; for Moltke was plagued by Austro–Hungarian demands to put more troops in the East. Nor did the legislature and the War Ministry adequately fund the Schlieffen plan. Supplies of weapons and equipment were inadequate. At the same time, Kaiser **Wilhelm II** persisted in saber rattling and posturing, which threatened to bring on war prematurely. Then there was Moltke himself. He was an intelligent and capable officer, but he did not have the strength of character to stand up to the government or the kaiser. The Schlieffen plan, as a result, was compromised and diluted.

With the outbreak of World War I on August 1, 1914, Moltke found himself incapable of directing his armies efficiently. The Schlieffen plan was carried out—to a point—but then Moltke altered the path of advance for General **Alexander von Kluck**'s First Army. Some thirty miles east of Paris, von Kluck was ordered to turn, thereby presenting a flank to the French and giving the Allies an opportunity to counterattack on the Marne during September 4–10. The German advance into and through France had been sweeping and speedy. Moltke's interference destroyed Germany's single chance for a quick victory. The opposing armies dug in, and Europe was devastated by four years of static, unprecedentedly brutal trench warfare.

Moltke was quite rightly held responsible for the failure of the Schlieffen plan in the West and was relieved of all executive responsibility by General Erich von Falkenhayn on September 14. He retained the title of chief of the General Staff for two until November 3 and was then assigned to the post of deputy chief of staff at home—a distinctly non-critical war role. Moltke fell into a profound depression and was dead within two years.

Moltke was a tragic figure. Highly intelligent, he lacked the personality and the combat experience necessary to direct massive movement of troops. Thanks to him, Germany lost the opportunity to win World War I within its first month.

MOLTKE, Count Helmuth Karl Bernhard von

BORN: October 26, 1800

DIED: April 24, 1891

SERVICE: First Turko–Egyptian War (1838–41); Schleswig-Holstein War (1864); Six Weeks' War (1866); Franco–Prussian War (1870–71)

Extraordinarily successful chief of the Prussian and German General Staff (1858–88); architect of victories over Denmark (1864), Austria (1866), and France (1871)

Moltke was born to a financially troubled aristocratic family in Parchim, Mecklenburg and was sent to the Royal Cadet Corps in Copenhagen for his education. He joined a Danish infantry regiment when he graduated, and, despite chronic ill health as well as a chronic shortage of funds, he determined to become a Prussian officer. In 1821, he was commissioned a lieutenant in the Leibgrenadier Regiment and proved such a promising officer that he was enrolled in the prestigious *Kriegsakademie*, which he attended from 1823 to 1826.

Moltke was an intelligent young man, who briefly turned to writing as a means of supplementing the meager salary of a junior officer. He published a novel in 1827 and a historical study of the Polish insurrection of 1830–31. His military career advanced steadily, and he was appointed to the Prussian general staff in 1833. Two years later, he was dispatched to Turkey and assigned to learn the language and customs and to advise the sultan on matters of military modernization. In part due to this experience, Moltke himself became an expert on modernization and would later make brilliant use of the telegraph and railroads.

Moltke, eager for action, violated his instructions by joining the Turkish service in order to fight in the Turkish sultan's war against Mehmet Ali of Egypt. He fought in the battle of Nezib, Syria on June 24, 1839 and saved the day for the Turks by maneuvering their artillery out of a hopeless disaster. After he returned to Germany late in 1839, Moltke published an account of his Turkish adventures in 1841. Three years later, he wrote an influential and forward-looking essay on railways, with particular attention to their value in war. In 1845, he published a history of the Russo–Turkish campaign of 1828–29. Also in 1845, Moltke was appointed to the coveted post of aide-de-camp to Prince Henry of Prussia. Henry died in July of the following year, and Moltke rejoined the General Staff.

Promoted to colonel in 1851, Moltke was appointed aide-de-camp to Prince Frederick William—the future Kaiser Frederick III—and traveled to England, Russia, and France. On October 7, 1857, he was named chief of staff and was admitted into the inner circle of Chancellor Otto von Bismarck and War Minister Albrecht von Roon. This triumvirate began the transformation of the Prussian army into Europe's most efficient and formidable war machine. Moltke promoted his vision of the role of the railway and telegraph in military operations. He also worked to expand the army and, equally important, to modify the system of command required to control larger forces deployed over greater areas. By 1859, he had radically reorganized the General Staff into four departments, including the East, German, and West departments—and a Railways Department.

With the outbreak of the Second Danish War in April 1864, Moltke became chief of staff to Karl Gustav von Wrangel's Austro–Prussian army. During April 25–June 25, Moltke took charge of the occupation of Jutland, which compelled Denmark to capitulate and seek terms on August 1, 1864. This triumph catapulted Moltke into a position of special favor and confidence with King William of Prussia. Recognizing the necessity of rapid movement in the war he planned to wage on Austria, William authorized Moltke to command Prussian force directly, without consulting himself of the war minister. For his part, Moltke devised a plan that relied heavily on rapid rail deployment of numbers of troops.

The opening invasion of Bohemia on June 16, 1866 was, in fact, plagued by interference from Bismarck, and Moltke did not always receive adequate compliance from subordinates. Nevertheless, he achieved enough of his plan's objectives to deal a decisive defeat to the Austro–Saxon army under Benedik at Königgrätz/Sadowa on July 3.

Moltke did not bask in his triumph. While the war had been successful for Prussia, he saw the weaknesses of his army. In the four years prior to the Franco–Prussian War of 1870–71, Moltke molded and remolded his forces, fashioning them into a far more tightly integrated unit with a General Staff that was much more capable—and compliant.

Moltke's first attempt to encircle the French army on August 2, 1870 failed. However, the consistently flexible and rapidly deployed Prussian army soon overwhelmed the stodgy French forces in any case. Prussian–German armies boxed in the French forces at Metz, then outflanked the army led to the relief of Metz by Marie Edmé MacMahon. MacMahon was surrounded at Sedan and suffered an ignominious defeat on September 1–2, 1870. After the collapse of Sedan, Moltke laid siege to Paris from October 19, 1870 to January 26, 1871 and readily picked off other French forces.

In June 1871, the triumphant Moltke was promoted to field marshal. Following the victory over France, he turned his attention to planning for the contingency of simultaneous war with France and

Russia—in effect evolving the scenario of World War I. He formulated a Russia-first strategy by 1879.

With advancing age, Moltke gradually transferred many of his duties to Count Alfred von Waldersee and, after Kaiser **Wilhelm II** ascended to the throne (June 15, 1888), Moltke retired (August 10, 1888).

Moltke was a great strategist and thoroughgoing organizer. He was perceptive and utterly free from the backward military prejudices that retarded developments in France. He brought warfare into a new age.

MONCEY, Bon Adrien Jeannot de, Duke of Conegliano

BORN: July 31, 1754

DIED: April 20, 1840

SERVICE: French Revolutionary Wars (1792–99); Napoleonic Wars (1800–15)

*Moncey was one of **Napoleon's** most respected commanders*

Moncey was born at Palisse, the son of an advocate. He entered the Conti Regiment as an enlisted man in 1769, but his father objected to having a son as a soldier. After six months' service, he purchased his son's release and set him to studying law. Moncey soon reenlisted in the Champagne Regiment and served with it through 1773, when he himself obtained discharge and voluntarily returned to the law. He practiced for a time, but was soon bored and joined the Gendarmes Anglais in 1774. It was 1779 before he obtained an officer's commission, as sublieutenant in the Nassau-Siegen Dragoon Regiment. When the unit was reformed into a light infantry battalion in 1791, Moncey became a captain, and three years later was promoted to major of the battalion when the dragoons were sent to join the Army of the Pyrenees.

Moncey's rise in 1794 was as rapid as his progress in his early military career had been slow. During the year, he was promoted to general of brigade, general of division, and, finally, commander in chief of the Army of the Pyrenees. Moncey achieved great success in the Pyrenees, pushing the Spanish back at the Battle of Bidassoa River (separating France and Spain at Fuenterrabia) in February 1794, then defeating them again at San Marcial (near Fuenterrabia) on August 1. He took the fortress of Fuenterrabia the following day, capturing two hundred guns and the flags of five units. Moncey went on to capture the fortress of San Sebastian on August 4, thereby taking another 148 guns, in addition to a substantial powder magazine. During October 15–17, he dealt a decisive blow to the Spanish at Orbaiceta.

Following the conclusion of a treaty at Basel in June 1795, Moncey was transferred from Spain to command of the army at Brest on September 1. The following year, he assumed command of the Eleventh Military Division at Bayonne and then the Fifteenth Division at Lyons. However, he was suspended from duty after the coup d'etat of Eighteenth Fructidor (September 4, 1797). Suspected of royalist sympathies, he was kept idle until 1800. In that year, he was assigned a command under **Jean Victor Moreau** in the Army of the Rhine, which was then in Switzerland. In 1801, he moved to the Army of the Reserve in Italy and fought in the Marengo campaign.

Napoleon appointed Moncey inspector general of the Gendarmerie in 1801, and he followed Napoleon into the Netherlands campaign. Awarded his marshal's baton in May 1804, he was subsequently elevated to grand officer of the Legion of Honor.

Moncey was sent to Spain in 1808 to command III Corps and was also created Duke of Conegliano, in July. He fought at Tudela on November 23 and mounted the second siege of Saragossa from December 20, 1808 to February 20, 1809, when he was transferred to command of the Army of Reserve of the North (in September).

After being elevated to the office of grand dignitary of the Order of the Iron Crown, Moncey was sent to the Pyrenees as commander in chief of the Army of Reserve in 1813, but he saw no action. On January 8, 1814, Moncey became major general of the National Guard at Paris and directed the defense of the capital until its capitulation on March 30.

After Napoleon's abdication, Moncey pledged himself to the Bourbons and was again named inspector general of the gendarmerie on May 13. He was appointed minister of state on June 2 and, two days later, was made a chevalier of the Order of St. Louis and a peer of France. When Napoleon returned from exile on Elba, Moncey was created a peer of the Empire, but he did not participate in the military campaigns of the Hundred Days.

After Napoleon's final defeat and exile, Moncey was commanded to serve on the court martial for treason of Marshal **Michel Ney.** When he refused, he was dismissed from his offices and from the military, and he was imprisoned for three months. He was not restored to favor until 1816. In 1823, he was given command of IV Corps in the Duke of Angouleme's Army of the Pyrenees and captured Catalonia. In December 1833, he was appointed governor of the Invalides and was given the honor of receiving Napoleon's ashes from St. Helena in 1840.

Moncey was perhaps the least colorful of Napoleon's marshals, always preparing for battle with meticulous care and ensuring that he was positioned for maximum effectiveness. He showed great regard for his troops, who reciprocated by showing him absolute loyalty. While he handled the opposing Spanish armies with dispatch, he was never cruel to the Spanish populace, a quality that made it that much easier to maintain what he had gained in the Spanish campaigns.

MONCK, George, First Duke of Albemarle

BORN: December 8, 1608

DIED: January 3, 1670

SERVICE: Anglo–French War (1626–30); Thirty Years' War (1618–48); Bishops' Wars (1639–40); First Civil War (1642–46); Second Civil War (1648–49); Third Civil War (1650–51); First Anglo–Dutch War (1652–54); Second Anglo–Dutch War (1665–67)

Distinguished British general and admiral whose sound judgment and great skill achieved much long-term success

Monck was born at Potheridge, Devon and first saw action as a volunteer in a force commanded by his kinsman Sir Richard Greenville under the Duke of Buckingham against Cadiz during September–October 1625. During 1627–28, he also participated in an expedition to the Ile de Ré. Both enterprises failed miserably. Nevertheless, Monck continued to pursue a military career, serving under Lord Vere in the Netherlands and later winning singular distinction at the year-long siege of Breda from October 1636 to October of the following year.

Promoted to lieutenant colonel of an English infantry regiment, he fought in the Bishops' Wars of 1639–40, then went to Ireland in 1642 at the head of an infantry regiment.

In 1643, during the First Civil War, he returned to England to aid Charles I, but was taken prisoner at Nantwich on January 25, 1644. Monck languished for the next two years in the Tower of London and used his confinement to write a book, *Observations on Political and Military Affairs*.

Following his release at the end of the war in 1646, he was appointed a major general with a command in Ulster during 1646–49. Confronted by an alliance among royalists, Irish rebels, and Ulster Scots, Monck had no choice but to conclude an armistice in the winter of 1649. Parliament censured him, but voted no further penalty, judging that what he had done had been motivated by military necessity. He did, however, have to relinquish command of his regiment.

Oliver Cromwell chose Monck to fight for him during the Third Civil War, assigning him to command an infantry regiment in Scotland. He fought with great distinction at Dunbar on September 3, 1650, so impressing Cromwell that, when the Lord Protector led his principal forces south, he left Monck behind with 8,000 men and charged him with the pacification of Scotland. Monck attacked swiftly and brutally at Dundee on September 1, 1651, and followed this with a series of equally efficient operations. His campaign was interrupted by illness late in the year, and he left the inhospitable climate of the north for the curative waters of bath.

On November 26, 1652, Monck accepted an appointment as the fourth General at Sea and, on

December 10, 1652, fought at Dungeness, During February 18–20, 1653, he fought at Portland, scoring a triumph off the North Foreland; he sunk nine of Admiral Maarten van Tromp's ships on June 11, 1653. On August 10, he again defeated Tromp, at Scheveningen.

Late in 1653, Cromwell ordered Monck back to Scotland as commander in chief of an army force there. Monck waged a lightning campaign in the Highlands and had soon quelled the incipient rebellion.

Following the death of Cromwell on September 3, 1658, Monck reluctantly acquiesced to the overthrow of the Protectorate, but he threw in his lot with Parliament after Major General John Lambert dissolved the Rump Parliament by force in October 1659. After purging his Scottish army of disloyal personnel, Monck established headquarters at Coldstream, Scotland, on December 8, 1659 and marched southward, crossing the border into England on January 1, 1660. After easily defeating Lambert, he marched into York on January 11 and London on February 3, 1660. After persuading the Rump Parliament to dissolve of its own accord in March 1660 and hold new elections, Monck contacted the exiled King Charles II. The new parliament invited Charles II to return, and Monck was at Dover to welcome him on May 25, 1660. The grateful monarch created Monck duke of Albemarle and made him a Knight of the Garter, additionally awarding him a generous pension. Charles appointed Monck captain general—effectively commander in chief of the army—on August 3, 1660, as well as master of the horse and lord lieutenant of Ireland. Monck's Scottish army was disbanded, but his own regiment was retained as the Lord General's Regiment of Foot Guards, a unit subsequently called the Coldstream Guards.

When war with the Netherlands broke out in 1666, Monck took command at sea and barely held off Admiral Michel de Ruyter's superior fleet during June 11–14, 1666. The timely arrival of a squadron under Prince **Rupert** on June 13 saved the day; then Monck sortied from the Thames, taking De Ruyter unawares. On July 25, Monck broke the Dutch blockade in St. James's Fight. De Ruyter countered the following year, during June 17–22, with the Thames-Medway raid, to which Monck responded well, but not fully enough to satisfy his critics. His reputation sullied by the incident, he resigned and retired.

Monck was a versatile commander on land and on sea. His response to the De Ruyter raid was entirely appropriate, but it was not the high note on which he deserved to end his illustrious career.

MONMOUTH, James Scott, Duke of

BORN: April 9, 1649

DIED: July 15, 1685

SERVICE: War of the Grand Alliance (1672–78); Covenanter Rising (1679); Monmouth's Rebellion (1685)

Claimant to the English throne who led an ill-conceived and unsuccessful rebellion against* King James II *in 1685

Monmouth was born in Rotterdam to the exiled King Charles II of England and his mistress of that period, Lucy Walters. He was raised by Charles's mother, Queen Henrietta Maria, and was subsequently brought to England by John Lord Crofts, his guardian, in 1662. Charles II, by this time reinstalled on the throne, welcomed his natural son and arranged a marriage for him with Anne Scott, Countess of Buccleuch. Knighted by the king—as Sir James Scott—he was created Duke of Monmouth on February 14, 1663. By virtue of his marriage, he also became Duke of Buccleuch in the Scots peerage.

In 1668, Monmouth was appointed captain of the First Troop of Horse Guards and in this position became highly visible. A handsome man who cut a dashing figure, he was charismatic and soon attracted a substantial following. He became a central figure in the party opposed to Charles's brother, the powerful James, Duke of York.

In April 1670, Monmouth became a member of the English Privy Council and, four years later, was admitted to the Scots Privy Council as well. During this period, he commanded the English army serving on the continent against the Dutch in the War of the Grand Alliance. He replaced Buckingham as Master of the Horse in April 1674 and four years later was named captain-general of all British forces. In this capacity, he commanded the royalist army that defeated the rebel "Covenanter" army at the Battle of Bothwell Bridge on June 22, 1679. His generous treatment of the rebels after this victory won him universal admiration.

In the years following the Covenanter Rebellion, Monmouth became involved in the "popish plot"—the fictitious conspiracy in which Jesuits were supposedly planning the assassination of King Charles II in order to bring his Roman Catholic brother, the Duke of York (later King James II), to the throne. Monmouth used the popish plot to consolidate support for all those, like himself, who opposed the Duke of York's succession to the throne. After the revelation of the confused Rye House Plot of 1683—an alleged Whig conspiracy to assassinate or mount a rebellion against Charles II—Monmouth was banished from court. He fled to the Netherlands in 1684, but when Charles died the following year—on February 6, 1685—Monmouth allowed himself to be persuaded to mount an expedition to England to claim the throne. He landed at Lyme Regis, Dorset, with an "army" of 89 followers on June 11, 1685, but rapidly accumulated a substantial force made up of commoners.

Monmouth marched into Taunton and proclaimed himself king on June 20. After considerable—and fatal—delay, he advanced on Bristol, where he met heavy royalist resistance and had to withdraw to Bridgewater. From this base, he led his makeshift army in a surprise nighttime raid on the royalist encampment at Sedgemoor on July 6. The well-trained royalist troops quickly rallied and repulsed Monmouth. They then took the offensive in a concentrated counterattack that annihilated Monmouth's army. Monmouth himself fled the field and sought to evade capture, but he was soon run to ground and imprisoned in London. Following a brief trial, he was found guilty of treason and was beheaded at the Tower of London.

Monmouth had more charisma than judgment. Although he was a dashing officer and an able commander, his expedition to England was doomed from the beginning.

MONTCALM, Louis-Joseph de Montcalm-Gozon, Marquis of

BORN: February 28, 1712

DIED: September 14, 1759

SERVICE: War of the Polish Succession (1733–38); War of the AustrianSuccession (1740–48); French and Indian War (1754–63)

French general best known for his service in the French and Indian War and for his defeat and death at the defense of Quebec

Montcalm was born at Candiac, near Nîmes, France, to an ancient aristocratic family. He was commissioned an ensign in the Hainaut Regiment at the age of twelve and was a soldier by fifteen. In 1733 he saw his first combat action, fighting the Austrians in the War of the Polish Succession. He temporarily left service in 1735, when he inherited his father's title and property, but returned to active duty in 1749 with France's entry into the War of the Austrian Succession.

Wounded at the defense of Prague during July 27–December 26, 1742, he was promoted to colonel of the Auxerrois Regiment in 1743 and fought with magnificent gallantry at the battle of Piacenza, during which most of his regiment was cut to shreds. He himself was wounded no fewer than five times by the sword and was ultimately made a prisoner of war on June 16, 1746.

After he was released in a prisoner exchange, Montcalm was promoted to brigadier general and was appointed to command of a cavalry regiment in 1748. In January 1756, Montcalm was appointed commander of French forces in North America and was promoted to major general. He arrived in Quebec on May 13 and was soon dismayed to discover that he had no authority over colonial troops or the militia, which answered to the governor of New France, the Marquis de Vaudreuil. Montcalm, a moody and irascible individual, argued bitterly with Vaudreuil, and military operations suffered as a result. Yet Montcalm repeatedly demonstrated his skill as a general, first by capturing Oswego, New York, and thereby securing control of Lake Ontario on August 14, then by taking Fort William Henry at the southern tip of Lake George on August 9, 1757. He also held off and repulsed an attack by 15,000 Troops against his garrison of 3,800 at Fort Ticonderoga on July 8, 1758.

Promoted lieutenant general, he was finally given authority over Vaudreuil in military matters, and he turned his attention to the defense of Quebec in 1759. He was able to hold off the British general **James Wolfe** for two months before he was taken by surprise after Wolfe's men ascended to the Plains of Abraham and attacked on September 13. In this battle, Montcalm made several uncharacteristic errors and was unable to muster and coordinate the full force of his defenders. Vaudreuil also failed to provide needed artillery support. The battle, so long in coming, was over in less than half an hour. Quebec fell to the British, and Montcalm lay dying of a wound. He was unaware that General Wolfe, the victor, had likewise suffered a mortal wound. Montcalm died the day after the battle.

Montcalm was an extraordinarily courageous and skillful commander, who suffered from depression and was given to fits of temper. Crossed by Vaudreuil, he obstinately refused to patch their relationship up. As a result, his military position suffered in North America.

MONTGOMERY, Sir Bernard Law, First Viscount Montgomery of Alamein

BORN: November 17, 1887

DIED: March 25, 1976

SERVICE: World War I (1914–18); World War II (1939–45)

British field marshal who was one of the great Allied commanders of World War II

London-born Montgomery was the son of a clergyman and was raised in Tasmania, although he returned to England for schooling at St. Paul's School. In 1906, he enrolled in Sandhurst and graduated in 1908 with a commission in the Royal Warwickshire Regiment. His regiment was sent into the field at the outbreak of World War I in August 1914, and Montgomery fought at Le Cateau on August 24–25. He distinguished himself at the first Battle of Ypres in

Sir Bernard Law Montgomery. Courtesy National Archives.

October–November, suffering a severe wound but earning the DSO.

Following his injury, Montgomery was sent back to England for training duties in Britain during 1915, but he returned to the front in France at the beginning of 1916 and was brigade major for 104th Brigade at the Somme from June 24 to November 13. He was promoted to staff duty for the Thirty-Third Division at Arras during April 9–15, 1917 and then for IX Corps at Passchendaele from July 31 to November 10. By the end of the war, he was a staff officer with the Forty-Seventh Division.

Montgomery served briefly in the Allied army of occupation after the war, then attended the staff college at Camberley in 1921. He served in various posts in Britain until 1926, when he was appointed an instructor at Camberley. In 1929, he rewrote the army's *Infantry Training Manual*, then was posted for three years at Jerusalem, Alexandria, and Poona, commanding a regiment from 1930 to 1933. He was appointed chief instructor at Quetta (India—now Pakistan) Staff College and served there from 1934 to 1937. Returning to England, he took command of the Ninth Brigade at Portsmouth until October 1938, when he was named commander of the Eighth Division in Palestine. Through August 1939, he was involved in suppressing Arab terrorism.

Recalled to Europe at the outbreak of World War II, Montgomery was commander of Third Division in II Corps during the failed offensive in Flanders. Montgomery won distinction in his management of the retreat to Dunkirk, fighting a brilliant and life-saving rearguard action during May–June 1940. Knighted for his services in this campaign, he was named to replace Sir **Claude Auchinleck** as commander of V Corps in July, then transferred to command XII Corps in April 1941. By November, he was in command of the Southeastern Army.

Montgomery played a role in planning the Dieppe raid of August 1942—another unsuccessful attempt at an early offensive—but before the operation was executed, he was participating in the Operation Torch landings in North Africa. He was slated to command First Army, but, following the death of General W. H. E. Gott in August, he was given command of Eight Army in Egypt. He found himself immediately under attack by the "Desert Fox," **Erwin Rommel,** whose August 31–September 7 offensive at Alam Halfa he repulsed. Seizing the initiative, he conducted his own offensive at El Alamein during October 23–November 4 and achieved a victory. Later in November, Montgomery was promoted to general and continued to dog Rommel's Afrika Korps to the border of Tunisia from November 5, 1942 to January 1943. However, the "Desert Fox" at last eluded Montgomery, and the British general found himself on the defensive at Medenine on March 6, 1943. Some days later, Montgomery, was attacked at Mareth, on March 20, but then outflanked the German position during March 27–April 7.

Montgomery continued to lead the Eighth Army through the balance of the Tunisian campaign, which ended on May 13. From here, he participated in the invasion of Sicily, pushing the Germans out of their positions around Mount Etna during July 9–August 17. He then went on to capture the airfields at Foggia during September 3–27, but was stopped at the Sangro River at the end of the year.

Bogged down, Montgomery was grateful for his recall to Britain, where he was assigned command of the Twenty-First Army Group in preparation for the invasion of France. Montgomery had overall charge of ground forces during the Normandy ("D-Day") landing and invasion, beginning on June 6, 1944. He relinquished overall command to the supreme Allied commander, **Dwight D. Eisenhower,** on September 1, the very day he was promoted to field marshal.

Montgomery conceived Operation Market-Garden to push the war to a quick conclusion. But the operation failed when the Allies suffered defeat at Arnhem during September 17–26. This failure resulted in some loss of confidence on the part of the Allies, Montgomery was temporarily shifted to a secondary role in the final months of the war in Europe. However, he partially redeemed himself when he took command of the northern end of the American line during the Battle of the Bulge (December 16, 1944–January 15, 1945), greatly helping to restore the American position. In the process, however, he offended certain American commanders. Nevertheless, Montgomery was assigned to plan and direct the British crossing of the Rhine at Wesel on March 23, 1945, and from here he pushed into northern Germany. It was Montgomery who accepted the surrender of German forces in the Netherlands, Denmark, and, on May 4, northwestern Germany.

Montgomery was named to command British occupation forces in Germany in May 1945 and was created Viscount Montgomery of Alamein in January of the following year. In June 1946, he became successor to Lord Alanbrooke as chief of the Imperial General Staff, but was not popular. He soon left this post to become chairman of the Western European Union commanders in chief in 1948. This evolved into the military arm of NATO, and led to Montgomery's appointment as first commander of NATO forces in Europe and Eisenhower's deputy supreme commander. Montgomery held this post from March 1951 until his retirement from the army in September 1958.

Montgomery was very popular with troops, who affectionately dubbed him Monty. He was careful and thorough—U.S. general **George S. Patton** and other Americans thought him *too* careful—yet he possessed a genuine ability to inspire with his leadership. He did not work very well with his American allies, of whom he had a low opinion, which he often aired. His opinion of himself was never in doubt: He believed that he was nothing less than the world's greatest living general.

MONTMORENCY, Anne, Duke of

BORN: March 15, 1493

DIED: November 12, 1567

SERVICE: War of the Holy League (1511–14); Swiss War (1511–14); First Hapsburg–Valois War (1521–26); Second Hapsburg–Valois War (1536–38); Fifth Hapsburg–Valois War (1547–59); Huguenot Wars (1560–92)

Constable of France who wielded great power during the reigns of Francis I, Henry II, and Charles IX

Montmorency was born at Chantilly, to Guillaume de Montmorency and Anne Pot and saw his first action at Ravenna on April 11, 1512. He fought at Marignano on September 13–14, 1515, and in 1519 was sent to England as a hostage. While in England, he witnessed the celebrated Battle of the Field of the Cloth of Gold in 1520.

After returning to France, Montmorency fought at the defense of Mézières in 1521, then, the following year, raised an army in Switzerland and led it into Italy, where he took Novara and Vigevano. He was in command of this army at La Bicocca, where he lost control of his troops, who advanced against his orders and were defeated. Montmorency was wounded in this action of April 27.

Emperor Francis I elevated Montmorency to the post of Marshal of France on August 5 and dispatched him once again to Italy, where he served during 1523–24. It is not certain whether he fought at the Sesia on April 30, 1524, but he did participate in the general French retreat into Provence, where he found himself fighting a delaying action against the rebellious Constable de Bourbon and the Marquis of Pescara in the summer of 1524.

Montmorency returned again to Italy late in the fall, and, on February 24, 1525, fought alongside Francis I at the Battle of Pavia. Taken prisoner, Montmorency was subsequently released to negotiate the Treaty of Madrid, which ended the first Hapsburg–Valois War. Subsequent to this, he was appointed grand master of France and governor of Languedoc. He was also elevated to the post of King's Chief Minister, entrusted with enormous power, including the conduct of war, the running of the royal household, the management of public works, and the conduct of foreign affairs. Montmorency soon acquired a reputation for great energy, but was also resented for his autocratic bearing, his dictatorial manner, and his arrogance.

Montmorency was instrumental in concluding the Peace of Cambrai in 1529, and when the Second Hapsburg–Valois War broke out in 1536, Montmorency forced the withdrawal of Emperor Charles V from Provence through a scorched-earth policy that removed all provisions from the region. Montmorency attained the office of Constable of France on February 10, 1538, but soon made many enemies, fell out of royal favor, and retired in 1541.

With the coronation of King Henry II in 1547, Montmorency was recalled to court and the following year was dispatched to Bordeaux to put down a rebellion there. He crushed it with unmitigated ferocity and was created Duke of Montmorency and peer of France in July 1551.

On August 10, 1557, Montmorency was riding to the relief of Saint Quentin, besieged by a combined English–Spanish army during the Fifth Hapsburg–Valois War, when his column was met by a Spanish cavalry charge as it was crossing the Somme. Montmorency's unit was cut to shreds, and he himself was captured. He remained a prisoner for the balance of the war, and was released only after a treaty was concluded on April 3, 1559.

Montmorency was a rigid Catholic who opposed Marie de Medici's toleration of the Huguenots (French Protestants) during the early 1560s. He plunged into the Huguenot wars and fought at Dreux on December 19, 1562, defeating the Grand **Condé.** Both commanders—he and Condé—were taken prisoner in the battle.

Following the Peace of Amboise in March 1563, which ended one phase of the Huguenot Wars, Montmorency and his son François led a combined Catholic–Huguenot army against the English and recovered Le Havre after a July 28, 1563 battle. When the Huguenot conflict resumed, Montmorency again defeated Condé at Saint-Denis on November 10, 1567. The victory came at a great price, however. Montmorency received a fatal wound and died two days after the battle.

Montmorency was indefatigable and his bravery was without bounds. An inspiring figure, he was nonetheless a poor leader, given to arrogance and an unwillingness to practice anything resembling diplomacy. He frequently alienated king and court.

MONTROSE, James Graham, Marquess of

BORN: 1612
DIED: May 21, 1650
SERVICE: First Bishops' War (1639); Second Bishops' War (1640); First English Civil War (1642–46); Third English Civil War (1650–51)

Scots general who achieved spectacular victories in Scotland for King Charles I of Great Britain during two of the English civil wars

Montrose was son of John, Fourth Earl of Montrose, and became the Fifth Earl after his father's death in 1626. He attended St. Andrews University and, in 1637, sided with the party of resistance to Charles I during the religious upheaval that attended the introduction of a new book of canons in 1635 and a new prayer book in 1637, both of which compromised Scots Presbyterianism and elevated British Anglicanism. Montrose subscribed to the National Covenant of February 28, 1638, which rejected the efforts of King Charles I and the Archbishop of Canterbury, William Laud, to force the Scottish church to conform to English liturgical practice. However, when the dispute came to armed rebellion, Montrose helped to suppress anti-Covenant forces around Aberdeen and among the Gordons in the spring of 1639.

Montrose opposed the anti-royalist policies of the Earl of Argyll during the Scots parliament of August–October 1639, yet he fought Charles I during the Second Bishops' War and achieved a splendid victory in the fighting at Newburn during August 1640. Following the Treaty of Ripon of November 1640, Montrose fell under suspicion and was imprisoned by his rival Argyll during June–August 1641.

After his release, Montrose tried desperately to keep Scotland out of the civil war in England, but failed. After a Scottish army invaded England in January 1644, he accepted Charles I's creation of him as a marquess and also accepted appointment as lieutenant general for Scotland in February 1644. Thus Montrose sided with the English king against the Scots rebels.

In April 1644, Montrose invaded Scotland with a force of 2,000 and took Dumfries. Argyll, with superior numbers, soon forced him out of the town. Montrose withdrew from Scotland, then reentered in August, engaging and defeating an Irish force led by Alasdair Macolla MacDonald on the twenty-eighth. Next, Montrose's 3,000 troops routed the 7,000-man army of Lord Elcho at Tippermuir on September 1. After this, on September 13, he annihilated Lord Burleigh's force of 2,500 outside Aberdeen.

Moving into the Grampians, Montrose raided the lands of the Campbells while consistently evading the pursuing forces of Argyll from September 25, 1644 to February 17, 1645. Then, on February 19, he turned on Argyll, took a stand, and destroyed Argyll's army of Campbells—even though Argyll enjoyed a 3,000- to 2,000-man advantage. Montrose next took Dundee on April 4 and, with a force of 1,700, beat Sir John Hurry's army of 4,000 at Auldearn on May 8. Next came the defeat of William Baillie's army at Alford on July 2, then, at Kilsyth, the coup de grace against the combined armies of Baillie and Argyll on August 15, 1645.

With the defeat of Baillie and Argyll, Glasgow and Edinburgh fell to Montrose. He convened a parliament at Glasgow on October 20—but, at this moment of victory, Charles I summoned Montrose south after the decisive Royalist defeat at Naseby on June 14, 1645. As he marched, Montrose lost his Highland clansmen, who decided to abandon his cause. Left with fewer than a thousand soldiers, Montrose was surprised by a night attack at Philiphaugh by General David Leslie during September 12–13, 1645. Routed, Montrose fled into the Highlands with fewer than a hundred horsemen. He set about raising a guerrilla army consisting of a few hundred men, but soon saw the futility of resistance. On May 31, 1646, he surrendered and set sail for Norway. Later, in 1647, he sought support for the royalists from Henrietta Maria in Paris—in vain.

Charles II restored Montrose to the lieutenancy of Scotland in 1650 and he returned from the continent, landing in the Orkneys in March. He raised an army of 1,200, then made the crossing to the Scottish mainland, where he attempted to gather support from among the clans. Failing, he went into hopeless battle at Carbiesdale on April 27, 1650 and was defeated. Fleeing into the hills, Montrose found refuge with Neil Macleod of Assynt, who betrayed him to the Covenanters. The cruelest blow came when Charles II, hoping to placate the Covenanters, refused to support Montrose. Sentenced to death in Edinburgh on May 20, 1650, he was hanged the following day.

Montrose was a great general, who, operating with very limited forces and almost always outnumbered, nevertheless brought most of Scotland to the Royalist cause. Charles II's betrayal of him was one of the great personal tragedies of the English civil wars.

MOREAU, Jean Victor Marie

BORN: February 14, 1763

DIED: September 2, 1813

SERVICE: French Revolutionary Wars (1782–99); Napoleonic Wars (1800–15)

Important general of the French Revolutionary Wars, who later became an opponent of Napoleon

Moreau was born at Morlaix, Brittany, and joined the National Guard at Rennes, where, in 1789, he raised an artillery company, taking charge of it as captain. After transferring to the infantry in 1791, he was elected lieutenant colonel of the First Battalion, Ille-et-Vilaine Volunteer Regiment. He led this unit in the Army of the North during 1792–93, fighting with signal distinction at Neerwinden on March 18, 1793 and earning promotion to general of brigade on December 20. On April 14, 1794, he was named general of division and early in March of the following year took temporary command of the Army of the North until, on March 14, he was assigned to command the Army of the Rhine and Moselle.

Moreau served with distinction in the German campaign of 1796, but was suspended from command because he was suspected of Royalist sympathies. When the Royalist coup d'etat of Eighteenth Fructidor (September 4, 1797) failed, however, Moreau was recalled to duty on September 9. Given provisional command of the Army of Italy after it suffered defeat at Magnano on April 5, 1799, he stepped down when Joubert arrived, but was returned to command of the army after Joubert died on August 15.

Moreau returned to France on September 21, 1799 to assist **Napoleon** in the coup d'etat of Eighteenth Brumaire (November 9–10) and for this was given command of the armies of the Rhine and Helvetia. He dealt the Austrians a harsh blow at Hohenlinden on December 3, 1800, prompting Austria to withdraw from the Second Coalition.

Moreau turned against Napoleon by joining a Royalist plot against him. Presumably, Moreau was motivated by nothing more than his own vanity in this. Arrested on April 15, 1804, Moreau was sentenced to two years' imprisonment, which Napoleon commuted to exile for life. Moreau immigrated to the United States, where he took up residence at Morrisville, Pennsylvania. In 1813, French Royalists and representatives of the Russian czar **Alexander I** invited him to return to Europe and joined the Allied coalition against Napoleon. Moreau met with them in Prague on August 17, 1813, and served as the czar's military adviser during the campaign in Germany. He received a fatal wound at the Battle of Dresden on August 27 and died on September 2. Burial was at St. Petersburg, Russia.

Moreau was a complex personality. Supremely courageous and adored by his troops, he was an excellent tactician, though, as Napoleon put it, "an absolute stranger to strategy." However, he could not bring himself to play a subordinate role under Napoleon. With more ambition than political savvy, he betrayed Napoleon and came to grief as a result.

MORGAN, John Hunt

BORN: June 1, 1825

DIED: September 4, 1864

SERVICE: U.S.–Mexican War (1846–48); Civil War (1861–65)

Confederate cavalry general who led raids in the border states and in Indiana and Ohio

A native of Huntsville, Alabama, Morgan was raised near Lexington, Kentucky. He joined the cavalry at the outbreak of the U.S.–Mexican War in 1846 and served in that conflict. As the nation approached civil war, Morgan organized a Kentucky militia company, the Lexington Rifles, in 1857. He joined the provisional Confederate army in September 1861 and was appointed captain of cavalry.

Serving under General **Braxton Bragg,** he commenced the series of raids, in the spring of 1862, that made him famous in the South and feared in the North. He conducted operations in Tennessee and Kentucky, destroying railroad tracks and cutting down telegraph lines. He also looted supplies and took Union prisoners, most notably at Hartsville on December 7, when he rounded up 1,700 Union troops.

Promoted to brigadier general, he led 2,400 men—on his own authority—in a raid on Indiana during June 1863. After creating havoc there, he was chased into Ohio, where he skirted Cincinnati and brought much panic to the area. In this operation, Morgan penetrated farther north than any other Confederate commander in the war. The Indiana–Ohio raid was costly, since most of Morgan's men were taken prisoner at Buffington Island, on the Ohio River on July 19. Morgan evaded capture, but was subsequently taken at Salineville on July 26. Imprisoned in Columbus, Ohio, he made a daring escape in November, returned to the South, and was given command of the Department of Southwestern Virginia early in 1864.

Morgan returned to Kentucky to conduct more raids, but lost control of his men, who merely pillaged at will. Morgan was accused by Confederate authorities of banditry. While these charges were under investigation, he made one more foray into Tennessee, but was killed by Union troops in a surprise attack at Greenville.

Morgan was a consummate cavalryman and a highly enterprising guerrilla leader. His early raids were quite effective at disrupting Union operations in Kentucky and Tennessee. Beginning with the Indiana operation, however, he showed increasingly poor judgment, and his command degenerated into a band of hooligans.

MORTIER, Edouard Adolphe Casimir Joseph, Duc de Treviso

BORN: February 13, 1768

DIED: July 28, 1835

SERVICE: French Revolutionary Wars (1792–99); Napoleonic Wars (1800–15)

One of* Napoleon's *marshals; also served as prime minister and minister of war during the reign of King Louis-Philippe

Born at Cateau-Cambrésis, he was the son of Antony Charles Joseph Mortier, who served in the States-General of May 1789 as Third State Deputy. Mortier joined the National Guard of Dunkirk and then of Le Cateau in 1789. In 1791, he was elected captain of the First Volunteer Battalion, Army of the North, and fought well at Jemappes on November 6 and at the siege of Namur during November 6–December 2, 1792. The following year, he participated in battles at Neerwinden (March 18) and Hondschoote (September

8), earning an appointment as adjutant general. After fighting at Fleurus on June 26, 1794, he was transferred to the Army of Sambre-et-Meuse, where he distinguished himself in crossing the Rhine at Maastricht during September 22–November 4, 1794.

Offered promotion to general of brigade in 1797, Mortier demurred and did not accept the promotion until 1799. He fought at the Battle of Zurich during June 4–7, 1799 and was promoted to general of division shortly afterward. **Napoleon** personally appointed Mortier to command the Seventeenth Military Division in Paris in 1800. Three years later, he was dispatched to occupy Hanover and accepted the surrender of the Hanoverian army in July. The following year, in 1804, Mortier was elevated to command of the artillery and sailors of the Consular Guard with the rank of colonel general. He was also created Grand Officer of the Legion of Honor and, in May, received his marshal's baton.

In 1805, Mortier became commander of the infantry at the Imperial Guard. In the campaign of that year, he commanded a provisional corps charged with supporting Marshal **Joachim Murat**'s advance guard. He was engaging Russian General **Mikhail Kutuzov**'s rearguard on November 11 at Dijrrenstein when he was in turn attacked by Doctorov. The fighting was punishing in its intensity, and two thirds of his leading division fell. Nevertheless, Mortier held his ground against vastly superior numbers until he was relieved by General Dupont's division.

After a distinguished performance at the battle of Austerlitz on December 2, Mortier was given command of V Corps, then, from October 1806 to June 1807, he had charge of VIII Corps, which he used to push the Prussian and Swedish armies out of Hanover and Pomerania.

Mortier was summoned back to the Grande Armée and took charge of the left wing at Friedland on June 14, 1807. The following year he was created Duke of Treviso, and in October 1808 resumed command of V Corps in Spain, where he fought at Somosierra on December 20 and then at the siege of Saragossa from December 20, 1808 to February 20, 1809. He fought at Arzobispo (August 4, 1809) and Ocaña, where he was wounded. Later, during January 26–March 9, 1811, he participated in the siege of Badajox.

Recalled to France in May 1811, Mortier prepared for the invasion of Russia. He was assigned to command the Young Guard and, after the Battle of Borodino on September 7, he was named military governor of Moscow. When the French army retreated from Moscow, Mortier fought at Krasnoye on November 17 and at the Berezina River on November 27–29. In January 1813, he assumed command of what remained of the battered Imperial Guard, then resumed command of the Young Guard in Germany, fighting directly under Napoleon at Lützen (May 2), Bautzen (May 20–21), Dresden (August 26–27), and Leipzig (October 16–19).

During the campaign for France in 1814, Mortier was given command of the Old Guard, which handled superbly at Montmirail (February 11), Craonne (March 7), and Laon (March 9–10). But he was defeated (with General **Auguste Marmont**) at La Fère-Champenoise on March 25 and at Montmartre—just outside Paris—on March 30. It was he and Marmont who opened negotiations with the Allies on the day following the Montmartre defeat.

When the Bourbons returned to power, Mortier was created a peer of France, a chevalier of the Order of St. Louis, and was appointed governor of the Sixteenth Military Division. When Napoleon returned from Elba, Mortier escorted Louis XVIII to the Belgian border, then joined Napoleon for the Hundred Days. He took command of the cavalry of the Guard on June 8, 1815, but a few days later, on June 15, he suffered a crippling attack of sciatica and was compelled to turn over his command.

After Napoleon's final exile, Mortier was ordered to sit on Marshal **Michel Ney**'s court-martial for treason. He refused, declaring himself incompetent to judge a fellow marshal. Briefly out of favor with the Bourbons as a result of this, he was restored to command in 1816 and given charge of the Fifteenth Military Division. In 1820, he was made commander of the Order of St. Louis, and in 1829 governor of the Fourteenth Military Division. From 1830 to 1832, he served as ambassador to Russia, and in 1831 became a Grand Chancellor of the Legion of Honor. He was King Louis Philippe's minister for war from November 18, 1834, to March 12, 1835. He was escorting the king during a parade of the National Guard on July 28, 1835, when he was killed by a bomb blast intended to assassinate Louis Philippe.

If Mortier was one of the less brilliant of Napoleon's marshals, he was among the most consistently reliable—a better subordinate than independent commander. Well liked by his own subordinates, he was universally respected by the other marshals—characteristically a discordant group.

Louis Mountbatten (Earl Mountbatten of Burma). Courtesy National Archives.

MOUNTBATTEN, Louis Francis Albert Victor Nicholas, Earl Mountbatten of Burma

BORN: June 25, 1900

DIED: August 27, 1979

SERVICE: World War I (1914–18); World War II (1939–45)

British admiral, statesman, last viceroy of India; commander in Southeast Asia during World War II

Mountbatten was born at Frogmore House, Windsor, to Prince Louis of Battenberg (later Lord of Milford Haven) and Princess Victoria of Hesse-Darmstadt (a granddaughter of Queen Victoria). Mountbatten's father changed the family name during World War I in response to public hostility to Germans.

Mountbatten entered the Royal Navy as a cadet at the Osbourne Naval Training College, which he attended from May 1913 to November 1914, then continued at the Royal Naval College, Devonport, graduating at the top of his class in June 1916. During World War I, he saw service as a midshipman aboard Admiral David Beatty's flagship H.M.S. *Lion* from July 1916 to January 1917 and H.M.S. *Queen Elizabeth* from February 1918 to July. Promoted to lieutenant, he transferred to P-boat (coastal torpedo boat) service in August and served aboard P-boats until the end of the war, November 1918.

Following the armistice, Mountbatten spent a year at Cambridge University, then toured Australia, Japan, and India with the Prince of Wales (later Edward VIII) beginning in 1920, and married Edwina Ashley on his return to Britain in 1922. Later, during World War II, Lady Edwina would become almost as famous as Mountbatten through her morale-lifting presence at Southeast Asian fronts.

In 1923, Mountbatten sailed aboard H.M.S. *Revenge* (1923), then enrolled in an advanced signals course, graduating, again, first in his class in July 1925. With his signals training, Mountbatten served as assistant fleet wireless officer in the Mediterranean during 1927–28 and became fleet wireless officer in 1931. Promoted to captain in 1932, he continued to serve as Mediterranean Fleet wireless officer through 1933, when he was briefly given command of the destroyer H.M.S. *Daring*.

From 1936 to 1938, he became naval aide-de-camp to Edward VIII and George VI and, on the eve of World War II, in June 1939, was assigned command of the destroyer H. M. S. *Kelly*, then under construction. After overseeing the completion of the vessel, Mountbatten sailed aboard her as commander of the Fifth Destroyer Flotilla, consisting of *Kelly* and *Kingston*, on September 20, 1939. Despite some incidents of questionable seamanship aboard *Kelly* (involving near capsize in heavy weather and a narrowly missed collision on another occasion), Mountbatten performed with great distinction in the evacuation of Namsos following the ill-fated offensive in Norway during June 1940. During the evacuation of Crete, *Kelly* was sunk by German dive bombers on May 23, 1941.

Following the loss of *Kelly*, Mountbatten was made captain of the aircraft carrier H.M.S. *Illustrious*, which was being repaired in the U.S. during October 1941. While overseeing the work, Mountbatten made many valuable U.S. contacts and greatly impressed U.S. naval leaders. Unlike some other British military leaders, Mountbatten early on formed a respectful opinion of American military leaders.

In April 1942, **Winston Churchill** recalled Mountbatten to England to serve as director of Combined Operations. Eager to take offensive action in the war, Mountbatten was among the key advocates and planners of the raid on Dieppe (August 18, 1942), which resulted in disastrous defeat. Despite this loss, Mountbatten was unshaken and became determined to improve the British military's amphibious capabilities. He built up the amphibious-capable force, so that by April 1943, Combined Operations consisted of some 2,600 landing craft and 50,000 personnel. He turned his attention to creating technological improvements to facilitate amphibious operations, including "mulberries" (towed harbors) and the PLUTO system (Pipe-Line Under the Ocean), both of which were important in the "D-Day" Normandy invasion of 1944.

In the August 1943 Quebec conference, Mountbatten was elevated above other more senior officers to become Supreme Allied Commander for Southeast Asia—a very difficult assignment in a corner of the war that was chronically undermanned and poorly supplied. He directed Allied operations in Burma and the Indian Ocean, and rapidly exhibited a genius for managing disparate resources and conflicting personalities toward a common goal. The assignment called for great flexibility and the ability to improvise. He developed a particularly close working relationship with General **Sir William Slim,** and, together, the two commanders led the liberation of Burma during late 1944 through August 1945. It was war waged on a shoestring, against a fanatically dedicated enemy, and in conditions of terrain and climate that were often more lethal than any human-made weapons.

If Mountbatten had held command in a neglected front, he had to manage, after the war, the daunting task of accepting Japanese surrender and reestablishing colonial authority in places that were now increasingly nationalistic in spirit. Indeed, Mountbatten executed his duties despite a growing conviction that the time for colonial rule had passed. Mountbatten also saw to the speedy and humane liberation of Allied POWs, who had been abused and starved by their Japanese captors. His postwar authority extended to Indochina and Indonesia during September 1945–46, and from March 24 to August 15, 1947, he served as the last British viceroy of India. In this post, he oversaw the British withdrawal from India and the inauguration of independence for India and Pakistan.

Created a viscount in 1946, he was made an earl in 1947. In 1950, he was appointed fourth sea lord, serving until 1952, when he became commander in chief of the Mediterranean Fleet. In 1954, he was made first sea lord and served in this post until 1959. Promoted to admiral of the fleet in 1956, he was named chief of the United Kingdom Defence Staff and chairman of the Chiefs of Staff Committee in July 1959. He served in these posts until July 1965. In 1965, he became governor and, in 1974, lord lieutenant of the Isle of Wight.

Mountbatten fell victim to Irish Republican Army (IRA) terrorists in 1979. He and his teenaged grandson Nicholas and a local Irish boy were killed when an IRA bomb exploded aboard Mountbatten's yacht.

Mountbatten combined many of the qualities of traditional British imperial aristocracy with a forward-looking tolerance for the realities of a changing world. While he could be vain and arrogant, greatly enamored of war's pomp and circumstance, he was also a highly flexible and innovative military thinker, who fostered effective cooperation among diverse interests.

MURAT, Joachim, King of Naples, Duke of Cleve and Berg

BORN: March 25, 1767

DIED: October 13, 1815

SERVICE: French Revolutionary Wars (1792–99); Napoleonic Wars (1800–15)

French cavalry officer who became* Napoleon's *most celebrated marshal and subsequently was made king of Naples

Murat was born at La Bastide-Fortunière (now named for him, Labastide-Murat), Quercy, Gascony. His father, a farmer and innkeeper, marked the boy for a career as a priest and sent him to Cahors and Toulouse for study. The high-spirited young Murat soon ran up formidable debts and, mostly to evade his creditors, he enlisted in the mounted chasseurs on February 23, 1787. After brief service in Louis XVI's Constitutional Guard in 1792, he joined the Twelfth Chasseurs and was commissioned a lieutenant in October. During the Revolutionary Wars, he saw service in the Army of the North (1792–93) and gained promotion to major in 1793.

A radical Jacobin during the Revolution, Murat was nearly purged after the coup d'etat of Ninth Thermidor, but the Committee of Public Safety quickly reinstated him, and **Napoleon** Bonaparte chose him to transport forty guns from the artillery park at Sablons to Paris to defend the Convention against the Royalist insurrection of Thirteenth Vendemiare (October 5, 1795). Murat arrived with the artillery just in time to drive back the Royalists with what Napoleon called a "whiff of grapeshot." This act solidified Napoleon's position in the new government and, for his role in Vendemiare, Murat became Napoleon's senior aide.

Murat served with Napoleon in the Italian campaign of 1796–97 and was promoted to general of brigade on May 10, 1796. Given a cavalry brigade command under General **Michel Ney**, he performed valiantly at the Tagliamento in March 1797. Murat truly came into his own as a military leader during the Egyptian and Syrian campaigns of 1798–99. He led a dragoon brigade at Alexandria and the battle of the Pyramids. At the first battle of Aboukir, on July 25, 1799, he was severely wounded in the jaw by a lance while leading a charge to take Aboukir castle. Despite his injury, he not only took the castle, but captured the Turkish general Mustapha Pasha. Napoleon promoted him, on the field, to general of division.

After returning to France, Murat hastened to aid Napoleon in the coup d'etat of 18-19th Brumaire (November 9–10, 1799). For this service, Napoleon, now First Consul, made him commander of the Consular Guard. Murat wed Napoleon's youngest sister, Caroline, in January 1800—thereby confirming his high stock in Napoleon's estimation.

In the second Italian campaign, Murat was given command of the cavalry of the Army of the Reserve. He acquitted himself well at Marengo on June 14, 1800 and was presented with a saber of honor. Napoleon then dispatched him to Tuscany, where he liberated the Papal States from control by Naples and forced upon the Neapolitan king the Armistice of Foligno in February 1801.

Murat was given various posts in Italy. Then, in January 1804, he was named governor of Paris and, later in the year, was made Marshal of the Empire. In February 1805, he was titled Prince and made "grand admiral." Once again on the battlefield, Murat commanded the cavalry reserve during the Ulm campaign and used the cavalry to great effect in pursuing the Austro–Russian armies through the Danube Valley. However, he broke off the pursuit to occupy Vienna. This outraged Napoleon, who was further angered when Murat accepted Prince Bagration's armistice proposal at Hollabrunn on November 15, 1805.

The following month, Murat redressed his wrongs in Napoleon's eyes by collaborating with **Jean Lannes** on the capture of the key Danube bridgehead at Spitz and, at the battle of Austerlitz (December 2, 1805), exploiting the Austro–Russian retreat and turning it into a rout.

Napoleon created Murat Grand Duke of Cleve and Berg on March 15, 1806. Then, on October 14, of that year, at Jena, Murat deployed his cavalry in support of Lannes's attack on the center. He occupied Erfurt and Prenzlau, boxing in General **Gebhard Blücher** at

Lübeck. Blücher surrendered on November 7, 1806, and Murat went on to occupy Warsaw. He next fought at Golymin on December 26, 1806, and then at Eylau on February 7–8, 1807. It was at Eylau that he led perhaps the greatest cavalry charge in modern military history, hurling eighty squadrons—10,000 men—against the Russian center. Breaking through the Russian line, he charged over a seventy-gun battery, brought confusion and panic that broke the Russians' formation, then he re-formed into a single column, wheeled about, and charged back through the Russian center. He pursued the retreating Russians and fought at Guttstadt on June 9 and Heilsberg on June 10. After this, he directed the siege of Königsberg (June 11–16, 1807).

Murat was sent to Spain, titled Imperial lieutenant, on February 20, 1808. Hoping to be crowned king of Spain, he had to abandon this ambition when he was forced to suppress the Dos de Mayo uprising in Madrid on May 2, 1808—a brutal act that earned him the wrath of the Spanish people. At this time, too, Murat fell ill and left Spain on June 15. He, in effect, swapped nations with Napoleon's brother Joseph, assuming Joseph's title of King of Naples on August 1, 1808, while Joseph went to Spain for a brief reign as king.

Murat wanted to expand his modest kingdom and, in 1809, attacked the British positions in Sicily. The expedition failed. For the next three years, Murat presided over Naples, then joined Napoleon for the invasion of Russia in 1812. He was assigned command of the advance guard and fought at Ostronovo, on July 25–26, Smolensk, on August 17–19, and Borodino, on September 7. Entering Moscow on September 14, he suffered a surprise attack at Vinkovo on October 18 and was defeated by General **Mikhail Kutuzov.**

Beginning in December, Murat took charge of the long, costly retreat from Russia. At Elbing, he turned over what was left of the army to Eugène de Beauharnais, viceroy of Italy, and returned to Naples on January 18, 1813. With his kingdom imperiled, he opened negotiations with the British and Austrians in Sicily. When negotiations failed, he rejoined Napoleon in the midst of the German campaign.

Murat fought at Dresden (August 26–27) and at Leipzig (October 16–19, 1813), then returned to Naples and renewed negotiations. This time, he reached an agreement, pledging on January 26, 1814, to furnish 30,000 men to fight *against* France. However, after Napoleon's abdication, Murat's new allies decided that they could not trust him. In response, Murat attempted to return to Napoleon's fold after the Emperor landed in France, having left his first exile. Murat's strategy was to win Italy's independence from Austria, thereby gaining Napoleon's favor. Despite a few initial successes, however, Murat was defeated by the Austrians at Tolentino on May 2, 1815. He fled Naples for France, but was rebuffed by Napoleon, who refused to see him. Now, without a country, he appealed to England for asylum—and was again rebuffed. He set sail for Corsica on August 22. There he raised a small army with the hope of winning back his Neapolitan throne. The army was wiped out at Pizzo, and Murat, arrested, was tried for treason. Found guilty, he was executed by firing squad on October 13.

No marshal was more ambitious than Murat. Unfortunately for him, his grasp of strategy was poor. He was not a great independent commander. Where he excelled was in cavalry tactics, and in this he served Napoleon well—until his overweening ambition induced him to betray the Emperor and led him into a series of politically desperate and inept acts.

MURRAY, Lord George

BORN: October 4, 1694

DIED: October 11, 1760

SERVICE: War of the Spanish Succession (1701–14); Jacobite risings of 1715 and 1719; The Forty-Five (1745–46)

Brilliant Scots Jacobite general who fought for Charles Edward, the Young Pretender, Stuart claimant to the English throne, in the Jacobite rebellion of 1745–46

Murray was born at Huntingtower, near Perth, and joined the English army in Flanders toward the end of the War of the Spanish Succession in 1712. After this experience, he returned to Scotland and threw his support behind the Jacobite Rebellion of 1715. He fought at Sheriffmuir on November 13, but the rising petered out under inept leadership. Four years later, he joined the renewed uprising and suffered a wound at Glen Shiel on June 10, 1719. He was forced to seek refuge in Rotterdam during the fall of 1719 and did not return to Scotland until 1724, when he was granted a pardon.

Murray settled in Tullibardine upon his return and lived quietly for years. Just prior to the rebellion known as the Forty-Five, the Duke of Perth called on him on behalf of the Young Pretender, Prince Charles Edward Stuart. Murray demurred, but ultimately accepted a commission in the army of Bonnie Prince Charlie. Although he demonstrated little enthusiasm for the cause, he out-generaled Sir John Cope and achieved a major victory over the English at Prestonpans on September 21, 1745.

Murray drew the line at invading England, and when Prince Charles took Carlisle in November, he resigned his commission—only to allow himself to be persuaded to take up command again after the army reached Derby on December 4. Here he was met by two British armies, and he quickly beat a retreat. In January 1746, Murray arrived with his force at Stirling and covered Bonnie Prince Charlie's siege of Stirling Castle. In the process, Murray defeated an army led by Henry Hawley at Falkirk in a battle of January 17.

Despite the victories, Murray's Jacobites were wracked by disease. Troops also deserted the cause at an alarming rate. With remnant of his forces, Murray withdrew into the Highlands, seeking time to recoup and avoid a premature confrontation with the army of the **Duke of Cumberland.** At last, on April 16, Murray and Prince Charles led a 5,000-man army by night to Culloden Moor, near Inverness. They had planned on surprising the Duke of Cumberland—but, in fact, he was ready for them with some 9,000 men. The attacking Jacobites were easily repulsed and routed with heavy losses. Murray re-formed a mere 1,500 men at Ruthven, but Prince Charles had given up, and what was left of his army dispersed. Seeing that the situation was quite hopeless, Murray fled to the continent. He died in the Netherlands in 1760.

Murray was a very able leader, but, in the end, was overwhelmed by superior numbers.

MUSSOLINI, Benito

BORN: July 29, 1883

DIED: April 28, 1945

SERVICE: World War I (1914–18); Fascist March on Rome (1922); Conquest of Ethipoia (1935–36); Spanish Civil War (1936–39); World War II (1939–45)

Italian dictator and founder of European fascism, who led Italy to defeat in World War II

Benito Mussolini was the son of a blacksmith with strong socialist and anti-church beliefs. He grew into a singularly spirited, even unruly youth, avidly imbibing his father's beliefs, which he embellished with the romantic, even mystical tendencies of his mother, who readily persuaded her son that he was destined for great things. Mussolini was a voracious reader, who consumed the works of such political philosophers as Louis Auguste Blanqui, Friedrich Wilhelm Nietszche, Georges Sorel, and, most significantly, **Niccoló Machiavelli.**

Mussolini enrolled in the Salesian college of Faenza and then the normal school, from which he obtained a teaching certificate. At age eighteen, Mussolini obtained a post as a provincial schoolteacher and also traveled, living for several years in Switzerland and the Austrian Trentino. With broadened experience, he gave up teaching for socialist journalism, becoming editor of the Milan Socialist party newspaper *Avanti!* in 1912. At this stage in his development, the socialist Mussolini was a pacifist, producing article after article arguing against Italy's entry into World War I. Then, in the most momentous decision of his life, he suddenly abandoned the socialist party line and urged Italy's entry into the war on the side of the Allies. The party responded to this change of heart by expelling Mussolini, who then set up his own newspaper in Milan, *Il popolo d'Italia*. In this vehicle he developed and broadcast the message of what became the Fascist movement. First, however, Mussolini enlisted in the Italian army as a private in 1915, serving until he was wounded in the buttocks by trench mortar fragments early in 1917.

After recovering from his wounds, Mussolini resumed publication of his newspaper, and, on March 23, 1919, both encouraged and inspired by the grandiose poet, novelist, patriot, and adventurer Gabriele d'Annunzio, he and other war veterans founded in Milan a revolutionary nationalistic group they called the Fasci di Combattimento. Its name was derived from the Italian word *fascio*, "bundle" or "bunch," which suggested unity, and the symbol of the *fasces*, a bundle of rods bound together around an ax with the blade protruding, the ancient Roman emblem of power.

Fascism quickly evolved from the its origins in leftwing socialism to become a radical rightwing nationalism, even though many of Mussolini's early speeches were more radically pro-labor and anti-church than anything the socialist left would have found acceptable. But what captured the public's imagination and support was the nationalist message suffused in hazy visions of ancient imperial Roman grandeur. Not only was the Fascist message popular with the common man, it found in the influential d'Annunzio a powerful exponent, and he and Mussolini soon gained the backing of large landowners in the lower Po valley, leading industrialists, and senior army officers. To underscore and enforce his program, Mussolini created squads of thugs, the Blackshirts, who waged what amounted to a street-level civil war against all opponents: Socialists, Communists, Catholics, and Liberals. By 1922, Mussolini, having moved far from socialism, enjoyed the support of the rich and powerful, as well as the masses. On October 28, he led a Fascist march on Rome, intimidating King Victor Emmanuel III into forming a coalition government with his party. Like the dictators of ancient Rome, Mussolini obtained dictatorial powers set to last one year, and he used that period to do nothing less than refashion Italy's economic structure, slashing government expenses for public services, reducing taxes on industry to encourage production, and centralizing as well as consolidating government bureaucracy. These measures revitalized Italy's lackluster economy, and it became a commonplace to summarize Mussolini's achievement by simply observing that he "made the trains run on time."

During his first dictatorial year, Mussolini also replaced the king's guard with his own Fascist *squadisti* and the Orva, a secret police force. He greatly increased Italy's prestige in foreign affairs when he responded to the murder of some Italian officials at the hands of bandits on the Greek–Albanian border by demanding a huge indemnity from the Greek government, then bombarding and seizing the Greek island of Corfu. He also negotiated an agreement with Yugoslavia to obtain Italian possession of the long-contested Fiume.

At first, Mussolini studiously avoided attacking labor, though he did not hesitate in the brutal suppression of the strikes that traditionally crippled the country's industry. In 1924, Mussolini even relinquished—ostensibly—his dictatorial powers and called for new elections. However, he had taken care to secure legislation guaranteeing a two-thirds parliamentary majority for his party regardless of the outcome of the popular vote. Among the handful of Socialists elected that year despite Fascist domination was Giacomo Matteotti, who made a series of withering speeches in opposition to Mussolini and the Fascists, exposing political outrages that ranged from acts of intimidation and violence, to misuse of public funds, to murder. Shortly after these speeches, Matteotti's own murdered body was found, and a protracted parliamentary crisis ensued. Emboldened, the opposition press attacked Mussolini and his followers.

Mussolini responded by dropping all pretext of democracy. He imposed a single-party dictatorship and a policy of strict censorship. He authorized his henchmen to terrorize all opponents, and one liberal editor was even beaten to death. It was now that Mussolini moved openly against labor, solidifying his power base among Italian capitalists by abolishing free trade unions. He also secured the backing of the Catholic church by negotiating the Lateran Treaty of 1929, by which the Vatican was established under the absolute temporal sovereignty of the pope.

An absolute dictator, Mussolini was now called *Il Duce*, the Leader, and he prosecuted an aggressive foreign policy during the balance of the 1930s. Seizing as a pretext a clash over a disputed zone on the Italian Somaliland border, he invaded Ethiopia during

1935–36 without a declaration of war, unleashing aerial bombardment and poison gas on the civilian population. On May 9, 1936, Italy annexed the African nation. During this period, Mussolini also assisted Generalissimo **Francisco Franco** in the Spanish Civil War, and developed a fateful alliance with **Adolf Hitler**'s Germany during 1936–39.

In April 1939, Mussolini sent his armies to occupy Albania, but stayed out of World War II until June 1940, when the fall of France was in the offing and Germany seemed invincible. Hitler embraced Mussolini as a mentor and admired colleague in conquest, but he soon had reason to regret the alliance. Conquering a relatively primitive Ethiopian army was one thing, but against more sophisticated powers, Mussolini's army suffered one disaster after another in Greece and North Africa. By the middle of the war, the popular tide turned inexorably against the dictator, and the leaders of his own party abandoned him. King Victor Emmanuel dismissed Mussolini as premier on July 25, 1943, ordering his arrest. But Hitler sent a rescue force on September 12 and installed Mussolini as his puppet in northern Italy, territory that had yet to be taken by the advancing Allies.

By the spring of 1945, Allied forces were closing in on Mussolini. In April, he and his mistress, Clara Petacci, fled, only to be captured by Italian partisans at Lake Como. The couple was summarily executed by firing squad on April 28, and their half-naked bodies were strung up in a public square in Milan, exposed to public shame and desecration.

N

NAPIER, Charles James

BORN: August 10, 1782

DIED: August 29, 1853

SERVICE: Napoleonic Wars (1800–15); War of 1812 (1812–15); Conquest of Sind (1842–43)

Distinguished British general of the Napoleonic era

London-born Napier was commissioned an ensign (1794) in the 33d Foot, under Lieutenant Colonel Arthur Wellesley, who would become the Duke of **Wellington** (1794). Napier did not begin his term of active duty until 1799, when he saw service in Ireland. In 1801, he transferred to the Ninety-Fifth Regiment of Foot and made captain in 1803, when he was appointed aide-de-camp to General Henry Edward Fox. He served in this post until 1805, when he was promoted to major in the Fiftieth Foot and, in September 1808, sent to Portugal, where he participated in General Sir John Moore's invasion of Spain during September 1808–January 1809. During the Battle of Corunna, he was wounded and then taken prisoner on January 16, 1809, but was soon released.

In January 1810, he fought under Wellington in Spain and was severely wounded in the face at the Battle of Busacco on September 27. Napier did not return to duty until 1811, when he fought at Fuentes de Onoro on May 5. On June 27, he was promoted to lieutenant colonel with command of the 102nd Regiment of Foot—though he spent an extended leave in England before actually assuming command of the regiment in January 1812. Napier led the 102nd to Bermuda, where he conducted operations along the East Coast of the United States during the War of 1812. He fought at Craney Island, Virginia, and at Ocracoke, North Carolina in 1813.

In September 1813, Napier transferred back into the Fiftieth Foot, but Napoleon had abdicated and gone into his first exile before Napier reached Europe in April 1814. Napier enrolled for study at a military college at Farnham in September 1814, then volunteered for service under Wellington in Belgium when Napoleon returned from Elba. Although he did not see action at Waterloo, he did participate in the balance of the campaign, and he was part of the Allied entry into Paris.

Napier resumed his studies at Farnham late in 1817 and accepted appointment as Inspecting Field Officer for the Ionian Islands in May 1818. Appointed Resident for Cephalonia in March 1822, he proved an able civil governor. During this period, he befriended the poet George Gordon, Lord Byron, who was active in the Greek independence struggle. Napier volunteered his professional advice to the Greek freedom fighters during 1823–25, and he even attempted to enter the Greek service. However, because he was unwilling to renounce his British army commission, he was unable formally to join the patriots.

Napier briefly returned to England, where he married an aging widow, Elizabeth Kelly, in April 1827, then returned to Cephalonia until February 1830, when his wife's failing health necessitated his return to England. Elizabeth died in July 1833, and Napier took up residence in Normandy, where he attended to a pair of daughters born out of wedlock in Cephalonia. He remarried—another widow, Frances Alcock—in 1835 and two years later was promoted to major general.

In April 1838, Napier was recalled to active duty as commander in the Northern District in England, and in June 1841, he accepted a command in the Bombay presidency. He set sail for India and assumed his duties at Poona in December. In August 1842, he also assumed command of British troops in the Baluchi province of the British protectorate of Sind. Napier put down a rebellion there, defeating some 20,000 Baluchi warriors with a mere 2,200 British troops on February 16, 1843. He scored a second decisive victory at Hyderabad on March 24, and, with that, resistance in Sind came to an end.

Napier was made a Knight Grand Commander of the Bath (G.C.B.), and Sind was annexed to the Empire. Despite his honors, Napier was embroiled in political struggles—largely the result of his grudge-holding temperament—and wearily resigned his offices in July 1847. Early in 1849, Wellington appointed him Commander-in-Chief for India, and, once again, Napier left England, arriving in Calcutta on May 6, 1849. Within a short time, he fell into a dispute with India's governor-general, Lord Dalhousie, over Napier's refusal to implement an ordered pay cut for sepoys on the Northwest Frontier. Dalhousie also objected to Napier's incorporating a native Gurkha mercenary unit into the Indian Army. At last, a frustrated Napier resigned in November 1850 and returned to England in March 1851. He retired in order to write on military subjects, but soon fell ill.

Napier was an extremely competent commander and a very able civil administrator. He was contentious, however, and earned many enemies during his long career.

NAPOLEON I
(Napoleon Bonaparte)

BORN: August 15, 1769

DIED: May 5, 1821

SERVICE: French Revolutionary Wars (1792–99); Napoleonic Wars (1800–15)

French Emperor; the single most famous conqueror-commander in modern military history

Napoleon I reshaped Europe, commanding hatred and adoration in his own time and creating a personal aura of fascination that endures to this day. Napoleon's early

Napoleon in his study. From Charlotte M. Yonge, Pictorial History of the World's Great Nations, *1882.*

life offered little hint of the momentous role he was to play in the world. Born in Ajaccio, Corsica, the second surviving son of Carlo and Marie-Letizia Buonaparte, he attended a French military school at Brienne-le-Chateau from April 1779 to October 1784, but was isolated and spurned as a foreigner—and a provincial bumpkin. Young Napoleon grew reclusive, throwing himself into his studies, yet graduating no better than near the bottom of his class: number forty-two out of fifty-eight. After further study at the Military Academy in Paris, he was commissioned a second lieutenant of artillery on September 1, 1785, and was assigned to La Fère Artillery Regiment.

During this early period, Napoleon read the works of the military theorist J. P. du Teil and, beginning in 1789, was also caught up in a Corsican nationalist movement. He was transferred to the Artillery Régiment du Grenoble in February 1791, attaining promotion to first lieutenant, at which time, he became active in the Jacobin Club of Grenoble, traveled to Corsica, and successfully lobbied for election as lieutenant colonel of the Ajaccio Volunteers on April 1, 1792. Torn between loyalty to Corsica and to France, Napoleon fell into a dispute with the anti-French Corsican nationalist Pasquale Paoli and fled with his family to Marseille on June 10, 1793.

When the Revolt of Midi (July) broke out, Napoleon joined in the fray on the side of the Republicans and was appointed commander of artillery in the Army of Carteaux. On September 16, 1793, he participated in the successful siege of Toulon, a royalist stronghold that had welcomed a counterrevolutionary British fleet. By December 17, the British had been driven out, and Toulon fell to the Republicans within five days. In recognition of his able command of the artillery, Napoleon was promoted to brigadier general and subsequently named artillery commander of the French Army of Italy in February 1794. Following the overthrow of Maximilien Robespierre in July 1794, however, Napoleon was imprisoned from August 6 to September 14, 1794, then released and offered artillery command of the Army of the West. He declined and was assigned instead to the Bureau Topographique, shortly thereafter achieving appointment as second in command of the Army of the Interior. In that capacity, he brought to an end the Parisian rebellion of 13 Vendémiaire (October 5, 1795), dispersing the insurrectionists with what he termed a "whiff of grapeshot," thereby saving the ongoing constitutional Convention. The Directory—the new governing body created by the Convention— rewarded Napoleon with full command of the Army of the Interior. At this time, in March 1796, Napoleon married Josephine de Beauharnais, the somewhat notorious widow of a republican general.

Following the revolution, France was beset by many enemies, and, immediately after he assumed command of the Army of the Interior, Napoleon moved against Piedmontese and Austrian forces, bringing about an armistice with the Piedmontese by the end of April after defeating them at Modovi on April 21, obtaining the cession of Savoy and Nice to France. He moved next against the Austrians, defeating them at Lodi on May 10, then entered Milan on May 15. After driving the Austrian forces out of Lombardy during May and June, he attacked Mantua—last Austrian stronghold in the region—which fell in February 1797 after a protracted siege. When Napoleon began an advance on Vienna, the Austrians sued for peace, and it was Napoleon who personally negotiated the Treaty of Campo Formio on October 17, 1797, which ended the War of the First Coalition—first of the French revolutionary wars.

Napoleon reshaped Italian politics by creating the Cisalpine Republic and establishing several puppet governments in Italy. Moreover, he pillaged Italian art collections to help finance French military operations. He was hailed as a hero by the Directory, which drew up plans to send Napoleon next to invade England, but the general proposed an alternative grand strategy: Invade Egypt to establish a staging area for an invasion of British India. He sailed on May 19, 1798, with 35,000 troops bound for Alexandria. On the way to Egypt, he took Malta, managing to evade the British fleet under the command of **Horatio Nelson,** then occupied Alexandria and Cairo. Napoleon pledged to preserve Islamic law, but he set about modernizing the secular government during September 1798–February 1799. In the meantime, however, Nelson destroyed the French fleet at Aboukir Bay on August 1, 1798, cutting Napoleon off from France. Undaunted, Napoleon continued the Egyptian campaign, and when the Ottoman Turks declared war on France in February 1799, he moved to head off an invasion of Egypt by striking preemptively at Syria. Turkish troops under British command checked his advance at Acre during March 15–May 17, and, stalled, the French army was suddenly swept by plague endemic to the region. Napoleon had little choice other than to bring his army back to Cairo in June. On July 25, he defeated an Anglo–Turkish invasion attempt to retake Aboukir.

While Napoleon was prosecuting the campaign in the Middle East, the situation in Europe reached a crisis, as French forces were suffering defeat at the hands of the Second Coalition. Napoleon set off for France on August 24, 1799, arriving in Paris on October 14, where he played a key role in the coup d'etat of November 9

(18 Brumaire) against the Directory. Appointed commander of the Paris garrison, he was made one of three consuls in a new Consulate. In February 1800, under the Constitution of the Year VIII, he was elected first consul, with power to appoint members of the council of state, government officials, and judges. From this position, Napoleon was able to acquire dictatorial power, and he radically centralized the government, effectively bringing it under his personal control. He was the right man at the right time. Racked by years of revolutionary terror on the one hand and general lawlessness on the other, always menaced by a still-powerful royalist faction, the people readily invested authority in one strong man. At this time Napoleon oversaw the creation of the Napoleonic Code, a supremely rational set of civil laws. Napoleon also concluded the 1801 Concordat with Pope Pius VII, reestablishing Roman Catholicism as the state religion. He next restructured the French national debt, setting the tottering French economy on a sound footing. From this new basis, he launched programs to develop industry and the educational system, initiating a series of public works inspired by examples of Roman imperial splendor.

While he was pursuing these domestic reforms, Napoleon continued to establish France as a world power, defeating the Austrians at the Battle of Marengo on June 14, 1800, bringing about the Treaty of Luneville (February 9, 1801) and initiating a brief interval of peace with all of Europe, including England, which signed the Treaty of Amiens (March 27, 1802). At this time, on August 2, 1802, a plebiscite created Napoleon first consul for life.

With his authority affirmed, Napoleon began to reshape the face of Europe. In the Netherlands he created the Batavian Republic and in Switzerland the Helvetic Republic. He annexed Savoy-Piedmont, then took the first step toward abolishing the decrepit Holy Roman Empire, instituting the Imperial Recess of 1803, which consolidated free cities and minor states dominated by the Holy Roman Empire. He also launched a campaign to recover the island nation of Haiti, which had successfully rebelled against French colonial domination. By May 1803, these acts of renewed aggression, coupled with a refusal to grant trade concessions to Britain, reignited war between the two nations.

As Napoleon was outfitting an army of 170,000 to invade England, a British-backed assassination scheme was discovered, sending the French Senate into a panic. Petitioned to establish a hereditary dynasty, the first consul seized the opportunity, and, on December 2, 1804, as Pope Pius VII looked on, he crowned himself emperor. In a stroke, he abolished the republic for which he had fought, instantly creating a royal court populated by former republicans and royalists alike. Nor was the emperor content to create a dynasty in France alone. He ultimately installed members of his family on the thrones of Naples, Holland, Westphalia, and Spain.

In 1809, Napoleon divorced Josephine on the grounds that she had failed to bear a male heir. On April 2, 1810, he married Marie Louise, daughter of the Austrian emperor, and a son was born within a year. In the meantime, he designed a plan to draw the British fleets away from England so that he could finally execute his long-cherished invasion. This strategy collapsed, however, and Austria now prepared to renew war. On May 26, 1805, Napoleon was crowned king of Italy, and during July through September maneuvered against the Austrians led by General Karl Mack von Leiberich, encircling and defeating them at Ulm during September 25–October 20, 1805. As usual, however, the triumph on land was offset by defeat at sea when Admiral Nelson annihilated most of the French ships at the Battle of Trafalgar on October 21, thereby saving England from all threat of invasion.

Napoleon nevertheless advanced on Vienna, which he took on November 13, then continued into Moravia, where the Russian army under Marshal **Mikhail Kutuzov** offered battle at Austerlitz. Napoleon's complete victory here on December 2 was his greatest single military triumph, and by the end of the month, Austria signed the Treaty of Pressburg, relinquishing Venice and Dalmatia to Napoleon's Kingdom of Italy. On July 12, 1806, Napoleon formally and finally abolished the Holy Roman Empire, organizing in its stead the Confederation of the Rhine, a French protectorate of German states. Now wishing to smooth relations with England, Napoleon offered to return Hanover to British control. That served only to provoke war with Prussia in September.

Under Prussian direction, the Fourth Coalition against Napoleon was created. Its armies were badly beaten at the battles of Jena and Auerstadt (both on October 14, 1806). Following this, Napoleon met the Russian army at Eylau on February 8, 1807, which resulted in a draw, and then at Friedland, on June 14, 1807, a clear victory for the emperor, which resulted in Czar **Alexander I**'s agreeing to the Treaties of Tilsit in July 1807. These created the French-controlled Grand Duchies of Warsaw and the Kingdom of Westphalia.

This marked the zenith of Napoleon's sway over Europe, and he controlled more of the continent than anyone had before him. Yet he was not satisfied. Unable to defeat England by military means, he decided to apply an economic stranglehold in 1806–07 with the Continental System, a blockade of British trade. This high-handed measure produced universal unrest throughout Europe, and Portugal immediately announced that it would not participate in the blockade. Napoleon responded with the Peninsular War to bring Portugal into line. This served to provoke civil unrest in Spain and the ultimate abdication of King Charles IV and his son Ferdinand VII during May 5–6, 1808, as well as a revolution against Napoleon's chosen successor to the Spanish throne, Joseph Bonaparte.

With Napoleon engaged on the peninsula, Austria formed the Fifth Coalition, beginning a war that produced early Coalition victories, but that ended in a decisive French triumph at the Battle of Wagram on July 5–6, 1809. Napoleon married Marie-Louise following the July 12 armistice with Austria, which, by the Treaty of Schonbrunn (October 14, 1809), relinquished Illyria and Galicia.

Despite this major victory, Spain and Portugal were slipping from Napoleon's grasp, and now Russia also refused to participate in the Continental System. An overextended Napoleon invaded that country on June 23–24, 1812, the Russian armies steadily retreating before the emperor's advance, actually drawing him deeper and deeper into that vast land. Victorious at Borodino on September 7, 1812, he arrived in Moscow within a week. But Czar **Alexander I** refused to surrender, and Russian partisans set fire to the city. Under attack by freshly reinforced Russian armies, and with the cruel Russian winter closing in, Napoleon began a massive retreat that degenerated into full-scale disaster. Considering the conditions he faced, the retreat was extraordinary; for Napoleon managed to preserve himself and the core of his Grand Army. Nevertheless, by December, much of his force was destroyed.

In the aftermath of this terrible defeat, the Prussians abandoned their short-lived alliance with the French to form a Sixth Coalition against France, consisting of Prussia, Russia, Britain, and Sweden. In Paris, however, the emperor built a new army, with which he defeated Coalition forces at Lützen on May 2, 1813, and at Bautzen on May 20–21, bringing about a brief armistice. In August, Austria joined the Sixth Coalition, and Napoleon defeated Austrian troops at Dresden during August 26–27. Grossly outnumbered, however, the French were in turn defeated at Leipzig on October 16–19, 1813. Napoleon retreated across the Rhine, but refused to give up any conquered territory.

The next year, Coalition armies invaded France itself, but the emperor prevailed against each attempt to penetrate to Paris until his marshals mutinied, and the capital fell on March 31, 1814. A few days later, on April 6, Napoleon abdicated in favor of his son. The allies rejected this, forcing Napoleon to abdicate unconditionally on April 11. He was exiled to the British-controlled island of Elba.

In 1815, Napoleon escaped exile and returned to France, landing at Cannes on March 1. The Bourbon monarch, Louis XVIII fled in terror, and Napoleon occupied Paris on March 20. He communicated his peaceful intentions to the Congress of Vienna, which spurned him. Fearing a combined attack by Russian and Austrian armies, Napoleon resolved to make a preemptive strike in order to divide and destroy Prussian and Anglo–Dutch armies in Belgium. Napoleon prevailed against the Austrians at Ligny on June 16 and against the British at Quatre Bras on the same day, but he was defeated at what might be the most famous battle in modern European history—Waterloo—by the Duke of **Wellington** reinforced by Austrians under **Gebhard von Blücher** on June 18, 1815.

Napoleon returned to Paris, abdicated for the second time on June 23, and was again exiled, this time to the island of Saint Helena. There he composed his memoirs and grew increasingly ill. On May 5, 1821, Napoleon died. Some believe the cause was cancer of the stomach, while others have theorized that he died of gradual arsenic poisoning, which may have been the result of a deliberate assassination scheme or due to overmedication with the arsenic-based drugs widely used during the period.

NARSES

BORN: ca. 478

DIED: 573

SERVICE: Gothic War (535–554)

Byzantine general who cleared Italy of the Goths and other invaders, thereby restoring Imperial authority

Narses was born in Persarmenia and was a court eunuch, who became a favorite of Theodora, wife of Emperor **Justinian.** Through her influence, Narses became commander of the eunuch bodyguard and, ultimately, grand chamberlain. He performed invaluable service during the Nika riots of January 532, skillfully using bribery to win over the "Blue" faction and thereby keep Justinian in power. He then led an army to Alexandria to enforce the installation of the Imperial patriarchal candidate, Theodosius, and to crush rioting that had erupted after a disputed election in 535.

Narses assisted **Belisarius** in his Italian campaigns during 538, but soon fell out with Belisarius and returned to Constantinople in 539.

During the summer of 551, Narses fought Huns as well as Gepid and Lombard raiders in the Balkans. In the fall, Justinian sent him to Italy with 20,000 men against the Goths. Marching on Rome, he defeated Totila's army at Taginae, decimating Totila's forces in a June 552 battle. He then dealt with the Gothic army under Teias, beating it Monte Lacteria in 553.

When he was faced with invasion by a superior Frankish force, he took advantage of his army's superior mobility and was able to destroy the heavy infantry phalanx of Buccelin at Casilinum in 554. With the invaders cleared, he governed the country and restored Imperial authority. He was recalled to Constantinople by Justin II in 567.

NAVARRO, Pedro, Count of Oliveto

BORN: ca. 1460

DIED: 1528

SERVICE: Turko-Venetian War (1499–1503); Neapolitan War (1501–03); North African expeditions (1508–11); War of the Holy League (1511–14); First Hapsburg-Valois War (1521–26); Second Hapsburg-Valois War (1527–30)

Spanish general, who also served the French; famed as a military engineer

Very little is known about Navarro's early life other than that he was born probably at Garde in the Roncal valley and fought for a time in the service of Cardinal Juan de Aragon before 1485. By the 1490s, he was working in Italy as a *condottiere*—a mercenary soldier—fighting against the Barbary pirates. In 1499, he was hired by Gonzalo de Cordoba and fought against the Turks, participating in the siege and capture of the Turkish fortress of Cephalonia in 1500. Navarro breached the fortress with mines, and this action was greatly admired at the time as a magnificent feat of military engineering.

Still in the employ of Gonzalo, Navarro marched to Naples, then defended Canosa (1502) and Taranto (1503) against the French. At Cerignola he designed and oversaw field fortifications that were instrumental in Gonzalo's victory over the Duke of Nemours on April 28. Navarro also figured in the Spanish victory at the Garigliano River on December 29, after which Navarro was created Count of Oliveto.

After returning for a period to Spain in 1507, Navarro followed Francisco Jiménez on an invasion of North Africa. He designed and built a floating artillery battery, which was used to help capture Veléz de la Gomera in 1508. His battery also was used in the taking of Mazalquivir and Oran in 1509.

In 1510, Navarro commanded Spanish forces marching to the conquest of Bougie, Algiers, Tunis, Tlemcen, and Tripoli. He broke off this campaign in 1510, when his services were required in Italy to checkmate renewed aggression from the French. In the service of Viceroy Ramón de Cardona, he designed lightweight carts to carry light field guns that could be easily maneuvered into position and that were extremely effective in breaking up enemy lines and formations. Despite these weapons, however, the Spanish–Papal army was defeated by the French at Ravenna on April 11, 1512), and Navarro was captured. When Spain's King Ferdinand V declined to ransom him, Navarro offered his services to Francis I of France, who eagerly accepted.

In the French service, Navarro accompanied the armies of Francis against Milan during 1515–16. At La Bicocca, on April 27, 1522, Navarro was captured by the Spanish, who did not release him until the signing of the Treaty of Madrid on January 14, 1526. Once again, Navarro returned to the French service—and, once again, he was taken prisoner in Italy in 1527. He died the following year, while he was being held in Castel Nuovo, a Neapolitan prison.

NELSON, Horatio, Viscount

BORN: September 29, 1758

DIED: October 21, 1805

SERVICE: American Revolution (1775–83); French Revolutionary Wars (1792–99); Napoleonic Wars (1800–15)

British admiral in the French Revolutionary and Napoleonic wars; perhaps the greatest naval officer to sail for Britain

Nelson was born at Burnham Thorpe in Norfolk, England, the son of a clergyman. After schooling at Norwich, Downham, and North Walsham, he joined the Royal Navy in 1770 as a midshipman, serving aboard on H.M.S. *Raissonable* under his uncle, Captain Maurice Suckling. In 1771, he transferred to a new ship, H.M.S. *Triumph,* also under the command of Suckling. During 1771–74, he sailed to the West Indies and on an Arctic expedition, then served aboard H.M.S. *Seahorse* on an East Indian cruise during 1774–76. He served briefly aboard the frigate H.M.S. *Worcester,* then secured a promotion to lieutenant assigned to H.M.S. *Lowestoft* in 1777 in the West Indies.

Nelson's first command was H.M.S. *Badger,* followed by promotion to captain, with command of the frigate *Hinchbrook* in 1779. During 1780, while conducting operations off Central America, against San Juan del Norte, Nicaragua, Nelson contracted a jungle fever and, desperately ill, had to return to England.

Upon his recovery, Nelson performed convoy escort duty in command of H.M.S. *Albemarle,* then returned to the West Indies. Following the conclusion of the American Revolution, he was given command of the frigate H.M.S. *Boreas,* again in the West Indies.

Nelson made a stormy marriage with Frances Nisbet (March 11, 1787) and eagerly put out to sea after war broke out with France. He was made captain of the sixty-four-gun H.M.S. *Agamemnon* on February 7, 1793. While serving in the Mediterranean, Nelson met the beautiful Emma, Lady Hamilton, wife of the British ambassador in Naples, and he began a long and celebrated liaison with her.

In May 1794, he served on and near Corsica and around Calvi during June–August. In battle here, he lost his right eye. He fought under Admiral William Hotham at Genoa on March 14, 1795, then sailed along the Italian Riviera, where he operated against the French lines of communication. With the entry of Spain into the war as an ally of France, Nelson was dispatched with two frigates to evacuate the British garrison from Elba during December 1796–January 1797.

Nelson was elevated to commodore by Admiral Sir John Jervis, and Nelson did not disappoint him. He was instrumental in Jervis's victory off Cape St. Vincent on February 14, 1797. Through brilliant and daring seamanship, he cut off the escape route of a portion of the Spanish fleet and captured two vessels in the process. Promoted to rear admiral, Nelson was also knighted.

During the spring of 1797, Nelson was assigned to blockade Cadiz. On July 24, 1797, he attacked Santa Cruz de Tenerife, failing to achieve his objective and receiving a severe wound, which resulted in the loss of his right arm. By April 1798, Nelson had recovered from his injury and rejoined Jervis off Gibraltar. Jervis dispatched him with a small squadron to probe French naval and land-based operations at Toulon on May 2. He was delayed by a May 20 storm, which damaged his flagship, H.M.S. *Vanguard*. By the time repairs had been effected, the French had left Toulon. Joined by Commodore Sir Ernest Charles Troubridge, who brought ten additional ships of the line, Nelson scoured the Mediterranean until, on August 1, 1798, he located the French fleet at Aboukir Bay, the fleet having just convoyed **Napoleon** Bonaparte's army to Egypt. Nelson wasted no time in pressing an attack, taking the French totally by surprise. The titanic battle was fierce. Troubridge lost H.M.S. *Culloden* when it ran aground, and 900 British sailors became casualties—including Nelson, who was again wounded. However, the French fleet was crushed. Only two of thirteen ships-of-the-line and two of four frigates survived. This great victory meant Britain's control of the Mediterranean, and it doomed Napoleon's army in Egypt.

Following Aboukir Bay, Nelson was created Baron Nelson of the Nile. On September 22, 1798, he arrived in Naples to support Neapolitan operations against the French. The campaign faltering, he evacuated the royal family to Palermo late in January 1799, then returned to blockade Naples and Malta. After the French had been ousted from Naples, Nelson returned the royal family in June 1799. Nelson placed the highest priority on these operations, and when Jervis's successor, Admiral Keith, ordered Nelson to Minorca while he was in the midst of the evacuation, Nelson ignored the order. He understood that the British base there was in no danger. Nevertheless, Nelson was reprimanded, and shortly after he joined Keith at Livorno on January 20, 1800, he was ordered back to England. He traveled overland, in company with Lady Hamilton—and her husband.

After promotion to vice admiral on January 1, 1801, Nelson was appointed second in command in Admiral Sir Hyde Parker's Baltic Fleet. During the Battle of Copenhagen on April 2, he once again ignored a superior's command, refusing to heed Parker's order to break off action. In an instantly famous gesture, he put the spyglass to his blind eye, remarking that *he* could see no signal from Parker. The result of Nelson's disobedience was a victory that beat the Danes into a much sought-after armistice.

Nelson succeeded to command of the Baltic Fleet after Parker was recalled in May. However, hostilities there ended in June, and Nelson, exhausted and broken in health, returned to England. He returned to sea in 1801, but was repulsed at Boulogne when he attempted to destroy invasion barges there on August 15. Following the Peace of Amiens—from March 27, 1802 to May 16, 1803—Nelson lived in Surrey with the Hamiltons. In June 1803, he was recalled to duty after war with France resumed. Given command of the Mediterranean Fleet, he arrived off Toulon and set up a blockade. Under cover of bad weather, however, the French fleet under Admiral Pierre Charles Villeneuve escaped on March 30, 1805. Nelson sailed after it, mistakenly believing Villeneuve had made for the West Indies. From April to August, he sailed to the West Indies and back, finally locating the French fleet at Cadiz in October. On the eighteenth, Villeneuve sortied to enter the Mediterranean in order to support Napoleon's Italian campaign. Nelson, lying in wait, engaged him off Cape Trafalgar on October 21, 1805. Thus commenced the most famous battle in modern naval history. Directing the combat from his flagship H.M.S. *Victory*, Nelson conducted a highly unorthodox departure from standard line-ahead attack tactics, leading a column of twelve ships toward the center of the combined Franco–Spanish fleet while, to the south, Vice Admiral Cuthbert Collingwood led fifteen ships. The two-column attack brought utter confusion among the French: one of their vessels was sunk outright, and eight French as well as nine Spanish ships were captured.

The battle was marked by sparse British casualties—among these few, however, was Horatio Nelson, cut down and mortally wounded by a sharpshooter's musket ball fired from Villeneuve's flagship *Bucentaure*. In great pain, he was taken below decks on the *Victory*, living long enough to see his victory.

Nelson is perhaps the most celebrated combat sailor in history. He possessed every quality of seamanship and military leadership. Although supremely confident, he was always flexible and encouraged contributions from his trusted officers. While he himself was supremely capable of independent action and often scorned for his egotism and arrogance, his tactics—as at Trafalgar—characteristically hinged upon close cooperation and coordination with key subordinates.

NERO, Gaius Claudius

ACTIVE: 214–199 B.C.

SERVICE: Second Punic War (219–202 B.C.); Second Macedonian War (200–196 B.C.)

Roman consul and general who fought Hannibal Barca

Nothing is known of Nero's birth and early life. He first enters history as the object of censure for having failed to heed the orders of **Marcus Claudius Marcellus** at Nola in 214. He appears next as praetor and propraetor, serving under the consuls at the siege of Capua during 212–211. He was given independent command during 211–210 in Spain. He briefly outgeneraled Hasdrubal, **Hannibal Barca**'s brother, but, ultimately, Hasdrubal evaded him and escaped capture in 210.

Nero next served under Marcellus at Canusium in 209 and, two years later, was elected consul. From this office, he launched a campaign against Hannibal in Campania and enjoyed a degree of success. When he learned that his old adversary Hasdrubal was marching to Italy with reinforcements to from Spain early in 207, Nero hastened northward with 6,000 handpicked troops in an impressive sustained forced march. He reached his fellow consul, Marcus Livius Salinator, in time to destroy Hasdrubal's army at the River Metaurus. This was the signal victory of the Second Punic War, for which both Nero and Salinator were elevated to censors in 204. Unfortunately, the two censors despised one another to such a degree that their hatred became a public scandal.

Nero served on the Roman embassy to Philip V of Macedon, prelude to the Second Macedonian War (201–200). Presumably, Nero fought in the war, but the details of his role are not known. Nor are the date and cause of his death.

Nero provoked strong reactions from colleagues, some of whom considered him a brilliant commander, while others faulted his tendency to impulsiveness and rash decision making. With **Scipio Africanus,** Nero was the only military match for Hannibal Barca.

NEY, Michel, Prince de la Moskova, Duc of Elchingen

BORN: January 10, 1769

DIED: December 7, 1815

SERVICE: French Revolutionary Wars (1792–99); Napoleonic Wars (1800–15)

Most celebrated and controversial of the marshals serving under Napoleon I

People who know nothing else about **Napoleon**'s armies know the name of Marshal Michel Ney. The most dashing and illustrious of the Emperor's commanders, Ney was also the most controversial. Vigorous, aggressive, but moody and arrogant, he was loved by his men but distrusted by his colleagues and regarded as insubordinate by his superiors. An early supporter of Napoleon and fiercely loyal to him, he nevertheless disliked the emperor personally. He led the other marshals in demanding Napoleon's first abdication in 1814, but declared for him upon his return from exile in Elba. A brilliant corps commander and tactician, Ney was an indifferent strategist who performed poorly when asked to command more than a single corps. Among these contradictions Napoleon himself identified Ney's one wholly unambiguous quality: he is, the emperor declared, "the bravest of the brave."

Son of a barrel cooper who had served in the Seven Years' War, Ney was born Saarlouis, Alsace, and was educated for a civil service position as an overseer of mines and forges, but quickly enlisted (February 12, 1787) in a regiment of hussars. He rose steadily through the ranks, serving as aide-de-camp to two generals (Lamarche and Colaud) and given a cavalry command of 500 men under General **Jean-Baptiste Kléber,** where he made a name for himself in the war with Austria. Promoted to *général de brigade* after a brilliant performance during the Battle of Altenkirchen (June 4, 1796), he became *général de division* after his role in the fall of Würzbourg (July 15, 1799). Instrumental in the victory at Hohenlinden (December 3, 1800), he came to the attention of Napoleon, who appointed him *Maréchal d'Empire* (May 19, 1804), making him twelfth in command seniority. (Though his military prowess certainly merited the appointment, his 1802 marriage to Agalée Louise Auguiée, protégée of the future Empress Josephine, must also have helped.)

Ney was a key figure in a series of victories against the Third and Fourth Coalitions (Austria, Great Britain, Sweden, Russia, and subsequently Prussia), including Elchingen (October 14, 1805), which led to the surrender of General Karl Mack's Austrian army at Ulm (October 20, 1805); followed by Jena-Auerstadt (October 14, 1806), which neutralized the Prussian army; Magdeburg (November 8, 1806); then Eylau (February 8, 1807) and Friedland (June 14, 1807), which led to the capitulation of Prussia and Russia in the Treaties of Tilsit (July 7–9, 1807).

Ney's conduct earned him the title of duc d'Elchingen, and he was sent to fight in Spain and Portugal. There, however, the French juggernaut faltered, and Ney, so recently in high favor, became the object of Napoleon's ire. When Marshal **Jean Lannes** defeated General Castaños at Tudela (November 23, 1808), Ney was unjustly blamed for failing to cut off the Spaniard's retreat. Ney did lay successful siege to Almeida and Ciudad Rodrigo (July 10, 1810), but in the process was insubordinate to his commander, **André Massena.** Ney also showed his tactical skill and personal courage to great advantage in rear guard actions against the Duke of **Wellington**—a performance that presaged his conduct of the retreat from Moscow—but relations between him and Massena steadily deteriorated and, early in 1811, Massena relieved the marshal of his command.

The next year, Napoleon appointed Ney to head III Corps in the Russian campaign. He was instrumental in brilliant victories at Smolensk (August 17, 1812) and Borodino (September 7, 1812), but it was in defeat, during the long, heartbreaking retreat from Moscow, that Ney's greatness as a tactician and leader of men was most fully realized. Through the bitter Russian winter, Ney, musket in hand, fought a seemingly endless series of rear guard actions with an ever-diminishing body of men and, in doing so, was largely responsible for saving as much of Napoleon's army as could be saved.

When Napoleon sent his troops into Saxony following the Russian debacle, Ney was appalled—"the machine no longer has either strength or cohesion: we need peace to reorganize everything"—but nevertheless served as III Corps commander, winning victories at Lützen (May 2, 1813), Bautzen (May 20–21), and Dresden (August 26–27), but suffering a defeat so disastrous at Dennewitz—in which he lost 24,000 men—just before Napoleon's even costlier defeat at Leipzig (October 16–19), that he contemplated suicide.

After recovering emotionally from Dennewitz and physically from a wound received at Leipzig, Ney participated in the defense of France, and, when the campaign on French soil at last proved hopeless, it was Ney who spoke for the other marshals in demanding the abdication of the emperor.

With the restoration of the Bourbons, Ney was appointed a member of the Council of War and given a cavalry command. But the contentious soldier soon chafed under the Old Regime and, on Napoleon's return from Elba, joined him. He fought at Quatre-Bras (June 16, 1815) and Waterloo (June 18, 1815), where he had four horses shot out from under him.

After Napoleon's final defeat, Ney, apparently too stubborn to go into self-imposed exile, took refuge in an isolated chateau near Aurillac. There, on August 3, 1815, he was arrested. When, during his trial for treason, one of his lawyers protested that the accused was not, in fact, a French citizen because Saarlouis, his birthplace, was no longer French, Ney rose up and declared: "I am French, I shall know how to die like a Frenchman!" He was executed before a firing squad on December 7, 1815.

NICHOLAS (Nikolai Nikolaevich Romanov), Grand Duke

BORN: November 18, 1856

DIED: January 6, 1929

SERVICE: Russo–Turkish War (1877–78); World War I (1914–18)

Russian general—one of the few of great ability during the period

Born in St. Petersburg, Nicholas was the son of Grand Duke Nikolai Nikolaevich, younger brother of Czar **Alexander II.** He joined the army in 1872 and served as a staff officer under his father, who was army commander in chief during the Russo–Turkish War. Following this war, Nicholas transferred to the Guard Hussar Regiment, which he came to command by 1884. A decade later, in 1895, he was appointed inspector general of cavalry and, during the ten years he held that post, did much to reform and modernize the service.

In 1905, Nicholas became commander of the St. Petersburg Military District, a post from which he continued to advocate and introduce reform and modernization. He was one of a handful of Russian generals who studied the many errors made during the Russo–Japanese War and worked to redress these faults. From 1905 to 1908, he served as president of the Imperial Committee of National Defense, but was dismissed by V. A. Sukhomlinov when he was appointed minister of War in 1909. The war minister was responsible for making plans to prepare the nation for what seemed like imminent war with Germany and Austro–Hungary, but Sukhomlinov stubbornly refused to consult with Nicholas, and the result, at the outbreak of World War I, was a series of disastrous losses for the Russians. Yielding to a public outcry—as well as the advice of government officials—the czar appointed Grand Duke Nicholas commander in chief of the army.

With the war already under way, Nicholas faced an impossible task. On the whole, the Russian army was obsolescent, ill-equipped, poorly trained, and ineptly led. Nevertheless, Nicholas eked out some success, and it was largely due to his leadership that the army recovered from the devastating losses of the Battle of Tannenberg. However, in the wake of the successful Austro–German offenses of April–September 1915, ousted Nicholas and assumed command of the army himself. Thus the fate of the Russian army—and that of the Romanov dynasty—was sealed.

As to the Grand Duke, he was dispatched to the Caucasus on September 24 and enjoyed moderate success in Armenia during 1916. He successfully met a Turkish counteroffensive during June–August 1916, albeit with very heavy losses.

Immediately before his abdication, the Czar Nicholas II reinstated the Grand Duke as commander in chief, but he held the post for no more than a day before he was dismissed by Prince G. E. Lvov (March 1917). Within a matter of months, the Revolutionary government of Russia would make a "separate peace" with Germany by acceding to the humiliating demands of the Treaty of Brest-Litovsk.

Nicholas retired to the Crimea during the balance of 1917 and through 1919, then moved to France, where he lived the rest of his life in exile from the Bolshevik regime.

Grand Duke Nicholas was one of very few competent, let alone intelligent, commanders in Russia during World War I. His relief by Czar Nicholas II set the seal on disaster for the Russian army.

NIMITZ, Chester William

BORN: February 24, 1885

DIED: February 20, 1966

SERVICE: World War I (1917–18); World War II (1941–45)

Commander of the U.S. Pacific Fleet during World War II

Nimitz was a native Texan, born in Fredericksburg. He enrolled in the U.S. Naval Academy in 1901 and graduated in 1905. Two years later, he was commissioned an ensign while serving on the China station and then served on the submarine *Plunger*. Promoted to lieutenant in 1910, he was given command of the submarine *Skipjack* as well as the Atlantic Submarine Flotilla in 1912. During 1913, he toured Germany and Belgium, studying diesel engines. Subsequently, he directed construction of the U.S. Navy's first diesel ship engine.

In 1916, Nimitz was promoted to lieutenant commander and, after U.S. entry into World War I, was appointed chief of staff to the commander of the Atlantic Fleet's submarine division. Following a series of miscellaneous assignments after the war, Nimitz was promoted to commander in 1921 and, from 1922 to 1923, attended the Naval War College. From 1923 to 1925, he was attached to the staff of the commander in chief, Battle Fleet, then, during 1925–26, served on staff of the commander in chief, U.S. Fleet. Following this assignment, he organized the first training division for naval reserve officers at the University of California and administered this program from 1926 to 1929. During this period, he was promoted to captain (1927), and in 1929, he was assigned command of Submarine Division 20, serving in this capacity through 1931.

Nimitz was given command of a surface vessel, the cruiser U.S.S. *Augusta* (CA-31) in 1933 and sailed as her skipper until 1935, when he was named assistant chief of the Bureau of Navigation. In 1938, promoted to rear admiral, he left the bureau to command a cruiser division and then a battleship division, returning to the Bureau of Navigation in June 1939 as its chief.

After Admiral **Husband E. Kimmel** resigned (December 17, 1941) following the disastrous December 7 Japanese surprise attack on Pearl Harbor, Nimitz, promoted to admiral, was named on December 31 to replace him as commander in chief of the Pacific Fleet. Nimitz plunged into the reorganization of all defenses in the Hawaiian Islands and directed the rebuilding of the partially shattered Pacific fleet. On March 30, 1942, he was given unified command of all U.S. naval, sea, and air forces in the Pacific Ocean Area.

Nimitz took advantage of superb naval intelligence to checkmate Japanese operations against Port Moresby at the battle of the Coral Sea on May 7–8, 1942, then at Midway during June 2–6, 1942. Both battles were crucial early victories for the U.S. Navy—and Midway was a turning point in the Pacific war. In addition to these battles, Nimitz played a key role in creating overall Allied "island-hopping" strategy for the Pacific. Although he personally directed the campaign in the Gilbert Islands (November 20–23, 1943) and in the Marshalls (January 31–February 23, 1944), Nimitz delegated specific tactical authority to key subordinates, with whom he worked brilliantly and whom he was always able to inspire.

Nimitz oversaw the advance into the Marianas during June 14–August 10, 1944 and then into the Palaus during September 15–November 25. At this time, he also linked up with General **Douglas MacArthur**'s forces from New Guinea and co-ordinated the invasion of Leyte in the U.S. return to the Philippines on October 20, 1944.

Nimitz was promoted to the newly created rank of fleet admiral—the naval equivalent of general of the army—on December 15, 1944, then went on to direct the capture of Iwo Jima during February 19–March 24, 1945. Iwo Jima was strategically critical because it was the best available air base between Saipan (some seven hundred miles to the south, which U.S. forces had already taken) and Tokyo. With the taking of Iwo Jima, the U.S. was clearly closing in on the Japanese homeland.

Next came the campaign to take Okinawa—April 1–June 21, 1945—followed by operations against Japan itself during January 1945 to August, when the Japanese surrendered. The surrender took place aboard Nimitz's flagship, U.S.S. *Missouri*, on September 2, 1945.

Following the war, Nimitz served as chief of naval operations from December 15, 1945 to December 15, 1947 and was then appointed special assistant to the Secretary of the Navy during 1948–49. He was a United Nations commissioner for Kashmir from 1949 to 1951 and wrote (with E. B. Potter) an important history of warfare at sea, *Sea Power: a Naval History*, published in 1960.

Nimitz bore tremendous responsibility during World War II as overall commander of Pacific operations. Like **Dwight David Eisenhower** in Europe, Nimitz was especially adept at working well with subordinates, whom he both inspired and trusted. He was a great naval strategist.

NOGI, Maresuke

BORN: 1843
DIED: July 30, 1912
SERVICE: Restoration War (1868); Satsuma Rebellion (1877); Sino–Japanese War (1894–95); Russo–Japanese War (1904–05)

Japanese general; hero of the Russo–Japanese War

Nogi was the son of a Choshu clan samurai, born in Edo (Tokyo). Like his father, he followed a military career, fighting first in the Restoration (Boshin) War as a Choshu samurai against the Tokugawa shogunate from January to June 1868. He made the transition from samurai to military officer when he was commissioned a major in the new Imperial army in 1871 and saw action in the defense of Kumamoto Castle against rebels led by Takamori Saigo during the Satsuma Rebellion February–September, 1877. Despite the humiliation of losing his unit's flag to the enemy during one battle of the rebellion, Nogi was promoted to major general in 1885 and went abroad, to Germany during 1885–86, to study European military practice.

During the Sino–Japanese War, he commanded a brigade at the siege of Port Arthur from October 24 to November 19, 1894 and at the battle of Yingkow on March 9, 1895. During the Russo–Japanese War, he was commander of the Third Army and directed the Japanese siege against Russia's "impregnable" fortress at Port Arthur from June 22, 1904 to January 2, 1905. The siege became infamous for its six-month duration and its terrible cost—more than 100,000 Japanese lives, including those of Nogi's two sons. However, Port Arthur finally collapsed, and Nogi could claim a bitter victory in the first siege of twentieth-century warfare.

Nogi completed the siege of Port Arthur in time to unite with the main Japanese army at the Battle of Mukden, where he succeeded in turning the right flank of the Russian forces, an action that ultimately led to the Russian army's withdrawal to Tieling during February 21–March 10, 1905.

As Mukden demonstrated, Nogi was a capable tactician. His Samurai code also motivated the kind of victory-or-death resolve evident in the Port Arthur siege. Western commanders criticized Port Arthur as suicidal. However, in retrospect, it seems a foreshadow both of World War I trench warfare and of the Japanese conduct of World War II. The ultimate measure of Nogi's devotion to his country and, in particular, to its monarch, came on July 30, 1912, when Nogi, together with his wife, committed suicide on the death of the Meiji emperor. Nogi's Tokyo house is venerated as the Nogi Shrine, and Nogi is considered a national hero.

NORSTAD, Lauris

BORN: March 24, 1907
DIED: September 12, 1988
SERVICE: World War II (1941–45)

U.S. Air Force general who planned the two atomic bomb missions against Japan in World War II and, during the Cold War, supreme Allied commander of NATO

Born in Minneapolis and raised in Red Wing, Minnesota, Norstad graduated from West Point in 1930 with a cavalry commission. Almost immediately, in 1931, he transferred to the Air Corps. From 1933 to 1936, he served as commander of the Eighteenth Pursuit Group, based in Hawaii, then transferred to the staff of Ninth Bombardment Group at Mitchell Field, on Long Island, New York, where he served until 1938. From 1938 to the end of 1939, he attended the Air Corps Tactical School and was assigned to intelligence staff duty with GHQ Air Force at Langley Field, Virginia, from 1940 to 1942. He served a brief stint on the staff of General **"Hap" Arnold,** the commanding general of Army Air Forces, then became assistant chief of staff for the Twelfth Air Force in August 1942.

Norstad accompanied the Twelfth to North Africa in October 1942 and was promoted to brigadier general in March of the following year. He was appointed director of operations for Mediterranean Allied Air Forces, serving in this capacity until August 1944, when he was returned to the United States as chief of staff to the Twentieth Air Force. The Twentieth was a special bombing unit that reported directly to General Arnold, and Norstad was given the responsibility of planning the top secret atomic bomb missions against the Japanese cities of Hiroshima and Nagasaki. He was promoted to major general in June 1945, and the atomic missions were carried out in August.

After the war, in June 1946, Norstad became director of the Operations Division (OPD) of the Army General Staff and was a key participant in the discussions between the army and navy that laid the foundation for the National Security Act, which Norstad helped to draft with Admiral Forrest P. Sherman in 1947. Among the results of the act was the creation of an independent Air Force in September 1947, and Norstad, promoted to lieutenant general, was named acting vice chief of staff for the new branch. He was also appointed commander of U.S. Air Forces, Europe in October 1950, as well as commander, Allied Air Forces in central Europe, in April 1951.

In July 1952, Norstad was promoted to general and, in July 1953, was named to the post of air deputy under army General **Matthew B. Ridgway,** Supreme Allied commander of NATO. After General Alfred M. Gruenther succeeded Ridgway, Norstad remained in his post, then succeeded Gruenther as commander of all NATO forces in Europe in November 1956.

Norstad retired in January 1963 and began a highly successful career in business, as president of Owens-Corning Corporation and, subsequently, as its CEO and, finally, chairman.

NURHACHI

BORN: 1559

DIED: 1626

SERVICE: Manchu conquest of China (1618–58)

A chieftain of the Chien-chou Juchen; one of the founders of the Manchu (Ch'ing) dynasty

The son of a Juchen tribal chieftain, Nurhachi assumed his father's position and title in 1586, after his father died. He was determined to expand his realm and increase the importance and influence of the Juchen. Through strategic marriage, diplomacy, and aggressive warfare, Nurhachi consolidated much of Manchuria under his direct control. In the process, he built up a cavalry that was without peer in China. The army was organized around mounted archers, and operated according to the "banner system," which was established by 1601. Nurhachi conscripted all the young men of his realm into military companies under the command of hereditary officers. These units were then grouped into regiments, and the regiments were grouped into banners. In this way, the military organization was made to supercede and replace clan and tribal allegiances.

Nurhachi pressed his expansion further by suppressing all competing Juchen tribes, but he could not subjugate the Yehe, who, with Ming aid, held out against him. Hurhachi proclaimed the Later Chin empire in 1616, then, two years after this, promulgated a list of grievances against the Ming. In effect, this was a declaration of war, and Nurhachi took the offensive. He captured the city of Fu-shun in the spring of 1618, then soundly defeated an army of 90,000 led by Yang Hao—even though it outnumbered his own forces by a factor of four.

In September 1619, with the Ming in defeat, Nurhachi finally succeeded in subduing the Yehe. He captured Liaoyang and Mukden in May 1621, and in 1625, proclaimed Mukden his capital. From here, he launched a massive expedition against the city of Ningyu, near the Great Wall in 1626. The Ming general Yüan Ch'ung-huan fired upon Nurhachi's army with cannon that had been cast by Jesuit missionaries. Nurhachi suffered his first major defeat and, afterward, fell into a severe depression from which he never recovered. He was dead within little more than half a year.

Nurhachi was obviously a magnetic leader and certainly a great commander. He also ruled with a considerable sense of justice and order. Almost single-handedly, he transformed loosely connected nomadic tribes into a cohesive society built around a supremely well-organized army.

O

ODA, Nohunaga

BORN: 1534
DIED: June 21,1582
SERVICE: Unification Wars (1550–1615)

Japanese warrior who overthrew the Ashikaga shogunate, thereby ending a long period of feudal wars by unifying half of Japan's provinces under his dictatorship

Oda was born into a *daimyo* (a term for Japan's largest and most powerful landholding magnates) family in Owari province, near Nagoya. From childhood, Oda resolved to become supremely powerful. As a youth, he began to fight his neighbors as well as his relatives, even driving his older brother out of the province during 1557–60. Having staked out his territory, Oda was threatened by an invasion under the leadership of Yoshimoto Imagawa. Oda had only 2,000 men with which to oppose Imagawa's force of 25,000. Oda used the element of surprise and attacked at Okehazarna on June 22, 1560, quickly killing Yoshimoto and throwing his army into confusion. The massive force scattered and evaporated, leaving Oda's domain intact.

Oda's magnificent victory prompted a series of powerful alliances, most importantly one with the ex-hostage of the Imagawa, a *daimyo* named Ieyasu Tokugawa. Within a few years, Oda effectively consolidated his dominion over central Honshu by means of alliances and marriages.

Oda attacked the Saito clan, overwhelming their hilltop fortress of Inabayama in 1567 and thereby destroying them in a single stroke. He next turned on the lords of Ise—the region surrounding Yokkaichi—and Omi—surrounding Biwa Ko—in 1568, defeating both of these groups. From these victories, Oda advanced on Kyoto and threw his support behind the former shogun Yoshiaki Ashikaga, whom he restored to the shogunate on November 9.

At this point, Oda's considerable ambitions expanded beyond merely making himself powerful. He saw an opportunity of unifying all of Japan. He attacked Echizen—the area around Fukui—but suffered a reversal in 1570. Ieyasu came to his aid, and, together, they advanced northward to defeat all of his major opponents in a massive battle at Anegawa on July 22, 1570. When the monastery on Mount Hiei outside Kyoto resisted him, he responded by annihilating it in 1571, thereby neutralizing Buddhist political power in Japan for a long time.

Oda turned against Yoshiaki Ashikaga in 1573, whom he had earlier restored to the shogunate, after he learned that Ashikaga had conspired with his rival **Shingen Takeda.** Oda also waged war on the Ikko-Ikki sectarians in 1574, but did not succeed in wiping them out.

Despite many successes, Oda was still faced with the cavalry of the Takeda clan in 1575. They held Nagashino Castle under siege. Oda provoked the Takeda into an open attack, then decimated their charging number with massed musket fire at the Battle of Nagashino in June. Oda failed, however, to capture the army's leader, Katsuyori Takeda.

Following the victory against the Takeda, Oda returned to battling the Ikko-Ikki, once again laying siege to their fortress at Hoganji in 1579. During April of the following year, the fortress surrendered. In 1582, Oda again confronted the Takeda, at Temmoku San, achieving victory yet again and, this time, killing Katsuyori Takeda.

At the height of his power—having seized thirty of Japan's sixty-eight provinces—Oda was set upon in his Kyoto palace later in 1582 by a force under the command of Mitsuhide Akechi. Rather than allow himself to be taken prisoner, Oda committed suicide.

ODAENATHUS, Septimius

DIED: 267

SERVICE: Persian War (258–264); Roman Civil War (262)

Prince of the Palmyra, a Roman colony in what is now Syria; stopped the Sasanian Persians from conquering the eastern provinces of the Roman Empire

Odaenathus was born in Palmyra on the Syrian frontier. An Arab, he had been Romanized, serving as chief of the Palmyrenes (from 251), and subsequently granted consular rank by Rome (in 258). After Emperor Valerian fell in 260, Odaenathus supported Valerian's son Gallienus and, in his service, led Palmyrian forces against the Persians as they returned from the sack of Antioch in 261. Odaenathus' victory against the Persians, which reaped a rich harvest of captured booty, prompted Rome to give him command of an army, which he led into Persia proper and conducted raids.

In 262, Odaenathus was recalled from Persia to deal with rebellion at home. He defeated the usurper Quietus at Emesa (Horns) in 262, after which Rome made him viceroy of all the East. Almost immediately after this—from 262 to 264—Odaenathus set off into Persia again, penetrating Persia as well as Armenia. From the Persians, he retook Nisibis (Nusaybin) and Carrhae. He captured Ctesiphon on two occasions. The Persian ruler came to terms with Odaenathus in 264.

In 267, Odaenathus was making his way to Cappadocia (present-day central Turkey) to fight the Goths when he was assassinated by his nephew Maeonius.

Odaenathus was one of Rome's best governors—though he clearly aimed at transforming Palmyra into an independent Levantine state. He was an extremely efficient general and, judging by the pace of his victories, tireless.

OKU, Yasukata

BORN: 1846

DIED: 1930

SERVICE: Satsuma Rebellion (1877); Sino–Japanese War (1894–95); Russo–Japanese War (1904–05)

Japanese field marshal who was a hero of the Russo–Japanese War

Oku was born at Kokura, near Kitakyushu, and was a samurai of the Kokura clan. In 1871, he joined the new Imperial army, first winning distinction during the Satsuma Rebellion of February–September 1877, after which he rose rapidly, becoming a major general in May 1885.

During 1893–94, Oku made a military tour of Europe, then returned to command the Fifth Division during the Sino–Japanese War. He achieved his greatest fame during the Russo–Japanese War, landing at Laiotung during May 5–19, 1904 and commencing the siege of Port Arthur with the capture of Nan Shan on May 25 and occupation of Dairen.

During June 14–15, he attacked entrenched Russian positions at Telissu, dislodging them and pushing them back. He also fought at Liaoyang (August 25–September 3) and the Sha Ho (October 5–17). Oku took part in the decisive Battle of Mukden during February 21–March 10.

In 1907, Oku was created the equivalent of viscount and, in 1911, was promoted to field marshal rank. From 1906 to 1912, he served as the army's chief of staff, then spent his declining years on the reserve list.

Unlike other top officers of the era of the Russo–Japanese War, Oku was not well connected politically. His progress through the ranks to leadership of the Imperial Army was exclusively due to his prowess as an officer.

OLDENDORF, Jesse Bartlett

BORN: February 16, 1887
DIED: April 27, 1974
SERVICE: World War I (1917–18); World War II (1941–45)

World War II U.S. admiral operating chiefly in the Pacific

A native of Riverside, California, Oldendorf graduated from the Naval Academy as a passed midshipman in 1909 and in 1911 was commissioned an ensign. Shortly after U.S. entry into World War I, he was promoted to lieutenant and sailed in the North Atlantic on convoy escort duty through the end of the war. From August 1918 to March 1919, he was engineer officer aboard the armored cruiser *Seattle*, then was transferred to the captured German transport *Patricia* during March–July 1919. After promotion to lieutenant commander in July, he was assigned to shore duty before securing an appointment as aide and flag secretary on the staff of the commander the Special Service Squadron in 1921. The following year, he became aide to the commandant of the Mare Island Navy Yard in California.

From 1924 to May 1927, Oldendorf was back at sea, in command of the destroyer *Decatur*. Promoted to commander, he was sent to the Army as well as the Naval War Colleges from June 1928 to July 1930, then was assigned as navigation officer to U.S.S. *New York*, on which he sailed from August 1930 to June 1932, when he accepted an appointment as instructor in seamanship and navigation at the Naval Academy.

Oldendorf returned to sea as executive officer of U.S.S. *West Virginia* (BB-48) in June 1935, but was soon back ashore as head of the Naval Recruiting Section of the Bureau of Navigation Promoted to captain on March 1, 1938, he was given command of the cruiser U.S.S. *Houston* shortly after World War II began in Europe. In the fall of 1941, he joined the staff of the Naval War College and, after Pearl Harbor, was promoted to rear admiral.

Oldendorf was assigned several Caribbean commands from January 1942 to April 1943, then led a task force in the Atlantic Fleet from May to the end of 1943. In January 1944, he was transferred to the Pacific as commander of Cruiser Division 4, aboard the flagship, U.S.S. *Louisville*. He fought in operations against the Marshalls during January–February and Truk, during February 17–18. From June to August, he was part of operations against the Marianas. Next came Peleliu in September and support of the Leyte Gulf landings in the Philippines in October.

Oldendorf was assigned to command the Seventh Fleet's Fire Support Force, which consisted of six superannuated battleships, three heavy and five light cruisers, some twenty destroyers, and nearly forty PT boats. With this force, he checked a Japanese attack through Surigao Strait against U.S. shipping in Leyte Gulf. Handicapped by the obsolescence of his battleships—which had antiquated radar systems and lacked modern fire-control equipment—Oldendorf nevertheless laid an ambush for the Japanese force that resulted in the sinking of both Japanese battleships, two of four cruisers, and four of their eight destroyers in a night battle at Surigao Strait on October 24–25.

Promoted to vice admiral in December, Oldendorf was reassigned to command Battleship Squadron 1 and Battleship Division 4 supporting landings at Lingayen Gulf, Luzon on January 9. Briefly laid up with an injury, he commanded Task Force 95 at Okinawa from June to November. He was reassigned in November to command of the Eleventh Naval District, as well as the San Diego Naval Base (January 1946). He spent his last two years as a naval officer in command of Western Sea Frontier and of the "mothball" fleet moored at San Francisco. Upon his retirement from the Navy on September 1, 1948, he was promoted to admiral.

At Surigao, Oldendorf commanded the last surface action ever fought between battleships. His performance there was the magnificent work of an accomplished naval commander.

ORD, Edward Otho Cresap

BORN: October 18, 1818

DIED: July 22, 1883

SERVICE: Second Seminole War (1835–42); U.S.–Mexican War (1846–48); Indian campaigns (1850s); Civil War (1861–65)

U.S. general who saw long and distinguished service in the nineteenth century

Ord was born in Cumberland, Maryland, but was raised in Washington, D.C. He graduated from West Point in 1839 and was commissioned in the artillery. He saw his first action during the Second Seminole War and was promoted to first lieutenant for gallantry in 1841. In 1846, at the outbreak of the U.S.–Mexican War, Ord was sent to California as governor of Monterey. He was promoted to captain in 1850 and from 1855 to 1859, he fought Indians in Oregon and Washington.

Sent back east in 1859, he was posted to the Artillery School at Fort Monroe. With **Robert E. Lee,** he was sent to apprehend John Brown at Harpers Ferry in October 1859, then returned to the West. When the Civil War began, he was in San Francisco. Recalled to the east, he was promoted to brigadier general of volunteers and given command of a brigade in the Army of the Potomac in September 1861. In November, he was promoted to major of regulars and scored a brilliant defensive victory against **J. E. B. Stuart** at Dranesville, Virginia on December 20. For this action, he was breveted lieutenant colonel.

In May 1862, Ord was promoted to major general of volunteers and assigned to the Department of the Mississippi, where he commanded the left wing of General **Ulysses S. Grant**'s Army of the Tennessee during August–September. He was again breveted—to colonel—after the Union victory at Iuka on September 19. The following month, he made a vigorous thrust against General Earl van Dorn at the Hatchie River as van Dorn's troops withdrew from the battle of Corinth. Although the October 5 exchange mauled van Dorn's command, Ord suffered a serious wound and did not return to active duty until June 1863, when he commanded XIII Corps in the Army of the Tennessee. He had a major role in the siege and capture of Vicksburg during May 19–July 4.

In August, Ord was transferred to the Army of Western Louisiana, but was felled by illness in October and did not return to duty until March 1864. He was assigned to command VIII Corps in the Shenandoah Valley in July, then moved on to head up XVIII Corps during the siege of Richmond and Petersburg. On September 29, he effected the capture of Fort Harrison, south of Richmond, but was again badly wounded and out of action until January 1865, when he was assigned command of the Army of the James and the Department of North Carolina. In March he received a brevet to major general of regulars and fought alongside Grant in the final operations against Lee.

After the war, in December 1865, Ord was promoted to lieutenant colonel of regulars and, in July of the following year, was promoted to brigadier general. After leaving volunteer service in December 1866, he assumed command of the Department of Arkansas. In his later career, he commanded the departments of California, the Platte, and Texas, retiring from that post in December 1880. Ord was promoted to major general on the retired list in January 1881.

Ord was a solid commander, much respected by superiors and subordinates alike. A front-line officer, his serious wounds attest to his gallantry under fire.

OTTO I THE GREAT

BORN: October 23, 912

DIED: May 7, 973

SERVICE: Revolt of Thankmar (939); Nobles' Revolt (939–41); Invasion of France (942); Bavarian Revolt (944–47); Invasion of France (948); Invasion of Bohemia (950); First Italian Expedition (951–52); Ludolf's Rebellion (953–55); Wendish War (955); Second Italian Expedition (961–64); Third Italian Expedition (966–72)

Greatest of the Saxon kings of Germany; founded the Holy Roman Empire

Otto I created the medieval German monarchy and created anew the Holy Roman Empire, encompassing Germany, Italy, and Burgundy. The Germany into which he was born, on October 23, 912, was by no means a nation, but a collection of often-feuding entities. Otto was the son of Duke Henry, who later assumed the throne of Saxony as King Henry I the Fowler, and Otto, in 929, married Eadgyth, the daughter of King Edgar of England. Henry the Fowler nominated Otto as his heir, a choice confirmed by the German dukes meeting at Aachen in August 7, 936, following the death of Henry.

Otto immediately proved himself a strong ruler. He began by asserting his authority over the German dukes, who included Otto's half brother Thankmar and younger brother Henry. Many of them chafed under the new direction, and Thankmar allied with Dukes Eberhard of Bavaria and Eberhard of Franconia to lead a revolt against Otto in 939. Otto demonstrated his military prowess by dealing with the revolt swiftly and thoroughly. Henry, with Eberhard of Franconia and Giselbert of Lorraine, secured the support of Otto's brother-in-law King Louis IV of France to stage a much more considerable rebellion from 939 to 941. Otto met these combined forces on the field, achieving decisive victories at the battles of Xanten in 940 and Andernach in 941, killing Eberhard and Giselbert in battle. Henry had little choice but to submit to Otto's authority, and Otto not only pardoned him for the revolt, but also forgave him even after he discovered, in December 941, that Henry was behind a plot against his life. These remarkable acts of clemency were successful in winning Henry's enduring loyalty and put an end to his dreams of autonomy.

To Louis IV of France, Otto did not demonstrate similar leniency. He invaded France in 942, swiftly defeating the king's forces and forcing a peace agreement. Otto enjoyed relative tranquility for no more than two years, however; for in 944, Bavaria erupted in revolt, and Otto was defeated at the Battle of Wels by Duke Bertold. In 946–47, however, Otto applied a combination of military pressure and diplomacy to regain control over the region. A second invasion of France came in 948, but this time Otto came to aid his former enemy, brother-in-law Louis IV, who had been imprisoned by the rebellious Hugh the Great, Count of Paris. Otto met and defeated the forces of Hugh at Rheims, capturing that city, whereupon the count surrendered and restored Louis to the throne.

Otto turned east in 950 to invade Bohemia, compelling Duke Boleslav to accept his suzerainty. The next year, he answered a plea for help from Princess Adelaide of Burgundy, who had been imprisoned by Berengar, margrave of Ivrea. Otto invaded Italy in 951, riding gallantly to the rescue of Adelaide, who was the widow of Italy's King Lothair. After defeating rival claimants to the throne of Lombardy in 952, he crowned himself king of the Lombards, freed Adelaide, and then—in fairy-tale fashion—took her as his second wife. Berengar acknowledged his submission by paying Otto homage.

Still, the German monarch had little opportunity to savor his Italian triumph. This time, his son Ludolf was staging a revolt with Duke Conrad the Red of Lorraine, Archbishop Frederick of Mainz, and others. Otto rushed back to Germany in 953, only to be defeated and captured by the rebels. He effected an escape and took the offensive, attacked Mainz and Regensburg in 953–54, both of which held out. At this point, 50,000 to 100,000 Magyars seized upon the German unrest to raid Bavaria and Franconia. Conrad forged a treaty with the raiders in 954, helping them to cross the Rhine at Worms, then facilitating their entry into Lorraine. They crossed the Meuse River to ravage northeastern France, traveling through Rheims and Châlons into Burgundy, thence to Italy, through Lombardy, and then into the valleys of the Danube and the Drava. From here, in defiance of the treaty, a Magyar force moved against Bavaria, and Ludolf decided to surrender Regensburg to Otto early in 955, so that his father could fight the invasion.

Leading an army of 10,000, Otto arrived at Augsburg, which was besieged by 50,000 Magyars. On August 10, 955, the Magyars lifted the siege and turned on Otto at the Battle of Lechfeld. At first, they readily out-maneuvered the German, capturing his camp and driving one-third of his forces from the field. Conrad, however, having been betrayed by the Magyars, now joined Otto to repel the enveloping Magyars and ultimately to turn the tide of battle against them. By the end of the day, it was Otto who held the Magyar camp. But at the moment of victory, Conrad fell, and Otto alone pursued the retreating Magyars for three days, bringing the Great Magyar Raid to an end.

Still, menaces abounded. Fresh from the harrowing confrontation with the Magyars, Otto marched to the north, where he joined forces with Gero, the margrave of Brandenburg, to drive the Slavic Wends out of Germany at the Battle of Recknitz in October 955. A five-year interval of relative peace ensued; then, in 960, Otto moved against Slavic tribes between the middle Elbe and the middle Oder. From here, he answered the request of Pope John XII to fight Berengar—now Berengar II, king of Italy—who was subdued by the fall of 961 and submitted to vassalage under Otto. At Rome, the Pope crowned Otto Holy Roman Emperor on February 2, 962, but then, to Otto's surprise and dismay, commenced to treat with Berengar. Thus provoked, Otto moved to depose John XII and replace him with Leo VIII. Next, Otto took prisoner, then returned to Germany only after suppressing a Roman revolt against Leo, who had been deposed by Benedict V. Leo died before the revolt had been put down, so Otto appointed John XIII as his successor. Yet another revolt, in 965, drove this pontiff from Rome, which moved Otto to a third Italian invasion in 966.

Otto put down the latest Roman revolt, then marched to southern Italy, where he prevailed against the Saracens and Byzantines. He next arranged a marriage between his son Otto II—his chosen successor—and the Byzantine princess Theophano on April 14, 972. The Germany to which Otto now returned was at the very least a proto-nation, unified to a degree previously unknown. He had strengthened not only the German states, but the Holy Roman Empire. The new-found stability that resulted brought a fleeting interval of artistic and intellectual productivity known as the Ottonian Renaissance. Otto's death on May 7, 973, however, ushered in a five-year period of violent civil war, as Bavaria's Henry the Wrangler and Boleslav of Bohemia rebelled against Otto II, and the cultural gloom of the Dark Ages returned.

OUDINOT, Nicolas Charles, Duke of Reggio

BORN: April 25, 1767

DIED: September 13, 1847

SERVICE: French Revolutionary Wars (1792–99); Napoleonic Wars (1800–15); Intervention in Spain (1823)

Rose from the ranks during the Revolution to become a marshal of France

Born at Bar-le-Duc, the son of a brewer, Oudinot joined the Médoc Infantry Regiment as a private on June 2, 1784, leaving the service in 1787 to return to his family's brewery. At the commencement of the Revolution, on July 14, 1789, he was elected captain in the Third Battalion, Meuse Volunteers, and was soon promoted to lieutenant colonel (1791). While leading his unit during 1792–93 in the Meuse and the Vosges, he suffered a gunshot wound to the head. Upon his recovery, he was promoted to general of brigade, having distinguished himself at Kaiserslautern on May 23, 1794.

From 1794 to 1798, Oudinot fought in the armies of the Rhine and the Moselle. He was wounded at Mannheim in 1795 and, at Ulm, in October 1795, he was wounded six times and taken as a prisoner of war. Released and repatriated in January 1796, he fought at Ingolstädt in September and was deeply slashed by four saber cuts. He also sustained a bullet wound in the thigh.

Promoted to general of division under **André Masséna** in 1799, he was yet again wounded, this time in the chest, at Zurich. Recovering swiftly, he was made Masséna's chief of staff. After suffering two more wounds, he accompanied Masséna to Italy in 1800 and was entrusted with Masséna's dispatches to General **Louis Suchet.** He penetrated the British blockade of Genoa and got the messages through.

After the siege of Genoa, later in 1800, Oudinot served under General Guillaume Brune at Mincio. Three years later, Oudinot was appointed inspector general of infantry and cavalry, and in 1805, he was assigned to command Oudinot's Grenadiers, an elite unit attached to V Corps under General **Jean Lannes.**

Leading the grenadiers, he participated in triumphs at Wertingen (October 8) and Ulm (October 17). On November 16, he was badly wounded at Hollabrünn, but he recovered in time to command his unit at Austerlitz. On December 2, the grenadiers, held in reserve on the left flank, were instrumental in trapping Allied units against the Satschan Pond.

Exhausted and ailing from his many wounds, Oudinot was largely out of action during 1806. He returned to duty for the Polish campaign of 1807, seeing action at Ostrolenka in February and participating in the siege of Danzig from March 18 to May 27. During this operation, Oudinot broke his leg, but nevertheless led a division at Friedland on June 14.

Created Count of the Empire in July 1808, Oudinot was dispatched to Germany to organize reserves under Louis Davout. The following year, Oudinot fought with these reserves as the provisional II Corps under Lannes in the Battle of Landshut on April 21. At Aspern-Essling, during May 21–22, Oudinot sustained an arm wound, but nevertheless took over command of II Corps after Lannes was killed.

Oudinot fought at Wagram during July 5–6 and was wounded in the ear. On July 12, he was presented with his marshal's baton on the field. The next year, on April 14, 1810, Oudinot was created Duke of Reggio, then was dispatched to Holland for service 1810–12. He was recalled for the Russian campaign in 1812 and was given command of II Corps, which formed part of the army's left wing. He commenced the Battle of Polotsk on August 17, but was shot through the shoulder and had to relinquish command to **Laurent Saint-Cyr,** who brought the battle to a successful conclusion.

Recovering before the Battle of the Berezina on November 27–28, Oudinot was shot in the side. His horse bolted, dragging Oudinot through the snow until his aide-de-camp rescued him. While he was lying in an aid station waiting for transportation, Cossacks attacked, and Oudinot was badly injured by a falling beam.

Remarkably, Oudinot recovered by April 1813 and was given command of XII Corps, which he led at Bautzen during May 20–21. **Napoleon** then additionally assigned IV and VII Corps to him and sent him to capture Berlin. Repulsed at Grossbeeren on August 23, he was forced to withdraw. He was assigned two divisions of the Young Guard at Leipzig during October 16–19, then, in 1814, fought at Brienne (January 29) during the campaign for France. In this battle he was wounded in both legs, but quickly recovered in time to participate in the defense of La Rothiere on February 1, 1815.

Oudinot was wounded a final time at Arcis, during March 20–21. Directing the rear guard, he was shot in the chest, but the ball was deflected by the Grand Eagle of the Legion of Honor, which he wore.

Following Napoleon's first abdication, Oudinot served on a commission to negotiate peace with the Allies. Louis XVIII named him to command of the Corps of Royal Grenadiers. Oudinot pledged loyalty to the Bourbon king, and when Napoleon returned for the Hundred Days, Oudinot was effectively exiled to his estate. He refused to partake in the Waterloo campaign. Louis subsequently promoted him to command of the grenadiers and chasseurs of the Royal Guard and of the third military district. Appointed a minister of the state, a member of the Privy Council, and a peer of France, he also became commander in chief of the National Guard of Paris as well as major general of the Royal Guard in 1816.

Oudinot's final campaign came in 1823, when he led I Corps in an invasion of Spain. He took, occupied, and governed Madrid. In old age, Oudinot was honored with an appointment as governor of the Invalides in 1842. He held that post until his death.

Oudinot suffered no fewer than thirty-four battle wounds fighting for Napoleon. An inspiring leader, he was a poor tactician and strategist who had little aptitude for independent command. He was, however, a splendid subordinate.

P

PAPPENHEIM, Count Gottfried Heinrich

BORN: May 29, 1594
DIED: November 17, 1632
SERVICE: Thirty Years' War (1618–48)

German cavalry general conspicuous for his aggressive zeal in the Thirty Years' War

Pappenheim was born at Treuchtlingen and inherited his family's military tradition as well as the office of archmarshal of the Empire. However, he early on intended to become a diplomat rather than a soldier and studied law at the Universities of Altdorf and Tübigen. Suddenly, he converted from the Lutheran religion of his family and became a Catholic. He also bolted from the university and launched into a military career, immediately manifesting the audacity and magnetism that characaterized his combat leadership style. He distinguished himself at White Mountain, fighting for the Catholic League under Count **Johan Tilly.** Although he suffered a serious wound in the battle of November 8, 1620, he acquired a reputation for valor that brought him command of his own regiment of cuirassiers by 1623. With this unit, he campaigned in the Spanish service in Lombardy and the Grisons during 1624–26.

Pappenheim was recalled to Austria to put down a peasant revolt in Upper Austria. He set about this assignment with bloodthirsty abandon in the fall of 1626, not merely defeating the peasant army at Wolfsegg, Gmunden, and Vocklabrück, but annihilating it. He may have killed upward of 40,000 men.

In 1627, he was sent north to take Wolfenbüttel, which he readily did and for which he was created, in 1628, an Imperial Count. Under Tilly, he next led the storming of Magdeburg on May 27, 1631. He turned his troopers loose on the fallen town, which they sacked with a ferocity that made Pappenheim infamous.

Pappenheim next met **Gustav II Adolph** at the first Battle of Breitenfeld on September 13, but did not prevail against him. He did, however, effectively cover Tilly's retreat from the battle, saving many casualties. Following Breitenfeld, Pappenheim was given independent command in northern Germany. **Wallenstein** sent him marching toward Westphalia and the lower Rhine region, but he was recalled to the north when Gustavus again threatened. No sooner did he arrive on the field at Lützen than he was wounded—mortally—on November 16, 1632. He died the next day.

Pappenheim was unpredictable, arrogant, insubordinate, and gratuitously cruel. Yet he was so fearless and vigorous a commander, handling cavalry with great skill, that his superiors—civil and military alike—overlooked much that made him objectionable.

PATCH, Alexander McCarrell, Jr.

BORN: November 23, 1889
DIED: November 21, 1945
SERVICE: World War I (1917–18); World War II (1941–45)

U.S. Army general considered one of the great trainers of troops

Patch was the son of an army captain stationed at Fort Huachuca, Arizona Territory. Born there, Patch was raised back east, in Pennsylvania, where he attended a year of Lehigh University before securing an appointment to West Point in 1909. After graduating in 1913, Patch was commissioned a second lieutenant of infantry and served on the Mexican border during 1916–17. After promotion to captain on May 15, 1917, he shipped out to France, and was put in charge of the

Army Machine Gun School in France from April to October 1918. During this period, he also fought at Aisne-Marne (July 18–August 5), at the Saint-Mihiel salient (September 12–16), and in the Meuse-Argonne (September 26–November 11).

Patch served in Germany with the army of occupation through 1919, then returned to the United States, where he reverted from lieutenant colonel to his prewar rank of captain. He was soon promoted to major (on July 1, 1920) and served in a series of training and education functions through 1924. He graduated with distinction from the Command and General Staff School in 1925, and later graduated from the Army War College in 1932. During 1925–28 and 1932–36, he taught as professor of military science and tactics at Staunton Military Academy, Virginia.

On August 1, 1935, Patch was promoted to lieutenant colonel and, the following year, was appointed to the Infantry Board at Fort Benning, Georgia. He was instrumental in testing the new three-regiment "triangular" division that streamlined the U.S. Army and made mass movement more flexible. The triangular division would be the foundation of army organization in World War II.

Promoted to colonel, Patch was given command of the Forty-Seventh Infantry in August 1939, then was promoted to brigadier general, with command of the Infantry Replacement Center at Camp Croft, South Carolina on August 4, 1941. After Pearl Harbor, late in January 1942, Patch was sent to New Caledonia with the remnants of units left over from the "triangularization" of the Twenty-Sixth and Thirty-Third Divisions. He was promoted to major general on March 10, and his units became the core of the famed Americal Division—"American troops on New Caledonia"—which was activated on May 27, 1942. Patch took the new division to Guadalcanal, where he relieved the First Marine Division there on December 9. He oversaw mop-up operations on Guadalcanal from December 1942 to February 7, 1943. From January to April 1943, Patch was commander of XIV Corps, but was then summoned back to the United States to assume command of IV Corps area. In this assignment, he was intensively involved in troop training from April 1943 to March 1944, when he was sent to Sicily in command of the Seventh Army. He led the Seventh in Operation Anvil-Dragoon, the invasion of southern France, beginning on August 15. Three days into the operation, Patch was promoted to lieutenant general.

Patch's invasion proceeded with great rapidity, and, on September 11, he linked up with **George Patton**'s Third Army at Dijon. The Seventh Army joined General Jacob L. Devers's Sixth Army Group on August 15 for the advance east into Alsace. Patch took Strasbourg in November and was then part of the resistance to a German counteroffensive in January 1945 and the reduction of the so-called Colmar Pocket the next month.

Patch led the Seventh Army through southern Germany and into Austria, linking up with units from General **Mark W. Clark**'s Fifth Army at the Brenner Pass on May 4. He returned to the United States in June and was assigned command of the Fourth Army at Fort Sam Houston, Texas, where he resumed troop training. In October, he was also assigned to a special group formed to study postwar defense reorganization. He succumbed to pneumonia shortly after completing this assignment.

PATE, Randolph McCall

BORN: February 11, 1898

DIED: July 31, 1961

SERVICE: World War II (1941–45); Korean War (1950–53)

U.S. Marine Corps general who was instrumental in World War II battles including Guadalcanal, Iwo Jima, and Okinawa

A native of Port Royal, South Carolina, Pate began his military career as an army private in 1918. He left to attend Virginia Military Institute and graduated in 1921, after which he was commissioned a second lieutenant in the Marine Corps Reserve. In September 1921, he received an active commission in the Corps and was dispatched on expeditionary duty to Santo Domingo (Dominican Republic) during 1923–24. After promotion to first lieutenant in September 1926, he was assigned to the expeditionary force in China during 1927–29.

Promoted to captain in November 1934, then major in October 1938, he was named assistant chief of staff for supply, First Marine Division, in 1939. Shortly after U.S. entry into World War II, Pate was promoted to lieutenant colonel (January 1942) and served with the First Marine Division on Guadalcanal from August 19, 1942 to February 7, 1943. Promoted to colonel in December 1943, he was named deputy chief of staff of the V Amphibious Corps during the Iwo Jima campaign of February 19–March 24, 1945 and the taking of Okinawa (April 1–June 22, 1945).

After the end of the war, in January 1946, Pate was named director of the Division of Reserve at Marine Corps Headquarters. He also sat on the General Board, Navy Department, in Washington, D.C. in 1947, then was appointed chief of staff of the Marine Corps Schools, at Quantico, Virginia the following year. Promoted to brigadier general in September 1949, he was made director of the Marine Corps Educational Center in 1950.

In July 1951, Pate was attached to the Office of the Joint Chiefs of Staff, where he was deputy director of the Joint Staff for Logistics Plans. He served again as director of the Marine Corps Reserve beginning in November 1951 and was promoted to major general in August 1952. In September, he was given command of the Second Marine Division.

Pate fought a new war beginning in 1953, when he was sent to Korea in command of the First Marine Division during the closing stages of the conflict. He returned to the United States in May 1954 and, promoted to lieutenant general, was appointed assistant commandant of the Marine Corps and chief of staff. On January 1, 1956, he became commandant, with the rank of general. He retired three years later.

PATTON, George Smith

BORN: November 11, 1885
DIED: December 21, 1945
SERVICE: Punitive Expedition against Pancho Villa (1916–17); World War I (1917–18); World War II (1941–45)

Controversial and brilliant commander of the U.S. Third Army in World War II; after Eisenhower and MacArthur, probably the best known general of the war

A native Californian, born at San Gabriel, Patton was the son of a family with a strong military tradition. Wishing to gain admission to West Point, Patton spent a year at Virginia Military Institute (1904), and then secured his West Point appointment, graduating in 1909 after a difficult four years. A superb horseman, Patton was commissioned a second lieutenant in the cavalry and served well in a number of army posts. He excelled as an athlete and represented the army on the

George S. Patton. Courtesy National Archives.

U.S. pentathlon team at the 1912 Stockholm Olympics. He was honored by an appointment to study at Saumur, the prestigious French cavalry school, and he also graduated from the Mounted Service School at Fort Riley, Kansas, in 1913. From 1914 to 1916, he served as an instructor at the school. An expert swordsman, he wrote the army's saber manual.

In 1916, he was assigned to General **John J. Pershing**'s punitive expedition to Mexico in pursuit of Pancho Villa, at the conclusion of which, in 1917, he was promoted to captain. Patton worshiped Pershing as a model military officer and was thrilled to be appointed to his staff and sent with him to France in May 1917, after the U.S. entry into World War I.

In Europe, Patton became the first American officer to receive tank training, and he became an enthusiastic convert to the potential of mechanized warfare. He set up the AEF Tank School at Langres in November 1917. Promoted to temporary lieutenant colonel, then temporary colonel, he organized and led the First Tank Brigade in the assault on the Saint-Mihiel salient during September 12–17, 1918. He was wounded in this engagement, but quickly recovered and fought at Meuse-Argonne (September 26–November 11).

Like many other officers, he reverted to his prewar rank on his return to the United States following the Armistice. However, he was soon promoted to major (1919) and given command of the 304th Tank Brigade (really only a battalion), at Fort Meade, Maryland, where he perfected armored tactics during 1919–21.

During 1921–22, Patton was posted with the Third Cavalry Regiment at Fort Myer, Virginia, and graduated at the top of his class from the Command and General Staff School in 1923. He served on the general staff from 1923 to 1927, then as chief of cavalry from 1928 to 1931, when he left to attend the Army War College. Patton was made executive officer of the Third Cavalry and promoted to lieutenant colonel in 1934, and he returned to service on the general staff in 1935, with a promotion to colonel coming in 1937.

From December 1938 to July 1940, Patton commanded the Third Cavalry, and then took command of the Second Armored Brigade during July–November 1940. Promoted to temporary brigadier general on October 2, 1940, he became acting commanding general of the Second Armored Division in November. The appointment was made permanent on April 4, 1941, and Patton was also promoted to temporary major general. He led his division in the massive war maneuvers conducted in Tennessee during the summer and fall of 1941.

Shortly after U.S. entry into World War II, Patton was named commander of I Armored Corps (January 15, 1942), and he was then assigned to command the Desert Training Center during March 26–July 30, 1942 in preparation for combat in North Africa. Patton was instrumental in the final planning for Operation Torch (July 30–August 21), the conquest of North Africa, and he commanded the Western Task Force in landings there on November 8, 1942. He was named to replace General Lloyd R. Fredendall as commander of II Corps on March 3, 1943 after Fredendall encountered disaster against Panzer general **Erwin Rommel** at the Kasserine Pass. Patton was promoted to temporary lieutenant general on March 12, only to be relieved of command three days later after a dispute with his British colleagues. Patton's arrogance and outspoken, impulsive nature would plague him throughout the war. Friction with Allies, the press, and with superiors would become chronic.

Soon after this incident, Patton was given command of I Armored Corps, which became the Seventh Army on July 10. Patton led it with extraordinary drive during the invasion of Sicily from July 10 through August 17. On the 16th, while visiting wounded soldiers in a field hospital, Patton encountered a soldier suffering from battle fatigue. He accused the man of cowardice and slapped him in the face. The public and Patton's superiors were scandalized by this incident, and Patton was sent to England in disgrace on January 22, 1944.

Patton publicly apologized for the incident, but he was temporarily sidelined in England. The planners of the Normandy invasion used Patton's presence in England to mislead the Germans into thinking that he was going to lead an invading army to Calais—not Normandy. Once the D-Day invasion was actually put into motion, Patton was given command of the newly formed Third Army and arrived with it in France on July 6.

It was in the operations that followed that Patton became one of great heroes of World War II. He led the Third Army during the breakout from Normandy and the lightning advance across France through the summer of 1944, collecting retroactive promotions to brigadier general and to major general in the process. He liberated town after town, delivering ruinous blows to the retreating German army.

When the Germans launched their desperate surprise offensive in the German Ardennes (December 16, 1944–January 1945)—the so-called Battle of the Bulge—Patton performed a tactical miracle by wheeling the entire Third Army, exhausted from months of forced marching and battle, 90 degrees north and launching a bold counterattack into the southern flank of the German penetration. By this action, he relieved Bastogne on December 26, 1944, ended the Battle of the Bulge, and then poised his army for the final push to the Rhine.

After encountering stiff resistance during January–March 1945, the Third Army crossed the Rhine at Oppenheim on March 22 and advanced into central Germany and northern Bavaria by April. Units of the Third reached Linz on May 5 and Pilsen, Czechoslovakia on May 6—even before the Germans surrendered.

With the war in Europe won, the indiscreet expression of political opinions once again got Patton into trouble. He outspokenly criticized the Soviet allies and, even worse, as military governor of Bavaria, he opposed de-Nazification policies because (he said) they left the conquered territories without qualified officials to maintain order. The Allied command yielded to public and diplomatic pressure and relieved Patton from command of the Third Army and from the governorship of Bavaria.

Although Patton desperately wanted to be sent to the Pacific to fight the war against Japan, he was appointed commander of the Fifteenth Army—essentially a "paper army," an administrative unit set up to collect records and compile a history of the war. On December 9, 1945, he broke his neck in an otherwise trivial automobile accident near Mannheim. Paralyzed from the neck down, he was hospitalized. Pulmonary edema and congestive heart failure developed, and he died on December 21.

Few modern American commanders have caused as much controversy as George S. Patton. Known to his men—not affectionately—as "Old Blood and Guts," he was a combination of flamboyance and spit-and-polish. A deeply religious man who wrote poetry and was a keen student of history, he was also incurably vulgar, liberally punctuating public and private discourse with obscenity. Although he was not a great strategist, he was a brilliant tactician, whose push through France was nothing short of a military masterpiece.

PAULUS, Friedrich von

BORN: September 23, 1890

DIED: February 1, 1957

SERVICE: World War I (1914–18); World War II (1939–45)

Nazi German field marshal on the Eastern Front; his total defeat at Stalingrad (now Volgograd) was a key turning point in World War II

Paulus was born at Breitenau in Hesse. His father was a school administrator. Young Paulus's ambition was to become an officer in the Imperial Navy, but he was unable to secure a cadetship. He turned instead to law, which he studied at Marburg University before abandoning his books to enter the 111th ("Markgraf Ludwig's Third Baden") Infantry Regiment as an officer cadet in February 1910. When World War I began in August 1914, he fought with his regiment (as part of the Seventh Army) at the Battle for the Frontiers during August–September 1914. He also served at Arras following the Battle of the Marne during September–October. In November, an attack of illness forced his return home until 1915, when he returned to active duty with an assignment in the *Alpenkorps*—the mountain troops—as a staff officer. He fought in Macedonia during 1915 and was promoted to *oberleutnant* and he and the *Alpenkorps* were sent to the Western Front. During June 23–30, 1916, Paulus fought at Fleury, and from February to November 1916, he was present at the Battle of Verdun.

Friederich von Paulus. Courtesy National Archives.

During April 9–17, 1918, he took part in **Erich Ludendorff**'s attack on the Lys and in the defense against the British Somme-Lys offensive as well as the Battle of Saint-Quentin during August 22–September 4.

Paulus ended World War I as a captain in the *Alpenkorps* and remained in the small *Reichswehr*—the 100,000-man army permitted by the Treaty of Versailles. He was a company commander in the Thirteenth Infantry Regiment at Stuttgart from 1919 to 1921, and then served in a number of staff posts through 1934, when he was given command of a motorized battalion. In 1935, he became chief of staff for the new Panzer (armored) headquarters. Four years later, he was promoted to *generalmajor* and was appointed chief of staff for Tenth Army under General Walther von Reichenau. He was with the Tenth in the blitzkrieg invasion of Poland during September 1–October 5, 1939 and remained with the army after it was redesignated the Sixth Army in Holland and Belgium during May 10–28, 1940.

Paulus was one of the key planners of Operation Sea Lion, the never-executed invasion of Britain. He was made Deputy Chief of Staff, Operations Section, *oberkommando des heeres* (OKH), a post from which he participated in the creation of Operation Barbarossa—the invasion of the Soviet Union. Paulus, after visiting General **Erwin Rommel** in North Africa in April 1941, reported that the "Desert Fox" would have to be restrained, lest his requirement for reinforcements weaken Barbarossa.

In January 1942, Paulus replaced Reichenau as commander of Sixth Army and repulsed a Soviet offensive at the First Battle of Khar'kov in February. He then led the Sixth Army toward the Volga during June 28–August 23 and approached Stalingrad in late August. Uniting with elements of the Fourth Panzer Army, he began to slog out the battle for the city. But it was here that the Red Army had decided to make its strongest stand, and Paulus was unable to advance. He was plagued by tenuous and overlong supply lines—and by the worst Russian winter in decades.

In November, the Red Army initiated a counteroffensive, which encircled battered Sixth Army during November 19–23. Paulus became wholly dependent on **Hermann Göring**'s promise to supply the isolated and starving Sixth Army by air. The promised support never materialized to any significant degree. On January 15, 1945, **Adolf Hitler** promoted Paulus *to feldmarschall* and awarded him the Oak Leaves grade of the Knights Cross. Nine days later, the last of the German wounded were evacuated by air. On February 2, Field Marshal Paulus surrendered Sixth Army's 91,000 survivors to the Soviets.

For Paulus, the war had ended. He was held under house arrest in Moscow until 1953, when he was released with 6,000 other long-incarcerated survivors of the Battle of Stalingrad. The Soviets permitted him to reside exclusively in East Germany. He developed a degenerative neuromuscular disorder soon after his release and died at a Dresden clinic two years later.

Paulus's reputation was forever destroyed by Stalingrad. While it is true that he failed to react in time to maintain land communications, there was probably little that could have been done in any case to prevent encirclement. He had the misfortune to be in a hopeless situation.

PERICLES

BORN: 495? B.C.

DIED: 429 B.C.

SERVICE: First Peloponnesian War (460–445 B.C.); Second ("Great") Peloponnesian War (432–404 B.C.)

Strategos (general) under whose rule Athens won dominance over other Greek city-states

Pericles was the most celebrated of Athenian statesmen, but remains an obscure figure and had no contemporary biographer. It is known, of course, that he was a champion of democracy, yet he relentlessly employed the means of tyranny to force that form of government on city-states subordinate to Athens. While Pericles presided over the zenith of Hellas' greatness, opening Athenian democracy to the ordinary citizen, building the temples and statues of the Acropolis, and creating the Athenian empire, his unyielding drive to impose his nation's will on others stirred the discontent and rebellion that eventually led to the fall of Athens and, ultimately, the decline of Greece.

The son of Zanthippus, a prominent statesman, and Agariste, a member of the influential Alcmaeonid family, Pericles first came to public attention in 472 B.C., when he provided and trained the chorus for Aeschylus' play The Persians. He was elected *strategos* (general) in 458 B.C., a key policy-setting office that carried a one-year term, and Pericles was frequently reelected over the next thirty years. Despite his noble lineage, Pericles was a supporter of democracy in a time of almost universal tyranny—though his personal political style was as haughty, reserved, and aristocratic as that of any tyrant.

The same may be said of his dealings with other city-states. The Delian League was a confederation of city-states formed to fight Xerxes and the Persians. Athens collected money from many of the member states to maintain an army and build a navy. These forces were soon dominated by Athenians, and the other states of the confederation grew to resent paying tribute to maintain forces that often served to impose the Athenian model of government on them. Moreover, a portion of the tribute money was diverted to finance Pericles' program of sculpture-raising and temple-building. Worse, under the system of *cleruchy*, Athenian colonists were granted lands abroad while retaining the rights and privileges of Athenian citizenship. It became clear that Athens was building an empire of tributary states.

In a celebrated speech recorded by the historian Thuycidides, Pericles declared that all subjects of Athens "confess that she is worthy to rule them." This high-handed assertion finally proved untrue. While Pericles realized his ambition to make Athens preeminent among Greek states, and while the oppression Athens wrought upon those states did at least enforce peace, stability, and the rule of law, he also lived to see Sparta lead the powers of the Peloponnesus in a revolt against Athens (the Peloponnesian War). While Athens was under siege during the conflict, a plague swept the city. Amid the turmoil, Pericles and those associated with him were censured: his mistress Aspasia and his teacher Anaxagoras were accused of impiety; his protégé, the great sculptor Phidias, was charged with embezzlement; and Pericles himself was briefly deposed from office. A few weeks later, in a fit of repentance, the Athenian people restored Pericles to office, investing him with even greater power. However, weakened by a bout with the plague, he died within a year.

PERRY, Matthew Calbraith

BORN: April 10, 1794

DIED: April 3, 1858

SERVICE: War of 1812 (1812–15); Algerine War (1815); U.S.–Mexican War (1846–48)

U.S. naval officer; headed an 1853–54 expedition that forced Japan to open trade and diplomatic relations with the West

Born at Rocky Brook, Rhode Island, Matthew C. Perry was the younger brother of **Oliver Hazard Perry,** who would become the hero of the Battle of Lake Erie in the War of 1812. Matthew entered the navy as a midshipman in 1809 and served aboard the twelve-gun schooner *Revenge,* commanded by his brother, during 1809–10. He then transferred to the U.S.S. *President,* on which he served until 1812, when he was commissioned a lieutenant under **Stephen Decatur** aboard U.S.S. *United States*. He sailed on this vessel through the War of 1812 and, afterward, during Decatur's 1815–16 Mediterranean expedition against the Barbary pirates.

During 1820–24, Perry sailed on small vessels as part of the antislavery patrol. He was also stationed in the West Indies in operations against pirates. In 1822, Perry assisted in transporting free blacks to Africa, where they founded Monrovia, the capital of Liberia.

During 1825–26, Perry served as first lieutenant and, at times, acting commander of the seventy-four-gun U.S.S. *South Carolina* on patrol in the Mediterranean. He represented the United States in preliminary negotiations with the Turks on a first U.S.–Turkish treaty in 1826. Promoted to master commandant in March 1826, he returned to the United States, first as second officer and then as commandant of the New York Navy Yard. He was instrumental in promoting the adoption of steam power for the U.S. Navy.

Promoted to captain in February 1837 and commodore in June 1841, he collaborated with Navy Secretary George Bancroft in drafting the first curriculum for the new U.S. Naval Academy in 1845. With the outbreak of war with Mexico in 1846, he served blockade duty along the east coast of Mexico. He also sailed in coastal operations against various Mexican towns and led an expedition up the Tabasco River during October 23–26. Perry assumed overall command of the squadron assigned to assist General **Winfield Scott**'s landing at and capture of Veracruz during March 9–27, 1847. This accomplished, he led expeditions up the Vinazco (April 18–22) and Tabasco (June 14–22) rivers with the objective of destroying fortifications and supply dumps.

After the war, Perry was appointed general superintending agent of mail steamers, During 1848–52, he oversaw and directed their construction and operation from his headquarters in New York. In 1852, he accepted a mission to open Japan, first ensuring that he would be given a sufficiently impressive array of ships to persuade the Japanese of U.S. naval power. He sailed aboard the steam frigate *Mississippi* in November 1852 to Okinawa, where he joined three other vessels in May 1853. The ships entered Tokyo Bay on July 8, and Perry presented representatives of the Japanese government with a letter from President Millard Fillmore on the 14th.

From Japan, Perry sailed to China, returning in the spring with a larger squadron. He concluded the Treaty of Kanagawa after three weeks of negotiations during March 3–31, 1854. The agreement established U.S.–Japanese diplomatic relations and secured permission for American vessels to call at two ports. It also assured aid for shipwrecked U.S. sailors. Beyond this, by opening up relations between the United States and Japan, America earned a prestigious place among the nations of the West. The Perry expedition was an important step toward making the United States a recognized world power.

After the mission to Japan, Perry returned to the U. S. and an assignment on the Naval Efficiency Board. He also collaborated with Francis L. Hawks on a book about the Japan expedition.

Perry, moody and unfriendly, lived in the shadow of his more "heroic" older brother. Nevertheless, he was considered the foremost U.S. naval officer of his generation, and his mission to Japan was a milestone for the development of the United States as a world power.

PERRY, Oliver Hazard

BORN: August 23, 1785

DIED: August 23, 1819

SERVICE: Tripolitan War (1801–05); War of 1812 (1812–15)

Naval hero of the Battle of Lake Erie, a turning point in the War of 1812

Born at Rocky Brook, Rhode Island, Oliver H. Perry was the older brother of **Matthew C. Perry.** He joined the navy in youth, sailing as a midshipman under his father, Christopher R. Perry, captain of the twenty-eight-gun frigate, U.S.S. *General Greene* in 1799. He made lieutenant in 1802 and served in the Mediterranean against the Tripolitan corsairs during 1802–03.

During 1804–06, Perry was captain of the twelve-gun schooner *Nautilus* in the Mediterranean. On his return to the United States, he supervised the building of gunboats to augment the fledgling U.S. Navy. He also sailed on enforcement patrols pursuant to President Thomas Jefferson's Embargo Act during 1807–09.

In 1811, he was in command of the twelve-gun U.S.S. *Revenge*, which ran aground in a dense fog in Newport Harbor. Perry was court-martialed on a charge of negligence. Acquitted, he was given command of the Newport gunboat flotilla when the War of 1812 commenced.

Perry was sent to Lake Erie to serve under Commodore Isaac Chauncey on February 17, 1813. He arrived at Presque Isle (modern Erie), Pennsylvania, and discovered that there were no American ships on the lake. Perry was faced with a dearth of supplies and a shortage of officers and sailors. Nevertheless, he set to work building gunships and, by the early summer of 1813, had completed nine vessels mounting fifty-four guns.

On May 27, Perry and Chauncey assisted then-Colonel **Winfield Scott** in the capture of Fort George (Ontario). Perry then returned to his gunboats, slipping them out of Presque Isle harbor during August 1–4. He engaged the British blockading fleet on Lake Erie on September 10, leading the attack from his twenty-gun flagship *Lawrence*—named for naval hero **James Lawrence,** recently slain in battle. His dying words—"Don't Give Up the Ship"—were emblazoned on the *Lawrence*'s banner. It was a gesture typical of the charismatic Perry.

The *Lawrence* fared poorly. While the rest of the American Lake Erie flotilla lagged behind, the British vessels pounded Perry's flagship. Perry left *Lawrence*, rowed back to the rest of the flotilla in a small boat, and, through dint of leadership, brought them into the action. The result was a stunning defeat for the British fleet. Perry sent a message to General **William Henry Harrison,** who was waiting to commence the land attack: "We have met the enemy and they are ours. Two ships, two brigs, one schooner, one sloop." Perry then went ashore to serve under Harrison in the ensuing Battle of the Thames River. He even led a charge on October 5. The victory at the Battle of the Thames turned the tide of the War of 1812, which had been, up to that point, a losing proposition for the Americans.

Perry was promoted to captain and was given the thanks of Congress in January 1814. Following the war, he commanded the forty-four-gun frigate *Java* in the Mediterranean during 1816–17. During this service, he struck a Marine captain, an act for which he was subsequently court-martialed and reprimanded.

In 1819, he took command of a diplomatic mission to the new republic of Venezuela. While in South America, on the Orinoco River, he contracted yellow fever. He died from it, at sea, off the coast of Trinidad.

PERSEUS

BORN: ca. 213 B.C.

DIED: ca. 165 B.C.

SERVICE: Third Macedonian War (172–167 B.C.)

Last king of Macedonia; his attempts to dominate Greece brought defeat at the hands of the Romans

Elder son of King Philip V of Macedon, the teenaged Perseus fought alongside his father against the Romans at Cynoscephalae in 197 B.C. Perseus plotted against his younger brother Demetrius, of whom he was jealous, and persuaded his father to execute Demetrius in 181 B.C. On Philip's death in 179 B.C., Perseus assumed the throne and launched a military campaign to push Macedonian influence into Thrace, Illyria (the Yugoslav region), and Greece.

In response to a treaty Perseus concluded with Boeotia (east-central Greece) in 174 B.C., Eumenes II of Pergamum engineered a breach between Perseus and Rome. Perseus managed to hold the Roman forces along the frontier south of Mount Olympus during 171–168 B.C., but General Genthius of Illyria, resentful of the secondary role Perseus had assigned to him, exposed the Macedonian flank. The Romans attacked, and Perseus was forced to fight at Pydna against L. Aemilius Paulus. His army was in a bad position, its phalanx hampered by rough terrain. The result was a broken advance, and the Romans were able to defeat Perseus on June 22, 168 B.C.

Perseus fled to Samothrace, where he was taken prisoner and transported to Rome in 167 B.C. Paraded in the triumph of Paulus, he was sent into exile at Alba Fucens, where he died within two years.

Ruthless but able as a military strategist and as a diplomat, Perseus had the misfortune of being betrayed into premature battle. However, it is also likely that Rome would have, sooner or later, acted to check Perseus' expansionism.

PERSHING, John Joseph

BORN: September 13, 1860

DIED: July 15, 1948

SERVICE: Indian Wars (1880s–90); Spanish–American War (1898); Philippine service (1899–1903; 1906–14); Punitive expedition against Pancho Villa (1916–17); World War I (1914–18)

American general who led the American Expeditionary Force (AEF) into World War I

Pershing was born in Laclede, Missouri, and was raised on a farm. He worked as a schoolteacher from 1878 to 1882, when he obtained an appointment to West Point. After graduating in 1886, he was commissioned a second lieutenant in the Sixth Cavalry Regiment and served in the West during the late phase of the Indian Wars. He helped round up fugitives from the massacre at Wounded Knee (December 28, 1890).

From 1891 to 1895, he taught as the University of Nebraska as commandant of cadets. Promoted to first lieutenant in 1892, he took time out to earn a law degree (awarded June 1898), then saw service with the Tenth Cavalry—a black regiment of "Buffalo Soldiers" commanded by white officers. It was from this assignment that Pershing earned his nickname: Black Jack. He served with the Tenth from October 1895 to October 1896, when he was appointed aide to General **Nelson A. Miles.**

From June 1897 to April 1898, Pershing returned to West Point as an instructor in tactics, then returned to the Tenth Cavalry as its quartermaster. During the Spanish–American War, he fought at El Caney-San Juan Hill (July 1–3, 1898), but was felled by malaria, contracted while in Cuba. He was briefly assigned quiet duty at the War Department toward the end of his convalescence in August 1898. He was next appointed chief of Bureau of Insular Affairs, serving from September 1898 to August 1899.

In 1899, Pershing requested a posting to the Philippines, where he served on northern Mindanao. From December 1899 to May 1903, he was engaged in the pacification of the Moros, who resisted American authority. He returned to Washington and staff duty shortly before his marriage to Helen Francis Warren on January 25, 1905. From March 5, 1905 to September of 1906, Pershing was stationed in Japan as a military attaché and observer in the Russo–Japanese War. The assignment turned out to be Pershing's great opportunity, for it brought him into contact with President Theodore Roosevelt, who, greatly impressed with Captain Pershing, caused his promotion—in one step—to brigadier general on September 20, 1906.

From December 1906 to June 1908, Pershing commanded a brigade at Fort McKinley, near Manila, then accepted an appointment as military commander of Moro Province in the Philippines in November 1909. He served in this post until early 1914, continually conducting small-scale operations against recalcitrant Moro rebels.

In April 1914, Pershing was assigned command of Eighth Brigade in San Francisco, but was almost immediately dispatched to the Mexican border during the civil war in Mexico. On August 27, 1915, while Pershing was in Texas, on the Mexican border, a fire swept through his family's quarters in the Presidio at San Francisco. Pershing was devastated to learn that his wife and three of their daughters had perished in the fire. Francisco "Pancho" Villa, whom Pershing would soon be assigned to capture, was among the many who wrote or wired their condolences to the general.

On March 9, 1916, Villa raided Columbus, New Mexico, leaving seventeen U.S. citizens dead. President Woodrow Wilson ordered Pershing to invade Mexico and capture Villa. Commanding a force of 4,800, he chased Villa and his men fruitlessly for ten months until he was ordered home on January 27, 1917.

On May 12, 1917, Pershing was named commander of the American Expeditionary Force (AEF), which was to be sent to Europe to fight in World War I. Pershing arrived in France well in advance of the force, on June 23, and directed the massive American buildup from August 1917 through October 1918. Apart from fighting Germans, his greatest challenge was preserving the independence of the AEF. It was unacceptable to him—and to the American people—for an American army to be commanded by French generals. With great difficulty, Pershing prevailed. He then went on to conduct three major offensives: Aisne-Marne (July 25–August 2, 1918), Saint-Mihiel (September 12–17), and the Meuse-Argonne (September 26–November 11). All were costly, but effective. The Americans had arrived at a time of Allied exhaustion. Their presence turned the tide of the war.

On his triumphal return to the United States, Pershing was promoted to general of the armies—and judiciously avoided offers to enter politics. He remained in the military as Army chief of staff (appointed on July 21, 1921) and held that post until his retirement on September 13, 1924. He published his memoirs in 1931.

Pershing nearly succumbed to a debilitating illness in 1938, but recovered—though never completely. By 1941, his health had deteriorated so that he took up residence in Walter Reed Hospital. He passed his final years there.

Personally cold and distant, Pershing was, in fact, an impressive leader. General **George Patton,** who had served as Pershing's aide during World War I, thought him the finest general the American military had ever produced.

PÉTAIN, Henri Philippe

BORN: 1856

DIED: 1951

SERVICE: World War I (1914–18); World War II (1939–45)

French field marshal and World War I hero, who "sold out" France to the Nazis and, during World War II, was titular head of the Vichy government

Pétain was born on April 24, 1856, in Cauchy-à-la-Tour, Pas de Calais, France. Pétain was the son of a peasant family, but he early on exhibited great promise and was enrolled at the École de Saint-Cyr, the French military academy, where he excelled. He was commissioned an officer of the *chasseurs alpins*—mountain troops—in 1876, but in the tradition-bound French military was promoted slowly. It was 1900 before he made major and was given a battalion-level command. Six years later, his abilities as a theorist were finally acknowledged, and he was appointed to the École de Guerre. However, Pétain was conservative and methodical, his strategic thinking always oriented toward defense and centered on the exploitation of artillery firepower. This put him at odds with the prevailing war policies of **Ferdinand Foch** and Grandmaison, advocates of nothing less than the all-out attack. Once again, promotion was delayed, and, at the outbreak of World War I in August 1914, Pétain was no more than a colonel in command of the Thirty-Third Regiment.

As a colonel, he quickly distinguished himself, achieving promotion to brigadier general within the first month of the war. Brilliant performance at the Battle of the Marne (September 4–10) earned him promotion to general of division, and by October 25 he was in command of XXXIII Corps in Artois. Again, Pétain performed superbly, this time during the Arras offensive of May 9–16, 1915, and was given command of the Second Army in June. In February 1916, when the all-important fortress of Verdun was menaced, it was Pétain to whom France turned, and he uttered the phrase that made him a popular hero of the war: "Ils ne passeront pas!" ("They shall not pass!")

Pétain successfully defended Verdun—perhaps too successfully, for, once again, his superiors became concerned that he was too committed to defensive strategy. He was promoted to command of Army Group Center, so that his subordinate, the hyper-aggressive and brilliantly dashing Robert Nivelle could be put in command of the Verdun sector. From this position, Nivelle was then jumped ahead of Pétain, but, in April 1917, he failed disastrously with his Chemin-des-Dames offensive, and the next month, Pétain was called in to relieve Nivelle and to assume supreme command of all the French armies.

Pétain's first crisis was a mass mutiny of the war-weary ranks. He vigorously prosecuted the ringleaders of the mutiny, even as he sought to act swiftly to address the soldiers' grievances, enacting reforms to humanize the French army and improve the treatment of the common soldier. With the mutiny quelled, Pétain conducted the closing months of the war with consummate skill, stressing preparedness and reasonable objectives rather than enforcing the long-cherished but often suicidal policy of all-out attack. At war's end, he was made marshal of France in recognition of his service to his country, and in 1920 he was appointed vice president of the Supreme War Council. In 1922, Pétain was made inspector general of the army, then served as minister of war during the brief government of Gaston Doumergue in February–November 1934. Now in civilian politics, he became an outspoken critic of civilian politicians, whom he derided for failing to lead or govern.

As the 1930s wore on, Pétain became increasingly disdainful of liberalism and turned to advocacy of autocratic government. Appointed ambassador to Spain in March 1939, he was recalled to France in May 1940 as his nation faced defeat at the hands of the Nazis. By this time Marshal Pétain was an old man, and some suggested that he was well on his way to senility, but French President Albert Lebrun was desperate and called upon this hero of Verdun to save France as he had saved her during World War I. He asked Pétain to form a new government, and on June 22, 1940, as titular head of the French government, Pétain negotiated surrender to the Germans.

Pétain was given emergency dictatorial powers, and although he attempted to retain as much independence from German domination as possible—even dismissing, in December 1940, Pierre Laval, the wholeheartedly collaborationist foreign minister who had been instrumental in bringing Pétain to power in the first place—Pétain was destined to be remembered as the man who cravenly sold his nation to the Nazis. For their part, the Germans steadily pressured Pétain, who repeatedly backed down, recalling Laval to power in 1942. In November 1942, the Germans occupied Vichy, the seat of the Pétain government, and in August 1943, they arrested Pétain himself, eventually imprisoning him in Germany.

After the war, Pétain was returned to France in April 1945, where he stood trial for treason, was found guilty, and was sentenced to death. **Charles de Gaulle,** wartime leader of the Free French forces and now provisional president of the republic, chose to remember the Pétain under whom he had served during World War I. He commuted the sentence pronounced upon his former commander to life imprisonment and Pétain was remanded to a fortress on Île d'Yeu. Subsequently, he fell ill and was transferred to a villa at Port-Joinville, where he died on July 23, 1951.

As a military man, Pétain favored defense. He sought always to minimize the loss of life, and it was this commitment that had motivated his actions at Vichy. Indeed, he was a patriot, who wanted to spare the French people, devastated by one war, from the further agonies of a second. But Pétain's Vichy government itself collaborated in the persecution of French citizens, including the deportation, forced labor, and death of thousands of French Jews. Moreover, Pétain's government sacrificed French national identity and honor. As one of the marshal's aides put it, Pétain thought "too much about the French and not enough about France."

PETER I THE GREAT

BORN: June 6, 1672

DIED: February 8, 1725

SERVICE: Revolt of the Streltsi (1682–84); Azov expedition (1695–86); Great Northern War (1700–21); Russo–Turkish War (1710–12)

Russian ruler who expanded the nation into an empire

Born in Moscow to Czar Feodor III Alakseevich and his second wife, Natalia Kirillovna Naryshkina, Peter was proclaimed czar when he was only ten years old, in 1682, after the death of his father. A few months later, however, the *streltsi* (militia musketeers) revolted, compelling the boy to share the throne with his semi-imbecile brother Ivan V under the regency of their sister Sophia. It was during this early period that Peter became enamored of the wider world beyond Russia, the world of European culture, learning, fashion—and military practices. While he was living outside Moscow in the suburb of Prebrazhenskoye, he formed two elite regiments of guards and plotted with his guardian, Prince Boris Golitsyn, to overthrow Sophia. Golitsyn moved successfully against Sophia in August 1689, and Peter seized the government.

In personality and appearance, the young man was commanding and powerful: six feet, six inches in height, he was energetic and decisive. He moved quickly to consolidate his newly won power during the period of 1689 to 1696, undertaking military expeditions to the White Sea during 1694–95 and liberating the Azov region from Turkish control during 1695–96. Following this conquest, he indulged his childhood wish to travel to Europe and from March 1697 to August 1698, toured Germany, Britain, and the Netherlands, greedily taking in examples of the most advanced technology and science as well as European fashion and style.

Another *streltsi* revolt forced Peter to cut short his tour, and he returned to Russia to squelch this latest threat with savage fury during the course of the summer of 1698. With the streltsi crushed, Peter instituted sweeping social reforms inspired by his European sojourn. He commanded the court nobility—the boyars—to shave their traditional beards and to don European-style dress. He imported European learning into Russia and abandoned the traditional Julian calendar for the new Gregorian calendar, which was just then gaining currency in the nations of Europe (this transformation was not generally adopted, however, until after the Russian Revolutions of 1917).

During this period, Peter also remodeled his army, introducing Prussian-style conscription and drafting some 32,000 commoners. Using his new forces, he allied himself with Augustus II of Poland and Saxony in November 1699 and attacked the Swedes at Livonia in August 1700. Unfortunately, Peter's military ambitions far outstripped his skill as a commander. He was badly defeated by Swedish king **Charles XII** at the Battle of Narva on November 30, 1700. Undeterred by this setback, Peter resumed the transformation of Russia into a European-style industrial and military power, establishing factories, arms manufacturers, and military schools. He developed a system of internal transport and fostered the growth of a shipbuilding industry in order to open up commerce with the rest of the world. He also had the good sense to learn from his defeat at Narva to surround himself with able military advisors and secured the leadership of Field Marshal Count Sheremetev in an invasion of Ingria, which was held by the Swedes. On January 9, 1702, Peter and Shermetev defeated a Swedish army at Erestfer and then, on July 29, at Hummelshof in Livonia. Peter occupied the valley of the Neva River in December 1702, and he founded the city of Saint Petersburg on May 16, 1703.

The new city was built on a frozen marsh along the Gulf of Finland, and it became the object of Peter's pride and joy, an urban monument designed according to the latest models of European neoclassicism and intended to attract the best of European culture and learning. Peter called the new city his "window on the West."

From June 12 to August 21, 1704, Peter besieged Narva, this time successfully. However, his ally, Augustus II, had surrendered to Charles XII of Sweden, and Peter, now holding Narva, proposed peace with the Swedish monarch. Charles scornfully rejected the tender and invaded Russia during 1707–08, pushing the overextended Peter as far as the central Ukraine, where he holed up at Poltava during the winter and spring of 1708–09. At last, on July 8, 1709, Peter massed his forces against the Swedish army, defeating Charles at the Battle of Poltava, all but destroying the opposing army.

Peter turned next to the south, moving against Turkish Moldavia in March 1711. He was outmaneuvered by the Turks, who hemmed him in at the River Pruth. Thus stymied, he negotiated a settlement on July 21, 1711. Three years later, Peter again directed his efforts against the Swedes, planning with Admiral Feodor Apraskin a devastating attack on the Swedish fleet near Hangö in the Baltic on July 7, 1714. Through this master stroke, Peter gained control of the Baltic Sea, acquiring Livonia, Estonia, Ingria, and southern Karelia, all of which were ceded to Russia by treaty on August 30, 1721.

With his empire greatly expanded, his court Europeanized, and the fractious nobility more unified than they had ever before been, Peter I discarded the title of "czar" for the western title of "emperor" on November 2, 1721. He spent the next two years campaigning against Persia on his empire's Asian frontier, occupying Derbent (in Dagestan), Rasht, and Baku during 1722–23.

Peter the Great died on February 8, 1725, in Saint Petersburg from complications of a cold he had caught after he helped rescue soldiers who had fallen into the frozen Neva River.

PHILIP II

BORN: 382 B.C.

DIED: 336 B.C.

SERVICE: Wars of expansion (359–339 B.C.); Third Sacred War (355–346 B.C.); Fourth Sacred War (339–338 B.C.)

***King of Macedonia and father of* Alexander III the Great**

Philip II conquered Greece, established the League of Corinth, built a formidable army, and began the Macedonian invasion of Persia, which was completed by his more famous son.

Philip was the son of the Macedonian king Amyntas III. Born in 382 B.C., Philip was sent at age fifteen to Thebes as a hostage, where he remained for three years, learning the art of warfare from the Greek general Epaminondas. In 359 B.C., after his brother Perdiccas was killed in battle, Philip II assumed the Macedonian throne, though it is unclear whether he did so as regent for Perdiccas's son or in his own right.

His reign saw the unification of the principalities of upper Madedonia and the creation of a professional army. The so-called "Companions"—members of the landed nobility—were pledged to him in service as cavalry troops. In addition, Philip organized the free peasants and shepherds into an infantry corps. Thus constituted, the army also benefited from all Philip had learned about the latest tactics and battle equipment. As infantry equipment, he introduced the *sarissa*—a pike nearly 50 percent longer than the Greek spear—and, for seiges, he brought the torsion catapult.

In 358 B.C., Philip used his new army to invade Paeonia, defeating the Illyrians there in 356 B.C., then invaded Thrace, capturing the silver and gold mines of Mount Pangeios. The capture of this prize—mines that produced a thousand talents each year—provoked the Athenians to battle Philip for the next decade. A year after the invasion of Thrace, Philip entered southern Thessaly and in 348 B.C. destroyed Chalcidian Olynthos, a triumph that brought him election as president of the Thessalian League (about 352 B.C.). For the next century and a half, this confederation united Thessaly and Macedonia.

In 348 B.C., Philip annexed Chalcidice and enslaved the population of Olynthus and other people of the region. Peace with Athens finally came in 346 B.C., and over the next three years he consolidated control over Greece, mainly through diplomatic rather than military means. Athens broke the peace in 340 B.C., but Philip quickly countered this at the Battle of Chaeronea. He next consolidated his hold on territory stretching from the Hellespont to Thermopylae, and in 337 B.C. he convened the Greek states (except for Sparta) to form the Corinthian League, of which Philip was named commander and president. Using the League, Philip increased the size of his army by requiring the states to supply troops and ships according to a quota system. He next abrogated the democratic constitution of Thebes and stationed a Macedonian garrison in the region—though he was careful not to provoke Athens, whose support he needed for a new campaign: large-scale war against Persia.

Philip prepared his army for war, and the combined Greek and Macedonian forces crossed the Hellespont in 336 B.C. In the meantime, however, Philip's wife and jealous Macedonian nobles plotted against him, and Philip was stabbed to death by one Pausanias, a young Macedonian noble, during the wedding of his daughter Cleopatra to King Alexander of Epirus. The conquest of Persia would be left to his son, Alexander the Great.

PHILIP II AUGUSTUS

BORN: August 21, 1165

DIED: July 14, 1223

SERVICE: Wars with England (1187–90, 1194–99, 1202–08); Third Crusade (1189–92); War with England, Germany, and Flanders (1213–14)

Perhaps the greatest monarch of his time; ruled France from 1179 until 1223, reestablishing local sovereignty over French lands formerly held by England

Philip II was born in Paris, the son of King Louis VII, who installed the young man as king during his own lifetime and immediately set off a power struggle among several French provinces and England which strove to control the fourteen-year-old monarch. When Henry II of England turned over all English territories in France (except Normandy) to his son John, the king's older son, **Richard the Lion Hearted,** rebelled. With Philip's assistance, Richard acquired the lands in question for himself and appointed Philip as his feudal lord.

In the meantime, Jerusalem had fallen to the armies of Saladin in 1187, and Philip joined forces with **Frederick Barbarossa** and Richard to form the major thrust of the Third Crusade. After the combined forces captured the city of Acre, Philip returned home, leaving Richard to attempt to capture Jerusalem. Failing in this, Richard set out for England, but was captured in Austria and held for ransom, during which time Philip became an ally of John in an attempt to wrest Normandy from Richard.

Upon Richard's death in 1199, John became king of England, and again, Philip turned from ally to enemy. Finally, in 1202, Philip declared all English holdings in France to be void and returned to the French crown, thereby ending a conflict that had raged for years. King John's attempt to repossess his lost French real estate ended in a battle fought at Bouvines on July 27, 1214, at which Philip defeated Otto IV of Germany and his combined army of English, German, and Flemish knights.

Philip was also known for his internal reforms, most especially, the awarding of high-ranking positions in the govenment to those who were qualified to hold them, rather than to those claiming them by hereditary right. His plans to make Paris a leader among European cities culminated in the construction of a defensive wall around the town, the paving of the city streets with stone, and the improvement of the University of Paris.

The last years of Philip's reign were peaceful, and when the monarch died at Mantes, France, in 1223, he left behind a kingdom that was one of the most powerful in Europe.

PHILOPOEMEN

BORN: ca. 252 B.C.

DIED: 182 B.C.

SERVICE: Spartan War (223–221 B.C.); Wars against Crete (221–211 B.C.); First Macedonian War (215–205 B.C.); War against Nabis (195–192 B.C.); Revolt of Messene (182 B.C.)

Greek general of the Achaean League who restored the league's military efficiency

Born at Megalopolis in Arcadia—in the central Peloponnese—Philopoemen enters history in 223 B.C., when he led the evacuation of his native city when it fell under attack from Cleomenes III of Sparta. From 223 B.C. to 221 B.C., he served under Antigonus III of Macedonia in the army of an alliance formed to combat Sparta. Philopoemen demonstrated his capacity for initiative and independent command when, in 222, he did not wait for orders to attack the Spartans at Sellasia. His independent action assured the defeat of Cleomenes—and garnered the gratitude of Antigonus.

From 221 to 210 B.C., Philopoemen earned a formidable reputation as a mercenary on Crete. In 210 B.C., he returned to Achaea, where he was elected cavalry general of the Achaean League. He set about thoroughly reorganizing and retraining the cavalry, shaping them into an effective fighting force. The very next year, he led them against the Aetolians and defeated them on the borders of Elis. Following this victory, Philopoemen was elevated from general of cavalry to *strategus*—general-in-chief and military dictator—of the Achaean League.

As *strategus*, Philopoemen made even more sweeping changes in the Achaean League's army. He modernized their weaponry, using patterns developed in Macedonia. He also retrained the army. In 207 B.C., he launched his reequipped and retrained force against a Spartan army under Machanidas at the second battle of Mantinea. Philopemen emerged victorious and was elected *strategus* again in 206, 204, and 201 B.C.

In 202 B.C., Philopoemen defeated Nabis of Sparta at Messene, then, the following year, beat Nabis again—this time at sea, in the battle of Tegea. He declined to fight in the Second Macedonian War against Rome and returned to Crete, where he lived quietly from 200 to 197 B.C. Some time after this, he returned to Achaea and was again elected *strategus*. He defeated Nabis' Spartans at Gythium in 193 B.C., but was enjoined by the Roman general T. Quinctius Flamininus from capturing Sparta. The following year, however Nabis was assassinated, and Philopemen annexed Sparta, Messene, and Elis to Achaea.

Four more times—in 191, 189, 187, and 183 B.C. — Philopoemen was elected *strategus* of Achaea. He treated the conquered Sparta so harshly that Rome (which established a protectorate over Sparta) intervened with censure in 198 B.C. Philopoemen was slain in a minor battle with Messene rebels in 182 B.C.

Philopemen was a military genius who was certainly the best Greek commander after **Alexander III the Great.** He was less successful as a diplomat, repeatedly antagonizing Rome—to the detriment of Achaean interests.

PHRAATES IV

REIGNED: ca. 37–2 B.C.

SERVICE: War with Rome (37 B.C.); War against Media (34 B.C.); Parthian Civil War (32–29 B.C.)

Parthian ruler who fought Rome's Marc Antony

After murdering his father, Phraates became king of Parthia and continually waged war against Rome during his stormy reign. As the second son of Orodes II, Phraates was not in the line of succession until his older brother, Pacorus died around 36 B.C., quite possibly the victim of foul play instigated by Phraates. Devastated by the death of his favorite son and heir, Orodes fell desperately ill, yet lingered and clung to life. Impatient, Phraates killed his father and seized the throne.

He began his reign by eliminating all opposition, beginning with any of his nobles who criticized or spoke out against him, executing some and exiling many others. In 37 B.C., learning of the planned Roman invasion of Parthia, Phraates assigned one of his generals, Monaeses, to pose as a traitor in order to infiltrate the camp of Marc Antony and discover the Roman commander's plans. The ruse failed, and Monaeses returned to do the best he could to lead the Parthian armies in the field—something Phraates, despite his bellicosity, never personally did.

Driving through Armenia into Media Atropatene, Marc Antony attacked Parthia, but was repulsed, retreating with heavy casualties. In 34 B.C., however, the king of Media, who was Phraates's vassal, betrayed his liege by making an alliance with Marc Antony, permitting the Romans to occupy Media. Ultimately, however, the Roman withdrew his forces, and an enraged Phraates ordered his army into Media to exact a savage revenge on the people of the treacherous vassal.

In the meantime, Parthia was racked by civil unrest, and Tiridates, one of Phraates's generals in the war against Marc Antony, revolted during late 32 B.C. By the summer of 31 B.C., his forces had succeeded in evicting Phraates from Parthia, forcing him to seek refuge among the Saka nomads. Phraates secured aid from the Scythians in 30 B.C., and marched against Tiridates, who, in turn, fled to Rome, but with Phraates's son as hostage. Hoping to redeem the boy, Phraates made peace with the Roman emperor Augustus, formally recognizing Armenia and Osroene as Roman dependencies. Augustus promised to secure the return of Phraates's son on condition that Roman prisoners and standards be released. Phraates agreed, but once he had his son back, he refused to release the prisoners and standards. Accordingly, in 29 B.C., Augustus invaded Parthia, and Phraates at last yielded, freeing the prisoners and restoring the standards.

In token of restored faith and friendship, Augustus sent Phraates a slave girl, Thea Urania Musa, as a concubine. She persuaded Phraates to send four of his own sons to Rome as hostages, a move that secured the succession of her own son to the throne. This accomplished, Musa poisoned Phraates in 2 B.C.

PICCOLOMINI, Prince Ottavio

BORN: November 11, 1599

DIED: August 11, 1656

SERVICE: Thirty Years' War (1618–48); War of the Mantuan Succession (1628–31)

Italian general and diplomat in the service of the Holy Roman Empire; one of* Albrecht von Wallenstein's *most trusted lieutenants

Born in Florence, Piccolomini first joined the Spanish army in 1616, then moved to the Austrian Imperial army—the army of the Holy Roman Empire. He served in Italy, Bohemia, Hungary, and northern Germany, and became the leading subordinate of **Wallenstein** after 1627. During the War of the Mantuan Succession, he commanded Wallenstein's bodyguard, but was soon given independent command during 1628–31.

Piccolomini fought at Lützen on November 16, 1632, demonstrating almost superhuman endurance. Repeatedly charging the enemy, he took seven wounds. His zeal notwithstanding, the Imperial forces narrowly lost the contest. In 1633, Wallenstein assigned Piccolomini key commands in Bohemia and Silesia.

Piccolomini owed much to Wallenstein, but he became increasingly angry and disillusioned when Wallenstein promoted others above him. Piccolomini turned against Wallenstein and played a leading role, with the Austrian general Matthias von Gallas, in the conspiracy that overthrew and, ultimately, assassinated Wallenstein (February 25, 1634). Following this, the emperor rewarded Piccolomini with an estate in Bohemia, but denied to him what he most wanted: to succeed to Wallenstein's command. When that went to Gallas, Piccolomini left the Imperial army and returned to the service of Spain in 1635.

On behalf of Spain, Piccolomini fought the French in the Netherlands, dealing a severe blow to the numerically superior army of the Marquis de Feuquiéres at Thionville on June 7, 1639. For this, King Philip IV bestowed on Piccolomini the duchy of Amalfi.

In the early 1640s, Piccolomini returned to the Imperial army, again in the hope of succeeding Gallas. This did not occur. Instead, he served under Archduke Leopold William, and was present at the Austrian defeat in the second battle of Breitenfeld on November 2, 1642. After this, discouraged, he transferred again to the Spanish army. Again, the Spanish crown rewarded him well for his service, but, when the Holy Roman emperor Ferdinand III at last named Piccolomini commander in chief of Imperial forces, he returned to Austria and commenced the final campaign of the Thirty Years' War, leading an army to the relief of Prague in May 1648.

After the war, he served as head of the imperial delegation to the Congress of Nürnberg in 1649, which negotiated issues left unsettled by the Peace of Westphalia that had been concluded in 1648. His diplomatic skill proved as remarkable as his military prowess, and Ferdinand III created Piccolomini a Prince of the Empire in 1650.

Piccolomini was a remarkable man—brave and skilled as a field cavalry commander, and also brilliant as a strategist. He was an adroit diplomat, whose negotiations at Nürnberg preserved the Empire from suffering under what might otherwise have been a very harsh peace.

PICKETT, George Edward

BORN: January 28, 1825

DIED: July 30, 1875

SERVICE: U.S.–Mexican War (1846–48); Civil War (1861–65)

Confederate general who organized perhaps the single most famous action of the Civil War: Pickett's Charge at Gettysburg

A native of Richmond, Virginia, Pickett was born into the family of a prosperous planter. He took a stab at the study of law, but secured an appointment to West Point in 1842—and graduated, last in his class, in 1846. He was commissioned a second lieutenant in the infantry and served under General **Winfield Scott** during the U.S.–Mexican War. He fought at Veracruz during March 9–27, 1847 and subsequently was breveted to first lieutenant for Contreras and Churubusco (August 18–20). He also performed valiantly in the storming of Chapultepec on September 13. Foreshadowing the famous Civil War charge named for him, he was the first man over the parapet, and it was he who tore down the Mexican flag to replace it with the colors of his regiment. This action won him a brevet to captain.

Following the war, Pickett was assigned to garrison duty in Texas. His brevet to captain was made permanent in 1855, and in 1859 he was transferred to service in the Northwest. He commanded a post on San Juan Island from 1859 to 1861, during the boundary dispute with Great Britain.

With the commencement of the Civil War, Pickett returned to Richmond, where he resigned from the U.S. Army to enter Confederate service in June. He was commissioned a colonel on July 23 and was promoted to brigadier general on January 14, 1862.

George Pickett. Courtesy Library of Congress.

Pickett fought bravely at Williamsburg (May 4–5), at Fair Oaks/Seven Pines (May 31–June 1), and at Gaines's Mill during the so-called Seven Days (June 27). He received a severe wound at Gaines's Mill, which put him out of service for a matter of months. Promoted to major general, he commanded a division in **General James Longstreet**'s Corps at Fredericksburg on December 13.

Pickett's division arrived late at Gettysburg, but **Robert E. Lee** used it as the core of a daring—and ill-advised—attack on the Union line at Cemetery Ridge. The attack, "Pickett's Charge," came on July 3, 1863, and resulted in terrible casualties for Pickett's division. Pickett fixed the blame on Lee—who accepted it.

Following Pickett's Charge, the general, characteristically flamboyant, was a changed man, somber and sober. He was appointed to command the Department of Virginia and North Carolina. He made an attempt to retake New Bern, North Carolina, in February 1864, but was repulsed. After he was checked by General Benjamin Butler at Bermuda Hundred near Petersburg in April, he was replaced by P. G. T. Beauregard.

Pickett rejoined what was left of his old division at Cold Harbor in June. He also fought at Petersburg during June 15–18. On March 29–31, 1865, he managed to retard General **Philip H. Sheridan**'s offensive at Dinwiddie Court House, but was swept away by Sheridan at Five Forks on April 1. He surrendered with Lee at Appomattox on April 9.

Pickett was offered a commission as brigadier general in the Egyptian Army, but turned it down. Shaken, perhaps even broken by his experience of the war, Pickett spent the remainder of his life as an insurance agent.

Pickett was not a great soldier, though he was a brave one, animated by romantic and chivalric ideals that were as anachronistic as they were shallow.

PIZARRO, Francisco

BORN: ca. 1475

DIED: June 26, 1541

SERVICE: Conquest of Peru (1531–48)

Spanish conqueror of the Inca empire in Peru

Born in Trujillo, Extremadura, Castile, the son of a Spanish hidalgo, Pizarro was given no formal education, no patrimony, and was forced as a youth to earn his living herding swine. When his relative **Hernán Cortés** voyaged to Hispaniola in the New World, Pizarro, like many of the other disaffected or disenfranchised young men who called themselves conquistadors, seized the opportunity to make his fortune. He joined the expedition. In 1510, Pizarro was a member of a party exploring the Gulf of Uraba in northern Colombia, and he subsequently served as the great Balboa's lieutenant on the 1513 expedition that discovered the Pacific Ocean.

Spanish exploration, conquest, and colonization, albeit authorized by the crown, was largely driven by individual entrepreneurship. In 1522, Pizarro formed a partnership with another conquistador, Diego de Almagro, and a priest, Hernán de Luque, to undertake a series of expeditions deep into South America. The first expedition took them as far as the San Juan River in Colombia before the hardships of the jungle turned them back. A second expedition in 1526–28 was even harder than the first, and the party was beset by starvation, disease, and mutiny. Yet this exploration also yielded riches. Reaching the Santa River in Peru, Pizarro and his men returned to Panama loaded down with gold, cloth, and llamas.

Pizarro's success roused the jealousy of Panama's royal governor, who denied permission for further exploration. In response, the conquistador took his case to Spain in 1528, and he and the king drew up an agreement giving the crown all of Peru, its people, and its treasure. For his part, Pizarro was to be a knight of Santiago and proclaimed governor and captain general of all the lands he conquered. A greedy man, Pizarro minimized the rewards available to his partners, who greatly resented the arrangement.

In June 1530, Pizarro, several of his brothers, and 180 men sailed from Panama for the coast of the Gulf of Guayaquil. They raided and razed the settlement of Tumbes, Peru, then marched inland, meeting with Atahualpa, ruler of the Incas, at Cajamarca on November 15, 1532. Rather than negotiate with the leader or confront his military forces directly, Pizarro simply took him captive, then demanded a huge ransom for his release. Atahualpa's people collected and rendered the ransom, whereupon the conquistador ordered Atahualpa's death by strangulation on August 29, 1533. Within a year, the Pizarro expedition subdued and captured Cuzco, the Inca capital. Atahualpa's son, Manco Capac, launched a desperate campaign to recover Cuzco, but he was defeated by Almagro during 1536–37. But, following this victory, Almagro's bitterness over the inequitable division of spoils prompted him to attack Pizarro. The conquistador bested his rival, took him prisoner, and executed him as a mutineer.

Raised a swineherd, Pizarro was now a member of the Spanish nobility, but, unlike Cortés in Mexico, Pizarro barely savored his triumphs. Instead, he spent the rest of his life in efforts to consolidate all that he had nominally conquered. He apportioned lands as well as Indian serfs among his loyal men, created new settlements, and established agricultural operations, yet most of these proved marginal and unprofitable operations, and the Native peoples were continually rebellious. Even worse, the execution of Almagro hardly disposed of all those among his own men who opposed him. Pizarro founded Lima in 1535, making it the capital of Peru, but dissent grew steadily. Finally, on June 26, 1541, partisans of Almagro assassinated the conquistador in his own capital city.

POMPEY THE GREAT (Gnaeus Pompeius Magnus)

BORN: September 29, 106 B.C.

DIED: September 28, 48 B.C.

SERVICE: Social War (91–88 B.C.); Civil War (88–82 B.C.); Sertorian War (80–72 B.C.); Mithridatic War (75–65 B.C.); Third Servile War (73–71 B.C.); Pirate War (67 B.C.); Civil War (50–44 B.C.)

One of the First Triumvirate of Rome; early in his career, he was a great general

Born into the senatorial nobility of Rome, Pompey spent his early years developing his military and diplomatic skills under the tutelage of his father, a Roman consul. When civil war broke out between the followers of **Lucius Sulla** and **Gaius Marius,** Pompey's father sided with Marius's faction. Following his father's death, however, Pompey distanced himself from Marius and concentrated on developing his own power base. After gaining control of three legions through mutiny, Pompey joined Sulla in fighting Marius and driving him from Rome.

Pompey next secured the Senate's permission to campaign for the recovery of Sicily and Africa from Marius, completing the mission in two offensives during 81 B.C. Marius's troops called him Sulla's Butcher because he ruthlessly slaughtered any and all who surrendered themselves to him.

After defeating Marius's troops, Pompey marched on Rome and demanded that Sulla give him an independent command. Sulla complied, but soon abdicated, leaving Lepidus as consul in 78 B.C. Pompey supported Lepidus until he attempted to establish a brutal dictatorship. Thereupon, Pompey joined forces with those opposed to Lepidus and overthrew him.

Pompey's power increased after he defeated Lepidus. Again he refused to disband his army and demanded proconsular authority in Spain to help defeat the Marius factions there. After straining his resources to the maximum, Pompey was victorious in Spain and this time was more generous in his dealings with the vanquished. He again returned to Rome with his army, this time to put down a slave revolt led by Spartacus. With the aid of Marcus Crassus, he crushed the rebellion and now looked forward to a consulship. The Senate named both Pompey and Crassus consul in 70 B.C., and the pair quickly set about reversing many of Sulla's policies that had led to the corruption of the Republic.

Pompey now stayed close to Rome in order to build a following and increase his popularity, which had been growing considerably ever since his first defeat of Marius's forces. When pirate raids off the Mediterranean coast became a menace in 67 B.C., Pompey received carte blanche to deal with the problem. Within three months, he defeated the pirates.

In the East, the Pontus king Mithradates waged war in Asia Minor. In 66 B.C., the Senate, though wary of Pompey's growing popularity, gave him even more power—command of the entire East—in order to deal with Mithradates. Within the year, Pompey had subjugated Mithradates and added most of Pontus to the Rome's tributaries. For good measure, Pompey laid siege to Jerusalem in 62 B.C., taking it after three months. When he returned to Rome, he was the most celebrated general of the age.

The Senate, now extremely jealous of Pompey's power and popularity, refused him the traditional allotment of lands he asked for himself and his soldiers. The Senate was also wary of Crassus and of **Julius Caesar,** each of whom had his own personal agenda for power. The three, realizing they needed each other if they were to gain anything, secretly formed the First Triumvirate in 59 B.C. The alliance proved successful almost immediately, with each of the three receiving the offices and rewards they sought.

They renewed the alliance in 56 B.C., but the situation soon deteriorated as each triumvir wanted more than the others. Pompey and Crassus were renamed consuls in 55 B.C., but Crassus died in battle two years later. With Caesar gaining personal wealth and fame in Gaul, Pompey essentially dominated Rome. He came to control the Senate and was able to get much of what he wanted through careful diplomacy or outright skullduggery. In 52 B.C. he was named sole consul, essentially the most powerful man in the Roman world.

Pompey now felt that a showdown between him and Caesar was imminent. He demanded that Caesar surrender his powers and his army, which Caesar refused to do. At this impasse, Pompey had the Senate issue the "last decree" on January 7, 49 B.C., effectively declaring war on Caesar. Caesar responded by crossing the Rubicon, his point of no return in his march against Pompey in southern Italy.

The two met in minor battles across southern Europe and eventually fought the climactic Battle of Pharsalus in Thessaly on August 9, 48 B.C. Caesar smashed Pompey's forces, and the consul fled to Egypt, hoping to secure refuge from King Ptolemy. Fearing the wrath of the victorious Caesar, Ptolemy had Pompey executed on September 28, 48 B.C.

POPÉ

BORN: ca. 1630

DIED: ca. 1690

SERVICE: Pueblo Revolt (1680–92)

A Tewa medicine man who planned and executed a mass rebellion among the Pueblos of the American Southwest against the Spanish colonizers

Popé was a respected medicine man of the Tewa pueblo at present-day San Juan, New Mexico (north of Santa Fe). By the 1670s, the pueblos had endured nearly a half-century of persecution, exploitation, and general cruelty at the hands of the Spanish colonial government. Desperate, the Indians of the pueblo made an alliance with their hereditary enemies, the Apaches (whose very name was derived from a Zuni word for "enemy"), and terrorized the Spanish. Colonial governor Antonio de Oterrmín determined in 1675 to put a stop to the resistance by arresting the influential medicine men in forty-seven pueblos. The governor hanged three and imprisoned the rest, among them Popé.

After some years, Popé was released. Far from having been broken in spirit, he was determined to throw off the Spanish yoke once and for all. He decided to unite the far-flung pueblos in one great revolt. In doing this, he faced a formidable problem of politics as well as logistics. To begin with, decisions among the pueblos were customarily a product of debate, discussion, and mutual agreement. None of the pueblo towns was likely to act without securing the unanimous consent of its council, and the Tewa medicine man had first to persuade the council of each pueblo to commit that pueblo and to act in concert with the others. In this, he was remarkably successful. Not only did he persuade all but a few of the most remote settlements along the Rio Grande to participate, but he managed to do so while preserving absolute secrecy.

Next came the logistical problem of coordinating the moment of the attack. Popé devised an ingenious scheme whereby runners were dispatched to the various towns, each bearing a knotted cord designed so that the last knot would be untied in each pueblo on the day set for the revolt: August 13, 1680. Popé was ruthless in his enforcement of secrecy. He even killed his brother-in-law because he suspected him of treachery. Despite his vigilance, word did leak, and Popé was forced to launch the attack three days early, on the tenth.

Despite the sudden change in plan, the rebellion was devastatingly successful. The missions at Taos, Pecos, and Acoma were burned to the ground and the priests killed, their bodies heaped upon the hated altars. Lesser missions fell one by one, as did the surrounding haciendas, which were destroyed along with their inhabitants. By August 14 or 15, leading five hundred warriors, Popé advanced against the colonial capital of Santa Fe. The settlement was defended by a small garrison of fifty, but it harbored some one thousand settlers. The garrison was also equipped with a brass cannon. After four days of fierce fighting, Santa Fe fell, and Popé installed himself in the palace Governor Oterrmín had evacuated at the last possible moment. In all, some four to five hundred settlers had been killed, together with twenty-one out of thirty-three missionaries assigned to the region. About 2,500 Spaniards fled downriver in terror, many as far as present-day El Paso, Texas. They abandoned all that they had owned.

As the Spanish had sought to wipe out all manifestations of the Indians' "pagan" culture, so Popé now directed his followers in an orgy of destruction across a region extending from Taos to Isleta.

The pleasures of liberation and revenge were short-lived for the Indians. Popé set himself up as a tyrant oppressive as any Spaniard had been. Until his death about 1690, he taxed and plundered his people relentlessly. Against marauding Utes and Apaches he showed none of the military brilliance that had guided the rebellion. This raiding, combined with internal strife and general famine, reduced the population of the pueblos from some 30,000 at the outbreak of the revolt to about 9,000 in 1692, two years after the dictator's death, when Spanish governor Don Diego de Vargas exploited the dissension and weakness of the pueblos to lay siege to Santa Fe and retake it. Within four years, all of the pueblos were once again firmly under Spanish colonial domination.

PORTER, David

BORN: 1780

DIED: March 3, 1843

SERVICE: Quasi War with France (1798–1800); Tripolitan War (1801–05); War of 1812 (1812–15)

Distinguished officer during the early years of the U.S. Navy

A native Bostonian, Porter was the son of a Revolutionary War naval commander, also named David. He got his sea legs accompanying his father on a voyage to the West Indies in 1796. Shortly after this, he was impressed—seized for duty—by a British vessel, but managed to escape and was subsequently commissioned a midshipman aboard the thirty-eight-gun *Constellation* on January 16, 1798.

During the Quasi War with France, Porter acquitted himself with conspicuous gallantry in an action against the frigate *Insurgente*, resulting in the ship's capture. On February 9, 1799, he was given command of the vessel as a prize. Later in the year, on October 9, he was promoted to lieutenant and sent to the Mediterranean in command of the twelve-gun schooner *Enterprise*.

Porter next saw action during the Tripolitan War, sailing aboard the thirty-eight-gun frigate U.S.S. *New York*. Subsequently transferring to the **Philadelphia,** he was taken prisoner when the vessel ran aground off Tripoli on October 31, 1803, and he was not released until the conclusion of the war in 1805. On his release, Porter was promoted to master commandant (commander) in April 1806.

From 1808 to 1810, Porter commanded the naval station at New Orleans and was then given command of U.S.S. *Essex*, a thirty-two-gun frigate, in July 1811. Promoted to captain after the outbreak of the War of 1812, he was sent to raid British commerce in the South Atlantic. His accumulated bag amounted to nine prizes, among them H.M. sloop of war *Alert*, which he captured near Bermuda on August 13. Later, during 1813–14, he harassed British whaling in the South Pacific, to the tune of $2.5 million in damages, and took many prizes.

On February 28, 1814, the thirty-six-gun H.M.S. *Phoebe* and the twenty-gun *Cherub* engaged Porter off Valparaiso. Although he drove off *Cherub*, he was finally compelled to surrender to *Phoebe*, which carried longer-range firepower. He was imprisoned until late in 1814 and released in time to take part in naval operations in the Chesapeake during the Washington–Baltimore campaign.

After the war, Porter served on the Naval Board beginning in 1815, But by 1823, he longed for the sea again and assumed command of the West Indies Squadron in operations against pirates. This assignment is historically important because Porter commanded the first steamship used in combat, the three-gun *Sea Gull.* When one of Porter's officers pursued pirates onto shore in Puerto Rico, Spanish authorities arrested the officer and imprisoned *him* on charges of piracy. An outraged Porter threatened bombardment if the officer was not released. He also extracted an apology. Later, the Spanish lodged an official protest, and Porter was recalled, court-martialed, and suspended from duty for six months in 1825. Disgusted, Porter resigned his commission and accepted an appointment as commander in chief of the Mexican Navy—which was then at war with Spain—in August 1826.

Porter served the Mexicans well, but disdained the politics that plagued his office and, in 1829, left Mexico to return to the United States. He was commissioned by President **Andrew Jackson** to go to Algiers as U.S. consul in 1830. The following year, he became chargé d'affaires in Constantinople (Istanbul), and, ten years later, in 1841 was made U.S. minister to Turkey. He succumbed to a bout of Yellow Fever while serving in this office.

Porter was one of the U.S. Navy's great sailors, always resourceful and daring—with the seamanship to back it up. It was Porter who introduced the idea of opening up trade with Japan—a proposal later acted upon when **Matthew C. Perry** was sent on a mission there.

PORTER, David Dixon

BORN: June 8, 1813

DIED: February 3, 1891

SERVICE: U.S.–Mexican War (1846–48); Civil War (1861–65)

American naval officer in the U.S.–Mexican War and the Civil war

Porter was born in the Delaware River town of Chester, Pennsylvania, the third of ten children of **David Porter** and the foster brother of **David G. Farragut.** His land-based education was meager, but he learned early on the ways of the sea, accompanying his father to the West Indies aboard the frigate *John Adams* in 1824. The young man did not begin his naval career proper in the U.S. Navy, however, but started as a midshipman in the Mexican Navy after his father accepted an appointment as commander in chief of that force in August 1826. Young Porter was captured by the Spanish in 1828, but was soon released and, in February 1829, was commissioned in the U.S. Navy as a midshipman.

He began his service in the Mediterranean and was subsequently attached to the Coast Survey during 1836–45. He was promoted to lieutenant in February 1845 and the following year was assigned to recruiting duty in New Orleans. When war with Mexico broke out, he accepted a first lieutenant's slot on the steamer *Spitfire* in February 1847 and participated in the naval bombardment of Veracruz led by **Matthew C. Perry** in March.

Porter earned distinction in the second Tabasco campaign from June 14 to June 22, when he led a detachment of seventy men in storming the main fort there. He was then given command of *Spitfire* for the balance of the war.

Following the war, Porter returned to the Coast Survey and the Naval Observatory in a shore assignment. When no navy sea duty was forthcoming, he took a leave of absence to command passenger and cargo steamers during 1849–55, then returned to the navy as captain of U.S.S. *Supply*. Among his more interesting assignments was the transport of camels as part of an army experiment in using them in the desert Southwest.

At the verge of the Civil War, Porter was again languishing in shore assignments and was contemplating resigning his commission. He was then given command of U.S.S. *Powhatan* for a mission to relieve Fort Pickens, off Pensacola, Florida, in April 1861. Promoted to commander, he was assigned to blockade duty on the Gulf coast, then planned a major naval offensive against New Orleans during November 1861–April 1862. Porter led a flotilla of twenty mortar boats in support of the capture of New Orleans on April 27. He was also crucial to the taking of Forts St. Philip and Jackson two days later.

Later in 1862, Porter sailed under Farragut in operations against Vicksburg, then, in October, he was jumped in rank to acting rear admiral—over some eighty senior officers—and was assigned to open a Mississippi river shipyard at Cairo, Illinois. There Porter built a river fleet of some eighty vessels—the Mississippi Squadron—which he used in support of operations against Arkansas Post during January 10–11, 1863. During the night of April 16–17, he made a daring dash under the Vicksburg batteries and, on April 29, sailed past the guns at Grand Gulf (below Vicksburg) in order to cover **Ulysses S. Grant**'s crossing of the Mississippi on April 30–May 1. In this operation, he sailed up the Red River during May 4–7, then sent gunboats from the Mississippi up the Yazoo River, which forced the Confederates to abandon three unfinished rams and a shipyard on May 13. On July 3, 1863, after the fall of Vicksburg, Porter received the thanks of Congress.

Following Vicksburg, Porter sailed in support of General Nathaniel P. Banks's failed Red River campaign during March 12–May 13, 1864—nearly losing his ships in the process. He was next appointed commander of the North Atlantic Blockading Squadron in October 1864, and, during December 24–25, mounted a massive bombardment of Fort Fisher, North Carolina, with 120 ships. General Benjamin Butler failed to coordinate the land attack with him, however, and Porter was forced to withdraw. He made a second attempt during January 13–15, 1865 when General Alfred H. Terry replaced Butler, and this time Terry was able to attack and storm the fort, taking it after a costly land battle. In April 1865, Porter led a gunboat squadron up the James River in Virginia, an act that forced Confederate admiral Raphael Semmes to scuttle his fleet.

Immediately after the war, Porter was appointed superintendent of the U.S. Naval Academy, in which post he served until 1869. Promoted to vice admiral in July 1866, he was named adviser to the secretary of Navy in 1869. By the 1870s, after the death of David Farragut, Porter was the de facto commander in chief of the navy and was promoted to admiral. He was named to head up the Board of Inspections in 1877. Porter continued on active duty until his death in 1891.

Witty, well-liked, bold, enterprising, and, as a naval commander during the Civil War, second only to Farragut, Porter was a major and formative figure for the United States Navy.

POTEMKIN, Prince Gregori Aleksandrovich

BORN: September 24, 1739

DIED: October 16, 1791

SERVICE: Russo–Turkish Wars (1768–74, 1787–92)

Russian field marshal and statesman who reformed and modernized the Russian army; for a time, he was the most powerful man in the Russian empire

Born in Chizevo in Byelorussia, Potemkin was of Polish extraction, the son of a noble. He was schooled at Moscow University, then entered the Horse Guards in 1755. He was a key participant in the coup d'etat that brought Catherine II the Great to power. She rewarded him with a small estate in 1762. After he distinguished himself during the first of Catherine's several wars with Turkey, during October 1768–July 1774, he became empress's fifth favorite and was elevated to field marshal. He was made commander in chief of the army and governor-general of the Ukraine. He also became Catherine's lover, and even after that relationship ceased in 1776, he remained very much within her inner circle.

As commander in chief of the army, Potemkin revamped and modernized Russia's forces. He planned the conquest of the Crimea in 1776, and formulated many other schemes for colonization and expansion—many of which were fantastically grandiose. His military reforms extended construction of a Black Sea Fleet that included fifteen ships of the line and twenty-five smaller vessels launched during 1784–87. To support the fleet, he established an arsenal at Kherson in 1778 and a great harbor at Sevastopol in 1784.

Despite his reforms, the army performed poorly at the outbreak of war with the Turks in 1787. Potemkin nearly resigned, but the tide of the war turned by its second year, and he was emboldened to invade Moldavia, where he took Ochakov on December 17, 1788 and then Bendery early the next year. During 1790, he concluded operations against the Turks, then returned to St. Petersburg, intending to oust the empress's last favorite, Platon Zubov. Catherine quickly sent him back to Jassy to negotiate peace terms with the Turks. He died en route.

Potemkin was a visionary who also had considerable ability as an administrator. His army reforms were valuable, and although the fleet he hastily built was hardly state of the art, it served Catherine well against the Turks.

POWELL, Colin Luther

BORN: April 5, 1937

SERVICE: Vietnam War (1965–73); Persian Gulf (Kuwait) War (1991)

American general; chairman of the Joint Chiefs of Staff during the Persian Gulf War

Powell was born in New York City, the son of Jamaican immigrants, and graduated from the City University of New York in 1958. He had enrolled in an ROTC program at the university and was commissioned a second lieutenant in the army upon graduation. His first assignment was as a platoon leader and company commander in Germany during 1959–62. He was sent to South Vietnam during 1962–63 as part of a contingent of U.S. military advisers to the Army of the Republic of Vietnam (ARVN). He returned to South Vietnam during 1968–70, as an infantry battalion executive officer and assistant chief of staff (G-3) with the Twenty-Third (Americal) Infantry Division.

In 1971, Powell earned an M.B.A. degree from George Washington University and was also honored by selection as a White House Fellow. He was made special assistant to the deputy director of the Office of the President, serving in that capacity during 1972–73.

Promoted to lieutenant colonel, Powell returned to field command during 1973–75, as commander of the First Battalion, Thirty-Second Infantry in South Korea. He returned to the U.S. to attend the National War College, from which he graduated in June 1976. Promoted to colonel, he was assigned command of the Second Brigade, 101st Airborne Division (Air Assault) at Fort Campbell, Kentucky during 1976–77, then returned to Washington, D.C., where he was assigned to the Office of the Secretary of Defense. Powell served for a brief time as executive assistant to the Secretary of Energy in 1979, then was appointed senior military assistant to the deputy secretary of defense, a post in which he served from 1979 to 1981.

Powell returned to field command again in 1981 as assistant division commander, Fourth Mechanized Infantry Division, at Fort Carson, Colorado, returning to Washington in 1983 as senior military adviser to Secretary of Defense Caspar Weinberger until 1985. The following year, he was put in command of U.S. V Corps in West Germany, returning to in 1987 as deputy assistant for National Security Affairs to President Ronald Reagan and, subsequently, assistant for National Security Affairs for Presidents Reagan and George Bush, from December 1987 to January 1989.

On October 1, 1989, President Bush selected Powell to serve as chairman of the Joint Chiefs of Staff as well as commander in chief, U.S. Army Forces Command. He was the first African American to hold this position.

The first major action he directed in this latter post was the December 1989 Operation Just Cause, an expedition into Panama to apprehend the nation's dictator, Manuel Noriega, and bring him back to the United States to stand trial on drug trafficking charges. From August 1990 to March 1991, Powell directed U.S. participation in United Nations operations against Iraqi dictator **Saddam Hussein,** including during the Persian

Gulf (Kuwait) War of January 17–February 28, 1991. This massive operation gained Powell a great deal of public exposure and approval. The highest-ranking African American in the U.S. armed forces, Powell was seen as an intelligent and politically attractive figure. Many Americans hoped that he would declare himself a candidate for president when he retired from the military in 1993. He chose not to do so, but left the possibility open for the future.

PTOLEMY I SOTER

BORN: ca. 367 B.C.

DIED: 283 B.C.

SERVICE: Consolidation War (336–335 B.C.); Conquest of Persia (334–330 B.C.); Invasion of Central Asia (329–328 B.C.); Invasion of India (328–327 B.C.); Wars of the Diadochi (323–281 B.C.)

Macedonian general under ***Alexander the Great;*** *became ruler of Egypt and founder of the Ptolemaic dynasty*

Ptolemy was the son of a Macedonian noble and, as a child, was the boon companion of **Alexander III the Great,** at whose side he first fought in the Consolidation War. Later, Ptolemy served Alexander during his campaigns in Persia, Central Asia, and India. He fought at such major battles as Granicus (May 334 B.C.), Issus (333 B.C.), Arbela/Gaugamela (October 1, 331 B.C.), and the Hydaspes (May 326 B.C.).

Immediately after Alexander died in June 323 B.C., Ptolemy became satrap of Egypt. He set about establishing himself as ruler, and he acted quickly to expand his domain. His first step was to annex neighboring Cyrenaica. He also secured diversion of Alexander's funeral cortege through Egypt, and he even lodged the dead King's body at Memphis in 322 B.C. This garnered much prestige for Ptolemy, but it also provoked the regent Perdiccas, who invaded Egypt. Ptolemy checkmated Perdiccas by suborning his troops, who turned against their leader, assassinating Perdiccas in the Sinai in 321 B.C.

Ptolemy united with Cassander and Lysimachus in an alliance against Antigonus I Monopthalmus in 316 B.C. during the long Wars of the Diadochi, the power struggle that followed Alexander's death. The allies gained Cyprus about 313 B.C. At Gaza, Ptolemy bested the son of Antigonus, Demetrius, in battle, and then made a favorable peace with the father in 311 B.C.

The power contest continued, as Ptolemy returned to the Wars of the Diadochi. He launched a series of seaborne raids on the coasts of Lycia and Caria (in the southwest corner of modern Turkey) in 309. He expanded combat into the Aegean, taking Corinth, Sicyon, and Megara by 308 B.C.; however, in 306 B.C., his fleet, commanded by his brother Menelaus, was mauled off Salamis in 306 B.C., and Ptolemy lost Cyprus to Demetrius.

During the winter of 306–305 B.C., Ptolemy successfully repulsed an invasion led by Antigonus. He next marched to the relief of Rhodes, under siege by Demetrius, in 304 B.C. His rescue of Rhodes earned him the by-name "Soter"—Savior. After forming a new coalition against Antigonus, Ptolemy next invaded Palestine. Hearing that Antigonus had defeated Lysimachus in Anatolia, he withdrew in 302 B.C. The news was false, and, in 301 B.C., when Ptolemy learned that Antigonus had been defeated at Ipsus, he rushed to occupy Palestine. Unfortunately, this came too late to please Ptolemy's allies, who awarded Palestine to Seleucus. The result was a prolonged struggle between the Ptolemies and the Seleucids for control of Palestine.

The aging Ptolemy abdicated in favor of his younger son Ptolemy II Philadelphus in 285 B.C. He died two years later.

Ptolemy was the most skillful general among the Diadochi—the "successors" who partitioned Alexander's empire after his death. He was also highly effective at building a viable state in Egypt.

PULASKI, Count Kazimierz (Casimir)

BORN: March 4, 1747

DIED: February 11, 1779

SERVICE: War of the Confederation of Bar (1768–72); American Revolution (1775–83)

Polish general best known for his service to America during the American Revolution

Pulaski was born at Winiary in Masovia, near Warsaw, the second son of a man who would later help found the Confederation of Bar, the league of Polish nobles and gentry that was created to defend the privileges of the Roman Catholic Church and, in particular, the independence of Poland from Russia. Young Pulaski would fight in the war triggered by the confederation, distinguishing himself at the defense of Berdičhev in the spring of 1768, then conducting a brilliant campaign against Russians invading near the Turkish frontier. Pulaski also directed the defense of Zwaniec and Okopy Swietej Trojcy in 1769. On September 10, 1770, he took and occupied the fortified monastery of Czestochowa and used it as a base from which he launched guerrilla operations that ranged from the Carpathian Mountains to Poznan. When the Russians attacked the monastery, Pulaski directed a valiant defense that gained him renown well beyond the borders of Poland.

Following the invasion of Poland by Prussian and Austrian troops, which brought about the First Partition of Poland, Pulaski fled to Saxony on May 31, 1772. From there, he traveled to Paris, then to Turkey, where he undertook a campaign to raise troops to fight Russia. But the war-weary Ottomans were now eager to make peace with the Russians, and Pulaski returned to Paris, having exhausted his personal funds, while, at home, the Russians had confiscated his estates.

In Paris, during December 1776, Pulaski met Benjamin Franklin, who had been dispatched to Europe by the Continental Congress to garner support for the American cause. Franklin realized that Pulaski's record as a freedom fighter, his skill at guerrilla warfare, and his prowess cavalryman would be valuable assets to **George Washington**'s army. He persuade Pulaski to sail for America in the spring of 1777. He arrived in Philadelphia and presented Franklin's letter of recommendation to Washington in June.

Pulaski first saw action at the Battle of Brandywine on September 11 and was afterward (September 15) promoted to brigadier general and appointed chief of cavalry. Pulaski did much to train the American cavalry forces, but relations between him and the Continental Army deteriorated. He insisted on high rank and authority, and he refused to make any effort to learn English. He also refused to obey Washington's orders. By March 1778, the situation had become acute, and Pulaski resigned as chief of cavalry, turning his efforts to raising his own mixed force of cavalry and infantry: Pulaski's Legion.

In its first engagement with the British, the unit met with defeat in a surprise attack at Little Egg Harbor on October 4–5. But the legion performed well in the upper Delaware River Valley. In February 1779, Washington ordered the unit to the Carolinas. Shortly after arriving at Charleston, the Pulaski Legion was mauled by the British during May 11–12. Pulaski persevered, however, and fought valiantly through the summer and fall. On October 9, while leading a cavalry charge during the siege of Savannah, he was mortally wounded. He lingered, in great pain, until February 11, 1779, when he died while being transported aboard the brig *Wasp* to Charleston.

As a Polish freedom fighter and patriot, Pulaski enjoyed more success than he did in America. Nevertheless, he was a gallant soldier, and his presence in the American Revolution—despite his arrogance—was an inspiration. His work with training the Continental Army was also of significant value.

PYRRHUS

BORN: ca. 319 B.C.

DIED: 272 B.C.

SERVICE: War against Demetrius (295–284 B.C.); War with Rome (281–272 , 275–275 B.C.); War with Carthage in Sicily (278–276 B.C.)

King of Epirus, whose name become synonymous with self-defeating victory

Pyrrhus would be remembered to history as a rather obscure ruler of Epirus (northwestern Greece) had he not provided the occasion for the phrase *pyrrhic victory* to describe a triumph so costly that it was self-defeating. He was a relative of **Alexander III the Great** of Macedonia and ascended the Epriote throne in 307 B.C., when he was only twelve years old. The youth's early reign was tumultuous, and he sought to bolster his position by forming an alliance with Demetrius I Poliorcetes. This notwithstanding, Pyrrhus was driven from power by a revolt and was compelled to flee for his life into Asia, where he joined Demetrius and the senescent Antigonus I Doson to prosecute one of the many struggles that constituted the Wars of the "Diadochi"—the feuding successors to Alexander the Great. At the Battle of Ipsus, in Asia Minor, Pyrrhus's two allies were defeated by Seleucus and Lysimachus. Demetrius escaped, however, and established control over western Asia Minor. Antigonus was killed, and Pyrrhus, captured, was sent as a hostage to the court of **Ptolemy I Soter** in accordance with the terms of a treaty between the Egyptian ruler and Demetrius.

Within a short time, Pyrrhus and Ptolemy formed a friendship and an alliance, and Ptolemy agreed to help him regain the Epriote throne in 297 B.C. Thus restored, Pyrrhus was at first content to share power with his kinsman Neoptolemus II, but within a very brief period, he arranged for his assassination in 296 B.C. and assumed sole rule. Pyrrhus skillfully exploited the chaotic political situation of Greece and Macedonia to expand his kingdom. Demetrius, his former ally, having seized Macedonia in 294 B.C., became the object of ten years of military operations. In 286 B.C., with the aid of Ptolemy and Lysimachus (another of the Diadochi), Pyrrhus finally succeeded in overthrowing Demitrius, who was deserted by his army and fled to Asia Minor. He threw himself on the mercy of Seleucus, who imprisoned him for the remainder of his life.

Pyrrhus left Greece for Italy in 280 B.C. in answer to a plea from the Greek colony of Tarentum (modern Taranto) for relief from Roman attack. Pyrrhus led 25,000 men and twenty elephants, twice defeating the Romans, most famously at the Battle of Heraclea in 280 B.C. He was congratulated on the victory, which had, in fact, very nearly destroyed his own army. Pyrrhus reportedly replied, "One more such victory and I shall be lost." Indeed, despite his triumph, the Romans refused to negotiate with Pyrrhus. He pushed toward Rome itself, but then withdrew to southern Italy to raise additional troops. At Asculum in 279 B.C. he won another "pyrrhic victory," also suffering a severe wound.

Pyrrhus, King of Epirus. From Charlotte M. Yonge, Pictorial History of the World's Great Nations, *1882.*

Pyrrhus turned away from Rome and advanced on Sicily instead in order to relieve Syracuse from Carthaginian siege in 278 B.C.

During 278–276 B.C., Pyrrhus enjoyed limited success and failed to dislodge the Carthaginians from their strongholds. Worse, Carthage allied itself with Rome, and Pyrrhus rushed back to Italy in 276 B.C. to confront Roman forces once again. The Battle of Beneventum in 275 B.C. ended in Pyrrhus's defeat when the Roman general M. Curius Dentatus drove Pyrrhus's own elephants back upon his lines, creating a panic that broke the commander's lines.

Pyrrhus marched back to Epirus, where he met and defeated Antigonus II Gonatus at Thessalonica in 274 B.C. Having secured Macedonia, he failed to consolidate his gains there, but turned instead to Greece, launching an attack on Sparta in a bid to restore his ally Cleonymus to the Spartan throne. Pyrrhus was killed in a minor nighttime skirmish on the streets of Argos in 272 B.C.

Q

QUANTRILL, William Clarke

BORN: July 31, 1837
DIED: May 1865
SERVICE: Civil War (1861–65)

Infamous Confederate guerrilla leader and outlaw

Born in Canal Dover (present-day Dover), Ohio, Quantrill became infamous during the Civil War as the leader of Confederate irregulars in the border states of Missouri and Kansas. He served under General Sterling Price at the siege of Lexington in September 1861 and was present at the capture of Independence, Missouri on August 11, 1862. He was nominally a captain in the Confederate army, but his conduct was hardly military. He often crossed the line separating militarily justifiable guerrilla action and cold-blooded outlawry.

On August 21, 1863, Quantrill and four hundred followers raided the abolitionist stronghold of Lawrence, Kansas, killing more than 150 unarmed civilians and putting most of the town to the torch. On October 6, 1863, after they surprised and defeated a Union detachment at Baxter Springs, Kansas, Quantrill and his raiders executed their prisoners, combatants and noncombatants alike.

Quantrill was run to ground by federal troops in Kentucky in May 1865. He was fatally wounded.

That Quantrill's men were in essence outlaws is attested to by the names of some of his alumni, which include Frank and Jesse James, and Bob and Cole Younger.

R

RABIN, Yitzhak

BORN: March 1, 1922

DIED: November 4, 1995

SERVICE: Israeli War of Independence (1948–49); Six-Day War (1967)

Israeli guerrilla fighter, general, and prime minister

A native of Palestine (now Israel), Rabin was born in Jerusalem to the family of Russian immigrants. In 1941, he joined the Haganah, Zionist underground army. Two years later, he became part of its elite strike force, the Palmach, and, during the War of Independence, he commanded the Harel Brigade within the Palmach. His theater of operations was the Jerusalem corridor from April 1948 to January 1949.

In 1953, Rabin attended the British staff college at Camberley, then served as commander in chief of Northern Command from 1956 to 1959. He did not see action during the Sinai War of October–November 1956. During 1959–60, Rabin was head of the army's Manpower Branch, then became deputy chief of staff and head of the general staff branch from 1960 to 1964, when he was appointed chief of staff.

Rabin was chief of staff during the Six-Day War of 1967 and played a key role in creating the strategy that was so thoroughly successful during June 5–10. After resigning from the army the following year, Rabin was appointed ambassador to the United States, serving from 1968 to 1973, then gaining election to the Israeli parliament, the Knesset, in December 1973. The following year, he succeeded Golda Meir as prime minister—the first native-born Israeli to hold that office.

Rabin resigned as prime minister in December 1976, remaining as leader of a caretaker government until Menachem Begin came to office. In April 1977, Begin also resigned as head of the Labor Party and spent some years in retirement. In 1984, Begin became Defense Minister under the National Unity government and served in this capacity to 1990.

In February 1992, Rabin was once again elected prime minister and, in a gesture toward peace with the Palestine Liberation Organization (PLO), put a freeze on new Israeli settlements in the occupied territories—the region claimed by the PLO as a Palestinian homeland. Rabin opened secret negotiations with the PLO, which culminated in the Israel–PLO accords of September 1993. Israel agreed to recognize the PLO and also gradually to implement limited self-rule for Palestinians in the West Bank and Gaza Strip. In 1994, Rabin's peace efforts were recognized by the Nobel committee, which awarded him that year's Peace Prize. On November 4 of the following year, Rabin was assassinated by a Jewish extremist who objected to the peace accords.

RAEDER, Erich

BORN: April 14, 1876

DIED: September 6, 1960

SERVICE: World War I (1914–18); World War II (1939–45)

German admiral who masterminded the covert post-World War I expansion of the navy and who directed all-out submarine warfare during World War II

Raeder was born in Wandsbek, a suburb of Hamburg, and joined the navy in 1894. Commissioned an ensign in 1897, he served during World War I primarily at the fleet and staff level, but also directed mine-laying operations and raids along the British coast.

A committed career officer, Raeder remained with the much-reduced German navy after World War I and, using materials in the naval archives, created and published a standard history of cruiser warfare. He was promoted to vice admiral in 1925, then to admiral three years later, when he was appointed chief of the Naval Command on October 1, 1928. From this powerful post, Raeder supervised the expansion and modernization of the German navy, transforming it from the coastal defense fleet allowed by the provisions of the punitive Treaty of Versailles to a full-scale blue-water force.

Adolf Hitler and the Nazis admired Raeder's work, and he was promoted to *generaladmiral*, a new rank created expressly for him in 1935. As commander in chief of the navy, he drew up a grand plan for naval expansion. His Plan Z called for a navy of six battleships, three battle cruisers, two aircraft carriers, and a massive force of cruisers and destroyers, all to be completed by 1944 or 1945. Ironically, the exigencies of World War II, which Hitler started in September 1939 by invading Poland, caused the plans to be shelved. Promoted to *grossadmiral* on April 1, 1939, Raeder turned from an emphasis on surface combat to submarine warfare—although he opposed total reliance on submarines.

Raeder had sharp strategic differences with his colleagues and, worse for him, with the Nazi leadership. Although he remained loyal to Hitler personally, he bitterly opposed the two-front war. Other differences developed and became increasingly strident until, on January 30, 1943, he was forced to resign in favor of **Karl Dönitz.**

Although Raeder was out of the war by the beginning of 1943, he was tried by the international military tribunal that convened at Nuremberg after the war. Found guilty of war crimes, relating to his use of unrestricted submarine warfare, on October 1, 1946, Raeder was sentenced to life imprisonment. Like most of the Nazi war criminals sentenced to lengthy prison terms, Raeder was released early. He was paroled on September 26, 1955.

Raeder did much to develop Germany's navy in despite of the provisions of the Versailles treaty. He also did much to develop the German submarine fleet, although, ironically, he protested against the wholesale conversion of the Nazi navy to a submarine fleet. He supported Hitler, but frequently disagreed with him. Dismissed by the Nazi hierarchy, he nevertheless was made to share Nazi guilt when it came to trial for war crimes.

RICHARD I THE LION-HEARTED

BORN: September 8, 1157

DIED: April 6, 1199

SERVICE: Revolt against Henry II (1173–74); Third Crusade (1190–92); War with Philip II of France (1194–99)

English king and Crusade commander

Richard was perhaps the most narrowly focused monarch in English history, his sole ambition being the "liberation" of Jerusalem. His dedication to the crusade earned him his by-name, "Lion-Hearted." He was the son of Henry II, king of England, and Eleanor of Aquitaine, born in Oxford. He was the third son and, therefore, not in line for his father's kingdom or lands, but he was set to inherit his mother's realm, and he was named duke of Aquitaine in 1172. Richard quickly demonstrated himself adept at the military as well as the political arts, combining both to curb the power of the aristocracy in Aquitaine.

When his three brothers rebelled against their father, Richard joined in the conspiracy, having learned that Henry intended to replace him in Aquitaine with his older brother John. There ensued, during 1173–74, a twisted struggle for power, as the four brothers and Henry repeatedly shifted alliances among themselves. In the end, Henry triumphed, having put down the rebellion against him. He twice invaded Aquitaine before Richard submitted, and the father pardoned his son in 1174.

With this episode behind him, Richard turned to the brutal suppression of revolts among the feudal barons in his duchy. The barons enlisted the aid of Richard's brothers Henry III and Geoffrey of Brittany to drive Richard from Aquitaine, the barons' revolt nevertheless collapsed when Henry III died in 1183, leaving Richard as heir to the throne.

Now that Richard was in line to receive Normandy and Anjou as well as England, his father pressed him to yield Aquitaine to his youngest brother, John. Richard refused to relinquish the realm of his birth and aligned himself with the French king Philip II against the Henry, who was compelled to acknowledge Richard as heir and supreme power in the region. This last war of father against son was concluded two days before Henry II's death on July 6, 1189.

Having schemed and fought to become king, Richard actually cared little about domestic policy. His single all-consuming goal was to "liberate" the Holy Land from the infidels, and, accordingly, he assembled the champions of Christendom. To finance his crusade, he put up for sale all that lay within his preorgative to offer, including government offices and clerical investiture. Richard even sold the king of Scotland his feudal obligation for 10,000 marks, thereby relieving him of fealty to himself. Thus financed, Richard set out for Sicily in 1190.

Richard began by alienating his German allies when he captured Messina, Sicily, on October 4; for the Germans had had their own designs on the town. After Messina, Richard conquered Cyprus and then Joppa, before reaching Acre in July 1191. The Christian coalition he led took Acre and advanced to within a few miles of Jerusalem itself by December 1191. It was there, so close to its goal, that the coalition, badly weakened by the controversy at Messina, began to fall apart. After two attempts to take the city, the Christian alliance fell apart completely, and Richard returned from the crusade without having liberated Jerusalem, although he had obtained pilgrimage rights to the Holy Land.

Worse, upon his return late in 1192, Richard was taken prisoner by his enemy Duke Leopold of Austria, who handed him over to Henry IV, the Germanic Holy Roman Emperor. Henry, still angry over the Messina debacle, held Richard hostage demanding an outrageous ransom of 150,000 marks that took two years to raise.

Immediately after his return to England, Richard moved to Normandy to save his lands from Philip II and a state of chronic warfare ensued. Although Richard made peace with Philip, he was fatally wounded on April 6, 1199, while besieging the castle of the Vicomte of Limoges in a dispute over the ownership of a treasure of gold discovered by a local peasant.

RICHARD III

BORN: October 2, 1452
DIED: August 22, 1485
SERVICE: Wars of the Roses (1455–85)

Last of England's Yorkist kings; ruthless in suppression of opponents

Richard was born at Fotheringhay Castle, Northamptonshire, during the long and bitter conflict between the houses of York and Lancaster, which contended for the English throne. Richard was the son of the Duke of York, who was killed by Lancastrian forces at the Battle of Wakefield in December 1460. The youth's older brother Edward deposed King Henry VI in 1461, but was himself deposed in turn by Richard Neville, earl of Warwick, nine years later. Richard commanded forces in two important Yorkist victories—at Barnet and Tewksbury—which paved the way to Edward's restoration. Richard also conspired in the murder of the deposed king, Henry VI, on May 21, 1471.

The death of Edward in 1483 left his twelve-year-old son, Edward V as heir, with Richard as regent. The family of Edward IV's widow, the Woodvilles, resolved to bring the new king to London, with an army, in order to establish themselves in power and oust the regent. In response, Richard united his forces with those of the Duke of Buckingham, intercepted Edward V, and persuaded him have the conspirators arrested. In the meantime, in 1483, Richard ruthlessly purged many of his closest advisors, believing they, too, had plotted against the regency. In June of the same year, London clerics declared Edward IV's marriage illegal and his children, therefore, illegitimate. It has never been clear whether Richard schemed to obtain this declaration or merely exploited his good fortune. In either case, he obtained the endorsement of an assembly of lords and commoners on June 25 and was himself crowned king on July 6, 1483.

To secure his tenuous hold on the throne, Richard hastily rounded up all those he considered a threat or potential threat, particularly among the Woodvilles. Not content with imprisonment, he executed these rivals. In the meantime, the "little princes," Edward V and his nine-year-old brother, were incarcerated in the Tower of London. In August, they disappeared, and it was later discovered that the two had been murdered. Although it is generally assumed that Richard authorized the murders, no evidence was ever discovered to link him incontrovertibly to the deaths.

When Henry Tudor rose in opposition to Richard in 1485, Richard's rivals flocked to Henry, hoping for a restoration of the house of Lancaster. Henry landed in Wales in early August and marched east, where he was met by the Yorkist forces under Richard at the Battle of Bosworth Field on August 22, 1485. Richard resolved to make short work of Henry and personally—if foolishly—led a valiant charge. He was slain and, at the fall of their leader, the Yorkists fled the field, leaving Henry victorious to establish the house of Tudor.

As to Richard, his corpse was stripped, tied to a horse, and dragged in ridicule. History has treated the monarch in similar fashion; for he was an unpopular ruler and, doubtless, a genuinely ruthless tyrant, bent on achieving and maintaining power by any means. Yet it must be observed that his reputation was deliberately blackened throughout the entire subsequent Tudor epoch, most effectively by William Shakespeare, who created him as one of literature's greatest, most complex, and most deeply sinister villains in the play *Richard III*.

RICHELIEU, Armand Jean du Plessis, Cardinal et Duc de

BORN: September 9, 1585

DIED: December 4, 1642

SERVICE: Thirty Years' War (1618–48); Huguenot War (1621–29); Anglo-French War (1626–30); War of the Mantuan Succession (1628–31)

Principal minister to Louis XIII; guided the nation to a dominant place in European affairs

Born Armand Jean du Plessis in Paris, the future Cardinal and Duke of Richelieu was the youngest son of a poor but noble family from Poitou. He was groomed for a military career, but, after an older brother died, he hastily sought ordination in order to secure his family's traditional benefice, the bishopric of Luçon. Henry IV nominated him as bishop in 1606, and he was ordained in Rome the following year. The clergy of Poitou elected him to the Estates General, and he traveled to Paris in 1614 for its session. He remained afterward, becoming secretary of state for foreign affairs in 1616 through the favor of Marie de Medici, the queen mother and regent. When King Louis XIII overthrew his mother's authority in 1617, Richelieu was removed as secretary, but remained in general favor with the court through his role as peacemaker between the king and his mother.

Richelieu became a cardinal in 1622 and was made chief of the royal council in 1624. His title was changed to first minister in 1628. As first minister, Richelieu developed a single-minded policy of developing the absolute authority of the crown and, with it, the preeminance of France in European affairs. To achieve these ends, he was willing to sacrifice whatever was necessary, including morality, religion, and ordinary law. When a program of warfare was called for, he was ruthless in his taxation of the lower classes—and it is to this willingness to exploit the common folk that Richelieu owes his historical reputation as a crafty and heartless archvillain.

Richelieu was faced with treacherous choices in matters of foreign affairs and religion. It was essential to the rise of France that the European hegemony of Spain and the Austrian Hapsburgs be broken. Accordingly, he initially supported the Protestant cause in the Thirty Years' War, desiring an alliance with the Protestant states aligned against Spain and Austria. However, at home, the French Protestants—the Huguenots—were becoming increasingly powerful, and their independence posed a threat to the absolute rule of the monarchy. Richelieu temporarily reconciled France with Spain and directed his attention to the Huguenots, successfully campaigning against their forces at La Rochelle and Languedoc, resulting in the Peace of Alès, which deprived them of military and political privileges while guaranteeing toleration of their religion. The policy of toleration alienated the devout Catholic party and others among the nobility, including Marie de Medici and the king's brother, Gaston d'Orleans. Richelieu found himself beset by numerous conspiracies to remove him from office, but, with consummate craft and the unwavering support of Louis XIII, he retained his power.

It was not the court and nobles alone who wished to see the overthrow of the first minister. In the tortuous course of the Thirty Years' War (1618–48), Richelieu carefully engineered a series of complex alliances with Sweden, the Netherlands, Denmark, Saxe-Weimar, Saxony, and Brandenburg, so that, on May 21, 1635, he could declare war on Spain. Success against the Spanish was at first slow in coming, but Richelieu was prepared to expend whatever sums were necessary to prosecute the war. To finance the struggle, he levied extraordinarily heavy taxes against those least able to pay, the lower classes. The result was substantial French gains in the war against Spain, but also the perpetual alienation of the lower classes, which, in the provinces, settled into a chronic state of revolt. To cope with the disorder in the provinces, Richelieu appointed and ruthlessly employed commisioners—*intendants*—to oversee them.

Richelieu's interests did extend beyond politics and the amassing of power. A brilliant man, he had an intense interest in literature and theology and became founder of the Académie Française in 1635. However, his overriding preoccupation was with foreign affairs, to the unfortunate exclusion of domestic reform. Richelieu must be regarded as one of the founders of modern France, but, in the process, he was also one of the architects of the social inequities that Louis XIV began to reform but that, in the course of the eighteenth century, intensified and made violent revolution inevitable. Cardinal Richelieu, exhausted by overwork, died in his Parisian palace in 1642.

RIDGWAY, Matthew Bunker

BORN: March 3, 1895

DIED: July 26, 1993

SERVICE: World War I (1917–18); World War II (1941–45); Korean War (1950–53)

U.S. Army general who planned and execute the world's first airborne assault—on Sicily, during World War II

Ridgway was born at Fort Monroe, Virginia and graduated from West Point in 1917. He was immediately sent to a U.S.–Mexican border post with the Third Infantry. By August 1918, he had been promoted to acting captain and given command of the regimental headquarters company. Toward the end of World War I, in September 1918, he returned to West Point as instructor in French and Spanish, and, promoted to the permanent rank of captain, he was sent to Infantry School, from which he graduated in 1925.

From 1925 to 1930, Ridgway served variously in China, Texas, Nicaragua, the Canal Zone, and the Philippines. He returned to the U.S., where he graduated from the Infantry School advanced course in 1930 and, in 1932, was promoted to major. After graduating from the Command and General Staff School in 1935, he served on the staffs of VI Corps and Second Army in Chicago, then attended the Army War College, graduating in 1937. Ridgway served briefly with the Fourth Army based in San Francisco, then joined the War Plans Division of the War Department in September 1939. After promotion to lieutenant colonel (July 1940), temporary colonel (December 1941), and temporary brigadier general (January 1942), he was appointed assistant division commander of the Eighty-Second Infantry Division based in Louisiana in March 1942. Ridgway directed the unit's conversion to the Eighty-Second Airborne Division by August, and, promoted to temporary major general, he accompanied the division to the Mediterranean early in 1943.

Ridgway planned and executed the first-ever airborne assault. A portion of the division parachuted into Sicily during July 9–10, 1943. Ridgway then led elements of the division into combat around Salerno on September 13, then, just before dawn on June 6, 1944, parachuted into France with his troops in support of the Normandy (D-Day) invasion.

Ridgway's command was expanded to the XVIII Airborne Corps (the combined Eighty-Second and 101st Airborne divisions) in August. He led the airborne contingent of the failed Operation Market-Garden masterminded by **Bernard Law Montgomery** in September. Ridgway's corps was far more successful at the Battle of the Bulge, where it was instrumental in checking the last great German offensive of World War II during December 16, 1944–January 15, 1945.

From January to April 1945, Ridgway served in the Rhineland and Ruhr campaigns. At the end of the war, he was promoted to lieutenant general in June 1945 and was given command of the Mediterranean region from November 1945 to January 1946. Following the war, he was appointed to the United Nations Military Staff Committee in January 1946 and subsequently directed the Caribbean Defense Command from July 1948 to August 1949, when he was appointed deputy chief of staff of the Army.

Ridgway took over command of the Eighth Army in Korea shortly after the death of General Walton H. Walker and the great Chinese counteroffensive at the end of 1950, which pushed the U.S.–U.N. forces relentlessly southward. Taking over a badly beaten and demoralized force, Ridgway managed to bring the Chinese counteroffensive to a standstill seventy-five miles south of Seoul. With the arrival of reinforcements, he was able to commence his own counteroffensive, taking advantage of the fact that Chinese lines of communication and supply were stretched to the breaking point. The Ridgway counteroffensive began on January 25, 1951, and he had liberated Seoul by March 14–15. On March 30, U.S. and U.N. troops advanced north of the Thirty-Fifth Parallel, halting the drive on April 22. At this point, President Harry S. Truman concluded his dispute with General **Douglas MacArthur** over goals and strategy by firing MacArthur and naming Ridgway to replace him as U.N. commander and commander in chief Far East on April 11.

In June, Ridgway commenced negotiations with the North Koreans, which continued after he left in May 1952 to succeed General **Dwight David Eisenhower** as NATO supreme Allied commander Europe. With this appointment came a promotion to general. The following year, Ridgway returned to the States as army chief of staff and was immediately embroiled in the red-baiting accusations of Senator Joseph McCarthy, who charged that the army was riddled with Communists and "fellow travelers." Ridgway parried McCarthy's thrusts masterfully and survived—along with the army—to see McCarthy suffer censure from his Senate colleagues for his reckless behavior.

In the Cold War climate of the world after World War II and Korea, in which armed conflict was often on a local scale, Ridgway became concerned that the United States relied too heavily on strategic—nuclear—weapons programs at the expense of developing and maintaining tactical—conventional—warfare capability. In this, at the end of his career, he came into increasingly sharp conflict with U.S. Secretary of Defense Charles Wilson, a proponent of reducing conventional forces and allocating increased funding to nuclear weapons development.

Ridgway stepped down in June 1955 and retired from the army. In 1956, he published his memoirs, *Soldier*, and entered into private business in an executive capacity.

Ridgway was an excellent, intelligent commander, at home behind a desk or in the field with his troops. He was a pioneering force in airborne assault, and he was one of comparatively few proponents of tactical capability at a time when defense policy was enamored of nuclear weapons and massive retaliation.

ROCHAMBEAU, Jean Baptiste Donatien de Virneur, Count of

BORN: July 1, 1725

DIED: May 10, 1807

SERVICE: War of the Austrian Succession (1740–48); Seven Years' War (1756–63); American Revolution (1775–83); Wars of the French Revolution (1792–99)

Marshal of France who aided the Patriot cause during the American Revolution

Born at Vendôme, Rochambeau was the number three son of a noble family and was marked out for a career in the Church. After the death of his elder brother, however, he decided to become a military officer instead and obtained an ensign's commission in the Saint Simon Cavalry Regiment in 1742 at the height of the War of the Austrian Succession. He quickly distinguished himself in actions in Bohemia, Bavaria, and the Rhineland during 1743–47 and was encouraged to purchase the colonelcy of the La Marche Infantry Regiment in March 1747. He was in command of this unit at Lauffeldt, where he was severely wounded on July 2, 1747. It was spring of the following year before he was sufficiently recovered to return to duty, and he participated in the siege of Maastricht from April 15 to May 7, 1748.

With the end of the war on October 18, 1748, Rochambeau returned to his family's estate in Vendôme, wed, and lived the quietly opulent life of the rural gentry. He still remained on duty, however, serving from time to time in garrison posts. The outbreak of the Seven Years' War in 1756 brought him back into battle, beginning with the siege of Minorca from April 12 to May 28, 1756. He fought next, with great distinction, at Crefeld (June 23, 1758) and at Clostercamp (October 16, 1760). Promoted to brigadier general, he was appointed inspector of cavalry in 1761. Rochambeau used his office to modernize the cavalry and introduce a higher degree of organization and professional discipline. He also revised outmoded tactics.

In 1764, Rochambeau was assigned command of a military district in southern France, then, in 1776, was named governor of Villefranche-en-Roussillon, near Prades. When King Louis XVI finally decided to commit French troops to the American Revolution in 1780, Rochambeau, now holding the rank of lieutenant general, was chosen to command the North American expeditionary force. He arrived at Newport, Rhode Island, at the head of 5,500 men, on July 11, 1780.

He did not leap immediately into battle, however, but encamped near Newport from July 1780 to June 1781, awaiting reinforcements—that never arrived. Finally, during June–July 1781, he aided **George Washington** in blockading British-held New York. Together, and in anticipation of naval support from French admiral **de Grasse,** Washington and Rochambeau coordinated a shift to the South to fight General **Charles Cornwallis** in Virginia. Rochambeau and Washington laid siege to Cornwallis at Yorktown on September 28, 1781, boxing in the British general's army and finally forcing its surrender on October 19, 1781. When Cornwallis's representative attempted to offer surrender to Rochambeau, the Frenchman gallantly directed him to Washington.

The Battle of Yorktown essentially brought the Revolution to an end, and, after wintering his army in Virginia, Rochambeau accepted the thanks of Congress, returned to Rhode Island, and sailed for France in the fall of 1782. He arrived at Saint-Nazaire in February 1783 and, from 1784 to 1788, commanded the military district of Picardy. He was transferred to the Alsace district in 1789 and was named a Marshal of France in December 1791.

Assigned command of the Army of the North in the spring of 1792, he failed to check the Austrian advance and summarily resigned his post. Arrested and imprisoned during the Reign of Terror in 1793, he was soon but was released and allowed to retire. **Napoleon** honored him as the great soldier he had been.

Rochambeau was a skilled, if not inspired commander. His innate sense of diplomacy and generosity of spirit are evident in his behavior during the British surrender at Yorktown.

RODNEY, George Brydges, Baron

BORN: February 1718

DIED: May 24, 1792

SERVICE: War of the Austrian Succession (1740–48); Seven Years' War (1756–63); American Revolution (1775–83)

British admiral who performed brilliantly against the French, Spanish, and Dutch

Harrow-educated Rodney volunteered for service aboard H.M.S. *Sunderland* on June 21, 1732. He transferred to the *Dolphin* and was promoted to lieutenant in 1739, then secured an appointment as post captain at Plymouth in 1742. Given command later that year of the sixty-gun *Eagle*, he participated in the victory at Finistere on October 14, 1747.

He was sent to North America as commodore and as governor and commander in chief of Newfoundland in 1749. Returning to England in 1751, he stood for election to Parliament and became the member from Saltash. Six years later, in 1757, he commanded the seventy-four-gun *Dublin* in the campaign to take Louisbourg, Newfoundland, from the French during May 30–July 27, 1758. At the conclusion of this action, he returned to Europe and engaged in raiding along the Norman coast during 1759–60. This operation was highly effective. Rodney destroyed or captured a large number of French ships massed for an invasion of England.

Rodney was assigned command of the Leeward Islands station in October 1761 and the following year, on February 12, took Martinique from the French. Shortly afterward, St. Lucia, Grenada, and St. Vincent all fell to Rodney, who then joined forces with Admiral Sir George Pocock to take Havana from Spain during June 20–August 10.

Rodney returned to England in 1763, triumphant and wealthy, and accepted a promotion to "vice admiral of the blue" as well as the thanks of Parliament. He took command of the Jamaica station during 1771–74, then found himself in the awkward position of fugitive from debtor's prison. For he had quickly squandered his considerable fortune. He remained self-exiled in France until 1778, when a friend helped him discharge his debts. After his return to England, he was appointed commander in chief of the Leeward Islands station in 1779, but had to detour en route to relieve besieged Gibraltar.

In January 1780, he captured a Spanish convoy of merchantmen together with their six escort vessels off Cape Finisterre, Spain. In a spectacular nighttime combat dubbed the Moonlight Battle, he defeated Don Juan de Langara's Spanish squadron off Cape St. Vincent, capturing six ships and sinking one out of thirteen engaged.

After the Midnight Battle, Rodney sailed on to the successful relief of Gibraltar, then proceeded to the West Indies, where he fought a drawn battle with a flotilla under the command of French admiral Luc Urbain du Bouexic, Count of Guichen on April 17. In May, he fought Guichen twice more, then sailed with half his vessels to New York to checkmate an impending American–French attack during September–December.

From New York, Rodney went on to capture the Dutch island of St. Eustatius on February 3, 1781. He took a leave of absence to straighten out a clutch of lawsuits relating to booty seized there. With Rodney away, French admiral **François J. P de Grasse** was able to push Admiral Samuel Hood's squadron away from Fort Royal (Martinique) on April 29. Rodney returned to command on February 19, 1782, and in April he caught de Grasse near Martinique and brilliantly broke through his line at the Saints on April 12. Rodney sunk one French vessel and captured five, including the French admiral's flagship.

The eleven-hour battle was the last sea battle of the American Revolution. It proved that, although the British had been defeated on land, Britannia still ruled the waves. Rodney was hailed as a war hero when he returned to England and was made a baron and given a pension for the balance of his life.

Rodney was a fine admiral, whose most spectacular victory brought an illustrious career to a glorious close.

ROMMEL, Erwin Johannes Eugen

BORN: November 15, 1891
DIED: October 14, 1944
SERVICE: World War I (1914–18); World War II (1939–45)

The Third Reich's "Desert Fox"; commander of the Afrika Korps and Germany's finest armored tactician

Born in Heidenheim, Württemberg, Rommel was the son of a schoolteacher. He joined the German Army in 1910 as an officer-aspirant in the 124th Infantry Regiment. In January 1912, he received his second lieutenant's commission and was assigned to a field artillery regiment in March 1914, shortly before the outbreak of World War I. He rejoined the 124th in August, when the war began and fought in the invasion of France. He was twice wounded, and, after recovering from the second wound, he transferred to the Württemberg Mountain Battalion, with which he served in the Vosges. He followed the mountain battalion to the Romanian Front and served there and in Italy with distinction during 1917–18. What gained Rommel particular notice was his flair for tactics, which he demonstrated by a bold infiltration during the Battle of Caporetto.

Rommel ended the war as a staff officer, then rejoined the 124th Regiment in December 1918, after the Armistice. In the much-reduced postwar army, Rommel commanded an internal security company in 1919–21, then took charge of a company in the Thirteenth Infantry Regiment headquartered at Stuttgart. In October 1929, Rommel was appointed an instructor at the Infantry School in Dresden, During this period he wrote a book that became a standard text on infantry tactics, *Infanterie greift an* (*Infantry Attacks*), which was published in 1937.

Rommel was given command of a *jaeger* battalion of the Seventeenth Infantry Regiment in October 1935, then, in 1938, was assigned to command the War Academy at Wiener Neustadt. During the prelude to World War II—**Adolf Hitler**'s annexation of the Czech Sudetenland and the occupation of Prague (1938–39)—Rommel was honored with assignment to

Erwin Rommel. Courtesy National Archives.

command the *Führerbegleitbataillon*, Hitler's personal bodyguard. Hitler had great confidence in Rommel and appointed him his chief for personal security during the Polish campaign. After this, he was given command of the Seventh Panzer Division in February 1940 and was a key participant in the rapid invasion of France during May–June.

Following the French campaign, Rommel was dispatched to Libya to command the Afrika Korps beginning in February 1941. He quickly earned legendary renown as the "Desert Fox," who twice pushed British forces back across the Egyptian–Cyrenaican frontier in a spectacular series of large-scale armored battles. After the British surrendered at Tobruk in June 1942, Rommel was promoted to field marshal.

But his fortunes began to turn. Running up against **Bernard Law Montgomery,** he was soundly defeated at El Alamein during October–November 1942. In March 1943, he was recalled to Europe, where he was assigned to command Army Group B in northern Italy.

Subsequently, he became commander of German forces in the Low Countries and northern France (January 1944) and oversaw reinforcement of the "Atlantic Wall" defenses in anticipation of an Allied invasion. It was Rommel's Army Group B that bore the brunt of the Normand "D-Day" invasion in June 1944.

Rommel failed to check the invasion on the beaches—in large part because Hitler refused Rommel's request for armored forces in the early stages of the fighting. He then fought fiercely against an Allied breakout, but ultimately failed. Then, on July 17, 1944, he was wounded in an Allied air attack. Shortly after this, he was implicated—but never formally accused—in a failed attempt to assassinate Hitler. He was given the choice of standing trial or ending his own life. On October 14, 1944, he took cyanide.

Rommel was a legendary tactician and, therefore, one of the great generals of World War II. He turned against Hitler because of Hitler's failures as a military strategist rather than because of moral qualms. However, popular legend has tended to paint Rommel as a noble warrior, an anti-Nazi, and even a kind of knight-errant. This is a romantic distortion. He was, however, admired for his skill—by Germans and Allies alike. The Nazis preserved the illusion that he had died a hero of the Reich, claiming he had succumbed to war wounds. He was given a full state funeral.

RUNDSTEDT (Karl Rudolf), Gerd von

BORN: December 12, 1875

DIED: February 24, 1953

SERVICE: World War I (1914–18); World War II (1939–45)

German field marshal; among* Adolf Hitler's *most capable military commanders

Born at Aschersleben into a venerable Prussian military family, Rundstedt attended the Oranienstein Cadet School from 1888 to 1891 and graduated from the Main Cadet School at Gross Lichterfelde in 1893. He was commissioned a second lieutenant that year in the Thirty-Third Infantry Regiment. In 1902, the promising young officer was sent to the *Kriegsakaillemie*—the German army's war college—from which he graduated with distinction, earning him a promotion to captain on the General Staff in 1909.

By the outbreak of World War I, Rundstedt was chief operations officer of the Twenty-Second Reserve Division, then took over command of the division for a time after its regular commander was wounded at the battle of the Marne during September 5–10, 1914. Promoted to major in November 1914, Rundstedt was assigned to a number of staff posts throughout the war, culminating in an appointment as chief of staff of the XV Corps in November 1918.

Rundstedt was one of the elite corps of four thousand officers selected to lead the limited army, the *Reischwehr*, of postwar Germany. Promoted to lieutenant colonel, he served as chief of staff of the Third Cavalry Division beginning in October 1920. In 1923, he was promoted to colonel and, two years later, was assigned command of the Eighteenth Infantry Regiment.

In November 1928, Rundstedt was promoted to *generalmajor* and given command of the Second Cavalry Division. He was promoted once again, to *generalleutnant*, in November 1929, then to *general der infanterie* with command of First Army Command in October 1932.

Rundstedt was instrumental in the secret rearmament of Germany, then retired from active service with the rank of *generaloberst* in October 1938, only to be recalled to active duty on June 1, 1939. He was assigned to command Army Group South during the lightning campaign in the Low Countries and France from May 10 to June 25, 1940. After his triumph in France, Rundstedt was promoted to *generafeldmarschall* on August 19, 1940.

The field marshal commanded Army Group South during Operation Barbarossa, the invasion of the Soviet Union. He advanced through the Ukraine with devastating results and reach the Don River on December 1. Advanced in age, he resigned at this point, only to be recalled yet again in March 1942 as commander in chief, west (OB West) and commander of Army Group B. He was charged with responsibility for making preparations against an Allied invasion of Western Europe. He and **Erwin Rommel** disagreed on how to employ the vast mobile reserve force. Shortly after the "D-Day" Normandy invasion, Rundstedt was relieved of command (on July 6, 1944); however, he was sufficiently respected to merit appointment by Hitler to the Court of Honor that tried the officers implicated in the July 20 plot to assassinate the Führer.

As the situation on the Western front rapidly deteriorated. Rundstedt was recalled to duty yet again, as commander of Army Group B on September 5, 1944. In this capacity, he directed the Ardennes offensive—the "Battle of the Bulge"—during September 16, 1944–January 16, 1945, the last German offensive of the war. He also commanded the last-ditch and quite doomed defense of the Rhineland during January 1–March 10, 1945.

Rundstedt was personally dismissed from command by Hitler on March 9, but was captured by American forces in May. Held in England from 1945 to 1948, he was released to a quiet retirement in Hanover.

Rundstedt was an anachronism in the Nazi army, a holdover from the traditional Prussian officer corps. He did not share enthusiastically in Nazi politics and was, for that reason, always suspect. In fact, he was a dedicated military professional, determined to serve his country regardless of its politics.

RUPERT, Prince, Count Palatine of the Rhine, Duke of Bavaria

BORN: November 28?, 1619

DIED: November 29, 1682

SERVICE: Thirty Years' War (1618–48); First English Civil War (1642–46); Naval War against Parliament (1647–53); Second Anglo–Dutch War (1665–67); Third Anglo–Dutch War (1672–74)

Anglo–German Royalist general and admiral during the English Civil War

Rupert was born in Prague, the third son of Frederick V, Elector Palatine, and Elizabeth, daughter of King James I of England. After the battle of White Mountain ended his father's brief reign on November 8, 1620, Rupert was raised in relative poverty in his father's court-of-exile at the Hague.

In 1630, he was present with Prince Frederick Henry of Orange at the siege of Rheinsberg, then joined the Prince's Life Guards and fought at the siege of Tirlemont in 1635. He traveled to the court of his uncle, Charles I of England, and stayed from 1635 to 1637, when he fought under **George Monck** at the capture of Breda on October 20, 1637. He was then assigned command of a cavalry regiment under his brother, Charles Louis, who made an ill-advised expedition to recover the Palatinate. Rupert was taken prisoner by the Imperial forces at Vlotho and was held from 1638 to 1641.

Released by Emperor Ferdinand III in December 1641, he returned to England at the outbreak of the Civil War and was made King Charles's general of the horse. He soon demonstrated prodigious talent, at Powick Bridge, at Edgehill (October 23), and at Birmingham and Lichfield early in April 1643. He triumphed at Chalgrove Field, near Dorchester, on June 18, then led his troops to join Sir Ralph Hopton in the taking of Bristol on July 26, 1643. Next, he took the offensive against Essex's army at Stow on the Wold on September 1, and attacked again at Newbury on September 19.

Created Duke of Cumberland and Earl of Holderness on January 24, 1644, he marched to the relief of Newark on March 20, 1644, forcing the surrender of the opposing army. On June 1, he united with the army of Lord George Goring, and, together, they marched north to relieve York and aid the Earl of Newcastle.

At Marston Moor on July 2, 1644, he fought the Scots–Parliamentary armies of Lord Leven, Sir Thomas Fairfax, and the Earl of Manchester. Rupert's army was badly outnumbered, and was defeated—very nearly destroyed—after a brutal combat. He fell back on Chester with the few thousand cavalrymen remaining to him.

On October 26, he fought the second Battle of Newbury to a draw, and the following month was appointed lieutenant general of all the king's armies. However, the following year Charles acted contrary to Rupert's counsel and sought battle with the more numerous and far better trained New Model Army of **Fairfax** and **Cromwell.** The Battle of Naseby, on June 4, 1645, was a disaster. Rupert fought brilliantly, but the Royalist forces were consistently outmaneuvered and out-gunned. The king's army was nearly destroyed.

Following Naseby, Rupert reasoned that the Royalist cause was untenable militarily. He advised negotiation, but was rebuffed by Charles. Loyal to the king, he defended Bristol during August 18–September 10, then surrendered.

King Charles dismissed Rupert in September 1645, and Rupert left England for France, where he became a *marechal de camp*. Wounded during the siege of La Bassée in 1647, he was quickly reconciled with Charles I and returned to England to take command of the king's fleet in the summer of 1648. During 1649, he privateered until Admiral **Robert Blake** drove him from the Mediterranean. Fleeing Blake, he made for the Azores and the West Indies, but, during 1651–52, he lost most of his small fleet in actions throughout the Caribbean.

Rupert returned to Europe in the spring of 1653, but, once again, Charles was discouraged with Rupert's results and disappointed in him. Rupert settled for a time in Germany, where he pursued science and invention. He perfected the mezzotint printing process during 1653–60.

With the Restoration, Rupert returned to England, where Charles II welcomed him wholeheartedly and appointed him to a naval command in May 1665, at the outbreak of Second Anglo–Dutch War. He was in charge of the van at Lowestoft on June 3, 1665, and, although he was wounded, his squadron arrived in good time to relieve Monck's beleaguered fleet on June 13, 1666 during the famous Four Days' Battle of June 11–14. When Admiral Michel de Ruyter, drove them into the Thames, Monck and Rupert sallied forth to drive off de Ruyter at the battle of the North Foreland on July 25, 1666.

Rupert commanded shore defenses during the Dutch raid on the Medway River (June 17–22, 1667) and successfully repulsed an attack on Upnor Castle, near Rochester. When war broke out anew during March–May 1672, Rupert rushed to reinforce the fleet under James Duke of York at Sole Bay on June 7, 1672. He subsequently took James's place as commander in chief of the fleet after the Test Act disqualified James for command in 1673. Rupert had mixed success leading the Anglo–French fleet against the Dutch. He never received support from the allies, and was unable to attain any decisive victories. With the end of the war in 1674, Rupert retired, succumbing to a fever eight years later.

Rupert's loyalty and bravery were boundless. He was an excellent cavalry leader, though he often acted rashly. He was the Royalist's most talented military leader.

S

SAINT-CYR, Laurent Gouvion, Count (Marquis of)

BORN: April 13, 1764

DIED: March 17, 1830

SERVICE: French Revolutionary Wars (1792–99); Napoleonic Wars (1800–15)

French Marshal of the Empire under ***Napoleon***

Saint-Cyr was born at Toul, the son of a tanner. He was studying art in Paris at the outbreak of the Revolutionary Wars and eagerly volunteered, in 1792, for service in the chasseurs of Paris. Enormously popular and charismatic, he rose through the ranks with lightning speed. The year he volunteered, he was made captain and given an appointment to the regimental staff. The following year, he was promoted to major, followed by another promotion, to adjutant general. In 1794, he became a colonel, general of brigade, and general of division.

Saint-Cyr fought in the first three campaigns of the Army of the Rhine. He then transferred to the Army of the Rhine and Moselle in 1795, and was a key participant in the siege of Mainz on October 29. The success of this operation resulted in his being given corps command of the left wing of the army under General **Jean Victor Marie Moreau** in 1796.

Saint-Cyr next served in Rome—in 1798—under General Brune, then, the following year, transferred to the Army of the Danube. With this force, he fought at Stockach on March 25 under General Jean-Baptiste Jourdan. Moving on to the Army of Italy, he fought under Joubert and Championnet, leading the right wing of the army at Novi on August 15, 1799. He performed with great distinction at this battle and was presented with a saber of honor and appointment as governor of Genoa.

In 1800, Saint-Cyr was assigned to command the center of the Army of the Rhine under Moreau. He hastened to the front at Engen on May 3 and was instrumental in achieving victory there. He next distinguished himself for valor at Biberach on May 9, defeating the Austrian army. For this, he was named a councillor of state by **Napoleon** Bonaparte. The following year, 1801, Saint-Cyr commanded the army that invaded Portugal.

From late 1801 to the middle of 1803, Saint-Cyr served as French ambassador in Madrid, then was sent in command of an army of observation in Naples. A lieutenant general in 1803, he was promoted to colonel general of cuirassiers the following year.

In 1805, Saint-Cyr was given command of the First Corps of Reserve at the camps at Boulogne. Later in 1805, he fought in Italy against Archduke Charles and captured an entire corps at Castelfranco di Veneto on November 24. He served in General **Nicolas Soult**'s IV Corps at Heilsberg on June 10, acquitting himself with conspicuous bravery.

Created a count in 1808, Saint-Cyr was dispatched to Catalonia in command of a corps. Napoleon's orders were simple and direct: "Preserve Barcelona for me." Saint-Cyr carried out his emperor's orders, taking the town of Rosas during November 7–December 5, then emerging victorious from the battle of Cardedeu on December 16. This relieved Barcelona.

After winning two additional battles—Molins del Rey (December 21) and Igualada (February 17, 1809—he besieged Gerona beginning on June 6. His patience in this operation exceeded that of Napoleon, who summarily relieved Saint-Cyr on October 5. Outraged, Saint-Cyr left his post before his replacement arrived. Napoleon responded by pronouncing Saint-Cyr in disgrace and exiling him to his estates. It was 1811 before he returned to the emperor's good graces. He was reappointed as councillor of state, and, in 1812, was given command of VI Corps in the Russian campaign.

Saint-Cyr fought at Polotsk, on August 17–18, and took over command when Marshal **Nicolas Oudinot** suffered one of his many battle woundings. Saint-Cyr was also wounded in battle, but remained on the scene and succeeded in driving off General Wittgenstein. His victory here effectively canceled out the vestiges of his earlier disgrace. Napoleon presented Saint-Cyr with his marshal's baton on August 27, 1812.

On October 18, at Polotsk, Saint-Cyr was hit in the foot by a musket ball. Severely injured, he had to be carried from the field. Defeated at the second battle of Polotsk during October 18–20, he relinquished command because of his wound. He returned to duty in 1813, when he served under Eugene Beauharnais in Germany. In March, however, he was felled by typhus, which made him sufficiently ill that he had once again to relinquish his command.

After his recovery later in 1813, Saint-Cyr was given command of XIV Corps and named governor of Dresden. At the Battle of Dresden, Saint-Cyr commanded the center directly under Napoleon during August 26–27. When Napoleon turned his attention to Leipzig, Saint-Cyr remained in Dresden to hold the town. However, he finally capitulated to an Allied siege in November. He was marched off, at the head of 33,000 men, to Bohemia, a prisoner of war. It was not until June 1814 that Saint-Cyr was released and returned to France. There Louis XVIII made him a peer of France.

Saint-Cyr did not fight at Waterloo, and, after Napoleon's final exile and the return of Louis, he was appointed Minister for War, governor of the twelfth military district, and member of the Privy Council—all in 1815. The following year, he was made Grand Cross of the Order of St. Louis and was appointed governor of the fifth military division. For a brief period during 1817, he was also Minister of the Marine and Minister of the Colonies, then was reappointed Minister for War, serving from September 1817 to November 1819.

Saint-Cyr proved to be a superb military administrator, introducing many reforms in the French army, which helped to rebuild the force in the post-Napoleonic era. Created a marquis in 1817, he retired two years later and composed a history of his Rhine campaigns, which was published in 1829, a year before his death. Saint-Cyr was a brilliant field commander and a gallant soldier.

SALADIN (Salah Ad-din Yusuf Ibn Ayyub, "Righteous of the Faith")

BORN: ca. 1138

DIED: March 4, 1193

SERVICE: Egyptian War (1164–68); Conquest of Syria (1174–86); Holy War (1164–92)

*Turkish–Egyptian general and ruler who opposed **Richard I the Lion-Hearted** during the Crusades*

Born in Takrit, Mesopotamia, Saladin was of Kurdish ancestry and, as a young man, accompanied his uncle on an expedition to free Egypt from the Frankish domination begun by the First Crusade. He served briefly as the governor of Alexandria, and when his uncle died, about 1169, he became the vizier of Egypt and the commander of Syrian troops. Within two years, he had established himself as the sole ruler of Egypt.

For the next decade, Saladin's primary goal was to unite the various countries of the Middle East under a single Islamic banner, something he achieved with a high degree of success. The Islamic Empire spread over thousands of square miles—from Gibralter in the west to beyond the Indus River in the east, and from Turkey in the north to the tip of the Arabian Peninsula in the south—yet this vast realm was racked by dispute and distrust. Saladin did much to reconcile the disparate factions of Islam and make the empire stronger that it had ever been. This progress toward unification was achieved only after much warfare with his Islamic brethren, but it was also accompanied by profound cultural developments and a growth in learning, as Saladin fashioned his realm into a haven for religious scholars and teachers, to whom he assigned the responsibility for instilling in the masses the tenets of Mohammadism.

By 1187, Saladin had sufficiently consolidated the Islamic Empire that he deemed himself ready to wage war on the Christian states established by the Crusaders during the preceding century. Jerusalem itself had been occupied for eighty-eight years, and one of Saladin's top priorities was to re-establish Islamic hegemony there. In northern Palestine, on July 4, 1187, he led a highly trained army against a poorly equipped Frankish force near Tiberius. Within hours, Saladin had routed the Latin army and, within the span of three months, was able to retake from the Crusaders the cities of Acre, Toron, Beirut, Sidon, Nazareth, Caesarea, Nabulus, Jaffa, Ascalon, and, of course, Jerusalem.

Although the Crusaders had freely persecuted those whom they had conquered, Saladin treated the vanquished with justice and courtesy. This did nothing to stop **Richard I the Lion-Hearted** of England, **Philip II Augustus** of France, and **Frederick I** of Germany from organizing the Third Crusade, which misfired when Frederick was drowned on the journey east, and Philip, after reoccupying Acre, returned home. Richard concluded a peace with Saladin; then he, too, returned to Europe. Following the Third Crusade, Saladin still held sway over most of the Islamic Empire, and Jerusalem remained in his hands.

Saladin was a consummate diplomat and military strategist. His unification of Islamic nations into a single military power strengthened the culture and religion of the Mohammedan world. He failed to purge all Crusader influence from the Holy Land, but he sharply curbed it. Perhaps most remarkably was how all of this was accomplished with much humanity and generosity toward enemies. Saladin died in Damascus on March 4, 1193, shortly after the end of the Third Crusade.

SAMORI TOURE

BORN: ca. 1835

DIED: June 2, 1900

SERVICE: Conquest of Kankan (1879); Wars with the French (1883–85); War with Sikasso (1887); War with the French (1889–98)

Dioula war chief, ruler, conqueror of Mali

Samori Toure led the Dioula resistence to French colonial domination of Guinea, Mali, and the Ivory Coast. He was the son of a merchant family and born near Sanakoro, Guinea. Trained as a soldier, he took the title of *almani*, or prayer leader in 1868 and that year proclaimed himself chief of Bissandougou. He organized the *sofa*, a well-equipped and well-trained army. For the next several years, well into the 1870s, Samori Toure consolidated his power, expanding the Bissandougou into a kingdom subject to his absolute rule and maintained by an army financed through exorbitant tribute payments extorted from the many tribes under his control. By the opening of the 1880s, Samori Toure controlled what amounted to a full-fledged state along the eastern side of the upper Niger and bounded by Burkina Faso (in Upper Volta) to Fouta Djallon (in northwestern Guinea).

In 1883, when French colonial forces invaded and occupied Bamako (in present-day Mali), Samori Toure mounted an offensive, engaging the French troops many times between 1884 and 1885 before he signed the Treaty of Bissandougou in 1886, establishing the Niger River as the western limit of his domain and agreeing to French "protection"—a transparent code word for the acceptance of colonial domination.

Samori very quickly violated the provisions of the protectorate by attacking Sikasso (in present-day Mali), seeking to expand his kingdom. In response, French forces aided the local tribes in repelling Samori's attacks during 1887–88, and Samori Toure repudiated the Treaty of Bissandougou, declaring all-out war against the colonial forces, beginning about 1891. The French relentlessly pushed Samori Toure from position to position, until he retreated to Dabakala (in the present-day

Ivory Coast), where he set up a guerrilla base in 1893. From this stronghold, he launched terrorist raids against the countryside for the next three years.

By 1896, Samori Toure was able to coalesce his guerrilla activities into a full-scale invasion of Burkina Fasso, penetrating the districts of Lobi and Sanufo (between the Léraba and Black Volta rivers). He established a new stronghold at Darsalami during the early summer of 1897, then, in July and August, moved against Noumodara, laying siege to that settlement and ultimately destroying it. Retreating before the approach of French colonial forces, Samori Toure terrorized and razed Kong (on the Ivory Coast) toward the end of 1897 and Bondoukou (on the border of the Ivory Coast and Ghana during mid 1898.

French forces finally captured Samori Toure at Guelemou, Ivory Coast. He was tried and exiled to Ndjole, an island in the Ogooué River in Gabon. There he died on June 2, 1900.

SAMSONOV, Aleksandr Vasilievich

BORN: 1859

DIED: August 30, 1914

SERVICE: Russo–Turkish War (1877–78); Russo–Japanese War (1904–05); World War I (1914–18)

Russian general in command at the disastrous battle of Tannenberg

Samsonov was a graduate of the cavalry school at St. Petersburg and saw his first combat service during the Russo–Turkish War of 1877–78. A solid career officer, he graduated from the staff academy and joined the General Staff in 1884. Promoted to colonel, he was assigned as commandant of the St. Petersburg Cavalry School from 1896 to 1904.

Samsonov was promoted to brigadier general in 1902 and commanded the Ussuri mounted brigade of the Siberian Cossack Cavalry Division during the Russo–Japanese War. He was made *ataman* of the Don Cossacks in 1909, and from 1910 to 1914, he was governor general and commander in chief of troops in Turkestan.

At the outbreak of World War I in August 1914, Samsonov was made commander of the Second Army along the River Narew in Poland. He was new to the territory and to his command, but was nevertheless ordered to invade East Prussia from the south. Unfortunately, he was assigned to work in coordination with the First Army, invading from the east, which was commanded by General Rennenkampf, a bitter personal rival. The two generals hated one another more than any foreign enemy and did little to communicate, let alone cooperate. Moreover, neither of the Russian forces was adequately equipped or staffed. Officers were few, and the enlisted ranks were poorly trained.

Samsonov obeyed orders from the high command to advance deep into East Prussia, even though Rennenkampf, with whom he was to attack in concert, had halted after the Battle of Gumbinnen on August 20. As a result of this incoordination, Samsonov was surprised by the German Eighth Army during August 24–25. The Germans virtually annihilated Samsonov's entire army at Tannenberg during August 26–31.

When Samsonov realized just how badly he had lost, he cut off his communications with the rear and rode to the fore in order to take personal command. It was too late. During the disorderly retreat that came on August 30, he apparently committed suicide. In a war of terrible carnage and notable blunders, Samsonov presided over one of the very worst. It was a blow from which the Russian army never really recovered.

SAMUDRAGUPTA

DIED: 375

SERVICE: Expansion of the Gupta Empire (335–75)

Son of Chandragupta I; led a series of wars that greatly expanded the territory he inherited; came to rule a vast portion of modern-day India

Upon the death of Chandragupta I, his son Samudragupta ruled a territory that extended from Magadha to Allahabab. When still new to power, he plotted to expand his holdings and accordingly attacked a host of lesser kings in the Upper Ganges valley, killing them one by one. From his base near present-day Delhi, Samudragupta spared but reduced to vassalage the kings of Samatata (eastern Bengal), Davaka (Nowgong in Assam), Kamarupa (western Assam), Nepal, Kartripura (Garhwal and Jalandhar), and numerous tribal leaders in the eatern and central Punjab, Malwa, and western India as well as the chiefs of the Kushans and Sakas. After his successes, he ventured into the Deccan, where he defeated and captured an impressive catalogue of rulers. Once his captors paid him handsome tributes, he restored them to their places of power.

Through his many conquests, Samudragupta expanded his holdings to include vast tracts extending south from the Himalyas to Narmada, and west from the Brahmaputra to the Jumna and Chambal. In regions beyond his direct control, kings in east Bengal, Assan, Nepal, the eastern portion of Punjab, and several tribes of Rajasthan paid tribute to him. In the course of expanding his domain, Samudragupta killed a total of nine kings and subjugated twelve others.

Samudragupta brought a measure of culture and benevolence to the regions he conquered. Most scholars agree that in the early days of the Gupta dynasty Indian society adopted its characteristic Hindu religion and culture, which endures to this day.

SANTA ANNA, Antonio Lopez de

BORN: February 21, 1794

DIED: June 21, 1876

SERVICE: Second War for Mexican Indepence (1821); Texan War for Independence (1835–36); U.S.–Mexican War (1846–48)

Mexican general and dictator, who took the Alamo and who was defeated in the U.S.–Mexican War

Antonio Lopez de Santa Anna seized opportunity, changed allegiances and positions when it suited his own ambitions, and sought, always, to enrich himself. Much of the political turbulence that plagued Mexico throughout the nineteenth century—and even into the twentieth century—resulted from Santa Anna's antics.

Born in Jalapa, Veracruz, the son of a minor colonial official, Antonio Lopez de Santa Anna joined the Spanish army in Mexico as a young man, eventually rising to the rank of captain. Even his early forays into politics revealed a duplicitous agility that allowed him to champion both sides of most issues. In a time characterized by intense political confusion and intrigue, Santa Anna first lent his support to Augustin de Iturbide in the war for Mexican Independence during 1821; two years later, he helped overthrow Iturbide. In 1828, he supported Vicente Guerrero in his bid for the presidency; later, Santa Anna participated in his overthrow.

Santa Anna became a hero to the Mexican people in 1829 when he fought Spanish forces in their attempt to regain Mexico. The "Hero of Tampico" rode to the presidency in 1833 on a wave of popularity and promises of a federalist government. In three years in office, however, Santa Anna actually solidified an absolutist state.

Santa Anna's rule was short lived. When Texas declared its independence from Mexico in 1836, the dictator answered. He marched his army to the Alamo, attacked the garrison, and slaughtered its 183 men. Texans, galvanized by "Remember the Alamo," met Santa Anna at the San Jacinto River, where he was defeated by forces under **Sam Houston** on April 21, 1836. Santa Anna's pleas for his life were successful, and the Texas commander sent him to Washington, D.C. for a dressing-down at the hands of President **Andrew Jackson.** Following his defeat and subsequent humiliation, Santa Anna returned to Mexico, stepped down as president, and entered into retirement.

He did not stay out of military and political affairs for long. Two years after his defeat, French forces landed at Veracruz to claim reparation for injuries to the French citizens living in Mexico. Santa Anna headed an expedition to challange the French troops. By the time he arrived, the French fleet was already in the process of departing, but there was a skirmish and Santa Anna was wounded.

The incident at Vera Cruz cost Santa Anna a leg, but he regained his prestige. He became dictator of Mexico for a few months in 1839, while the duly elected president was away. Two years later, he led a revolt against the government and seized power again, but in 1845 he was driven into exile in Cuba.

When war broke out in 1846 between the United States and Mexico following the U.S. annexation of California and Texas, Santa Anna craftily devised a way to get back into the fight. Promising to help restore peace, he convinced U.S. President James K. Polk to send him to Mexico on board an American ship. Once Santa Anna arrived, however, he catapaulted himself to the head of the Mexican Army. Santa Anna's war was a disaster for Mexico. It suffered defeat after defeat and ultimately ceded to the United States New Mexico, Arizona, Nevada, Utah, and California. Santa Anna retired in 1847—even before the war was officially concluded by the Treaty of Guadalupe Hidalgo—and went to live first in Jamaica, then in New Granada.

Recalled to Mexico as dictator between 1854 and 1855, Santa Anna was again ousted and exiled to Nassau. He offered his old gambit yet again when Napoleon III installed Archduke Maximilian of Austria as emperor of Mexico in 1863. Incredibly, Santa Anna appealed to the United States to support him in an attempt to oust the emperor—even as he offered Maximilian his support against the Mexican nationalists. Both sides, however, knew full well that Santa Anna was untrustworthy and declined his proposals.

Santa Anna remained in exile until 1874, when, impoverished and blind, he was allowed to return to Mexico. He died in Mexico City on June 21, 1876.

SAXE, Hermann Maurice, Count

BORN: 1696

DIED: November 30, 1750

SERVICE: Great Northern War (1700–21); War of the Spanish Succession (1701–14); Austro-Turkish War (1716–18); War of the Polish Succession (1733–38); War of the Austrian Succession (1740–48)

Born in Germany, Saxe was a highly effective marshal of France

Saxon-born Saxe was the natural child of the elector of Saxony Augustus II the Strong, who was also king of Poland. Augustus cared for his illegitimate son and found him a berth as an ensign in a Saxon infantry regiment when he was twelve. He first saw action under **Marlborough** and **Eugene, Prince of Savoy-Carignan** in Flanders during 1709. His experience at the Battle of Maiplaquet on September 11, 1709 profoundly affected him. He was attracted to battle, though repelled by the magnitude of carnage.

Titled "Count of Saxony" by his father in 1711, Saxe participated in campaigns against the Swedes in Pomerania (now northern Poland) during 1711–12, then served at the siege of Stralsund in 1712. He was appointed a colonel of cavalry in this year, but stepped down from active service the following year, when he married a youthful heiress. After a period of moneyed domestic bliss, Saxe put himself in the Austrian service in 1717 and fought once again under the command of Prince Eugene, this time at the siege of Belgrade from June 29 to August 18.

Saxe served with the Austrians until the Treaty of Passarowitz of July 21, 1718, when he returned to Dresden. Having squandered his bride's fortune, he prevailed upon his father to purchase for him the colonelcy of a German regiment in French service in 1719. He commanded the unit with such skill that he earned promotion to *marechal de camp* in 1720 and, from then until 1725, he enjoyed the pleasures of the French court—while simultaneously acquiring the tools of a military commander and becoming expert in the arts of war.

Succeeding to the vacant duchy of Courland in 1725, he reigned only until 1727, when he met with popular opposition and returned to Paris. On February 1, 1733, the death of Augustus II, Saxe's father, gave rise to the War of the Polish Succession, in which Saxe fought in the French army against Eugene, his former comrade-at-arms and mentor, and against his own legitimate brother, Frederick Augustus II (known as the Weak). Saxe commanded a covering force during the Duke of Berwick's siege of Philippsburg from May 25 to July 27, 1734, successfully repulsing all of Prince Eugene's relief attempts.

Promoted to lieutenant general in 1736, Saxe became a favorite of Louis XV's famed mistress, Madame Pompadour. When the War of the Austrian Succession broke out in 1740, Saxe served under Marshal François M. de Broglie's force in the invasion of Upper Austria during July–September 1741 and Bohemia, during October–November. Saxe triumphed by taking Prague in a nighttime action on November 19. He then won universal admiration for his ability to maintain order among the ranks of the victors, who were accustomed to despoiling and plundering.

The next spring, Saxe took the fortress of Eger after an efficiently executed siege during April 7–20, 1742. However, he was unable to resist Austrian relief forces and withdrew. At this juncture, he obtained a leave of absence to attempt to regain his control over the duchy of Courland. Failing in this, he returned to duty and was welcomed back with a promotion to the rank of marshal of France on March 26, 1743.

Saxe became involved in a scheme in support of Bonnie Prince Charlie—Prince Charles Edward Stuart—in Scotland during 1743–44. However, the French fleet was wrecked in a storm off Dunkirk, and the aid plans never materialized. The cause of Bonnie Prince Charlie and the Jacobite Rebellion was ultimately lost.

When France declared war on Austria in April 1744, Saxe was given command of the principal army in Flanders, but King Louis XV's illness put a stop to French operations until the spring of 1745. Saxe now rushed into Flanders to lay siege against Tournai during April 25–June 9 and did battle with the Duke of **Cumberland**'s British army, which came to Tournai's relief. Saxe defeated Cumberland at the splendid Battle of Fontenoy on May 11, 1745, directing a combined-arms counterattack that drove the Allies back. Saxe directed the battle personally, despite the fact that he had been stricken with dropsy (edema). Despite his infirmity, Saxe pressed the counterstrike and captured Tournai on June 19, 1745. During August 9–14, he laid siege to Ostend, which fell, thereby yielding to Saxe most of Flanders.

In 1746, Saxe went up against Charles of Lorraine, after taking Brussels (February 7–20, 1746) and Antwerp (May 31). He faced Charles at Raucoux, defeating this very able opponent on October 11. This left the way clear for an invasion of Holland. Met by an Allied army under Cumberland and the Prince of Orange, Saxe attacked at Lauffeld, breaking through the opposing army's center after a bloody battle on July 2, 1747. Costly though it was, the victory isolated Maastricht, to which Saxe successfully laid siege during April 15–May 7, 1740. Following this triumph, Saxe was promoted to marshal-general and retired soon afterward at that rank. He lived his last years at the Chateau de Chambord, given to him by Louis XV.

Saxe was a spectacular soldier, whose personality was in every respect bigger than life. He died in his estate, after having entertained eight actresses. His death certificate attributed the cause of his demise to "une surfeit des femmes."

SCHARNHORST, Count Gerhard Johann David von

BORN: November 12, 1755

DIED: June 8, 1813

SERVICE: French Revolutionary Wars (1792–99); Napoleonic Wars (1800–15)

Hannover-born Prussian general whose military reforms were essential to the reorganization of the army

Born at Bordenau, near Hannover, Scharnhorst was commissioned in the Hannoverian army in 1788 and became an artillery instructor. He collaborated on an officers' handbook as well as a military pocketbook for use in the field, which were published in 1792. On September 8, 1793, he fought under the Duke of York in the Netherlands at Hondschoote, followed by Menin during April 27–30, 1794. After this battle, he wrote and published a study of the defense of Menin and a work on the reasons for the French success in the Revolutionary Wars.

Scharnhorst was a well-known figure in military circles when he was promoted to major and transferred to the staff of the Hannoverian army, on which he served from 1796 to 1801, when he joined the Prussian army with the rank of lieutenant colonel. Ennobled by the King of Prussia, Scharnhorst accepted an appointment as an instructor at the War Academy in Berlin. Among his most illustrious students was the great theoretician of war, **Karl von Clausewitz.** Scharnhorst also acted as tutor to the Crown Prince during 1802–04.

From March 1804 to 1805, Scharnhorst served as deputy quartermaster general and commander of the Third Brigade. From 1805 to 1806, he was chief of staff to the Duke of Brunswick. He fought at Auerstädt on October 14, 1806 and was wounded. The following month, he was captured at Ratkau, but was speedily exchanged in time to serve with the Prussians at Eylau during February 7–8, 1807.

Promoted to major general in July, he was appointed minister of war and chief of the general staff on March 1, 1808. From these posts, he worked to reform and rebuild the Prussian army in order to make it a truly efficient national force. He had been appointed to head the Army Reform Commission after the 1807 Peace of Tilsit, by the terms of which **Napoleon** compelled Prussia to limit the size of its army to 42,000 men. Working with fellow military reformer **August von Gneisenau,** Scharnhorst created the "shrinkage system"—the *Krümpersystem*—by which new army recruits were very rapidly trained, then sent into the reserves (a body not counted in the 42,000-man limit), so that more men could be trained. The "shrinkage system" actually increased Prussia's pool of trained soldiers and officers, yet kept the size of the active army within treaty limits. (This strategem would also serve German military planners as a model for stealthy rearmament, in defiance of the provisions of the Treaty of Versailles, after World War I.) During this period, Scharnhorst also worked to overhaul officer education, which not only improved the level of leadership throughout the Prussian army, but enabled the development of the modern general staff system.

Pursuant to Napoleon's edict of September 26, 1810, barring foreigners from service in the Prussian army, the Hannoverian Scharnhorst had to step down and leave Prussia. He entered semiretirement, taking the time to write a manual on firearms. However, in 1812, he was recalled to active duty as chief of staff to **Gebhard Blücher.** He fought at Lützen on May 2, 1813 and was wounded there. Subsequently, he was sent to Prague to negotiate for Austria's entry into the war, but he failed to recover from his wound. His health deteriorated, and he died on June 8, 1813.

Scharnhorst was a fine staff officer, whose writings were in part responsible for the reform of the Prussian army into the extraordinarily efficient war machine that it became later in the nineteenth century.

SCHLIEFFEN, Count Alfred von

BORN: February 28, 1833

DIED: January 4, 1913

SERVICE: Austro–Prussian War (1866); Franco–Prussian War (1870–71)

Prussian–German general who drew up the plan for European war that became the basis for Germany's opening offensive in World War I

Berlin-born, Schlieffen was the son of a Prussian army officer, but was initially destined for a career in law. However, he decided on a military career and joined the Second Guard Uhlans Regiment as a one-year volunteer in 1853. Discovering a passion for military life, he transferred to the regular service and was commissioned a second lieutenant in December 1854. Marked out as a promising young officer, he was tapped for enrollment at the *Kriegsakademie* (war college) in 1858 and graduated in 1861, immediately joining the General Staff.

Schlieffen saw service on the staff of Prince Albert of Prussia during the Austro–Prussian (Six Weeks') War and fought at Königgrätz on July 3, 1866. Next, he served on the staff of XIII Corps under **Frederick II** of Mecklenburg-Schwerin during the Franco–Prussian War.

From 1876 to 1884, Schlieffen held command of the First Guard Uhlans Regiment, then was assigned as head of the military history section of the general staff. He was promoted to *generalmajor* in 1886, then to *generalleutnant* in 1888, when he was also appointed both quartermaster general as well as deputy chief of the general staff. On February 9, 1891, he succeeded Alfred von Waldersee as chief of the general staff and served in that post until his retirement on January 1, 1906. In 1893, he was promoted to general of cavalry.

During his tenure as chief of the general staff, Schlieffen formulated and refined plans for simultaneous war with France and Russia. He concluded that it was not possible to defeat Russia quickly, so his plan called for the concentration of no less than seven-eighths of the German army against France, which could be invaded through a lightning advance through Belgium, followed by envelopment of the left wing of the French army. The decisive battle, as Schlieffen envisioned it, would take place just outside of Paris and would result in the destruction of the French army. With the fall of France, a larger portion of the army could be redirected to the Russian front.

Even after his retirement, Schlieffen labored at refining his plan down to the last detail. He worked through to the day of his death, and the so-called Schlieffen Plan became German policy in the opening weeks of World War I—even though the plan had been compromised and diluted by **Helmut von Moltke.**

Schlieffen was a bold thinker, but also a rigid one. The Schlieffen plan did not allow for the evolution of political realities in the early twentieth century.

SCHWARZKOPF, H(erbert) Norman

BORN: August 22, 1934

SERVICE: Vietnam War (1965–73); Grenada Invasion (1983); Persian Gulf (Kuwait) War (1990–91)

U.S. Army general; commander in chief, U.S. Central Command during the Persian Gulf (Kuwait) War

As a U.S. Army four-star general and Commander in Chief, U.S. Central Command, Schwarzkopf successfully led operations Desert Shield and Desert Storm during the so-called Persian Gulf War against Iraq. He was born in Trenton, New Jersey, August 22, 1934, the son of a World War I army officer who commanded the New Jersey State Police from 1921 to 1936 before returning to the army for duty in the Middle East during World War II. H. Norman Schwarzkopf was commissioned second lieutenant after graduating from West Point (1956) and steadily rose through the grades to four-star general.

Schwarzkopf had numerous staff strategic and personnel management assignments, but always relished field command. He served two tours of duty in Vietnam—early in the war, as adviser to the Vietnamese Airborne (1965–66) and, later, as battalion commander in the Americal Division (1969–70). He eagerly accepted a brigade command in Alaska, an assignment few coveted, rather than man a desk (1974–76). After Vietnam, Schwarzkopf saw action as deputy commander of U.S. forces in the invasion of Grenada (1983). He gained tremendous public recognition as director of operations Desert Shield and Desert Storm (August 2, 1990–April 10, 1991) during the so-called Persian Gulf War fought against Iraq by the United States and a coalition of twenty-eight nations pursuant to a U.N. resolution demanding the withdrawal of Iraqi forces from neighboring Kuwait.

The resolution mandated Iraq's withdrawal by January 15, 1991. In anticipation of that date, the U.S. and coalition nations built up a combined land, sea, and air presence in Saudi Arabia beginning early in August 1990. The build-up, code named Operation Desert Shield, became Operation Desert Storm on January 17, when Iraq failed to withdraw and U.S. President George Bush, after consultation with the coalition leaders, ordered air strikes against Bagdhad and other Iraqi targets. Schwarzkopf commanded a combined U.S.-coalition force of 530,000 troops opposed to 545,000 Iraqis, whose weaponry was vastly inferior to that of the U.S. and coalition forces. After a month of unremitting air attacks against Iraq, Schwarzkopf launched a massive ground assault, which achieved victory in approximately one hundred hours. A cease-fire was called on February 23, 1991, and formalized on April 10, 1991. U.S. and coalition losses were 149 killed, 238 wounded, 81 missing, and 13 taken prisoner (they were subsequently released); Iraqi losses were estimated in excess of eighty thousand men, with overwhelming loss of materiel.

Schwarzkopf, who frequently appeared on television press conferences during Operation Desert Storm, impressed the American public with his forthrightness, military skill, and soft-spoken humanity. "Any soldier worth his salt," Schwarzkopf declared, "should be anti-war. And still there are things worth fighting for."

SCIPIO AFRICANUS MAJOR
(Publius Cornelius Scipio)

BORN: 235 or 236 B.C.

DIED: 184 or 183 B.C.

SERVICE: Second Punic War (219–202 B.C.); Syrian War with Antiochus III (192–188 B.C.)

*Classical Roman commander who defeated **Hannibal Barca** in the Second Punic War*

Born into one of Rome's great patrician families, Publius Cornelius Scipio—called Scipio Africanus after his decisive defeat of Hannibal Barca's Carthaginian forces in Africa—made the most of his position and connections to become one of ancient Rome's greatest military leaders and, later, an important political figure. His father, also named Publius Cornelius Scipio, was a general and consul of 218 B.C. Young Scipio married into another prominent Roman family when he took as his bride Aemilia, daughter of Aemilius Paullus, the consul of 216 B.C. who fell at the Battle of Cannae (216 B.C.).

As a teenager, Scipio assumed his first command, of cavalry at the Battle of Ticinus River (218 B.C.). He accompanied his father, who had been sent to campaign against Hannibal in Spain while an army was being raised to invade Africa and attack Carthage itself. Hannibal moved so swiftly, however, that P. Scipio (father) had to fight a delaying action in Italy itself, among the tributaries of the River Po, until the army that was being mustered for action in Africa could join him there instead. At the Ticinus, according to (possibly credible) legend, Scipio saved his wounded father, who had been cut off by the enemy.

The next recorded appearance of Scipio was during the aftermath of the disastrous Battle of Cannae (216 B.C.), in which Hannibal, whose forces numbered 42,000, killed some 60,000 Roman soldiers (including consul Aemilius Paullus) out of an army of 80,000. At Canusium, Scipio rallied the panic-stricken and doom-saying survivors of the battle, forging them into a new fighting force. Riding on his popularity, Scipio was elected aedile (magistrate) in 213 B.C. but soon left civil life to return to military command when his father and uncle, Gnaeus, were defeated and killed in Spain in 211 B.C. The Romans originally sent **Gaius Claudius Nero,** who had successfully retaken from Hannibal the key city of Capua, to replace the fallen brothers. But in 210 B.C., the People of Rome granted twenty-five-year-old Scipio unprecedented proconsular powers and sent him to replace Nero.

Scipio quickly reestablished Roman control north of the Ebro River, from which he, commanding 27,500 men, made a surprise attack on the capital city Carthago Nova (New Carthage, present-day Cartagena), laid siege by land and sea, and took the town (209 B.C.). Scipio next defeated Hannibal's brother Hasdrubal Barca at the Battle of Baecula, near present-day Cordova (208 B.C.). By 206 B.C., after Scipio led a force of 48,000 to victory against 70,000 Carthaginians under Mago and Hasdrubal Gisco at the Battle of Ilipa (or Silpia), Spain was securely in Roman hands.

Returned to Rome, Scipio was elevated to consul for 205 B.C. and (despite political opposition from the ultraconservative **Fabius Maximus** and his adherents) set about raising a magnificent volunteer army of 30,000 men, which left the Sicilian port of Lilybaeum in 204 B.C., bound for an invasion of Africa and a strike at Carthage itself. After landing near the North African city of Utica (near present Tunis), Scipio fought the Battle of the Tower of Agathocles (204 B.C.), pillaged the countryside, and then laid siege to Utica. At some time during the action preceding the siege, Hannibal's brother Hanno was killed. However, Hasdrubal Gisco and King Syphax of the Masaesyles approached with a large army, forcing Scipio to break off the forty-day-old siege and take up a fortified position near the coast. An armistice was agreed to, during which Scipio participated in peace negotiations—only to make a surprise nighttime attack on the Carthaginian camp, which he burned (203 B.C.), butchering the enemy army as it ran, unarmed, from the flames. Hasdrubal Gisco and Syphax soon raised another army, however, and fought Scipio at the Battle of Bagbrades (203 B.C.).

Scipio achieved total victory and captured Syphax. The Carthaginian Senate sued for peace—even as it summoned Hannibal and his brother Mago to return from Italy. Mago, who had been wounded in Liguria, died en route, but Hannibal was able to build an army around the 18,000 men he had brought with him from Italy. With 45,000 infantry and 3,000 cavalry, Hannibal marched from Carthage in an effort to draw the Roman forces away from that city. Scipio, who had been joined by forces under the command of the Numidian tribal leader Massinissa, pursued with 34,000 infantry and 9,000 cavalry. Following an inconclusive parley between Hannibal and Scipio, the Battle of Zama (202 B.C.) commenced. Half of Hannibal's men were raw recruits, whereas Scipio commanded battle-tested veterans. Scipio also deployed his forces brilliantly, so that the elephants with which the Carthaginians attacked could be herded to slaughter. Cavalry, which had played a crucial role in Hannibal's earlier victories, was again decisive. But, this time, Scipio commanded three times Hannibal's number, and the latter's horsemen were swept from the field. After the battle, twenty thousand Carthaginians lay dead; the survivors, including Hannibal, fled to Carthage. Roman losses were trivial by comparison: 1,500 dead, 4,000 wounded. Carthage again sued for peace—this time in earnest—forfeiting its warships and elephants, agreeing to establish Massinissa as King of Numidia in place of the Carthaginian ally Syphax, and paying a heavy tribute to Rome.

Scipio Africanus returned to Rome in triumph, was elected Censor for 199 B.C., and named *princeps senatus* (leader of the Senate). He was elected consul for 194 B.C., but was unsuccessful in his attempt to convince Rome to maintain a presence in Greece in order to check the ambitious Antiochus III of Syria. Scipio's very charisma and popularity earned him the opposition of Roman conservatives, including the powerful elder Cato. Disgusted with Roman politics and in ill health, Scipio Africanus retired to his estate at Liternum, where he died.

SCOTT, Winfield

BORN: June 13, 1786

DIED: May 29, 1866

SERVICE: War of 1812 (1812–15); Seminole War (1835–41); Aroostook War (1838); U.S.–Mexican War (1846–48); Civil War (1861–65)

American general; hero of the War of 1812, U.S.–Mexican War, and general-in-chief of the Union army at the start of the Civil War

Scott was born near Petersburg, Virginia and was educated briefly at the College of William and Mary in 1805, then apprenticed himself to a lawyer. In 1807, he enlisted in a local cavalry troop during the patriotic fervor that followed the *Chesapeake–Leopard* affair, which raised the issue of British impressment of American sailors. He was commissioned a captain of light artillery and dispatched to New Orleans in May 1808, but fell afoul of his commander, the thoroughly corrupt General James Wilkinson, the following year and was suspended from 1809 to 1810. He returned to service in New Orleans during 1811–12, but then returned to Washington, D.C., where he obtained promotion to lieutenant colonel in July 1812.

Joining the army of General Stephen van Rensselaer at Niagara in October 1812, he commanded a small detachment of volunteers at Queenston on October 13, where he was captured. Soon exchanged, he was promoted to colonel and made adjutant to Henry Dearborn in March 1813. He planned and led a successful attack on Fort George, Ontario, on May 27, suffering a wound in the battle. He was given command of the captured fort, then joined Wilkinson's advance on Montreal in October and fought at Chrysler's Farm (near Cornwall, Ontario) on November 11, 1813.

Scott was promoted to brigadier general in March 1814 and devoted himself to training a brigade in Buffalo. Under General Jacob Brown, he led his brigade into Canada on July 3, achieving a victory at Chippewa on July 5, 1814. At the major battle of Lundy's Lane on July 25, 1814, he was twice wounded and became a national hero. Breveted to major general, he was appointed to head the board of inquiry on General William H. Winder's conduct in the ignominious defeat at Bladensburg, Maryland, and he was also one of the authors of the U.S. Army's first standard drill book.

Appointed to command of the army's Northern Department in 1815, following the war, he twice visited Europe to observe military developments there. He was disappointed when he failed to be promoted to commanding general following the death of Jacob Brown in February 1828, and he tendered his resignation, which was refused. Instead, he was made commander of Eastern Division in 1829.

Scott was called on to lead a force during the Black Hawk War of 1832, but many of his troops were stricken in camp by a cholera epidemic. Delayed, Scott did not arrive in Wisconsin until Chief Black Hawk and his followers had been neutralized. However, he was instrumental in negotiating the Treaty of Fort Armstrong with the Sauk and Fox tribes on September 21, 1832.

Late in 1832, President **Andrew Jackson** prudently sent Scott to South Carolina to observe the military situation there during the Nullification Crisis (a prelude to civil war, in which South Carolina defied a federal tariff, claiming that it was unconstitutional). Scott saw to the reinforcement of the federal garrisons around Charleston, yet did so in a subtle manner that avoided stirring local passions.

In 1836, Scott was dispatched to Florida to fight the Seminoles, who were resisting removal to Indian Territory (modern Oklahoma), as prescribed by treaties most of the tribe repudiated. Scott was plagued by ill-trained and downright insubordinate troops, as well as a lack of supplies. Soon relieved of command and replaced by General Thomas S. Jesup, he was brought up on charges before a board of inquiry, which, however, cleared him in 1837. Following this, during January–May 1838, he was sent to Maine during the border dispute with Britain known as the Aroostook War. He exercised admirable diplomacy and thereby averted a shooting war.

During the late summer and early fall of 1838, Scott was assigned to oversee the forcible removal of the Cherokees from Georgia, South Carolina, and Tennessee to Indian Territory. Through no fault of Scott, the operation was marked by corruption and brutality and has been recorded in the Native American memory as the Trail of Tears.

Scott was sent back to Maine as the border dispute flared up again in January 1839. He parleyed with the lieutenant governor of New Brunswick and, once again, avoided a shooting war.

On July 5, 1841, Scott was at last appointed commanding general of the army. With the outbreak of war with Mexico, Scott at first supported General **Zachary Taylor**'s position in Texas and northern Mexico, but he soon became convinced that something more aggressive was called for and proposed an invasion of Mexico City itself. President James Polk approved Scott's plan and replaced Taylor with Scott, who carefully laid plans for a landing at Veracruz. Accomplished on March 9, 1847, the Veracruz assault was the first amphibious operation ever undertaken by the U.S. Army.

The city of Veracruz fell on March 27, after a brief siege. Scott marched inland, trouncing General **Santa Anna**'s much larger army at Cerro Gordo on April 18 and occupying Puebla on May 15. He resupplied and reinforced his army there, then set off for Mexico City on August 7. In a bold maneuver that no less an international figure than the Duke of **Wellington** pronounced foolhardy, Scott severed his line of supply so that he could rapidly circle south. Thus positioned, he beat Santa Anna at Contreras and Churubusco during August 19–20. Scott pushed Santa Anna back into Mexico City.

Scott suspended operations during peace negotiations, which commenced on August 25, but broke down on September 6, whereupon Scott dealt the Mexican forces a blow at Molino del Rey on September 8 and Chapultepec on September 13. The following day, he took Mexico City. Scott set up there as military governor until he returned to the United States on April 22, 1848. Accused of exceeding his authority in the course of the invasion, Scott was quickly cleared of misconduct. He was, in fact, a national hero, who narrowly lost the Whig presidential nomination to Zachary Taylor in 1848. Scott ran for presidency under the banner of a breakaway Whig faction in 1852, but lost by a wide margin to Democrat Franklin Pierce.

Breveted to lieutenant general in February 1855, Scott was sent to negotiate during an Anglo–American boundary crisis over possession of San Juan Island in Puget Sound during September 1859.

As the clouds of civil war gathered on the horizon, Scott returned to the east and advised on preparing for the conflict. His counsel fell on deaf ears, as the administration of James Buchanan marked time. In January 1861, Scott transferred his headquarters to Washington, D.C., and, when war began, he supervised the opening engagements. His boldest stroke was the so-called Anaconda, a proposed naval blockade of Confederate ports. However, Scott was aged, obese, and infirm—called, half derisively, half affectionately, "Old Fuss and Feathers." He retired from the army on November 1, 1861 and was replaced as general in chief by **George B. McClellan.**

Scott was a brave and vigorous soldier, who was strong on strategy and quite capable in the area of diplomacy. Although he left the army early in the Civil War, his Anaconda Plan and his grand strategy for taking the Mississippi Valley were eventually adopted by the Union.

SEVERUS (Lucius Septimius Severus Pertinax)

BORN: April 11, 146

DIED: February 4, 211

SERVICE: Wars of Succession (193–197); Parthian War (195–202); Caledonian War (208–210)

Roman despot; founded a dynasty that remained in power until 235 and transformed the Roman Empire into a military monarchy

A member of the Roman Senate in about 173, Septimius Severus became consul in 190. By the time the Emperor Commodus was assassinated on December 31, 192, Severus was governor of Upper Pannonia and in command of the largest army on the Danube. When Pertinax, the successor to Emperor Commodus, was murdered in March 193, the title was auctioned off to Marcus Didius Julianus. Severus seized his chance, claiming to be Pertinax's avenger. He proclaimed himself emperor with the support of his troops on April 13. As Severus led legionnaires on Rome, the Senate sentenced Didius Julianus to death, and he was assassinated on June 1.

Severus immediately undertook the reorganization of the military. He disbanded the Praetorian Guard, placing in its stead troops from the legions and doubling its strength. He also increased the size of the regular army, including those units assigned to defend Rome against external invasion and internal plotting. He raised the number of men permanently under arms to over thirty thousand, including thirty-three legions. He improved the lot of the common soldier by raising his pay and allowing him, for the first time in the history of the Empire, to marry.

After improving his ability to make war, he promptly made it, advancing Gaius Pescennius Niger, who had proclaimed himself emperor in the East. At the Battle of Issus in 194, Severus's legions crushed Niger, and Severus acted swiftly to pacify the region, splitting it in two to discourage further rebellion.

He turned next to Britain to attack Clodius Albinus, the governor whom Severus has previously designated as heir in a vain effort to secure his loyalty. Severus openly rebuked Albinus by naming his own eldest son heir. In response, Albinus's troops proclaimed their leader emperor, and he advanced with them into Gaul, preparing to march on Rome. Severus faced him at the Battle of Lugdunum (present-day Lyon, France) in 197 and easily defeated the usurper.

Returning to Rome, Severus destroyed Albinus's supporters, executing twenty-nine senators and countless knights. He next set about punishing those on the frontier who had refused to become allies. He attacked the Parthians and captured their capital in 198; he also annexed Mesopotamia to the Empire, the second such annexation in thirty years. His later efforts to push into present-day southern Iraq ended in defeat. Twice he was turned back.

After falling ill in 211, Severus called his two sons to his deathbed and urged them to favor the military. "Be good to the soldiers," he advised, "and take heed of no one else!"

SHAKA (CHAKA, TSHAKA)

BORN: ca. 1787

DIED: September 22, 1828

SERVICE: Later Wars of Dingiswayo (1810–18); Wars of Zulu Expansion (1818–28)

Zulu ruler who transformed the Zulus from a clan of 1,500 to a nation of a quarter million

Ruler the Zulu Empire from 1816 until his death in 1828, Shaka transformed a minor African tribe into one of the most formidable armies of warriors on the African continent.

Shaka was born to a Zulu chieftan and a princess, both of whom belonged to the Langeni clan. Such a close relationship between husband and wife was foreign to Zulu custom, and Shaka's parents separated after a few years of marriage. Shaka grew up without a father; when he was about fifteen years old, he and his mother, Nandi, were driven out of the Langeni village and found a home among the Dietsheni, a subclan of the Mtetwa. A few years later, Shaka was called to military service for his adopted clan. He performed his duties brilliantly.

In 1816, Shaka's natural father died, and his adoptive father, a chief among the Mtetwa, sent the young man to become chief of the Zulu, at the time, a small and relatively unimportant clan of no more than ifteen hundred members. As the new chief, Shaka transformed the Zulu army from a poorly armed and inadequately trained mob into a disciplined, polished, and well-equipped professional fighting machine feared by everyone with whom it came into contact. Traditionally, the warriors of southern Africa were armed with a rather small shield and a long spear made for throwing. Shaka adopted a large shield that covered the entire front of the body and reduced the size of the spear so that it could be used for jabbing in hand-to-hand combat. Traditionally as well, native troops faced each other from a distance, threw their spears, jeered at each other, and called the fight off, usually with no clear-cut winner. When Shaka had completed his overhaul of the Zulu army, his men were trained to get in close, protect themselves with the large shield, and slash with the short, powerful spear.

Shaka also instituted a regimental concept in his army. He troops were quartered, according to regiment, in seperate *kraals*, or villages, distributed across the countryside. He developed a tactic of offense in which his army was divided into four elements, shaped like a bull. The central element, or "chest," was sent to engage the enemy, while the "horns" surrounded the foe and attacked from behind. The "loins," a reserve element, backed up the "chest," if needed.

Shaka's military innovations were overwhelming. His army conquered sub-clans and clans, and as his soldiers won battle after battle, the ranks of the Zulu army grew. Shaka killed anyone who stood in his way. In 1817, when he eliminated Dingiswayo—his protector who had made him the chief of the Zulus in the first place—Shaka's path to a complete takeover of the Zulu Nation was clear.

In the following ten years, Shaka led his vast armies in the annihilation of tribe after tribe. More than two million people had been killed or displaced, leaving much of southern Africa depopulated when the Boers made their "Great Trek" in the 1830s.

When Shaka's mother, Nandi, died in 1827, the event sent her son on a redoubled rampage of killing that lasted for a year. More than seven thousand Zulus fell victim to Shaka's grief. Shaka ordered no crops to be planted, and he forbade the drinking of milk, a staple food among the Zulus. Pregnant women were killed along with their husbands.

In 1828, Shaka committed his exhausted army to a series of campaigns. In September of that year, his half brothers, Dingane and Mhlangana, weary of Shaka's relentless drive and the violent reprisals that had followed his mother's death, assassinated him.

SHAPUR I

DIED: 270 or 273

SERVICE: War with Rome (241–244); Second war with Rome (258–261); Wars with Odenathus of Palmyra (261–266)

Iranian ruler, expanded Sasanian Empire

At the time Shapur I took over from his father Ardashir I, his empire was expanding. Continuing the empire's wars with Rome, in 244 Shapur I defeated the Roman emperor Gordian and extracted a huge ransom from Philip the Arab, Gordian's successor. Twelve years later, the Roman Empire still coveted Armenia and Syria, but Shapur I successfully repelled all Roman advances.

After conquering Dura-Europos, on the Euphrates River, and Antioch, the capital of Syria, Shapur I removed the defeated populations of the defeated towns—many of them Christians—to Iran. Once his captives were settled, Shapur I forced them to work on massive construction projects, including palaces, bridges, dams, and entire cities.

Beginning in 258, the Roman emperor Valerian moved against Shapur's sieges of Carrhae and Edessa. Shapur's forces captured the emperor at Edessa in 260, however, imprisoning him for life at Gundeshapur, a city built expressly to house prisoners of war.

After the success of his sweeping conquests, Shapur I named himself King of Iran and Aniran, meaning "non-Iran." At the time of his death, his Sasanian Empire comprised Mesopotamia, the Transcaucasus, Oman, and the Kushan territory. Through his shahrabs, Shapur I ruled over fifteen royal cities, and his sons reigned over Mesene, Armenia, and Gilan. The expanded Sasanian Empire contained many former Roman outposts, thus, disrupting the patterns of trade in Euphrates region and slowing down the spread of Roman Christianity. Ruthless in conquest, Shapur was nevertheless tolerant of Christianity, Judaism, and other religions, but he actively promoted Zoroastrianism within his empire.

SHARON, Ariel

BORN: 1928

SERVICE: Israeli War of Independence (1948–49); Sinai War (1956); Six Day War (1967); October War (1973); Israeli Invasion of Lebanon (1982)

Important Israeli general and politician whose reputation was marred when he was held indirectly responsible for massacres at two Palestinian refugee camps in Israeli-occupied Beirut

Born in Kefar Malal, Palestine—now part of Israel—Sharon joined the Haganah, the Jewish underground army, about 1936. In 1947, he was an instructor of the Jewish Settlement Police and, the following year, fought, at the outbreak of the War of Independence, as platoon commander in the Haganah's Alexandroni (Third) Brigade. He quickly rose to brigade intelligence officer and then company commander by early 1949. Commanding the Third Brigade reconnaisance unit during 1949–50, he moved on to become intelligence officer for Central Command and Northern Command during 1951–52.

After study at Hebrew University in Jerusalem during 1952–53, Sharon took command of Unit 101, a special force whose mission was retaliation against terrorist raids into Israel. He led this elite group until 1956, when he took command of the 202nd Paratroop

Brigade during the Sinai War. His motorized advance, during October 29–31, to relieve an advance battalion parachuted in to seize Mitla Pass became legendary, as did his instant turnaround to the south to join Ninth Brigade and Israeli naval forces at Sharm-el Sheikh at the southern tip of the Sinai during November 1–3.

In 1957, Sharon traveled to Britain for study at the Staff College, Camberley. He returned to Israel in 1958 and was appointed training director for the General Staff, then head of the Infantry School from 1958 to 1959. In 1962, he was assigned command of an armored brigade, gaining promotion in 1964 to the post of chief of staff for Northern Command. In 1966, he was appointed director of Training Services for the entire Israeli Defense Force, then, in 1967, commanded one of the three armored division task forces in the Israeli Sinai offensive (June 5–9).

After destroying Egyptian positions around Abu Ageila on June 5, Sharon headed south to Naichi, then west to the Mitla Pass, where he joined forces from General Avraham Yoffe's division on June 8. From 1969 to 1970, he was involved in the War of Attrition with Egypt.

Retiring in July 1973, Sharon was soon recalled at the outbreak of the October War on October 6, 1973. He commanded an armored division in the Sinai and was instrumental in checking an Egyptian offensive there during October 7–8. He then fought the celebrated Battle of Chinese Farm during October 16–17, then led an attack across the Suez Canal into Egypt during October 17–19. He fell back under stiff Egyptian resistance during October 19–22.

After retiring a second time, Sharon entered politics, helping to create the conservative Likud Party, then gaining election to the Knesset (parliament) in 1973–74 and periodically thereafter. In addition, he has held numerous government posts, including adviser to the Prime Minister (1975–77), Minister of Agriculture in Charge of Settlements (1977–81), and Minister of Defense (1981–83). As Defense minister, Sharon directed the Israeli invasion of southern Lebanon during June 6–13, 1982 and the siege of Beirut during June 15–August 12. During this operation, Lebanese Christian extremists perpetrated massacres at the Sabra and Shetila refugee camps (September 1982). An inquiry found Sharon indirectly responsible for the carnage, and he was replaced as defense minister.

Subsequently, Sharon became a minister without portfolio from 1983 to 1984, then Minister of Trade and Industry in 1984 and a leading figure of the Likud Party.

SHERIDAN, Philip Henry

BORN: March 6, 1831

DIED: August 5, 1888

SERVICE: Civil War (1861–65); Indian Wars (1865–91)

Union general in the American Civil War and in the Indian Wars

Born in Albany, New York, Sheridan graduated from West Point in 1853 and first saw service with the First Infantry Regiment in Texas and the Fourth Infantry in Oregon, fighting the Indians. Promoted to first lieutenant in March 1861, he rose to captain in May and fought in the Corinth (Mississippi) campaign of 1862.

Sheridan won appointment as colonel of the Second Michigan Cavalry in May 1962 and, after a daring raid on Booneville, Mississippi, on July 1, he was promoted to brigadier general of volunteers. As commander of the Eleventh Division of the Army of the Ohio, he fought extremely well at Perryville (October 8) and at Stones River (December 31, 1862–January 3, 1863). Promoted to major general of volunteers in 1863, he served with the Army of the Cumberland in the Tullahoma campaign of 1863, then led the XX Corps to support General George Henry Thomas at Chickamauga during September 19–20. He fought the rearguard for Thomas's retreat as well.

During November 24–25, he charged up Missionary Ridge at the battle of Chattanooga and, by April 1864, had gained an appointment as commander of Cavalry Corps, Army of the Potomac. In this assignment, he fought under **Ulysses S. Grant** in the Battle of the Wilderness (May 5–6) and at Spotsylvania Court House (May 8–18). His raid against rebel lines of supply and communications during Spotsylvania resulted in the defeat of **J. E. B. Stuart** at Todd's Tavern on

May 7 and, again, at Yellow Tavern on May 11. Assigned to destroy rail lines near Charlottesville, Virginia, he engaged Confederate units at Haw's Shop on May 28 and at Trevilian Station on June 11–12.

Sheridan was assigned command of Union forces in the Shenandoah Valley in August and conducted his finest campaign of the war through the valley. He defeated the Confederate army at Winchester on September 19 and at Fisher's Hill on September 22, then cut a swath through the Shenandoah to make it useless to the enemy.

Sheridan was promoted to brigadier general of regulars in September. That month, while he was away from his army, it was surprised at Cedar Creek on October 19. Hearing of the attack, Sheridan galloped twenty miles to the battle—an action immortalized as Sheridan's Ride—and managed to regroup and rally his troops, who succeeded in repulsing the enemy. In November, he was promoted to major general of regulars.

Sheridan received the thanks of Congress in February 1865, then went on to raid Petersburg during February 27–March 24. Joining up with Grant, he turned **Robert E. Lee**'s flank at Five Forks on April 1, then engaged Lee's rear guard at Sayler's Creek on April 6. This way, he blocked Lee's avenue of retreat at Appomattox Court House, and there, on April 9, Lee surrendered the Army of Northern Virginia—effectively ending the Civil War.

Named commander of the Military Division of the Gulf in May 1865, Sheridan was later appointed commander of the Fifth Military District in March 1867, to which was added the vast Department of the Missouri in September. He initiated a campaign against the Indian tribes of the Washita Valley in Oklahoma during 1868–69. In March 1869, Sheridan was promoted to lieutenant general with command of the Division of the Missouri. He was detached from this post during the Franco–Prussian War of 1870–71, and sent to Europe as an observer and as a liaison officer with the Prussians.

During 1876–77, Sheridan directed the campaign against the Southern Plains Indians, then became commander of the Military Divisions of the West and Southwest in 1878. In November 1883, he replaced his old friend and mentor **William Tecumseh Sherman** as commanding general of the Army. He was promoted to general two months before his death.

Sheridan was a fine cavalry officer. An excellent tactician, he was no strategist. But he believed fully in Sherman's policy of total warfare, especially when it came to fighting Indians. He developed a series of winter campaigns precisely because combat was more brutal and effective against the Indians in the winter months. Sheridan was greatly loved by his troops, who affectionately called him "Little Phil," and he was respected by his superiors. A man of blunt-spoken wit, he summed up one aspect of his western service by quipping, "If I owned both Hell and Texas, I'd rent out Texas and live in Hell."

SHERMAN, William Tecumseh

BORN: February 8, 1820

DIED: February 14, 1891

SERVICE: Second Seminole War (1835–43); U.S.–Mexican War (1846–48); Civil War (1861–65)

Union general in the American Civil War; brought the concept of "total war" to the conflict

A native of Lancaster, Ohio, Sherman was the son of an Ohio Supreme Court judge. He obtained an appointment to West Point, graduating in 1840 with a commission in the artillery. He was dispatched to Florida to serve against the Seminoles and soon gained promotion to first lieutenant (November 1841). With the outbreak of the U.S.–Mexican War, Sherman was assigned to the staff of **Stephen Watts Kearny,** but was disappointed that he saw no battle action during 1846–47. Instead, he served most of the war as an administrative officer in California until that territory joined the Union in 1848.

Sherman became a commissary captain in September 1850 and, feeling profound dissatisfaction with his career, he resigned his commission to start a building firm in 1853. His business failed, and he moved to Leavenworth, Kansas, where he set up as a lawyer in 1857. In this enterprise he also failed, but

William Tecumseh Sherman. Courtesy Library of Congress.

finally found a satisfying position as superintendent of the newly established Alexandria Military Academy, which later became the Louisiana Military Academy, and, finally, Louisiana State University. He served in this post from October 1859 until January 1861, when it became clear that the country was about to fight a civil war.

Sherman returned to duty as colonel of the Thirteenth Infantry in May 1861, then was appointed to a brigade command at First Bull Run (July 21). In August, he became a brigadier general of volunteers and was then given command of Union forces in Kentucky in October. He faltered badly in this assignment and seemed on the verge of a nervous collapse. However, he was transferred to the Western Department in November and pulled himself together. In February 1862, he was assigned as a division commander in the Army of the Tennessee, commanded by **Ulysses S. Grant,** and fought with great distinction at Shiloh during April 6–7.

Shiloh wholly redeemed Sherman after his problems in Kentucky, and, in May, he was promoted to major general of volunteers. He fought under Grant in the Corinth (Mississippi) campaign during May–June and began operations against the fortified Mississippi River town of Vicksburg. However, he suffered a defeat at Chickasaw Bluffs on December 29 and stopped his advance. Transferred to command of XV Corps, Army of the Mississippi, he took Arkansas Post on January 11, 1863, then transferred with XV Corps to the Army of the Tennessee. In this capacity, he returned to Vicksburg to support Grant's siege of the town during January 1863–July 1864, He took Jackson, Mississippi on May 14.

Promoted to brigadier general of regulars in July, Sherman rushed to the relief of William Rosecrans at Chattanooga, then succeeded Grant as commander of the Army of the Tennessee in October. He played a strong supporting role in coordinating with George Henry Thomas's Army of the Cumberland, and he was in command of Union left at Chattanooga during November 24–25. In December, he marched to the relief of **Ambrose Burnside,** who, with characteristic ineptitude, had gotten himself besieged in Knoxville.

Named commander of the Military Division of the Mississippi in March 1864, he now controlled the Armies of the Cumberland, the Tennessee, and the Ohio. He consolidated these forces—some 100,000 men—in a spectacular drive toward Atlanta, which he coordinated with Grant's advance on Richmond. Sherman advanced one hundred miles in seventy-four days, pushing the army of **Joseph E. Johnston** before him, fighting battles at Dalton, Georgia, on May 8–12, at Resaca on May 15–16, at New Hope Church on May 24–28, and at Dallas on May 25–28, closing inexorably on Atlanta. Although he suffered a reverse at Kenesaw Mountain on June 27, he beat John Bell Hood at the Battle of Peachtree Creek, just outside of Atlanta, on July 20. Northwest of the city, he defeated Ezra Church on July 28, and, on September 2, occupied the city.

Atlanta was razed by a fire, which Sherman's troops may or may not have started. Whoever actually set the blaze, the fire was a fitting prelude to the devastation that followed. An advocate of "total war"—the policy of carrying combat to the civilian population, rather than confining it to a contest of armies—Sherman now commenced his famed—or infamous—"March to the Sea" on November 16. His soldiers looted and destroyed all that lay before them, clear through to the coast. Sherman's army occupied Savannah on December 21. This accomplished, Sherman turned to the Carolinas, beginning a drive on February 1, 1865 that culminated in the capture and burning of Columbia, South Carolina on February 17. The burning was blamed on Sherman's troops, though it is more likely that retreating rebel troops set the blazes.

Sherman successfully repulsed a surprise attack by Johnston at Bentonville, North Carolina, during March 19–20. This threat neutralized, he captured Raleigh on April 13, then, two weeks later, on April 26, he received Johnston's surrender near Durham Station, North Carolina. With Grant's acceptance of **Robert E. Lee**'s surrender at Appomattox (on April 9), this victory marked the end of the Civil War.

After the war, Sherman was appointed commander of the Division of the Missouri in June and was promoted to lieutenant general of regulars in July 1866. From his headquarters in Chicago, he directed much of the strategy and policy during the Indian Wars—though he participated in no battles. In November 1869, he became commanding general of the army and received a promotion to general. He held this often largely ceremonial post until his retirement in 1884.

Sherman published a fine memoir in 1875, and he created a still-important army officer training center at Fort Leavenworth, Kansas in 1884. But he may be best known to most people as the general who declared that "war is hell" (in a speech in Columbus, Ohio on August 11, 1880).

A tremendously popular and much lionized figure, Sherman seemed certain to be chosen by the Republicans as their candidate for president in 1884. But Sherman, who had a strong professional dislike of politicians, declined the nomination.

Sherman was among the most intelligent and ruthless of the Union's commanders. He understood that the essence of war is destruction, and his most sweeping action in the war was to destroy the South's ability to continue fighting. Sherman was also a thoroughgoing professional soldier, who did much to improve army training. His men were intensely loyal to the man they called "Uncle Billy."

SHORT, Walter Campbell

BORN: March 30, 1880

DIED: September 3, 1949

SERVICE: Punitive Expedition against Pancho Villa (1916–17); World War I (1917–18); World War II (1941–45)

American general whose inadequate defenses made Pearl Harbor vulnerable to Japanese surprise attack on December 7, 1941

Short was born in Fillmore, Illinois and graduated from the University of Illinois in 1901. He joined the army with a commission as a second lieutenant of infantry in March 1902 and was posted to the Southwest. After receiving a promotion to first lieutenant, he was sent to the Philippines in 1907, then to a post in Alaska.

In 1913, Short was attached to the Musketry School staff at Fort Sill, Oklahoma. In 1916, he was assigned to the Punitive Expedition against Pancho Villa under General **John J. Pershing,** and that same year he was promoted to captain.

With the U.S. entry in World War I, Short was ordered to France with the First Division in June 1917. He served in various staff assignments with Pershing's headquarters. Appointed chief of staff of the Third Army, he fought at Aisne-Marne during July 18–August 5, 1918; Saint-Mihiel, September 12–16; and Meuse-Argonne, September 26–November 11.

Short was promoted to temporary colonel, then returned to the United States, where he reverted to captain, but was soon promoted to major in 1920. He was assigned as an instructor at the General Staff School in Fort Leavenworth during 1919–21, and he wrote a textbook, *Employment of Machine Guns*, which was published in 1922. The following year, Short was promoted to lieutenant colonel, and in 1925, he graduated from the Army War College.

Short served with the Sixty-Fifth Infantry on Puerto Rico from 1925 to 1928, then returned to Fort

Leavenworth as an instructor until 1930. He was next assigned to the Bureau of Indian Affairs, when that agency was part of the War Department, during 1930–34. After promotion to colonel, Short was assigned to command the Sixth Infantry at Jefferson Barracks in St. Louis. He left this post in 1936 to become assistant commandant of the Infantry School at Fort Benning, Georgia, and was promoted to brigadier general in December. He commanded a brigade of the First Division in New York in 1937, then the entire First Division from July 1938 to October 1940.

During the maneuvers of 1940, Short was given command of a provisional corps and was promoted to major general to command of I Corps at the end of the year. He was assigned command of the Hawaiian Department in January 1941 and promoted to temporary lieutenant general in February.

Short made many mistakes while commanding army forces at Pearl Harbor, including failure to communicate regularly with his navy counterpart, Admiral **Husband E. Kimmel,** and failure to gather and analyze intelligence adequately. He also made the aircraft parked on Hickam Field easy targets for Japanese bombers by massing them together, wingtip to wingtip. His purpose was to make them easier to guard from possible saboteurs—but the tactic was clearly misguided. Almost immediately after the December 7, 1941 attack on Pearl Harbor, Short was removed from command. He was retired on February 28, 1942 after a presidential commission (the Roberts Commission) found him guilty of poor judgment and dereliction of duty. Later, the joint Congressional Investigating Committee softened this judgment a bit, but still maintained that Short had made bad errors of judgment.

Short, whose administrative abilities were considerable, became an executive with the Ford Motor Company from 1942 to 1946, but was never in good health after Pearl Harbor. It was illness that forced his retirement from Ford.

There can be little doubt that Short lacked sufficient foresight and vigilance to defend Pearl Harbor adequately. He was an able and professional soldier, but insufficiently imaginative to avert disaster.

SIMS, William Sowden

BORN: October 15, 1858

DIED: September 28, 1936

SERVICE: Spanish–American War (1898); World War I (1917–18)

Admiral who directed U.S. naval operations in World War I

Canadian-born Sims (Port Hope, Ontario) moved with his family to Pennsylvania in 1872. He graduated from Annapolis in 1880 and served in the Atlantic and Pacific, as well as in Asian waters from then until 1897, when he became naval attaché in Paris and St. Petersburg through 1900. Sims was responsible for coordinating intelligence activities in Europe during the Spanish–American War with Spain during April–October 1898.

Serving aboard the battleship U.S.S. *Kentucky* (BB-6) in Asian waters (beginning November 1900), he learned the new British method of "continuous firing," which improved gunnery accuracy. Sims became an advocate of the practice and urged the U.S. Navy to adopt it. He took the bold step of writing to President Theodore Roosevelt about it in 1901. Far from being brought up on charges of insubordination for this act, Sims was made inspector of gunnery training and promoted to lieutenant commander in November 1902. He reduced firing time for large-caliber guns in the U.S. Navy from five minutes to thirty seconds, with an accompanying improvement in accuracy.

Sims was dispatched to Asia as an observer during the Russo–Japanese War to study the effect of modern guns on armored plate, as well as the effectiveness of battleships against smaller vessels. These studies and reports spanned 1904–1908.

Promoted to commander in July 1907, Sims became a naval aide to President Roosevelt in November and in 1909 was given command of the U.S.S. *Minnesota* (BB-22). Promoted to captain in March 1911, he was appointed instructor at the Naval War College and served until 1913, when he took command of the Atlantic Torpedo Flotilla through 1915. Sims formulated tactical doctrine for destroyers.

After commanding the new U.S.S. *Nevada* (BB-36) from November 1915 to January 1917, he was appointed president of the Naval War College and commander of the Second Naval District. Prior to U.S. entry into World War I, Sims was promoted to rear admiral and, upon the nation's entry into the war, he assumed command of all U.S. destroyers, tenders, and auxiliaries operating from British bases. Promoted to vice admiral in May, Sims was named Commander U.S. Naval Forces Operating in European Waters the following month.

Sims coordinated his efforts with the British Admiralty to protect Allied shipping. Through his work, the American navy became a significant presence in the war. Sims was promoted to admiral in December 1918, the month after Armistice, then returned to the States, where he reverted to rear admiral and took up once again his duties as president of the Naval War College in April 1919.

After the war, Sims was vociferous in his criticism of the U.S. Navy, which he declared incapable of fighting a major opponent. Sims succeeded in prompting a Congressional investigation, but he also engaged in a sharp feud with Navy Secretary Josephus Daniels during 1920–21. Although he retired in October 1922—as a rear admiral—he continued to voice his opinions on naval affairs.

Promoted to admiral on the retired list in June 1930, he ended his "retired" career by advocating the development of naval aviation, and even uttered the blasphemy that aircraft had rendered the battleship obsolete.

Sims was perhaps the most progressive naval officer—in any nation's navy—during the early twentieth century. His leadership during World War I was valuable, but his agitation for modernization after the war was perhaps even more important and doubtless put the U.S. Navy in a much stronger position going into World War II.

SITTING BULL (Tatanka Yotanka)

BORN: ca. 1831

DIED: December 15, 1890

SERVICE: Sioux War (1862–68); Sioux and Northern Cheyenne War (1875–77)

Hunkpapa Sioux chief and medicine man; perhaps the most famous Indian warrior–leader in American history

Sitting Bull was a member of the Hunkpapa tribe, a branch of the Teton Sioux. He was born on the Grand River in South Dakota, the son of a distinguished chief. From an early age, Sitting Bull acquired fame in his own right: as a hunter by age ten and as a warrior by fourteen. He became prominent in the Strong Heart warrior lodge by about 1856.

Sitting Bull's first major engagement against white settlers came in the Minnesota Sioux Uprising of 1862–63, which began with a major attack against settlers in New Ulm. On July 28, 1864, he fought General Alfred Sully at Killdeer Mountain, North Dakota, but agreed to peace through the Treaty of Fort Laramie in 1868.

Like many Indian–white agreements during this period, the peace was not destined to endure. When prospectors began to invade the Black Hills country—sacred to the Sioux—after gold was discovered there in 1874, Sitting Bull became chief of the war council of combined Sioux, Cheyenne, and Arapaho in Montana. He did not actually fight against **George Armstrong Custer** at the Battle of the Little Bighorn on June 25, 1876, but, as medicine man, he "made the medicine" that (according to the Indian warriors' belief) had made the triumph possible.

In the wake of Little Bighorn, Sitting Bull led most of the Hunkpapa to Canada in May 1877 in order to avoid reprisals. The tribe suffered from hunger and disease in their exile and, at length, Sitting Bull brought those who remained back to the United States, where he and 170 followers surrendered at Fort Buford, North Dakota, in July 1881. Sitting Bull was held at Fort Randall, South Dakota from 1881 to 1883 and was then placed at Standing Rock Reservation, North Dakota.

During this period, Sitting Bull became an advocate of traditional Sioux culture, which he struggled to maintain against the inexorable incursions of the whites. He did befriend and respect at least one white man, however: William "Buffalo Bill" Cody, who recruited him as a performer in his Wild West Show during 1885–86.

Back on the reservation, Sitting Bull supported the Native religious revival called the Ghost Dance movement during 1889–90. Government authorities saw this passionate religion, replete with ritual, as a threat and feared an uprising. An order went out for Sitting Bull's arrest. Native American reservation police personnel were dispatched to Sitting Bull's home on the Grand River, South Dakota. A scuffle and near riot broke out during the arrest, and the Hunkpapa leader was slain with two of his sons on December 15, 1890.

Sitting Bull led through a combination of religious and moral force. His fame was legendary not only among Sioux, but among a vast number of whites. He is perhaps the most famous of all Indian leaders.

SLIM, Sir William Joseph

BORN: August 6, 1891

DIED: December 14, 1970

SERVICE: World War I (1914–18); World War II (1939–45)

British field marshal who accomplished much with little in Burma during World War II

Born near Bristol, England, Slim, enrolled in King Edward's School in 1907 and participated in the Officer Training Corps (OTC) there. However, he did not enter the army directly from school, but taught school himself for a time (1909–10), then worked as a bank clerk. Although he was not a student, he continued OTC at Birmingham University in 1912 and joined the army at the outbreak of World War I. He was gazetted to second lieutenant in the Ninth Battalion of the Royal Warwickshire Regiment on August 22, 1914 and first distinguished himself by heading off a mutiny in December. From July 13 to August 8, 1915, he served with his regiment in the disastrous Gallipoli campaign, suffering a severe wound on August 8. It was January 1916 before he returned to duty. He conducted replacements to France in April, then, promoted to temporary captain with the Ninth Battalion, he conducted replacements to Mesopotamia in November 1916.

Slim fought during General **Sir Frederick Maude**'s advance on Baghdad during December 10, 1916–March 11, 1917, but found himself suddenly invalided to India because he had never been officially certified as fit for duty after his first wound. In India, he was temporarily attached to the general staff, then appointed a general staff officer III in November 1917. He was promoted to temporary major and GSO II in November 1918, after which, at his request, he was gazetted captain in the Indian Army on May 31, 1919. On March 27 of the following year, Slim was assigned as a captain to the 2/7 Gurkha Rifles. He had greatly admired the Gurkhas at Gallipoli, and now he served with them in the campaign against the Tochi Wazirs in October 1920.

Slim returned to England in 1924 and passed the necessary tests in May 1925 to enter the Staff College. He returned to India in 1926 and began work at the Staff College in Quetta. Returning again to England, he served at Army Headquarters from 1929 to 1933, then became an instructor at the Imperial Staff College, Camberley, serving there from 1934 through 1936. In 1937, he taught at the Imperial Defense College and was promoted to lieutenant colonel the following year, when he also returned to India. There he enrolled in the Senior Officer's School at Belgaum and assumed command of the 2/7 Gurkhas in Assam during 1938–39. He returned to Belgaum as commandant of the Senior Officer's School and was promoted to the local rank of brigadier general.

With the outbreak of World War II, Slim was given command of the Tenth Indian Brigade on September 23, 1939, which he prepared for mechanized desert warfare. He and his brigade were sent to East Africa as part of the Fifth Indian Division, arriving at Port Sudan on August 2, 1940. They were, however, repulsed at Gallabat, Sudan, on November 6, 1940, due to poor air support and faulty reconnaissance. After this setback, however, Slim was more succesful, but then suffered a wound late in January 1941.

After convalescing in India, Slim assumed command of the Tenth Indian Division in Iraq on May 15, 1941. He captured Baghdad early in June, then pushed into Vichy-controlled Syria, where he captured Deir-es-Zor, using the slightest of resources. On August 25, Slim invaded Iran in order to open supply lines to Russia. This accomplished, he returned to India in March 1942, where he took command of the newly created I Burma Corps (the Seventeenth Indian and the Burma Divisions) on March 13 at Prome. In this forlorn theater of the war, Slim conducted a determined delaying campaign against the Japanese, but it was a lost cause. Reinforcements were rare, and Chinese assistance under General **Joseph Stilwell** were scarce. Inevitably, Slim was driven out of Burma by the end of April.

After executing a retreat across the River Chindwin on Imphal in Manipur, Slim directed the defense of India from June to December 1942. This was severely complicated by the Indian nationalist movement's anti-British "Quit India" campaign during August–September.

Slim took command of the IV Corps as well I Burma Corps on April 14 and extricated the corps, which was pinned down at Arakan. He next replaced General Giffard at the head of the Eastern Army on October 16, 1943 and transformed this group into the Fourteenth Army. Undermanned and chronically undersupplied, the army felt that it had been abandoned and consequently suffered severe morale problems. Slim turned the situation around with extraordinary leadership, shaping the Fourteenth Army into a great fighting unit.

Slim led the Fourteenth Army in an operation that halted the Japanese offensives in Arakan during February 1944 and at Imphal-Kohima during March 6–July 15. He was in command of operations to liberate northern Burma during July–December 1944, finishing work begun by Stilwell and by Orde Wingate.

With the liberation of northern Burma accomplished, Slim directed the British offensive of December–March 1945 that took Mandalay on March 20, 1945. The capture of Mandalay was a masterpiece of planning for combat in hostile terrain and under very difficult logistical circumstances.

After Mandalay, Slim proceeded south along the Irrawaddy and Sittang valleys in order to take Rangoon before the May monsoons made movement impossible. He reached Pegu at the end of April and marched into Rangoon on May 2. Promoted to general on July 1, Slim became commander in chief of Southeast Asian ground forces on August 16. He conducted pacification operations in Malaysia and Indonesia during August–November 1945; even after the Japanese had surrendered, holdouts in these remote places continued to fight.

After World War II, Slim became commandant of the Imperial Defense College, serving from 1946 to 1948. He then became chief of the Imperial General Staff on November 1, 1948 and was promoted to field marshal on January 4, 1949.

Slim served as Governor-General of Australia from 1952 to 1960. He wrote an account of his war experiences in 1956, calling it, aptly, *Defeat into Victory*. After returning to England, Slim was named lieutenant governor deputy constable (1963–64), then governor and constable of Windsor Castle. He held the latter post from 1964 until his death in 1970.

Slim was one of the greatest—yet least familiar—commanders of World War II. His skill at operating on a logistical shoestring was phenomenal, and his leadership was always inspiring.

SMUTS, JAN CHRISTIAAN

BORN: September 24, 1870

DIED: September 11, 1950

SERVICE: Second Anglo–Boer War (1899–1902); Boer Uprising (1914); World War I (1914–18); World War II (1939–45)

South African (Afrikaner) general and prime minister who sought accommodation with the British

Born near Riebeeck West (near Malmesbury), Cape Colony, Smuts was raised as a farm boy, but was then sent to England for study at Christ's College, Cambridge, from which he graduated in 1894. He passed the bar a year after graduation and then returned to South Africa, setting up a law practice in Capetown.

Smuts had been a wholehearted supporter of Cecil Rhodes until the Jameson Raid (December 29, 1895–January 2, 1896), in which Rhodes authorized a violent attempt to overthrow the Afrikaners, sent Smuts into the political camp of Paul Kruger and the Afrikaaners.

Smuts moved to Johannesburg in 1897 and became state attorney for the Transvaal government in June 1898. When Pretoria fell to the British on June 5, 1900, Smuts became a general and fought his first battle at Diamond Hill on June 11–12. He next led the Boer offensive at Nooitgedacht/Magaliesburg on December 13, taking Modderfontein in Transvaal and killing its native inhabitants on January 31, 1901.

Smuts planned to invade Cape Colony in an effort to stir up rebellion among the Cape Colony Boers. He invaded on September 3 and defeated the Seventeenth (British) Lancers at Elands River Poort on September 17, but soon found himself in trouble when British reinforcements arrived and he received little of the anticipated support from the Cape Boers. He took a new tack the following year, laying siege to O'okiep in the western Cape from April 4 to May 3, 1902. After this operation, he participated in the peace conference at Vereeniging during May 15–31 and concluded the Treaty of Vereeniging on May 31, 1902.

Following the war, Smuts was appointed Colonial Secretary under Transvaal Prime Minister **Louis Botha** in March 1907. Smuts was one of the main authors of the constitution of the Union of South Africa, which was enacted on May 31, 1910. Seeking accommodation with the British, Smuts suppressed Christian de Wet's anti–British Boer uprising during September–December 1914, then led South African forces in the conquest of German Southwest Africa (modern Namibia) from February to July 9, 1915. Smuts was in charge of East African operations against the German guerilla-style general **Paul von Lettow–Vorbeck** during February 1916. Although Smuts succeeded in overrunning most of German East Africa (today's Tanzania) by 1917, he never decisively defeated the wily Lettow–Vorbeck, whose small army remained highly effective.

After attending an Imperial war conference in London in 1917, Smuts was appointed a cabinet minister and privy councillor in March 1917 and was invited to sign the Treaty of Versailles—even though he privately believed the treaty too punitive.

Smuts became South African prime minister upon Botha's death in August 1919, but he was never a popular leader, and his party lost the 1924 election, whereupon Smuts retired until 1934, when he became deputy prime minister to J. B. M. Hertzog. When Hertzog's government fell on September 5, 1939 over the issue of neutrality in World War II—Herzog supported neutrality; Smuts did not—Smuts became prime minster again and oversaw the South African war effort, which was primarily aimed at blocking Germany and Italy from conquering North Africa. **Winston Churchill** frequently consulted him, and Smuts was promoted to field marshal in the British army in 1941; however, he had little direct input into the conduct of the war.

At the conclusion of World War II, Smuts was present at the San Francisco Conference and was among the drafters of the United Nations charter in 1945. His later career in South African politics again revealed his marginal popularity. Defeated in the 1948 elections, he stepped down as prime minister, but retained a seat in parliament.

Smuts was a fine general, although he met his match in Lettow-Vorbeck. He was popular with the men he commanded, but less well liked by South African voters. His position on the British Commonwealth—which he supported—cost him the increasingly important nationalist vote.

SOULT, Nicolas Jean de Dieu, Duke of Dalmatia

BORN: March 29, 1769

DIED: November 26, 1851

SERVICE: French Revolutionary Wars (1792–99); Napoleonic Wars (1800–15)

Marshal General of France; as courageous on the battlefield as he was opportunistic in politics

Born at Saint-Amand-de-Bastide, France Soult was the son of a humble notary and joined the army as a private on April 16, 1785 when his father died. Soult possessed a remarkable personality. He was quiet and dutiful, yet also charismatic. In 1791, he was commissioned as an officer in the grenadiers, and two years later he made captain. His skills as a soldier were readily apparent, and Soult became an instructor specializing in infantry maneuvers. He was then assigned as adjutant to General Jean-Baptiste Jourdan in the Army of the Moselle. There he first saw action against the Austrians. He led an assault against the Austrian encampment at Marsthal in 1793 and was promoted at the beginning of the following year to colonel. Shortly afterward, on June 26, 1794, he was appointed chief of staff to **Francis Joseph Lefebvre** at Fleurus, and in this assignment he made a name for himself as an officer whose steadiness in a crisis and under fire was nothing short of phenomenal. In November, he was promoted to general of brigade, though he continued to serve as Lefebvre's chief of staff through 1797.

In April 1799, Soult was promoted to general of division and took over Lefebvre's division after Lefebvre was wounded. Soult fought at Zurich on June 4–7 and September 26, 1799, taking his direction from Marshal **André Masséna,** whom he served brilliantly as a subordinate. Promoted to lieutenant general in 1800, he followed Masséna into the Army of Italy and was first tested in his new role at the siege of Genoa, where he led a bold sortie against Monte Cretto on April 13. In this action, he took a musketball in the leg and was captured by the Austrians, who released him after **Napoleon**'s victory at Marengo on June 14, 1800. Still suffering from the effects of his wound, he returned to France to recuperate, after which he was made commander of the army in Piedmont. He set about vigorously squashing the insurrection at Aosta, after which he brought order to region not through terror, but through even-handed and compassionate rule.

In 1801, Soult took command of a "corps of observation," which he used to occupy Taranto and Otranto, and the following year, he was elevated to an appointment of one of the four colonel-generals of the Consular Guard. Soult trained this elite unit to such a high level that Napoleon made him commander in chief of the camp at Saint-Omer in 1801 and promoted him to marshal of the Empire on May 19, 1804.

At Ulm in 1805, Soult commanded IV Corps at Ulm, and at Austerlitz, on December 2, he was in command of the critical assault on the Pratzen Heights. This operation was executed with flawless vigor, as was his handling of the French right wing at Jena on October 14, 1806. During February 7–8, 1807, at Eylau, Soult was in command of the left and persuaded Napoleon not to order a retreat, but to make a stand. As a result, Soult emerged victorious after the Russians retired during the night.

Soult suffered a costly reverse at Heilsberg on June 10, 1807, but he took Konigsberg on June 16. The following year, the emperor created him Duke of Dalmatia and assigned him to command of II Corps. With this unit, Soult accompanied Napoleon to Spain and took charge of the pursuit of the English army to Corunna on January 16, 1809. After the British slipped away by sea, Soult invaded Portugal in hopes that he would be made king of that country. But Wellesley (later Duke of **Wellington**), soundly defeated him at Oporto on May 12. Soult later countered at Ocalia on November 19, smashing the Spanish Army of the Center and harvesting 15,000 prisoners. After this triumph, he invaded Andalusia, taking the fortress of Badajoz on March 11, 1811.

At this point, Soult began to look after his self-interest with considerable zeal. He declined to support Masséna against the lines of Torres Vedras and Marmont on the Douro, thereby giving himself ample time to despoil the wealth of Andalusia and amass for himself a fabulous fortune.

On May 16, 1811, Soult's efforts to raise the British siege of Badajoz were thwarted by his defeat at Albuera. When the French lost at Salamanca on July 22, 1812, Soult left Andalusia and withdrew to the north, occupying Madrid. He commenced a pursuit of Wellington, just then retiring toward Portugal after his own failed siege of Burgos during September 19–October 22, but was forced to break off when he was summoned to Germany to replace Jean-Baptiste Bessières—slain in battle—as commander of the Imperial Guard.

In Germany, Soult fought at Bautzen on May 20–21, 1813, but immediately after this was sent back to Spain, where he assumed command of what was left there of the French army. Beginning in June 1813 and extending well into 1814, he repeatedly engaged Wellington in the Pyrenees. Finally, on April 10, 1814, he was defeated at Toulouse. With this loss, the British invasion of France began.

Following Napoleon's first exile, Soult became minister for war under the Bourbons from December 3, 1814 to March 11, 1815. When Napoleon returned, however, he was made chief of staff of the Army of the North for the Waterloo campaign. Following that final defeat, Soult fled to Düsseldorf in January 1816 and remained in exile there until 1819, when the Bourbon king recalled him to France. The following year, he was restored to the rank of marshal, and, ten years later, he was once again named minister for war, serving in that post until 1834. In 1838, he traveled to London as the French representative at Queen Victoria's coronation. He came face to face with his old adversary Wellington. The Iron Duke good naturedly took him by the arm and exclaimed, "I have you at last!"

Soult spent his later years in dignified service to France. He was minister for foreign affairs in 1839 and also presided over the Council of Ministers during 1832–34 and 1840–47. From 1840 to 1845, he became war minister again, finally retiring on September 15, 1847. Just before this, he was named the fourth marshal-general in the history of France, thereby joining the ranks of three other greats, the Viscount of **Turenne,** Claude Villars, and the Count de **Saxe.**

Like most of Napoleon's marshals, Soult was more capable as a tactician than as a strategist. Cool in a crisis, he was also an excellent trainer of troops. Only once did he sacrifice military integrity to personal gain—when he plundered Andalusia to the detriment of the Spanish campaign.

STALIN, Josef

BORN: December 21, 1879

DIED: March 5, 1953

SERVICE: Russian Civil War (1917–22); World War II (1939–45)

Absolute dictator of the Union of Soviet Socialist Republics (USSR); leader in the Russian Civil War; led his nation through World War II

Born Josef Vissarionovich Djugashvili in Gori, a hill town in tsarist Georgia, Stalin was subjected to a brutal childhood at the hands of his alcoholic and abusive father, an impoverished shoemaker. After his father's death in a brawl, eleven-year-old Josef was indulged by his doting mother, who groomed him for the Orthodox priesthood. By the time he entered the Tiflis

Josef Stalin. Courtesy Library of Congress.

Theological Seminary at age fourteen, the youth's rebelliousness had earned him the nickname Koba, after a legendary Georgian bandit and rebel. He soon bridled under the harsh corporal discipline of the seminary and became involved in radical anti-tsarist political activity in 1898. In 1899, he abruptly left the seminary to become a full-time revolutionary organizer. Within three years, he was a member of the Georgian branch of the Social Democratic party, touring the Caucasus, stirring up laborers, and organizing strikes. Like his German contemporary, **Adolf Hitler,** young "Koba" came from a squalid and brutal family background and showed remarkably little early promise. Unlike Hitler, the aspiring Georgian revolutionary possessed neither personal magnetism nor appreciable rhetorical skills, but he was fearless, ruthless, and a brilliant organizer.

In 1903, the Social Democrats split into two groups, V. I. Lenin's radical Bolshevik ("majority") faction and the more moderate faction Lenin called the Mensheviks—the "minority"—which actually outnumbered the Bolsheviks. Stalin fell in with the radicals and grew close to Lenin. For the next decade, from 1903 until he was exiled to Siberia in 1913, Stalin worked to expand the party's power, organizing cell after cell across the nation, and financing the party's work by planning and executing daring robberies. His activities caused him to be arrested many times, but he always managed to escape. In 1912, Lenin elevated him to the Bolshevik Central Committee, the party's inner circle. Though taciturn and even inarticulate, Stalin became the first editor of *Pravda* ("Truth"), the Bolsheviks' official newspaper. It was during this period that he took the name Stalin—"Man of Steel."

In 1913, Stalin's ability to escape capture and punishment at last failed him, and he was sentenced by the tsar to Siberian exile. He endured through four fateful years, returning to Russia only after the overthrow of Nicholas II in March 1917. In the wake of the failure of the first Bolshevik attempt to seize power during the summer of 1917—and in the absence of Leon Trotsky, who had been arrested, and Lenin, who had gone into hiding—Stalin worked to reorganize the party, thereby playing a major role in its successful acquisition of power during the November Revolution. With Lenin's return from self-imposed exile, Stalin was given a succession of commissar posts, all the while working quietly to consolidate greater power. By 1922, he was named general secretary of the party's Central Committee, a position of tremendous influence from which Stalin controlled the apparatus and official personnel of most of the party. When Lenin died in 1924, Stalin promoted himself as the Communist leader's handpicked successor and ruthlessly exploited his position as general secretary to eliminate all who opposed him.

Stalin quickly demonstrated his ability and willingness to manipulate ideology in order to ensure his personal supremacy. He began by announcing a retreat from Lenin's ideal of world communist revolution by advocating "socialism in one country." He also proposed an economic program far more moderate than what others who had been close to Lenin envisioned. Trotsky, Lev Kamenev, and Grigory Zinoviev, party leftists, went on the offensive against Stalin and his moderate policies, but by 1928, Stalin consolidated the party's right wing in opposition to the left and managed to oust that faction's leadership. Having accomplished this, he performed an abrupt about face and summarily adopted radical leftist economic programs, including the wholesale collectivization of agriculture and greatly accelerated industrialization. Now he successfully attacked the party's right wing, led by Nikolai Bukharin. Within a year, opposition on the left *and* the right had been quashed, and Stalin emerged as absolute dictator of the Soviet Union.

Stalin sought to transform the Soviet Union from a primarily agricultural nation into a modern industrial power. To do this, he was willing to sacrifice human life on an unprecedented scale. Late in 1928 he expropriated the lands of the middle-class farmers ("kulaks"), "deporting" or killing those who offered resistance. Stalin's regime proposed a series of "five-year plans" by which collectivization and industrialization were to be achieved. His administration adhered rigorously to the plans, raising capital to finance industrialization by exporting grain and other produce despite a devastating famine that swept the Soviet Union in 1932. Millions who resisted were executed, and millions more starved to death. A 1988 estimate put the number of deaths that directly resulted from the forced collectivization of 1928–33 at 25 million.

During the period of the first five-year plan, opposition to Stalin mounted, and there was a short-lived peasant revolt, which the dictator easily crushed. When the Seventeenth Party Congress showed support for Sergei Kirov, a moderate and a potential rival, Stalin engineered his assassination in December 1934. Having disposed of Kirov, Stalin then used his murder as a pretext for arresting most of the party's highest-ranking officials as counter-revolutionary conspirators. From 1936 to 1938 Stalin conducted a long series of public trials in which party officials and many in the senior officer corps of the Red Army were convicted of outrageous crimes or acts of treason. The results of the purge were devastating. By 1939, 98 of the 139 central committee members elected in 1934 had been executed, and 1,108 of the 1,966 delegates to the Seventeenth Congress arrested. Worse, under KGB chief Lavrenti Beria, the secret police arrested, executed, exiled, or imprisoned millions of individuals in the general population. By the eve of World War II, Stalin had destroyed all serious opposition and had terrorized his nation into submission even as he built it into an industrial giant and created about himself a cult of personal worship.

As Adolf Hitler came to dominate more and more of Europe in the late 1930s, Stalin had no desire to oppose this ideological antithesis of communism. He did make attempts to reach agreements with the western democratic powers, but was rebuffed. So he turned instead to Hitler himself, concluding a Nazi–Soviet non-aggression pact on August 23, 1939. With the German invasion of Poland, Stalin moved to increase Soviet influence in the West by invading Finland on November 30, 1939. A short but costly war secured Finland's surrender on March 12, 1940. Then, on June 22, 1941, Hitler abrogated the non-aggression pact by invading the Soviet Union.

Early resistance to the invasion was poorly coordinated because Stalin's purges had stripped the Red Army of thousands in its senior officer corps. After an initial period of panic and disorganization, Stalin took personal command of the Red Army and organized an increasingly effective counter-force. He acted swiftly to move vital war industries east, into Siberia and Central Asia, just ahead of the advancing German armies. He rallied the Soviet people by appealing to patriotism and even disbanded the Communist International and officially rehabilitated the Orthodox Church. Despite the damage he himself had inflicted on his officer corps, Stalin identified dependable—even brilliant—commanders in Marshals **Georgi Zhukov** and Ivan Konev, supporting them in extremely costly but ultimately successful campaigns against the invaders.

By the middle of the war, Stalin had earned great prestige as a military leader and was in a strong negotiating position at the major allied conferences conducted in Tehran, Yalta, and Potsdam. By the end of the war, many Russians—and others—were willing to overlook the enormities of Stalin's regime, regarding him now as the savior of his nation.

Having triumphed over Hitler, however, Stalin quickly instituted a new regime of terror and repression at home, imposing more taxes on peasants, and announcing fresh discoveries of sabotage and conspiracy. Hard on the heels of victory in World War II, Stalin aggressively expanded the Soviet sphere of influence into neighboring regions, setting up puppet regimes and client states among the nations of the Balkans and eastern Europe, creating what **Winston Churchill** called an Iron Curtain separating the communist realm from the rest of the world, and touching off nearly four decades of "Cold War" between Soviet-aligned nations and the West.

In 1953, Stalin declared that he had discovered a plot among the Kremlin's corps of physicians, and the Soviet people trembled on the brink of what seemed an imminent and inevitable new round of blood purges. But dictator died of a massive cerebral hemorrhage before he could inaugurate these.

Stalin was an incredibly cruel and incredibly complex leader. His political paranoia caused him to purge his army of some of his best officers, making the nation that much more vulnerable to Nazi aggression during World War II. However, he also built the Soviet Union into a military superpower, whose only rival after World War II was the United States.

STEUBEN, Baron Friedrich Wilhelm von

BORN: September 17, 1730

DIED: November 28, 1794

Principal wars: Seven Years' War (1756–63); American Revolution (1775–83)

German officer who did much to train the Continental Army

Steuben was born in Magdeburg, Prussia, the son of a distinguished military family, which, however, had no hereditary right to the ennobling "von" Steuben's grandfather had taken it upon himself to adopt. In 1747, Steuben entered the Prussian army and served with it during the Seven Years' War, both as a field officer in the infantry and as a staff officer. He was assigned confidential diplomatic missions and was then attached to the staff of **Frederick the Great** from 1757 to 1763, when he resigned (with the rank of captain) to become chamberlain at the petty court of Hohenzollern-Hechingen. He served in this capacity until 1771, then journeyed to France with the prince to raise money for the court during 1771–75. When this came to nothing, Steuben tried to enter the French, Austrian, and Baden armies, but was admitted to none. At last, he returned to France in 1777 to enter the American service, and, bearing a recommendation from Benjamin Franklin to **George Washington,** he sailed from Marseilles on September 26 and arrived in Portsmouth, New Hampshire on December 1.

On February 5, the Continental Congress accepted Steuben's offer to serve as a volunteer, and he hastened to join Washington at Valley Forge, Pennsylvania on February 23. Steuben set about training Washington's regulars, beginning with a model company of one hundred. His methods spread throughout the Continental Army quickly, despite the fact that Steuben knew no English and had to use interpreters to make himself understood. His charisma transcended the language barrier, and he was highly successful at training and motivating troops.

Appointed major general and inspector general by Congress in May, he participated in the Battle of Monmouth on June 28, then, in 1778–79, wrote *Regulations for the Order and Discipline of the Troops of the United States*, a manual that served the American army for three decades.

Steuben went to Virginia in the spring of 1780 to raise troops and supplies and to organize local defenses against the depredations of turncoat Benedict Arnold along the coast during the fall of 1780 and into the spring of the following year. Steuben fell ill and turned over his command to the **Marquis de Lafayette** on June 19, but recovered in time to join the army in September just prior to Yorktown, He took command of a division at Yorktown from September 28 to October 19.

At the conclusion of the Revolution, Steuben tried with little success to secure the surrender of British posts on the frontier during the spring and summer of 1783. He left the Continental Army on March 24, of the following year and lived a lavish retirement in upstate New York. Soon weighed down by debt, he secured a well-deserved pension from the government in June 1790 and was saved from destitution. From his retirement, he remained active in U.S. military affairs and was instrumental in urging the creation of a national military academy. He also advocated a standing militia to supplement a small professional army.

Steuben was enthusiastic and dedicated. He trained the Continental Army to a high level and fashioned it into a more effective fighting force.

STILWELL, Joseph Warren

BORN: March 19, 1883

DIED: October 12, 1946

SERVICE: Moro pirate suppression (1902–06); World War I (1917–18); World War II (1941–45)

Called "Vinegar Joe" because of his acidic and irascible temperament; a U.S. general who fought vigorously on the China–Burma–India front in World War II

A native of Palatka, Florida, Stilwell was raised in Yonkers, New York. He graduated from West Point in June 1904 and was commissioned a second lieutenant in the infantry. His request for duty in the Philippines was granted, and he was posted to the Twelfth Infantry Regiment there. He saw action on Samar against the rebel Puljanes during February–April 1905.

He returned to West Point as a foreign language instructor in February 1906, and the assignment stretched to four years—March 1906–January 1911—of teaching history and tactics in addition to coaching athletics. In January 11, he transferred back to the Philippines and was promoted to first lieutenant in March. During November–December, he visited China for the first time.

He returned again to West Point as a language instructor from 1913 to 1916 and was promoted to captain in September 1916. Named brigade adjutant in the Eightieth Division after promotion to temporary major in July 1917, he shipped out to France in January 1918, where he was appointed a staff intelligence officer. During March 20–April 28, he was attached to the French XVII Corps near Verdun, then received an assignment as deputy chief of staff for intelligence under General Joseph T. Dickman in IV Corps during the Meuse-Argonne campaign (September 26–November 11).

Stilwell was promoted to temporary lieutenant colonel on September 11, 1918, then temporary colonel in October. After the Armistice, he served in occupation duties until May 1919. Not wishing to return to the States, he requested an assignment to China as a language officer and served there from August 6, 1919 to July of 1923. During this period, Stilwell made an ally of the warlord **Feng Yu-hsiang.**

Back in the U.S., Stilwell attended Infantry School at Fort Benning from 1923 to 1924, followed by the Command and General Staff School at Fort Leavenworth from 1925 to 1926. He was then returned to China in command of a battalion of the Fifteenth Infantry at Tientsin in August 1926. There he met **George C. Marshall.** Promoted to lieutenant colonel in March 1928, he became head of the tactical section of the Infantry School in July 1929, largely through the influence of Marshall.

Stilwell left the Infantry School in 1933 to become training officer for the IX Corps reserves from 1933 to 1935. After this, he was appointed military attaché to China and was promoted to colonel on August 1, 1935. During his assignment in China, Stilwell became intimately familiar with vast regions of the country, and he was a keen observer of the developing Sino–Japanese War. He returned to the United States in 1939 and received his promotion to brigadier general while en route on May 4, 1939.

Assigned command of the Third Brigade, Second Division in September 1939, Stilwell played a major role in the Third Army's maneuvers of January 1940, then commanded the "red" forces in the massive Louisiana–Texas maneuvers of May 1940. He gained attention for his ability to move troops and in unexpected ways. Following the maneuvers, he was assigned to command on July 1, 1940 the newly created Seventh Division stationed at Fort Ord, California and, in September, was promoted to temporary major general. He was moved up to command of III Corps in July 1941.

After Pearl Harbor, Stilwell was promoted to lieutenant general and appointed commanding general of U.S. Army forces in the China–Burma–India (CBI) Theater in January 1942. He set up a headquarters at Chungking (Chongqing), China, where he made an ally of **Chiang Kai-shek,** who agreed to his commanding Chinese forces in Burma on March 6, 1942. Stilwell arrived in Burma on March 11 with a single Chinese division. Subsequently, he raised eight more.

Like his British counterpart in the CBI theater, General **William Slim,** Stilwell suffered from chronic shortages of supply and reinforcements. He also had to contend with conflicting orders from Chiang Kai-shek. Ultimately, he was forced to withdraw from Burma to India during May 11–30. After the disappointment in Burma, Stilwell turned his attention to training and equipping the three Chinese divisions in India and to creating "the Hump," an airlift chain to supply Kunming during January–February 1943.

In July 1943, Stilwell was appointed deputy supreme Allied commander in the CBI under Lord **Louis Mountbatten.** Stilwell advocated the Salween–Myitkyina Mogaung offensive of March–August 1944, which was a hard-won success that ended in the capture of Myitkyina on August 3 and the subsequent liberation of all northern Burma. Following this, Stilwell was promoted to temporary general, but he could not maintain a viable working relationship with Chiang Kai-shek, and, at Chiang's request, President Franklin D. Roosevelt recalled him on October 19, 1944.

Back in the United States, Stilwell was named commander of Army Ground Forces on January 23, 1945 and was decorated with the Legion of Merit and the Oak Leaf cluster of the DSM on February 10, 1945. Dispatched to Okinawa, he took command of the Tenth Army there on June 23 and was among those present at the Japanese surrender ceremony in Tokyo Bay on September 2, 1945.

After the war, Stilwell became president of the War Equipment Board and then commander of the Sixth Army and the Western Defense Command in January 1946, succumbing later in the year to stomach cancer.

As his nickname testifies, Stilwell was irascible and acidic. He was also a no-nonsense soldier, who performed miracles in the CBI theater with very little in the way of support or supplies. He was one of the commanders Lord Louis Mountbatten most admired.

STUART, James Ewell Brown (Jeb)

BORN: February 6, 1833
DIED: May 12, 1864
SERVICE: Civil War (1861–65)

Confederate general who was an expert in cavalry reconnaissance

Born in Patrick County, Virginia, Stuart graduated from West Point in 1854 and was commissioned in the Mounted Rifles. He served in Texas and Kansas with the First Cavalry through 1860 and was **Robert E. Lee**'s aide-de-camp during the operation against John Brown at Harper's Ferry in October 1859. Stuart was promoted to captain in April 1861, but resigned from the U.S. Army the following month to join the provisional Confederate forces and was commissioned both as a lieutenant colonel of Virginia infantry and captain of cavalry. In July, he was promoted to colonel of the First Virginia Cavalry and immediately distinguished himself at the first Battle of Bull Run on July 21. Promoted to brigadier general in September, he fought against **Edward Ord** at Dranesville on December 20 and suffered a defeat. He fared better at Williamsburg on May 5, 1862 and fought brilliantly at Fair Oaks on May 31.

Against the slow-moving **George B. McClellan,** he demonstrated superior tactical ability, running 1,200 cavalrymen around McClellan's Army of the Potomac during June 12–15, then harrying the Union commander's rear during the Seven Days (June 26–July 2).

Promoted to major general with command of the cavalry of the Army of Northern Virginia in July, Stuart raided Union general John Pope's headquarters at Catlett's Station, near Warrenton, Virginia, on August 22, then hit his supply depot at Manassas Junction on the twenty-seventh. He coordinated with **Thomas "Stonewall" Jackson** at Second Bull Run on August 29–30, then fought with considerable distinction at South Mountain, Maryland, on September 14 and at Antietam, which followed on the seventeenth.

In October, Stuart led a daring cavalry raid into Pennsylvania, taking the town of Chambersburg. He then commanded the right flank at Fredericksburg on December 13. At Chancellorsville, on May 2, 1863, he showed himself a master of cavalry reconnaissance, probing General **Joseph Hooker**'s position and exposing the vulnerability of Hooker's right flank.

With the wounding of Jackson and Hill, Stuart assumed temporary command of II Corps and fought General Alfred Pleasanton to a standoff at Brandy Station on June 9. At Gettysburg, Stuart failed Robert E. Lee by raiding into Pennsylvania rather than performing the reconnaissance vital to the battle. Lee was thus deprived of information as to the movement of the Union Army—a deficiency that put Lee at a fatal disadvantage. At Gettysburg, Stuart did not engage federal cavalry until July 3.

On October 19, Stuart routed Union cavalry at Backband Mills, near Warrenton, Virginia. The following spring, he engaged **Philip Sheridan** at Todd's Tavern on May 7, 1864 and at Yellow Tavern on May 11. Here he was mortally wounded and died the following day. Called "Lee's eyes and ears," Stuart raised cavalry reconnaissance to a high level of art and of daring.

His flamboyance was highly attractive and inspirational, although it caused irreparable harm at Gettysburg.

STUYVESANT, Peter

BORN: ca.1592

DIED: Februray 1672

SERVICE: Dutch–Portuguese Wars (ca. 1620–25); Wars against the Indians (1650–59); War with New Sweden (1655); Esopus War (1659); Peach War (1655); British Conquest of New Netherland (1664)

Last and most important governor of New Netherland

The son of a Calvinist minister, Peter Stuyvesant was born in Holland and entered its University of Franeker in 1629. He was expelled after only a year, for reasons unknown. Something of an adventurer, Stuyvesant joined the Dutch West India Company in 1635 and traveled to the New World. In 1638, he became the chief commercial officer for Curaçao and within five years, was made its governor. It was a position that cost him his leg, in an ill-fated attack against the Portuguese-held St. Martin. Popular accounts often illustrate Stuyvesant stumping about on his wooden leg, which contributed to a likely accurate image of him as a quick-tempered and irascible autocrat.

In 1645, Stuyvesant volunteered to serve as governor of New Netherland and arrived in what would later become New York on Christmas Day. As a devout Calvinist, Stuyvesant immediately set restrictions on the sale of alcohol and decreed strict observance of the Sabbath. He vigorously persecuted non-Calvinists, particularly Quakers and Lutherans, whom he feared would be most likely to instigate revolt.

Stuyvesant did attempt to establish an honest and efficient administration, including a limited public works campaign of improving roads, repairing fences, constructing a wharf on the East River, and building a defensive wall on the northern edge of New Amsterdam along what would later be known as Wall Street. His greatest loyalty, however, was to the Dutch West India Company, and this, combined with his despotism, prompted considerable protest among his subjects. Colonists, particularly the burghers of New Amsterdam (later New York), clamored for more self-government. To appease them, Stuyvesant instituted the Board of Nine Men in September 1647 as an advisory council—but he reserved the right to appoint the members. Because of his high-handed and autocratic ways untempered by the board, the citizens of New Amsterdam appealed directly to the West India Company. In February 1653, the company directed Stuyvesant to grant independent municipal control to New Amsterdam.

Beyond his rule of New Netherland, Stuyvesant had mixed success in dealing with the colonies of other European powers. In 1655, he invaded the Delaware Valley and forced the surrender of New Sweden. He was not as fortunate in dealing with New England, however. In 1650, he was compelled, by the Treaty of Hartford, to cede the entire Connecticut Valley to English control.

Although, of all the colonial powers, the Dutch often enjoyed the most peaceful and profitable relations with the Indians, Stuyvesant was at times as autocratic and brutal to Native Americans as he appeared to be to his own subjects. In 1655, the so-called Peach War broke out when a Dutch farmer killed a Delaware Indian woman for picking peaches in his orchard. The Delaware sought revenge against the white settlers, and in response Suyvesant called out the militia in a punitive campaign. Three years later, Stuyvesant sought to end conflict between the Esopus Indians and Dutch settlers by recruiting Mohawks to terrorize the tribe. Violence escalated, and in 1659 the Dutch settlement of Wiltwyck (modern Kingston, New York) was attacked. Stuyvesant called for a parley, a delegation of Esopus chiefs responded by traveling to the town for a conference. Stuyvesant ordered the chiefs killed as they slept following the first day of talks. Predictably, the murders brought Indian reprisals, to which Stuyvesant responded by taking as hostages the children of the Esopus and other tribes. When the Esopus refused to yield all of their children as directed, Stuyvesant sold those hostages he held into the West Indian slave trade.

Continued strife between Stuyvesant and the British reached an end on September 8, 1664, when British warships sailed against the colony and the Dutch residents simply refused to defend themselves. Without any supporters, Stuyvesant had no choice but to surrender, having at least secured trading rights for the West India Company. He retired to his farm—the Bouwerie (through which Bowery Street now passes)—on Manhattan Island. He died there in February 1672.

SULEIMAN I THE MAGNIFICENT

BORN: ca. 1494

DIED: September 5, 1566

SERVICE: Hungarian War (1521–33); Conquest of Rhodes (1522); Persian War (1526–55); Venetian War (1537–39); Austrian War (1537–44); War with Austria in Hungary (1566)

Ruled the Ottoman Empire, greatly enlarging it while also advancing its art and architecture

Suleiman was the only son born to Sultan Selim I. As such, the young man was given a taste of administrative matters early on. While his grandfather, Sultan Bayezid II, ruled the Ottoman Empire, Suleiman was appointed governor of Kaffa, in the Crimea. Later, during his father's reign, young Suleiman became governor of Manisa in Asia Minor.

Suleiman became sultan in 1520 and almost immediately began a series of campaigns against the Christian nations that surrounded the Mediterranean Sea. He took Belgrade in 1521 and Rhodes in 1522. Suleiman defeated the Hungarian king Louis II at Mohacs in August 1526 and installed a vassal, John, in his place. When John died in 1540, Suleiman committed vast resources to the final conquest of the country. By 1562, after years of fighting, no progress had been made, and Suleiman agreed to a peace.

In the meantime, as early as 1534, Suleiman had turned his eyes eastward and began a series of wars against Persia. Following up on his father's earlier efforts, Suleiman captured eastern Asia Minor and Iraq.

Suleiman I. From Stanley Lane-Poole, The Story of Nations: Turkey, *1888.*

However, two additional protracted eastern campaigns failed to gain for Suleiman all of the territory that he wanted.

In addition to his considerable skills as commander of land forces, Suleiman also became famous for the amassing a navy of considerable size and strength. In 1538, the sultan's ships defeated a combined force sent by Venice and Spain, and in 1551, his navy attacked and took the city of Tripoli. The empire's ships sailed far and wide, even reaching distant India.

In the course of his military operations in Europe, Suleiman demonstrated great cunning in pitting Protestant factions against the Pope and his followers. Although he was not particularly interested in Western religion, Suleiman realized that it was to his benefit to keep Europe divided in matters of religion, with the Protestant nations in direct conflict with the rulers of the Ottoman's arch-rival Habsburg Empire. In Hungary, for example, Suleiman allowed Calvinism to flourish, and, so liberal was his treatment of Protestantism—when it served his purposes—that some churchmen looked upon the sultan as their protector.

Suleiman strengthened the Ottoman Empire, raising it to a position of leadership in Eastern Europe, the Mediterranian basin, and the Near East. He also brought Ottoman culture to a height that surpassed anything in Europe of the day. In addition to Suleiman's massive building programs, which produced scores of mosques, bridges, fortresses, and other public works, his administrative efforts included the conversion of Constantinople into Istanbul, making it the center of Ottoman government. However, as Suleiman grew older, his empire was shaken by disputes between his two sons. Neither of them had inherited his father's zeal for conquest, nor his diplomatic acumen. Accordingly, after Suleiman fell in battle during a September 1566 assault against a Hungarian fortress, the Ottoman Empire entered a long period of decline under a series of weak and even degenerate sultans.

SULLA, Lucius Cornelius

BORN: ca. 138 B.C.

DIED: 78 B.C.

SERVICE: Jugurthine War (112–105 B.C.); Cimbri and Teutones War (105–101 B.C.); Social War (91–88 B.C.); First Mithridatic War (89–84 B.C.); Roman Civil War (88–82 B.C.)

Won the first major civil war in Roman history; became dictator

Lucius Cornelius Sulla rose to the dictatorship of Rome, instituting a harsh regime of largely self-serving reforms. Sulla was born into a patrician family of little distinction and, as a foppish and effete young man, exhibited nothing to foretell his rise. The military tribune Gaius Marius put Sulla on his staff as quaestor in 107 B.C., and Sulla gained an immediate reputation by engineering the capture of the Numidian king Jugurtha in 105, thereby bringing the Jugurthine Wars to a successful conclusion. Although Marius retained Sulla's services in his campaigns against the Cimbri and the Teutons during 105–101 B.C., he became increasingly jealous of the young man's triumphs.

In 93 B.C. Sulla became praetor and, the following year, governor, of Cilicia (the southern portion of modern Turkey). In this post he installed Ariobarzanes as king of Cappadocia. From 91 to 88 B.C., Sulla was called back to Italy to take command of forces prosecuting the Social War (war against the Allies or Socii) and successfully laid siege against Pompeii in 89 B.C. His growing reputation earned him election as consul for 88 B.C., and he obtained the command against the powerful Mithridates VI of Pontus, the conqueror of Anatolia. M. Sulpicius Rufus, a partisan of Marius, attempted to replace Sulla with Marius, an action that prompted Sulla to march on Rome with six legions. He took the city and drove Marius into exile, then continued his program of military triumph unimpeded. In 86, Sulla invaded Greece and captured Athens. He performed brilliantly against the superior numbers of a Pontic army led by Archelaus at Chaeronea in the summer of 86 B.C.

The successor to Marius, Lucius Cornelius Cinna, made an attempt to prosecute Sulla for treason, but Sulla sailed east, defeated Archelaus again at Orchomenus in 85 B.C., then invaded Asia, taking it upon himself to conclude a hasty treaty with Mithridates—notwithstanding the presence of another Roman legion under G. Flavius Fimbria. Flushed with victory, Sulla returned to Italy, landing at Brundisium (modern Brindisi) in 83 B.C., where he joined forces with Marcus Licinius Crassus and **Pompey the Great** for a full-scale advance on Rome. In the spring of 82 B.C., he and his allies defeated the Marians and their allies at the Battle of the Colline Gate and captured Rome. Much as Marius had done before him in 87 B.C., Sulla and his allies embarked on an orgy of massacres and proscriptions in order to purge the capital of their enemies. Soon afterward, Sulla was made dictator (82–81 B.C.), then dictator and consul (80 B.C.).

Sulla introduced a number of sweeping reforms, strengthening the Senate—and making it impossible for outsiders to challenge the powerful families who controlled that body—and revamping the criminal courts. In an effort to thwart incipient opposition, he greatly reduced the power and authority of the tribunes. He also secured, by force, the settlement of his many veterans in Italy.

Soon after achieving his ends, Sulla retired from public life, retreating to his estate in Campania in 79 B.C., where he died a year later.

For all the bloodshed associated with them, Sulla's system of reforms did not appreciably outlive him. His personal example, more than his reforms, determined the shape of the Rome to come. He failed to inject new life into the traditional oligarchic republic. Instead, he provided a model of absolute personal leadership that inspired the likes of the egocentric and ambitious Pompey and **Julius Caesar** to overturn republicanism altogether in favor of rule by emperor.

T

TAKEDA, Shingen (Harunobu)

BORN: 1521

DIED: 1573

SERVICE: Unification Wars (1550–1615); War against Kenshin Uesigi (1547–73)

Among the most violent of the many warlords who contended for power during Japan's "Warring States" period

Takeda's given name was Harunobu, and he was the oldest son of the ruler, Katsuyori Takeda, whom Takeda deposed in 1541 to prevent his younger brother from inheriting his father's domains. Takeda continued his father's aggression against other warlords, expanding his territory in the provinces of Shinano and Hida. By 1547, he faced Kenshin Uesugi, warlord of Echigo province, in battle for the first time. The struggle between Takeda and Uesugi continued for the next thirty years, ending only with Takeda's death in 1573. The conflict involved annual battles fought in the vicinity of Kawanakajima from 1553 to 1564. Characteristically, each battle ended in a futile, but bloody, draw. In one, some eight thousand samurai perished, and Takeda himself was severely wounded in the 1561 engagement.

In 1567, Takeda joined in an alliance with the powerful general and statesman Ieyasu Tokugawa to attack Ujizama Imagawa, who held sway over Suraga province. The house of Imagawa, which had been weakening, fell to Takeda. But even he proved unable to hold the province, which was subject to unrelenting attack from the Hojo clan.

Following his experience in Suraga, Takeda became a Buddhist priest, taking the name of Shingen. He did not, however, renounce his holdings nor his warlike ways. In 1572, he turned on his ally of a few years before, Ieyasu Tokugawa, who had joined forces with **Nobunaga Oda,** a powerful warrior well on his way to consolidating enough power to make him warlord of a unified Japan. Takeda soundly defeated Ieyasu Tokugawa at Mikatagahara in 1572. Takeda, however, failed to capitalize on his victory. It is possible that he had taken a musketball to the head during the siege of Ieyasu's stronghold and lost heart for the fight. He died—perhaps from this wound—the next year.

TAMERLANE (Timur, Timur the Lane)

BORN: 1336

DIED: January 19, 1405

SERVICE: Conquest of Transoxiana (1364–70); War with Khwarizm (1370–80); Conquest of Persia (1381–87); Wars with Toktamish (1385–86, 1388–95); Invasion of Russia (1390–91); Conquest of Mesopotamia and Georgia (1393–95); Invasion of India (1398–99); Invasion of Syria (1400); Invasion of Anatolia (1402)

Amir of Samarkand, conqueror and plunderer of Mesopotamia, Persia, Afghanistan, and much of India

Born at Kesh in 1336, most likely the son of a Tartar chieftain, Tamerlane became vizier to Khan Tughlak Timor of Kashgar in 1361. He left the post to join his brother Amir Hussain in an expedition to conquer Transoxiana. The brothers spent six years raiding the territory, beginning their rampages in 1364. In 1369, Tamerlane succeeded his father as amir of Samarkand, then battled the khans of Khwarizm and Jatah for the next ten years. In 1381, Tamerlane invaded Persia, capturing Herat. The next year, he began several years of raids into Khurasan and eastern Persia. During 1386–87, Fars, Armenia, Azerbaijan, and Iraq fell to him in succession.

His former ally Toktamish invaded Samarkand in 1385–86 and again in 1388 and 1389. All three times, Tamerlane defeated Toktamish, and in 1390–91 pushed him back into Russia. He was unable to capitalize on his victories, however, because he was called back to Persia to put down a rebellion mounted there by Shah Mansur at Shiraz in 1392. Tamerlane defeated the shah and went on to reconquer Armenia, Azerbaijan, and Fars during 1393–94. In 1393, Tamerlane captured Baghdad, and within two years held Mesopotamia and Georgia.

In 1395, Toktamish invaded yet again, provoking the worst of Tamerlane's wrath in response. Tamerlane defeated the upstart at the Battle of Terek late in the year. He followed his victory with a rampage through southern Russia and the Ukraine, visiting great destruction and misery in a relentless punitive raid. Taking no prisoners, he slaughtered all of the Mongols he encountered.

Tamerlane interrupted his orgy of killing by returning again to Persia to suppress rebellion there in 1396–97. A year later, he launched a massive cavalry invasion of India. Tamerlane cut a broad swath of destruction on his way to Delhi, just as he had through Russia. At the outskirts of Delhi, he met and routed the army of Sultan Mahmud Tughluq and entered the city on December 18, 1398. He turned his troops loose upon its citizens for little more than two weeks, killing tens of thousands, looting all that was worth taking, then began a march back toward Samarkand. While he did not seek to occupy India or add it to his empire, he did gut it mercilessly, thereby undermining the Delhi sultanate, which soon decayed and fell.

Following the Indian expedition, Tamerlane invaded Syria, annihilating the Mameluke army at the Battle of Aleppo on October 30, 1400. After sacking Aleppo and Damascus, he returned to rebellious Baghdad, which he utterly destroyed in 1401 as retribution for revolt. Though awash in blood, Tamerlane was unsated and moved on in 1402 to invade Anatolia, smashing the army of Sultan Bayazid I at Ankara on July 20, 1402. When he captured Smyrna from the Knights of Rhodes, he collected tribute from the Sultan of Egypt as well as from John I, emperor of the Byzantines.

Undefeated, Tamerlane returned to Samarkand in 1404 and set about planning a large-scale invasion of China. He was stricken with illness and died on January 19, 1405, before putting his plans into operation.

TAYLOR, Zachary

BORN: November 24, 1784

DIED: July 9, 1850

SERVICE: War against Tecumseh (1811); War of 1812 (1812–15); Black Hawk War (1832); Second Seminole War (1835–42); U.S.–Mexican War (1846–48)

U.S. general; hero of the War of 1812 and U.S.–Mexican War; president

Taylor was born in Orange County, Virginia, but was raised near Louisville, Kentucky. After a spotty education, he entered the militia as a short-term volunteer in 1806, then joined the army as a first lieutenant in the Seventh Infantry in March 1808. Two years later, he was promoted to captain, and in 1811, he served under General **William Henry Harrison** against **Tecumseh**'s band of Shawnee and other Indians of the Old Northwest.

With the outbreak of the War of 1812, Taylor made a valiant defense of Fort Harrison on the Wabash River on September 4, 1812, which resulted in a brevet promotion to major. By 1814, the promotion was confirmed.

Taylor continued to serve on the frontier, and commanded an expedition down the Mississippi to assert federal authority on the frontier. However, in the postwar rush to demobilize, Taylor was reduced in rank to captain in June 1815. Indignant, he resigned, but was restored to the rank of major through the intervention of his friend President James Madison. From 1817 to 1819, he served garrison duty with the Third Infantry at Green Bay, Wisconsin Territory, then assumed command of Fort Winnebago. Promoted to lieutenant colonel in April 1819, he transferred to Louisiana in 1822. There he built Fort Jesup, near Natchitoches.

From 1829 to 1832, Taylor was at Fort Snelling (present-day St. Paul, Minnesota), where he served as Indian superintendent, then moved into the Wisconsin–Illinois area to fight Chief Black Hawk. He fought with distinction at the decisive battle of Bad Axe River on August 2, 1832, which ended the conflict.

In July 1837, Taylor was dispatched with a force to fight the Seminoles, who were resisting "removal" to Indian Territory (modern Oklahoma). After he managed to lure the Seminoles into open battle at Lake Okeechobee, he scored a signal victory on December 25. In a war characterized by guerrilla encounters, a field battle like this was a rarity.

Breveted to brigadier general in 1838, Taylor could not lure the Indians into open conflict again and, therefore, was unable to end the war decisively. In April 1840, he was relieved, having himself requested this action. He served briefly in Louisiana, then was named commander of the Second Department, Western Division, headquartered at Fort Smith, Arkansas in May 1841. In May 1844, as war clouds gathered over Texas, Taylor put together a combat force at Fort Jesup and was ordered to march to Texas when the republic was annexed to the United States in June 1845. With a force of four thousand, he established his headquarters at Corpus Christi in October.

Taylor and his army were dispatched to the Rio Grande during February 1846 and occupied territory claimed by Mexico. Hostilities commenced near Fort Brown on April 25. On May 8, at Palo Alto, he defeated a vastly superior Mexican force, then defeated the Mexicans again on the next day at Resaca de Ia Palma. Taylor occupied Matamoros on May 18.

Breveted major general and elevated to command of the Army of the Rio Grande in July, Taylor crossed the Rio Grande and invaded Mexico with 6,200 men. He attacked Monterrey, but agreed to an armistice on September 24. President James Polk repudiated the armistice and ordered Taylor to press his advance more deeply into Mexico. Following orders, Taylor captured Saltillo in November, but then sent most of his regular troops—four thousand strong—and many volunteers to General **Winfield Scott,** whose army was invading central Mexico during the winter of 1846–47.

Taylor used his remaining troops—4,600 green volunteers—to continue his own advance. He confronted a 15,000-man army under General **Santa Anna** at Buena Vista on February 21. The Mexican commander demanded his surrender. Taylor refused and launched into battle, repulsing the Mexicans in a bloody two-day battle during February 21–22.

Buena Vista made Taylor a national hero, who became the Whig candidate for president in June 1848. He won handily and took office in March 1849. He died of heatstroke the following year—ironically, after an Independence Day celebration. Vice President Millard Fillmore served out the balance of his term.

TECUMSEH

BORN: ca. 1768

DIED: October 5, 1813

SERVICE: American Revolution (1775–83); Northwest Indian Wars (1790–95; 1811); War of 1812 (1812–15)

Shawnee chief who organized among several tribes powerful resistance to white settlement

Tecumseh was the child of a Shawnee father and a Creek mother. It is believed that he was born in the Shawnee village of Old Piqua, which occupied the location of present-day Springfield, Ohio. After his father fell in Lord Dunmore's War on October 10, 1774, Tecumseh was raised by Chief Blackfish.

Tecumseh early on acquired a reputation as a great warrior. He fought on the side of the British during the later years of the American Revolution, from about 1780 to 1783, then organized Shawnee resistance during the Indian Wars of the Old Northwest from September 1790 to August 1795. General **Anthony ("Mad Anthony") Wayne** defeated him decisively at Fallen Timbers on August 20, 1794, a battle that ended the Indian Wars in the Old Northwest and brought relative peace to this unsettled region until the eve of the War of 1812.

In August 1795, Tecumseh refused to sign the Treaty of Greenville (Indiana), which ceded (in return for cash) much Indian land to the federal government. Instead he moved to what was then Indian territory along the Wabash River. There, with his brother

Fanciful depiction of the death of Tecumseh. Courtesy Library of Congress.

Tenskwatawa—known as "the Prophet"—he worked for several years to organize an alliance or union among several tribes. His object was to save Indian land and Indian culture, and this meant organizing armed resistance to white incursions. Tecumseh, a brilliant orator, spread his gospel throughout Iowa, New York, and, with little success, into the Southeast.

While Tecumseh was absent form his Wabash River headquarters village, his brother violated Tecumseh's instructions to refrain from violence at this point. The Prophet attacked a force under Ohio territorial governor **William Henry Harrison** at Tippecanoe on November 7, 1811. The Prophet was not only defeated, but disgraced—since he did not participate in the fighting and since he had told his followers that his "medicine" would protect them and ensure victory. The headquarters village was burned, and the union Tecumseh had created was left in tatters. When Tecumseh returned, he led the remnants of his followers to Canada, where they joined with British army forces at the outbreak of the War of 1812 in June 1812.

Tecumseh commanded a mixed force of whites and Indians at the Battle of Detroit in August and scored a victory at nearby Maguada on August 9. British authorities commissioned him a brigadier general, and he was assigned to fight under General Henry Proctor during the sieges of Forts Meigs (Maumee, Ohio) and Stephenson (Fremont, Ohio) during April–August 1813. Tecumseh was far bolder—and a better tactician—than the cautious Proctor, and the two never worked well together. Tecumseh led the rearguard that covered Proctor's withdrawal into Canada during September, and then he led his Indian warriors alongside Proctor's regulars against William Henry Harrison at the Battle of the Thames on October 5, 1813. Once again, coordination with Proctor was poor, and Harrison defeated the British–Indian forces. Tecumseh was slain in this battle.

With the death of Tecumseh died one of the last great hopes of forming a politically and militarily effective union of Indian tribes.

TIGLATH-PILESER III

ACTIVE: 745–727 B.C.
SERVICE: Conquests in the Middle East (745–727 B.C.)

Assyrian king; returned Assyria to its former pre-eminence in the Middle East; established an imperial government that would last a hundred years; made far-reaching conquests, including Babylonia, Syria, and Palestine

In 745 B.C., Tiglath-pileser III, son of Adad-nirari III, fought his brother for the Assyrian throne. The former governor of Calah, Tiglath-pileser was a masterful administrator who knew how to make the best use of his governors. He assigned his strong and loyal governors to the larger Assyrian provinces, which had been vying for independence. In outlying regions, he installed Assyrian officials directly accountable to him. By 738 B.C., Assyrian government had numbered some eighty provinces.

During Tiglath-pileser's reign, the Assyrian army drastically changed in character. Previously, it had been composed of Assyrian peasants called to military service during time of crisis and released once the crisis abated. Tiglath-pileser created a standing army composed primarily of Assyrians, but significantly supplemented by foreign mercenaries and troops of vassal states.

With his empire reorganized and a strong standing army in place, Tiglath-pileser set out to conquer his surrounding enemies. First he struck against Zamua, then the Medes and the Puqudu, northeast of Baghdad. These regions he annexed to Assyrian provinces. To maintain control, Tiglath-pileser instituted a policy of relocating conquered subjects. When the king had reason to doubt their loyalty, the residents of defeated regions were resettled elsewhere. Between 742 and 741 B.C., tens of thousands of conquered people were relocated.

In 743 B.C., Tiglath-pileser attacked and defeated Urartu and its allies, the neo-Hittites and Aramaeans. For three years, he laid siege to the city of Arpad, which ultimately fell in 741 B.C. With the defeat of Arpad, Tiglath-pileser was able to command tribute from Damascus, Tyre, Cilicia, and other important cities.

During the 730s, the Assyrian king conquered Syria and Palestine and ousted the Chaldean chief Ukin-zer, who had seized the Babylonian throne in 734 B.C.. In 729 or 728 B.C., Tiglath-pileser claimed the throne himself, becoming King Pulu, or Pul, of Babylon. He died shortly afterward.

TIGRANES

BORN: ca. 140 B.C.

DIED: 55 B.C.

SERVICE: War with Rome (93–92 B.C.); War with Seleucia (83 B.C.); Wars in Syria, Cilicia, and Egypt (78–70 B.C.); War with Rome (69–66 B.C.); Third Mithdridatic War (75–65 B.C.)

Built Armenia into the most powerful kingdom of western Asia

The origin of Tigranes is somewhat obscure. Through Artavasdes, who was either his uncle or his father, he was part of the dynastic line founded by Artaxias. In his youth, he was taken hostage by the Parthian monarch Mithridates II, to whom he was forced to cede terrritories corresponding to present-day central Iran in exchange for his freedom. After assuming the throne in 95 or 94 B.C., Tigranes set about augmenting his kingdom.

He began by attacking and deposing the ruler of Sophene (Diyarbakir region), annexing his kingdom in 93 B.C. He allied himself with the Pontine monarch Mithridates VI Eupator by marrying his daughter Cleopatra (not to be confused with her far more famous namesake, who lived from 69–30 B.C.). Thus allied, he invaded Cappadocia (modern-day central Turkey), but was driven out by the forces of the Roman governor **Sulla** during 93–92 B.C. Although he next voiced his support for his Pontine ally's first war against Rome during 89–84 B.C., he did not send troops, but instead waged his own war against Parthia, which was reeling after the death of Tigranes's former captor, Mithridates II.

From 88 to 84 B.C., Tigranes conqured much of Media, northern Mesopotamia, and Atropatene (modern Azerbaijan, Iran). He also dominated Gordyene, Adiabene, and Osroene. In 83 B.C., he invaded Syria, defeating the Seleudcids.

With the death of the Roman governor Sulla, Tigranes again moved against Cappadocia, this time successfully, occupying much of it by 78 B.C. He staged a second Syrian invasion, then raided Cilicia (in modern southern Turkey), destroying Soli, a city near Tarsus. Emboldened by success after success, Tigranes moved against Cleopatra Selene of Egypt.

In conjunction with his military triumphs, Tigranes set about reshaping the realms he now dominated. He began by building a new capital city, Tigranocerta, east of the Tigris River. He transplanted a large Greek population to Armenia and moved Arabs into Mesopotamia in an effort to stabilize his conquests and consolidate his power. This did not go unnoticed by the Roman empire, which feared an erosion of its influence. With a small army, Lucius Licinius Lucullus invaded Tigranes's kingdom and attacked the new capital. The Battle of Tigranocerta, fought on October 6, 69 B.C., resulted in the defeat of Tigranes, whose forces, effective against any number of semibarbarian peoples, were no match for Roman soldiers. Lucullus attacked next at Artaxata in September of 68 B.C., defeating Tigranes again, but then was recalled to Rome.

Tigranes and his ally Mithridates VI, duly chastened, welcomed the respite. But Tigranes's hot-headed son, also called Tigranes, rebelled against his father. The son fled to Parthia, then returned with an army of Parthians in 67 B.C., intending to resist the Romans. Tigranes the elder offered a bounty for the death or capture of his own son, then surrendered to the forces of **Gnaeus Pompey** when he invaded Armenia in 66 B.C.

Having secured control of Armenia, Pompey acted with great moderation, permitting the now-aged Tigranes senior to remain on the throne. Pompey sent the younger Tigranes to Rome, where he was out of harm's way and could do no harm himself. When Tigranes senior died ten years later, another son, Artavasdes, was permitted to inherit the throne.

TILLY, Count Johan Tserclaes

BORN: February 1559

DIED: April 30, 1632

SERVICE: *Eighty Years' War (1567–48); Franco–Spanish War (1589–98); Thirty Years' War (1618–48)*

Flemish general in the Bavarian service who was principal general for the Catholic League in the Thirty Years' War

Tilly was born at Castle Tilly in Brabant, Spanish Netherlands and was exposed to the rigors of a Jesuit education. He served under the Duke of Parma as a cadet beginning about 1574 and later participated in the lengthy siege of Antwerp, which the duke capture on August 17, 1585. Tilly fought at Ivry on March 10, 1590, then was appointed governor of the Meuse River town of Dun and of Villefranche in Lorraine from 1591 to 1594.

Tilly next went to work for the Holy Roman Empire, fighting in Hungary during 1600–1608 to expel the Turks. He was appointed colonel of a Walloon regiment in the Austrian (Holy Roman) service in 1602, and by 1604 had become a general of artillery. The next year, he was promoted to field marshal. In 1610, Tilly put himself in service to **Maximilian I** of Bavaria, leader of the Catholic League, and he commanded an advance into Bohemia in July 1620 in order to fight the Bohemian Protestants. He met their forces in the crucial Battle of White Mountain on November 8, 1620, and emerged victorious, but was beaten two years later by Ernst von Mansfeld at Mingolsheim on April 27, 1622.

Tilly joined a Spanish force commanded by Gonzales de Cordoba and, with him, defeated George Frederick of Baden-Durlach at Wimpfen on May 6, 1622. He then attacked Christian of Brunswick's poorly trained army at Höchst on June 20. This victory earned him creation as a count; however, he could not check Christian's advance to join forces with Mansfeld. Nothing daunted, Tilly invaded and took Heidelberg in the culmination of an eleven-week siege on September 19. Then, on August 6, 1623, he met and again defeated Christian, at Stadtlohn, and went on to another triumph, over another Christian—King Christian IV of Denmark—at Lutter during August 24–27, 1626. On May 20, 1627, Tilly led the storming of fortified Magdeburg, then turned his troops loose to sack the town.

Tilly met defeat at the hands of **Gustav II Adolph** of Sweden at Breitenfeld on September 17, 1631 and had to withdraw into Bavaria. Fighting a holding action at River Lech, he was severely wounded on April 16, 1632, and succumbed within two weeks.

Tilly was the outstanding general of the Thirty Years' War. His ultimate defeat was less his fault than that of his uncooperative ally, **Wallenstein;** it is also true that Gustav II Adolph performed brilliantly.

TITO, Josip Broz

BORN: May 7, 1892

DIED: May 4, 1980

SERVICE: World War I (1914–18); World War II (1939–45)

As marshal of Yugoslavia, led partisan resistance to* Adolf Hitler *in World War II; after the war, was first Communist national leader openly to defy the Soviet Union

Born Josip Broz in Kumrovec, near Zagreb, Tito was one of fifteen children in a peasant family. At the age of thirteen, he moved to the town of Sisak to become a locksmith's apprentice, after which he traveled about central Europe as a metalworker. Broz joined the metalworkers trade union, which led him to membership in the Social Democratic Party of Croatia.

The outbreak of World War I in 1914 interrupted his socialist activism, and he enlisted in the Twenty-Fifth Regiment of Zagreb, which marched against the Serbs in August 1914. Broz was accused of disseminating antiwar propaganda and was imprisoned, only to be released in January 1915 after the charges were dropped. He was sent back to his regiment on the Carpathian front and was decorated for bravery. The Twenty-Fifth was then transferred to the Bukovina front, where it saw heavy action. Broz was severely wounded in hand-to-hand combat and taken prisoner by the Russians.

Josip Broz Tito. Courtesy National Archives.

Broz was laboring as a prisoner of war in the Ural Mountains when the Bolshevik Revolution took place in 1917, He made his way to Siberia, where he sided with the Bolsheviks and fought in the Red Guard during the Russian Civil War. Finally making his way home again in 1920, Tito now considered himself a Communist—albeit a moderate in the degree of his convictions.

He joined the Communist Party of Yugoslavia (CPY) and was promptly arrested in 1923, tried, and acquitted. He began working in a shipyard in Croatia, but was arrested again in 1925 and sentenced to seven months probation. Government harassment only strengthened his resolve, and he now began a climb in the Communist hierarchy, gaining membership in the Zagreb Committee of the CPY in 1927, and in 1928 becoming deputy of the Politburo of the Central Committee of the CPY, as well as secretary general of the Croatian and Slavonian committees. In August, 1928, he was again arrested. Declaring that the tribunal had no right to judge him, Broz was sentenced to five years imprisonment.

After his release in 1934, Broz traveled throughout Europe on behalf of the party. For reasons of security on these trips, he adopted the codename Tito, which he would use thereafter. Returning to Moscow in 1935, Tito worked in the Balkan section of the Comintern, the organization of international Communism. In August 1936, Tito was named organizational secretary of the CPY Politburo.

In 1937, **Josef Stalin** began his infamous purges, and prominent Yugoslav Communists were among the first to be "liquidated." Some eight hundred Yugoslavs disappeared in the Soviet Union, and only a high-ranking Comintern official saved the CPY from total dissolution. Tito not only escaped the purge, but by the end of 1937, he was named secretary general by the Executive Council of the Comintern. He returned to Yugoslavia to reorganize the Communist Party and was formally named secretary general of the CPY in October 1940.

The international communist movement was stunned when Stalin concluded a non-aggression pact with **Adolf Hitler**'s Germany, betraying Communism's unshakable opposition to Fascism. Yugoslavia itself was officially neutral in the war that had just begun. But when the pro-Axis leader Prince Paul was overthrown in a coup, Hitler promptly invaded Yugoslavia in reprisal. Even under German occupation, the Yugoslavs at first remained passive. This changed dramatically in June 1941, after Hitler launched Operation Barbarossa and, in violation of the non-aggression pact, invaded the Soviet Union.

Tito now led his followers in a well-coordinated campaign of sabotage and resistance against the occupying Germans. His movement was so successful that, in the summer of 1942, he was able to organize an offensive into Bosnia and Croatia, forcing the Germans to commit substantial numbers of troops to stop the partisans. Despite the counteroffensive, the partisans held their own, and by December 1943, Tito announced a provisional government in Yugoslavia, with himself as president, secretary of defense, and marshal of the armed forces.

With the defeat of Nazi Germany in May 1945, Tito went about the business of establishing his government on a permanent basis. He formed a party oligarchy, the Politburo, and held Soviet-style elections in November. In the manner of Lenin and Stalin, he promulgated a Five Year Plan for economic recovery and development, and was greatly aided in these efforts by his universal popularity as a war hero and patriot.

It was these very qualities that Stalin feared and resented, especially as it became clear that Tito would not allow Yugoslavia to become a Soviet puppet. Tito's stance as a Yugoslav first and a Communist second—in bold defiance of Stalin—further enhanced the dictator's popularity. Indeed, the word *Titoism* was coined to describe the opposition by Communist satellites to Soviet domination.

Tito was a twentieth-century example of what the eighteenth and nineteenth centuries would have called an "enlightened despot." His grip on Yugoslavia was absolute, yet, within a framework of totalitarianism, he granted considerable constitutional leeway. He was a Communist, yet he refused to align his nation with the Soviet bloc, and he permitted a substantial degree of free enterprise that made Yugoslavia one of the richest of Eastern European nations. Moreover, he brought three decades of stability to a country traditionally racked by violence, dissension, and civil war.

When Tito died three days before his eighty-eighth birthday, he was a living legend in Yugoslavia and around the world. His opposition to Stalin and the mother Soviet state made him more appealing to the democratic West. His longevity allowed him to travel Eastern Europe and Asia speaking on communism and Titoism to nonaligned countries and his firsthand observations of some of the most important events in the history of the world.

TOJO, Hideki

BORN: December 30, 1884

DIED: December 23, 1948

SERVICE: Second Sino–Japanese War (1937–45); World War II (1941–45)

Japan's prime minister and military commander during World War II

During the period of World War II, the only man the people of the Allied nations hated as much as **Adolf Hitler** was probably Hideki Tojo. Yet, unlike Hitler, a genuine—if evil—political leader, Tojo was little more than the bureaucratic head of the military regime that dominated Japanese empire. Not a popular leader to his countrymen, he nonetheless gained a reputation among his fellow bureaucrats as "The Razor"—a hyperefficient bureaucrat who knew how to get things done.

Tojo was born into a traditionally military family on December 30, 1884, in Iwate prefecture. His father, Eikyo Tojo, was a general, and Hideki Tojo was expected to attend the military academy, to graduate—as he did in 1905—and to pursue a career in the army.

Tojo distinguished himself in much of his military positions, but no one called his career glorious. He engaged in combat only once, in August 1937, when he directed operations against the Chinese in Chahar (in the vicinity of Zhangjiakou). Prior to this one campaign, he held regimental staff assignments from 1905 to 1909. In 1915, he graduated from the Army Staff College and served in Berlin as assistant military attaché from 1919 to 1921. During the years in Germany, he attained the rank of major, becoming a resident Japanese officer from 1921 to 1922.

Tojo's diplomatic service brought him into the inner circles of government, and he contributed to the military's efforts to take control of national policy and administration. Promoted to lieutenant colonel in 1924, Tojo became chief of the Army Ministry's important Mobilization Plans Bureau, where he had a hand in directing Japan's preparations for war.

After a promotion to colonel in 1929, Tojo received a regimental command, then assumed the position of chief of the Army General Staff's Organization and Mobilization Section. He served in this capacity from 1931 to 1933, when he was promoted to major general and made deputy commandant of the Military Academy.

Tojo continued his orderly rise in the military hierarchy when he was given command of an infantry brigade in 1934–35, then of the Kwantung Army Gendarmerie, a post in which military leaders of the highest rank were traditionally groomed. He held this position until 1937, gaining promotion to lieutenant general in 1936. From 1937 to 1938, he was chief of staff of the Kwantung Army.

Following his years in China, Tojo became vice minister of the army and chief of Army Air Headquarters on the eve of World War II (1938–41). From this position, Tojo emerged as a leading spokesman for the most aggressive, pro-Axis faction of the army, even as pro-Axis supporters gained dominance in government affairs. Promoted to general in 1941, he was named as prime minister. Even members of the government who objected to military domination approved Tojo's selection as a way of staving off an outright military coup.

Tojo established a hard line in international and military affairs, propelling his nation into war and steadily expanding the scope of that war. During 1941 to 1944, he was often given wide-ranging dictatorial powers in foreign as well as domestic affairs. For most of the war, Prime Minister Tojo also seved as chief of the Army General Staff. In this position, he directed military operations with ruthless but indifferent skill. For example, he never really managed to develop a long-term strategy. After Saipan fell to Allied forces on July 12, 1944, a coalition of Japanese statesmen exerted their influence to force Tojo's removal as military head. Shaken by his country's rapidly deteriorating military situation, Tojo bowed to the coalition without a struggle. His attempted suicide after the surrender of Japan in 1945 failed, and he was taken into custody by Allied forces. A tribunal convicted him as a war criminal, and he was hanged in 1948.

TRAJAN (Marcus Ulpius Traianus)

BORN: 53

DIED: 117

SERVICE: Dacian Wars (101–107); Eatern War (113–117)

Expanded the Roman Empire to its greatest extent; following his death, the long decline of Rome began

One of the greatest military emperors of Rome, Trajan was born not in Italy, but in Spain, in Italica, a town near present-day Seville. His father was an able general in the Roman army, and Trajan served as a military tribune from 78 to 88, sometimes under the command of his father in Syria. In 88, Emperor Domitian gave him a legion to command in Spain, then quickly summoned him with it to Germany to fight Antonius Saturninus during 88–89. His success earned him a consulship in 91, and in 96 he was made governor of Upper Germany by the emperor Nerva, who had succeeded to the throne after the assassination of Domitian.

Nerva, as ineffectual a ruler as Domitian had been tyrannical, was despised by his own Praetorian Guard, who mutinied against him in 97. On October 27, to stave off outright revolt, the emperor named as his successor Trajan, by now enormously popular with the military. Nerva died late in January 98, but Trajan delayed coming to Rome to assume the purple until he had completed an inspection tour of the Rhine and Danube frontiers.

Trajan's first major military expedition was against Decabulus, ruler of the Dacians, a warlike tribe who lived in the area corresponding to modern Hungary and Rumania. The Dacians had long terrorized their neighbors by extorting tribute payments, and in 101–102 Trajan defeated them, but, shortly afterward, Decabulus rebelled, sending raiding parties into Moesia (northern Bulgaria). Trajan pushed these forces out of Moesia and, after a difficult campaign spanning 105–106, took the Dacian capital of Sarmizgetusa. He annexed Dacia to the empire in 107, and Decabulus committed suicide.

Even while he was prosecuting the Dacian campaigns, Trajan moved his forces into Arabia Petraea—the Sinai, Negev, and part of Jordan—in 106, taking the capital city of Petra and annexing the region to the empire. Chosroes of Parthia, a region roughly corresponding to modern Iran, put a puppet ruler on the throne of Armenia in 113. Trajan responded by invading Armenia as well as Mesopotamia during 113–114, then captured the Parthian capital city of Ctesiphon, whence he sailed down the Tigris River to the Persian Gulf during 114–115. Having lost much of his territory, Chosroes raised a new army and staged a vigorous counterattack that drove Roman occupying forces out of northern Mesopotamia. Trajan responded with equal vigor in 116, pushing Chosroes out of the region.

At this triumphal juncture, however, the emperor fell ill. Early in 117, he designated Publius Aelius Hadrianus—**Hadrian**—as his successor, and he embarked on a desperate journey back to Rome. His condition worsened, however, and Trajan died at Selinus in Ciclicia (part of modern Turkey) on August 8, 117. Almost immediately, the greatly enlarged empire began to shrink, as Hadrian relinquished the hard-won Mesopotamian and Assyrian conquests to Parthia. Whereas Trajan had begun his reign by subduing the barbarian tribes that demanded tribute, Hadrian intiated a policy of appeasement, which resulted in further reductions of the empire. The twenty-year reign of Trajan was indeed the highwater mark of the Roman Empire.

TURENNE, Henri de la Tour d'Auvergne, Viscount of

BORN: September 11, 1611

DIED: July 27, 1675

SERVICE: Thirty Years' War (1618–48); Franco–Spanish War (1635–59); the Fronde (1648–53); War of Devolution (1665–68); Dutch War (1672–78)

Marshal of France; one of the greatest military leaders during the reign of Louis XIV

Born at Sedan to Henri, Duke of Bouillon, and Elizabeth of Nassau, Turenne was raised as a Protestant. Fascinated from childhood by military history, he entered the army of his uncle, Maurice of Nassau, as a private when his father died in 1625. He saw his first action the very next year, at the siege of Bois-le-Duc and performed with such dash and brilliance that he was quickly promoted, becoming a captain in 1627.

In 1630, Turenne entered the French service as an infantry regimental commander, and fought against Spain during 1630–34. In 1634, he directed the assault on the fortress of La Motte, for the success of which he was promoted to *marechal-de-camp*.

After fighting in the failed Rhineland campaign of Cardinal la Vallette in 1635, he was part of the reinforcing unit at the siege of Breisach on August 5, 1638. He then served in Italy under the Duke of Harcourt during 1639–41 and fought with distinction at Casale on April 29, 1640, then at Turin. In 1643, Cardinal Mazarin elevated him to marshal of France and charged him with the task of rebuilding and reorganizaing the Army of Germany after it was crushed at the Battle of Tuttlingen on November 24, 1643. After crossing the Rhine in May 1644, Turenne invaded the Black Forest, but, finding Baron Franz von Mercy's force numerically superior, he fell back on Breisach. where he joined forces with **Condé**'s Army of Champagne. He and Condé conducted a spirited counteroffensive against Mercy, then achieved victory at the Battle of Freiburg during August 3–9, 1644.

Turenne led the Army of Germany into Franconia in 1645, only to be hit with a surprise attack by Mercy at Mergentheim on May 2, 1645. He withdrew quickly into Hesse and there regrouped. Subsequently, Condé arrived and they jointly invaded Bavaria, meeting Mercy at Allerheim on August 3, 1645. Mercy was not only defeated, he fell in the battle. However, forces of the Archduke Leopold pushed Condé and Turenne back Philippsburg. Condé returned to France, and Turenne bided his time for a year before joined forces with Karl Gustav von Wrangel's Swedish army on August 10, 1646 near Giessen.

Wrangel and Turenne devastated the Bavarian countryside during September–November, until Bavaria agreed to the Truce of Ulm on March 14, 1647. After squashing a revolt among his Weimarian troops, then beating back a Spanish assault on Luxembourg in May 1647, Turenne again linked up with Wrangel in a second sweep through Bavaria. Fighting in the rear guard, Turenne defeated General Melander at Zusmarshausen on May 17, 1648. Melander was killed in the battle.

During the Fronde, Turenne threw in his lot with the rebel *Parlement de Paris* during January–February 1649, but soon found himself seeking refuge in the Netherlands. In the second Fronde rebellion, he assumed leadership of the rebel forces, and, allied with the Spanish, he led an army to the relief of Rethel. However, his force was nearly crushed at Champ Blane—near Rethel—in a battle of October 15, 1650, and Turenne returned to Paris on February 15, 1651, after Condé was released.

Turenne decided to support Louis XIV during Condé's revolt and the outbreak of the Third (Spanish) Fronde during September–November 1651. Now he warred against his former ally Condé throughout the Loire valley, defeating him at Gien on April 7, 1652 and at Porte de St. Antoine/St. Denis, outside Paris, on July 5. For all practical purposes, this brought an end to the Third Fronde. The following year, Turenne was able to halt Condé's invasion, and he went on the offensive, taking Stenay and defeated the Spanish at Arras on August 25, 1654. He captured three more towns before the end of 1654, and another three during 1655, but he suffered defeat at the hands of Condé on July 1656, when he laid siege against Valenciennes. Condé also bested him at Cambrai in 1657; however, he captured Mardyck (near Dunkirk) with the aid of English allies. After this victory, he laid siege to Dunkirk during May 1658, scoring a massive triumph against a relief force commanded by Don John of Austria and Condé at the Battle of the Dunes on June 14, 1658. This battle was the culmination of the Franco–Spanish War.

Following the successful conclusion of the war, Louis XIV named Turenne marshal general on April 4, 1660, which made Turenne, in effect, supreme commander of the army. He fought under the nominal command of Louis XIV in the invasion of Flanders during the War of Devolution in the spring of 1667 and found further favor with the king when, upon the death of his wife the next year, he converted from Protestantism to the Catholic faith.

Turenne commanded the French army on the left bank of the Rhine in the first campaign of the War of the Triple Alliance in 1672. He managed this modest force with great skill, readily outgeneraling Prince Raimundo Montecuccoli and **Frederick William** of Brandenburg. The Brandenburger was so badly defeated that, compelled to assent to the Treaty of Vassem on June 6, 1673, he withdrew from the anti-French alliance. Unfortunately for Turenne, however, he was denied reinforcements and could not long hold his position. He retreated back across the Rhine during the fall.

In 1674, Turenne hastened to the Alsace, which was threatened by the substantially superior forces of Charles of Lorraine and the Austrian–Imperial general Alexandre Bournonville. Once again, Turenne crossed the Rhine, this time to take the initiative against Lorraine on June 16, 1674. Successful in this, he next captured Strasbourg on September 24, then eked out a narrow and costly victory over Bournonville at Enzheim on October 4, 1674.

Turenne remained on the move. He maneuvered skillfully in a flank march, regrouped and rallied his troops, then launched a highly unconventional surprise winter campaign against the Allied armies in December. He bested Bournonville at Mulhouse on December 29, then hit him again at Turckheim on January 5, 1675. This double blow smashed the opposing army, thereby saving Alsace.

Turenne checked Montecuccoli outside of Strasbourg and took the offensive against him at Sasbach (near Strasbourg) on July 27, 1675. Turenne was riding out ahead of his troops on a reconnaissance mission when he was struck by a cannonball and killed. It was an irreparable loss to the French army.

Turenne was a universal soldier: certainly the greatest tactician of his time and one of its best strategists. Moreover, he was personally courageous and an inspiring leader, who was greatly loved by his troops.

V

VANDEGRIFT, Alexander Archer

BORN: March 13, 1887

DIED: May 5, 1973

SERVICE: Veracruz Expedition (1914); Occupation of Haiti (1915–34); World War II (1941–45)

U.S. Marine Corps general famous for his fierce defense of Guadalcanal in World War II

Born in Charlottesville, Virginia, Vandegrift was educated at the University of Virginia, which he left after two years, in 1908, to enlist in the Marine Corps. Commissioned a second lieutenant on January 22, 1909, he was assigned primarily to duty posts in the Caribbean, most notably U.S. interventionist operations in Nicaragua during August–November 1912. He also participated in occupation of Veracruz, Mexico, in April, and the occupation of Haiti from July to November 1915. Vandegrift returned to Haiti as a Marine officer in the Haitian Constabulary, during 1916–18 and 1919–23.

From 1923 to 1926, Vandegrift served as a staff officer at Marine headquarters in Quantico, Virginia. He also attended field officer's school during this period, graduating in 1926. During 1927–28. he was dispatched to China as part of an expeditionary force under General Smedley D. Butler, then served in Washington, D.C. on the Federal Coordinating Service—essentially a procurement position—from 1928 to 1933, when he became personnel officer for East Coast Expeditionary Force commander General Charles Lyman. During this period, through 1935, Vandegrift was instrumental in composing the *Tentative Manual of Landing Operations*, which became the foundation for tactics used in amphibious assaults in the Pacific during World War II.

From 1935 to 1937, Vandegrift was back in China, where he was assigned to command the U.S. Marine guard at the Peking (Beijing) embassy. Returning to Washington in 1937, he served as secretary and, subsequently, assistant to Marine Corps commandant General Thomas Holcomb and was promoted (in 1940) to brigadier general.

Just before Pearl Harbor, Vandegrift was assigned to the First Marine Division at New River, North Carolina (November 1941), then assumed command of the division in March 1942. In June, he and his unit shipped out to the South Pacific, and he led the First in landings at Tulagi and Guadalcanal on August 7, 1942. These were the first U.S. amphibious assaults of the war.

Tulagi fell quickly, but Guadalcanal became a nightmare of unceasing Japanese resistance and spotty support. Vandegrift's Marines held Henderson Field and successfully repulsed repeated Japanese attacks from August to December, when the Second Marine and Twenty-Fifth U.S. Army Infantry Divisions came to their relief. Vandegrift and the First Marines regrouped and recuperated in Brisbane, Australia, then planned the invasion of Bougainville in November 1943. After this, Vandegrift prepared to return to the States as the new Marine Corps commandant, but he instead had to assume command of I Marine Corps when its commander, General Charles Barrett, died. He led I Corps in its landings at Empress Augusta Bay (Bougainville) on November 11.

In January 1944, Vandegrift did return to the United States and was appointed USMC commandant. In April 1945, he was promoted to general—the first active-duty marine to achieve that rank. By the end of the war, he had overseen the expansion of the USMC to half a million men, only to direct its demobilization after the war. He retired on December 31, 1947.

Vandegrift's tenacious direction of the Guadalcanal campaign, in the face of fierce resistance and under terrible conditions of climate and supply, became legendary and did much to create the mystique of the Marines in the Pacific.

VANDENBURG, Hoyt Sanford

BORN: January 24, 1899

DIED: April 2, 1954

SERVICE: World War II (1941–45); Korean War (1950–53)

One of the prime architects of the U.S. Air Force

Milwaukee-born Vandenberg graduated from West Point in 1923 and then attended flying school at Brooks and Kelly Fields in Texas. After earning his wings, he was assigned to the Third Attack Group in 1924. Three years later, he became a flight instructor at the Air Corps Primary Flying School, March Field, Riverside, California. Promoted to first lieutenant in August 1928, he joined the Sixth Pursuit Squadron at Wheeler Field, Hawaii in May of the following year. In November, he was given command of the squadron, then returned to the mainland in September 1931 as an instructor at Randolph Field, Texas. He attended the Air Corps Tactical School, graduating in 1935, then went on to the Command and General Staff School at Fort Leavenworth, from which he graduated in 1936. After graduation, he returned to the Tactical School as an instructor.

In 1939, Vandenberg graduated from the Army War College and took up an assignment at the war plans division of Air Corps Headquarters under General **Henry H. "Hap" Arnold** (the officer often called the "father" of the Air Force). Vandenberg was promoted to major in July 1941, lieutenant colonel in November, and colonel on January 27, 1942. In August, he became operations and training officer (A-3) in the Air Staff, as well as chief of staff of the Twelfth Air Force under General **James H. Doolittle.** In this position, he was instrumental in planning the U.S. air component of the invasion of North Africa during November 1942. Following this mission, he was promoted to brigadier general in December.

Vandenberg became chief of staff of the Northwest Africa Strategic Air Force in March 1943 and flew with his men many missions over Tunisia, Sicily, Sardinia, and southern Italy from March through August, when he was recalled to Washington as deputy chief of the Air Staff.

In September, Vandenburg was a member of delegation to the Soviet Union to arrange bases for shuttle-bombing missions against eastern European targets. After successfully negotiating the use of the bases, he returned to Air Corps Headquarters in Washington, where he formulated plans for the air component of the invasion of France.

Vandenburg was promoted to major general in March 1944 and was assigned to **Dwight D. Eisenhower**'s staff as deputy to Air Vice-Marshal Sir Trafford Leigh-Mallory, who was commander of the Allied Expeditionary Air Force. After the "D-Day" invasion of Normandy, Vandenburg was assigned to command the Ninth Air Force, which provided air support for General **Omar N. Bradley**'s Twelfth Army Group in July. Vandenburg delivered the required support and was instrumental in Bradley's advance across Europe. He was promoted to lieutenant general in March, then returned to Washington at the conclusion of the war in Europe.

Vandenburg became assistant chief of staff for operations for the Army Air Forces in July 1945 and, after victory in the Pacific ended World War II, he became chief of the army general staff's Intelligence Division in January 1946. In June, he was made director of the Central Intelligence Group—precursor to the CIA. He returned to military aviation in September 1947 as vice chief of staff of Air Force, which had just been created as an independent service arm. In July 1948, he succeeded General Carl Spaatz as Air Force chief of staff, and Vandenburg served during the early years of the Cold War. Most significantly, he held responsibility for overseeing the planning and execution of the Berlin Airlift from June 1948 to May 1949 and directed air operations during the Korean War, from June 1950 to

July 1953. Stricken with cancer, he retired from the Air Force in June 1953 and died a year later.

With "Hap" Arnold, Vandenburg was one of the builders of the U.S. Air Force. He was an extraordinarily efficient and perceptive officer, who executed the mission of ground support in World War II with brilliance. During the Cold War, he championed the development of the Air Force as the nation's chief deterrent to attack.

VAN FLEET, James Alward

BORN: March 19, 1892

DIED: September 23, 1992

SERVICE: World War I (1917–18); World War II (1941–45); Greek Civil War (1946–49); Korean War (1950–53)

American general; controversial and flamboyant combat commander

Born in Coytesville, New Jersey, Van Fleet was a graduate of West Point (1915) and was commissioned in the infantry. He served overseas in World War I, in the Sixth Division, as commander of the Seventeenth Machinegun Battalion (July 1918). Van Fleet fought in the Meuse-Argonne offensive during September 26–November 11 and was wounded a few days before the Armistice. He remained in Germany on occupation duty until June 1919, when he returned to the United States as an ROTC instructor at the university level during 1920–25.

From 1925 to 1927, he commanded a battalion of the Forty-Second Infantry in Panama, then returned to the States to teach at the Infantry School in Fort Benning, Georgia. He left this post in 1928 to take the advanced course at the Infantry School, which he completed in 1929. From 1929 to 1933, he taught ROTC at the University of Florida, and from 1933 to 1935 was stationed with the Fifth Infantry in Maine. Promoted to lieutenant colonel, he was sent to San Diego to instruct reservists in 1936.

Van Fleet served with the Twenty-Ninth Infantry at Fort Benning from 1939 until the outbreak of World War II in 1941, when he was given command of the Eighth Infantry, Fourth Infantry Division. He led this unit through July 1944, and landed with it on Utah Beach during the Normandy invasion on June 6, 1944. He performed brilliantly in the capture of Cherbourg during June 22–27, a campaign for which he was highly decorated and accorded a field promotion to brigadier general.

Second in command of the Second Infantry Division, he participated in the siege of Brest during August–September, then became commander of the Ninetieth Infantry Division under **George S. Patton** in the Third Army in October. Promoted to major general in November, he fought at Metz and then in the Battle of the Bulge from December 16, 1944 to January 15, 1945. Given command of XIII Corps in England during February–March 1945, he subsequently led III Corps, First Army in the breakout from the Remagen bridgehead and throughout the campaign into central Germany during March and May.

Van Fleet remained in command of III Corps during the occupation, then returned to the States with III Corps early in 1946. He was named chief of the Second Service Command headquartered in New York in February, then in June was appointed deputy commander of the First Army. Assigned as deputy chief of staff of the European Command in December 1947, he was promoted to lieutenant general and was tapped by Harry S. Truman to lead the military advisory mission to Greece and Turkey in February 1948. Van Fleet directed the training, organization, and tactical deployment of Greek government forces against communist insurgents during the Greek Civil War from 1946 to 1949. After the war, he remained in Greece and Turkey through 1951, when he was sent to Korea as commander of the Eighth Army, replacing, on April 11, 1951, General **Matthew B. Ridgway.**

Van Fleet was highly effective at checking the communist offensive in action from April 22 to May 20, and then he commenced a counteroffensive to retake ground lost by the U.N. However, he was frequently frustrated by U.S. insistence on limiting the war during truce negotiations. He was relieved late in the war, in February 1953, by General Maxwell D. Taylor, and he retired from the army the following month. However, he served as a civilian consultant on guerrilla warfare to

Secretary of Defense Robert S. McNamara during 1961–62.

VERCINGETORIX

DIED: ca. 45 B.C.
SERVICE: Caesar's Conquest of Gaul (58–51)

***Gallic chieftain famed for rebelling against* Caesar**

History records few facts about Vercingetorix's origins other than that he was the son of a king of the Arverni tribe in central Gaul. Early in the winter of 53 B.C., the Roman army was spread throughout northern Gaul, and Julius Caesar himself was in Italy. In a plan to confront the Roman troops, Vercingetorix rallied followers to his cause at Cenabum (modern Orléans) and gathered a large army. The chieftain trained his army in the tactics of the Roman Legion, so that, when Caesar returned to Gaul from Italy, he found himself confronted by a very formidable force. Caesar's first task was to break through Gallic opposition in order to join the main body of his legions in the north. Break through, he did, and, with his united forces, retook Cenabum.

Vercingetorix's defeat made him rethink his tactics. Soon he abandoned the Roman battle procedures he had instilled in his troops for the tactics of guerilla warfare, His methods of harassment and an unsparing scorched earth policy worked hardship not only on Roman legions but also on the Gallic people themselves.

Despite Vercingetorix's new measures of resistance, Caesar laid siege to the rebel stronghold of Avaricum (modern Bourges) beginning in February 52. Vercingetorix unsuccessfully attempted to relieve the fortress, but it fell to Caesar's forces in March. The Roman leader, however, faced stronger opposition when he led a seige against the Arverni captial of Gergovia. The people of Gergovia repelled the seige, and Rome's legions could not long support themselves in a country ravaged by Vercingetorix's scorched earth measures. During April and May of 52 B.C., Caesar withdrew his forces from central Gaul.

Encouraged by his triumph, Vercingetorix once again assembled a large army, numbering some 80,000 infantrymen and 15,000 cavalry troops, in an attempt to keep the legions from leaving the valley of the Saone and entering the valley of the Seine. Roman cavalry, however, handily defeated Vercingetorix's mounted troops at the Vingeanne River, Vercingetorix pulled back rather than engage in a major battle. In July, he was trapped by Roman forces in Alesia, a commanding hilltop fortress. Caesar led 50,000 troops in siege, forcing the Gallic leader and his army to take refuge within the walls of Alesia. As Caesar settled in for a protracted siege, Vercingetorix appealed to allied tribes for support. They obliged, as best they could. The reinforcements, however, lacked strong leadership, and Caesar was able to drive off the relief forces.

Throughout the summer and into early autumn, Caesar patiently maintained the siege, denying Vercingetorix food and supplies. Each attempt to break out failed, and each successive attempt was weaker than the previous one. His troops facing certain starvation, Vercingetorix surrendered and was taken into capitivity, in which he was held for more than a decade before he was at last put to death about 45 B.C.

VESPASIAN (Titus Flavius Vespasianus)

BORN: November 17 or 18, A.D. 9
DIED: June 4, 79
SERVICE: Conquest of Britain (43–60); Judean Revolt (66–70); Year of the Four Emperors (68–69)

Emperor of Rome who brought a degree of peace and prosperity to a war-weary empire

Vespasian was born near Reate (modern Rieti), Italy. He embarked on a military career at a young age, serving first in Thrace, then as quaestor in Crete and Cyrenaica. Popular and able, he became aedile and then praetor. He served with distinction in the German provinces during the 30s, then commanded a legion in Britain under Aulus Plautius. During 43–44, he conquered the Isle of Wight and advanced with his Second

Legion to the region of modern-day Somersetshire. After serving a brief term as consul in 51, he became governor of the African provinces in 63. Vespasian accompanied **Nero** on his Greek tour, then, late in 66, was dispatched to Palestine to put down the Jewish rebellion in Judaea. There he confronted determined, even fanatical resistance, but by 69 had opened the siege of Jerusalem.

While Vespasian was occupied in the Middle East, Nero died by his own hand in 68, and Servius Sulpicius Galba was proclaimed emperor and recognized as such by most of the empire. The Praetorian Guard, always a volatile force, turned on Servius, however, and murdered him in order to bring to the throne early in 69 their favored candidate, Aulus Vitellius, who had served as commander of Roman forces in Germany. In July 69, the army in Egypt and his own forces in Judaea proclaimed Vespasian emperor. Vitellius, in Rome, had the support of the German legions, but Vespasian garnered the backing not only of the Middle Eastern troops, but those in the Balkans and in Illyria. With loyalties thus aligned, Vespasian entrusted command of the Middle Eastern forces to his son Titus and made his way back to Italy while another ally, Antonius Primus, used the legions of the Danube region to invade Italy and engage Vitellius, whom he defeated at the second Battle of Bedriacum in October 69. When Vespasian arrived in Rome in December 69, his rival emperor was dead and the city itself was held by Primus's troops. Thus ended the "Year of the Four Emperors."

Vespasian assumed the throne just as his son Titus effected the capture of Jerusalem in 70. This same year, Vespasian's forces put down the Rhineland revolt of Claudius Civilis. With these matters successfully resolved, Vespasian turned to a program of sorely needed domestic reform. Nero's extravagance had created a dangerous treasury deficit, which Vespasian quickly reduced and eliminated through a combination of economy measures and increased taxation. He broadened Rome's tax base in part by granting Roman citizenship to selected towns and provinces. After annexing portions of Anatolia and Germany, he reinforced garrisons in Great Britan, especially Wales and Scotland. Although he practiced prudent economies, Vespasian embarked on an ambitious program of construction, the grandest product of which was the Roman Colosseum.

Vespasian reintroduced Rome to something it had lacked for many years: good government conducted for the public benefit. Concerned to bring lasting stability to Rome, Vespasian established the Flavian dynasty by naming his son Titus as his successor. He gave him the strategic appointment of prefect of the notoriously dangerous Praetorian Guard. The strategy worked. When Vespasian died—of natural causes—on June 4, 79, having brought prosperity and peace to the empire, Titus succeeded him with the full support of the praetorians and the acclamation of the Senate. Titus and, after him, Domitian had learned well from their father. Like him, they ruled wisely, honestly, and equitably.

VICTOR-PERRIN, Claude, Duke of Belluno

BORN: December 7, 1764

DIED: March 1, 1841

SERVICE: French Revolutionary Wars (1792–99); Napoleonic Wars (1800–15)

French Marshal of the Empire; one of* Napoleon*'s leading lieutenants

Victor-Perrin was born in Lamarche, Lorraine, and was the son of a notary. He was a teenager when he joined the artillery as a drummer boy in 1781. Working his way through the enlisted ranks, he then purchased his discharge in order to join the National Guard as an elite grenadier in 1791. With the outbreak of the revolution, he joined the volunteers and rose meteorically, achieving election as lieutenant colonel in the Second Battalion of the Bouches-du-Rhône Volunteers on September 15, 1792. With this unit, he joined the Army of Italy, but was soon severely wounded in the December 17 attack on the fort of "Little Gibraltar" during the siege of Toulon.

Promoted to provisional general of brigade on December 20, while he recovered, Victor-Perrin transferred to the Army of the East Pyrenees during 1794–95, but returned to the Army of Italy in 1796 and fought at Dego during April 14–15, 1796. He then fought at Borghetto (May 30), Lonato (August 2–5), Roveredo (September 4), and San Giorgio (September 15). He was again wounded in this last engagement.

Promoted to provisional general of division, Victor-Perrin distinguished himself at Rivoli on January 14, 1797, then at La Favorita (January 16) and at Ancona (February 9). Following these engagements **Napoleon** Bonaparte confirmed his promotion to general of division and entrusted him with command of the Twelfth Division on March 10. He fought on in Italy, where he was wounded for a third time, at the Trebbia during June 17–19, 1799. The following year, Victor–Perrin participated in the battles of Montebello (June 9, 1800) and Marengo (June 14). He was then assigned to command the French occupation forces in the Netherlands during 1800–04.

Napoleon appointed Victor-Perrin captain-general of Louisiana in 1803, but the emperor sold the territory to the United States before Victor-Perrin assumed the position. Instead, he was posted to Denmark as Napoleon's minister there in 1805.

On October 10, 1806, Victor-Perrin was appointed chief of staff under Marshal **Jean Lannes** at the Battle of Saalfeld. He also served under Lannes at Jena on October 14, at Spandau on October 25, and at Pułtusk on December 26. The following January, he received command of X Corps, which he led against the Prussians at Stettin and was captured. He was exchanged the following month and replaced the wounded General Jean Baptiste Bernadotte as commander of I Corps on June 5. In this command, he performed brilliantly at Friedland on June 14 and was created a Marshal of the Empire on July 13. He was also appointed governor of Prussia and Berlin. In September 1808, he was additionally created Duke of Belluno.

During 1808, Victor-Perrin was sent to Spain with I Corps and there defeated the British at Espinosa during November 10–11. On November 30, he fought at the Somosierra Pass, then devastated the Spanish forces at Medellin on March 29, 1809. Coming up against Arthur Wellesley (subsequently the Duke of **Wellington**) at Talavera on July 28, however, he was badly defeated. And he was bested once again, by Sir Thomas Graham, at Barrosa on March 5, 1811.

Victor-Perrin laid a very long and ultimately unsuccessful siege against Cadiz from February 5, 1810 to August 24, 1812, when he was recalled to France and assigned to command IX Corps, which was in Germany preparing to invade Russia. Victor-Perrin distinguished himself most during the Russian campaign in the retreat from Moscow, when he was in command of the rearguard at the crossing of the Berezina River during November 27-29. The following year, he was back in Germany, in command of II Corps, and performed with distinction at Dresden on August 26–27 and at Leipzig during October 16–19. Next, during the campaign for France, he fought bravely at Erienne on January 29, 1814 and at La Rothiere on February 1, but his troops, exhausted from ceaseless battle, failed to take and hold the bridge at Montereau on February 17. Napoleon, in consternation, summarily relieved him from command—only to restore him to command of a pair of divisions of the Imperial Guard. He these until he was gravely wounded at Craonne on March 7.

After Napoleon's first exile and during the Hundred Days, Victor-Perrin remained loyal to the Bourbons. He was created a peer of France and appointed major general of the Royal Guard in 1815. He was among the marshals who consented to sit on the court-martial trial of fellow marshal **Michel Ney.** He voted for execution.

Appointed governor of the Sixteenth Military Division in 1816, Victor-Perrin became minister for war and commandant of the Sixteenth, Seventeenth, and Nineteenth Military Divisions in 1821. He was further appointed minister of State and elevated to membership in the Privy Council in 1828, when he was also appointed to the Superior Council of War. However, following the Revolution of 1830, he fell from grace with the Bourbons.

Victor-Perrin was one of Napoleon's more dependable marshals. Brave and inspiring as a leader, he was a capable tactician. As a strategist, he left much to be desired, as was evident in his Spanish tenure.

WAINWRIGHT, Jonathan Mayhew IV

BORN: August 23, 1883

DIED: September 2, 1953

SERVICE: World War I (1917–18); World War II (1941–45)

U.S. general best known for his valiant—but hopeless—defense of the Philippines at the start of World War II

A native of Walla Walla, Washington, Wainwright graduated from West Point and became a second lieutenant in the cavalry in 1906. He saw service with the First Cavalry in Texas, and went with the First to the Philippines. There he fought against Moro pirates on Job Island during 1908–10. Promoted to first lieutenant in 1912, he graduated from the Mounted Service School four years later and was also promoted to captain.

Upon U.S. entry into World War I, Wainwright was promoted to temporary major of field artillery, and served first as an instructor at the officers' training camp in Plattsburgh, New York during 1917. He shipped out for France with the Seventy-Sixth Division in February 1918 and served on detached service with the British near Ypres, Belgium. He was then posted as assistant chief of staff for operations (G-3) in the U.S. Eighty-Second Division and served with the unit in the Saint-Mihiel offensive of September 12–16 and at Meuse-Argonne during September 26–November 11.

After the Armistice, Wainwright remained in Germany with Third Army on occupation duty until October 1920, when he returned to the United States and reverted to his permanent rank of captain. However, Wainwright was quickly promoted to major and assigned as an instructor at the Cavalry School, Fort Riley, Kansas, where he served until 1921, when he became a general staff officer with Third Infantry Division.

From 1921 to 1923, Wainwright served in the War Department, then was assigned to the Third Cavalry until 1925, when he returned to the War Department. He was promoted to lieutenant colonel in 1929 and, in 1931, graduated from the Command and General Staff School at Fort Leavenworth. Next, he graduated from the Army War College in 1934 and the following year was promoted to colonel while serving as commandant of the Cavalry School. He left the school in 1936 to assume command of the Third Cavalry. In 1938, promoted to the temporary rank of brigadier general, he was assigned to command the First Cavalry Brigade.

In September 1940, Wainwright was promoted to major general and sent to the Philippines to command the Philippine Division. He served as senior field commander under General **Douglas MacArthur.**

When the Japanese attacked and invaded the Philippines at the start of World War II in the Pacific, the brunt of the defense fell on Wainwright. There was no hope of reinforcement and no hope of resupply. His mission, as commander of the Northern Luzon Force (Eleventh, Twenty-First, Seventy-First, and Ninety-First Filipino Divisions, and the U.S. Twenty-Sixth Cavalry Regiment), was to delay the Japanese, who had landed at Lingayen Gulf during December 22–31, so that the American and Filipino forces could fall back to Bataan and take a stand there. The defense of northern Bataan fell back under a first assault during January 10–25, and Wainwright and MacArthur repulsed a second assault during January 26–February 23.

Wainwright was promoted to lieutenant general and made commander of U.S. Forces in the Philippines after President Franklin Roosevelt ordered MacArthur to flee to Australia in March 1942. Remarkably,

Wainwright conducted his defense through early April 1942. His obstinate and skilled action had cost the Japanese dearly, but Wainwright and his men also paid a terrible price. The U.S.–Filipino forces on Bataan surrendered on April 9, and a massive Japanese assault on Corregidor forced Wainwright to surrender all forces on May 6. The general and his men were sent to POW camps in the Philippines, Taiwan, and finally Manchuria, and were treated with terrible inhumanity by their Japanese captors. Wainwright did much to sustain the morale and honor not only of himself, but of his imprisoned command.

Whereas a Japanese commander in Wainwright's situation would have been disgraced by surrender, Wainwright, after his liberation by Russian troops in Manchuria in August 1945, was hailed as a war hero. The emaciated Wainwright was accorded the honor of attending the Japanese surrender ceremonies in Tokyo Bay aboard U.S.S. *Missouri* on September 2. He commented with characteristic reserve, good humor, and self-irony: "The last surrender I attended the shoe was on the other foot."

Wainwright was awarded the Medal of Honor and assigned to command the Fourth Army in Texas in January 1946, but retired the following year.

Wainwright was not only a fine commander, whose resourcefulness, skill, and courage made the Japanese victory in the Philippines extremely costly, his example served as a symbol of the gallantry of which U.S. forces were capable.

WALKER, William

BORN: May 8, 1824

DIED: 1860

SERVICE: Invasion of Mexico (1853); Conquest of Nicaragua (1855–56)

American adventurer and filibuster in Mexico and Nicaragua

Called in the popular press the "gray-eyed man of destiny," William Walker was an example of a political-military adventurer or "filibuster." He was born into a wealthy Nashville, Tennessee, family. His father was a prominent insurance company executive, and his mother was a recently converted member of the newly formed denomination called the Church of Christ. At the age of fourteen, Walker became one of the youngest men ever to graduate from the University of Nashville. His parents wanted him to pursue a career in the ministry of the Church of Christ, but an interest in medicine and a friendship with a local doctor convinced the young man that his future lay in Nashville as a physician. In 1843, after graduating from the University of Pennsylvania Medical School, he returned to his hometown, one of the youngest qualified medical doctors in the United States.

Walker quickly became bored with his fledgling medical practice, and sailed overseas to enroll in the University of Edinburgh, following this with a tour across much of western Europe. He then returned to Nashville, but was soon off again, this time to New Orleans, where he embarked on a new profession, law, gaining admission to the Louisiana bar after a brief period of study. Next, in 1848, he embraced a third occupation, journalism, and subsequently moved to California, where he became the editor of a San Francisco newspaper.

It was during his California days that the restless Walker became obsessed with the idea of "liberating" the Mexican state of Sonora. Accordingly, in November 1853, he assembled a rag-tag army of forty-five like-minded men and, quite simply, invaded Mexico. Within six months he had established his own "republic" of Sonora and engineered his election as president, only to surrender to American authorities at the international border. Walker was tried for violation of U.S. neutrality laws, but the public was sympathetic to his filibustering, and the jury handed up a "not guilty" verdict. Apparently feeling vindicated, Walker formulated a new plan: the invasion of Nicaragua.

At the time, that nation was indeed ripe for revolution. Civil unrest and a heedless power struggle among several leaders was tearing the country apart. Quick to take advantage of the situation, Walker obtained a contract from the *de facto* government of Nicaragua in 1854, allowing him to bring some three hundred colonists to settle a land grant of 50,000 acres. In return, Walker and his American colonists would be liable for military service, for which they would receive monthly compensation. To avoid U.S. prosecution, Walker legitimated his operation by submitting the entire scheme to review by the U. S. Attorney at San Francisco and by the commander of the Pacific Division of the U.S. Army, neither of whom could find anything particularly illegal in the paperwork presented.

Walker entered Nicaragua with considerably fewer than the three hundred colonists he had intended to bring. Only fifty-six men arrived with him at Realejo on June 1, 1855, and they were all immediately absorbed into the Nicaraguan army. After Walker led his small contingent to several victories against anti-government insurgents, the gray-eyed man of destiny was proclaimed a national hero and, incredibly enough, elected to the presidency of Nicaragua in June 1856. Conflicts with American business leaders who were interested in building a canal across his country, as well as difficulties with the U.S. government over neutrality laws combined with the unceasing internal strife in Nicaragua to prompt Walker to resign the presidency in May 1857.

Walker attempted to recapture Nicaragua several times during the late 1850s, but in 1860, while he was trying one more time, Walker was captured by British authorities, who turned him over to the Honduran army. The Hondurans executed him at Trujillo, Honduras. He was thirty-six years old.

WALLENSTEIN, Albert Eusebius von, Duke of Friedland and Mecklenburg, Prince of Sagan

BORN: September 24, 1583

DIED: February 25, 1634

SERVICE: Austro–Turkish War (1593–1606); War with Venice (1615–17); Siege of Gradisca (1617–19); Thirty Years' War (1618–48)

Bohemian–Austrian commanding general of the Holy Roman armies

Wallenstein was born in Hermanice, Bohemia, and was educated by Jesuits after he was orphaned. He also enrolled in the University of Altdorl, but was expelled because of bad conduct in 1599. He then studied at Bologna and Padua, and took in much of southern and western Europe. Finally, his drifting ended as he joined the army of Emperor Rudolf II of Hungary. In his first engagement, the siege of Gran, during September 19–October 10, 1604, he gained a degree of fame.

Wallenstein returned to Bohemia and married a wealthy widow, whose Moravian estates he inherited after her death in 1614. He used a portion of his wealth to raise a cavalry force of two hundred men for the siege of Gradisca in the War with Venice during 1615–17, then, at the outbreak of the Thirty Years' War, he turned down offers of command from the Bohemian rebels against the Holy Roman Empire and went to Vienna, where he funded the creation of a cuirassier regiment under his command.

Leading his cuirassiers, he defeated the anti-Hapsburg rebel Gabriel Bethlén and profited by the acquisition of even vaster estates in Moravia as well as land in Bohemia after the Protestant defeat at White Mountain on November 8, 1620. Wallenstein became an Imperial count palatine in 1622, then a prince in 1623, and wed another woman of wealth, Isabella Katharina Harrach, in 1623.

During 1623–24, he continued to wage war against against Bethlén in Moravia and was created Duke of Friedland in 1624. As the Danes were about to intervene in the Thirty Years' War, Hapsburg emperor Ferdinand II asked Wallenstein to raise an army in April 1625, and Wallenstein took his new force into the field, where he defeated Ernst von Mansfeld—a Catholic mercenary in the Protestant service—at Dessau Bridge on April 25. Following this triumph, Wallenstein marched north to join forces with **Johan Tilly** to drive the Danish king Christian IV into Jutland, thereby forcing the withdrawal of Denmark from the war in 1627.

For his triumphs over Denmark, Wallenstein was made Duke—and, later, Prince—of Sagan, and he was additionally rewarded with the duchy of Mecklenburg in 1628. But Wallenstein's effort to pacify northern Germany stumbled over resistance from the Baltic city of Stralsund during February 24–August 4, 1629. Nevertheless, Wallenstein's many gains and honors had made him practically independent. He could afford to support his army on his own. He remained nominally loyal to the Holy Roman Empire, but, in fact, he obeyed only those Imperial orders that suited him. At length, Ferdinand dismissed his general on August 1630, and Wallenstein retired to his duchy of Friedland.

It proved a brief retirement. When Sweden's **Gustav II Adolph** invaded Germany and scored a victory over the great Tilly at the first Battle of Breitenfeld on September 17, 1631, Ferdinand recalled his erstwhile commander. Wallenstein swung into action, pushing the Saxons out of Bohemia and checking the advance of the Swedish army as it neared Nurnberg Johan during July–August 1632. Next, he repulsed Gustav II Adolph's attack on his encampment at Alte Veste during September 3–4, 1632.

Late in September, Wallenstein advanced northward to invade Saxony, but Gustavus met him in battle at Lützen. Although Gustavus died in combat, Wallenstein was forced to retreat on November 16, 1632 and settled into winter quarters in Bohemia. Come spring, he campaigned in desultory fashion even as he prepared to defect from the service of the Holy Roman Empire and foist a negotiated peace on Ferdinand. The emperor checkmated Wallenstein, however, by plotting his removal from command on January 24, 1634. Next, he charged him with high treason on February 18. In response, Wallenstein fled from Pilsen to Eger, where he intended to go over to the Swedish side. However, officers at Eger who were loyal to Ferdinand, assassinated Wallenstein, and thus ignominiously died one of the two or three great generals to emerge form the Thirty Years' War.

WASHINGTON, George

BORN: February 22, 1732
DIED: December 14, 1799
SERVICE: French and Indian War (1754–63); American Revolution (1775–83)

American general who commanded the Continental Army during the Revolution; served as the first U.S. president

Washington was born in Westmoreland County, Virginia, and, following the death of his father on April 12, 1743, he was raised by his eldest brother, Lawrence. Washington had an indifferent education, but learned the surveyor's trade and became an assistant surveyor in 1748. From 1749 to 1751, he was surveyor for Culpeper County.

With the death of his brother Lawrence in July 1752—and that of Lawrence's daughter Sarah two months later—Washington came into a substantial inheritance, the crown jewel of which was Mount Vernon, a fine Virginia plantation estate. Now a solid property holder, Washington was appointed adjutant for southern Virginia and given the militia rank of major in November 1752. He was additionally named adjutant of Northern Neck—between the Rappahannock and the Potomac rivers—and the Eastern Shore in 1753.

On the eve of the French and Indian war, Virginia governor Robert Dinwiddie sent Washington to assess French military activity in the Ohio Valley and, indeed, to warn the French against trespassing on British colonial ground. The expedition, which spanned October 31, 1753 to January 16, 1754, pressed from coastal Williamsburg, Virginia, through the wilderness to Fort Duquesne, on the site of modern Pittsburgh,

Pennsylvania). Washington emerged victorious in an initial skirmish with a small French party, but French troops out of Fort Duquesne drove him back to a hastily constructed stockade Washington had dubbed Fort Necessity. After almost half of his men were killed, Washington surrendered to the French on July 3, 1754.

Washington and his men were well treated by the French, who let them return to Virginia, where, despite his defeat, Washington was greeted as a hero. He was given the dubious assignment of aide-de-camp to British general **Edward Braddock,** and he was promoted to the local rank of colonel in March 1755.

Washington and his Virginia volunteers marched with Braddock on a full-scale mission to retake Fort Duquesne during May–July. On July 9, 1755, just outside of the fort, Braddock and his troops, including Washington, were set upon by a combined French and Indian ambush. The British regulars panicked, but Washington's Virginians fought well. Nevertheless, all were overwhelmed, and the battle resulted in extremely heavy casualties for the colonials. Braddock himself was slain in the battle.

The defeat outside of Fort Duquesne was recognized as Braddock's fault, not Washington's. The young militia colonel was greeted as a hero and given command of all Virginia forces in August. He set about supervising the construction of a series of frontier forts during 1755–57.

During July–November 1758, Washington accompanied another British general, John Forbes—considerably more able than Braddock—in an arduous but successful expedition to Fort Duquesne. Forbes retook the fort for the British—albeit it lay in ruins, the French having blown part of it up after withdrawing from it.

Washington was elected to the House of Burgesses in 1758. Upon his resignation from the militia, he was promoted to honorary brigadier general. During this period, on January 6, 1759, Washington married Martha Dandridge Custis, a wealthy widow. He served as justice of the peace for Fairfax County for many years—from 1760 to 1774—and prospered as a plantation owner.

As the clouds of Revolution gathered, Washington became a key member of the first Virginia provincial congress in August 1774. He was chosen as one of seven Virginia delegates to the First Continental Congress in September 1774 and served as a member of the Second Congress in May 1775. When war broke out in earnest, the Congress asked Washington to serve as general-in-chief of the Continental forces. He accepted the challenge.

On June 15, 1775, he directed the blockade of Boston, then formally took command of the army at Cambridge, Massachusetts on July 3, 1775. The army was poorly disciplined, poorly equipped, and disorganized to the point of disintegration. Washington reshaped it and, through his inspiring leadership, injected it with patriotic fervor.

Washington scored an important victory at Boston when he successfully fought to gain Dorchester Heights on March 4, 1776, so that he could position cannon there in order to fire against the British. This maneuver forced the British commander, **Richard Howe,** to withdraw from Boston. Far less successful was Washington's defense of New York City during August–November 1776. Even in losing New York, however, Washington proved himself a capable commander, drawing the British far from supply as he withdrew across the Hudson and into New Jersey and, in November, Pennsylvania. Suddenly, on December 25–26, 1776, he recrossed the Delaware and caught unawares the British-employed Hessian troops at Trenton, New Jersey.

This signal victory not only cost the British substantially, but greatly elevated the morale of the Continental Army and others involved in the rebellion. Washington next emerged victorious at the Battle of Princeton, where he managed to rout three British regiments on January 3, 1777 before retiring into winter quarters at Morristown, New Jersey, on January 6.

Washington soon learned that British general **John Burgoyne,** planning to cut off New England from the other colonies, intended a large-scale invasion of the Hudson Valley from Canada. Washington saw the principal weakness of Burgoyne's plan as an absence of naval support. Accordingly, he dispatched many of his best troops and ablest officers north to check Burgoyne while he remained in New Jersey to engage Howe. Washington bargained for time, parrying through May and July 1777 Howe's attempts to force a battle. When a fight finally came at Brandywine Creek in Pennsylvania on September 11, 1777, Washington was defeated, but withdrew intact and regrouped for a surprise counterattack at Germantown on October 4. Hobbled by bad weather and a plan too complex for relatively inexperienced soldiers, the Germantown campaign failed. However, his subordinates fared far better in the north, with General Burgoyne suffering defeat at the hands of General **Horatio Gates** at the Battle of Saratoga on October 17.

The forces gathered around Washington were demoralized as they went into winter quarters at Valley Forge, Pennsylvania. The winter was harsh—though not unusually so—but the ragged, threadbare condition of the troops' clothing and their meager supplies brought great hardship. It is a measure of Washington's character and ability to lead that he managed to hold the Continental army together during November 1777–April 1778. During this period, he was faced with a crisis of leadership, the so-called Conway Cabal, a scheme to replace him in command with General Gates. Washington successfully suppressed the movement during November–December 1777.

Washington engaged General Sir Henry Clinton in a battle at Monmouth, New Jersey, that resulted in a costly draw on June 28, 1778. Gates, now in command of southern forces, was badly beaten at Camden, South Carolina, on August 16, 1780, and Congress appealed to Washington to save the South. On October 22, 1780, he appointed General **Nathanael Greene** to command in the region, and the Americans enjoyed greater success. In the meantime, the French, implacable enemies of Great Britain, had joined forces with the Americans, and Washington worked with French general **Jean Baptiste Rochamheau** to plan the Yorktown (Virginia) campaign executed during August 21–September 26. The campaign succeeded brilliantly, forcing the surrender of General **Charles Cornwallis** and his army on October 19, 1781.

Although major cities and geographical areas remained under British control, the defeat of Cornwallis meant that the Revolution was drawing to a successful conclusion. In November 1781, Washington returned north to reinstitute the blockade of British-held New York and continued to prosecute the war elsewhere. The British did not evacuate the city of New York until November 25, 1783, well after the definitive treaties of the Peace of Paris had been concluded on September 3, 1783 (preliminary accords had been signed the previous year). In an eloquent and touchingly simple ceremony, Washington took leave of his troops at Fraunces Tavern on December 4, then returned to Mount Vernon and resigned his commission.

That Washington made no move to become a revolution-borne dictator attested to the nobility of his character and his commitment to the ideals of democracy. He actively worked to unite the disparate colonies into a single government, and he was elected president of the Constitutional Convention in Philadelphia in 1787. In February 1789, he was unanimously elected as the first president of the United States and was inaugurated on April 30. He was reelected to a second term in December 1792, but declined a third, thereby establishing a tradition of a two-term limit that was unbroken until the four-term tenure of Franklin D. Roosevelt in the mid-twentieth century. (A constitutional amendment now limits presidential tenure to two terms.)

The two-term tradition was hardly the only precedent Washington set. In addition to guiding the nation successfully through its first eight years, he essentially created everything that the office of president would be, establishing a high and highly democratic standard.

Washington finally retired to his beloved Mount Vernon in March 1797, although he briefly served as commander in chief during a crisis with France, which looked as if it would flare into war in 1798. In December 1799, Washington took ill with severe laryngitis and died.

Washington combined qualities of character and natural military ability to prevail against great odds during the American Revolution. More than on any other single figure, the success of the Revolution, both militarily and politically, depended on him. He was a selfless and rational leader of the highest moral principles, who rejected attempts to make of him a military dictator.

WAYNE, Anthony

BORN: January 1, 1745

DIED: December 15, 1796

SERVICE: American Revolution (1775–83); War with the Indians of the Old Northwest (Little Turtle's War) (1790–94)

An important American general in the Revolution, "Mad Anthony" Wayne brought relative peace to the Old Northwest after the war

Wayne was born in Waynesboro, Chester County, Pennsylvania and made his early living as a surveyor and land agent in Nova Scotia, beginning in 1765. He returned to Pennsylvania in 1767 to take over his father's prosperous tannery, which he ran until 1774, when he was elected to the Pennsylvania legislature and served on the proto-revolutionary Committee of Public Safety. In September 1775, he resigned his legislature seat to raise a regiment of volunteers and, on January 3, 1776, he was formally commissioned colonel of the Fourth Pennsylvania Battalion.

"Mad Anthony" Wayne. *Courtesy Library of Congress.*

Wayne participated in the unsuccessful invasion of Canada early in the Revolution and was wounded at Trois Rivières on June 8, 1776. Assigned to command Fort Ticonderoga, he was promoted to brigadier general on February 21, 1777, then commanded the Pennsylvania Line at Brandywine on September 11.

Wayne suffered a bad defeat during a surprise night attack on his encampment at Paoli, Pennsylvania on September 21. This misfortune cast him under a cloud, and he requested a court-martial on the matter, which acquitted him of wrongdoing. Putting this defeat behind him, Wayne fought bravely at the Battle of Germantown on October 4, sustaining a minor wound in this action. He wintered with the army at Valley Forge during 1777–78, then participated in the triumph at Monmouth on June 28, 1778, leading the initial attack and successfully defending the center against the British counterattack.

But it was for the daring and spectacular night attack on Stony Point, New York, on July 16, 1779, that earned Wayne his early reputation. His famous sobriquet—"Mad Anthony"—was born at about this time as well, though it had little to do with the action at Stony Point. A neighbor of Wayne's deserted from the Continental Army. Arrested, he told the authorities to contact Wayne, who would vouch for him. Not only did General Wayne refuse to help the deserter, he denied knowing him at all. "He must be mad," the deserter responded incredulously. The epithet stuck.

Wayne was commanding operations along the lower Hudson River in 1780 when he suffered defeat at Bull's Ferry during July 20–21. Early the following year, he responded immediately to news of Benedict Arnold's treachery by rushing to defend imperiled West Point. Through force of personality, Wayne succeeded in suppressing a mutiny in the Pennsylvania Line during January 1–10, 1781, but he was less successful later that year in Virginia, when he was defeated at Green Spring on July 6. He did, however, manage to save his army from annihilation by the greatly superior British force.

Wayne participated in the victory at Yorktown during May–October 17, 1781, then, fighting under General **Nathanael Greene,** he led an effective expedition into Georgia against British-allied Creek and Cherokee Indians during January–July 1782.

Wayne was breveted to the rank of major general on September 30, 1783, but retired from the army at the end of the Revolution to take up a quiet life as a farmer. He was active in politics as Chester County's representative to the Pennsylvania General Assembly during 1784–85, then was elected by Georgians—familiar with exploits against the Creeks and Cherokees—representative to Congress on March 4, 1791. He did not serve, however, because his seat was declared vacant on March 21 on the grounds of election fraud as well as failure to meet residency requirements.

In any case, Wayne's life as a farmer did not long endure. On March 5, 1792, he was appointed major general and commander in chief of the army in the Northwest (essentially Ohio and the upper Midwest) and assigned to fight the Indians of that region. The Shawnee chiefs Little Turtle and Blue Jacket had dealt severe defeats to two previous commanders, Joshua Harmar and Arthur St. Clair—inexperienced commanders who led poorly trained units—and Wayne was determined not to let this happen again. He patiently and thoroughly prepared his troops, maneuvering them into an advantageous position. On August 20, 1794, he met Little Turtle at the Battle of Fallen Timbers and soundly defeated him. The result was the Treaty of Greenville, by which the Shawnee ceded vast tracts of land to the government, thereby opening most of the Ohio country to settlement and bringing to the Old Northwest some fourteen years of relative calm.

As a result of Wayne's victory over the Indians, British traders in the Ohio region lost their allies and were compelled to surrender to Wayne the British forts on the Great Lakes during 1796. Unfortunately, the general did not have long to bask in his triumphs. He succumbed suddenly to illness at the end of 1796.

Although his famous sobriquet was not born in battle, it aptly described his ferocity in combat, which was a subject of awe and wonder to his men. But Wayne was not an impetuous commander. As his campaign against Little Turtle demonstrated, he was a careful and patient planner, one of the U.S. Army's first thoroughly professional officers.

WELLINGTON, Arthur Wellesley, First Duke of

BORN: May 1(?), 1769

DIED: September 14, 1852

SERVICE: Fourth Mysore War (1799); Napoleonic Wars (1800–15); Second Maratha War (1803–05)

"The Iron Duke"; British field marshal who defeated* Napoleon *at Waterloo, thereby ending the Napoleonic Wars

Wellington was born Arthur Wellesley in Dublin, Ireland, the fifth son of the First Earl Mornington. He was sent off to Eton for an education he valued all of his life. (He once declared that the Battle of Waterloo had been won "on the playing fields of Eton.") Wellington also briefly attended a French military school at Angers.

On March 7, 1787, he was commissioned an ensign in the Sevety-Third Regiment and, on September 30, 1793, purchased the rank of lieutenant colonel in command of the Thirty-Third Regiment of Foot. During this period, he also served in the Irish Parliament for the family borough of Trim.

During 1793–95, Wellesley led the Thirty-Third in the Netherlands campaign of the Duke of York and enjoyed great success, distinguishing himself and his unit at Hondschoote on September 8, 1793. He was in charge of the rearguard brigade during the campaign in Flanders and the late fall withdrawal toward Hanover. (late autumn 1794). Wellington was exposed to the blunders of his superiors in Flanders and resigned from the army in some disgust. He was appointed military undersecretary to the Lord Lieutenant of Ireland, Lord Camden, in May 1795. However, he resumed command of his regiment in 1796, taking it to India, where his elder brother, Richard Wellesley, Second Earl Mornington, had been appointed governor general.

In India, Wellesley quickly made a name for himself as commander of one of two columns sent against Tipu Sahib, Sultan of Mysore. He participated in the storming of Seringapatam, where Tipu was killed in May 1799, and, after this, Wellesley was appointed governor of Seringapatam. Next, during the Second Maratha War, he led British and allied forces in the Deccan campaign of 1803, in which he was responsible for the capture of Poona on March 20, 1803, and Ahmadnagar on August 11. Engaging the Maratha army led by Daulat Bao Sindhia and the Raja of Besar at Assaye, he emerged victorious after a very bloody battle on September 23. He bested Sindhia again at Argaon on November 29, then stormed Gawilgarh on December 15. This brought the war to a successful conclusion and made a hero of Wellesley.

Wellesley returned to England in September 1805 and plunged into the Napoleonic Wars as brigade commander in the abortive Hanover expedition that December. Wellesley then withdrew from the military for a time, gaining election to Parliament as member for Rye in April 1806, sitting for St. Michael's, Cornwall in 1807, and becoming Irish Secretary in March. However, by the summer of 1807, the battlefield beckoned again, and Wellesley joined the expeditionary force sent to Denmark during July–September. He was instrumental in the defeat of a Danish force at Kjoge on August 29 and was afterward dispatched to Portugal with a small force during July–August 1808. Here he defeated **Jean Junot** at Vimeiro on August 21, but was frustrated by the arrival of Sir Hew Dalrymple, whose overly cautious approach to command prevented Wellesley from capitalizing on his victory. Discouraged, he resigned from the army and returned to England to take up his post as Irish Secretary once again.

His absence from the military did not last long. In April 1809, he was appointed to command the remaining British forces in Portugal and, after obtaining reinforcements, he engaged and defeated **Nicolas Soult** at Oporto on May 12. Following this, he led an invasion of Spain in June. There he defeated Marshal **Claude Victor-Perrin** and King Joseph at Talavera during July 27–28 and pushed them back to Madrid. For this triumph, Wellesley was created "Viscount Wellington" in September 1809. Wellington's victory was short lived, however, as he was soon forced to fall back on Lisbon because of the enemy's superior numbers.

Wellington took up a strong defensive position, building the three fortified "lines of Torres Vedras" during the winter of 1809–10. He successfully defended against Marshal **André Masséna**'s advance during July–October 1810, then halted the French advance at Bussaco on September 27. Following this action, he pulled back to the safety of the lines of Torres Vedras. For a year, from December 1810 to December 1811, Wellington fought along the Portuguese frontier. He eked out a victory over Masséna—who commanded a larger force—at Fuentes de Onoro on May 5, 1811, then took the offensive early in 1812. Wellington laid siege to Ciudad Rodrigo during January 7–20 and Badajoz during March 17–April 9. This paved the way to Madrid.

Elevated to Earl of Wellington in February, the general dealt a severe blow to the army of **Auguste Marmont** at Salamanca on July 22, then pushed on to Madrid, which fell to him on August 12. Created a marquess in October, he failed in his siege of Burgos during September 19–October 21, and had to fall back on Ciudad Rodrigo in November. Nevertheless, Wellington was named commander in chief of allied forces in northern Spain in March 1813 and summarily retook Madrid on May 17. He disposed of the smaller army of King Joseph at Vitoria on June 21, then hammered away at Soult in the Battle of the Pyrenees during July 25–August 2. Wellington helped capture San Sebastián after a siege that spanned July 9 to September 7. He then entered France. Here he scored minor triumphs at Bidassoa and La Rhune during October 7–9, then won a series of actions around Bayonne during December 9–13.

Wellington was promoted to field marshal in 1813 and enjoyed victory over Marshal Soult at Orthez on February 27, 1814 and pushed him out of Toulouse on April 10.

In May 1814, the ever-victorious Wellington was made a Duke and, with Napoleon's first exile, was appointed British ambassador to the court of the restored Bourbon king, Louis XVIII. He served in this post from August 1814 to January 1815, then assumed Lord Castlereagh's place at the Congress of Vienna during January–February.

When Napoleon's returned from Elba in March 1815, commencing the Hundred Days, the indefatigable Wellington assumed command of an Anglo–Dutch army in Flanders. In concert with a Prussian army under **Gebhard von Blücher,** Wellington prepared to invade France. He was caught by surprise when Napoleon's army invaded Belgium. Wellington scrambled to repulse Marshal **Michel Ney**'s jabs at Quatre Bras. Although he succeeded in fending Ney off, he suffered severe losses during the June 16 battle.

Wellington fought a holding action at Waterloo, awaiting the arrival of the Prussians under Blücher. Once united, the allies dealt Napoleon his final defeat on June 18.

During 1815–18, Wellington was in charge of the allied army of occupation in northern France. Returning to England, Wellington immersed himself in politics through 1846. He was appointed Master-General of the Ordnance in 1818, and he served as British ambassador to the Congress of Vienna from 1822 to 1826, when he was appointed ambassador to Russia. The following year, he returned to England to accept appointment as commander in chief of the army after the death of the Duke of York in January 1827.

Wellington's steadfast conservatism prompted him to resign all of his offices in April 1827 when the famed liberal George Canning became prime minister. Wellington himself was appointed prime minister in January 1828, but his ultraconservative resistance to parliamentary reform brought about his ouster on November 15, 1830. He next served as foreign secretary to Robert Peel from November 1834 to April 1835 and was again appointed commander in chief in 1842. Wellington retired from government after the defeat of Peel in June 1846.

In 1848, Wellington performed his last military service, organizing resistance against the London Chartists, members of a British working-class movement for parliamentary reform.

Wellington was an extraordinary strategist, who was a worthy match for Napoleon and his marshals. In addition, he was preternaturally calm in battle and a man of dauntless courage.

WESTMORELAND, William Childs

BORN: March 26, 1914

SERVICE: World War II (1941–45); Korean War (1950–53); Vietnam War (1964–72)

U.S. general best known as the principal U.S. commander in the Vietnam War

A native of Spartanburg, South Carolina, Westmoreland enrolled for a year at the South Carolina military academy known as The Citadel, then entered West Point, graduating in 1936. He was commissioned a second lieutenant in the artillery and saw service in Oklahoma, Hawaii, and North Carolina before U.S. entry into World War II. Promoted to major in April 1942, he commanded an artillery battalion in Tunisia from November 8, 1942 to May 13, 1943. Westmoreland distinguished himself in the disastrous battle of Kasserine Pass during February 14–22, 1943. Subsequently, he commanded the unit in Sicily from July 9 to August 17.

Westmoreland participated in the Normandy "D-Day" invasion as part of Ninth Infantry Division, hitting the beach on June 10, 1944. The following month, he was promoted to colonel and made divisional chief of staff, serving with the Ninth in its advance across northern France and into Germany to the Elbe River during July 1944 to May 1945.

After Germany's surrender, Westmoreland commanded the Sixtieth Infantry Regiment in occupation duty into early 1946, then transferred to the Seventy-First Division through March 1946. Next, after undergoing parachute training, Westmoreland assumed command of the 504th Parachute Infantry Regiment, then was appointed chief of staff of the Eighty-Second Airborne Division, a post he held from 1947 to 1950.

In 1950, Westmoreland was assigned as an instructor at the Command and General Staff School and at the Army War College, leaving in August 1952 to command the 187th Airborne Regimental Combat Team in Korea. He was promoted to brigadier general in November and served on the General Staff from 1953 to 1958, becoming secretary of the General Staff in December 1956, after promotion to major general.

In 1958, Westmoreland was made commander of the famed 101st Airborne Division—the "Screaming Eagles"—at Fort Campbell, Kentucky–Tennessee. He served in this post through July 1960, when he returned to West Point as the academy's superintendent. During these Cold War years, he oversaw the expansion of the cadet program, approximately doubling the number of students by the time he left in 1963—as a lieutenant general—to resume Airborne command, now as commander of the XVIII Airborne Corps.

In June 1964, Westmoreland began what was undoubtedly his most fateful assignment, taking over from General Paul D. Harkin as commander of the U.S. Military Assistance Command in Vietnam (MACV). Promoted to general in August, he served in Vietnam through mid 1968 and oversaw the massive escalation of U.S. involvement in the war, from a relatively small cadre of military "advisers" to half a million U.S., South Vietnamese, and Allied troops. Westmoreland pursued a strategy of attrition against an enemy that was willing to suffer and die in great number. As the war ground on, U.S. popular and political support for the effort waned. During January–April 1968, the Viet Cong launched the infamous Tet Offensive, a massive attack on many fronts. U.S. forces sustained substantial losses; however, they dealt the communists a much harder blow. In a military sense, Westmoreland had presided over a victory. The Tet Offensive proved incredibly costly to the Viet Cong. Yet the communists refused to quit, and the Tet Offensive became a moral victory for the North. In the U.S. antiwar protest became epidemic, and president Lyndon Johnson took the first steps to reduce American participation in the war. He effectively promoted Westmoreland out of command, bringing him home as army chief of staff and replacing him in Vietnam with General **Creighton W. Abrams, Jr.** Westmoreland served as chief of staff until his retirement in July 1972. However, in retirement, he remained a fierce advocate of the U.S. role in Vietnam, publishing a memoir, *A Soldier Reports*, in 1974 and winning a 1986 libel case against CBS News over its reports that he and his staff had deliberately falsified enemy strength and casualty estimates during the Vietnam War.

Westmoreland was an able career commander charged with an impossible task in Vietnam, conducting a war on behalf of a corrupt South Vietnamese government and on the shaky foundation of diminishing political support in the United States. Unable to obtain a sufficient number of troops to occupy and hold territory in Vietnam, he had to fight a grim war of attrition. Unfortunately, Westmoreland discovered, he was fighting an enemy apparently willing endlessly to endure attrition.

WEYLER Y NICOLAU, Valeriano

BORN: 1838

DIED: October 20, 1930

SERVICE: Cuban Ten Years' War (1868–78); Second Carlist War (1873–76); Cuban Revolution (1895–98)

Spanish general, scourge of nineteenth-century Cuban rebels

After joining the Spanish military in his youth, Valeriano Weyler y Nicolau first fought against Cuban rebels in 1868–72. After his tour of duty in Cuba, Weyler returned to Spain to fight against the Carlists—the Spanish Bourbon traditionalists—in their attempts to restore a Bourbon monarchy. Following his success in helping repress the Carlist uprising, Weyler was sent to the Canary Islands in 1878, where he was named captain general (military governor), serving until 1883. An appointment in the Balearic Islands followed, after which he was assigned to suppress rebel uprisings in the the Philippines.

In January 1896, Weyler was sent back to Cuba to put down yet another Cuban revolt. He replaced an ineffective Martinez Campos as commander-in-chief and was determined to handle the rebellion as a straightforward war in which force was to be met with overwhelming force. The Cuban rebels had achieved a fearsome reputation for savagery, and Weyler intended to meet their threats with savagery of his own, including the institution of concentration camps to imprison captured rebels.

Weyler had badly misjudged the Cuban rebellion as a "simple" military operation. His uncompromising harshness might have led to military success, but it also incited Americans' outrage over Spanish imperialism and sparked sensational accounts of Spanish brutality in the "Yellow" press of William Randolph Hearst and Joseph Pulitzer. In 1898, the United States declared war against Spain.

By the the time of the United States' entry into the war, however, Weyler had pretty much defeated the rebels and killed their leader, Maceo, in December 1896. In fact, he had been relieved of his duties in October 1897 because of the criticism, not the least from the United States, levelled against his prosecution of the war. Back in Spain, Weyler held a succession of government appointments and became army commander-in-chief in 1921. In 1926, he participated in an abortive attempt to overthrow the regime of Primo de Rivera. Weyler died on October 20, 1930.

WILHELM II

BORN: January 27, 1859

DIED: June 4, 1941

SERVICE: World War I (1914–18)

The last kaiser (emperor) of Germany; his bellicose and expansionist were instrumental in triggering World War I

A member of the reigning Hohenzollern family that had ruled German politics for centuries, Wilhelm II was the son of Crown Prince Frederick Wilhelm, later Kaiser Frederick III, and grandson of Kaiser Wilhelm I. The predominant influence in his youth, was his mother, Victoria, daughter of Queen Victoria of England. Victoria tried to raise Wilhelm as an English gentlemen and a liberal, contrary to the prototype of the traditional Prussian ruler, who was militant, firm, frugal, and conservative. Caught between his mother and his heritage, Wilhelm became neither a true liberal nor conservative.

When Wilhelm I died in March 1888, Frederick Wilhelm was already dying of cancer. He reigned for a mere ninety-nine days before succumbing, and Wilhelm II became kaiser on June 15, 1888. Chancellor Otto von Bismarck looked favorably on his accession, feeling that Wilhelm was a more apt successor to his conservative grandfather than his liberal father. Yet Wilhelm and Bismarck soon clashed as the kaiser assumed prerogatives that had been the chancellor's. At length, Wilhelm, whose ambition was to restore the monarchy to the degree of absolute authority **Frederick the Great** had enjoyed. He dismissed Bismarck in March 1890 and attempted to rule through handpicked puppet chancellors.

Yet Wilhelm II proved unable to dominate men the way Bismarck had done, and he was further unable to institute lasting and meaningful policy. He wanted the world and Germany to believe he ran the country, but, in fact, he was quite incapable of doing so. Wilhelm fancied himself adept at foreign policy, feeling that central European politics still revolved around Germany as it had a century before around Brandenburg-Prussia. In 1896, he sent a telegram to South African President Paul Kruger, congratulating him on his suppression of the British-led Jameson raid. He further insulted the British by secretly challenging their supremacy on the seas. Wilhelm let his ministers talk him into not renewing Bismarck's 1887 Reinsurance Treaty with Russia. In consequence, Russia promptly aligned itself with France, Germany's bitterest enemy. Finally, the kaiser added insult to injury by giving a tactless interview to *The Daily Telegram*—on British soil no less—declaring the majority of German citizens were adamantly anti-English.

As kaiser, Wilhelm had the power to appoint the chancellor, who ran the civil government, but the military was not under the civil government. It was run at the discretion of the kaiser and his general staff. Wilhelm effectively relinquished control of the military to his generals, who began to plan and arm or global conflict, developing the famed **Schlieffen** Plan, first developed in 1904. Germany was not the only country to have large-scale war contingency plans—France, for example, developed Plan 17—but Wilhelm, essentially a non-belligerent, failed to curtail a military bloodlust that soon got out of hand.

When Austro-Hungarian Archduke Francis Ferdinand was assassinated in June 1914, Wilhelm pushed for the punishment of Serbia by Austria-Hungary in the hope of solidifying that ailing ally. He failed to realize that, because of the Triple Entente and the Triple Alliance, any conflict would quickly escalate to engulf all of Europe. When the interlocking treaties were activated, the kaiser backed away, even as World War I began, and the Schlieffen Plan was being put into effect with Germany's advance toward Paris.

Yielding before the juggernaut of world war, Wilhelm increasingly relinquished control of the country to his two senior generals. By early 1918, after four ruinous years of slaughter, it was apparent that Germany could not win the war. The chancellor, Prince Max of Baden, announced Wilhelm's abdication without even consulting him, and General **von Hindenburg** informed the kaiser that the army would no longer support him. Accordingly Wilhelm fled to the Netherlands.

Thus the kaiser left Germany to a humiliating peace. His own life was spared because the Dutch government declined to extradite the kaiser to the Allies, who intended to try him as the author of the war. Wilhelm remained in Dutch exile until his death in the early years of World War II.

WILLIAM I THE CONQUEROR

BORN: ca. 1028

DIED: September 9, 1087

SERVICE: Norman Baronial Revolt (1035–47); Wars against Henry I of France (1053–58); Norman Conquest of England (1066–77)

Norman ruler and conqueror of England

Called the Conqueror after his momentous victory over the Anglo-Saxon forces at the Battle of Hastings in 1066, William I brought a semblance of centralized order to feudal England by establishing the tradition of primacy in determining royal succession. Born in the city of Falaise, the bastard son of the Duke of Normandy, Robert I, William had no claim to his father's title, but Robert I died without legitimate issue in 1035 and named William as his heir. The appointment did not sit well with some Normans, and William was the victim of several attempted murders. Although the young duke remained unharmed, his guardians did not fare as well. Four of them were murdered in the course of nine years.

In 1046, William's cousin, Guy of Burgundy, determined to remove the duke from his throne. Guy might have succeeded except for the support William received from his feudal overlord, Henry I, King of France. With the king's aid, William defeated Guy at the Battle of Val-es-Dunes the following year, and in his victory, William secured his rule over Normandy. Indeed, his firm handling of potential usurpers left a strong impression on the aristocracy, and he maintained a very good relationship with them that was virtually unique for the time and that enabled him to make full use of all available resources.

The force of William's rule and personality soon prompted King Henry I of France to reconsider his alliance with the duke of Normandy. As William became more successful he also became more of a potential rival to the King. The king, thus, abrogated his previously amicable ties with the Norman monarch and used William's consanguinous marriage to Mathilda, daughter of the Count of Flanders, as an excuse for waging war against the duke. Accordingly, Henry allied himself with Geoffrey Martel, Count of Anjou, and attacked. William twice defeated the allies, at the Battle of Mortemer in 1054 and again at the Battle of Varaville in 1058. Strengthened by his victories, William had gained the feudal independence that allowed him to turn his attention to more distant lands.

William turned his eyes westward where King Edward the Confessor ruled over a chaotic state. In return for Norman support, Edward named William heir to the English throne. William strengthened his claim to the throne in a deal with Harold Godwin, brother of Edward's wife. Harold was shipwrecked in Normandy, and in exchange for his freedom, he pledged to support William's claim as successor to Edward.

William I the Conqueror, depicted in an engraving after the Bayeux Tapestry. From J. N. Larned, A History of England for the Use of Schools and Academies, *1900.*

In 1060, both Henry I and Geoffrey Martel died, leaving William dominant in northern Europe. William quickly marched into Anjou and conquered Maine before the Anjouan succession could be determined. William now felt that he could make his move on England.

In the meantime, however, Edward, on his deathbed at the beginning of 1066, designated Harold as heir to the throne. Harold abrogated his oath to William and assumed the throne. William raised an army and landed at the town of Pevensey on the southern tip of England in September 1066. He advanced on Hastings, where he met and defeated Harold's forces on October 15 in a battle matched in importance only by **Henry V**'s upset victory at Agincourt 350 years later.

After his victory, William moved northward, terrorized the English countryside, and finally took London in December with little opposition. He was crowned William I, King of England, in Westminster Abbey on Christmas Day, 1066. There were many uprisings in the wake of William's accession. These he put down swiftly, effectively, and brutally. By 1071, the seeds of rebellion had been trampled.

William ruled England and Normandy concurrently, frequently journeying back and forth between the two, managing the effective administration of both. Shortly before his death in 1087, William designated his first son, Robert, as the heir to the Duchy of Normandy, and his third—and favorite—Rufus, as heir to the English throne. Rufus reigned as William II and established the law of primacy henceforth to govern the succession of English monarchs.

WOLFE, James

BORN: January 2, 1759

DIED: September 13, 1759

SERVICE: War of the Austrian Succession (1740–48); Jacobite Rebellion (1745–46); Seven Years' War-French and Indian War (1754–63)

British general most famous for defeating Montcalm at the decisive Battle of Quebec during the French and Indian War

Born at Westerham, Kent, Wolfe was plagued in childhood by ill health, but nevertheless was determined to follow in his father's footsteps and become a military officer. After education at Westerham and Greenwich, he was commissioned into his father's old Marine regiment as a second lieutenant on November 3, 1741, then transferred to the Twelfth Regiment of Foot as an ensign. He went to Flanders and fought at Dettingen as acting regimental adjutant on June 27, 1743. Promoted to brigade major, he fought at Falkirk on January 17, 1746, and at Culloden on April 16 in the suppression of the Jacobite Rebellion. Following the English victory at Culloden, he returned to the Netherlands and the War of the Austrian Succession. On July 2, 1747, he was wounded at Lauffeld.

During 1748–49, Wolfe was a major with the Twentieth Foot and subsequently served garrison duty in southern Scotland from 1749 to 1757, when he was appointed quartermaster general for an abortive British expedition against Rochefort-sur-Mer, France, during September 20–30, 1757. Despite the failure of the mission, Wolfe distinguished himself and earned a brevet to colonel.

Wolfe was next dispatched to North America to fight in the phase of the Seven Years' War known as the French and Indian War. he was given command of a brigade in the army of Sir Jeffrey Amherst and marched in the expedition against the French-held port of Louisbourg, Nova Scotia during June 30–July 27, 1758. Successful in this venture, he returned to England, where Prime Minister William Pitt fitted him out with nine thousand men for an expedition against Quebec.

On this return to North America, Wolfe was commissioned a local major general in February 1759 and plotted out a strategy to take the strongly defended city of Quebec. He tried to storm the city, but was readily repulsed with great loss. Finally, after repeated frustration, he made a landing with five battalions just above the city at an area known as the Plains of Abraham. From this height, he took the French commander, the Marquis de **Montcalm,** totally by surprise. The usually coolly skilled Montcalm made a number of uncharacteristic blunders as he threw together a counterattack, which was quickly squashed.

The battle was over in a quarter of an hour. In the end, Quebec fell to the British—effectively marking an end to French power in North America (though the war dragged on for another three years)—but both Wolfe and his adversary, Montcalm, lay mortally wounded. Wolfe survived long enough to see the victory of September 13, 1759.

Wolfe's early death elevated him to semilegendary status as a figure of great romantic dash and daring. He was, in fact, a moody young man, who never quite overcame his tendency to chronic illness and chronic melancholia. Had he lived longer, however, he might well have been one of the great British generals.

WOOD, Leonard

BORN: October 9, 1860

DIED: August 7, 1927

SERVICE: Apache War (1885–86); Spanish–American War (1898); World War I (1917–18)

U.S. general who served as an army medical officer and became chief of staff of the army

Wood was born in Winchester: New Hampshire and graduated from Harvard Medical School in 1884. After practicing briefly in Boston, he joined the army as acting assistant surgeon with the rank of second lieutenant in June 1885. Promoted to assistant surgeon—with the rank of first lieutenant—in January 1886, he participated in the expedition against the Apache leader **Geronimo** during May through September.

Promoted to captain in January 1891, he was posted to Washington, D.C., where he was appointed White House physician to President Grover Cleveland in 1895. He also served for a time during the administration of William McKinley. While Wood was on the White House staff, he became a close personal friend of Theodore Roosevelt, who was at the time assistant secretary of the navy. With Roosevelt, he organized the First Volunteer Cavalry—the famed Rough Riders—for service in the Spanish–American War in May 1898.

Wood was appointed colonel of volunteers in command of the Rough Riders and led them to Cuba in June. He was soon promoted to command of the Second Brigade—which included the First Volunteers and the First and Fiftieth Cavalry regiments—and he personally led the Second Brigade at the make-or-break Battle of San Juan Hill on July 1. He was also in command during subsequent actions, culminating in the Capture of Santiago de Cuba on July 17. At this time, he was promoted to brigadier general and was named military governor of Santiago on July 18. In this post, the doctor in Wood came to the fore, as he worked tirelessly to improve sanitation, as well as to restore order. In October, his appointment as governor was extended to the entire province.

Wood was promoted to major general in December 1898, but reverted to brigadier general after the war, in April 1899; however, when he became governor of all of Cuba, he was reinstated as a major general. Wood worked vigorously to improve the Cuban infrastructure by building roads, schools, and communications. He collaborated closely with Major William C. Gorgas to attack the Yellow Fever, which was endemic to the region. Wood also created a constitution and legal code for Cuba, then stepped down as governor when the newly elected president, Tomás Estrada Palma, took office on May 20, 1902.

Wood was appointed governor of the rebellious Moro province of the Philippines in March 1903, and the commander of the Department of the Philippines at Manila in April 1906. Returning to the United States in November 1908, he became commander of the Department of the East and, in April 1910, was appointed chief of staff for the army. His tenure was a stormy one, as he worked tirelessly to reform and shake up the sleepy peacetime army.

A proponent of preparedness after World War I broke out in Europe in 1914, Wood created civilian training camps in Plattsburgh, New York, without the cooperation or consent of the War Department. While this alienated the department, it did help to ensure that the United States army was at least on the road to preparedness when the nation entered World War I in 1917. Wood himself was passed over for major command during the war, however, and served during part of 1918 as commander of the Eighty-Ninth Infantry Division at Camp Funston (now part of Fort Riley), Kansas. After the war, in January 1919, he was appointed commander of the Central Department, headquartered in Chicago.

Wood sought the nomination as Republican candidate for president in 1920, but lost to Warren G. Harding, who subsequently sent him to the Philippines as governor general. He served there until his death from a brain tumor in 1927.

The speed with which he advanced, coupled with his aggressive and abrasive character, caused much resentment among fellow officers. However, his reform work left a lasting mark on the U.S. Army, most notably in the enduring role of the General Staff as the guiding body in matters of policy and strategy. His work at Plattsburgh led ultimately to the creation of the ROTC—and was a key factor in achieving preparedness for World War I. Although he had relatively little combat experience, Wood was an important figure in the formation of the modern U.S. Army.

X

XERXES I

BORN: ca. 519 B.C.

DIED: 465 B.C.

SERVICE: Persian War with Greece (480 B.C.–448 B.C.)

Persian ruler, successor to Darius I

Xerxes I, king of Persia, is best known for his unsuccessful invasion of Greece. His defeat at a naval battle at Salamis marked the beginning of the decline of the Achamenid Empire. With the death of his father, Darius I, in 486 B.C., a thirty-five-year-old Xerxes I inherited the throne of the Achamenid Empire. One of his first decisions as ruler was to regain control of Egypt, which, for two years, had been in the hands of a usurper. In 484 B.C., Xerxes attacked the Nile Delta and swiftly suppressed the Egyptian rebellion. On the heels of his victory, he learned of another revolt, this one in Babylon. Xerxes sent his son-in-law to the region to quell the rebellion. Persian forces tore down Babylon's fortresses, pillaged temples, and destroyed the statue of Marduk, the Babylonian god.

Once firmly in control of his empire, Xerxes was persuaded by his brother-in-law Mardonius to attack the Greeks in retaliation for the humiliating defeat they had dealt Darius I at Marathon in 490 B.C. For three years, Xerxes prepared for war. He gathered some 360,000 troops from across the realm, built a navy of seven to eight hundred ships, dug a channel across the Isthmus of Actium, and built boat bridges to be used at the Hellespont.

Xerxes himself led his army from Sardis in 481 B.C. Tolerating no deviation from his battle plans, he became enraged when storms destroyed his carefully constructed boat bridges. He ordered his troops to punish the sea by whipping it with lashes.

Despite the mishap at the Hellespont, Xerxes triumphed at Thermopylae in August 480 B.C., and moved on to occupy Attica and Athens. By the end of September, however, his navy suffered a stunning defeat at Salamis, and he was forced to retreat, leaving Mardonius in Thessaly. When Mardonius was killed the following year at Plataea, the entire Persian army withdrew.

For the remainder of his life, Xerxes turned his attention to monumental construction projects. In his capital city of Persepolis, he built a huge terrace of the Apadana, completed his father's palace, and then built one of his own, including a building called the Harem, which may have served as his treasury. Construction was also begun on the Hall of a Hundred Columns—the throne room.

During the last years of his life, Xerxes was drawn into various palace intrigues. Persuaded by the queen that his brother's family meant to do him in, he had them killed preemptively. Nevertheless, in 465 B.C., he and his eldest son were apparently murdered by a court cabal that included Xerxes' trusted minister Atabanus. Another of the king's sons, Artaxerxes I, succeeded Xerxes to power.

Y

YAMAMOTO, Isoroku

BORN: April 4, 1884
DIED: April 18, 1943
SERVICE: Russo–Japanese War (1904–05); World War II (1941–45)

The Japanese admiral who planned and executed the December 7, 1941, surprise attack on the U.S. naval and air base at Pearl Harbor, Hawaii

Born Isoroku Takano in Nugata prefecture, the future admiral was adopted by the Yamamoto family, whose name he subsequently took. After graduating from the naval academy in 1904, saw action at the battle of Tsushima during the Russo–Japanese War and was wounded on May 26, 1905, losing two fingers from his left hand. His injury nearly resulted in his dismissal from the navy.

Following the war, Yamamoto served aboard several vessels, and graduated from the Torpedo School in 1908. As a promising young officer, he was enrolled in the Naval Staff College, from which he graduated in 1911 and, that same year, graduated from the Naval Gunnery School. Appointed an instructor there, he was promoted to lieutenant commander in 1915, then graduated from the Naval Staff College's senior course the following year.

Yamamoto was appointed a staff officer with the Second Fleet and was sent to the United States for study at Harvard University from 1919 to 1921. He then returned to Japan as an instructor at the Naval War College from 1921 to 1923, when he was promoted captain and sent on a tour of inspection and observation to the United States and Europe as an admiral's aide.

In 1924, Yamamoto was made deputy commander of Kasumiga Ura Naval Air Station, leaving that post the following year to become naval attaché in Washington, D.C. from 1925 to 1928. Returning to Japan, he was made captain of the aircraft carrier *Akagi*, but left this command in 1929 when he was promoted to rear admiral and, a year later, appointed chief of the Technological Division of the Navy Technological Department. In 1933, he was assigned to command the First Naval Air Division, then, promoted vice admiral in 1934, he became head of the Japanese delegation to the London Naval Conference of 1934–35.

Yamamoto was personally opposed to official Japanese insistence on naval parity with Britain and the United States. Nevertheless, acting on orders, he negotiated strongly and rejected any further extension of the tonnage ratios established by the Washington Naval Treaty on 1922. Later, he played a key role in Japan's abrogation of the treaty arrangements.

Made chief of Naval Aviation Headquarters in 1935, Yamamoto championed the use of the aircraft carrier as the principal offensive weapon of the navy. He then served as Navy Minister from 1936 to 1939, using his position to attempt to moderate the extreme militarism of the a government on the verge of war. This was ultimately to no avail. In the meantime, Yamamoto accepted a concurrent reappointment as chief of Naval Aviation Headquarters in 1938, then left both of his positions to become commander of the Combined Fleet in 1939 and, in 1940, commander of First Fleet as well.

Yamamoto had chief responsibility for making preparations for war against Britain and the United States. With regard to the U.S., he approached this task fatalistically, in the full belief that American industrial power and population would make it almost impossible

Isoroku Yamamoto. Courtesy National Archives.

for Japan to win a war—especially a long war—against the nation. When his efforts to head off such a war appeared hopeless, he took the next best step and planned a surprise attack on the American fleet and naval base at Pearl Harbor, hoping that a devastating attack would bring about a negotiated peace with the United States. The actual attack was executed on December 7, 1941, under the command of Vice Admiral Nagumo Chuichi.

The attack on Pearl Harbor achieved the object of surprise, and it did cause severe damage to the U.S. fleet: the battleship *Arizona* was completely destroyed, and the *Oklahoma* capsized. The battleships *California*, *Nevada*, and *West Virginia* sank in the shallow water of the harbor. Three other battleships, three cruisers, three destroyers, and numerous other vessels were also damaged. Some 180 aircraft, including B-17s, were destroyed, and more than 2,300 soldiers, sailors, and marines were killed, with an additional 1,100 wounded. Japanese losses were minimal: perhaps as many as sixty aircraft (though half that number is more likely), five midget submarines, one or two fleet submarines, and fewer than one hundred men. Crippling as it was, the attack could have been worse. The three U.S. aircraft carriers of the Pacific Fleet were not present at Pearl Harbor during the attack, and of the eight battleships hit, all but the *Arizona* and *Oklahoma* were ultimately returned to service. The attack did not destroy the vital oil storage facilities on the island.

Most of all, Yamamoto had badly misread the climate of American response to the attack. Far from bringing the nation to the negotiating table, it propelled the United States into World War II on December 8 and, as Yamamoto had feared, spelled the doom of the Japanese cause.

Yamamoto next planned lightning naval campaigns that captured the East Indies during January–March 1942 and that achieved success in the Indian Ocean during April 2–9, 1942. However, he met defeat against the U.S. Navy at the Battle of Midway on June 4, which began to turn the tide of the war in the Pacific against his nation. The defeat may be ascribed in part to Yamamoto's overly complex plan and unclear objectives. Moreover, the Midway debacle seems to have shaken Yamamoto's confidence, as he managed the subsequent Solomons campaign poorly, refusing to commit his resources aggressively.

Unknown to Yamamoto and other Japanese war leaders, the United States had broken the chief Japanese military codes even before the war had begun. An intercepted and decoded message revealed that Yamamoto was to fly to tour Japanese installations on Shortland Island on April 18, 1943. U.S. fighter aircraft were dispatched to intercept and shoot down his flight, and he crashed and perished near Bougainville.

During his brief World War II career, Yamamoto masterminded Japan's greatest naval successes and its first decisive defeat. His death was a blow from which the Japanese navy never recovered.

YAMASHITA, Tomoyuki

BORN: November 8, 1885
DIED: February 23, 1946
SERVICE: World War II (1941–45)

The "Tiger of Malaya"; Japanese general of World War II famous for his successful attacks on Malaya and Singapore and infamous for war crimes associated with his later defense of Manila

Born in Kochi prefecture, Yamashita graduated from the Military Academy in 1906 and was commissioned as an infantry officer. He was sent to the Staff College, from which he graduated in 1916 and was assigned to the German Section of the Intelligence Division of the Army General Staff two years later. He served as resident officer in Bern, Switzerland during 1919–21 and, subsequently, in Germany, during 1921–22.

After promotion to major in 1922 and to lieutenant colonel in 1925, Yamashita became military attaché in Vienna and, concurrently, in Budapest, serving in these posts from 1927 to 1929, when he returned to Japan as a colonel assigned to the Military Research Division, Central Ordnance Bureau. In 1930, Yamashita assumed regimental command, then, in 1932, was appointed chief of the Military Affairs Section in the Army Ministry. He was appointed chief of the Military Research Section in the Army Ministry's Military Research Bureau in 1935.

Yamashita was politically active as a member of General Sadao Araki's ultranationalistic Kodo-ha (Imperial Way) faction. He initially supported the revolt of young Koda-ha officers on February 21, 1936 and acted as liaison between them and the army central command. However, he soon turned against the faction and abandoned their cause. Following the rebellion, he was sent to Korea to command a brigade during 1936–37 and was promoted to lieutenant general in 1937.

Appointed chief of staff for the North China Area Army in 1937, he served in this post until 1939, when he took command of the Fourth Division, then, in 1940, became inspector general of Army Aviation and chief of the Military Aviation Observation Mission to Germany and Italy. He was assigned to command the Kwantung Defense Army in 1941, then transferred to command of Twenty-Fifth Army in November. It was at the head of this force that he led the invasion of Malaya during December 8–10 and directed the Japanese campaign down the Malay peninsula, which swept away the British Commonwealth defenders. Outnumbered by a factor of two, the "Tiger of Malaya" nevertheless pushed the British back to Singapore, where they surrendered on February 15, 1942.

Despite Yamashita's successes, the military dictator **Hideki Tojo,** long a rival and enemy, ordered Yamashita's transfer to command of the First Area Army in Manchuria, a backwater of World War II by July 1942. Promoted to general in 1943, he was returned to the principal theater of the war as commander of the Fourteenth Area Army and was assigned to defend the Philippines and northern Luzon. He reached Manila barely a week before U.S. forces landed on Leyte on October 20, 1944. He therefore had little effect on the landings, but did mount a fierce and well-executed defense of Luzon. Nevertheless, by February–April, his army had withdrawn into the mountains of northeastern Luzon, and in September 1945, he surrendered.

Yamashita was tried for the war crimes of Japanese troops defending Manila in early 1945. Although he bore no direct responsibility for his troops' excesses, the tribunal ruled that, as overall commander, he was ultimately responsible. He was convicted and, on February 23, 1946, executed. Today, many believe that the execution was unjust, since, in the desperate circumstances of the Manila defense, Yamashita had little control of his Manila garrison troops after he moved his headquarters into the mountains.

Yamashita was an extraordinarily energetic commander, whose victory over the British in Malaya was one of the boldest and most effectively conducted operations of World War II.

YANG CHIEN

DIED: 604
SERVICE: Rise of the Sui (581–600)

Unified China, creating a powerful army and a well-administered central government; he continued construction of the Great Wall and started building the Grand Canal

Little is known of Yang Chien's early life beyond the fact that he was of mixed Chinese and Hsien-pi heritage and served the Hsien-pi Northern Chou dynasty as a military leader in northern China. Yang distinguished himself as a commander and was named Duke Sui. He went on to lead Chou forces against Northern Ch'i, effectively reuniting northern China under the Chou in 577. Having consolidated Chou dominion over northern China, Yang seized on the death of the Chou emperor to place on the throne a seventeen-year-old youth in 580, securing for himself an appointment as regent. Barely a year later, Yang compelled the boy to abdicate, whereupon Yang assumed the throne and established the Sui dynasty. To secure his power, the new emperor subsequently arranged for the death of the teenager he had displaced.

Yang Chien mobilized forces against Turkic tribes west of his realm, campaigning against them from 582 to 603. While he did not succeed in utterly subjugating these people, he did extend Chinese influence deeply into central Asia. He moved, at first unsuccessfully, against Korea, managing by 589 to dominate that country's northern region. His greatest campaign, however, was conducted against the Ch'en dynasty of southern China. Massing a vast army—Chinese historical tradition puts its number at half a million—he defeated the Ch'en in 589, thereby unifying northern and southern China for the first time since the end of the Han dynasty three centuries before.

Yang Chien was not only a military genius, but an extraordinarily skilled administrator. He reintroduced Han governmental policies and ushered in a return to Confucianism. To combat the deadly cycle of harvest and famine that ravaged the country, he caused large, centralized granaries to be built and he introduced a just and productive system of taxes based on harvest yield. Remorseless in his grasping for power, he was himself killed under obscure circumstances. Historians suspect that he was murdered by his own son, Yang Ti, in 604.

YEN HSI-SHAN

BORN: 1882
DIED: 1960
SERVICE: Chinese Revolution (1911–17); Chinese Civil War (1917–26); Northern Expedition (1926–28); Second Sino-Japanese War (1937–45)

One of the more successful of the many modern-day warlords who contended for power in China

Yen Hsi-shan was born in Shansi province and was educated at the Taiyuan Military Academy, which trained him for a career in the Taiyuan New Model Army, which he entered in 1904. Four years later, he was sent to Japan, to study at its advanced military academy. There, during 1908–10, he met the young military leaders of China's growing nationalist movement, which was bent on overthrowing the late Ch'ing (Manchu) empire. Yen was installed as a regimental commander of the Taiyuan garrison in October 1911, when the Chinese Revolution broke out. He led his troops in support of the revolutionaries, eventually casting his lot with the revolutionary faction of Yüan Shih-k'ai.

Yen's genius lay in his ability to navigate the treacherous waters of revolutionary China. His allegiance to Yüan gained him the military governorship of Shansi, which he ruled with an autocratic hand—though he also instituted a far-reaching program of modernization and industrialization unknown under the emperors. By 1915, it was clear to Yen that Yüan was losing his power base, and he quickly shifted allegiance to the Anhwei faction led by Premier Tuan Ch'i-jui. This move gained him the civil governorship of Shansi, a strong foundation on which to consolidate and build additional power. When, in his turn, Tuan faltered and fell in 1918, Yen skillfully shifted his apparent allegiance between Wu P'ei-fu and **Feng Yü-hsiang.** Finally, in 1925, he joined Chang Tso-lin, military governor of Fengtien, to drive Wu out of north-central China. Then, with Wu disposed of, he quickly shifted alliance to the KMT (Kuomintang, or Nationalist Party) and was put in command of the Nationalist Third Army Group. From Septmeber 1927 to June 1928, he joined in a successful assault against Peking (Beijing).

The nationalists rewarded his services to the cause by making him Interior Minister and commander of the Peking (Beijing)-Tientsin (Tianjin) garrison. With even more than his usual agility, however, Yen turned his back on the KMT, took neither post, and joined in the revolt of Feng Yü-hsiang and Wang Ching-wei in 1930. When this proved abortive, Yen suffered exile to Dairen (Luda), but returned to Shansi a scant six months later, in 1931, and contented himself with governing the province. This was no small accomplishment, since the overwhelming majority of Chinese warlords failed to command sustained loyalty and were subjected to almost continual revolt. Yen, though stern and dictatorial, did exhibit concern for the people he governed. He also commanded great loyalty from his military subordinates. But he did not depend on progressive programs and personal loyalty alone. He was among the first of the Chinese revolutionary leaders to develop and employ the methods of political indoctrination that **Mao Tse-tung** would later bring to a high state of evolution in forging the People's Republic of China.

When the Japanese invaded China during July 1937, Yen put up a stout resistance that resulted in an uneasy stalemate, during which he maneuvered among the sharply contending forces of the Japanese, the KMT, and the CCP (Chinese Communist Party), pitting one against the other in a sustained and highly successful effort to retain his own power. Remarkably, after the Japaense defeat in World War II, Yen retained the military services of four divisions of the Japanese army to keep the CCP out of Taiyuan from 1945 to 1949. When it fell at last in April 1949, Yen quietly relinquished power, retired, and died of natural causes in 1960.

Z

ZHUKOV, Georgi Konstantinovich

BORN: December 1, 1896

DIED: June 18, 1974

SERVICE: World War I (1914–18); Russian Civil War (1918–22); Manchurian border conflict (1938–39); World War II (1939–45)

Marshal of the Soviet Union and the best Soviet commander during World War II

Zhukov was born to an obscure peasant family in the village of Strelkovka, about sixty miles east of Moscow. In 1908, he was apprenticed to a fur trader and pursued that vocation until 1915, when he was drafted into the Imperial Russian Army. He was quickly promoted from private to noncommissioned officer and fought in various cavalry units, including, most notably, the Novgorod Dragoons. He distinguished himself at the front and was awarded two Orders of St. George for bravery.

With the outbreak of the Civil War, Zhukov threw in his lot with the Red Army in October 1918 and was given command of a cavalry squadron in the First Cavalry Army. Zhukov graduated from a junior officers' military school in 1920, then, after the Civil War, he moved on to an intermediate-level cavalry officer course, which he completed in 1925. After this, he studied military science at a higher level in a clandestine *Kriegsakademie* (war college) in Germany as part of the secret military collaboration that took place between the Soviet Union and the Weimar Republic in the late 1920s. Returning to the USSR, Zhukov studied at the Frunze Military Academy from 1928 to 1931.

Appointed deputy commander of the Byelorussian Military District in 1938, he appeared doomed to the fate that befell many other top-level Soviet military officers during the wholesale purges of a paranoiac **Josef Stalin.** That he escaped imprisonment and execution was apparently due to a providential administrative accident.

Heading up the Soviet First Army Group, he defeated the Japanese Sixth Army at the Khalka River near Nomonhan during July–August 1939, after which he was appointed deputy commander (1939), then commander (1940) of the Kiev Military District. When Nazi Germany invaded the Soviet Union during Operation Barbarossa in June 1941, Zhukov was rushed to the front to help defend Smolensk in August. When that campaign failed, he organized the defense of Leningrad (present-day St. Petersburg) as commander of the Leningrad Front (Leningrad Army Group) during September–October 1941. Next, he directed the Western Front, which—at great cost—defeated the German effort to capture Moscow during 1941–42.

Zhukov was a key staff and field officer throughout the balance of the war, participating in every major operation, including the costly but decisive defense of Stalingrad during 1942–43—the turning point of the war in the USSR. He also directed the Battle of Kursk in July 1943, the Byelorussian offensive during the summer of 1944, and the advance into Germany and the capture of Berlin in 1945. It was Zhukov who accepted the surrender of Nazi Germany on behalf of the Soviet Union on May 8, 1945, and it was he who headed the military administration of the Soviet Zone of occupied Germany from May 1945 to March 1946.

Following World War II, the immensely popular Zhukov was assigned a series of obscure regional commands—most notably the Odessa Military District—by Stalin, who was jealous of his renown and feared that he would become too powerful. On Stalin's death in 1953, Zhukov was elevated to deputy minister of defense in March. He supported Nikita Khrushchev in his opposition to the chairman of the Council of

Ministers, Georgy Malenkov, who sought a reduction in military spending. After Khrushchev forced Malenkov to resign and replaced him with Nikolay Bulganin in February 1955, Zhukov succeeded Bulganin as minister of defense. He was also elected an alternate member of the Presidium.

As minister of defense, Zhukov undertook vigorous programs to introduce greater professionalism into the Soviet armed forces. This meant reducing the Communist party's role in military affairs and elevating non-political militarily qualified officers to positions of greater power. The result was a falling out with Khrushchev, who was now Soviet premier. However, Zhukov temporarily redeemed himself in Khrushchev's eyes by his effort to keep the premier in power when a majority of the Presidium (the so-called "anti-party" group) tried to oust Khrushchev. Zhukov ordered aircraft to transport members of the Central Committee from far-flung regions of the country to Moscow in order to restore the political balance in Khrushchev's favor in June 1957. Khrushchev promoted Zhukov to full membership in the Presidium in July, but still objected to Zhukov's efforts to replace party officials with military officers in the administration of the armed forces. Finally, on October 26, 1957, Zhukov was formally dismissed as minister of defense and, a week later, was removed from his party posts.

Zhukov retired into obscurity until Khrushchev himself fell from power in October 1964. Two years later, Zhukov was awarded the Order of Lenin and was allowed to publish his autobiography in 1969.

Zhukov was a man of great courage and fortitude, who possessed enormous tactical and strategical talent. It is sad that, in the postwar political climate of the Soviet Union, he was little rewarded for his extraordinary leadership against the Nazis.

Suggested Reading

Abd el-Kader

Danziger, Raphael. *Abd-al Qadir and the Algerians: Resistance to the French Internal Consolidation* (New York, 1977).

Agricola, Gnaeus Julius

Tacitus, *Agricola* (Cambridge, Ma., 1935).

Agrippa, Marcus Vipsanius

Reinhold, Meyer. *Marcus Agrippa* (New York, 1933).

Alanbrooke, Sir Alan Francis Brooke

Bryant, Sir Arthur. *The Turn of the Tide, 1939–43* (London, 1957).

——. *Triumph in the West, 1943–1946* (London, 1959).

Alexander I, Czar of Russia

Troyat, Henri. *Alexander of Russia* (New York, 1982).

Alexander III the Great

Plutarch. "Alexander," in *Parallel Lives* (New York, 1992).

Tarn, William Woodthorpe. *Alexander the Great*, 2 vols. (Cambridge, 1948).

Allen, Ethan

Jellison, Charles A. *Ethan Allen: Frontier Rebel* (Syracuse, N.Y., 1969).

Allenby, Edmund Henry Hynman, First Viscount

Wavell, Archibald P. *Allenby: A Study in Greatness* (London, 1943).

Anson, George, Lord

Pack, S. W. C. *Admiral Lord Anson* (London, 1960).

Arnold, Henry Harley ("Hap")

Coffey, Thomas M. *HAP: The Story of the U.S. Air Force and the Man Who Built It* (New York, 1982).

Atatürk (Mustafa Kemal, Kemal Atatürk)

Kinross, Lord. *Ataturk* (New York, 1964).

Attila the Hun

Thompson, E.A. *A History of Attila and the Huns* (Oxford, 1948).

Bainbridge, William

Guttridge, Leonard F., and Jay D. Smith. *The Commodores* (New York, 1969).

Barry, John

Clark, William Bell. *Gallant John Barry* (New York, 1938).

Berthier, Louis Alexandre

Delderfield, R. F. *Napoleon's Marshals* (Philadelphia, 1966).

Blake, Robert

Beardon, R. H. *Robert Blake* (N. p., 1935).

Curtiss, C. D. *Blake: General-at-Sea* (N.p., 1934).

Blücher, Gebhard Leberecht von

Warner, Peter. *Napoleon's Enemies* (London, 1976).

Bolívar, Simón

Bushnell, David. *The Liberator: Simón Bolívar* (New York, 1970).

Botha, Louis

Farwell, Byron. *The Great Anglo-Boer War* (New York, 1977).

Pakenham, Thomas. *The Boer War* (New York, 1979).

Braddock, Edward
McCardell, Lee. *Ill-Starred General: Braddock of the Coldstream Guards* (Pittsburgh, 1958).

Bradley, Omar Nelson
Bradley, Omar N. *A Soldier's Story* (New York, 1951).

Bragg, Braxton
McWhiney, Grady. *Braxton Bragg and Confederate Defeat* (New York, 1969).

Burgoyne, John
Huddleston, F. J. *Gentleman Johnny Burgoyne* (New York, 1927).

Burnside, Ambrose Everett
Ballon, Daniel R. *The Military Service of Major General Ambrose Everett Burnside in the Civil War* (Providence, 1914).

Caesar, Gaius Julius
Caesar, Gaius Julius. *War Commenataries* (John Warrington, ed. and tr.; London, 1953).

Dupuy, Trevor N., *The Military Life of Julius Caesar, Imperator* (New York, 1969).

Cassius Longinus, Gaius
Syme, Ronald. *The Roman Revolution* (Oxford, 1939).

Cato, Marcus Porcius Cato ("Cato the Elder")
Plutarch. "Cato," in *Parallel Lives* (New York, 1992).

Cetshwayo
Morris, Donald R. *The Washing of the Spears* (New York, 1965).

Champlain, Samuel de
Morison, Samuel Eliot. *Samuel Champalin: Father of New France* (Boston, 1972).

Charlemagne (Karl Der Grosse, Carolus Magnus)
Lamb, Harold. *Charlemagne* (New York, 1954).

Charles XII
Hatton, Ragnhild Marie. *Charles XII of Sweden* (New York, 1968).

Chiang Kai-shek
Hsuing, S. I. *The Life of Chiang Kai-shek* (London, 1948).

Tuchman, Barbara. *Stilwell and the American Experience in China, 1911–45* (New York, 1970).

Churchill, Sir Winston Spencer
Dupuy, Trevor N. *The Military Life of Winston Churchill of Britain* (New York, 1970).

Gilbert, Martin. *Winston Churchill: The Challenge of War* (London, 1971).

Clark, George Rogers
Bakeless, John E. *Background to Glory: The Life of George Rogers Clark* (Philadelphia, 1957).

Clark, Mark Wayne
Clark, Mark W. *Calculated Risk* (New York, 1950).

——. *From the Danube to the Yalu* (New York, 1954).

Clausewitz, Karl Maria von
Parkinson, Roger. *Carl von Clausewitz* (London, 1971).

Cochise
Sweeney, Edwin R. *Cochise: Chiricahua Apache Chief* (Norman, Okla., 1991).

Condé, Louis II de Bourbon, Prince of
Godfrey, Eviline C. *The Great Condé: A Life of Louis II de Bourbon* (London, 1915).

Constantine I the Great
Smith, John Holland. *Constantine the Great* (New York, 1971).

Cornwallis, Charles, First Marquess
Wickwire, Franklin B. *Cornwallis: The American Adventure* (Boston, 1970).

Cortés, Hérnan
Madariaga, Salvador de. *Hernán Cortéz, Conqueror of Mexico* (1941; reprint ed., New York, 1979).

Cromwell, Oliver
Fraser, Antonia. *Oliver Cromwell* (London, 1973).

Hill, Christopher. *God's Englishman: Oliver Cromwell and the English Revolution* (London, 1973).

Crook, George
Crook, George. *General George Crook: His Autobiography* (Norman, Okla., 1960).

Cumberland, William Augustus, Duke of
Prebble, John. *Culloden* (New York, 1962).

Custer, George Armstrong
Hutton, Paul A., *The Custer Reader* (Lincoln, Nebr., 1992).

Utley, Robert M. *Cavalier in Buckskin: George Armstrong Custer and the Western Military Frontier* (Norman, Okla., 1988).

Dayan, Moshe
Dayan, Moshe, *The Story of My Life* (New York, 1976).

Decatur, Stephen
Lewis, Charles L. *The Romantic Decatur* (Philadelphia, 1937).

De Gaulle, Charles
Crozier, Brian. *De Gaulle* (New York, 1973).
Lacouture, Jean. *De Gaulle* (New York, 1966).

Dewey, George
Dewey, George. *Autobiography* (New York, 1913).

Diocletian (Gaius Aurelius Vaerius Diocletianus)
Barnes, T. D. *The New Empire of Diocletian and Constantine* (Cambridge, Mass., 1982).

Williams, Stephen. *Diocletian and the Roman Recovery* (New York, 1983).

Dönitz, Karl
Dönitz, Karl. *Memoirs: Ten Years and Twenty Days* (London, 1958).

Doolittle, James ("Jimmy") Harold
Doolittle, James ("Jimmy") Harold. *I Could Never Be So Lucky Again* (New York, 1991).

Edward I Longshanks
Salzman, Louis Francis. *Edward I* (London, 1968).

Edward III of Windsor
Hewitt, Herbert James. *The Organization of War Under Edward III, 1338–1362* (New York, 1966).

Edward, the Black Prince
Harvey, John. *The Black Prince and His Age* (New York, 1976).

Eisenhower, Dwight David
Eisenhower, Dwight David. *Crusade in Europe* (New York, 1948).

Eisenhower, Dwight David. *Mandate for Change* (New York, 1963).

Farragut, David Glasgow
Lewis, C. L. *David Glasgow Farragut* (Annapolis, 1941–43).

Feng, Yü-hsiang
Tuchman, Barbara. *Stilwell and the American Experience in China 1911–45* (New York, 1970).

Foch, Ferdinand
Foch, Ferdinand. *Memoirs* (Garden City, New York, 1931).

Forrest, Nathan Bedford
Wyeth, John Allen. *That Devil Forrest* (New York, 1958).

Franco (Bahamonde), Francisco (Paulino Hermenegildo Teódulo)
J. W. D. Trythall. *El Caudillo: A Political Biography of Franco* (New York, 1970).

Frederick I (Frederick Barbarossa, "Red Beard")
Munz, Peter. *Frederick Barbarossa* (Ithaca, N.Y., 1969).

Frederick II the Great
Dupuy, Trevor N. *The Military Life of Frederick the Great* (New York, 1969).

Genghis Khan ("Temujin")
Duopuy, Trevor N. *The Military Life of Genghis Khan* (New York, 1969),

Giap, Vo Nguyen
Stetler, Russell, ed. *The Military Art of People's War: Selected Writings of Vo Nguyen Giap* (New York, 1970).

Gordon, Charles George ("Chinese Gordon")
Lord Elton, *General Gordon* (London, 1954).

Göring, Hermann William
Manvell, Roger, and Henrich Fraenkel. *Göring* (New York, 1972).

Grant, Ulysses Simpson
Catton, Bruce. *Grant Moves South* (Boston, 1960).

——. *Grant Takes Command* (Boston, 1968).

Grant, Ulysses Simpson. *Personal Memoirs* (New York, 1885).

Grasse, Count François Joseph Paul de
Lewis, Charles Lee. *Admiral de Grasse and American Independence* (Annapolis, Md., 1945).

Gustav II Adolf (Gustavus Adolphus)
Dupuy, Trevor N. *The Military Life of Gustavus Adolphus* (New York, 1968).

Haig, Douglas, First Earl
Duff Cooper, Alfred. *Haig* (London, 1935–36).

Halsey, William Frederick, Jr.
Halsey, William Frederick, Jr., and J. Bryan III. *Admiral Halsey's Story* (New York, 1947).

Hannibal
Bradford, Ernle D.S. *Hannibal* (London, 1981).

Harrison, William Henry
Cleaves, Freeman. *Old Tippecanoe: William Henry Harrison and His Time* (New York, 1939; reprint ed., 1969).

Henry II
Salzman, Louis Francis. *Henry II* (1914; reprint ed, New York, 1967).

Henry V
Wylie, James Hamilton. *The Reign of Henry the Fifth* (London, 1914–24).

Hill, Ambrose Powell
Hassler, William Woods. *A. P. Hill: Lee's Forgotten General* (Richmond, Va., 1957).

Hindenburg, Paul Ludwig Hans von
Dupuy, Trevor N. *The Military Lives of Hindenburg and Ludendorff* (New York, 1970).

Hitler, Adolf
Bullock, Alan. *Hitler: A Study in Tyranny* (London, 1952).

Hitler, Adolf. *Mein Kampf* (Munich, 1925–27).

Wykes, Alan. *Hitler* (New York, 1970).

Ho Chi Minh
Lacoutre, Jean. *Ho Chi Minh: A Political Biography* (New York, 1968).

Hooker, Joseph
Herbert, Walter H. *Fighting Joe Hooker* (Indiapolis, 1944).

Houston, Sam(uel)
Marquis, James. *The Raven: A Biography of Sam Houston* (N. p., 1953).

Howard, Oliver Otis
Howard, Oliver Otis . *Autobiography* (New York, 1907).

Howe, Richard, Earl
Barrow, Sir John. *Life of Howe* (London, 1838).

Howe, Sir William, Second Earl
Anderson, Troyer. *The Command of the Howe Brothers During the American Revolution* (New York, 1936).

Hull, Isaac
Roosevelt, Theodore. *The Naval War of 1812* (New York, 1894).

Hussein, Saddam (Takriti)
Karsh, Efraim, and Inari Rautsi. *Saddam Hussein: A Political Biography* (New York, 1991).

Sciolino, Elaine. *The Outlaw State: Saddam Hussein's Quest for Power and the War in the Gulf* (New York, 1991).

Ivan III the Great (Ivan Vasilievich)
Fennell, John Lister. *Ivan the Great of Moscow* (London, 1961).

Jackson, Andrew
Remini, Robert. *Andrew Jackson and the Course of American Empire* (New York, 1977).

Jackson, Thomas Jonathan ("Stonewall")
Farwell, Byron. *Stonewall: A Biography of General Thomas J. Jackson* (New York, 1993).

Freeman, Douglas S. *Lee's Lieutenants: A Study in Command* (New York, 1942–46).

Joan of Arc (Jeanne d'Arc)
Sackville-West, Vita. *Saint Joan of Arc* (London, 1958).

Joffre, Joseph Jacques Césaire
Barnett, Correlli. *The Swordbearers* (New York, 1964).

Johnston, Albert Sidney
Roland, Charles P. *Albert Sidney Johnson: Soldier of Three Republics* (Austin, Tex., 1964).

Jones, John Paul
Morison, Samuel Eliot. *John Paul Jones: A Sailor's Biography* (Boston, 1959).

Chief Joseph
Josephy, Alvin. *The Nez Perce Indians and the Opening of the Northwest* (New Haven, Conn., 1965).

Joubert, Petrus Jacobus
Farwell, Byron. *The Great Anglo-Boer War* (New York, 1977).

Justinian I the Great
Ostrogorsky, George. *A History of the Byzantine State* (New Brunswick, N.J., 1969).

Kalb, Johann (Baron de Kalb)
Kapp, Friederich. *Life of John Kalb: Major-General in the Revolutionary Army* (New York, 1884).

Kearny, Philip
Kearny, Thomas. *General Philip Kearny* (New York, 1947).

Kearny, Stephen Watts

Clarke, Dwight L. *Stephen Watts Kearny: Soldier of the West* (Norman, Okla., 1961).

Keitel, Wilhelm

Keitel, Wilhelm. *Memoirs* (New York, 1966).

Kesselring, Albert von

Kesselring, Albert von. *Memoirs* (N.p., 1953).

Kitchener, Horatio Herbert, First Earl Kitchener of Khartoum

Magnus, Philip. *Kitchener: Portrait of an Imperialist* (London, 1958).

Kosciusko, Tadeusz Andrezj Bonawentura

Haiman, Miecislaus. *Kosciuszko: Leader and Exile* (New York, 1946).

Krueger, Walter

Krueger, Walter, *From Down Under to Nippon: The Story of the Sixth Army in World War II* (Washington, D.C., 1953).

Lafayette, Marie Joseph Paul Yves Roch Gilbert du Motier, Marquis de

Gottschalk, Louis. *Lafayette and the Close of the American Revolution* (Chicago, 1942).

——. *Lafayette between the French and American Revolutions* (Chicago, 1950).

——. *Lafayette Comes to America* (Chicago, 1935).

Lawrence, James

Poolman, Kenneth. *Guns off Cape Ann: The Story of the "Shannon" and the "Chesapeake"* (Chicago, 1961).

Lawrence, T[homas] E[dward]

Lawrence, T[homas] E[dward]. *The Seven Pillars of Wisdom* (London, 1935).

Lee, Henry ("Light Horse Harry")

Boyd, Thomas. *Light-Horse Harry Lee* (New York, 1931).

Lee, Robert Edward

Freeman, Douglas Southall. *Lee's Lieutenants* (New York, 1942–44).

——. *R. E. Lee: A Biography* (New York, 1949).

LeMay, Curtis Emerson

LeMay, Curtis Emerson, with MacKinlay Kantor. *Mission with LeMay* (Garden City, N.Y., 1965).

Lincoln, Abraham

Ballard, Colin R. *The Military Genius of Abraham Lincoln* (Cleveland, 1952).

Sandburg, Carl. *Abraham Lincoln: The War Years* (New York, 1939).

Longstreet, James

Sanger, Donald Bridgman. *General James Longstreet and the Civil War* (Chicago, 1937).

Ludendorff, Erich

Barnett, Correlli. *The Swordbearers* (New York, 1964).

MacArthur, Douglas

Clayton, James D. *The Years of MacArthur* (Boston, 1970–86).

MacArthur, Douglas. *Reminiscences* (New York, 1964).

Manchester, William. *American Caesar* (New York, 1976).

Machiavelli, Niccolò

Ridolfi, Roberto. *The Life of Niccolò Machiavelli* (London, 1963).

Mackenzie, Ranald Slidell

Carter, Robert G. *On the Border with Mackenzie* (New York, 1961).

Mao Tse-tung

Mao Tse-tung. *Selected Military Writings* (Beijing, 1963).

——. *Selected Works* (London and New York, 1954–56).

Payne, Robert. *Mao Tse-tung* (New York, 1969).

Marcus Aurelius [Antoninus]

Birley, Anthony R. *Marcus Aurelius* (Boston, 1966).

Marcus Aurelius. *Meditations* (London, 1944).

Marion, Francis

Bass, Robert Duncan. *Swamp Fox: The Life and Campaigns of General Francis Marion* (New York, 1959).

Marlborough, John Churchill, First Duke of

Chandler, Daivd. *Marlborough as Military Commander* (New York, 1973).

Marshall, George Catlett

Pogue, Forrest C. *George C. Marshall: The Making of a General, 1880-1939* (New York, 1963).

——. *George C. Marshall: Ordeal and Hope, 1939–1942* (New York, 1966).

——. *George C. Marshall: Organizer of Victory, 1943–1945* (New York, 1973).

Masséna, André, Duke of Rivoli and Prince of Essling
Cornwall, James Marshall. *Marshal Masséna* (London, 1965).

Mazepa [Mazeppa], Ivan Stepanovich
Manning, Clarence. *Hetman of Ukraine: Ivcan Mazeppa* (New York, 1957).

McClellan, Georg Brinton
Hassler, Warren W., Jr. *General George B. McClellan: Shield of the Union* (Baton Rouge, 1957).

McClellan, George B. *McClellan's Own Story* (New York, 1887).

Meade, George Gordon
Cleaves, Freeman. *Meade of Gettysburg* (Norman, Okla., 1960).

Merrill, Frank Dow
Ogburn, Charlton. *The Marauders* (New York, 1959).

Miles, Nelson Appleton
Johnson, Virginia W. *The Unregimented General: A Biography of Nelson A. Miles* (Boston, 1962).

Utley, Robert M. *Frontier Regulars: The United States Army and the Indian, 1866–1890* (New York, 1973).

Mitchell, William ("Billy")
Hurley, Col. Alfred F. *Billy Mitchell: Crusader for Air Power* (Bloomington, Ind., 1975).

Mohammed II the Conqueror
Critobulus (C. T. Riggs, tr.). *History of Mehmud the Conqueror by Kristovoulos* (Princeton, 1954).

Moltke, Count Helmuth Johannes Ludwig von
Dupuy, Trevor N. *A Genius for War* (Fairfax, Va., 1984).

Moltke, Count Helmuth Karl Bernhard von
Dupuy, Trevor N. *A Genius for War* (Fairfax, Va., 1984).

Moncey, Bon Adrien Jeannot de, Duke of Conegliano
Delderfield, R. F. *Napoleon's Marshals* (Philadelphia, 1962).

Monck, George, First Duke of Albemarle
Ashley, Maurice, P. *Cromwell's Generals* (London, 1954).

Monmouth, James Scott, Duke of
Fea, Allan. *King Monmouth* (London, 1902).

Fraser, Antonia. *Royal Charles: Charles II and His Times* (New York, 1980).

Montcalm, Louis-Joseph de Montcalm-Gozon, Marquis of
Stacey, Charles P. *Quebec, 1759: The Siege and the Battle* (New York, 1959).

Montgomery, Sir Bernard Law, First Viscount Montgomery of Alamein
Hamilton, Nigel. *Monty: The Making of a General, 1887–1942; The Master of the the Battlefield, 1942–1944; The Field Marshal, 1944–1976.* (London, 1981–86).

Morgan, John Hunt
Butler, Lorine. *John Morgan and His Men* (Philadelphia, 1960).

Mortier, Edouard Adolphe Casimir Joseph, Duc de Treviso
Delderfield, R. F. *Napoleon's Marshals* (Philadelphia, 1962).

Mountbatten, Louis Francis Albert Victor Nicholas, Earl Mountbatten of Burma
Ziegler, Philip. *Mountbatten* (London, 1985).

Murat, Joachim, King of Naples, Duke of Cleve and Berg
Delderfield, R. F. *Napoleon's Marshals* (Philadelphia, 1962).

Murray, Lord George
Preble, John. *Culloden* (Harmondsworth, 1967).

Mussolini, Benito
Collier, Richard. *Duce!* (New York, 1971).

Kirkpatrick, Ivone. *Mussolini: A Study in Power* (New York, 1964).

Napier, Charles James
Farwell, Byron. *Eminent Victorian Soldiers* (New York, 1985).

Napoleon I (Napoleon Bonaparte)
Chandler, David. *The Campaigns of Napoleon* (New York, 1966).

——. *Napoleon* (New York, 1973).

Markham, Felix. *Napoleon* (New York, 1964).

Nelson, Horatio, Viscount
Warner, Oliver. *Nelson* (London, 1975).

Ney, Michel, Prince de la Moskova, Duc of d'Elchingen
Delderfield, R. F. *Napoleon's Marshals* (Philadelphia, 1962).

Nimitz, Chester William
Nimitz, Chester William, and E. B. Potter. *Sea Power: A Naval History* (Annapolis, 1960).

Otto I the Great
Bäuml, F. H. *Medieval Civilization in Germany* (New York, 1968).

Oudinot, Nicolas Charles, Duke of Reggio
Delderfield, R. F. *Napoleon's Marshals* (Philadelphia, 1962).

Patton, George Smith
D'Este, Carlo. *Patton: A Genius for War* (New York, 1995).

Pericles
Burns, Andrew Robert. *Pericles and Athens* (New York, 1962).

Perry, Matthew Calbraith
Morison, Samuel Eliot. *"Old Bruin": Commodore Matthew Calbraith Perry, 1794–1858* (Boston, 1967).

Perry, Oliver Hazard
Dillon, Richard. *We Have Met the Enemy* (New York, 1978).

Pershing, John Joseph
Smythe, Donald. *Pershing: General of the Armies* (Bloomington, Ind., 1986).

Pétain, Henri Philippe
Barnett, Correlli. *The Swordbearers* (London, 1963).

Peter I the Great
Grey, Ian. *Peter the Great: Emperor of All Russia* (London, 1968).

Pickett, George Edward
Freeman, Douglas Southall. *Lee's Lieutenants* (New York: 1942–43).

Pizarro, Francisco
Shay, Frank. *The Incredible Pizarro: Conqueror of Peru* (New York, 1932).

Porter, David
Long, David F. *Nothing Too Daring: A Biography of Commodore David Porter* (Annapolis, 1970).

Porter, David Dixon
Anderson, Bern. *By Sea and By River: The Naval History of the Civil War* (New York, 1961).

Powell, Colin Luther
Powell, Colin Luther (with Joseph E. Persico). *My American Journey* (New York, 1995).

Pulaski, Count Kazimierz (Casimir)
Manning, Charles A. *Soldier of Liberty: Casimir Pulaski* (New York, 1945).

Pyrrhus
Cross, G. N. *Epirus* (London, 1932).

Quantrill, William Clarke
Nichols, Alice. *Bleeding Kansas* (New York, 1954).

Rabin, Yitzhak
Rabin, Yitzhak, *The Rabin Memoirs* (Berkeley, Calif., 1996)

Raeder, Erich
Raeder, Erich. *My Life* (Annapolis, 1960).

Richard I the Lion-Hearted
Gibb, Christopher. *Richard the Lion Heart and the Crusades* (New York, 1985).

Russell, Kate. *Richard the Lion Heart* (London, 1969).

Richard III
Ross, Charles D. *Richard III* (Berkeley, Calif., 1981).

Richelieu, Armand Jean du Plessis, Cardinal et Duc de
Bergin, Joseph. *Cardinal Richelieu* (New Haven, 1985).

Ridgway, Matthew Bunker
Ridgway, Matthew Bunker. *Soldier: The Memoirs of Matthew B. Ridgway* (New York, 1956).

Rochambeau, Jean Baptiste Donatien de Virneur, Count of
Whiridge, Arnold. *Rochambeau* (New York, 1965).

Rodney, George Brydges, Baron
McIntyre, Donald. *Admiral Rodney* (New York, 1963).

Rommel, Erwin Johannes Eugen
Douglas-Home, Charles. *Rommel* (New York, 1973).

Rundstedt, (Karl Rudolf) Gerd von
Keegan, John. *Rundstedt* (New York, 1974).

Saint-Cyr, Laurent Gouvion, Count (Marquis of)
Delderfield, R. F. *Napoleon's Marshals* (Philadelphia, 1962).

Saladin (Salah Ad-din Yusuf Ibn Ayyub, "Righteous of the Faith")
Hindley, Geoffrey. *Saladin* (New York, 1976).

Santa Anna, Antonio Lopez de
Eisenhower, John. *So Far from God* (New York, 1989).

Schlieffen, Count Alfred von
Dupuy, Trevor N. *A Genius for War: The German Army and General Staff, 1807–1945* (Fairfax, Va., 1985).

Schwarzkopf, H(erbert) Norman
Schwarzkopf, H(erbert) Norman. *It Doesn't Take a Hero: The Autobiography* (New York, 1992).

Scott, Winfield
Elliot, Charles Winslow. *Winfield Scott: The Soldier and the Man* (New York, 1937).

Shaka (Chaka, Tshaka)
Osei, Gabriel Kingsley. *Shaka the Great, King of the Zulus* (London, 1971).

Sheridan, Philip Henry
Hergesheimer, Joseph. *Sheridan: A Military Narrative* (Boston, 1931).

Sherman, William Tecumseh
Fellman, Michael. *Citizen Sherman* (New York, 1995).

Short, Walter Campbell
Prange, Gordon. *At Dawn We Slept* (New York, 1981).

Sims, William Sowden
Sims, William Sowden, and Burton J. Hendrick. *The Victory at Sea* (New York, 1920).

Sitting Bull (Tatanka Yotanka)
Utley, Robert M. *The Lance and the Shield: The Life and Times of Sitting Bull* (New York, 1993).

Smuts, Jan Christiaan
Smuts, Jan Christiaan. *Jan Christiaan Smuts* (London, 1952).

Soult, Nicolas Jean de Dieu, Duke of Dalmatia
Delderfeld, R. F. *Napoleon's Marshals* (Philadelphia, 1962).

Stalin, Josef
Bullock, Alan. *Hitler and Stalin: Parallel Lives* (New York, 1992).

McNeal, Robert H. *Stalin: Man and Ruler* (New York, 1988).

Steuben, Baron Friedrich Wilhelm von
Palmer, John. *General von Steuben* (New Haven, 1937).

Stilwell, Joseph Warren
Stilwell, Joseph Warren. *The Stilwell Papers* (New York, 1948).

Tuchman, Barbara. *Stilwell and the American Experience in China, 1911–1945* (New York, 1970).

Stuart, James Ewell Brown
Davis, Burke. *Jeb Stuart: The Last Cavalier* (New York, 1957).

Stuyvesant, Peter
Kessler, Henry, and Eugene Rachlis. *Peter Stuyvesant and His New York* (New York, 1959).

Suleiman I the Magnificent
Braudel, Fernand. *The Mediterranean and the Mediterranean World in the Age of Philip II* (New York, 1973).

Takeda, Shingen [Harunobu]
Dening, Walter. *The Life of Toyotomi Hideyoshi* (London, 1930).

Tamerlane (Timur, Timur the Lane)
Sokol, Edward D. *Tamerlane* (Lawrence, Kan., 1977).

Taylor, Zachary
Dyer, Brainerd. *Zachary Taylor* (Baton Rouge, 1946).

Tecumseh
Wilson, William E. *Shooting Star: The Story of Tecumseh* (New York, 1942).

Tilly, Count Johan Tserclaes
Wedgwood, Cicely V., *The Thirty Years War* (1938; reprint ed., Garden City, N.Y., 1961).

Tito, Josip Broz
Auty, Phyllis. *Tito* (New York, 1970).

Christian, Henry M., ed. *The Essential Tito* (New York, 1970).

Tojo, Hideki
Coox, Alvin. *Tojo* (New York, 1975).

Turenne, Henri de la Tour d'Auvergne, Viscount of
Weygand, Maxime. *Turenne* (Boston, 1930).

Vandegrift, Alexander Archer
Asprey, Robert B., Alexander Archer Vandegrift. *Once a Marine: The Memoirs of Alexander Archer Vandegrift, USMC* (New York, 1966).

Victor-Perrin, Claude, Duke of Belluno
Delderfield, R. F. *Napoleon's Marshals* (Phaildelphia, 1962).

Wainwright, Jonathan Mayhew IV
Wainwright, Jonathan Mayhew IV, and Robert Consdadine. *General Wainwright's Story* (1946; reprint ed., Westport, Conn., 1973).

Wallenstein, Albert Eusebius von, Duke of Friedland and Mecklenburg, Prince of Sagan
Mann, Golo. *Wallenstein: His Life Narrated* (New York, 1977).

Washington, George
Dupuy, Trevor N. *The Military Life of George Washington* (New York, 1969).

Freeman, Douglas Southall. *George Washington: A Biography* (New York, 1947–52).

Wayne, Anthony
Boyd, Thomas. *Mad Anthony Wayne* (New York, 1929).

Wellington, Arthur Wellesley, First Duke of
Longford, Elizabeth. *Wellington* (London, 1969–72).

Westmoreland, William Childs
Westmoreland, William Childs. *A Soldier Reports* (New York, 1974).

Wilhelm II
Tuchman, Barbara. *The Guns of August* (New York, 1962).

William I the Conqueror
Ashley, Maurice. *The Life and Times of William I* (London, 1973).

Walker, David. *William the Conqueror* (Oxford, 1968).

Yamamoto, Isoroku
Potter, John Deane. *Admiral of the Pacific: The Life of Yamamoto* (London, 1965).

Zhukov, Georgi Konstantinovich
Zhukov, Georgi Konstantinovich. *Marshal Zhukov's Greatest Battles* (London, 1969).